Graduate Texts in Mathematics 273

Graduate Texts in Mathematics

Graduate Texts in Mathematics bridge the gap between passive study and creative understanding, offering graduate-level introductions to advanced topics in mathematics. The volumes are carefully written as teaching aids and highlight characteristic features of the theory. Although these books are frequently used as textbooks in graduate courses, they are also suitable for individual study.

More information about this series at http://www.springer.com/series/136

Anatoly Fomenko • Dmitry Fuchs

Homotopical Topology

Second Edition

Anatoly Fomenko
Department of Mathematics
and Mechanics
Moscow State University
Moscow, Russia

Dmitry Fuchs
Department of Mathematics
University of California
Davis, CA, USA

ISSN 0072-5285 ISSN 2197-5612 (electronic)
Graduate Texts in Mathematics
ISBN 978-3-319-79490-7 ISBN 978-3-319-23488-5 (eBook)
DOI 10.1007/978-3-319-23488-5

Printed on acid-free paper

This Springer imprint is published by Springer Nature
The registered company is Springer International Publishing AG Switzerland

Preface

The book we offer to the reader was conceived as a comprehensive course of homotopical topology, starting with the most elementary notions, such as paths, homotopies, and products of spaces, and ending with the most advanced topics, such as the Adams spectral sequence and K-theory. The history of homotopical, or algebraic, topology is short but full of sharp turns and breathtaking events, and this book seeks to follow this history as it unfolded.

It is fair to say that homotopical topology began with *Analysis Situs* by Henri Poincaré (1895). Poincaré showed that global analytic properties of functions, vector fields, and differential forms are greatly influenced by homotopic properties of the relevant domains of definition. Poincaré's methods were developed, in the half-century that followed *Analysis Situs*, by a constellation of great topologists which included such figures as James Alexander, Heinz Hopf, Andrey Kolmogorov, Hassler Whitney, and Lev Pontryagin. Gradually, it became clear that the homotopical properties of domains which were singled out by Poincaré could be best understood in the form of groups and rings associated, in a homotopy invariant way, with a topological space which has to be "good" from the geometric point of view, like smooth manifolds or triangulations. Thus, *Analysis Situs* became algebraic topology.

Still the algebra used by the algebraic topology of that epoch was very elementary: It did not go far beyond the classification of finitely generated Abelian groups (with some exceptions, however, such as Van Kampen's theorem about fundamental groups). More advanced algebra (which, actually, was not developed by algebraists of that time and had to be started from scratch by topologists under the name "homological algebra," aka "abstract nonsense") invaded topology, which made algebraic topology truly algebraic. This happened in the late 1940s and early 1950s. The leading item of this new algebra appeared in the form of a "spectral sequence." The true role of a spectral sequence in topology was discovered mainly by Jean-Pierre Serre (who was greatly influenced by older representatives of the French school of mathematics, mostly by Henri Cartan, Armand Borel, and Jean Leray).

The impact of spectral sequences on algebraic topology was tremendous: Many major problems of topology, both solved and unsolved, became exercises for students.

The progress of the new algebraic topology was very impressive but short-lived: As early as in the late 1950s, the results became less and less interesting, and the proofs became more and more involved. The last big achievement of the algebraic topology which was started by Serre was the Adams spectral sequence, which, in a sense, absorbed all major notions and methods of contemporary algebraic topology. Using his spectral sequence, J. Frank Adams was able to prove the famous Frobenius conjecture (the dimension of a real division algebra must be 1, 2, 4, or 8); it was also used by René Thom in his seminal work, becoming the starting point of the so-called cobordism theory.

Reviving ailing algebraic topology required strong means, and such means were found in the newly developed K-theory. Created by J. Frank Adams, Michael Atiyah, Raoul Bott, and Friedrich Hirzebruch, K-theory (which may be regarded as a branch of the broader "algebraic K-theory") had applications which were unthinkable from the viewpoint of "classical" algebraic topology. It is sufficient to say that the Frobenius conjecture was reduced, via K-theory, to the following question: For which positive integers n is $3^n - 1$ divisible by 2^n? (Answer: for $n = 1, 2,$ and 4.)

Developing K-theory was more or less completed in the mid-1960s. Certainly, it was not the end of algebraic topology. Very important results were obtained later; some of them, belonging to Sergei Novikov, Victor Buchstaber, Alexander Mishchenko, James Becker, and Daniel Gottlieb, are discussed in the last chapter of this book. Many excellent mathematicians continue to work in algebraic topology. Still, one can say that, from the students' point of view, algebraic topology can now be seen as a completed domain, and it is possible to study it from the beginning to the end. (We can add that this is not only possible, but also highly advisable: Algebraic topology provides a necessary background for geometry, analysis, mathematical physics, etc.) This book is intended to help the reader achieve this goal.

The book consists of an introduction and six chapters. The introduction introduces the most often used topological spaces (from spheres to the Cayley projective plane) and major operations over topological spaces (products, bouquets, suspensions, etc.). The chapter titles are as follows: "Homotopy"; "Homology"; "Spectral Sequences of Fibrations"; "Cohomology Operations"; "The Adams Spectral Sequence"; and "K-theory and Other Extraordinary Theories." Chapters are divided into parts called "Lectures," which are numerated throughout the book from Lecture 1 to Lecture 44. Lectures are divided into sections numerated with Arabic numbers, and some sections are divided into subsections labeled with capital letters of the Roman alphabet. For example, Lecture 13 consists of Sects. 13.1, 13.2, 13.3, ...,13.11, and Sect. 13.8 consists of the subsections $A, B, \dots, E$ (which are referred to, in further parts of the book, as Sects. 13.8.A, 13.8.B, and so on).

To present this huge material in one volume of moderate size, we had to be very selective in presenting details of proofs. Many proofs in this book are algebraic, and they often involve routine verifications of independence of the result of a construction of some arbitrary choices within this construction, of exactness of this or that sequence, of group or ring axioms for this or that addition or multiplication. These verifications are necessary, but they often repeat each other, and if included in the book, they will only lead to excessively increasing the volume and irritating the reader, who will probably skip them. On the other hand, we did not want to follow some authors of books who skip details of proofs which are inconvenient, for this or that reason, for an honest presentation. We did our best to avoid this pattern: If a part of a proof is left to the reader as an exercise, then we are sure that this exercise should not be difficult for somebody who has consciously absorbed the preceding material.

As should be clear from the preceding sentences, the book contains many exercises; actually, there are approximately 500 of them. They are numerated within each lecture. They may be serve the usual purposes of exercises: Instructors can use them for homework and tests, while readers can solve them to check their understanding or to get some additional information. But at least some of them must be regarded as a necessary part of the course; in some (not very numerous) cases we will make references to exercises from preceding sections or lectures. We hope that the reader will appreciate this style. The most visible consequence of this approach to exercises is that they are not concentrated in one special section (which is common for many textbooks) but rather scattered throughout every lecture.

Being a part of geometry, homotopic topology requires, for its understanding, a lot of graphic material. Our book contains more than 100 drawings ("figures"), which are supposed to clarify definitions, theorems, or proofs. But the book also contains a chain of drawings that are pieces of art rather than rigorous mathematical figures. These pictures were drawn by A. Fomenko; some of them were displayed at various exhibitions. All of them are supposed to present not the rigorous mathematical meaning, but rather the spirit and emotional contents of notions and results of homotopical topology. They are located in the appropriate places in the book. A short explanation for these pictures can be found at the end of the book.

We owe our gratitude to many people. The first, mimeographed, version of the beginning of this book (which roughly corresponded to Chap. 1 and a considerable part of Chap. 2) was written in collaboration with Victor Gutenmacher; we are deeply grateful to him for his help. The idea of formally publishing this book was suggested to us by Sergei Novikov. We are grateful to him for this suggestion. Some improvements to the book were suggested by several students of the University of California; we are grateful to all of them, especially to Colin Hagemeyer. The whole idea of publishing this book under the auspices of Springer belonged to Boris Khesin and Anton Zorich; we thank them heartily. And the last but, maybe, the most important thanks go to the brilliant team of editors at Springer, especially to Eugene Ha and Jay Popham. It is the result of their work that the book looks as attractive as it does.

Moscow, Russia Anatoly Fomenko
Davis, CA, USA Dmitry Fuchs

Contents

Introduction: The Most Important Topological Spaces

There exists a longstanding tradition to begin a course of homotopic (or other) topology with an introductory lecture dedicated to the point set topology which studies topological spaces in a maximal generality. We violate this tradition, assuming that the reader either already has some knowledge of it, or is ready to experience small inconveniences stemming from an insufficient knowledge of it, or will look through some not too boring text (the first section of the book by Fuchs and Rokhlin [40] will do). Anyhow, we acquire the right to use without explanations terms like "Hausdorff space," or "compact space," or "countable base space," and so on, and also use (explicitly or implicitly) facts like "a bijective continuous map of a compact space onto a Hausdorff space is a homeomorphism," or "a compact subset of a Hausdorff space is closed." As to the introductory part of the book, we dedicate it not to the general notion of a topological space, but rather to creating a list of the most frequently used topological spaces which will serve as a source of examples and motivations, and also will participate in various geometric constructions. First, we will get acquainted with the most important, "classical" spaces, and then we will describe major constructions involving topological spaces, which will amplify our supply of topological spaces and will have a great importance of their own.

Lecture 1 Classical Spaces

1.1 Euclidean Spaces, Spheres, and Balls

The notations $\mathbb{R}^n$ and $\mathbb{C}^n$ will have the usual meaning. The spaces $\mathbb{C}^n$ and $\mathbb{R}^{2n}$ are identified by the correspondence $(x_1 + iy_1, \dots, x_n + iy_n) \leftrightarrow (x_1, y_1, \dots, x_n, y_n)$. The *sphere* S^n and the *ball* D^n are defined, respectively, as the unit sphere and the unit ball centered at the origin in spaces $\mathbb{R}^{n+1}$ and $\mathbb{R}^n$; thus, the sphere S^{n-1} is the boundary of the ball D^n. The symbol $\mathbb{R}^\infty$ always means the union (inductive limit) of the chain $\mathbb{R}^1 \subset \mathbb{R}^2 \subset \mathbb{R}^3 \subset \dots$; thus, $\mathbb{R}^\infty$ is the set of sequences $(x_1, x_2, x_3, \dots)$

of real numbers with only finitely many nonzero terms a_n. The topology in $\mathbb{R}^\infty$ is introduced by the rule: A set $F \subset \mathbb{R}^\infty$ is closed if and only if all the intersections $F \cap \mathbb{R}^n$ are closed in respective spaces $\mathbb{R}^n$. The symbols $\mathbb{C}^\infty, S^\infty, D^\infty$ have a similar sense.

EXERCISE 1. Show that a sequence

$$(a_1, 0, 0, \dots), (0, a_2, 0, \dots), \dots, (\underbrace{0, \dots, 0}_{n-1}, a_n, 0, \dots), \dots$$

has a limit if and only if it has only finitely many nonzero terms.

EXERCISE 2. Show that none of the spaces $\mathbb{R}^\infty, S^\infty, D^\infty$ is metrizable.

Remark. There are other definitions of $\mathbb{R}^\infty$ in the literature. For example, (1) the "Hilbert space" ℓ_2 is the set of all real sequences $(x_1, x_2, x_3, \dots)$ for which the series $\sum x_i^2$ converges; the topology is defined by the metric $d^2((x_1, x_2, x_3, \dots), (y_1, y_2, y_3, \dots)) = \sum (y_i - x_i)^2$; (2) the Tychonoff space T is the set of *all* real sequences $(x_1, x_2, x_3, \dots)$ with the base of topology formed by the sets $\{(x_1, x_2, x_3, \dots) \in T \mid (x_1, \dots, x_n) \in U\}$ for all n and all open $U \subset \mathbb{R}^n$ [a sequence $X_i = (x_{i1}, x_{i2}, x_{i3}, \dots)$ in T converges to $X = (x_1, x_2, x_3, \dots) \in T$ if and only if $\lim_{i \to \infty} x_{in} = x_n$ for every n].

EXERCISE 3. Which of the inclusion maps $\mathbb{R}^\infty \to \ell_2$, $\mathbb{R}^\infty \to T$, $\ell_2 \to T$ (if any) are continuous? Which of them (if any) are homeomorphisms onto their images?

EXERCISE 4. Is the space T metrizable?

EXERCISE 5. The *unit cube* of $\mathbb{R}^\infty, \ell_2, T$ is defined by the condition $0 \le x_i \le 1$ for $i = 1, 2, 3, \dots$. Which of these cubes (if any) are compact?

1.2 Real Projective Spaces

The *real n-dimensional projective space* $\mathbb{R}P^n$ is defined as the set of all straight lines in $\mathbb{R}^{n+1}$ passing through the origin equipped with the topology determined by the angular metric: The distance between two lines is defined as the angle between them.

EXERCISE 6. Prove that the real projective line is homeomorphic to the circle S^1.

The coordinates $(x_0, x_1, \dots, x_n)$ of the directing vector of the line (defined, obviously, up to a proportionality) are called the *homogeneous coordinates* of a point of a projective space; the common notation is $(x_0 : x_1 : \dots : x_n)$. The points with $x_i \ne 0$ form the *ith principal affine chart*. The correspondence $(x_0 : x_1 : \dots : x_n) \leftrightarrow (x_0/x_i, \dots, x_{i-1}/x_i.x_{i+1}/x_i, \dots, x_n/x_i)$ yields a homeomorphism of the affine chart onto $\mathbb{R}^n$ and equips the former with coordinates.

If we assign to a point of $S^n \subset \mathbb{R}^{n+1}$ a line passing through this point and the origin, we get a continuous map $S^n \to \mathbb{R}P^n$. This map sends two different points of S^n into the same point of $\mathbb{R}P^n$ if and only if these two points are antipodal (opposite). Thus, every point of $\mathbb{R}P^n$ has, with respect to this map, precisely two preimages (the map itself is a *twofold covering*; see Lecture 7). Having this map in mind, we say that $\mathbb{R}P^n$ is obtained from S^n by identifying all pairs of opposite points. (This statement has the following precise sense. Suppose that, in a topological space X, there is some chosen set of pairs of points, subject to identification. After the identification, there arise a set Y and a map $X \to Y$. We introduce a topology in Y declaring a subset of Y open if its inverse image in X is open; this is the weakest of all topologies with respect to which the map $X \to Y$ is continuous[1].) The upper hemisphere of S^n (composed of points with a nonnegative last coordinate) is canonically homeomorphic to the ball D^n (the homeomorphism is established by an orthogonal projection of the upper hemisphere onto the equatorial ball). The restriction of the last map $S^n \to \mathbb{R}P^n$ to the upper hemisphere is, therefore, a map $D^n \to \mathbb{R}P^n$. This map sends to the same point only opposite points of the boundary sphere $S^{n-1} \subset D^n$. Thus, $\mathbb{R}P^n$ may also be obtained from D^n by identifying all pairs of opposite points on the boundary sphere.

The infinite-dimensional real projective space $\mathbb{R}P^\infty$ may be defined by any of these three constructions. We can also put $\mathbb{R}P^\infty = \bigcup_i \mathbb{R}P^i$.

1.3 Complex and Quaternionic Projective Spaces

If in the definition of $\mathbb{R}P^n$, we replace $\mathbb{R}^{n+1}$ by $\mathbb{C}^{n+1}$ and real lines by complex lines, we will obtain a definition of a *complex projective space* $\mathbb{C}P^n$. (The angular metric still makes sense.)

Like $\mathbb{R}P^n$, the space $\mathbb{C}P^n$ is covered by $n + 1$ affine charts. If we assign to a point of $S^{2n+1} \subset \mathbb{C}^{n+1}$ a complex line passing through this point and the origin, we will obtain a continuous map $S^{2n+1} \to \mathbb{C}P^n$ which sends to the same point the whole circle $\{(z_0 w, \dots, z_n w)\}$, where $(z_0, \dots, z_n)$ is a fixed point of S^{2n+1} and w runs through the circle $|w| = 1$. One can say that $\mathbb{C}P^n$ is obtained from the sphere S^{2n+1} by collapsing each such circle to one point. If we restrict the map $S^{2n+1} \to \mathbb{C}P^n$ to the ball D^{2n} embedded into S^{2n+1} as the set of points $(z_0, \dots, z_n) \in S^{2n+1}$ whose last coordinate is real and nonnegative, we get a description of $\mathbb{C}P^n$ as obtained from D^{2n} by the same identification which is performed only on the boundary $\{z_n = 0\}$ of D^{2n}.

[1] In mathematics, the terms "weak topology" and "strong topology" do not have any commonly accepted meaning. We call a topology weaker if it has more open sets, that is, fewer limit points (for us, the weakest topology is the discrete topology). Informally speaking, we call a topology weak if the attraction forces between the points are weak. The opposite terminology considers points as repelling each other; from this point of view, the discrete topology is the strongest.

A similar construction is possible if the field $\mathbb{C}$ is further replaced by the algebra (skew field) $\mathbb{H}$ of quaternions. We get a definition of a *quaternionic projective space* $\mathbb{H}P^n$. One should notice, however, that , because of noncommutativity of the algebra $\mathbb{H}$, one has to distinguish between left and right lines. Considering $\mathbb{H}P^n$, one should choose one of these two possibilities and consider, say, left lines.

EXERCISE 7. Prove that projective lines $\mathbb{C}P^1$ and $\mathbb{H}P^1$ are homeomorphic, respectively, to S^2 and S^4.

There are also obvious definitions of $\mathbb{C}P^\infty$ and $\mathbb{H}P^\infty$.

1.4 Cayley Projective Plane

The reader may find it unfair that quaternionic projective spaces occupy in our list of classical spaces such an honorable place: next to spheres and balls. However, in reality, not only quaternionic projective spaces, but even such an exotic object as the Cayley projective plane are very important for topology.

Let us recall the definition of *Cayley numbers* or *octonions*. Suppose that in some space $\mathbb{R}^n$ two operations are defined: multiplication, $a, b \mapsto ab$, and conjugation, $a \mapsto \overline{a}$. Then we define similar operations in $\mathbb{R}^{2n} = \mathbb{R}^n \times \mathbb{R}^n$ by the formulas

$$(a,b)\cdot(c,d) = (ac - b\overline{d}, b\overline{c} + ad),\ \overline{(a,b)} = (\overline{a}, -b).$$

Starting from the usual multiplication and identical conjugation ($\overline{a} = a$) in $\mathbb{R}^1 = \mathbb{R}$, we get (bilinear) multiplications and conjugations in $\mathbb{R}^2, \mathbb{R}^4, \mathbb{R}^8, \mathbb{R}^{16}, \dots$. The multiplication in $\mathbb{R}^2$ is the usual multiplication of complex numbers. The multiplication in $\mathbb{R}^4$ is the multiplication of quaternions. It is bilinear, associative, and admits a unique division (that is, the equation $ax = b$ has a unique solution if $a \neq 0$) but is not commutative. The multiplication in $\mathbb{R}^8$ is still worse: Not only it is not commutative, but also it is not associative [although the associativity relations involving only two letters, such as $(ab)a = a(ba)$, $(ab)b = ab^2$, $(ab)a^{-1} = a(ba^{-1})$, etc., hold]. Still this multiplication possesses a unique division. The algebra $\mathbb{R}^8$ with this multiplication is called the *Cayley algebra* or *octonion algebra* and is denoted as $\mathbb{C}\mathbf{a}$. Much later in this book we will consider (and prove) the famous Frobenius conjecture: If the space $\mathbb{R}^n$ possesses a bilinear multiplication with a unique division, then $n = 1, 2, 4$, or 8. (By the way, it is not right that even in these dimensions any bilinear multiplication with unique division is isomorphic to one of multiplications described above; there are, for example, nonassociative bilinear multiplications with a unique division in $\mathbb{R}^4$.)

The nonassociativity of the Cayley multiplication impedes defining any lines in the space $\mathbb{C}\mathbf{a}^n$ with $n \geq 3$. Indeed, if we define a line ℓ_x through $x \in \mathbb{C}^n$ and the origin as the set $\{tx \mid t \in \mathbb{C}\mathbf{a}\}$, then the line ℓ_{t_0x} through a point of this line and the origin will not, in general, coincide with ℓ_x (see Fig. 1).

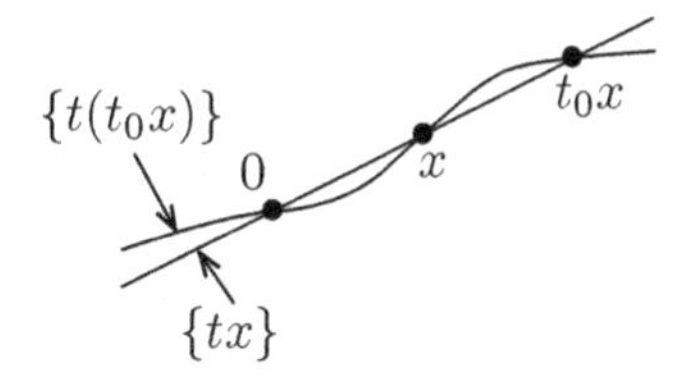

Fig. 1 Cayley lines in Cayley plane

For this reason, there is no satisfactory definition of a Cayley projective space (we will see later that spaces with expectable properties of Cayley projective spaces do not exist by purely topological reasons). Still, it remains possible to define a Cayley projective plane. For this purpose, we consider not whole lines in $\mathbb{C}\mathbf{a}^3$, but rather traces of these lines on the union T of three planes: $x = 1, y = 1, z = 1$ (x, y, and z denote the "Cayley coordinates" in $\mathbb{C}\mathbf{a}^3$). More precisely: A point $(a', b', c') \in \mathbb{C}\mathbf{a}^3 - 0$ is called collinear to the point $(a, b, c) \in \mathbb{C}\mathbf{a}^3 - 0$ if there exists a $t \in \mathbb{C}\mathbf{a} - 0$ such that $a' = ta, b' = tb, c' = tc$. The collinearity relation is reflective and symmetric but, in general, not transitive. However, it becomes transitive if we restrict ourselves to points in T. For example, if $(c, 1, d) = t(1, a, b)$ and $(e, f, 1) = u(c, 1, d)$, then $t = c = a^{-1} = db^{-1}$, $u = ec^{-1} = f = d^{-1}$, and $(tu)(1, a, b) = ((ec^{-1})c, (fa^{-1})a, (d^{-1}(db^{-1}))b) = (e, f, 1)$. Moreover, a point of any of the three planes is not collinear to any point of the same plane and collinear to no more than one point of any of the other two planes. The space obtained from T by identifying all collinear points is $\mathbb{C}\mathbf{a}P^2$. Each of the three planes in T is mapped into $\mathbb{C}\mathbf{a}P^2$ without folding; these three subsets of $\mathbb{C}\mathbf{a}P^2$ form a covering similar to the covering by affine charts.

EXERCISE 8. What will we get if we identify all pairs of collinear points in $\mathbb{C}\mathbf{a}^3 - 0$?

1.5 Grassmann Manifolds

This is a generalization of projective spaces. A *real Grassmann manifold* $G(n, k)$ is defined as the space of all k-dimensional subspaces of the space $\mathbb{R}^n$.[2] The topology in $G(n, k)$ may be described as induced by the embedding $G(n, k) \to \text{End}(\mathbb{R}^n)$ which assigns to a $P \in G(n, k)$ the orthogonal projection $\mathbb{R}^n \to P$ combined with the inclusion map $P \to \mathbb{R}^n$; a more convenient description of the same topology arises from a realization of $G(n, k)$ as a subspace of a projective space; see ahead.

[2] There exists another system of notation where the space which we denote as $G(n, k)$ is denoted as $G(n - k, k)$.

Obviously, $G(n,k) = G(n,n-k)$ and $G(n,1) = \mathbb{R}P^{n-1}$. There are also obvious embeddings $G(n,k) \to G(n+1,k)$ (arising from the inclusion $\mathbb{R}^n \subset \mathbb{R}^{n+1}$) and $G(n,k) \to G(n+1,k+1)$ (the space $\mathbb{R}^n$ and its k-dimensional subspaces are multiplied by a line).

An analog of an affine chart in a Grassmann manifold (which is a generalization of an affine chart in a projective space) is defined in the following way. Choose a sequence $1 \leq i_1 < \dots < i_k \leq n$ and consider the subset $G_{i_1,\dots,i_k}(n,k)$ of $G(n,k)$ composed of subspaces whose projection onto the space $\mathbb{R}^k_{i_1,\dots,i_k} \subset \mathbb{R}^n$ of i_1st, ..., i_kth coordinate axes is nondegenerate. If $P \in G(n,k)$ belongs to this part, then P may be considered a graph of a linear map of the space $\mathbb{R}^k_{i_1,\dots,i_k}$ into its orthogonal complement. Thus, the points of $G_{i_1,\dots,i_k}(n,k)$ are characterized by $k \times (n-k)$ matrices, that is, by sets of $k(n-k)$ real numbers. This defines a homeomorphism of $G_{i_1,\dots,i_k}(n,k)$ onto $\mathbb{R}^{k(n-k)}$ and yields a coordinate system in $G_{i_1,\dots,i_k}(n,k)$.

It is also possible to introduce a coordinate system in the whole space $G(n,k)$. For a $P \in G(n,k)$ choose a basis in $P \subset \mathbb{R}^n$. Let $(x_{i1},\dots,x_{in})$, $i = 1,\dots,k$, be coordinates of vectors of this basis. For $1 \leq j_1 < \dots < j_k \leq n$, put

$$\xi_{j_1,\dots,j_k}(P) = \det \begin{bmatrix} x_{1j_1} & \dots & x_{ij_k} \\ \dots & \dots & \dots \\ x_{kj_1} & \dots & x_{kj_k} \end{bmatrix}.$$

The numbers $\xi_{j_1,\dots,j_k}(P)$ are called *Plücker coordinates of* P; they are not all zero, and if we change the basis in P, they all will be multiplied by the same number (the determinant of the transition matrix to the new basis). Thus, Plücker coordinates of P may be regarded as homogeneous coordinates of a certain point of $\mathbb{R}P^{\binom{n}{k}-1}$. We get an embedding $G(n,k) \to \mathbb{R}P^{\binom{n}{k}-1}$. Certainly, the image of this embedding does not cover the whole space $\mathbb{R}P^{\binom{n}{k}-1}$; that is, there are some relations between the Plücker coordinates of a point in $G(n,k)$. For example, the six Plücker coordinates $\xi_{12}, \xi_{13}, \xi_{14}, \xi_{23}, \xi_{24}, \xi_{34}$ of a point in $G(4,2)$ satisfy the relation $\xi_{12}\xi_{34} - \xi_{13}\xi_{24} + \xi_{14}\xi_{23} = 0$, and no other relations. Thus, $G(4,2)$ is homeomorphic to a hypersurface in $\mathbb{R}P^5$ defined by the equation of degree 2 given above.

All this can be repeated, with obvious modifications, in the complex and quaternionic cases; the Grassmann manifolds arising are denoted as $\mathbb{C}G(n,k)$ and $\mathbb{H}G(n,k)$. One more version of Grassmann manifolds arises as the set of *oriented* k-dimensional subspaces of $\mathbb{R}^n$; the corresponding notation is $G_+(n,k)$.

There are obvious complex and quaternionic versions of the equalities $G(n,k) = G(n,n-k)$, $G(n,1) = \mathbb{R}P^{n-1}$; also, $G_+(n,k) = G_+(n,n-k)$ and $G_+(n,1) = S^{n-1}$. The embeddings $G(n,k) \to G(n+1,k)$ and $G(n,k) \to G(n+1,k+1)$ also have complex, quaternionic, and oriented analogs.

Notice also that there are Plücker coordinates in $\mathbb{C}G(n,k)$ and $G_+(n,k)$. In $\mathbb{C}G(n,k)$ they are defined up to a complex proportionality and yield an embedding $\mathbb{C}G(n,k) \to \mathbb{C}P^{\binom{n}{k}-1}$. In $G_+(n,k)$ they are defined up to a multiplication by positive numbers and give an embedding $G_+(n,k) \to S^{\binom{n}{k}-1}$.

Finally, the infinite-dimensional version of Grassmann manifolds is provided by the *Grassmann space* $G(\infty, k)$, which is the union of the chain $G(k+1, k) \subset G(k+2, k) \subset G(k+3, k) \subset \dots$, and $G(\infty, \infty)$, which is the union of the chain $G(\infty, k) \subset G(\infty, k+1) \subset G(\infty, k+2) \subset \dots$. There are also spaces $\mathbb{C}G(\infty, k)$, $\mathbb{C}G(\infty, \infty)$, $\mathbb{H}G(\infty, k)$, $\mathbb{H}G(\infty, \infty)$, $G_+(\infty, k)$, $G_+(\infty, \infty)$.

1.6 Flag Manifolds

This is a generalization of Grassmann manifolds. Let there be given a sequence of integers $1 \leq k_1 < \dots < k_s < n$. A flag of type $(k_1, \dots, k_s)$ in $\mathbb{R}^n$ is a chain $V_1 \subset \dots \subset V_s$ of subspaces of the space $\mathbb{R}^n$ such that $\dim V_i = k_i$. The set of flags has a natural topology [for example, as a subset of $G(n, k_1) \times \dots \times G(n, k_s)$] and becomes a "flag manifold" $F(n; k_1, \dots, k_s)$. The versions $\mathbb{C}F(n; k_1, \dots, k_s)$, $\mathbb{H}F(n; k_1, \dots, k_s)$, and $F_+(n; k_1, \dots, k_s)$ of this definition are obvious. The spaces $F(n; 1, 2, \dots, n-1)$, $\mathbb{C}F(n; 1, 2, \dots, n-1)$, $\mathbb{H}F(n; 1, 2, \dots, n-1)$, and $F_+(n; 1, 2, \dots, n-1)$ are called (understandably) manifolds of full flags.

1.7 Compact Classical Groups

The compact classical groups include the group $O(n)$ of orthogonal $n \times n$ matrices, the group $U(n)$ of unitary $n \times n$ matrices, the groups $SO(n)$ and $SU(n)$ of matrices from $O(n)$ and $U(n)$ with determinant 1, and the group $Sp(n)$ of quaternionic matrices of unitary transformations of $\mathbb{H}^n$.

Notice that the group $SO(2)$ of rotations of the plane around the origin is homeomorphic to a circle. The group $SO(3)$ is homeomorphic to $\mathbb{R}P^3$; the homeomorphism assigns to counterclockwise rotation by an angle $\alpha \leq \pi$ of $\mathbb{R}^3$ around an oriented axis ℓ a point of ℓ at the distance α/π from the origin (in the positive direction). Since the rotation by the angle π around an oriented axis is not different from the rotation by the angle π around the same axis with the opposite orientation, the image of this map is the unit ball in $\mathbb{R}^3$ with the opposite points on the boundary identified, that is, $\mathbb{R}P^3$. Another construction of (the same) homeomorphism $\mathbb{R}P^3 \to SO(3)$ assigns to a line $\ell \subset \mathbb{R}^4 = \mathbb{H}$ the transformation $p \mapsto qpq^{-1}$ of the space $\mathbb{R}^3$ of purely imaginary quaternions where $0 \neq q \in \ell$ (we leave the details to the reader).

The group $SU(2)$ is homeomorphic to S^3: It consists of matrices $\begin{bmatrix} \alpha & \beta \\ -\overline{\beta} & \overline{\alpha} \end{bmatrix}$, where $|\alpha|^2 + |\beta|^2 = 1$, that is, $(\alpha, \beta) \in S^3 \subset \mathbb{C}^2$. Finally, the groups $U(1)$ and $Sp(1)$, which are isomorphic, respectively, to the groups $SO(2)$ and $SU(2)$, are homeomorphic to S^1 and S^3.

1.8 Stiefel Manifolds

The spaces of orthonormal k-frames in $\mathbb{R}^n$ (topologized as a subset of $\mathbb{R}^n \times \cdots \times \mathbb{R}^n$) is called the *Stiefel manifold* and is denoted as $V(n,k)$. This space has complex and quaternionic analogs: $\mathbb{C}V(n,k)$ and $\mathbb{H}V(n,k)$. Stiefel manifolds generalize classical groups: $V(n,n) = O(n), \mathbb{C}V(n,n) = U(n), \mathbb{H}V(n,n) = Sp(n), V(n,n-1) = SO(n), \mathbb{C}V(n,n-1) = SU(n)$. Notice also that $V(n,1) = S^{n-1}, \mathbb{C}V(n,1) = S^{2n-1}, \mathbb{H}V(n,1) = S^{4n-1}$.

1.9 Classical Actions of Classical Groups in Classical Spaces

The action of the group $O(n)$ in $\mathbb{R}^n$ gives rise to its actions in S^{n-1}, D^n, $G(n,k)$, and $V(n,k)$. The subgroup $SO(n)$ of $O(n)$ acts also in $G_+(n,k)$. There is also an action of $O(k)$ in $V(n,k)$: The matrices from $O(k)$ are applied to the vectors of the frame. All these actions have complex and quaternionic analogs.

The actions of $O(n)$ in $S^{n-1}, G(n,k)$, and $V(n,k)$ are transitive. The same is true for the complex and quaternionic analogs of these actions, and also for the action of $SO(n)$ in $G_+(n,k)$. Thus, almost all classical spaces described above are homogeneous spaces of compact classical groups; that is, they can be described as quotient spaces of these groups over some subgroups. Here are these descriptions:

$$\begin{aligned}
S^{n-1} &= O(n)/O(n-1) = SO(n)/SO(n-1);\\
S^{2n-1} &= U(n)/U(n-1) = SU(n)/SU(n-1);\\
S^{4n-1} &= Sp(n)/Sp(n-1);\\
G(n,k) &= O(n)/O(k) \times O(n-k);\\
\mathbb{C}G(n,k) &= U(n)/U(k) \times U(n-k);\\
\mathbb{H}G(n,k) &= Sp(n)/Sp(k) \times Sp(n-k);\\
G_+(n,k) &= SO(n)/SO(k) \times SO(n-k);\\
V(n,k) &= O(n)/O(n-k) = (\text{if } n > k)\ SO(n)/SO(n-k);\\
\mathbb{C}V(n,k) &= U(n)/U(n-k) = (\text{if } n > k)\ SU(n)/SU(n-k);\\
\mathbb{H}V(n,k) &= Sp(n)/Sp(n-k).
\end{aligned}$$

($O(k) \times O(n-k)$ is a subgroup of $O(n)$ consisting of block diagonal matrices with $k \times k$ and $(n-k) \times (n-k)$ blocks; $U(k) \times U(n-k)$, etc., have a similar sense). Similarly, for flag manifolds,

$$F(n; k_1, \dots, k_s) = O(n)/O(k_1) \times O(k_2 - k_1) \times \cdots \times O(k_s - k_{s-1}) \times O(n-k_s),$$

etc. In particular, the manifold of full flags, $\mathbb{C}F(n; 1, 2, \dots, n-1)$, is the quotient space of the group $U(n)$ over its "maximal torus" $U(1) \times \cdots \times U(1)$.

The action of the group $O(k)$ in $V(n,k)$, as well as its complex and quaternionic analogs, are free. With respect to these actions,

$$V(n,k)/O(k) = G(n,k), \qquad V(n,k)/SO(k) = G_+(n,k),$$
$$\mathbb{C}V(n,k)/U(k) = \mathbb{C}G(n,k), \quad \mathbb{H}V(n,k)/Sp(k) = \mathbb{H}G(n,k).$$

1.10 Classical Surfaces

The most classical of classical surfaces are the two-dimensional sphere S^2, the projective plane $\mathbb{R}P^2$, and the Klein bottle, which, as is well known, can be realized in $\mathbb{R}^3$ only as a surface with self-intersection (see Fig. 2).

A similar self-intersecting surface representation, although not this broadly known, exists for a projective plane (it is called *Boy's surface*, by the name of the discoverer). It is shown as the top left drawing in Fig. 3. To make this drawing easier to understand, we show the sections of the surface by seven horizontal planes numerated from the top to the bottom (right drawing in Fig. 3). Notice that there is a saddle point between Sections 2 and 3, and a section by some horizontal plane has a triple self-intersection point. This triple point is also visible in Fig. 3. There is a theorem that a self-intersecting surface in space representing the projective plane must have at least one triple self-intersection point.

There is a surface with a more complicated singularity called a *cross cap* also representing the projective plane. It is shown as the bottom left drawing in Fig. 3.

We also count as classical all surfaces obtained from the sphere, the projective plane, and the Klein bottle by drilling some (finite) number of (small, round) holes and attaching some (finite) number of handles.

EXERCISE 9. Prove that the projective plane with one hole is homeomorphic to the Möbius band (thus, the Möbius band is a "classical surface").

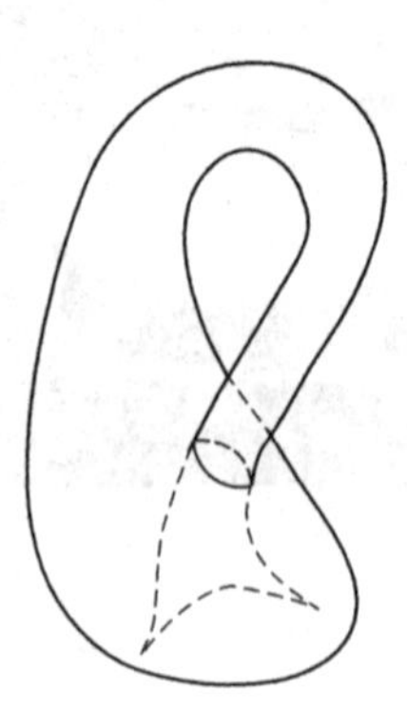

Fig. 2 Klein bottle

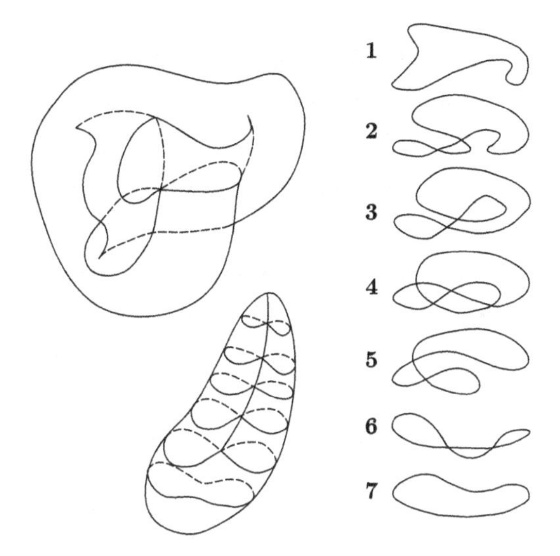

Fig. 3 Projective plane

EXERCISE 10. Prove that a surface which is obtained by joining two Klein bottles by a tube is homeomorphic to a Klein bottle with a handle.

EXERCISE 11. Prove that a surface which is obtained by joining a Klein bottle by a tube with a projective plane is homeomorphic to a projective plane with a handle.

EXERCISE 12. Prove that a surface which is obtained by joining two projective planes by a tube is homeomorphic to a Klein bottle.

EXERCISE 13. Deduce from Exercises 9–12 that a surface obtained by joining two classical surfaces by a tube is again a classical surface.

There is also a classical procedure of constructing classical surfaces from polygons (closed planar polygonal domains) by gluing together several pairs of sides. The procedure is as follows. We take a planar polygon, for example, a regular n-gon, and then form k pairs of $2k$ ($\leq n$) its sides. Furnish each of these $2k$ sides by an orientation (shown by an arrow). After that, we attach to each other the two sides of each pair in a way compatible with the orientations. (Sometimes, one can make these attachments using glue, but more often it can be done only mentally, as described in Sect. 1.2.)

EXERCISE 14. In Fig. 4, there are six polygons; some sides have numbers and arrows, and every number is repeated twice. Prove that after attaching the sides with equal numbers compatible with the arrows, we obtain the following classical surfaces: (a) an annulus (that is, a sphere with two holes); (b) a Möbius band; (c) a torus (that is, a sphere with one handle); (d) a Klein bottle; (e) a projective plane; (f) a sphere with two handles.

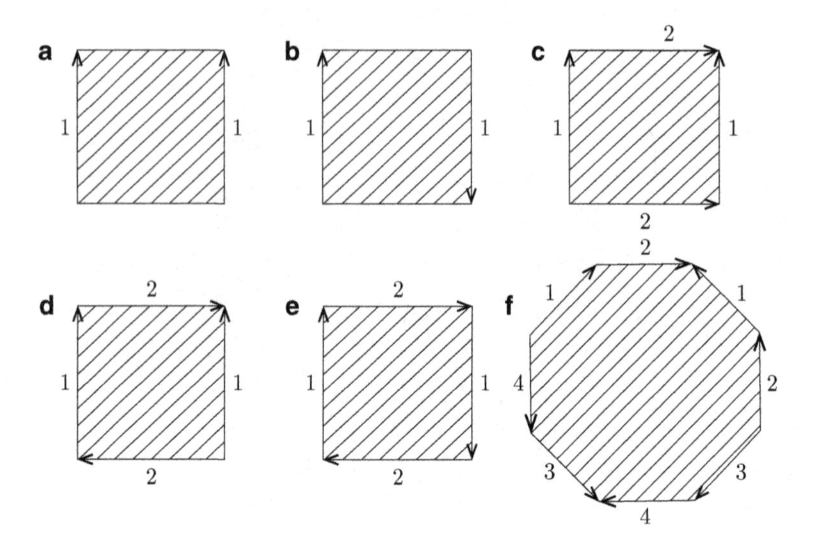

Fig. 4 For Exercise 14

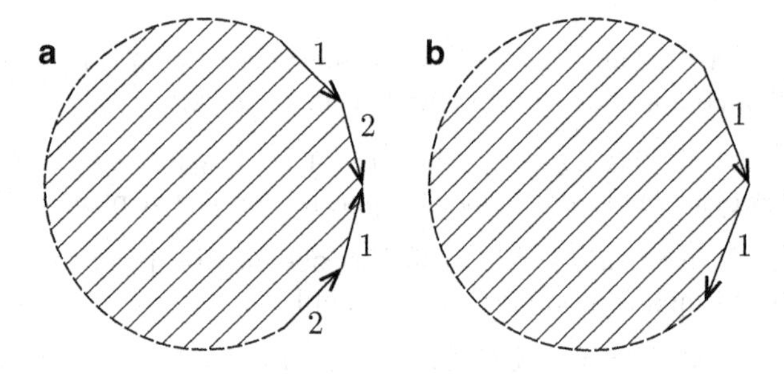

Fig. 5 For Exercises 15 and 16

EXERCISE 15. Show that if one inserts into a polygon four sides labeled and oriented as shown in Fig. 5a, then a handle is added to the surface.

EXERCISE 16. Show that if one inserts into a polygon two sides labeled and oriented as shown in Fig. 5b, then a projective plane joined to the surface by a tube is added (in other words, a hole is drilled in the surface, and a Möbius band is attached by its boundary to the boundary of the circle).

EXERCISE 17. Show that any classical surface can be obtained from a polygon by a procedure described above.

EXERCISE 18. Is it true that the procedure described above always yields a classical surface?

A torus (a sphere with one handle) can be constructed in $\mathbb{R}^3$ as a surface of revolution generated by revolving a circle around an axis in the plane of this circle but disjoint from it. The curves, which are the positions of the circle at intermediate moments of time, are called meridians of the torus, and the trajectories of points of the circle are called parallels of the torus. According to this, circular coordinates arise on the torus: The latitude is the angle measured counterclockwise from some fixed parallel (for example, from the longest one) and the longitude is the angle measured counterclockwise along a parallel from some fixed meridian (for example, from the initial position of the circle).

EXERCISE 19. Factorizing the torus by the relation $(\varphi, \psi) \sim (\varphi + \pi, \psi + \pi)$ (that is, by identifying points symmetric with respect to the symmetry center of the torus) provides a Klein bottle.

EXERCISE 20. Factorizing the torus by the relation $(\varphi, \psi) \sim (\psi, \varphi)$ provides a Möbius band.

In conclusion, we will calculate "the genus of complex curves" (it is useful to know the term "genus" as it is used in algebraic geometry: The genus of a sphere with handles is the number of handles).

EXERCISE 21. Prove that the subset of the complex projective plane $\mathbb{C}P^2$ consisting of points whose homogeneous coordinates satisfy the equation $x_0^n + x_1^n + x_2^n = 0$ is homeomorphic to the sphere with $\dfrac{(n-1)(n-2)}{2}$ handles.

If you cannot do this exercise now, you can return to it later.

Lecture 2 Basic Operations over Topological Spaces

2.1 Product Spaces

Recall that the product $X \times Y$ of two sets X and Y is the set of pairs (x, y) where $x \in X$ and $y \in Y$. If X and Y are topological spaces, then $X \times Y$ acquires a canonical topology: The base of open sets in $X \times Y$ is formed by products $U \times V$ where U is open in X and V is open in Y [so $\lim(x_i, y_i) = (x, y)$ if and only if $\lim x_i = x$ and $\lim y_i = y$]. The product of three or more topological spaces is defined in a similar way.

We have already encountered product spaces in Sect. 1.9: The subgroup of the group $O(n)$ denoted here as $O(k) \times O(n-k)$ as a topological space is the product of $O(k)$ and $O(n-k)$; the same is true for other products mentioned in Sect. 1.9.

Notice also that the torus is homeomorphic to the product $S^1 \times S^1$ (see its description at the end of Sect. 1.10). For this reason, the product $S^1 \times \cdots \times S^1$ is also called a torus (or an n-dimensional torus). Let us mention a less obvious product presentation: The Grassmann manifold $G_+(4, 2)$ is homeomorphic to $S^2 \times S^2$. Indeed, the Plücker coordinates $(\xi_{12}, \xi_{13}, \xi_{14}, \xi_{23}, \xi_{24}, \xi_{34})$ in $G_+(4, 2)$ (see Sect. 1.5) are defined up to a multiplication by a positive (because it is G_+) number, not all equal to 0, and satisfy the relation $\xi_{12}\xi_{34} - \xi_{13}\xi_{24} + \xi_{14}\xi_{23} = 0$. We can assume that the sum of the squares of these numbers is 1 (then we do not need to admit the multiplication by positive numbers). Make a coordinate change:

$$\xi_{12} = \frac{x_1 + x_4}{2}, \xi_{13} = \frac{x_2 + x_5}{2}, \xi_{14} = \frac{x_5 + x_6}{2},$$
$$\xi_{23} = \frac{x_5 - x_6}{2}, \xi_{24} = \frac{x_5 - x_2}{2}, \xi_{34} = \frac{x_1 - x_4}{2}.$$

Then our equation becomes

$$x_1^2 + x_2^2 + x_3^2 - x_4^2 - x_5^2 - x_6^2 = 0$$

and the condition "the sum of the squares is 1" becomes

$$x_1^2 + x_2^2 + x_3^2 + x_4^2 + x_5^2 + x_6^2 = 2.$$

Together, these equations show that

$$x_1^2 + x_2^2 + x_3^2 = 1 \text{ and } x_4^2 + x_5^2 + x_6^2 = 1,$$

which is the system of equations of $S^2 \times S^2 \subset \mathbb{R}^3 \times \mathbb{R}^3 = \mathbb{R}^6$.

EXERCISE 1. Show that the complex quadric, that is, the subspace of the complex projective space $\mathbb{C}P^3$ defined in the homogeneous coordinates $(z_0 : z_1 : z_2 : z_3)$ by the equation $z_0^2 + z_1^2 + z_2^2 + z_3^2 = 0$, is homeomorphic to $S^2 \times S^2$.

EXERCISE 2. Show that the group $SO(4)$ is homeomorphic to $S^3 \times SO(3)$, that is, to $S^3 \times \mathbb{R}P^3$.

Remark. One could expect that the group $SU(3)$ is homeomorphic to $S^5 \times S^3$; indeed, $SU(3)/SU(2) = S^5$. Thus, there is a mapping of $SU(3)$ onto S^5 such that the inverse image of every point of S^5 is homeomorphic to $SU(2)$, that is, to S^3, or, as people say, $SU(3)$ is fibered over S^5 with the fiber S^3. However, $SU(3)$ is not homeomorphic to $S^5 \times S^3$; we will not be able to explain this before Chap. 4.

Notice in conclusion that there are continuous *projections* of $X \times Y$ onto X and Y, and a continuous mapping of a third space, Z, into $X \times Y$ is the same as a pair of maps, $Z \to X$ and $Z \to Y$ (the composition of the map $Z \to X \times Y$ with the projections).

2.2 Cylinders, Cones, and Suspensions

For a topological space X and its subspace A, we denote as X/A the space obtained from X by collapsing A to a point; we call X/A the *quotient space* of X by A.

We always denote the segment $[0, 1]$ as I. The product $ZX = X \times I$ is called the *cylinder* over X; the subsets $X \times 0$ and $X \times 1$ of the cylinder (which are copies of X) are called its (*upper* and *lower*) *bases*. Smashing the upper base of the cylinder into one point gives the *cone* CX over X; thus, $CX = (X \times I)/(X \times 1)$ [sometimes, it is more convenient to define CX as $CX = (X \times I)/(X \times 0)$; see Exercise 11 ahead]. The base of the cylinder not affected by the factorization is called the base of the cone, and the point of CX obtained from the other base of the cylinder is called the *vertex* of the cone (Fig. 6).

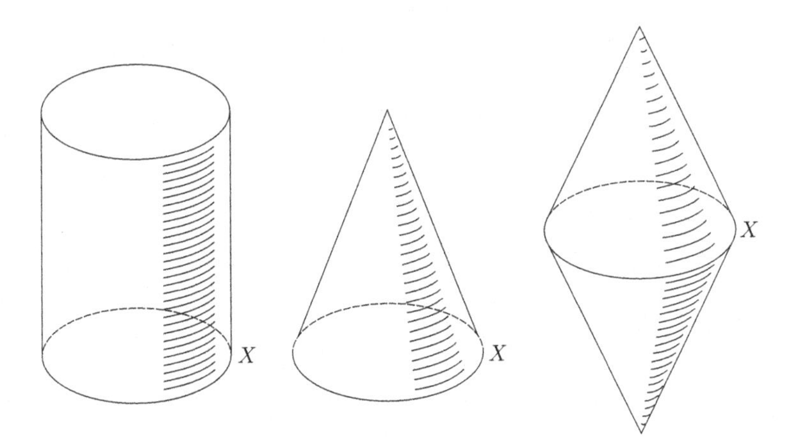

Fig. 6 Cylinder, cone, and suspension

If we further factorize the cone over its base, we get the *suspension* ΣX over X. The suspension has two vertices. The image of the "middle section" $\left(X \times \frac{1}{2}\right) \subset X \times I$ is called the base of the suspension. (This term can be justified by the fact that the suspension may be thought of as the union of two cones attached to each other by the bases; these bases become the base of the suspension.) The points of the cone and of the suspension will still be denoted as (x, t), $x \in X$, $0 \leq t \leq 1$ with the understanding that in the cone $(x', 1) = (x'', 1)$ for any $x', x'' \in X$, and in the suspension $(x', 1) = (x'', 1)$ and $(x', 0) = (x'', 0)$ for any $x', x'' \in X$. If we need to specify that a point (x, t) lies in the cone or in the suspension, we may write $(x, t)_C$ or $(x, t)_\Sigma$.

EXERCISE 3. Show that the cone and the suspension over S^n are, accordingly, D^{n+1} and S^{n+1}.

EXERCISE 4. Prove that no closed (that is, without holes) classical surface except S^2 is homeomorphic to a suspension over any other space.

2.3 Attachings: Cylinders and Cones of Maps

Let X, Y be topological spaces, let A be a subspace of Y, and let $\varphi: A \to X$ be a continuous map. Take the sum $X \coprod Y$ (that is, a space composed of X and Y as of two unrelated parts) and make an identification: We attach every point $a \in A \subset Y$ to $\varphi(a) \in X$. The resulting space is denoted as $X \cup_\varphi Y$, and the procedure for its construction described above is called attaching Y to X by means of the map φ.

We distinguish two special cases of this construction. Let $f: X \to Y$ be an arbitrary continuous map. The space obtained by attaching the cylinder $X \times I$ to Y by means of the map $X \times 0 = X \xrightarrow{f} Y$ is called the *cylinder of the map f* and is denoted as Cyl(f). The space obtained by attaching the cone CX to Y by means of the same map is called the *cone of the map f* and is denoted as Con(f). (See Fig. 7.) The cylinder of f contains both X and Y; the cone of f contains Y.

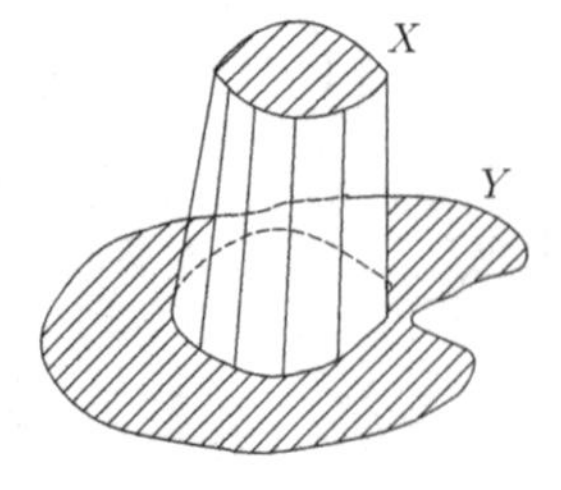

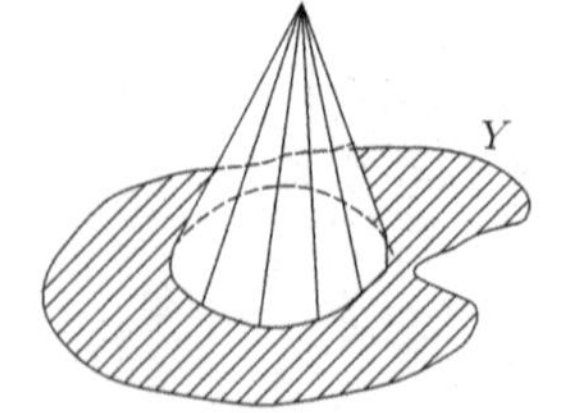

Fig. 7 Cylinder and cone of a map

EXERCISE 5. Show that the cone of a canonical map $S^n \to \mathbb{R}P^n$ (see Sect. 1.2) is homeomorphic to $\mathbb{R}P^{n+1}$. In particular, the cone of a double-rotation map of a circle onto itself [defined by the formula $(\cos\theta, \sin\theta) \mapsto (\cos 2\theta, \sin 2\theta)$, or, in complex coordinates, $z \mapsto z^2$] is homeomorphic to $\mathbb{R}P^2$.

EXERCISE 6. Formulate and prove the complex and quaternionic analogs of this statement.

2.4 Joins

The *join* $X * Y$ of topological spaces X and Y can be conveniently described as the union of segments joining every point of X with every point of Y.

EXERCISE 7. Show that the join of two (closed) segments containing two skew lines in $\mathbb{R}^3$ is a tetrahedron.

The formal definition of the join is as follows: We take the product $X \times Y \times I$ (we think of $x \times y \times I$ as a segment joining $x \in X$ with $y \in Y$) and then make a factorization: We glue together the points $(x, y', 0), (x, y'', 0)$ for every $x \in X$, $y', y'' \in Y$ and the points $(x', y, 1), (x'', y, 1)$ for every $x', x'' \in X, y \in Y$ (meaning that the segments joining x with y' and joining x with y'' have a common beginning, and the segments joining x' with y and joining x'' with y have a common endpoint). The "horizontal sections" $X \times Y \times t$ are copies of $X \times Y$ for $0 < t < 1$; the section $X \times Y \times 0$ is collapsed into X and the section $X \times Y \times 1$ is collapsed into Y. This "stack of cards" structure of the join is shown in Fig. 8.

EXERCISE 8. Show that the join of a space X and a one-point space (= the zero-dimensional ball D^0) is the same as the cone CX over X.

EXERCISE 9. Show that the join of a space X and a two-point space (= the zero-dimensional sphere S^0) is the same as the suspension ΣX over X.

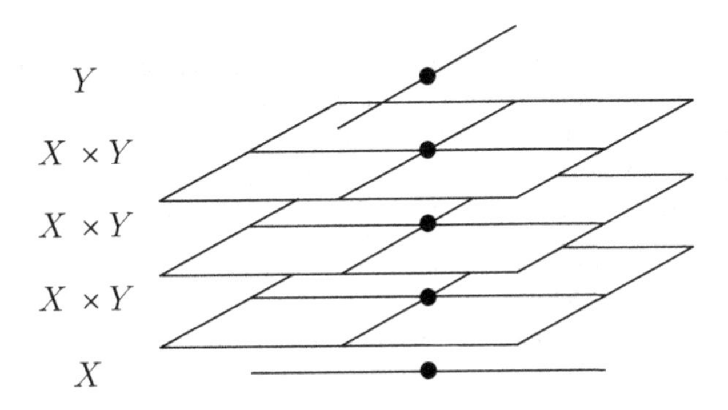

Fig. 8 The "horizontal sections" of a join

EXERCISE 10. Show that the join $S^m * S^n$ is homeomorphic to S^{m+n+1} (in view of Exercise 9, this is a generalization of Exercise 3).

Remark. The stack of cards structure of the join $S^m * S^n$ has an interesting relation to the geometry of the sphere S^{m+n+1}. Namely, for $0 \leq t \leq 1$, consider the subset

$$Q_t = \{(x_1, \ldots, x_{m+n+1}) \mid x_1^2 + \cdots + x_{m+1}^2 = t,\ x_{m+2}^2 + \ldots x_{m+n+2}^2 = 1 - t\}$$

of the sphere S^{m+n+1}. If $0 < t < 1$, then Q_t is the product of the spheres S^m and S^n (of radii $\sqrt{t}$ and $\sqrt{1-t}$). On the other hand, Q_0 is S^n and Q_1 is S^m. This construction is especially useful in the case when $m = n = 1$; it shows that the three-dimensional sphere S^3 is made out of a one-parameter family of tori and two circles. (We will return to it in Sect. 10.5.)

For sufficiently good spaces (say, Hausdorff and locally compact), the join operation is associative: The joins $(X * Y) * Z$ and $X * (Y * Z)$ are homeomorphic to the "triple join" $X * Y * Z$, which is defined as the union of triangles with vertices in X, Y, and Z[3]. (Formally, this triple join is defined as a result of an appropriate factorization in the product $X \times Y \times Z \times \Delta$ where Δ is a triangle.) For a better understanding of this matter, we can use another construction of the join (see the next exercise).

EXERCISE 11. For a space X, define a "height function" $h\colon CX \to [0, 1]$ by the formula $h(x, t)_C = 1 - t$ (so the height of the vertex is 0, and the height of the base is 1). Consider the alternative definition of the join: $X \widehat{*} Y = \{(\xi, \eta) \in CX \times CY \mid h(\xi) + h(\eta) = 1\}$. This new operation is obviously associative [an n-fold join $X_1 \widehat{*} \ldots \widehat{*} X_n$ is defined as $\{(\xi_1, \ldots, \xi_n) \in CX_1 \times \cdots \times CX_n \mid h(\xi_1) + \cdots + h(\xi_n) = 1\}$]. Prove that for good spaces (for example, for Hausdorff locally compact spaces) the operations $\widehat{*}$ and $*$ are the same. (This will imply the associativity of the usual join for good spaces.)

2.5 Mapping Spaces: Spaces of Paths and Loops

The set $C(X, Y)$ of all continuous maps of a space X into a space Y is furnished by *compact–open topology* (which can be thought of as the topology of uniform convergency on compact sets). The base of open sets of this topology consists of sets of the form $U(K, O)$, where K is a compact subset of X and O is an open subset of Y; the set $U(K, O)$ consists of continuous maps $f\colon X \to Y$ such that $f(K) \subset O$.

EXERCISE 12. If X is the one-point space, then $C(X, Y) = Y$; if X is a discrete space of n points, then $C(X, Y) = Y \times \cdots \times Y$ (n factors). [The last equality provides a reason for an alternate notation for the mapping space: $C(X, Y) = Y^X$.]

[3]For "bad spaces" this homeomorphism does not hold and the join operation is not associative.

Let X, Y, Z be topological spaces. The formula

$$\{f: C(X, Y)\} \mapsto \{(x, y) \mapsto [f(x)](y)\}$$

defines a map $C(X, C(Y, Z)) \to C(X \times Y, Z)$.

EXERCISE 13. Show that if the spaces X and Y are Hausdorff and locally compact, then this is a homeomorphism. [In this case we can write $(Z^Y)^X = Z^{X \times Y}$; this formula provides an additional justification for the notation Y^X and is called the *exponential law*.]

A *path* in the space X is defined as a continuous map $I \to X$. The points $s(0)$ and $s(1)$ are called the *beginning* and the *end* of the path $s: I \to X$. A path whose end coincides with the beginning is called a *loop*. The following subspaces of the space $E(X) = X^I$ of paths are considered: the space $E(X; x_0, x_1)$ of paths with the beginning $x_0 \in X$ and the end $x_1 \in X$; the space $E(X, x_0)$ of paths beginning at x_0 (with the end not fixed); the space $\Omega(X, x_0)$ of loops of X with the beginning (and end) x_0.

EXERCISE 14. Prove that the space $E(S^n; x_0, x_1)$ does not depend (up to a homeomorphism) on x_0 and x_1 (in particular, on these two points being the same or different). By what spaces can S^n be replaced in this exercise?

EXERCISE 15. Construct a natural (see the footnote on Exercise 18) homeomorphism between $C(X, E(Y; y_0, y_1))$ and the subspace of $C(\Sigma X, Y)$ consisting of maps taking the upper and lower vertices of ΣX, respectively, into y_0 and y_1.

By the way, a topological space, for which every two points can be joined by a path, is called *path connected*. This notion is slightly different from the notion of connectedness used in point set topology: A space is connected if it does not contain proper subsets which are both open and closed.

EXERCISE 16. Prove that every path connected space is connected, but the converse is false: Prove the first and find an example confirming the second (the fans of the function $\sin \frac{1}{x}$ will not experience any difficulty with such an example).

Still for the spaces which are mostly used in topology, like manifolds or CW complexes, the two notions of connectedness coincide. For this reason we sometimes will omit the prefix "path" and speak of connected spaces when we mean path connected spaces.

2.6 Operations over Base Point Spaces

Topologists often have to consider topological spaces with base points, that is, to assume that for every space a base point is selected and all maps considered take base points to base points; different choices of a base point in the same topological space yield different base point spaces. The transition to base point spaces leads to various modifications of operations considered above. Sometimes the modification consists only in a choice of a base point in the result of a construction. For example, the base point of the product $X \times Y$ of spaces X and Y with the base points x_0 and y_0 is chosen as (x_0, y_0). Sometimes, the modification affects the construction itself. For example, the cone over a space X with a base point x_0 is obtained from the usual cone CX by collapsing the segment $x_0 \times I$ to a point which is chosen for the base point of the modified cone; the latter may be denoted $C(X, x_0)$ (if it is not clear from the context that the construction involves a base point). Suspensions and joins are modified in a similar way (in the join the segment joining the base points is collapsed to a point), and the images of segments collapsed are taken for the base points; if necessary, the notations $\Sigma(X, x_0)$ and $(X, x_0) * (Y, y_0)$ are used. Cylinders and cones of maps (which are supposed to take base points into base points) are modified in a similar way.

EXERCISE 17. Show that the homeomorphisms $CS^n = D^{n+1}, \Sigma S^n = S^{n+1}, S^m * S^n = S^{m+n+1}$ hold after the modifications described above if we consider spheres and balls as spaces with base points [which is always assumed to be $(1, 0, \dots, 0)$].

The mapping space is reduced to the space of maps taking the base point into the base point; the base point of the mapping space is chosen as the constant map with the value at the base point. For a base point space $X = (X, x_0)$ the path space EX is defined as the space $E(X, x_0)$ of paths beginning at x_0, and the loop space ΩX is defined as the space $\Omega(X, x_0)$ of loops beginning (and ending) at x_0; the constant path and the constant loop become the base points of EX and ΩX.

EXERCISE 18. For base point spaces X, Y, construct a homeomorphism $\mathcal{C}(\Sigma X, Y) = \mathcal{C}(X, \Omega Y)$ that is natural with respect to X and Y.[4]

[4]The words "natural with respect to X and Y" may mean "defined for all X and Y in a unified way," but it is possible to attach to them a more formal sense. Namely, if X', Y' are other base point spaces, then for every (base point–preserving) map $\varphi: X' \to X$, $\psi: Y \to Y'$ there arises a commutative diagram

$$\begin{array}{ccc} \mathcal{C}(\Sigma X, Y) & \longrightarrow & \mathcal{C}(X, \Omega Y) \\ \downarrow & & \downarrow \\ \mathcal{C}(\Sigma X', Y') & \longrightarrow & \mathcal{C}(X', \Omega Y') \end{array}$$

where the horizontal arrows denote the homeomorphisms above, and vertical arrows denote maps induced by the given maps φ, ψ. [In detail: The left vertical arrow takes an $f: \Sigma X \to Y$ to the map $\Sigma X' \to Y'$ acting by the formula $(x', t)_\Sigma \mapsto \psi(f((\varphi(x'), t)_\Sigma))$; the right vertical arrow takes a $g: X \to \Omega Y$ into the map $X' \to \Omega Y'$ acting by the formula $x' \mapsto \{t \mapsto \psi(g(\varphi(x'))(t))\}$.] Actually, it is useful to keep in mind that all our constructions are "natural" in the sense that they can be applied not only to spaces, but also to maps which should act in an appropriate direction; in algebra, this phenomenon is described by the word *functor*.

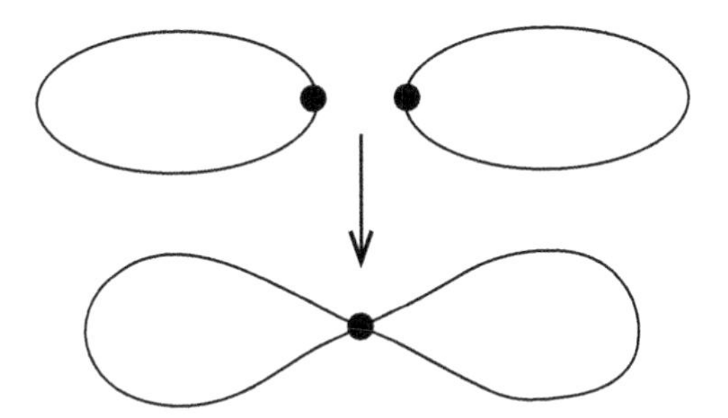

Fig. 9 The bouquet of two circles

In conclusion, we will describe two operations which exist only for base point spaces. The *bouquet* (or *wedge*) of two base point spaces, X and Y, is obtained from their disjoint union by merging their base points. For example, the bouquet of two circles is "the figure eight" (Fig. 9). The notation for the bouquet: $(X, x_0) \vee (Y, y_0)$ or $X \vee Y$.

Alternatively, one can define the bouquet $X \vee Y$ as the subspace of the product $X \times Y$ composed of points (x, y) for which $x = x_0$ or $y = y_0$. The quotient space (see Sect. 2.2) $X \# Y = (X \times Y)/(X \vee Y)$ is called the *smash product* or *tensor product*[5] of X and Y. The base points in $X \vee Y$ and $X \# Y$ are obvious.

EXERCISE 19. Show that $S^m \# S^n = S^{m+n}$.

EXERCISE 20. For a base point space X, construct a natural (with respect to X) homeomorphism $\Sigma X = X \# S^1$.

[5] In category theory, there exists a general notion of a tensor product; the definition of a smash product matches the definition of the tensor product for the category of base point topological spaces.

Chapter 1
Homotopy

Lecture 3 Homotopy and Homotopy Equivalence

3.1 The Definition of a Homotopy

Let X and Y be topological spaces. Continuous maps $f, g: X \to Y$ are called *homotopic* ($f \sim g$) if there exists a family of maps $h_t: X \to Y, t \in I$ such that (1) $h_0 = f, h_1 = g$; (2) the map $H: X \times I \to Y$, $H(x, t) = h_t(x)$, is continuous. [Condition (2) reflects the requirement that h_t depends "continuously" on t.] The map H (or, sometimes, the family h_t) is called a *homotopy* joining f and g.

It is obvious that the homotopy relation for maps is reflexive, symmetric, and transitive.

Example. All continuous maps of an arbitrary space X into the segment I are homotopic to each other: A homotopy $h_t: X \to I$ joining continuous maps $f, g: X \to I$ is defined by the formula $h_t(x) = (1 - t)f(x) + tg(x)$. Here I can be replaced by any convex subset of any space $\mathbb{R}^n$ or $\mathbb{R}^\infty$, in particular, by the whole spaces $\mathbb{R}^n$ or $\mathbb{R}^\infty$.

3.2 The Sets $\pi(X, Y)$

The equivalence classes for the homotopy relation in $C(X, Y)$ are called *homotopy classes*. The set of homotopy classes in $C(X, Y)$ is denoted as $\pi(X, Y)$.

Example 1. The set $\pi(X, I)$ consists (for every X) of one element.

Example 2. The set $\pi(*, Y)$ (where $*$ denotes a one-point space) is the set of path components (maximal path connected components) of Y.

A. Fomenko, D. Fuchs, *Homotopical Topology*, Graduate Texts in Mathematics 273,
DOI 10.1007/978-3-319-23488-5_1

Obviously, the set $\pi(X, Y)$ can be regarded as the set of path components of $C(X, Y)$.

Let X, X', Y, Y' be topological spaces, and let $\varphi: X' \to X$ and $\psi: Y \to Y'$ be continuous maps. Obviously, for continuous maps $f, g: X \to Y, f \sim g \Rightarrow f \circ \varphi \sim g \circ \varphi$ and $f \sim g \Rightarrow \psi \circ f \sim \psi \circ g$. Thus, the operations $\circ\varphi$ and $\psi\circ$ can be applied to homotopy classes of maps $X \to Y$, which gives the maps $\varphi^*: \pi(X, y) \to \pi(X', Y)$ and $\psi_*: \pi(X, Y) \to \pi(X, Y')$.

EXERCISE 1. Prove the relations $(\varphi_1 \circ \varphi_2)^* = \varphi_2^* \circ \varphi_1^*$, $(\psi_1 \circ \psi_2)_* = \psi_{1*} \circ \psi_{2*}$, and $\varphi^* \circ \psi_* = \psi_* \circ \varphi^*$ (we leave to the reader the work of determining the exact meaning of the notations in these equalities).

3.3 Homotopy Equivalence

We will give three definitions of this notion.

Definition 1. The spaces X, Y are called *homotopy equivalent* ($X \sim Y$) if there exist continuous maps $f: X \to Y$ and $g: Y \to X$ such that the compositions $g \circ f: X \to X$ and $f \circ g: Y \to Y$ are homotopic to the identity maps $\mathrm{id}_X: X \to X$ and $\mathrm{id}_Y: Y \to Y$.

In this situation, the maps f and g are called *homotopy equivalences homotopy inverse to each other.*

Remark. If the conditions $g \circ f \sim \mathrm{id}_X, f \circ g \sim \mathrm{id}_Y$ are replaced by conditions $g \circ f = \mathrm{id}_X, f \circ g = \mathrm{id}_Y$, then mutually homotopy inverse homotopy equivalences f, g become mutually inverse homeomorphisms. Having this in mind, we can say that homotopy equivalences are homotopy versions of homeomorphisms.

Definition 2. $X \sim Y$ if there exists a way to define for every space Z a bijective map $\alpha^Z: \pi(Y, Z) \to \pi(X, Z)$ such that for any continuous map $\psi: Z \to W$ the diagram

$$\begin{array}{ccc} \pi(X,Z) & \xleftarrow{\alpha^Z} & \pi(Y,Z) \\ \downarrow{\scriptstyle \psi_*} & & \downarrow{\scriptstyle \psi_*} \\ \pi(X,W) & \xleftarrow{\alpha^W} & \pi(Y,W) \end{array}$$

is commutative (that is, $\alpha^W \circ \psi_* = \psi_* \circ \alpha^Z$).

Definition 3. $X \sim Y$ if there exists a way to define for every space Z a bijective map $\beta_Z: \pi(Z, X) \to \pi(Z, Y)$ such that for any continuous map $\varphi: Z \to W$ the diagram

$$\begin{array}{ccc} \pi(Z,X) & \xrightarrow{\beta_Z} & \pi(Z,Y) \\ \uparrow \varphi^* & & \uparrow \varphi^* \\ \pi(W,X) & \xrightarrow{\beta_W} & \pi(W,Y) \end{array}$$

is commutative (that is, $\beta_Z \circ \varphi^* = \varphi^* \circ \beta_W$).

Theorem. *Definitions* 1, 2, *and* 3 *are equivalent.*

Proof. Let us prove the equivalence of Definitions 1 and 2. Assume that $X \sim Y$ in the sense of Definition 2. Then there is a bijection $\alpha^Y: \pi(X,Y) \to \pi(Y,Y)$, and we take for $f: X \to Y$ any representative of the homotopy class $(\alpha^Y)[\mathrm{id}_Y]$ (where the square brackets mean the transition from a map to its homotopy class). Also, there is a bijection $\alpha^X: \pi(Y,X) \to \pi(X,X)$, and we take for $g: Y \to X$ any representative of the homotopy class $(\alpha^X)^{-1}[\mathrm{id}_X]$. Consider the diagram in Definition 2 for ψ being $g: Y \to X$ and then for ψ being $f: X \to Y$:

$$\begin{array}{ccc} \pi(X,Y) & \xleftarrow{\alpha^Y} & \pi(Y,Y) \\ \downarrow g_* & & \downarrow g_* \\ \pi(X,X) & \xleftarrow{\alpha^X} & \pi(Y,X) \end{array}, \qquad \begin{array}{ccc} \pi(X,X) & \xleftarrow{\alpha^X} & \pi(Y,X) \\ \downarrow f_* & & \downarrow f_* \\ \pi(X,Y) & \xleftarrow{\alpha^Y} & \pi(Y,Y) \end{array}.$$

From the first diagram, $g_* \circ \alpha^Y = \alpha^X \circ g_*$. Apply this to $[\mathrm{id}_Y]$:

$$\begin{aligned} g_* \circ \alpha^Y[\mathrm{id}_Y] &= g_*[f] = [g \circ f], \\ \alpha^X \circ g_*[\mathrm{id}_Y] &= \alpha^X[g] = [\mathrm{id}_X]. \end{aligned}$$

Thus, $[g \circ f] = [\mathrm{id}_X]$; that is, $g \circ f \sim \mathrm{id}_X$. From the second diagram, $f_* \circ \alpha^X = \alpha^Y \circ f_*$, or $(\alpha^Y)^{-1} \circ f_* = f_* \circ (\alpha^X)^{-1}$. Apply the last equality to $[\mathrm{id}_X]$:

$$\begin{aligned} (\alpha^Y)^{-1} \circ f_*[\mathrm{id}_X] &= (\alpha^Y)^{-1}[f] = [\mathrm{id}_Y], \\ f_* \circ (\alpha^X)^{-1}[\mathrm{id}_X] &= f_*[g] = [f \circ g]. \end{aligned}$$

Thus, $[f \circ g] = [\mathrm{id}_Y]$; that is, $f \circ g \sim \mathrm{id}_Y$. We see that $X \sim Y$ in the sense of Definition 1.

Now let us assume that $X \sim Y$ in the sense of Definition 1. Then there exist continuous maps $f: X \to Y$, $g: Y \to X$ such that $g \circ f \sim \mathrm{id}_X$, $f \circ g \sim \mathrm{id}_Y$. For an arbitrary Z, let $\alpha^Z = f^*: \pi(Y,Z) \to \pi(X,Z)$. This is a bijection: The inverse map is g^*. Indeed, $g^* \circ f^* = (f \circ g)^* = (\mathrm{id}_Y)^* = \mathrm{id}_{\pi(Y,Z)}$ and $f^* \circ g^* = (g \circ f)^* = (\mathrm{id}_X)^* = \mathrm{id}_{\pi(X,Z)}$. Also, for any $\psi: Z \to W$ the diagram

$$\begin{array}{ccc} \pi(X,Z) & \xleftarrow{f^*} & \pi(Y,Z) \\ \downarrow \psi_* & & \downarrow \psi_* \\ \pi(X,W) & \xleftarrow{f^*} & \pi(Y,W) \end{array}$$

is commutative. Indeed, for an $h: Y \to Z$, $\psi_* \circ f^*[h] = \psi_*[h \circ f] = [\psi \circ h \circ f]$ and $f^* \circ \psi_*[h] = f^*[\psi \circ h] = [\psi \circ h \circ f]$ (a reader who did not skip Exercise 1 may be familiar with this argumentation). Thus, $X \sim Y$ in the sense of Definition 2.

The equivalence of Definitions 1 and 3 is checked precisely in the same way, and we leave it to the reader.

It is obvious that the relation of homotopy equivalence is reflexive, symmetric, and transitive. A class of homotopy equivalent spaces is called a *homotopy type*.

EXERCISE 2. Prove that a space that is homotopy equivalent to a path connected space is path connected.

An example of nonhomeomorphic homotopy equivalent spaces: X is a circle and Y is an annulus. One can take for $f: X \to Y$ the inclusion of X into Y as the outer boundary circle and put $g = f^{-1} \circ h: Y \to X$, where h is the radial projection of the annulus onto the outer boundary circle (see Fig. 10). The homotopy relations $g \circ f \sim \mathrm{id}_X, f \circ g \sim \mathrm{id}_Y$ are obvious.

A space X is called *contractible* if the identity map $\mathrm{id}_X: X \to X$ is homotopic to a constant map taking the whole space X to one point.

EXERCISE 3. Prove that a space is contractible if and only if it is homotopy equivalent to a one-point space.

EXERCISE 4. Prove that the cone over any (nonempty) space is contractible.

EXERCISE 5. Prove that the space $E(X, x_0)$ is contractible for any space X and any point $x_0 \in X$.

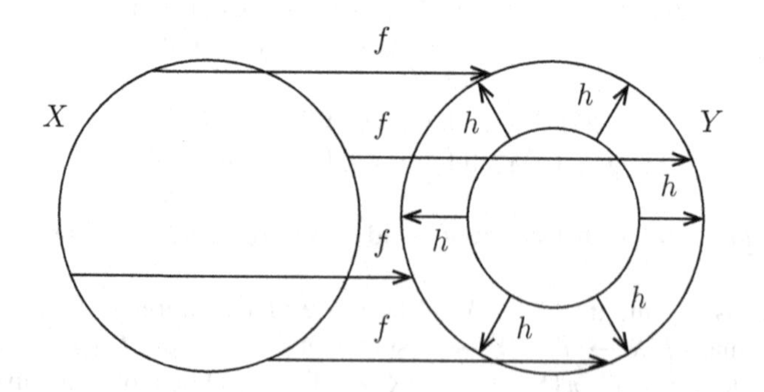

Fig. 10 A homotopy equivalence

EXERCISE 6. Prove that the cylinder of any continuous map $X \to Y$ is homotopy equivalent to Y.

EXERCISE 7. Prove that if $X \sim Y$, then $\Sigma X \sim \Sigma Y$.

EXERCISE 8. The previous statement is called the homotopy invariance of the operation of suspension. Prove that the operations of product, join, mapping spaces, path and loop spaces are homotopy invariant in a similar sense.

3.4 Retracts and Deformation Retracts

A subspace A of a space X is called a *retract* of X if there is a continuous map $r: X \to X$ ("retraction") such that $r(X) = A$ and $r(a) = a$ for every $a \in A$. For example, any point of a topological space is a retract of this space, but the union of the two endpoints of a segment is not a retract of this segment (the intermediate value theorem for continuous functions provides a reason for that). The boundary circle of a disk, and, more generally, $S^{n-1} \subset D^n$ are not retracts; but at the moment we do not have tools to prove that.

EXERCISE 9. Show that a retract of a path connected space is path connected.

EXERCISE 10. Prove that the bases of a cylinder are its retracts.

EXERCISE 11. Prove that the base of a cone CX is a retract of CX if and only if X is contractible.

If a retraction $r : X \to X$ of X onto A is homotopic to the identity $\mathrm{id}_X: X \to X$, then A is called a *deformation retract* of X. If a homotopy joining r with id_X may be made fixed on A [that is, $F_t(a) = a$ for all $t \in I, a \in A$], then A is called a *strong deformation retract* of X.

Obviously, a deformation retract of X is homotopy equivalent to X. Moreover, A is a deformation retract of X if and only if the inclusion map $A \to X$ is a homotopy equivalence (compare the example of a homotopy equivalence given above). Thus, the notion of a deformation retract is essentially not new for us. This cannot be stated regarding the notion of a strong deformation retract, but, as we will see later, the difference between deformation retracts and strong deformation retracts arises only in really pathological cases.

EXERCISE 12. A point is a deformation retract of a space X if and only if X is contractible.

EXERCISE 13. Show an example of a deformation retract which is not a strong deformation retract. (It is reasonable to regard this exercise as a sequel of the preceding exercise.)

In conclusion, we exhibit a pair of homotopy equivalent spaces of which neither is a deformation retract of the other one.

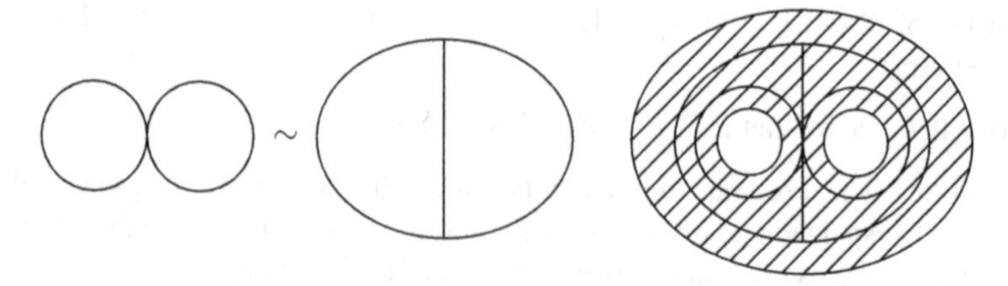

Fig. 11 Homotopy equivalence with no deformation retraction

The two spaces shown in Fig. 11 (a pair of mutually tangent circles and an ellipse with a diametrical segment) are homotopy equivalent since they both are deformation retracts of an elliptical domain with two circular holes; but neither of them is homeomorphic to a deformation retract of the other one.

3.5 An Example of a Homotopy Invariant: The Lusternik–Schnirelmann Category

We say that a subspace A of a topological space X is contractible *in* X if the inclusion map $A \to X$ is homotopic to a constant map $A \to X$. It is clear that if A is contractible (in our usual sense; see Sect. 3.3), then it is contractible in X, but the converse is not necessarily true. The minimal n (maybe, ∞) for which there exists a covering of X by n open subsets contractible in X is called the *(Lusternik–Schnirelmann) category* of X and is denoted as $\operatorname{cat} X$. If we replace in this definition the condition that the open sets from the covering are contractible in X by the condition that they are contractible, we will get a definition of a *strong category* of X, which is denoted as $\operatorname{cat}^s X$.

Theorem. *The category is homotopy invariant: If* $X \sim Y$, *then* $\operatorname{cat} X = \operatorname{cat} Y$.

Proof. Let $f: X \to Y$ and $g: Y \to X$ be mutually inverse homotopy equivalences, and let $h_t: X \to X$ be a homotopy such that $h_0 = \mathrm{id}_X$ and $h_1 = g \circ f$. Let $\{U_1, \dots, U_{\operatorname{cat} Y}\}$ be a covering of Y by open sets contractible in Y, and let $k_{i,t}: U_i \to Y$ be a homotopy with k_0 being the inclusion map of U_i into Y and k_1 being a constant map. Let $V_i = f^{-1}(U_i)$; the sets V_i form an open covering of X. Consider two homotopies $V_i \to X$: The first consists of maps $x \mapsto h_t(x)$, and the second consists of maps $x \mapsto g(k_{i,t}(f(x)))$ [this makes sense, since $f(x) \in U_i$]. The first homotopy joins the inclusion map $V_i \to X$ with the restriction map $(g \circ f)|_{V_i}$, and the second homotopy joins this restriction map with a constant map. Together they show that V_i is contractible in X. We see that $\operatorname{cat} X \leq \operatorname{cat} Y$ and a similar argumentation shows that $\operatorname{cat} Y \leq \operatorname{cat} X$; thus, $\operatorname{cat} X = \operatorname{cat} Y$.

EXERCISE 14. Prove that for any nonempty space X, $\operatorname{cat} \Sigma X \leq 2$. (Obviously, $\operatorname{cat} X = 1$ if and only of X is contractible.)

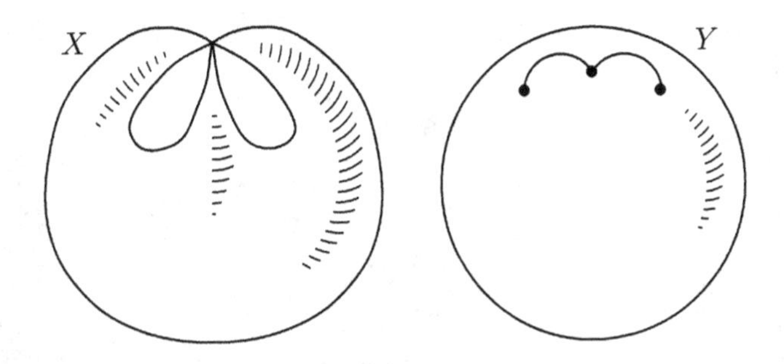

Fig. 12 An example for studying relations between categories and strong categories

Later on, we will be able to compute the category for a broad class of spaces.

Now, let us discuss the relations between the category and the strong category. It is obvious that the strong category is never less than the category.

Consider two spaces shown in Fig. 12.

The space X is obtained from the sphere S^2 by gluing together three points. The space Y is obtained from the sphere not by gluing together the three points, but rather by joining them by arcs attached to the sphere from the outer side. It is very easy to see that $X \sim Y$.

EXERCISE 15. Prove that $\operatorname{cat} X = \operatorname{cat} Y = \operatorname{cat}^s Y = 2$, but $\operatorname{cat}^s X = 3$.

This computation shows that the strong category does not need to be the same as the category, and also that the strong category is not homotopy invariant.

3.6 The Case of Base Point Spaces, Pairs, Triples, etc.

The definitions of a homotopy and a homotopy equivalence are modified in an obvious way for base point spaces. The set of (base point) homotopy classes of maps between base point spaces X and Y is also denoted as $\pi(X, Y)$, but, if necessary, the specific notation $\pi_b(X, Y)$ is used.

EXERCISE 16. Prove the base point homotopy invariance of the operations $\vee, \#, \Omega$ and also the base point versions of suspensions and joins.

A further generalization of the base point homotopy theory is a homotopy theory of pairs. A *pair* (X, A) is simply a topological space X with a distinguished subspace A. A map of a pair (X, A) into a pair (Y, B) is simply a continuous map $X \to Y$ taking A into B. Homotopies, homotopy equivalences, and so on are defined for pairs in the obvious way. Similar theories exist for triples (X, A, B) (where it is assumed that $X \supset A \supset B$), triads $(X; A, B)$ (where it is assumed that $X \supset A$, $X \supset B$), and so on.

Lecture 4 Natural Group Structures in the Sets $\pi(X, Y)$

Homotopy topology studies invariants of topological spaces and continuous maps which are discrete by their nature. Usually, these invariants have equal values on homotopy equivalent spaces and homotopy maps. The most usual procedure for constructing such invariants consists in a fixation of some space Y and then assigning to a topological space X the set $\pi(X, Y)$ or $\pi(Y, X)$ and to a continuous map $f: X \to X'$ or $f: X' \to X$ the map f^* or f_*. (Certainly, there are invariants of a completely different nature, like the Lusternik–Schnirelmann category—see Sect. 3.5.)

It is much easier to deal with such invariants if they possess some natural algebraic structure, most commonly a natural structure of a group. Before describing and studying these structures, we want to make a remark regarding the form of further exposition. We consider the invariants of two different kinds: $X \mapsto \pi(X, Y)$ and $X \mapsto \pi(Y, X)$ (for a fixed Y). Each of these kinds gives rise to a theory, and, for a long time, the two theories remain parallel or, better to say, dual. This duality is important for homotopy topology; it is called the *Eckmann–Hilton duality*. We will not explicitly describe it in this book, but, just to make it more visible, we will arrange the majority of this section in a two-column format, so that the dual statements will be written next to each other.

In this section, we assume that all spaces have base points and accordingly understand all maps, homotopies, homotopy equivalences, etc. We fix, once and forever, a space Y with a base point y_0.

Suppose that for every space X a group structure is introduced in the set $\pi_b(X, Y)$. This structure is called *natural* if for every continuous map $\varphi: X \to X'$, the map

$$\varphi^*: \pi_b(X', Y) \to \pi_b(X, Y)$$

is a homomorphism.

Suppose that for every space X a group structure is introduced in the set $\pi_b(Y, X)$. This structure is called *natural* if for every continuous map $\varphi: X \to X'$, the map

$$\varphi^*: \pi_b(Y, X) \to \pi_b(Y, X')$$

is a homomorphism.

Definition. The space Y is called an H-space if there are maps

$$\mu: Y \times Y \to Y$$

(multiplication) and

$$\nu: Y \to Y$$

(inversion) such that

Definition. The space Y is called an H'-space if there are maps

$$\mu: Y \times Y \to Y$$

(comultiplication) and

$$\nu: Y \to Y$$

(co-inversion) such that

(1) (homotopy unit). The compositions

$$Y \xrightarrow{j_1} Y \times Y \xrightarrow{\mu} Y$$
$$Y \xrightarrow{j_1} Y \times Y \xrightarrow{\mu} Y$$

where

$j_1(y) = (y, y_0)$, $j_2(y) = (y_0, y)$, are homotopic to $\mathrm{id}_Y: Y \to Y$.

(2) (homotopy associativity). The compositions
$Y \times (Y \times Y) \xrightarrow{\mathrm{id} \times \mu} Y \times Y \xrightarrow{\mu} Y$,
$(Y \times Y) \times Y \xrightarrow{\mu \times \mathrm{id}} Y \times Y \xrightarrow{\mu} Y$
are homotopic.

(3) (the property of the homotopy inversion). The maps

$$Y \longrightarrow Y \times Y \xrightarrow{\mu} Y$$
$$Y \longrightarrow Y \times Y \xrightarrow{\mu} Y$$

where the two left arrows mean, respectively, the maps $y \mapsto (y, \nu(y))$ and $y \mapsto (\nu(y), y)$, are homotopic to the constant map.

An important example of an H-space: the loop space ΩZ of an arbitrary space Z. The map

$$\mu: \Omega Z \times \Omega Z \to \Omega Z$$

(1) (homotopy co-unit). The compositions

$$Y \xrightarrow{\mu} Y \vee Y \xrightarrow{p_1} Y$$
$$Y \xrightarrow{\mu} Y \vee Y \xrightarrow{p_2} Y$$

where p_1 is the identity on the first Y and maps the second Y into y_0, and p_2 is the identity on the second Y and maps the first Y into y_0, are homotopic to $\mathrm{id}_Y: Y \to Y$.

(2) (homotopy co-associativity). The compositions
$Y \xrightarrow{\mu} Y \vee Y \xrightarrow{\mathrm{id} \vee \mu} Y \vee (Y \vee Y)$,
$Y \xrightarrow{\mu} Y \vee Y \xrightarrow{\mu \vee \mathrm{id}} (Y \vee Y) \vee Y$
are homotopic.

(3) (the property of the homotopy co-inversion). The maps

$$Y \xrightarrow{\mu} Y \vee Y \longrightarrow Y$$
$$Y \xrightarrow{\mu} Y \vee Y \longrightarrow Y$$

where the two right arrows mean, respectively, the map which is id on the first Y and ν on the second Y, and the same with id and ν swapped, are homotopic to the constant map.

An important example of an H'-space: the suspension ΣZ over an arbitrary space Z. The map

$$\mu: \Sigma Z \to \Sigma Z \vee \Sigma Z$$

is defined by the formula

$$[\mu(f, g)](t) = \begin{cases} f(2t), & \text{if } t \leq 1/2, \\ g(2t-1), & \text{if } t \geq 1/2, \end{cases}$$

that is, μ assigns to two loops a loop obtained by a successive passing these two loops:

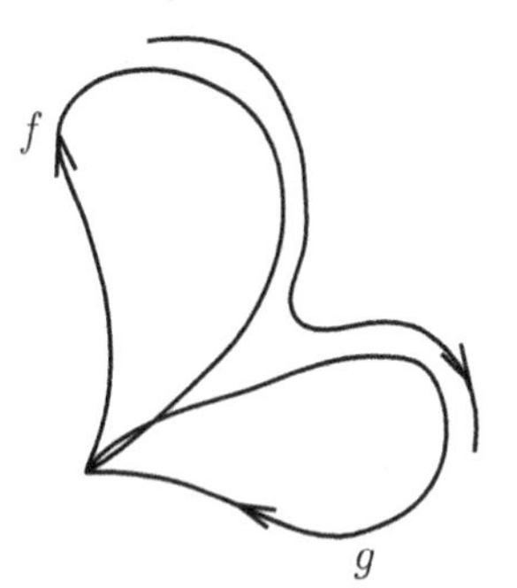

and the map $\nu: \Omega Z \to \Omega Z$ is defined by the formula

$$[\nu(f)](t) = f(1-t);$$

that is, ν assigns to a loop the same loop passed in the opposite direction.

Another important example of an H-space is a topological group.

(The dual statement is formulated in terms of *category*; the reader can try to conceive of an appropriate definition of this.)

Theorem. *The set $\pi_b(X, Y)$ possesses a natural (with respect to X) group structure if and only if Y is an H-space.*

is defined by the formula

$$\mu(z, t)_\Sigma = \begin{cases} (z, 2t)^I_\Sigma, & \text{if } t \leq 1/2 \\ (z, 2t-1)^{II}_\Sigma, & \text{if } t \geq 1/2, \end{cases}$$

where the Roman numerals show in which of the two ΣZs composing $\Sigma Z \vee \Sigma Z$ the point is taken:

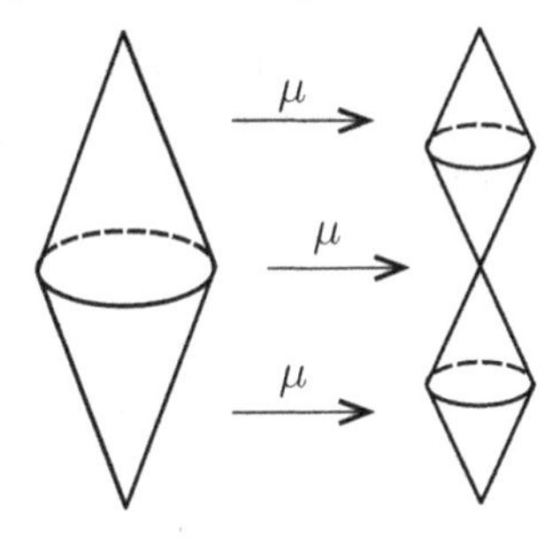

and the map $\nu: \Sigma Z \to \Sigma Z$ is defined by the formula

$$\nu(z, t)_\Sigma = (z, 1-t)_\Sigma;$$

that is, ν turns the suspension upside down.

(no dual notion)

EXERCISE 1. Prove that the Lusternik–Schnirelmann category of an arbitrary H'-space is ≤ 2.

Actually, the class of H'-spaces is very close to the class of spaces of category ≤ 2, but we will not go into the details of this statement.

Theorem. *The set $\pi_b(Y, X)$ possesses a natural (with respect to X) group structure if and only if Y is an H'-space.*

Proof of the Only If Statement.
Let the set $\pi_b(X, Y)$ have a natural with respect to X group structure. Take $X = Y \times Y$ and consider the classes $[p_1], [p_2] \in \pi_b(Y \times Y, Y)$ of the projections of $Y \times Y$ onto the factors.

Set $[\mu] = p_1 \cdot [p_2]$ (the multiplication is in the group $\pi_b(Y \times Y, Y)$) and choose an arbitrary map $\mu: Y \times Y \to Y$ of the class $[\mu]$. For $\nu: Y \to Y$ we will take an arbitrary map of a homotopy class $[\mathrm{id}_Y]^{-1}$ (the inversion is in the group $\pi_b(Y, Y)$).

Properties (1)–(3) of maps μ and ν are checked automatically. Let us prove, for example, that $\mu \circ j_1 \sim \mathrm{id}_Y$. The map $j_1: Y \to Y \times Y$ induces a map $j_1^*: \pi_b(Y \times Y, Y) \to \pi_b(Y, Y)$ which takes $[p_1]$ into $[p_1 \circ j_1]$ and $[p_2]$ into $[p_2 \circ j_1]$; but $p_1 \circ j_1 = \mathrm{id}_Y$ and $p_2 \circ j_1 = \mathrm{const}$.

Since the group structure in $\pi_b(X, Y)$ is natural, the map j_1^* takes products into products; hence, $[\mu \circ j_1] = j_1^*[\mu] = j_1^*([p_1] \cdot [p_2]) = j_1^*[p_1] j_1^*[p_2] = [p_1 \circ j_1] \cdot [p_2 \circ j_1] = [\mathrm{id}_Y] \cdot [\mathrm{const}] = [\mathrm{id}_Y]$. Thus, $\mu \circ j_1 \sim \mathrm{id}_Y$.

(We used the fact that the class of the constant map $\mathrm{const}: Y \mapsto y_0$ is the unit of the group $\pi_b(Y, Y)$. To prove that, we consider the one-point space pt and the (unique) map $Y \to \mathrm{pt}$. The homomorphism $\pi_b(\mathrm{pt}, Y) \to \pi_b(Y, Y)$ arising takes the unique element of the group $\pi_b(\mathrm{pt}, Y)$, on one hand, into the unit of the group $\pi_b(Y, Y)$, and, on the other hand, into the class of the constant map $Y \to Y$.)

Proof of the Only If Statement.
Let the set $\pi_b(Y, X)$ have a natural with respect to X group structure. Take $X = Y \vee Y$ and consider the classes $[\eta_1], [\eta_2] \in \pi_b(Y, Y \vee Y)$ of the natural embeddings of Y into $Y \vee Y$.

Set $[\mu] = \eta_1 \cdot [\eta_2]$ (the multiplication is in the group $\pi_b(Y, Y \vee Y)$) and choose an arbitrary map $\mu: Y \to Y \vee Y$ of the class $[\mu]$. For $\nu: Y \to Y$ we will take an arbitrary map of a homotopy class $[\mathrm{id}_Y]^{-1}$ (the inversion is in the group $\pi_b(Y, Y)$).

Properties (1)–(3) of maps μ and ν are checked automatically. Let us prove, for example, that $\pi_1 \circ \mu \sim \mathrm{id}_Y$. The map $\pi_1: Y \to Y \times Y$ induces a map $\pi_1*: \pi_b(Y, Y \vee Y) \to \pi_b(Y, Y)$, which takes $[\eta_1]$ into $[\pi_1 \circ \eta_1]$ and $[\eta_2]$ into $[\pi_1 \circ \eta_2]$; but $\pi_1 \circ \eta_1 = \mathrm{id}_Y$ and $\pi_1 \circ \eta_2 = \mathrm{const}$.

Since the group structure in $\pi_b(Y, X)$ is natural, the map π_{1*} takes products into products; hence, $[\pi_1 \circ \mu] = \pi_{1*}[\mu] = \pi_{1*}([\eta_1] \cdot [\eta_2]) = \pi_{1*}[\eta_1]\pi_{1*}[\eta_2] = [\pi_1 \circ \eta_1] \cdot [\pi_1 \circ \eta_2] = [\mathrm{id}_Y] \cdot [\mathrm{const}] = [\mathrm{id}_Y]$. Thus, $\pi_1 \circ \mu \sim \mathrm{id}_Y$.

(We used the fact that the class of the constant map $\mathrm{const}: Y \mapsto y_0$ is the unit of the group $\pi_b(Y, Y)$. To prove that, we consider the one-point space pt and the (unique) map $\mathrm{pt} \to Y$. The homomorphism $\pi_b(Y, \mathrm{pt}) \to \pi_b(Y, Y)$ arising takes the unique element of the group $\pi_b(Y, \mathrm{pt})$, on one hand, into the unit of the group $\pi_b(Y, Y)$, and, on the other hand, into the class of the constant map $Y \to Y$.)

Proof of the If Statement. Let Y be an H-space and X be an arbitrary (base point) space. The map $\mu: Y \times Y \to Y$ induces a map $\mu_*: \pi_b(X, Y \times Y) \to \pi_b(X, Y)$ which can be regarded, in view of the obvious equality $\pi_b(X, Y \times Y) = \pi_b(X, Y) \times \pi_b(X, Y)$, as a map

$$\mu_*: \pi_b(X, Y) \times \pi_b(X, Y) \to \pi_b(X, Y).$$

Furthermore, the map $\nu: Y \to Y$ gives rise to a map $\nu_*: \pi_b(X, Y) \to \pi_b(X, Y)$. Together, the multiplication μ_* and inversion ν_* determine in $\pi_b(X, Y)$ a natural with respect to X group structure. The verification of the details is left to the reader.

EXERCISE 2. Show that the group

$$\pi_b(X, \Omega\Omega Z)$$

is commutative.

Since for $n \geq 1$, $S^n = \Sigma S^{n-1}$, the set $\pi_b(S^n, X)$ is a group for any X (natural with respect to X). This group is called the nth homotopy group of X and is denoted as $\pi_n(X)$. Exercise 2 shows that this group is commutative for $n \geq 2$.

We will see ahead (in Lecture 15) that

$$H^i(S^n) = \begin{cases} 0 \text{ for } i \neq n, \\ \mathbb{Z} \text{ for } i = n. \end{cases}$$

Proof of the If Statement. Let Y be an H'-space and let X be an arbitrary (base point) space. The map $\mu: Y \to Y \vee Y$ induces a map $\mu^*: \pi_b(Y \vee Y, X) \to \pi_b(Y, X)$ which can be regarded, in view of the obvious equality $\pi_b(Y \vee Y, X) = \pi_b(Y, X) \times \pi_b(Y, X)$, as a map

$$\mu^*: \pi_b(Y, X) \times \pi_b(Y, X) \to \pi_b(Y, X).$$

Furthermore, the map $\nu: Y \to Y$ gives rise to a map $\nu^*: \pi_b(Y, X) \to \pi_b(Y, X)$. Together, the multiplication μ^* and inversion ν^* determine in $\pi_b(Y, X)$ a natural with respect to X group structure. The verification of the details is left to the reader.

EXERCISE 3. Show that the group

$$\pi_b(\Sigma\Sigma Z, X)$$

is commutative.

We will see ahead (in Lecture 11) that for any $n \geq 1$ there exists a (homotopically unique) space K_n (usually denoted as $K(\mathbb{Z}, n)$) such that

(1) $\pi_i(K_n) = \begin{cases} 0 \text{ for } i \neq n, \\ \mathbb{Z} \text{ for } i = n, \end{cases}$

(2) $K_n \sim \Omega K_{n+1}$.

Since $K_n \sim \Omega K_{n+1}$, the set $\pi_b(X, K_n)$ is a group for any X (natural with respect to X). This group is called the nth (integral) cohomology group of X and is denoted as $H^n(X)$ [or $H^n(X; \mathbb{Z})$]. Exercise 3 shows that this group is always commutative.

Decades ago, the computation of the homotopy groups of spheres seemed to the topologists a very important problem. This problem has not been solved yet (see some details in Chaps. 3–6).

The computation of the cohomology groups of the spaces K_n turned out to be a very important problem. This problem was solved in the 1950s, mostly in the works of H. Cartan, A. Borel, and J.-P. Serre (see the details in Chap. 3).

Lecture 5 CW Complexes

Homotopy topology almost never considers absolutely arbitrary spaces. Usually, the spaces studied are equipped with some additional structure, and, since the times of the founder of algebraic topology, Henri Poincaré, two kinds of structures have been considered. The structure of the first kind have their origin in analysis: differential, Riemannian, complex, symplectic, etc. We will deal with structures of this kind (see Lectures 17, 19, 30, 41–43), but not too often. Usually the structures of this kind are natural: The spaces considered have such a structure from the very beginning, and we do not need to construct it. The structures of the other, more important for our type, are combinatorial structures. This structure consists of representing a space as a union of more or less standard pieces, and then studying spaces is reduced to studying the mutual arrangement of these pieces.

In this lecture we consider the most important combinatorial structure: the so-called CW structure. Although we will prove in this lecture some properties of CW complexes (this is how spaces with these structure are called) which will justify the usefulness of the notion, its real role will show itself later, in the chapter entitled "Homology," where the CW structures will become a powerful computational mean. Still we cannot postpone the preliminary study of CW complexes until the homology chapter.

5.1 Basic Definitions

A *CW complex* is a Hausdorff space X with a fixed partition $X = \bigcup_{q=0}^{\infty} \bigcup_{i \in I_q} e_i^q$ of X into pairwise disjoint set (*cells*) e_i^q such that for every cell e_i^q there exists a continuous map $f_i^q \colon D^q \to X$ (a characteristic map of the cell e_i^q) whose restriction to $\operatorname{Int} D^q$ is a homeomorphism $\operatorname{Int} D^q \approx e_i^q$ whose restriction to $S^{q-1} = D^q - \operatorname{Int} D^q$ maps S^{q-1} into the union of cells of dimensions $< q$ (the *dimension* of the cell e_i^q, $\dim e_i^q$ is, by definition, q). The following two axioms are assumed satisfied.

(C) The boundary $\dot{e}_i^q = \overline{e}_i^q - e_i^q = f_i^q(S^{q-1})$ is contained in a finite union of cells.

(W) A set $F \subset X$ is closed if and only if for any cell e_i^q the intersection $F \cap \overline{e}_i^q$ is closed (in other words, $(f_i^q)^{-1}(F)$ is closed in D^q).

Remarks. (1) We assume characteristic maps *existing* but not *fixed.* If we need to consider a CW complex with characteristic maps selected, that is, we need to have them as a part of the structure, we will explicitly specify this. (2) The term "CW complex" is not universally used. People also say *cell spaces*, or a *CW decomposition.* (3) The notations (C) and (W) of the axioms are standard. They abbreviate the expressions "closure finite" and "weak topology".

EXERCISE 1. Prove that the topology described in axiom (W) is the weakest of all topologies with respect to which all characteristic maps are continuous.

A CW subcomplex of a CW complex X is a closed subset composed of whole cells. It is obvious that a CW subcomplex of a CW complex is a CW complex. The most important CW subcomplexes of a CW complex X are *skeletons*: The nth skeleton X^n or $\text{sk}_n X$ of X is the union of all cells e_i^q with $q \leq n$. By the way, sometimes people say "n-dimensional skeleton," but this is not right: The *dimension* of a CW complex is the supremum of dimensions of all its cells, and the dimension of the nth skeleton may be less than n. Another example of a CW subcomplex: the union of the nth skeleton and any set of $(n+1)$-dimensional cells.

Later on we will refer to pairs (X, A) in which X is a CW complex and A is a CW subcomplex of X as *CW pairs.*

A CW complex is called *finite* or *countable* if the set of cells is finite or countable. By the way, for finite CW complexes the axioms (C) and (W) are not needed: They are satisfied automatically.

EXERCISE 2. Prove that every point of a CW complex belongs to some finite CW subcomplex.

A CW complex is called *locally finite* if every point has a neighborhood which is contained in some finite CW subcomplex.

EXERCISE 3. Prove that every compact subset of a CW complex is contained in some finite CW subcomplex.

EXERCISE 4. Prove that a CW complex is finite (locally finite) if and only if it is compact (locally compact).

EXERCISE 5. Prove that a map of a CW complex into any topological space is continuous if and only if its restriction to every finite CW subcomplex is continuous.

EXERCISE 6. The same with the words "finite CW subcomplex" replaced by the word "skeleton."

A continuous map f of a CW complex X into a CW complex Y is called *cellular* if $f(\text{sk}_n X) \subset \text{sk}_n Y$ for every n. Notice that this definition, which is, as the reader will soon see, the most appropriate, gives to cellular maps a lot of freedom: A cell does not need to be mapped into a cell, but can be spread along several cells of the same or smaller (but not bigger!) dimensions.

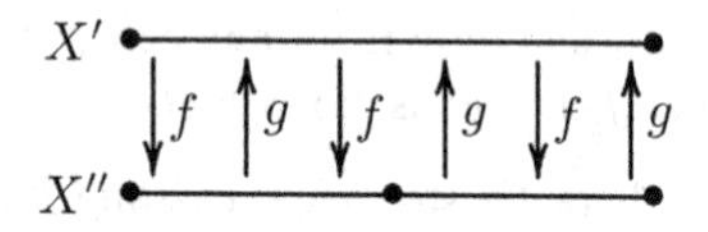

Fig. 13 For Exercise 7

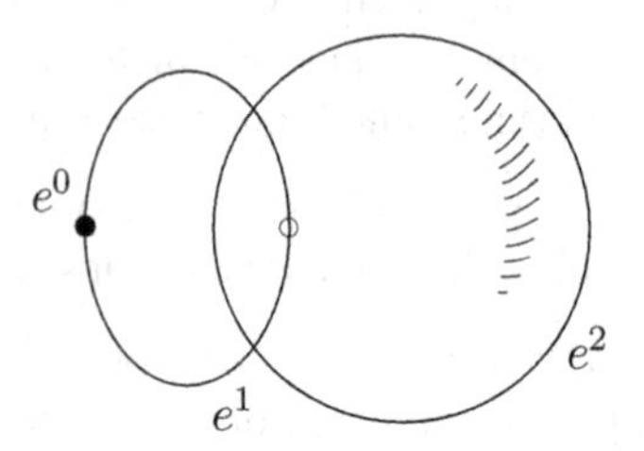

Fig. 14 The closure of a cell is not a CW subcomplex

EXERCISE 7. Let X' and X'' be the segment I decomposed into cells as shown in Fig. 13. Are the identity maps $f: X' \to X''$ and $g: X'' \to X'$ cellular? (Answer: yes for f, no for g.)

5.2 Comments to the Definition of a CW Complex

Remark 1. The closure of a cell does not need to be a CW subcomplex. Here is the example (Fig. 14). Let $X = S^1 \vee S^2$. We decompose it into three cells: e^0, e^1, e^2. For e^0 we take a point of S^1 which is not the base point. Then we put $e^1 = S^1 - e^0$, $e^2 = X - S^1$. Obviously, this is a CW decomposition, but $\overline{e}^2$ does not consist of whole cells; thus, it is not a CW subcomplex.

Remark 2. (W) does not imply (C). The decomposition of D^2 into $\operatorname{Int} D^2$ and all separate points of S^1 satisfies (W) (since $F \cap \overline{\operatorname{Int} D^2} = F$ for every F), but does not satisfy (C).

Remark 3. (C) does not imply (W). Take the infinite family $\{I_k \mid k = 1, 2, \dots\}$ of copies of the segment I and glue all the zero ends into one point. Topologize this set by the metric: The distance between $x \in I_k, y \in I_\ell$ is $x + y$ if $k \neq \ell$ and is $|y - x|$ if $k = \ell$. Consider the decomposition of the resulting space X into cells where every I_k is a union of three cells: 0, 1, and $\operatorname{Int} I$. The set $\{\frac{1}{k} \in I_k \mid k = 1, 2, \dots\} \subset X$ has a one-point, hence closed, intersection with every I_k but is not closed since it does not contain its limit point 0.

By the way, if a decomposition of a space into cells satisfies all the conditions listed in the beginning of Sect. 5.1 with the exception of Axiom (W) (as in the last example), we always can change (weaken) the topology, introducing it by Axiom (W). We will have to use this trick, called the "cellular weakening of topology," as soon as in Sect. 5.3.

EXERCISE 8. Prove that a CW complex is metrizable if and only if it is locally finite.

5.3 CW Structures and Constructions from Lecture 2

All the operations over topological spaces considered in Lecture 2, including the specific operations over base point spaces, and excluding the operation involving mapping spaces (like Ω and E), are defined in the CW setting. To begin with, the cylinder, cone, and suspension over CW complexes are, in a natural sense, CW complexes (for example, the cells of a suspension ΣX over a CW complex X are the two vertices and suspensions over cells of X with vertices removed). The cylinder and cone of a cellular map are also CW complexes (this appears to be our first justification of the definition of a cellular map); the same is true for the spaces of the form $X \cup_{\varphi} Y$ if φ is a cellular map of a CW subcomplex of Y into X, and, certainly, for the quotient space X/A of a CW complex X over a CW subcomplex A. But we encounter an unexpected obstacle when we try to introduce a CW structure into a product and, the more so, smash product or join of two CW complexes. Say, cells of the product of two CW complexes, $X \times Y$, are defined in the most natural way, as products of cells of X and Y, but there arises trouble with Axiom (W): It does not hold, in general. When topologists discovered this circumstance, they rushed to investigate it, and they proved a variety of theorems. We will refrain from discussing this matter, restricting ourselves to three exercises (see below) and the following remark. If the natural decomposition of $X \times Y$ into cells does not satisfy Axiom (W), we can apply the cellular weakening of topology [that is, redefine topology by Axiom (W)] and get a CW complex. We will define the latter as $X \times_w Y$. Luckily, it turns out that the replacement of space $X \times Y$ by $X \times_w Y$ does not spoil anything essential: The most important properties of the product remain true for this new operation. This allows us to forget the difference between $\times$ and $\times_w$, which we will do. The same can be said regarding joins and smash products.

EXERCISE 9. Show an example when $X \times_w Y \neq X \times Y$.

EXERCISE 10. Prove that if one of the CW complexes X, Y is locally finite, then $X \times_w Y = X \times Y$.

EXERCISE 11. Prove that if both CW complexes X, Y are locally countable, then $X \times_w Y = X \times Y$.

As to the mapping spaces, they are too big to have any hope of being decomposed into cells. Still, there is the following theorem proven by Milnor.

Theorem (Milnor [56]). *If X and Y are CW complexes, then the space Y^X is homotopy equivalent to a CW complex.*

(We will see ahead that to be homotopy equivalent to a CW complex is not bad at all. Anyway, Milnor dedicated the work cited above to a propaganda of this property.)

To finish our discussion of relations of CW complexes to constructions from Lecture 2, we will notice that every CW complex can be obtained by applying sufficiently many (sometimes, infinitely many) such constructions to the simplest spaces: to balls. Indeed, let $\{e^n_\alpha\}$ be the set of all n-dimensional cells of a CW complex X, and let $f^n_\alpha: D^n \to X$ be corresponding characteristic maps. Since $f^n_\alpha(S^{n-1}) \subset \mathrm{sk}_{n-1} X$, we can restrict f^n_α to a map $g^n_\alpha: S^{n-1} \to \mathrm{sk}_{n-1} X$ (the maps g^n_α are called *attaching maps*). Take the disjoint union $\mathcal{D} = \coprod_\alpha D^n_\alpha$ of n-dimensional balls, one for each n-dimensional cells of X, and put $\mathcal{S} = \coprod_\alpha S^{n-1}_\alpha \subset \mathcal{D}$. Then consider the map $g^n: \mathcal{S} \to X$, $g^n|_{S^{n-1}_\alpha} = g^n_\alpha$.

Obvious Lemma.

$$\mathrm{sk}_n X = (\mathrm{sk}_{n-1} X) \bigcup_{g^n} \mathcal{D}; \qquad (*)$$

that is, $\mathrm{sk}_n X$ *is obtained from* $\mathrm{sk}_{n-1} X$ *by attaching n-dimensional balls by means of attaching maps corresponding to all n=dimensional cells of X.*

The equality $(*)$ may be regarded as a step of a universal inductive procedure which allows us to construct an arbitrary CW complex from a discrete space ($\mathrm{sk}_0 X$ is discrete) or even an empty space ($\mathrm{sk}_{-1} X$ is empty) by successively attaching balls of growing dimensions. By the way, if the CW complex is infinite dimensional, then this inductive procedure includes a limit transition which is regulated by Axiom (W). Directly or indirectly, this inductive procedure creates a base for a proof of any statement about CW complexes: It allows us to reduce such a statement to the case of spheres or balls.

EXERCISE 12. Prove that a CW complex is path connected if and only if its first skeleton is path connected.

EXERCISE 13. Prove that a CW complex is path connected if and only if it is connected.

EXERCISE 14. Prove that a finite-dimensional CW complex can always be embedded into a Euclidean space of sufficiently large dimension.

5.4 CW Decompositions of Classical Spaces

A: Spheres and Balls

For a finite n, there are two canonical CW decompositions of the sphere S^n; they are shown for $n = 2$ in Fig. 15. The first consists of two cells: a point e^0 (for example, $(1, 0, \dots, 0)$) and the set $e^n = S^n - e^0$; a characteristic map $D^n \to S^n$ can be chosen like the usual making a sphere from a ball by gluing all points of the boundary sphere into one point:

$$(x_1, \dots, x_n) \mapsto \left(-\cos \pi\rho, x_1 \frac{\sin \pi\rho}{\rho} \dots, x_n \frac{\sin \pi\rho}{\rho}, \right)$$

where $\rho = \sqrt{x_1^2 + \dots + x_n^2}$ and $\dfrac{\sin \pi\rho}{\rho} = \pi$ for $\rho = 0$.

The other classical CW decomposition of S^n consists of $2n+2$ cells $e^0_\pm, \dots, e^n_\pm$, where $e^q_\pm = \{(x_1, \dots, x_{n+1}) \in S^n \mid x_{q+2} = \dots = x_{n+1} = 0, \pm x_{q+1} > 0\}$. Here we do not need to care about characteristic maps: Closures of all cells are obviously homeomorphic to balls (see Fig. 15).

Notice that both CW decompositions described above are obtained from the only possible cellular decomposition of S^0 (the two-point space) by the canonical cellular version of the suspension (see Sect. 5.3). In the first case, we use the base point version of suspension, and in the second case we take the usual suspension.

Certainly, there are a lot of other CW decompositions of the spheres. For example, S^n can be decomposed into $3^{n+1} - 1$ cells as the boundary of the $(n+1)$-dimensional cube, or into $2^{n+2} - 2$ cells as the boundary of the $(n+1)$-dimensional simplex (if you do not know what the simplex is, you will have to wait until Chap. 2).

All these CW decompositions, except the first one, work for S^∞.

A CW decomposition of the ball D^n may be obtained from any CW decomposition of the sphere S^{n-1} by adding one n-dimensional cell, namely $\operatorname{Int} D^n$. Thus, the smallest possible number of cells for D^n with $n \geq 1$ is 3. Notice, however, that no one of these CW decompositions will work for D^∞.

EXERCISE 15. Make up a CW decomposition for D^∞.

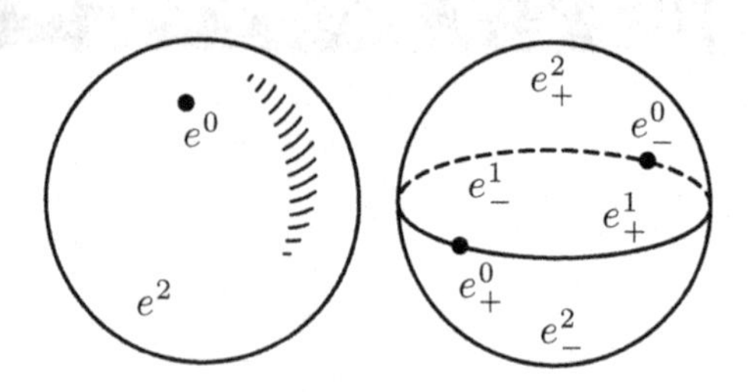

Fig. 15 Two CW decompositions of S^2

B: Projective Spaces

The identification of the antipodal points of the sphere S^n glues together the cells e^q_+, e^q_- of the above-described CW decomposition of S^n into $2n+2$ cells. This gives a decomposition of $\mathbb{R}P^n$ into $n+1$ cells e^q, one in every dimension from 0 to n. The other way of describing this CW decomposition of $\mathbb{R}P^n$ is provided by the formula

$$e^q = \{(x_0 : x_1 : \dots : x_n) \in \mathbb{R}P^n \mid x_q \neq 0,\ x_{q+1} = \dots = x_n = 0\}.$$

One more description is provided by the chain of inclusions

$$\emptyset = \mathbb{R}P^{-1} \subset \mathbb{R}P^0 \subset \mathbb{R}P^1 \subset \dots \subset \mathbb{R}P^n :$$

We set $e^q = \mathbb{R}P^q - \mathbb{R}^{q-1}$. A characteristic map for e^q may be chosen as the composition of the canonical projection $D^q \to \mathbb{R}P^q$ (see Sect. 1.2) and the inclusion $\mathbb{R}P^q \to \mathbb{R}P^n$. For $n = \infty$, this construction provides a CW decomposition of $\mathbb{R}P^\infty$ with one cell in every dimension.

The construction also has complex, quaternionic, and Cayley analogs. In the complex case, we get a CW decomposition of $\mathbb{C}P^n$ into $n+1$ cells $e^0, e^2, e^4, \dots, e^{2n}$ and also a CW decomposition of $\mathbb{C}P^\infty$ with one cell of every even dimension. In the quaternionic case, we get a CW decomposition of $\mathbb{H}P^n$ into $n+1$ cells $e^0, e^4, e^8, \dots, e^{4n}$ and also a CW decomposition of $\mathbb{H}P^\infty$ with one cell of every dimension divisible by 4. For the Cayley projective plane $\mathbb{C}\mathbf{a}P^2$, we get a CW decomposition into cells of dimensions 0, 8, and 16. For example, for $\mathbb{C}P^n$,

$$\begin{aligned} e^{2q} &= \{(z_0 : z_1 : \dots : z_n) \in \mathbb{C}P^n \mid z_q \neq 0, z_{q+1} = \dots = z_n = 0\} \\ &= \mathbb{C}P^q - \mathbb{C}P^{q-1} \end{aligned}$$

with characteristic maps $D^{2q} \to \mathbb{C}P^q \to \mathbb{C}P^n$, where the first arrow is the canonical projection (see Sect. 1.3) and the second arrow is the inclusion.

C: Grassmann Manifolds

The CW decomposition of the Grassmann manifold $G(n,k)$ described below is very important in topology (in particular, for the theory of characteristic classes; see Lecture 19 ahead) and also in algebra, algebraic geometry, and combinatorics. The cells of this decomposition are called *Schubert cells* (and the whole decomposition is called sometimes the Schubert decomposition).

Let $m_1, \dots, m_s$ be a finite (possibly, empty) nonincreasing sequence of positive integers less than or equal to k, where $s \leq n-k$. We denote as $e(m_1, \dots, m_s)$ the subset of $G(n,k)$ composed of all k-dimensional subspaces π of $\mathbb{R}^n$ such that, for $0 \leq j \leq n-k$,

$$\dim(\pi \cap \mathbb{R}^m) = m - j, \text{ if } k - m_j + j \leq m < k - m_{j+1} + (j+1),$$

where we put $m_0 = k$ and $m_j = 0$ for $s < j \leq n - k + 1$. It is clear that the sets $e(m_1, \ldots, m_s)$ are mutually disjoint and cover $G(n,k)$. For example, $G(4,2)$ is covered by six sets,

$$e(\emptyset), e(1), e(1,1), e(2), e(2,1), e(2,2),$$

which are composed of two-dimensional subspaces of $\mathbb{R}^4$ whose intersections with $\mathbb{R}^1, \mathbb{R}^2, \mathbb{R}^3$ have dimensions

$$(1,2,2), (1,1,2), (1,1,1), (0,1,2), (0,1,1), (0,0,1).$$

Differently, these six sets can be described the following way. Let

$$A = \{\pi = \mathbb{R}^2\}, B = \{\mathbb{R}^1 \subset \pi \subset \mathbb{R}^3\}, C = \{\mathbb{R}^1 \subset \pi\},$$
$$D = \{\pi \subset \mathbb{R}^3\}, E = \{\dim(\pi \cap \mathbb{R}^2) > 0\}.$$

Then

$$A \subset B \begin{matrix} \subset C \subset \\ \subset D \subset \end{matrix} E \subset G(4,2),$$

and

$$e(\emptyset) = A, e(1) = B - A, e(1,1) = C - B, e(2) = D - B,$$
$$e(2,1) = E - (C \cup D), e(2,2) = G(4,2) - E.$$

Let us provide a similar explanation in the general case.

Recall that the *Young diagram* of the sequence (partition) $m_1, \ldots, m_s$ is a drawing on a sheet of checked paper as shown in Fig. 16, left (the columns, from the left to the right, have the lengths $m_1, \ldots, m_s$). From the diagram in Fig. 16, left, we create a slant diagram in Fig. 16, right. The boldfaced polygonal line is a graph of a nondecreasing function d, and the condition in the definition of $e(m_1, \ldots, m_s)$ can be formulated as $\dim(\pi \cap \mathbb{R}^m) = d(m)$. This simple description of the set $e(m_1, \ldots, m_s)$ justifies its notation as $e(\Delta)$, where Δ is the notation for the Young diagram of the sequence $(m_1, \ldots, m_s)$. We will prove that the sets $e(\Delta)$ form a CW decomposition of $G(n,k)$ and thus the Schubert cells are labeled by Young diagrams contained in the rectangle $k \times (n-k)$; moreover, the dimension of the cell $e(\Delta)$ equals the number $|\Delta| = m_1 + \cdots + m_s$ of cells of the Young diagram Δ.

We begin with this computation of dimension.

Lemma. *The subspace $e(m_1, \ldots, m_s)$ is homeomorphic to $\mathbb{R}^{m_1 + \cdots + m_s}$.*

Proof. Redraw the picture in Fig. 16, right, as shown in Fig. 17 (that is, place the graph in Fig. 16, right, into the rectangle $k \times n$, then for every horizontal segment

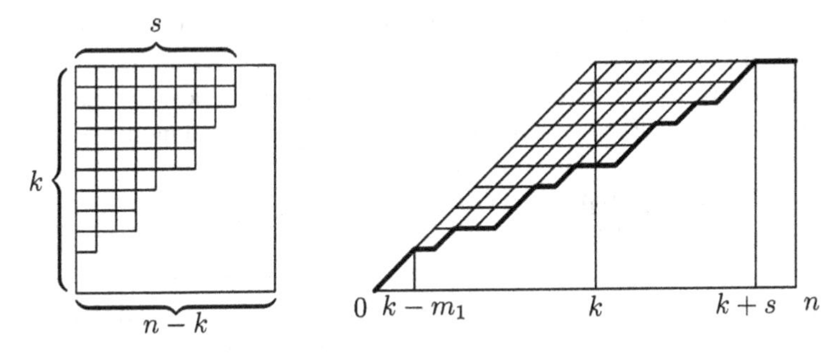

Fig. 16 Young diagram and slanted Young diagram

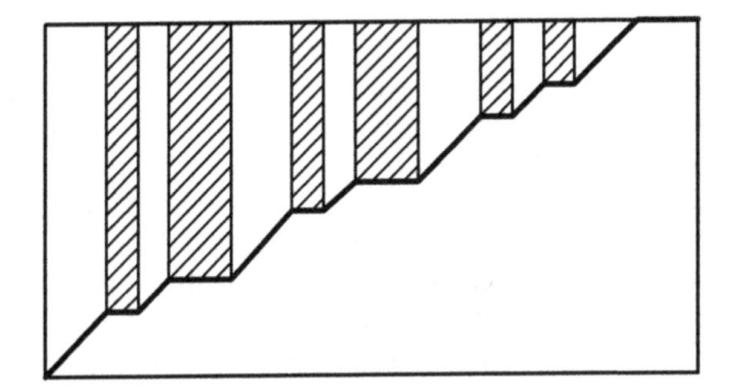

Fig. 17 Constructing a matrix from a Young diagram

of this graph construct a vertical strip with this segment as the lower base with the upper base on the upper side of the rectangle, and then shadow all the stripes).

Next, we make a $k \times n$ matrix out of the diagram of Fig. 17 in the following way. We place entries 1 on the slant intervals of the graph, arbitrary numbers (marked below as $*$) into the shadowed strips, and zeroes elsewhere. We obtain a matrix

$$\begin{bmatrix}
 & & * & & * & * & & & * & & * & * & & & * & & * & & 1 \\
 & & * & & * & * & & & * & & * & * & & & * & & * & 1 & \\
 & & * & & * & * & & & * & & * & * & & & * & 1 & & & \\
 & & * & & * & * & & & * & & * & * & & 1 & & & & & \\
 & & * & & * & * & & & * & & * & * & 1 & & & & & & \\
 & & * & & * & * & & & * & 1 & & & & & & & & & \\
 & & * & & * & * & & 1 & & & & & & & & & & & \\
 & & * & & * & * & 1 & & & & & & & & & & & & \\
 & & * & 1 & & & & & & & & & & & & & & & \\
 & 1 & & & & & & & & & & & & & & & & & \\
1 & & & & & & & & & & & & & & & & & &
\end{bmatrix}$$

The k rows of this matrix are linearly independent and form a basis of a k-dimensional subspace π of $\mathbb{R}^n$, and it is clear that this gives a bijection between matrices of this form and πs from $e(m_1, \ldots, m_s)$. These matrices are parametrized by values of entries marked as $*$, these values are arbitrary real numbers, and there are $m_1 + \cdots + m_s$ of them. This proves the lemma.

To prove that the decomposition $G(n,k) = \bigcup_{(m_1,\ldots,m_s)} e(m_1, \ldots, m_s)$, we need to extend the homeomorphism $\operatorname{Int} D^{m_1+\cdots+m_s} \approx \mathbb{R}^{m_1+\cdots+m_s} \to e(m_1, \ldots, m_s)$ of the lemma to a continuous map $D^{m_1+\cdots+m_s} \to G(n,k)$ [it is not hard to see that the boundary $\overline{e(m_1, \ldots, m_s)} - e(m_1, \ldots, m_s)$ is contained in the union of cells of dimensions $< m_1 + \cdots + m_s$]. There are explicit formulas for this map, but they are complicated, and we do not give them here. An interested reader can find them in the book [73] of J. T. Schwartz.

There is a remarkable property of Schubert cells: Embeddings of $G(n,k)$ to $G(n+1,k)$ and $G(n+1,n+1)$ map every cell $e(m_1, \ldots, m_s)$ onto a cell with the same notation. For this reason, the spaces $G(\infty,k)$ and $G(\infty,\infty)$ are decomposed into cells corresponding to Young diagrams: In the second case they correspond to all Young diagrams, while in the first case they correspond to Young diagrams contained in the infinite horizontal half-strip of height k.

Complex and quaternionic versions of Schubert cells are obvious: They have dimensions two and four times the dimensions in the real case. The Grassmann manifold $G_+(n,k)$ is decomposed into cells $e_\pm(m_1, \ldots, m_s)$ of the same dimension as $e(m_1, \ldots, m_s)$.

EXERCISE 16. The CW decompositions of $\mathbb{R}P^n = G(n+1,1)$, $\mathbb{C}P^n = \mathbb{C}G(n+1,1)$, $\mathbb{H}P^n = \mathbb{H}G(n+1,1)$ constructed above are particular cases of the Schubert decomposition.

D: Flag Manifolds

The flag manifolds have natural CW decompositions which generalize the Schubert decomposition of the Grassmann manifolds. This decomposition as well as its cells are also called Schubert. We will describe this decomposition only in the real case (the complex and quaternionic cases differ from the real case only by doubling and quadrupling of the dimensions of cells).

Schubert cells of a flag manifold are characterized by dimensions d_{ij} of intersections $V_i \cap \mathbb{R}^j$. The numbers d_{ij}, however, must satisfy several, rather inconvenient, conditions, and we prefer the following more reasonable definition.

The cells of the space $F(n; k_1, \ldots, k_s)$ correspond to sequences $m_1, \ldots, m_n$ of integers taking values $1, \ldots, s+1$ such that precisely $k_j - k_{j-1}$ of these numbers are equal to j ($j = 1, \ldots, s+1$; we put $k_0 = 0$ and $k_{s+1} = s$. The cell $e[m_1, \ldots, m_n]$ corresponding to the sequence $m_1, \ldots, m_n$ consists of those flags $V_1 \subset \cdots \subset V_s$ such that

$$\dim \frac{V_i \cap \mathbb{R}^j}{(V_{i-1} \cap R^j) + (V_i \cap \mathbb{R}^{j-1})} = \delta_{im_j} = \begin{cases} 0, \text{ if } i = m_j, \\ 1, \text{ if } i \neq m_j \end{cases}$$

(we put $V_0 = 0$ and $V_{s+1} = \mathbb{R}^n$), or, differently,

$$\dim(V_i \cap \mathbb{R}^j) = \text{card}\{p \leq i \mid k_p \leq j\}.$$

The dimension of the cell $e[m_1, \ldots, m_n]$ is equal to the number of pairs (i,j) for which $i < j, m_i > m_j$.

In particular, the manifold $F(n; 1, \ldots, n-1)$ of full flags is decomposed into the union of cells corresponding to usual permutations of numbers $1, \ldots, n$, and the dimension of a cell is equal to the number of inversions in a permutation.

If the flag manifold is the Grassmann manifold $G(n,k)$, then $s = 1$ and the sequence $m_1, \ldots, m_n$ consists of k ones and $n-k$ twos. Using this sequence, we construct an n-gon line starting at the point $(0, -k)$ and ending at the point $(n-k, 0)$ with all edges having the length 1, such that the ith edge is directed up if $m_i = 1$ and is directed right if $m_i = 2$. This line bounds (together with the coordinate axes) a Young diagram Δ, and it it is easy to see that $e[m_1, \ldots, m_n] = e(\Delta)$.

Notice in conclusion that the cells $e[m_1, \ldots, m_n]$ (as well as their complex and quaternionic analogs) may be described in pure algebraic terms: They are orbits of the group of lower triangular matrices with diagonal entries 1 in the flag manifold. Namely, the cell $e[m_1, \ldots, m_n]$ is the orbit of a flag whose ith space is spanned by the coordinate vectors whose numbers p satisfy the condition $m_p \leq i$.

E: Compact Classical Groups

They also have good CW decompositions. These decompositions are described (implicitly) in a classical work of Pontryagin [67].

F: Classical Surfaces

We already have CW decompositions of S^2 and $\mathbb{R}P^2$. For the other surfaces without holes, we can use their construction by gluing sides of a polygon (see Exercise 14 in Lecture 1). The interior of a polygon becomes a two-dimensional cell (and the projection of the polygon onto the surface becomes a characteristic map), the (open) sides become one-dimensional cells, and the vertices become zero-dimensional cells. The most common CW decomposition of every classical surface has one two-dimensional cell and one zero-dimensional cell. Also, a sphere with g handles has $2g$ one-dimensional cells (see Fig. 18 for $g = 2$), a projective plane with g handles

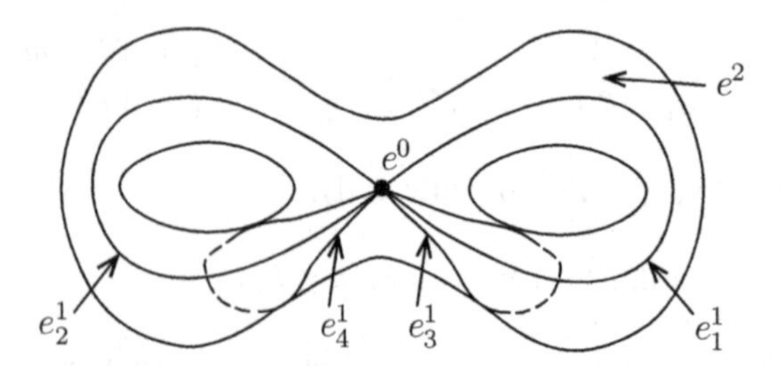

Fig. 18 A CW decomposition of a sphere with two handles

has $2g + 1$ one-dimensional cells, and a Klein bottle with g handles has $2g + 2$ one-dimensional cells.

EXERCISE 17. Construct CW decompositions of classical surfaces with holes with the minimal possible number of cells.

The rest of this lecture will be devoted to *homotopy properties of CW complexes*.

5.5 Borsuk's Theorem on Extension of Homotopies

Definition. A pair (X, A) is called a *Borsuk pair* if for every topological space Y, every continuous map $F: X \to Y$, and every homotopy $f_t: A \to Y$ such that $f_0 = F|_A$, there exists a homotopy $F_t: X \to Y$ such that $F_0 = F$ and $F_t|_A = f_t$.

Theorem (Borsuk). *Every CW pair is a Borsuk pair.*

Proof. Let (X, A) be a CW pair. We are given maps $\Phi: A \times I \to Y$ (this is the homotopy f_t) and $F: X \times 0 \to Y$ such that $F|_{A\times 0} = \Phi|_{A\times 0}$. To extend the homotopy f_t to a homotopy F_t we need to extend the map F to a map $F': X \times I \to Y$ such that $F'|_{A\times I} = \Phi$. We will construct this extension by induction with respect to dimension of cells. The first step of this induction is the extension of the map Φ to $(A \cup X^0) \times I$:

$$F'(x,t) = \begin{cases} F(x,0), & \text{if } x \text{ is a 0-dimensional cell of } X, x \notin A, \\ \Phi(x,t), & \text{if } x \in A. \end{cases}$$

Assume now that the map F' has been already defined on $(A \cup X^n) \times I$ and is equal to Φ on $A \times I$ and to F on $X \times 0$. Take an $(n+1)$-dimensional cell $e^{n+1} \subset X - A$. By assumption, F' is defined on the set $(\overline{e^{n+1}} - e^{n+1}) \times I$ (since the boundary $\dot{e}^{n+1} = \overline{e^{n+1}} - e^{n+1}$ is contained in X^n by definition of a CW complex). Let $f: D^{n+1} \to X$ be a characteristic map for the cell e^{n+1}. We want to extend the map F' to the interior of the cylinder $f(D^{n+1})$ from its side surface $f(S^n) \times I$ and the bottom base $f(D^{n+1}) \times 0$. But it is clear from the definition of a CW complex that it is the same as to extend the map $\psi = F' \circ f: (S^n \times I) \cup (D^{n+1} \times 0) \to Y$ to a continuous map $\psi': D^{n+1} \times I \to Y$.

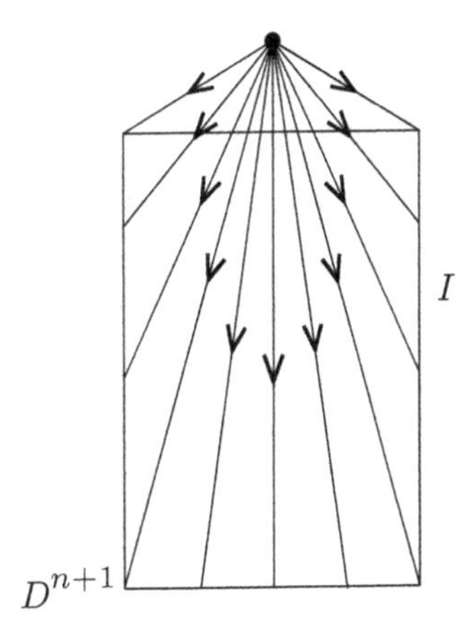

Fig. 19 The projection $\eta: D^{n+1} \times I \to (S^n \times I) \cup (D^{n+1} \times 0)$

Let $\eta: D^{n+1} \times I \to (S^n \times I) \cup (D^{n+1} \times 0)$ be the projection of the cylinder $D^{n+1} \times I$ from a point slightly above the upper base of the cylinder; it is the identity on $(S^n \times I) \cup (D^{n+1} \times 0)$ (Fig. 19).

We define the map ψ' as the composition

$$D^{n+1} \times I \xrightarrow{\eta} (S^n \times I) \cup (D^{n+1} \times 0) \xrightarrow{\psi} Y.$$

We can do this simultaneously for all $(n + 1)$-dimensional cells in $X - A$, and we get an extension of the map F' to $(A \cup X^{n+1}) \times I$.

In this way, skeleton after skeleton, we construct an extension of the map Φ to a map $F': X \times I \to Y$. Notice that if $X - A$ is infinite dimensional, then the construction will involve infinitely many steps. In this case, the continuity of the map F' obtained will follow from Axiom (W).

5.6 Corollaries from Borsuk's Theorem

Corollary 1. *Let (X, A) be a CW pair. If A is contractible, then $X/A \sim X$. More precisely: The projection $X \to X/A$ is a homotopy equivalence.*

Proof. Let p be the projection $X \to X/A$. Since A is contractible, there is a homotopy $f_t: A \to A$ such that $f_0 = \text{id}_A$ and $f_1 = \text{const}$. By Borsuk's theorem, there exists a homotopy $F_t: X \to X$ such that $F_0 = \text{id}_X$ and $F_t|_A = f_t$; in particular, $F_1(A)$ is a point. The latter means that F_1 "factorizes" through X/A; that is, there exists a (unique) continuous map $q: X/A \to X$ such that $F_1 = q \circ p$. Thus, $q \circ p \sim \text{id}_X$ (F_t is a homotopy).

Let us prove that $p \circ q \sim \text{id}_{X/A}$. Since $F_t|_A = f_t: A \to A$, we have $F_t(A) \subset A$, so F_t can be factorized to a map $h_t: X/A \to X/A$, which means that $p \circ F_t = h_t \circ p$.

Hence, $h_t \circ p$ is a homotopy between $p \circ F_0 = p \circ \mathrm{id}_X = p = \mathrm{id}_{X/A} \circ p$ and $p \circ F_1 = p \circ (q \circ p) = (p \circ q) \circ p$, so h_t is a homotopy between $\mathrm{id}_{X/A}$ and $p \circ q$.

Thus, p and q are mutually homotopy inverse, which completes the proof.

Corollary 2. *If (X, A) is a CW pair, then $X/A \sim X \cup CA$, where CA is a cone over A.*

Proof. $X/A = (X \cup CA)/CA \sim X \cup CA$. The latter follows from Corollary 1 applied to the CW complex $X \cup CA$ and its contractible CW subcomplex CA.

Remark. Both propositions may be regarded not as corollaries from Borsuk's theorem but as independent theorems, only the assumption of (X, A) being a CW pair should be replaced, in the first case, by the assumption that (X, A) is a Borsuk pair, and in the second case, by the assumption that $(X \cup CA, CA)$ is a Borsuk pair.

5.7 The Cellular Approximation Theorem

Theorem. *Every continuous map of one CW complex into another CW complex is homotopic to a cellular map.*

We will prove this theorem in the following, *relative* form.

Theorem. *Let f be a continuous map of a CW complex X into a CW complex Y such that the restriction $f|_A$ is cellular for some CW subcomplex A of X. Then there exists a cellular map $g: X \to Y$ such that $g|_A = f|_A$, and, moreover, g is A-homotopic to f.*

The expression "g is A-homotopic to f" (in formulas, $g \sim_A f$) means that there is a homotopy h_t between g and f which is fixed on A; that is, $f_t(x)$ does not depend on t for every $x \in A$. It is clear that if $g \sim_A f$, then $g|_A = f|_A$. Certainly, $g \sim_A f$ implies $g \sim f$, but not vice versa. For example, the maps $f, g: I \to S^1$, where f is the winding of the segment about the circle mapping both endpoints into the same point of the circle and g is a constant map, are homotopic, but not $(0 \cup 1)$-homotopic (strictly speaking, we will prove this only in Lecture 6).

Proof of Theorem. Assume that the map f has already been made cellular not only on all cells from A, but also on all cells from X of dimensions less than p. Take a p-dimensional cell $e^p \subset X - A$. Its image $f(e^p)$ has a nonempty intersection with only a finite set of cells of Y [this follows from the compactness of $f(\overline{e^p})$—see Exercise 3]. Of these cells of Y, choose a cell of a maximal dimension, say, ϵ^q, $\dim \epsilon^q = q$. If $q \leq p$, then we do not need to do anything with the cell e^p. If, however, $q > p$, we will need the following lemma.

Free-Point Lemma. *Let U be an open subset of $\mathbb{R}^p$ and $\varphi: U \to \mathrm{Int}\, D^q$ be such a continuous map that the set $V = \varphi^{-1}(d^q) \subset U$ where d^q is some closed ball in $\mathrm{Int}\, D^q$ is compact. If $q > p$, then there exists a continuous map $\psi: U \to \mathrm{Int}\, D^q$ coinciding with φ in the complement of V and such that its image does not cover the whole ball d^q.*

We will postpone the proof of this lemma (and a discussion of its geometric meaning) until the next section. For now, we restrict ourselves to the following obvious remark. The map ψ is automatically $(U-V)$-homotopic to φ: It is sufficient to take the "straight" homotopy joining φ and ψ when, for every $u \in U$, the point $\varphi(u)$ is moving to $\psi(u)$ at a constant speed along a straight interval joining $\varphi(u)$ and $\psi(u)$.

Now, let us finish the proof of the theorem. The free-point lemma implies that *the restriction $f_{A\cup X^{p-1}\cup e^p}$ is $(A \cup X^{p-1})$-homotopic to a map $f': A \cup X^{p-1} \cup e^p \to Y$ such that $f'(e^p)$ has nonempty intersections with the same cells as $f(e^p)$, but $f'(e^p)$ does not cover the whole cell ϵ^q.* Indeed, let $h: D^p \to X$ and $k: D^q \to Y$ be characteristic maps corresponding to the cells e^p and ϵ^q. Let $U = h^{-1}(f^{-1}(\epsilon^q) \cap e^q)$ and define a map $\varphi: U \to \operatorname{Int} D^q$ as a composition

$$\begin{array}{ccccccc} u & \stackrel{h}{\longmapsto} & x & \stackrel{f}{\longmapsto} & y & \stackrel{k^{-1}}{\longmapsto} & v = \varphi(u) \\ \Cap & & \Cap & & \Cap & & \Cap \\ U & & e^p \cup f^{-1}(e^q) & & e^q & & \text{Int } D^q \end{array}$$

Denote as d^q a closed concentric subball of the ball D^q. The set $V = \varphi^{-1}(d^q)$ is compact (because it is a closed subset of a closed ball D^p). Let $\psi: U \to \operatorname{Int} D^q$ be a map provided by the free-point lemma. We define the map f' as coinciding with f in the complement of $h(U)$ and as the composition

$$\begin{array}{ccccccc} x & \stackrel{h^{-1}}{\longmapsto} & u & \stackrel{\psi}{\longmapsto} & v & \stackrel{k}{\longmapsto} & y = f'(u) \\ \Cap & & \Cap & & \Cap & & \Cap \\ h(U) & & U & & \text{Int } D^q & & e^q \subset Y \end{array}$$

in $h(U)$. It is clear that the map f' is continuous [it coincides with f on the "buffer" set $h(U-V)$] and $(A \cup X^{p-1})$-homotopic [actually, even $(A \cup X^{p-1} \cup (e^p - h(V)))$-homotopic] to $f|_{A\cup X^{p-1}\cup e^p}$ [because $\varphi \sim_{(U-V)} \psi$]. It is also clear that $f'(e^p)$ does not cover ε^q.

It is very easy now to complete the proof. First, by Borsuk's theorem, we can extend our homotopy fixed on $A \cup X^{p-1}$ between $f|_{A\cup X^{p-1}\cup e^p}$ and f' to the whole space X, which lets us assume that the map f' with all necessary properties is defined on the whole space X. After that, we take a point $y_0 \in \epsilon^q$, not in $f(e^p)$, and apply to $f'|_{e^p}$ a "radial homotopy": If $x \in e^p - f^{-1}(\varepsilon^q)$, then $f'(x)$ does not move, but if $f'(x) \in \epsilon^q$, then $f'(x)$ is moving, at a constant speed, along a straight path going from y_0 through $f'(x)$ to the boundary of ϵ^q [more precisely, along the k-image of a straight interval in D^q starting at $k^{-1}(y_0)$ and going through $k^{-1}(f'(x))$ to the boundary sphere S^{q-1}]. We extend this homotopy to a homotopy of $f'|_A \cup X^{p-1} \cup e^p$ (fixed in the complement of e^p), and then, using Borsuk's theorem, to a homotopy of the whole map $f': X \to Y$. In this way, we reduce the number of q-dimensional cells hit by $f'(e^p)$ by one, and, repeating this procedure a necessary amount of times,

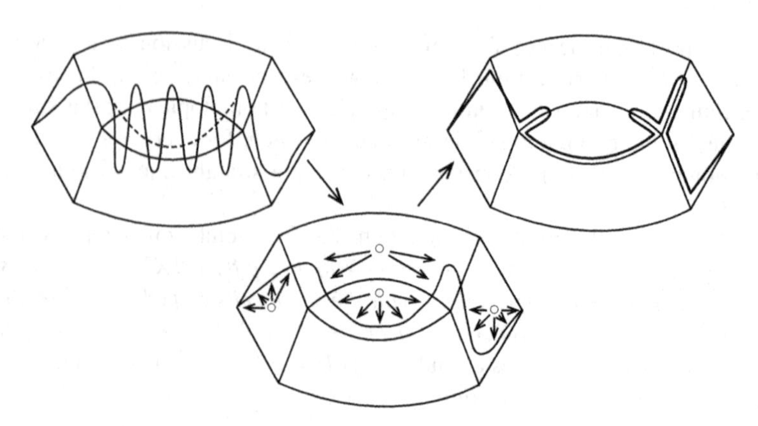

Fig. 20 The proof of the cellular approximation theorem

we get an $(A \cup X^{p-1})$-homotopy of f to a map cellular on $A \cup X^{p-1} \cup e^p$. The whole procedure is presented, schematically, in Fig. 20.

Notice now that the "correction" of the map f which we did for the cell e^p can be done simultaneously for all p-dimensional cells in $X - A$. We will arrive at a map cellular on $A \cup X^p$ and $(A \cup X^{p-1})$-homotopic to f.

To get an A-homotopy connecting f with a cellular map we need to perform this construction successively for $p = 0, 1, 2, \dots$. The number of steps may be infinite, but this is not a problem: We can perform the pth homotopy at the parameter interval $1-2^{-p} \leq t \leq 1-2^{-p-1}$. The continuity of the whole homotopy is secured by Axiom (W): For every cell e, the homotopy will be fixed starting with some $t_e < 1$.

This completes the proof of the theorem.

5.8 Fighting Chimeras: A Proof of the Free-Point Lemma

For a reader not hopelessly spoiled by popular mathematical literature, the very statement of the lemma looks awkward: How can a continuous image of a space of a smaller dimension cover a space of a bigger dimension? But everybody knows that there is the Peano curve which is propagandized not less than, say, the Klein bottle and which is a continuous map of an interval onto a square. Therefore, we have to prove the lemma, and it is especially difficult, because geometric intuition cannot help: It persistently repeats that it is not possible at all. Difficulties of this kind arise when a "rigorous" definition of this or that notion (in our current case, the ε-δ-definition of continuity) does not fully correspond to the initial intuitive image: We need to study the structure not of a real object, but rather of a chimera. But we have no choice: The lemma needs a proof.

There are two common ways of proving propositions like this: One needs to approximate the map φ by either a smooth map or a piecewise linear map. The first

way looks more natural, but it requires a familiarity with Sard's theorem, which is not covered by a standard university calculus course. We will need this theorem anyhow, but for now it is better to postpone a discussion of this matter. The second way is based on the notion of a *triangulation*. Recall that a *q-dimensional Euclidean simplex* is a subset of the space $\mathbb{R}^n$, $n \geq q$, which is a convex hull of a set of $q+1$ points not contained in one $(q-1)$-dimensional plane. (Euclidean simplices of dimensions 0, 1, 2, 3 are points, closed intervals, triangles, tetrahedra.) These $q+1$ points are called *vertices* of a simplex. Subsimplices, that is, convex hulls of nonempty subsets of the set of vertices, are called *faces* of the simplex. They are simplices of dimensions $\leq q$. A zero-dimensional face is a vertex. A remarkable property of simplices is that a linear map of a Euclidean simplex into an arbitrary space $\mathbb{R}^m$ is fully determined by its values at the vertices, while these values may

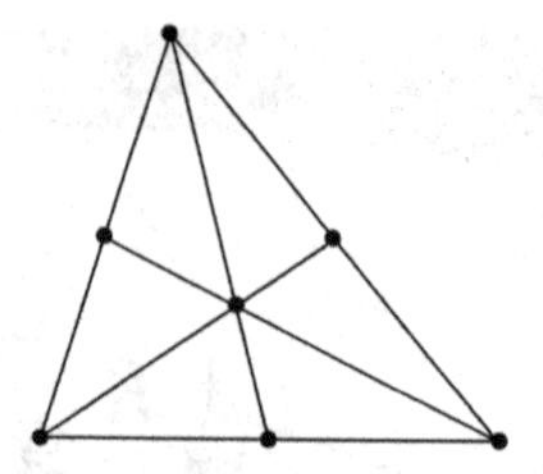

Fig. 21 The barycentric subdivision of a (two-dimensional) simplex

be absolutely arbitrary. A *finite triangulation* of a subset of a Euclidean space is a finite covering of this space by (different) Euclidean simplices such that any two of these simplices either are disjoint or meet each other at a whole face of both. It is convenient to include faces of simplices of a triangulation into the set of simplices of this triangulation.

The *barycentric subdivision* of a q-dimensional simplex consists in a partition of this simplex into $(q+1)!$ smaller q-dimensional simplices. The vertices of the new simplices are barycenters (mass centers) of faces of the old simplex (including the old simplex itself). A set $\{x_0, x_1, \dots, x_r\}$ of such centers is a set of vertices of an r-dimensional simplex of the barycentric subdivision if and only if they are centers of faces which form a chain of faces successively embedded into each other: x_i is a center of Γ_i and $\Gamma_0 \subset \Gamma_1 \subset \dots \subset \Gamma_r$. (See Fig. 21.)

Another description of the barycentric subdivision of a simplex is inductive: The barycentric subdivision of a zero-dimensional simplex is just this zero-dimensional simplex; to obtain the barycentric subdivision of a q-dimensional simplex, we take barycentric subdivisions of all its $q-1$-dimensional faces (they are compatible on $q-2$-dimensional faces) and add simplices which are cones over simplices of the subdivisions of faces with the vertex at the barycenter of the simplex. One more description can be made in terms of "barycentric coordinates": If $v_0, v_1, \dots, v_q$ are vertices of the given simplex, then every point of the simplex has the form $\sum t_i v_i$, where $t_0, t_1, \dots, t_q$ are nonnegative numbers with the sum 1; these numbers are uniquely defined and are called barycentric coordinates of the point. The $(q+1)!$ q-dimensional simplices of the barycentric subdivision correspond to permutations $(i_0, i_1, \dots, i_q)$ of $0, 1, \dots, q$ and are described by the inequalities $t_{i_0} \le t_{i_1} \le \dots \le t_{i_q}$.

The *barycentric subdivision of a triangulation* is a triangulation composed of simplices of barycentric subdivisions of the simplices of the triangulation (see Fig. 22).

Now turn to our map φ. First of all, consider in $d = d^q$ concentric balls d_1, d_2, d_3, d_4 of radii $\frac{\rho}{5}, \frac{2\rho}{5}, \frac{3\rho}{5}, \frac{4\rho}{5}$ where ρ is the radius of d. Then we consider a triangulated set $K \subset \mathbb{R}^q$ such that $\overline{V} \subset K \subset U$ (for example, take a big Euclidean simplex $\Delta \supset \overline{V}$ and apply the barycentric subdivision to Δ so many times that every simplex of the final subdivision that has a nonempty intersection with $\overline{V}$ is contained in U; the union of simplices with this property is K). Then

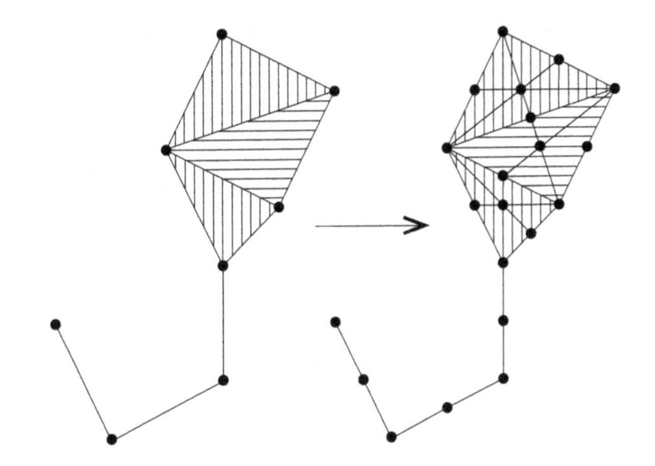

Fig. 22 The barycentric subdivision of a triangulation

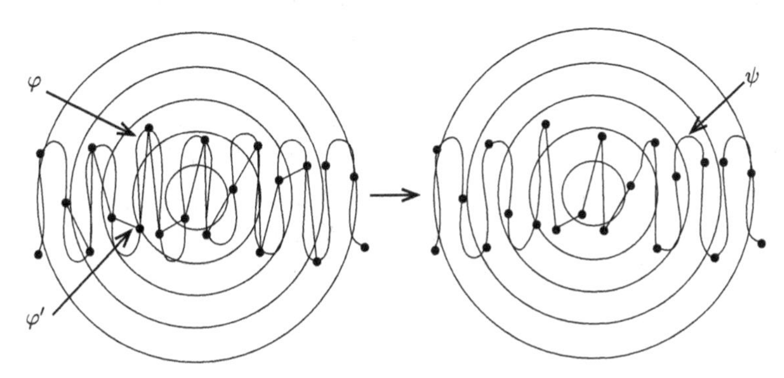

Fig. 23 Sewing φ and φ' into ψ

take a multiple barycentric subdivision of K such that for every simplex s of this subdivision, $\operatorname{diam}\varphi(s) < \frac{\rho}{5}$. Let L be the union of simplices of this triangulation of K whose φ-images hit d_4. Then $d_4 \cap \varphi(U) \subset \varphi(L) \subset d$. Let $\varphi': L \to d$, which coincides with φ at all vertices of L and is linear at every simplex of L. The maps $\varphi|_L$ and φ' are homotopic: They are connected by the straight-line homotopy $\varphi_t: L \to d$, $\varphi_0 = \varphi|_L, \varphi_1 = \varphi'$.

Now, we "sew" the maps φ and φ' into a single map $\psi: U \to \operatorname{Int} D^q$:

$$\psi(u) = \begin{cases} \varphi(u), & \text{if } \varphi(u) \notin d_3, \\ \varphi'(u), & \text{if } \varphi(u) \in d_2, \\ \varphi_{3-5\rho(u)}(u), & \text{if } \varphi(u) \in d_3 - d_2. \end{cases}$$

Here $\rho(u)$ is the distance from $\varphi(u)$ to the center of d. (See Fig. 23.)

The map ψ is continuous, it coincides with φ on $U - V$, and the intersection of its image with d_1 is contained in the union of a finite set of p-dimensional planes.

Thus, this image does not cover the whole ball d_1 (and hence does not cover the whole ball d). These completes the proof of the lemma.

5.9 First Applications of the Cellular Approximation Theorem

Theorem. *If X is a CW complex with one vertex (= zero-dimensional cell) and without other cells of dimensions $< q$ and Y is a CW complex of dimension $< q$, then every continuous map $Y \to X$ is homotopic to a constant map. The same statement holds in the base point case (it is convenient to assume that a base point of a CW complex is a zero-dimensional cell).*

This follows directly from the cellular approximation theorem, since the qth skeleton of Y is the whole Y, and the qth skeleton of X is one point.

In particular, if $m < q$, then $\pi(S^m, S^q) = \pi_b(S^m, S^q) = 0$ (that is, consists of one element).

Definition. A space X is called n-connected if for $q \leq n$ the set $\pi(S^q, X)$ consists of one element (that is, any two continuous maps $S^q \to X$ with $q \leq n$ are homotopic).

EXERCISE 18. Prove that each of the following two conditions is equivalent to n-connectedness. (1) For $q \leq n$, the set $\pi_b(S^q, X)$ consists of one element. (2) For $q \leq n$, every continuous map $S^q \to X$ can be extended to a continuous map $D^{q+1} \to X$.[1]

EXERCISE 19. Prove that 0-connectedness is the same as path connectedness.

The term 1-connected(ness) is usually replaced by the term "simply connected(ness)."

Theorem. *Let $n \geq 0$. An n-connected CW complex is homotopy equivalent to a CW complex with only one vertex and without cells of dimensions $1, 2, \ldots, n$. (In particular, every path connected CW complex is homotopy equivalent to a CW complex with only one vertex.)*

Proof. Choose in our CW complex X some vertex e_0 and join all other vertices, $e_1, e_2, e_3, \ldots$, with e_0 by paths s_1, s_2, s_3. This is possible since X is n-connected and hence path connected. (The paths may cross.) The cellular approximation theorem lets us assume that every path s_i lies in the first skeleton of X. For every i, attach a two-dimensional disk to X by s_i regarded as a map of the lower semicircle to X (see Fig. 24).

[1] Condition (2) makes sense for $n = -1$ and means that X is nonempty. Sometimes it is convenient to assume that (-1)-connected is the same as nonempty.

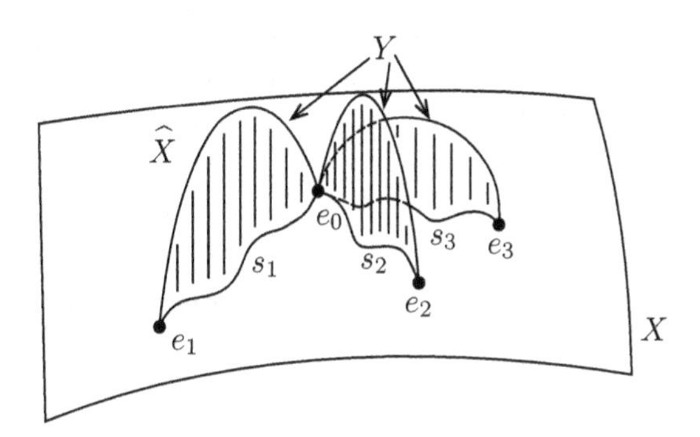

Fig. 24 Attaching disks to a path connected CW complex

We get a new CW complex $\widehat{X}$ which contains X and also cells e_i^2 and e_i^1 (interiors and upper semicircles of the disks attached). The boundaries of the cells e_i^2 are contained in the first skeleton because the paths s_i have this property.

It is clear that X is a deformation retract in $\widehat{X}$: Every attached disk can be smashed onto the corresponding path s_i.

Denote by Y the union of the closures of the cells e_i^1. Obviously, Y contains all the vertices of $\widehat{X}$ and is contractible. Hence, $\widehat{X}/Y$ has only one vertex and also $\widehat{X}/Y \sim \widehat{X} \sim X$.

The rest of the proof is quite similar. We assume that $X \sim X'$, where X' has only one vertex and has no cells of dimensions $1, 2, \ldots, k-1$, where $k \leq n$. In this case the closure of every k-dimensional cell is a k-dimensional sphere. Since X (and hence X') is n-connected, the inclusion of this sphere into X' can be extended to a continuous map $D^{k+1} \to X'$, and by the cellular approximation theorem, we can assume that the image of this map is contained in the $(k+1)$st skeleton of X'. Using this map, which we consider a map of the lower hemisphere of the sphere S^{k+1}, we attach a ball D^{k+2} to X', and we do this for every k-dimensional cell of X'. Thus, X' acquires, for every k-dimensional cell, two new cells: one $(k+1)$-dimensional, and one $(k+2)$-dimensional. The new CW complex $\widehat{X}'$ is homotopy equivalent to X' (which is a deformation retract of $\widehat{X}'$) and contains a contractible CW subcomplex Y', the union of all new $(k+1)$-dimensional cells, which contains the whole kth skeleton of $\widehat{X}'$. The quotient $\widehat{X}'/Y'$ is homotopy equivalent to X', and hence to X, and it does not have cells of dimensions $\leq k$, besides its only vertex. This induction proves our theorem.

Corollary. *If a CW complex X is n-connected, and a CW complex Y is n-dimensional, then the set $\pi(Y, X)$ consists of one element. The same is true for $\pi_b(Y, X)$ if X and Y have base points which are zero-dimensional cells.*

Remark. The procedure of killing k-dimensional cells used in the last proof includes attaching cells of dimension $k+2$. If our CW complex has dimension $n+1$, this could lead to increasing the dimension. However, as we will see in Chap. 2, an n-connected $n+1$-dimensional CW complex must be homotopy equivalent to a bouquet of $(n+1)$-dimensional spheres, so no increasing the dimension is actually taking place. In addition to that, in the important case when $n = 0$, there exists another proof of last theorem, which does not involve any attaching of cells: See Exercise 20.

EXERCISE 20. Prove that a connected CW complex X always contains a contractible one-dimensional CW subcomplex ("a tree") Y, which contains all vertices of X. Then X/Y is a CW complex with one vertex which is homotopy equivalent to X. (If the number of cells of X is infinite, then a proof of that requires Zorn's lemma.)

The last theorem has a relative version which requires a definition of an *n-connected pair*. A topological pair (X, A) is called n-connected if any continuous map $(D^k, S^{k-1}) \to (X, A)$ with $k \leq n$ is homotopic (as a map between pairs) to a map which sends the whole ball D^k into A.

EXERCISE 21. Make up equivalent definitions of an n-connected pair in the spirit of Exercise 18 and the interpretation of 0-connectedness and 1-connectedness in the spirit of Exercise 19.

EXERCISE 22. Prove that every n-connected CW pair (X, A) is homotopy equivalent to a CW pair (X', A') such that A' contains all cells of X' of dimensions $\leq n$ (that is, contains the nth skeleton of X').

Lecture 6 The Fundamental Group and Coverings

6.1 Definition of the Fundamental Group

The *fundamental group* of a space X with a base point is its first homotopy group, $\pi_1(X) = \pi_b(S^1, X)$. Since the definition in Lecture 4 was too general, we will repeat it now in our particular case.

Recall that a homotopy $s_t: I \to X$ of a path is always supposed to be fixed at the ends: $s_t(0)$ and $s_t(1)$ do not depend on t. Recall also that the product ss' of two paths, $s, s': I \to X$, is defined if $s(1) = s'(0)$, and in this case $(ss')(t) = s(2t)$ if $t \leq \frac{1}{2}$ and $ss'(t) = s'(2t-1)$ if $t \geq \frac{1}{2}$. A path s with equal ends, $s(0) = s(1)$, is called a loop.

We consider the set $\Omega(X, x_0)$ of loops of the space X with a base point x_0. The product is defined for every two loops from $\Omega(X, x_0)$. This multiplication gives rise to a multiplication in the set $\pi_1(X, x_0)$ of homotopy classes of loops (it is easy to check is that if $s \sim s'$ and $r \sim r'$, then $rs \sim r's'$). This multiplication determines a group structure in the set $\pi_1(X, x_0)$ (unlike the multiplication in the set of loops, where the group axioms hold only "up to a homotopy": $u(vw) \sim (uv)w$, and so on). The inverse to the class of a loop $s: I \to X$ is the class of the same loop passed in the

opposite direction; that is, the inverse of the class of the loop $s: I \to X$ is the class of the loop s', $s'(t) = s(1-t)$. The identity element in $\pi_1(X, x_0)$ is the class of the constant loop.

A continuous map $f: (X, x_0) \to (Y, y_0)$ induces, in the obvious way, a homomorphism $f_*: \pi_1(X, x_0) \to \pi_1(Y, y_0)$. It is clear that if maps f, g are (base point) homotopic, then the homomorphisms f_*, g_* coincide.

We finish this section with the following simple observation.

Theorem. *For any spaces with base points,* (X, x_0) *and* (Y, y_0),

$$\pi_1(X \times Y, (x_0, y_0)) \cong \pi_1(X, x_0) \times \pi(Y, y_0).$$

Proof. Indeed, a loop in $X \times Y$ with the beginning at (x_0, y_0) is the same as the pair of loops in X and Y with the beginnings at x_0 and y_0; the same for products and homotopies of loops.

6.2 Dependence of the Base Point

Theorem. *If the space* X *is path connected, then* $\pi_1(X, x_0) \approx \pi_1(X, x_1)$ *for any points* $x_0, x_1 \in X$.

Proof. Since X is path connected, there exists a path $u: I \to X$ such that $u(0) = x_0, u(1) = x_1$. We will construct a map $u_\#: \pi_1(X, x_0) \to \pi_1(X, x_1)$: For the class $[s] \in \pi_1(X, x_0)$ of a loop s, we put $u_\#[s] = [(u^{-1}s)u] \in \pi_1(X, x_1)$. [It is clear that if $s \sim s'$, then $(u^{-1}s)u \sim (u^{-1}s')u$.] It is clear also that if we replace the path u by a homotopic path (a homotopy of paths is defined, like a homotopy of loops, as a homotopy with ends fixed), then the map $u_\#$ remains the same. The map inverse to the map $u_\#$ is defined similarly: $u_\#^{-1}[r] = [(ur)u^{-1}]$.

It is easy to check that the map $u_\#$ is a homomorphism ($u_\#[ss'] = [(u^{-1}(ss'))u] = [(u^{-1}su)(u^{-1}s')u] = u_\#[s]u_\#[s']$), and hence an isomorphism.

Certainly, the isomorphism $u_\#$ depends on the path u: If we replace the path u by a path v not homotopic to u, we will get a different isomorphism. More precisely: If $r = v^{-1}u$, then for any loop $s \in \Omega(X, x_0)$, $v_\#[s] = [v^{-1}sv] = [v^{-1}uu^{-1}suu^{-1}v] = [r]u_\#[s][r]^{-1}$. We see that the isomorphism $v_\#$ is "conjugated" to the isomorphism $u_\#$ by means of $[r] \in \pi_1(X, x_1)$. Thus, if the fundamental group of a connected space X is commutative (and only in this case), the isomorphism between $\pi_1(X, x_0), \pi_1(X, x_1)$ is "canonical"; that is, it does not depend on the path joining x_0 and x_1. In this case we can speak of the group $\pi_1(X)$ without fixing a base point. Otherwise, we can speak of the fundamental group of X only as of an abstract group; that is, we can say that it is finite, or unipotent, or finitely generated, and so on, but we cannot, say, specify an element in it.

Remark. The statement $\pi_1(X)$ is trivial means precisely that X is simply connected.

EXERCISE 1. Prove that if $f: X \to Y$ is a homotopy equivalence, then $f_*: \pi_1(X, x_0) \to \pi_1(Y, f(x_0))$ is an isomorphism.

6.3 The First Computation: The Fundamental Group of a Circle

There are two major ways to compute the fundamental groups of a given space. The first one is based on the theory of covering, and we will consider it in this lecture. The main ingredient of second way is the so-called Van Kampen theorem; in the next lecture we will prove this theorem and apply it to the computation of fundamental groups of CW complexes.

We begin with the computation of the fundamental group of the circle. It can be regarded as the first application of the covering method; however, we will not explicitly mention coverings in this section: The definition will be given in Sect. 6.4.

Theorem. *The group $\pi_1(S^1)$ is isomorphic to the group $\mathbb{Z}$ of integers.*

Proof. For every point of the circle, we assign, in the usual way, a real number defined up to a summand of the form $2k\pi$. For the base point, we take 0. A loop $s: I \to S^1$ becomes a multivalued function on the segment $[0, 1]$ whose value at every point is defined up to a summand $2k\pi$ and whose value at 0 and 1 is the set $\{2k\pi\}$ itself. This function has a "univalent branch" $s^\#: [0, 1] \to \mathbb{R}$, which is a continuous function of the value $s^\#(t)$ of which at every point $t \in [0, 1]$ belongs to the set of values of the multivalued function s at t. This function will be unique if we require that $s^\#(0) = 0$ (see Fig. 25).

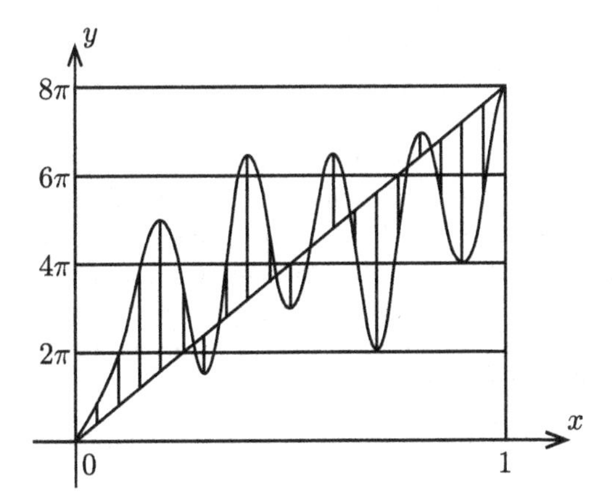

Fig. 25 The function $s^\#$ and its homotopy to a linear function

[For a pedant, we give the details of the construction of the function $s^\#$. Let n be so large that if $|t'' - t'| \leq \frac{1}{n}$, then the points $s(t'), s(t'') \in S^1$ are not diametrically opposite. Then for $0 \leq t \leq \frac{1}{n}$, we take for $s^\#(t)$ the value of $s(t)$ which differs from 0 by less than π; after this, for $\frac{1}{n} \leq t \leq \frac{2}{n}$ we take for $s^\#(t)$ the value of $s(t)$ which differs from $s^\#\left(\frac{1}{n}\right)$ by less than π; and so on.]

Point out two important properties of the function $s^\#$: (1) Its value at $t = 1$ is a multiple of 2π; (2) it depends on s continuously in the sense that if s_t is a homotopy, then $s_t^\#$ is a homotopy. Notice also that every continuous function h: $[0, 1] \to \mathbb{R}$ such that $h(0) = 0$ and $h(1)$ is a multiple of 2π is a function $s^\#$ for some loop $s: I \to S^1$.

To complete the proof, it remains to make the following four obvious remarks. First, the number $k = \frac{s^\#(1)}{2\pi}$ does not vary when we perform a homotopy of s, since a continuous variation of an integer is constant; thus, k depends only on an element of $\pi_1(S^1)$. Second, any k can be obtained in this way: One can take $h_k(t) = 2k\pi t$. Third, any two functions h with the same $h(1)$ are homotopic: A homotopy between $s^\#$ and $h_k(t) = 2k\pi t$ is shown in Fig. 25. Fourth, if $s^\#(1) = 2k\pi$ and $(s')^\#(1) = 2\ell\pi$, then $(ss')^\#(1) = 2(k + \ell)\pi$.

Remark. We are presenting this result as an example, but, actually, it is a crucially important theorem. Just imagine, for a moment, that $\pi_1(S^1)$ be 0. Then the fundamental groups of all spaces would be zeroes! Indeed, let $\sigma \in \pi_1(X, x_0)$ be represented by a loop $s: I \to X$. Then there is a map $f: S^1 \to X$ such that $s = f \circ h_1$, and $[s] = f_*[h_1] = 0$.

Our next goal is to generalize the last proof to a much broader context.

6.4 Coverings: Definition and Examples

Definition. We say that a path connected space T *covers* a path connected space X if there is a continuous map $p: T \to X$ such that every point $x \in X$ has a neighborhood U whose inverse image $p^{-1}(U) \subset T$ falls into a disjoint union of open sets $U_\alpha \subset T$ such that for every α, p maps homeomorphically U_α onto U. in this situation, the map $p: T \to X$ is called a *covering*. We will call open sets U with the described property of $p^{-1}(U)$ "properly covered open sets." Thus, properly covered open sets form an open covering of X.

Example 1. $p: \mathbb{R} \to S^1$, $p(t) = (\cos t, \sin t)$, or, if we interpret S^1 as a unit circle in $\mathbb{C}$, $p(t) = e^{it}$.

Example 2. $p: S^1 \to S^1$, $p(z) = z^k$, where k is a nonzero integer.

Example 3. $T = S^n, X = \mathbb{R}P^n, p: T \to X$ maps a point $x \in S^n$ to a line passing through 0 and x (compare with Sect. 1.2).

It is clear that if $p_1: T_1 \to X_1$ and $p_2: T_2 \to X_2$ are two coverings, then $p_1 \times p_2: T_1 \times T_2 \to X_1 \times X_2$ is also a covering. For example, the square of the covering in Example 1 is a covering of a torus by a plane, $\mathbb{R}^2 \to S^1 \times S^1$. A product of two coverings from Example 2 (with, possibly, different ks) is a covering of a torus by a torus.

EXERCISE 2. Prove that for any $g \geq 2$ a sphere with g handles can cover a sphere with two handles. (Think about which classical surfaces can cover other classical surfaces; however, at the moment, we do not have sufficient technical means to answer this questions. Such technical means will appear in Chap. 2.)

EXERCISE 3. Prove that for every $n \geq 2$ there exists a space homotopy equivalent to the bouquet of n circles which can cover the bouquet of two circles ("the figure-eight space").

6.5 Lifting[2] Paths and Homotopies

In many statements in this section, topological spaces considered are assumed "sufficiently good." Usually this means that the space is *locally path connected*; that is, for every point x and every neighborhood U of x there exists a neighborhood V of x such that $\overline{V} \subset U$ and any two points in V can be connected by a path in U. Sometimes we also require that the space is "semilocally simply connected," which means that for every neighborhood U of x there exists a neighborhood V of x such that $\overline{V} \subset U$ and every loop in V is homotopic to a constant loop *in the whole space*. These properties will be needed to check the continuity of some maps. They will usually routinely hold for spaces we will consider. For this reason, we will never specify the meaning of being sufficiently good in the statements, but sometimes (not always) we will explain in proofs, what and when is needed. However, for the first statement that follows, no assumption like this is needed.

Lifting Path Lemma. *Let $p: T \to X$ be a covering, $\widetilde{x}_0 \in T$, and $s: I \to X$ be a path such that $s(0) = x_0 = p(\widetilde{x}_0)$. Then there exists a unique path $\widetilde{s}: I \to T$ such that $\widetilde{s}(0) = \widetilde{x}_0$ and $p \circ \widetilde{s} = s$.*

Proof of Existence of $\widetilde{s}$. Choose an n such that for every k, $1 \leq k \leq n$, the set $s\left[\dfrac{k-1}{n}, \dfrac{k}{n}\right]$ is contained in some properly covered set, U_k. Assume that for some k, $0 \leq k < n$, there exists a map $\widetilde{s}_k: \left[0, \dfrac{k}{n}\right] \to T$ such that $\widetilde{s}_k(0) = \widetilde{x}_0$ and $p \circ \widetilde{s}_k =$

[2]Sometimes, the terminology of the theory of coverings is based on a visual presentation of a covering, in which T lies "above" X and the projection p is vertical and directed down. This is reflected not only in terminology, but also in many pictures in this section.

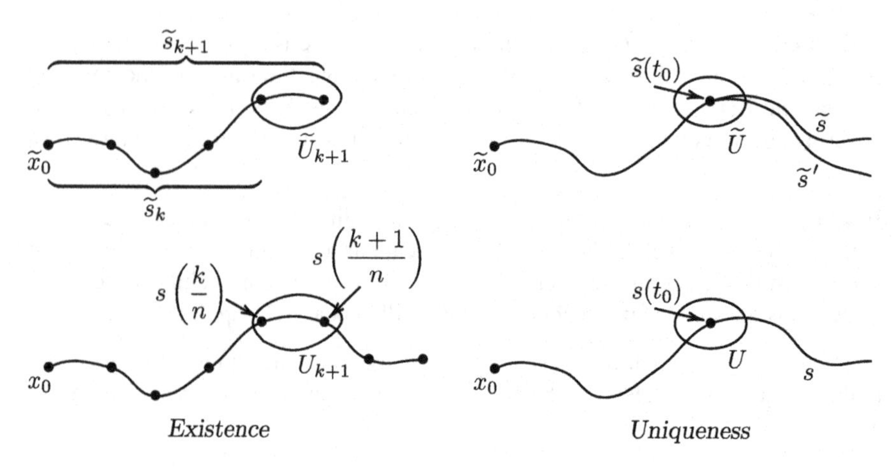

Fig. 26 Proof of the lifting path lemma

$s|_{[0,\frac{k}{n}]}$. For $k = 0$ this holds: $\widetilde{s}_0: 0 \mapsto \widetilde{x}_0$. Since $s\left(\dfrac{k}{n}\right) \in U_{k+1}$ and $p^{-1}(U_{k+1}) = \bigcup_\alpha U_{k+1,\alpha}$ as in the definition of a covering, there exists precisely one α such that $U_{k+1,\alpha} = \widetilde{U}_{k+1} \ni \widetilde{s}_k\left(\dfrac{k}{n}\right)$. Let $p_{k+1}: \widetilde{U}_{k+1} \to U_{k+1}$ be a restriction of p; it is a homeomorphism. Extend $\widetilde{s}_k: \left[0, \dfrac{k}{n}\right] \to T$ to $\widetilde{s}_{k+1}: \left[0, \dfrac{k+1}{n}\right] \to T$ by setting $\widetilde{s}_{k+1}|_{[\frac{k}{n},\frac{k+1}{n}]} = p_{k+1}^{-1} \circ s|_{[\frac{k}{n},\frac{k+1}{n}]}$. This completes the step of induction. The path $\widetilde{s}$ arises as $\widetilde{s}_n$.

Proof of Uniqueness of $\widetilde{s}$. Suppose that there are two different paths, $\widetilde{s}, \widetilde{s}': I \to T$, satisfying the requirements of the lemma. Let $t_0 = \inf\{t \mid \widetilde{s}(t) \neq \widetilde{s}'(t)\}$; then $\widetilde{s}(t_0) = \widetilde{s}'(t_0)$. Let $U \subset X$ be a properly covered neighborhood of $s(t_0)$. Then there is a neighborhood $\widetilde{U} \subset T$ of $\widetilde{s}(t_0) = \widetilde{s}'(t_0)$ which is mapped homeomorphically onto U by p. Since $\widetilde{U}$ is open and the maps $\widetilde{s}, \widetilde{s}'$ are continuous, there exists an $\varepsilon > 0$ such that $\widetilde{s}(t_0 - \varepsilon, t_0 + \varepsilon), \widetilde{s}'(t_0 - \varepsilon, t_0 + \varepsilon) \subset \widetilde{U}$. But since $p \circ \widetilde{s} = p \circ \widetilde{s}' (= s)$ and p maps $\widetilde{U}$ homeomorphically onto U, this means that $\widetilde{s}$ and $\widetilde{s}'$ coincide on $(t_0 - \varepsilon, t_0 + \varepsilon)$, which contradicts the definition of t_0.

Remark. This proof, like some other proofs below, does not use the connectedness of T. If we exclude the condition of the connectedness of T from the definition of the covering, we will get the definition of a "generalized covering," a notion that in many respects is not worse than that of a covering; in particular, the last lemma, as well as several upcoming theorems (such as the next theorem), actually holds for a generalized covering. Still, from the point of view of fundamental groups, we need coverings rather than generalized coverings.

Lifting Homotopy Theorem. *Let $p: T \to X$ be a covering, and let Z be a sufficiently good topological space. Let $f_t: Z \to X$ be a homotopy and $F: Z \to T$ be a continuous map such that $p \circ F = f_0$. Then there exists a unique homotopy $F_t: Z \to T$ such that $F_0 = F$ and $p \circ F_t = f_t$ for all t.*

Proof. For a $z \in Z$, consider the path $s_z: I \to X$, $s_z(t) = f_t(z)$. By the lifting path lemma, there exists a unique path $\widetilde{s}_z: I \to T$ such that $p \circ \widetilde{s}_z = s_z$ and $\widetilde{s}_z(0) = F(z)$. Put $F_t(z) = \widetilde{s}_z(t)$. We will prove that F_t is a homotopy. It is obvious that F_t satisfies the requirements of the theorem, and the uniqueness follows from the uniqueness in the lifting path lemma. It remains to prove the continuity of $\{F_t\}$. The proof uses the local path connectedness of Z.

We denote as $h: Z \times I \to X$ and $H: Z \times I \to T$ the maps $(z,t) \mapsto f_t(z)$ and $(z,t) \mapsto F_t(z)$. The continuity of h is given; we need to prove that H is also continuous. By construction, H is continuous on each segment $z_0 \times I$.

Let $E = \{(z,t) \mid H \text{ is continuous in a neighborhood of } (z,t)\} \subset Z \times I$. Obviously, E is open. First, remark that $E \supset Z \times 0$. Indeed, let $z_0 \in Z$ and U be a properly covered neighborhood of $h(z_0, 0)$. Let $\widetilde{U}$ be a neighborhood of $H(z_0, 0)$ homeomorphically mapped onto U by p. Since h is continuous, there exist neighborhoods $A \subset Z$ of z_0 and $J \subset I$ of 0 such that $h(A \times J) \subset U$. By local connectedness of Z, there exists a neighborhood $B \subset A$ of z_0 such that every two points in B can be joined by a path in A. We will show that $H(B \times J) \subset \widetilde{U}$, and hence $H|_{B \times J} = p^{-1} \circ h|_{B \times J}$ is continuous and $(z_0, 0) \in E$. For a $(z,t) \in B \times J$, consider a path $u: I \to Z \times I$ from $(z_0, 0)$ to (z,t) consisting of a path from (z_0) to $(z, 0)$ within $A \times 0$ and a path from $(z, 0)$ to (z,t) within $z \times [0,t]$. The path $H \circ u$ is continuous; hence, it coincides with $p^{-1} \circ h \circ u$ [both begin at $H(z_0, 0)$ and projected by p onto $h \circ u$], and hence $H(z,t) = (H \circ u)(1) \in V$.

Suppose now that H is discontinuous at some (z_0, t). Let $t_0 = \inf\{t \mid (z_0, t) \notin E\}$. By the remark above, $t_0 > 0$. Let $U \subset X$ be a properly covered neighborhood of $h(z_0, t_0)$, and let $\widetilde{U} \subset T$ be a neighborhood of $H(z_0, t_0)$ which is mapped homeomorphically by p onto U. Since h is continuous, there exist neighborhoods $A \subset Z$ of z_0 and $J \subset I$ of t_0 such that $h(A \times J) \subset U$. We choose successively: a $t_1 < t_0$ in J; neighborhoods $B \subset A$ of z_0 and $K \subset I$ of t_1 such that H is continuous in $B \times K$; a neighborhood $C \subset B$ of z_0 such that every two points in C can be connected by a path in B. We will show that $H(C \times J) \subset V$, which will mean, as before, that H is continuous in $C \times J$, in contradiction to the definition of t_0. For a $(z,t) \in C \times J$ take a path u from (z_0, t_0) to (z,t) consisting of three paths: from (z_0, t_0) to (z_0, t_1) in $z_0 \times [t_1, t_0]$; then from (z_0, t_1) to (z, t_1) in $B \times t_1$; and then from (z, t_1) to (z,t) in $z \times [t_1, t]$. The composition $H \circ u$ is continuous, covers $h \circ u: I \to U$, and hence $H \circ u(I) \subset V$; thus, $H(z,t) = h \circ u(1) \in V$. It is a pleasure to state that this boring proof is over.

Corollary. *If paths $s, s': I \to X$ are homotopic [and in particular $s'(0) = s(0)$ and $s'(1) = s(1)$], then the lifted paths $\widetilde{s}, \widetilde{s}': I \to T$ (meaning that $p \circ \widetilde{s} = s$ and $p \circ \widetilde{s}' = s'$) with $\widetilde{s}(0) = \widetilde{s}'(0)$ are also homotopic (in particular, $\widetilde{s}(1) = \widetilde{s}'(1)$).*

Proof. The lifting homotopy theorem yields a homotopy $\widetilde{s}_t: I \to T$ covering a homotopy $s_t: I \to X$ between the paths s, s' such that $\widetilde{s}_0 = \widetilde{s}$. Since the sets $p^{-1}(s(0)$ and $p^{-1}(s(1))$ are discrete and $s_t(0), s_t(1)$ do not depend on t, so $\widetilde{s}_t(0), \widetilde{s}_t(1)$ also do not depend on t. Thus, $\widetilde{s}_t$ is a path homotopy. Finally, $\widetilde{s}_1 = \widetilde{s}'$, since both paths cover s' and have the same beginning.

6.6 Coverings and Fundamental Groups

Theorem. *If $p: T \to X$ is a covering, then $p_*: \pi_1(T, \widetilde{x}_0) \to \pi_1(X, x_0)$ is a monomorphism (a one-to-one homomorphism).*

Proof. Let s be a loop in $\Omega(T, \widetilde{x}_0)$ and let $p_*[s] = 0$. The latter means that the loop $p \circ s \in \Omega(X, x_0)$ is homotopic to the constant loop $I \to x_0$. Hence, the loops s and $I \to \widetilde{x}_0$ cover homotopic loops $p \circ s$ and const, and hence, they are homotopic by the corollary in Sect. 6.5.

The subgroup $p_*\pi_1(T, \widetilde{x}_0) \approx \pi_1(T, \widetilde{x}_0)$ of $\pi_1(X, x_0)$ is called the *group of the covering*. If we change the point $\widetilde{x}_0$ without changing x_0, then the group of covering will be replaced by a conjugated group (the conjugation is performed by the homotopy class of the loop in X which is obtained by applying p to a path in T joining the two points $\widetilde{x}_0$; see Fig. 27).

For different points x_0, the groups of coverings are taken into each other by isomorphisms of the form $u_\#$ (see Sect. 6.2).

Our next goal is to show that the difference between the groups $\pi_1(T)$ and $\pi_1(X)$ is measured by the number of inverse images of a point of X in T. Namely, we will construct *a canonical bijection between the set $p^{-1}(x_0)$ and the set $\pi_1(X, x_0)/p_*\pi_1(T, \widetilde{x}_0)$.*

Consider a loop s in $\Omega(X, x_0)$. Lift it to a path $\widetilde{s}$ in T with the beginning $\widetilde{x}_0$. Let us assign to s the endpoint $\widetilde{s}(1) \in p^{-1}(x_0)$ of the path $\widetilde{s}$. It is obvious that this point depends only on the homotopy class $[s]$ of s: A homotopy of the loop s is lifted to a homotopy of the path $\widetilde{s}$, and this homotopy leaves the endpoint $\widetilde{s}(1)$

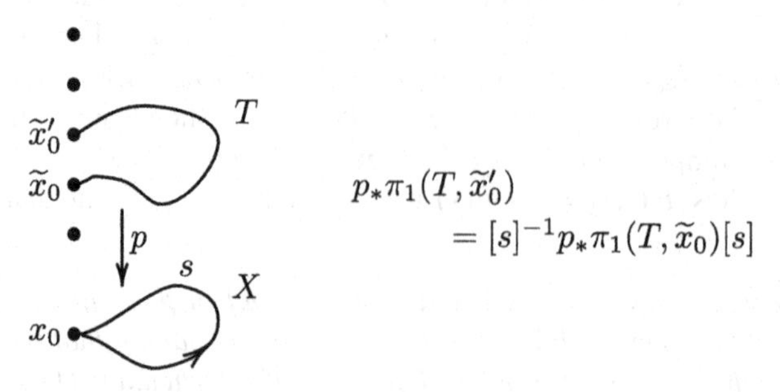

Fig. 27 Dependence of the group of the covering on $\widetilde{x}_0$

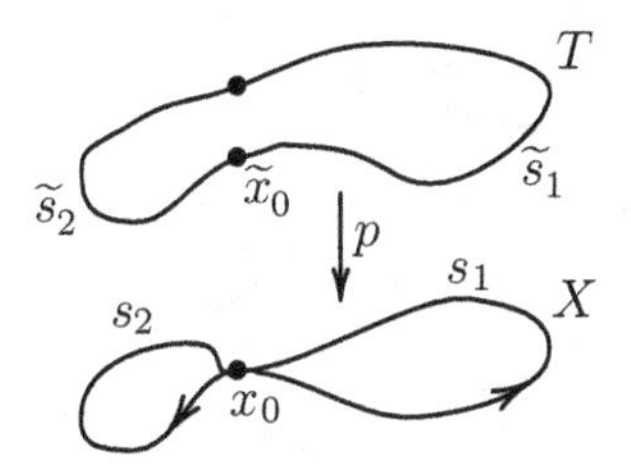

Fig. 28 The construction of a map $\pi_1(X, x_0)/p_*\pi_1(T, \widetilde{x}_0) \to p^{-1}(x_0)$, detail

unchanged, since it can vary only within the discrete set $p^{-1}(x_0)$. It is clear also that the loops $s_1, s_2 \in \Omega(X, x_0)$ determine the same point in $p^{-1}(x_0)$ if and only if the loop $s_1 s_2^{-1}$ is lifted to a loop in T (see Fig. 28), that is, if and only $[s_1 s_2^{-1}] = [s_1]\,[s_2]^{-1} \in p_*\pi_1(T, \widetilde{x}_0)$. Thus, our correspondence $s \mapsto \widetilde{s}(1)$ provides a one-to-one map $\pi_1(X, x_0)/p_*\pi_1(T, \widetilde{x}_0) \to p^{-1}(x_0)$, and this map is onto: Since T is path connected, every point $y \in p^{-1}(x_0)$ can be joined by a path u with this point, and u^{-1} is the lifting of the loop $s = p \circ u^{-1}$ to T with the beginning $\widetilde{x}_0$; thus, $[s] \mapsto y$.

This construction shows that the inverse images of different points x, y of the base X of a covering $p: T \to X$ have the same cardinality. This fact also can be easily deduced directly from the lifting path lemma. Indeed, let $s: I \to X$ be a path joining two points of X. We can lift this path to paths beginning at every point of $p^{-1}(x)$. The ends of these paths belong to $p^{-1}(y)$, and this results in a map $p^{-1}(x) \to p^{-1}(y)$. In a similar way, the path s^{-1} yields a map $p^{-1}(y) \to p^{-1}(x)$, and these maps are inverse to each other, because the products of the paths from the two collections cover the path ss^{-1} homotopic to zero.

If the cardinality of $p^{-1}(x)$ is n, we refer to the covering as an n-fold covering (These terms are often abbreviated to "finite covering," "infinite covering," "countable covering," and so on.).

6.7 Application: Noncommutativity of Fundamental Group

The results of Sect. 6.6 show that if a loop in X lifts to a path in T which is not a loop (the endpoints are different), then this loop in X is not homotopic to a constant (represents a nonidentity element of the fundamental group). Similarly, if two loops in X are lifted to two paths in T with the same beginning but different ends, then these two loops are not homotopic to each other. We can immediately apply this result to discovering nontrivial element of fundamental groups. In particular, we can show that fundamental groups are not necessarily commutative.

Theorem. *Let X be the "figure-eight space," $S^1 \vee S^1$, and let $i, j: S^1 \to X$ be the two natural embeddings of S^1 into $S^1 \vee S^1$. Let $\alpha, \beta \in \pi_1(X)$ be the images of the generator of $\pi_1(S^1) = \mathbb{Z}$ with respect to i_* and j_*. Then $\alpha\beta \neq \beta\alpha$.*

Proof. Consider a fivefold covering of the figure-eight space shown in Fig. 29. The loop $\alpha\beta\alpha^{-1}\beta^{-1}$ (we denote loops by the same letter as their homotopy classes) is

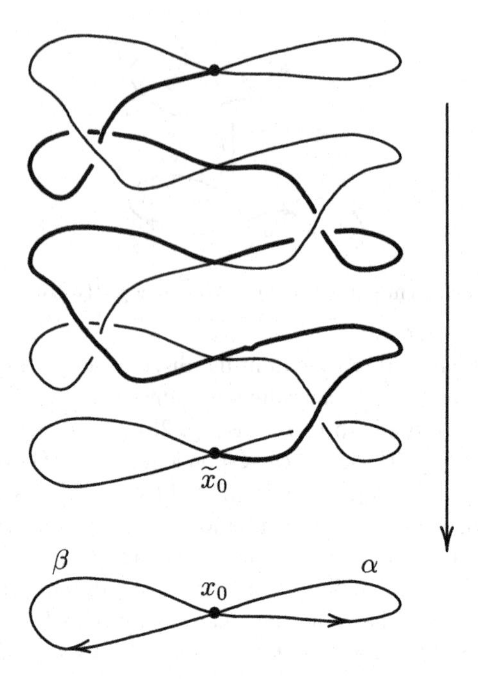

Fig. 29 A fivefold covering of the figure-eight space

lifted to a path shown in the figure by a thick line. Its end does not coincide with its beginning; thus, $\alpha\beta\alpha^{-1}\beta^{-1} \neq 1$.

Remark. This result has an importance comparable with that of the computation of $\pi_1(S^1)$ (see the remark in Sect. 6.2). The commutativity of the group $\pi_1(S^1 \vee S^1)$ would have implied the commutativity of the fundamental groups of all spaces. Indeed, any two loops s, t with the same beginning of any topological space X form a map $f: S^1 \vee S^1 \to X$, and $s = f \circ \alpha$, $t = f \circ \beta$; thus, $[\alpha]\,[\beta] = [\beta]\,[\alpha]$ would have implied $[s]\,[t] = [t]\,[s]$.

Notice in conclusion that the fact that the fundamental group is not always commutative makes it different from the majority of groups assigned in a homotopy invariant way to topological spaces. For this reason, the study of fundamental groups requires in many cases using specific, and not always standard, algebraic means. To avoid this, topologists often prefer to impose (with or without sufficient reasons) on the spaces considered the condition of simply connectedness, or, at least conditions which imply the commutativity of the fundamental group. One such condition is contained in the following exercise.

EXERCISE 4. Prove that if X is a topological group (not necessarily commutative!) or, at least, an H-space, then the fundamental group of X is commutative.

6.8 Deck Transformations

Definition. Let $p: T \to X$ be a covering. A *deck transformation* of this covering is a homeomorphism $f: T \to T$ such that $p \circ f = p$. [This condition means, in particular, that $f(p^{-1}(x)) = p^{-1}(x)$ for every $x \in X$.]

Proposition 1. *Let $y \in T$. A deck transformation $f: T \to T$ is fully determined by the image $f(y)$ of y. In particular, if a deck transformation f has a fixed point, then $f = \text{id}$.*

Proof. Let $z \in T$. Choose a path $s: I \to T$ joining y with z. Let $s': I \to T$ be a path obtained by lifting $p \circ s: I \to X$ with the beginning $f(y)$. The paths $f \circ s$ and s' have a common beginning and a common projection $p \circ s$ in X. Hence, $f \circ s = s'$ and $f(z) = f \circ s(1) = s'(1)$, which is determined by $f(y)$.

Proposition 2. *Let $y, y' \in T$ and $p(y) = p(y')$. A deck transformation $f: T \to T$ such that $f(y) = y'$ exists if and only if $p_*\pi_1(T, y) = p_*\pi_1(T, y')$. (The if part requires that X is sufficiently good.)*

Proof. Let there be a deck transformation $f: T \to T$ with $f(y) = y'$. Let $s: I \to T$ be a loop with the beginning y representing an arbitrarily chosen element α of $\pi_1(T, y)$. Consider the lifting $s': I \to T$ of the loop $p \circ s: I \to X$ with the beginning y'. Then $s' = f \circ s$ (because the two paths have a common beginning and a common projection in X). Hence, s' is a loop, and if $\beta \in \pi_1(T, y')$ is the class of s', then $p_*(\beta) = p_*(\alpha)$. This shows that $p_*\pi_1(T, y) \subset p_*\pi_1(T, y')$ and in the same way, using f^{-1} instead of f, we can prove that $p_*\pi_1(T, y') \subset p_*\pi_1(T, y)$. Thus, $p_*\pi_1(T, y) = p_*\pi_1(T, y')$.

Assume now that $p_*\pi_1(T, y) = p_*\pi_1(T, y')$. For a $z \in T$, choose a path s joining y with z. Lift the path $p \circ s: I \to X$ to a path $s': I \to T$ with the beginning y' and put $f(z) = s'(1)$. This point does not depend on the choice of s. Indeed, let s_1 be another path joining y with z. Since $p_*\pi_1(T, y) = p_*\pi_1(T, y')$, the loop $p \circ (ss_1^{-1})$ is covered by a loop with the beginning y'. But the latter is the product of paths covering $p \circ s$ and $p \circ s_1^{-1}$, which means that the lifting of the path $p \circ s_1$ with the beginning y' has the same end as the lifting of the path $p \circ s_1$. Thus, the map $s: T \to T$ is well defined. To check its continuity, we need the assumption that X is locally path connected (the proof is similar to that in Sect. 6.5; we leave the details to the reader). It is obvious that $p \circ f = p$ and that $f(y) = y'$ (for the latter we apply the construction to the constant path $s: I \to y$). The inverse map f^{-1} is constructed in the same way as f with y and y' swapped.

Theorem. *Let $p: T \to X$ be a covering, and let $\widetilde{x}_0 \in T, x_0 = p(\widetilde{x}_0)$. The group D of deck transformations of the covering p is isomorphic to the quotient of the normalizer $N = \{\gamma \in \pi_1(X, x_0) \mid \gamma p_*\pi_1(T, \widetilde{x}_0)\gamma^{-1} = p_*\pi_1(T, \widetilde{x}_0)\}$ of the group of covering $p_*\pi_1(T, \widetilde{x}_0)$ over this group.*

Proof. We already have a bijection between the sets $p^{-1}(x_0)$ and $\pi_1(X, x_0)/p_*\pi_1(T, \widetilde{x}_0)$ (Sect. 6.6). The orbit of $\widetilde{x}_0$ with respect to the action of D is a subset of $p^{-1}(x_0)$ which consists of cosets whose elements α satisfy the condition $\alpha p_* \pi_1(T, \widetilde{x}_0)\alpha^{-1} = p_* \pi_1(T, \widetilde{x}_0)$. These are cosets of $p_* \pi_1(T, \widetilde{x}_0)$ in the normalizer of $p_* \pi_1(T, \widetilde{x}_0)$.

6.9 Regular Coverings and Universal Coverings

Definition. A covering $p: T \to X$ is called *regular* if the group of covering $p_* \pi_1(T, \widetilde{x}_0)$ is a normal subgroup of $\pi_1(X, x_0)$ (we already know that this property does not depend on the choice of $\widetilde{x}_0 \in T$; see Sect. 6.6).

Equivalent Definition. A covering $p: T \to X$ is regular if the group of deck transformations acts transitively on $p^{-1}(x_0)$ (again, this property does not depend on the choice of $x_0 \in X$; the equivalence of the two definitions is contained in the theorem of Sect. 6.8).

Thus, for a regular covering $p: T \to X$, the orbits of the group D of deck transformations coincide with inverse images $p^{-1}(x_0)$ of points of X in T. This means that $X = T/D$, the orbit space of the group action (it is obvious that the orbit space topology is the same as topology of X. This also provides a new approach to a definition of regular coverings. Let T be a connected topological space with a discrete action of a group D (meaning that every point $y \in T$ has a neighborhood U such that sets $d \cdot U$ for all $d \in D$ are mutually disjoint). Then the projection $T \to T/D$ is a regular covering, and all regular coverings can be constructed in this way.

EXERCISE 5. Prove that any twofold covering is regular (this is equivalent to the well-known algebraic fact: For any group, any subgroup of index 2 is normal).

EXERCISE 6. Construct an example of an irregular threefold covering over the figure-eight space and over the sphere with two handles.

Definition. A covering $p: T \to X$ is called *universal* if the space T is simply connected.

Obviously, all universal coverings are regular.

Our observation above shows that for every point $x_0 \in X$ there is a one-to-one correspondence between $\pi_1(X, x_0)$ and $p^{-1}(x_0)$. Moreover, it is possible to give this correspondence an appearance of a group isomorphism. Namely, let there be a discrete action of a group D in a simply connected space T; then $\pi_1(T/D) \cong D$. Actually, the computation of $\pi_1(S^1)$ in Sect. 6.3 can be regarded as an application of this theorem (see the following examples).

Example 1. The covering $p: \mathbb{R} \to S^1$ (see Sect. 6.4) is a universal covering corresponding to the action of $\mathbb{Z}$ in $\mathbb{R}$, $(n \in \mathbb{Z}): x \mapsto x + 2n\pi$.

Example 2. The covering $p: S^n \to \mathbb{R}P^n$ (again see Sect. 6.4) is a universal covering with the group $\mathbb{Z}_2$ acting in S^n by the antipodal map. Thus, the fundamental group of $\mathbb{R}P^n$ is isomorphic to $\mathbb{Z}_2$.

Example 3. If Γ is a discrete subgroup of a topological group G, then there arises a regular covering $G \to G/\Gamma$. For example, there are many known discrete subgroups in the group $SO(3)$: the dihedral groups, the groups of symmetries of Platonic solids, and so forth. For each of these groups Γ there arises a regular covering $SO(3) \to SO(3)/\Gamma$; since $SO(3)$ is (canonically homeomorphic to $\mathbb{R}P^3$; see Sect. 1.7), we can combine this covering with the covering $S^3 \to \mathbb{R}P^3$ from the previous example and obtain a universal covering over $SO(3)/\Gamma$.

Example 4. Let X be the union of all lines $x = n$, $n \in \mathbb{Z}$ and $y = n$, $n \in Z$ (an infinite sheet of graph paper). The group $\mathbb{Z} \times \mathbb{Z}$ acts in X in the obvious way (and this action is discrete). The quotient $X/(\mathbb{Z} \times \mathbb{Z})$ is, obviously, the figure-eight space (the map $p: X \to S^1 \vee S^1$ maps every vertical segment $[(m, n), (m, n + 1)]$ onto the left S^1 and every horizontal segment $[(m, n), (m + 1, n)]$ onto the right S^1). Thus, we have a regular covering $X \to S^1 \vee S^1$.

Example 5. Figure 30 presents a space of a universal covering over $S^1 \vee S^1$ (we leave details to the reader).

(There is a similar construction of a universal covering over a bouquet of a set of circles.)

EXERCISE 7. Prove that every classical surface (Sect. 1.10) without holes, except S^2 and $\mathbb{R}P^2$ has a universal covering with the space homeomorphic to $\mathbb{R}^2$.

6.10 Lifting Maps

Theorem. *Let $p: T \to X$ be a covering, and let Z be a path connected space. Let $\widetilde{x}_0 \in T$, $x_0 = p(\widetilde{x}_0) \in X$, and $z_0 \in Z$ be base points and let $f: Z \to X$ be a continuous map such that $f(z_0) = x_0$. Then*

(1) *There exists no more than one continuous map $F: Z \to T$ such that $F(z_0) = \widetilde{x}_0$ and $p \circ F = f$.*
(2) *If the space Z is good enough, then the map F with the properties listed in* (1) *exists if and only if $f_* \pi_1(Z, z_0) \subset p_* \pi_1(T, \widetilde{x}_0)$.*

Proof. Let F', F'' be two such maps. Let $z \in Z$ be an arbitrary point, and let $s: I \to Z$ be a path joining z_0 with z. The paths $F' \circ s$, $F'' \circ s: I \to T$ both begin at $\widetilde{x}_0$ and both are projected by p into the path $f \circ s: I \to X$. Hence, they coincide, and $F'(z) = F' \circ s(1) = F'' \circ s(1) = F''(z)$, so $F' = F''$.

The same argumentation gives a clue to constructing the map F. For a point $z \in Z$, we take a path $s: I \to Z$ joining z_0 with z, then lift the path $f \circ s: I \to X$ to a path $\widetilde{s}: I \to T$ beginning at $\widetilde{x}_0$ and put $F(z) = \widetilde{s}(1)$. However, we need to verify that

Fig. 30 For Example 5: the space of a universal covering over the figure-eight space

this $\widetilde{s}(1)$ does not depend on the choice of s. For this, the inclusion $f_*\pi_1(Z, z_0) \subset p_*\pi_1(T, \widetilde{x}_0)$ provides a necessary and sufficient condition. Indeed, if s' is another path joining z_0 with z, then $s(s')^{-1}$ is a loop in Z, and the equality $\widetilde{s}'(1) = \widetilde{s}(1)$ means precisely that the loop $f \circ (s(s')^{-1})$ is lifted to a loop with its beginning at $\widetilde{x}_0$ in T. In other words, we need that $f_*[s(s')^{-1}] \in p_*\pi_1(T, \widetilde{x}_0)$. It remains to check that the map F constructed is continuous. This holds if Z is locally path connected; the proof is a replica of the proof of a similar statement in Sect. 6.5, and the reader, who prefers to do that, can recover it.

Corollary. *If the sufficiently good space Z is simply connected, then for every covering $p: T \to X$, the map p establishes a homeomorphism between base point mapping spaces $\mathcal{C}_b(Z, T)$ and $\mathcal{C}_b(Z, X)$ (the base points $z_0 \in Z, \widetilde{x}_0 \in T, x_0 \in X$ are as in the theorem). Subsequently, there arises a bijection $p_*: \pi_b(Z, T) \to \pi_b(Z, X)$.*

6.11 A Criterion of Equivalence of Coverings

Definition. Coverings $p_1: T_1 \to X$ and $p_2: T_2 \to X$ (with the same base) are called *equivalent* if there exists a homeomorphism $f: T_1 \to T_2$ such that the diagram

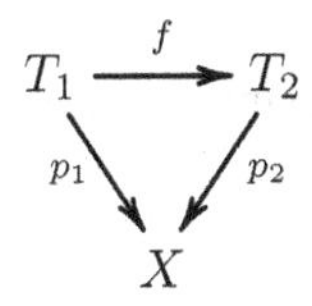

is commutative. Such a map f is called an *equivalence*.

Theorem. *Let $p_1: T_1 \to X$, $p_2: T_2 \to X$ be coverings and $x \in X, \widetilde{x}_1 \in T_1, \widetilde{x}_2 \in T_2$ be points such $p_1(\widetilde{x}_1) = p_2(\widetilde{x}_2) = x$. If the space X is good enough, then the coverings p_1, p_2 are equivalent if and only if the subgroups $(p_1)_*\pi_1(T_1, \widetilde{x}_1), (p_2)_*\pi_1(T_1, \widetilde{x}_2)$ of the group $\pi_1(X, x)$ are conjugated.*

In particular, two universal coverings over a sufficiently good space are always equivalent.

Proof of Theorem. In one direction (if the coverings are equivalent, then the groups are conjugated), this is, essentially, known to us (see Sect. 6.6). Let us prove the statement in the other direction. Let the subgroups $(p_1)_*\pi_1(T_1, \widetilde{x}_1), (p_2)_*\pi_1(T_1, \widetilde{x}_2)$ of the group $\pi_1(X, x)$ be conjugated. We can change, if necessary, the point $\widetilde{x}_2$ in such a way that the subgroups $(p_1)_*\pi_1(T_1, \widetilde{x}_1), (p_2)_*\pi_1(T_1, \widetilde{x}_2)$ will be equal, not just conjugated (see again Sect. 6.6). Then we apply the theorem from Sect. 6.10 to the map $p_1: T_1 \to X$ and the covering $p_2: T_2 \to X$, and get a continuous map $f: T_1 \to T_2$ such that $p_2 \circ f = p_1$. In the same way (swapping p_1 and p_2), we get a continuous map $g: T_2 \to T_1$ such that $p_1 \circ g = p_2$. Both maps are unique. This uniqueness implies the equalities $g \circ f = \mathrm{id}_{T_1}, f \circ g = \mathrm{id}_{T_2}$; indeed, both maps $g \circ f$ and id_{T_1} satisfy the conditions imposed on f in the case when $T_2 = T_1, p_2 = p_1$ and similarly for $f \circ g$ and id_{T_2}.

6.12 Existence, Classification, and Hierarchy of Coverings

Theorem. *Let X be a sufficiently good path connected space with a base point x_0. Then*

(1) *For every subgroup $H \subset \pi_1(X, x_0)$ there exists a unique, up to a base point equivalence, covering $p: (T, \widetilde{x}_0) \to (X, x_0)$ such that $p_*\pi_1(T, \widetilde{x}_0) = H$.*
(2) *Let H_1, H_2 be subgroups of $\pi_1(X, x_0)$ and let $H_1 \subset H_2$. Let $p_1: (T_1, \widetilde{x}_{10}) \to (X, x_0)$, $p_2: (T_2, \widetilde{x}_{20}) \to (X, x_0)$ be the coverings with $(p_i)_*\pi_1(T_i, \widetilde{x}_{i0}) = H_i$ existing and unique by the part (1). Then there exists a unique covering $q: (T_1, \widetilde{x}_{10}) \to (T_2, \widetilde{x}_{20})$ such that $p_2 \circ q = p_1$.*

Proof. The only thing which we still need to prove is the existence statement of part (1). Notice that this is the first (and last) case, when we need to use not only local connectedness, but also semilocal simply connectedness of the space X.

Define an equivalence relation in the set $E(X, x_0)$ of paths on X beginning at x_0: Two paths $s, s' \in E(X, x_0)$ are equivalent if $s(1) = s'(1)$ and the homotopy class of the loop $s(s')^{-1} \in \Omega(X, x_0)$ belongs to H. Let T be the set of equivalence classes. Define a topology in T in the following way. Let $V \subset U$ be open sets in X, and let $s \in E(X, x_0)$ be a path with $s(1) \in V$. Denote by $N(U, V, s)$ the set of equivalence classes of paths sw where $w: I \to X$ is a path with $w(0) = s(1), w(1) \in V$, and $w(I) \subset U$. The sets $N(U, V, s)$ form a base of topology in T (recall that a family $\mathcal{F}$ of subsets of some set Z is a base of a topology in Z if and only if for every $U, V \in \mathcal{F}$ and $z \in U \cap V$ there exists a $W \in \mathcal{F}$ such that $z \in W \subset U \cap V$; this condition obviously holds for $\mathcal{F} = \{N(U, V, s)\}$). Thus, T becomes a topological space.

Let $p: T \to X$ take the class of a path $s \in E(X, x_0)$ into $s(1) \in X$; obviously, it is well defined. It is clear also that p is continuous: For an open $V \subset X$, $p^{-1}(V) = \bigcup_{s(1)\in V} N(X, V, s)$.

For a path $s \in E(X, x_0)$ and $\tau \in I$, let $s_\tau: I \to X$ be a path defined by the formula $s_\tau(t) = s(\tau t)$, Obviously, the function $\tau \mapsto s_\tau$ defines a continuous map $I \to T$, and it is a path joining the class of s with the class $\widetilde{x_0} \in T$ of the constant path $I \to x_0$. In particular, T is path connected.

Let us prove now that p is a covering. Let $x \in X$, let U be a neighborhood of x such that every loop in U is homotopic to a constant loop in X, and let $V \subset U$ be a neighborhood of x such that every point in V can be joined with x by a path in U. We will prove that V is properly covered with respect to p. Let $y \in p^{-1}(x)$ be represented by a path $s \in E(X, x_0)$ [thus $s(1) = x$]. Then the neighborhood $N(U, V, s)$ of y is one-to-one projected by p onto V. Indeed, for an $x' \in V$, there exists a path w joining x with x' in U. Moreover, this path is unique up to a homotopy. Hence, there exists a unique $y' \in N(U, V, s)$ (the class of the path sw) such that $p(y') = x'$.

Finally, let us prove that the group of the covering $p: T \to X$ is H. We need to prove that a loop $s \in \Omega(X, x_0)$ is covered by a loop in $\Omega(T, \widetilde{x_0})$ if and only if the homotopy class of this loop belongs to H. But the loop s is covered by the loop $\widetilde{s} = \{\tau \mapsto s_\tau\}$, and $\widetilde{s}(1) = \widetilde{x_0}$ if and only if the path s and the constant path $I \to x_0$ belong to the same equivalence class, which means precisely that $[s] \in H$.

Corollary. *Any sufficiently good path connected space possesses a (unique) universal covering.*

Lecture 7 Van Kampen's Theorem and Fundamental Groups of CW Complexes

7.1 Van Kampen's Theorem

Let X be a topological space, and let $U_1, U_2 \subset X$ be open subsets such that $U_1 \cup U_2 = X$ and $U_1 \cap U_2 \neq \emptyset$. We assume also that X, U_1, U_2, and $U_1 \cap U_2$ are (path) connected. Let $x_0 \in U_1 \cap U_2$. Then the inclusion maps of U_1 and U_2 into X and of $U_1 \cap U_2$ into U_1 and U_2 induce homomorphisms of fundamental groups which form a commutative diagram

$$\begin{array}{ccccc} & & \pi_1(U_1, x_0) & & \\ & \overset{j_1}{\nearrow} & & \overset{i_1}{\searrow} & \\ \pi_1(U_1 \cap U_2, x_0) & & & & \pi_1(X, x_0) \\ & \underset{j_2}{\searrow} & & \underset{i_2}{\nearrow} & \\ & & \pi_1(U_2, x_0) & & \end{array}$$

Our goal is to reconstruct the group $\pi_1(X, x_0)$ (and homomorphisms i_1, i_2) from the groups $\pi_1(U_1, x_0), \pi_1(U_2, x_0), \pi_1(U_1 \cap U_2, x_0)$, and homomorphisms j_1, j_2. We formulate the result in terms of generators and relations.

Theorem (Van Kampen[3]). *Let A_i, R_i be systems of generators and relations for the groups $\pi_1(U_i, x_0)$ $(i = 1, 2)$. Let B be a system of generators for the group $\pi_1(U_1 \cap U_2, x_0)$. Then the group $\pi_1(X, x_0)$ is generated by the set $A_1 \coprod A_2$ with the set of relations $R_1 \coprod R_2 \coprod B$ where the relation corresponding to $b \in B$ is $j_1(b) = j_2(b)$, where, in turn, $j_1(b)$ is regarded as a word in A_1 and $j_2(b)$ is regarded as a word in A_2. The homomorphisms i_1, i_2 map generators from A_1, A_2 into the same elements of A_1, A_2 regarded as generators of $\pi_1(X, x_0)$.*

This theorem is covered by the following two propositions.

Proposition 1. *Every element of $\pi_1(X, x_0)$ may be presented as a product*

$$i_{k_1}(\alpha_1) \dots i_{k_N}(\alpha_N), \qquad (*)$$

where $k_s = 1$ or 2 and $\alpha_s \in \pi_1(U_{k_s}, x_0)$.

Proof. Let $\sigma: I \to X$ be a loop representing the chosen element of $\pi_1(X, x_0)$. A simple fact from analysis states that there exists an n such that for each $r = 1, \dots, n$, $\sigma\left(\left[\frac{r-1}{n}, \frac{r}{n}\right]\right)$ is contained in U_1 or U_2 (or both). Define $\sigma_r: I \to X$ by $\sigma_r(t) = \sigma\left(\frac{r-1+t}{n}\right)$. Then $\sigma = \sigma_1\sigma_2 \dots \sigma_n$.

[3]This theorem is often called Seifert–Van Kampen Theorem.

For each $r = 1, \dots n-1$, fix a path $\tau_r: I \to X$ such that (i) $\tau_r(0) = x_0$; (ii) $\tau_r(1) = \sigma\left(\frac{r}{n}\right)$; (iii) if $\sigma\left(\frac{r}{n}\right) \in U_i$, then $\tau_r(I) \subset U_i$ $(i = 1, 2)$. [Condition (iii) implies that if $\sigma\left(\frac{r}{n}\right) \in U_1 \cap U_2$, then $\tau_r(I) \subset U_1 \cap U_2$; the existence of such paths τ_r follows from the connectedness of U_1, U_2, and $U_1 \cap U_2$.]

We have

$$\sigma = \sigma_1\sigma_2 \dots \sigma_n \sim \sigma_1\tau_1^{-1}\tau_1\sigma_2\tau_2^{-1}\tau_2 \dots \tau_{n-1}^{-1}\tau_{n-1}\sigma_n.$$

The latter is the product of loops

$$\sigma_1\tau_1^{-1}, \tau_1\sigma_2\tau_2^{-1}, \tau_2\sigma_3\tau_3^{-1}, \dots, \tau_{n-2}\sigma_{n-1}\tau_{n-1}^{-1}, \tau_{n-1}\sigma_n,$$

each of which lies either in U_1, or in U_2. The proposition follows.

To state Proposition 2, we describe what we will call *admissible transformations* of expression (1). There are two sorts of admissible transformations.

(*Splitting/Merging*). If $k_s = k_{s+1}$, we can replace $i_{k_s}(\alpha_s)i_{k_{s+1}}(\alpha_{s+1})$ by $i_{k_s}(\alpha_s\alpha_{s+1})$; and vice versa: If $\alpha_s = \alpha_s'\alpha_s''$, then we can replace $i_{k_s}(\alpha_s)$ by $i_{k_s}(\alpha_s')i_{k_s}(\alpha_s'')$.

(*Renaming*). If $\alpha_s = j_{k_s}(\beta)$, then we can replace $i_{k_s}(\alpha_s)$ by $i_{k_s'}(\alpha_s')$, where $k_s' \neq k_s$ (that is, $k_s' = 3 - k_s$) and $\alpha_s' = j_{k_s'}(\beta)$.

Proposition 2. *The word* (1) *is equal to* $1 \in \pi_1(X, x_0)$ *if and only if it can be reduced by admissible transformations to the trivial word* ($i_1(1)$ *or* $i_2(1)$).

Proof. The *if* part of this proposition is obvious [admissible transformations do not change the product (1)]. Prove the *only if* part.

Suppose that a product (∗) is equal to $1 \in \pi_1(X, x_0)$. Let $\sigma_r: I \to U_{k_r} \subset X$ be a loop (of X) representing $i_{k_r}(\alpha_r)$, and let $\sigma: I \to X$ be the loop $\sigma\left(\frac{r+t}{n}\right) = \sigma_r(t)$ $(t \in I)$. Then $\sigma \sim$ const; let $S: I \times I \to X$ be a homotopy, that is, $S(t, 0) = S(t, 1) = S(1, t) = x_0$, $S(0, t) = \sigma(t)$. Choose a big m such that for all r, s, $S\left(\left[\frac{r}{m}, \frac{r+1}{m}\right] \times \left[\frac{s}{m}, \frac{s+1}{m}\right]\right) \subset U_1$ or U_2; it will be convenient to assume that $n \mid m$, that is, $m = \ell n$.

For each r, s between 0 and m, fix a path $\tau_{rs}: I \to X$ such that $\tau_{rs}(0) = x_0$, $\tau_{rs}(1) = S\left(\frac{r}{m}, \frac{s}{m}\right)$ and if $S\left(\frac{r}{m}, \frac{s}{m}\right)$ is contained in $U_1, U_2, U_1 \cap U_2$, or x_0, then so is $\tau_{rs}(I)$. Consider short "horizontal" and "vertical" paths

$$\alpha_{rs}: I \to X,\ \alpha_{rs}(t) = S\left(\frac{r-1+t}{m}, \frac{s}{m}\right),\ 1 \le r \le m,\ 0 \le s \le m.$$

$$\beta_{rs}: I \to X,\ \beta_{rs}(t) = S\left(\frac{r}{m}, \frac{s-1+t}{m}\right),\ 0 \le r \le m,\ 1 \le s \le m.$$

Then set

$$\alpha'_{rs} = \tau_{rs}\alpha_{rs}\tau^{-1}_{r+1,s},\ \beta'_{rs} = \tau_{rs}\beta_{rs}\tau^{-1}_{r,s+1}.$$

These are loops; each is contained in U_1 or U_2 (or both). We will transform the product loop $\sigma_1 \dots \sigma_m$; in $\pi_1(X, x_0)$, each of the factors represents an element of the form $i_k(\alpha)$, and we will see that, at the level of $\pi_1(X, x_0)$, our transformations will be admissible. For this purpose, we fix for the homotopy type of each loop $\alpha'_{rs}, \beta'_{rs}$ a representation as $i_1(\alpha)$, or $i_2(\alpha)$, or, if the loop is contained in $U_1 \cap U_2$, as $i_1(j_i(\beta)) = i_2(j_2(\beta))$ with the understanding that switching this last representations is an admissible transformation of *renaming*.

First, we replace each σ_r by the product $\alpha'_{(r-1)\ell+1,0}\alpha'_{(r-1)\ell+2,0}\dots\alpha'_{r\ell,0}$. As a result, our product is replaced by $\alpha'_{1,0}\alpha'_{2,0}\dots\alpha'_{m,0}$, and, in $\pi_1(X, x_0)$, this transition is a sequence of the *splitting* transformations (and, possibly *renaming*). Then multiply this product from the left by m trivial loops (which also may be regarded as a sequence of *splittings* and *renamings*):

$$\alpha'_{1,0}\alpha'_{2,0}\dots\alpha'_{m,0}\beta'_{m1}\beta'_{m2}\dots\beta'_{m,m}.$$

Using a sequence of admissible transformations, we join this product with

$$\beta'_{0,1}\beta'_{0,2}\dots\beta'_{0,m}\alpha'_{1,m}\alpha'_{2,m}\dots\alpha'_{m,m},$$

which is a product of trivial loops. An intermediate step is

$$\beta'_{01}\dots\beta'_{0s}\alpha'_{1s}\dots\alpha'_{rs}\beta'_{r,s+1}\alpha'_{r+1,s+1}\dots\alpha'_{m,s+1}\beta'_{m,s+2}\dots\beta'_{mm}$$

(if r or s is equal to 0 or m, some groups of factors may be missing). The steps are labeled with pairs (r, s) and are performed in the following order: $(m, 0) \to (m-1, 0) \to \dots \to (0, 0) \to (m-1, 1) \to (m-2, 1) \to \dots \to (0, 1) \to (m-1, 2) \to \dots \to (1, m) \to (0, m)$. One step consists in replacing $\alpha'_{r+1,s}\beta'_{r+1,s+1} \to \beta'_{r,s+1}\alpha'_{r+1,s+1}$. Assume that $S\left(\left[\frac{r}{m}, \frac{r+1}{m}\right] \times \left[\frac{s}{m}, \frac{s+1}{m}\right]\right) \subset U_1$ (the case of U_2 is absolutely similar). Then $\alpha'_{r+1,s}(I), \beta'_{r+1,s+1}(I), \beta'_{r,s+1}(I)$, and $\alpha'_{r+1,s+1}(I)$ are all contained in U_1 and the products are, obviously, homotopic in U_1. So our transition is made by (if needed) *renaming* the homotopy classes of $\alpha'_{r+1,s}, \beta'_{r+1,s+1}$; then *merging* these two classes; then *splitting* them into the product of homotopy classes of $\beta'_{r,s+1}, \alpha'_{r+1,s+1}$; then (if needed) renaming these classes. These procedure proves Proposition.

In these exercises, U_1 and U_2 are open subsets of a space X with $U_1 \cup U_2 = X$, and $x_0 \in U_1 \cap U_2$.

EXERCISE 1. Prove that if U_1 and U_2 are simply connected and $U_1 \cap U_2$ is path connected, then X is simply connected.

EXERCISE 2. Prove that if U_1 and U_2 are simply connected and $U_1 \cap U_2$ consists of two path components connected, then $\pi_1(X) \cong \mathbb{Z}$.

EXERCISE 3. Prove that if U_1 and U_2 are path connected and $U_1 \cap U_2$ is not path connected, then X is simply connected.

Terminological remarks. The operations over groups used in the statement and the proof of Van Kampen's theorem have standard names and notations in algebra. For groups G_1, G_2 with the sets of generators A_1, A_2 and the set of defining relations R_1, R_2 their *free product* $G_1 * G_2$ is defined as the group with the set of generators $A_1 \coprod A_2$ and the set of defining relations $R_1 \coprod R_2$. A more invariant (not depending on the sets of generators and relations) definition: $G_1 * G_2$ is the group of words

$$g_1 g_2 g_3 \dots g_n, \ g_k \in G_1 \text{ or } G_2$$

with obvious identifications (if g_k, g_{k+1} belong to the same group, then we have the right to replace them by their product; the inverse operation is also allowed) and group structure. Examples: $\mathbb{Z} * \mathbb{Z}$ is a free group with two generators; $\mathbb{Z} * \mathbb{Z} * \mathbb{Z}$ is a free group with three generators; and so on.

One more equivalent, axiomatic, definition: A group P given with monomorphisms $i_1: G_1 \to P$, $i_2: G_2 \to P$ is called a free product of G_1 and G_2 if for any homomorphisms $f_1: G_1 \to H, f_2: G_2 \to H$ there exists a unique homomorphism $f: P \to H$ such that $f_1 = f \circ i_1, f_2 = f \circ i_2$. The existence and uniqueness, up to a canonical isomorphism of a thusly defined free product are easily checked.)

There is a generalization of this notion. Let Γ be one more group, and let $\gamma_1: \Gamma_1 \to G_1$, $\gamma_2: \Gamma_2 \to G_2$ be homomorphisms. The *amalgamated product* $G_1 *_\Gamma G_2$ is defined as the set of words as before with one more admissible operation: If $g_k \in G_1$ and $g_k = \gamma_1(h)$, then we can replace g_k by $\gamma_2(h)$; and the same for $g_k \in G_2$. The axiomatic definition also can be modified to give the amalgamated product: A group H with given homomorphisms $i_1: G_1 \to P$, $i_2: G_2 \to P$ such that $i_1 \circ \gamma_1 = i_2 \circ \gamma_2$ is $G_1 *_\Gamma G_2$ if for any homomorphisms $f_1: G_1 \to H, f_2: G_2 \to H$ such that $f_1 \circ \gamma_1 = f_2 \circ \gamma_2$ there exists a unique homomorphism $f: P \to H$ such that $f_1 = f \circ i_1, f_2 = f \circ i_2$. (The notation $G_1 *_\Gamma G_2$ may be misleading since it does not specify γ_1 and γ_2; in some cases additional explanations may be necessary.)

Thus, Van Kampen's theorem states that

$$\pi_1(X, x_0) = \pi_1(U_1, x_0) *_{\pi_1(U_1 \cap U_2, x_0)} \pi_1(U_2, x_0).$$

EXERCISE 4. Prove that the group $\mathbb{Z}_2 * \mathbb{Z}_2$ has a (normal) subgroup of index 2 isomorphic to $\mathbb{Z}$.

EXERCISE 5. Prove that $SL(2, \mathbb{Z}) \cong \mathbb{Z}_4 *_{\mathbb{Z}_2} \mathbb{Z}_6$.

7.2 First Applications of Van Kampen's Theorem

Theorem. *For sufficiently good spaces X, Y with base points, $\pi_1(X \vee Y) = \pi_1(X) * \pi_1(Y)$.*

Proof. All we need from X and Y is that base points have base point contractible neighborhoods U and V in X and Y. Then, by Van Kampen's theorem, $\pi_1(X \vee Y) = \pi_1(X \vee V) *_{\pi_1(U \vee V)} \pi_1(U \vee Y)$, but X is a deformation retract of $X \vee V$, Y is a deformation retract of $U \vee Y$, and $U \vee V$ is contractible. Hence, $\pi_1(X \vee Y) = \pi_1(X) * \pi_1(Y)$.

Corollary. *The fundamental group of a bouquet of n circles is a free group with n generators.*

Remark. Actually, the fundamental group of a bouquet of any set of circles is a free group with generators corresponding to the circles, provided that the bouquet is endowed by the weak topology. Indeed, any loop of such a bouquet is contained in a finite subbouquet, and the same for a homotopy of loops.

EXERCISE 6. Prove that the suspension over any nonempty path connected space is simply connected. (See Exercise 1.)

EXERCISE 7. Prove that the join of two nonempty path connected spaces is simply connected,

EXERCISE 8. (A generalization of Exercise 7.) Prove that the join of two nonempty spaces, of which one is path connected, is simply connected.

7.3 A More Serious Application of Van Kampen's Theorem: Groups of Knots and Links

A *knot* is a closed nonself-intersecting smooth curve in $\mathbb{R}^3$. Knots K, K' are called isotopic if there exists a homotopy $h_t: \mathbb{R}^3 \to \mathbb{R}^3$ consisting of smooth homeomorphisms such that $h_0 = \mathrm{id}$ and $h_1(K) = K'$. An important invariant related to a knot is the fundamental group of its complement; sometimes it is briefly called *the group of a knot*. The following is obvious.

Theorem 1. *If the knots K and K' are isotopic, then $\pi_1(\mathbb{R}^3 - K) \cong \pi_1(\mathbb{R}^3 - K')$.*

The circle $x^2 + y^2 = 1$ in $\mathbb{R}^2 \subset \mathbb{R}^3$ is, by definition, unknotted. We say that K is an *unknot* if it is isotopic to this circle.

EXERCISE 9. If K is an unknot, then $\pi_1(\mathbb{R}^3 - K) \cong \mathbb{Z}$.

The following result is highly nontrivial.

Theorem 2. *If $\pi_1(\mathbb{R}^3 - K) \cong \mathbb{Z}$, then K is an unknot.*

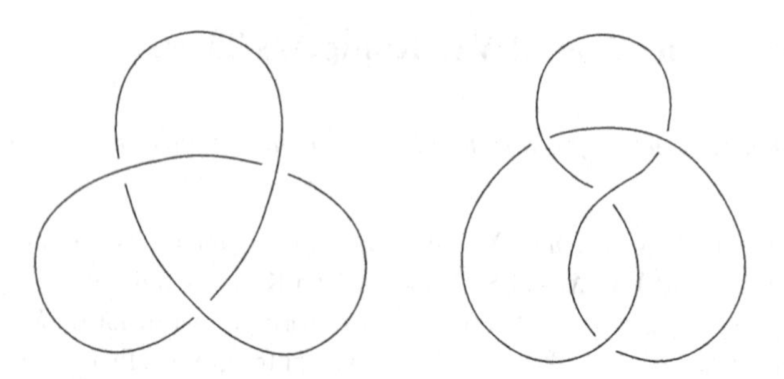

Fig. 31 Knot diagrams: trefoil knot and figure-eight knot

It could be expected that if $\pi_1(\mathbb{R}^3 - K) \cong \pi_1(\mathbb{R}^3 - K')$, then the knots K and K' are isotopic. However, it is wrong. The simplest example: The trefoil knot (see ahead) is known to be not isotopic to its mirror image, but the two knots (the trefoil and the mirror trefoil), certainly, share the fundamental group of the complement. There are more interesting examples, when $\pi_1(\mathbb{R}^3 - K) \cong \pi_1(\mathbb{R}^3 - K')$, but K is not isotopic to either K', or to the mirror image of K'. Still, the fundamental group of the complement is a very effective tool for distinguishing nonisotopic knots.

Knots are usually presented by *knot diagrams*, like the two shown in Fig. 31. These are projections of knots onto a plane; the knots are nonself-intersecting, the projections have self-intersections; the breaks in the curves in the diagram should indicate which of the two strands is above the other one in space. Every knot has a diagram, and a diagram determines the isotopy type of a knot. But isotopic knots may have diagrams that look very differently (see Exercise 10 ahead).

Transformations of a knot diagram which do not change the knot are called *Reidemeister moves*. They are described in the next exercise.

EXERCISE 10. (This exercise has nothing to do with a fundamental group; it is purely geometric. Still, it may be useful for some exercises ahead.) Prove that two knot diagrams represent the same knot (the same isotopy class of knots) if and only if they can be obtained from each other by a series of transformations called *Reidemeister moves*.

Move 1.

Move 2.

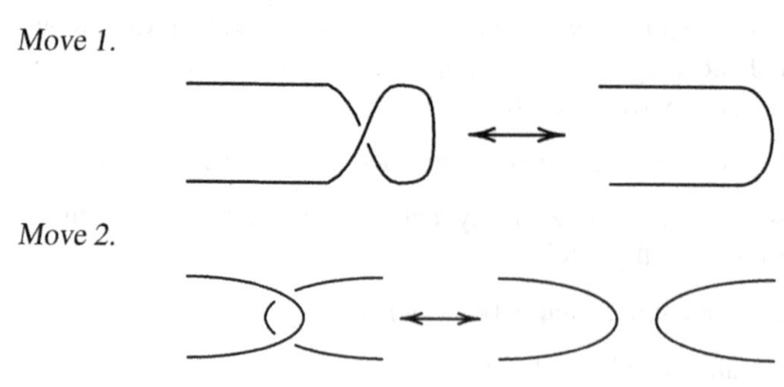

Move 3.

The knots in the preceding diagrams are called a *trefoil knot* (or simply a trefoil) and a *figure-eight knot*. For many reasons, they are usually considered the simplest knot and the second simplest knot.

Here we will develop a machinery for computing fundamental groups for complements of knots with given diagrams. We will begin with the case of a trefoil, but, as will be explained later, the construction is actually quite general.

Our knot K will consist of the diagram (several disjoint *curves*, that is, smooth curvilinear intervals in the plane) and the arcs (*gates*) joining the ends of the curves below the plane (our sheet of paper).

We apply Van Kampen's theorem. Let U be the intersection of $\mathbb{R}^3 - K$ with the half-space above the plane, and let V be the intersection of $\mathbb{R}^3 - K$ with the half-space below the plane. (To make U and V open, we should take slightly overlapping open half-spaces and make the knot solid, thicker than the width of the intersection of the half-spaces; we will not be this scrupulous.) Then $U \cap V$ is the complement to the diagram in the plane. Obviously, U is contractible, and thus $\pi_1(U) = \{1\}$. Furthermore, V is, essentially, a half-space minus the arcs, that is, the gates (our diagram, the engraving on the boundary, does not affect the homotopy type of V); thus, $\pi_1(V)$ is a free group "generated" by the gates. Finally, $U \cap V$ is a perforated plane; its fundamental group is a free group "generated" by the curves. Thus, $\pi_1(\mathbb{R}^3 - K)$ has a system of generators and relations where the generators correspond to the gates and the relations correspond to the curves. Obviously, there are equal numbers of gates and curves. We mark the gates as $a, b, \ldots$ and the curves as $A, B, \ldots$. To specify the generators in $\pi_1(V)$, we need to orient the gates; for this purpose, we simply fix an orientation of the knot and then take for the generators of $\pi_1(V)$ loops which go through the gates in the direction of the knot. For the trefoil diagram all this is done in Fig. 32.

Now, let K be the trefoil, as shown in the diagram. The group $\pi_1(\mathbb{R}^3 - K)$ is generated by a, b, and c. To find the relations, we need an explicit description of generators of the group $\pi_1(U \cap V)$, that is, of the complement to the diagram in the plane. Choose a point x_0 in this complement; for each of the components of the diagram (curves), take a disjoint from the diagram path from x_0 to a point near the curve (B on the diagram), and compose a loop from this path, a loop closely encircling the curve and following at the beginning the orientation of the knot, and the same path back to x_0 (see Fig. 32).

Obviously, loops like this for all the curves represent a system of generators for $\pi_1(U \cap V)$. Moreover, it is easy to express the classes of this loop in $\pi_1(V)$ in terms of generators $a, b, \ldots$. The loop in Fig. 32 obviously belongs to the class $cac^{-1}b^{-1}$; the similar loops around the other two curves correspond, similarly, to the classes $aba^{-1}c^{-1}$ and $bcb^{-1}a^{-1}$. Thus, the fundamental group of the complement to the trefoil is a group with three generators, a, b, c, and three relations,

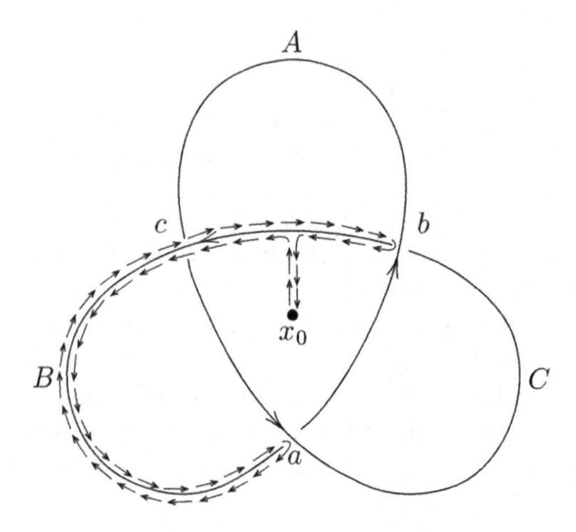

Fig. 32 Calculating the group of a trefoil

$$cac^{-1}b^{-1} = 1, \ aba^{-1}c^{-1} = 1, \ bcb^{-1}a^{-1} = 1,$$

that is,

$$a = bcb^{-1}, \ b = cac^{-1}, \ c = aba^{-1}.$$

The number of generators may be reduced: If we use the last relation to express c in terms of a and b, and plug these expressions into the first two relations, we get $a = baba^{-1}b^{-1}$ and $b = aba^{-1}aab^{-1}a^{-1}$, which are actually the same:

$$aba = bab.$$

We arrive at the following result.

Theorem 3. *The fundamental group of the complement to the trefoil is a group with two generators, a and b, and one relation: aba = bab.*

One can take for the generators $u = ab$ and $v = bab$; then the relation takes the form $u^3 = v^2$.

Theorem 4. *The fundamental group of the complement to the trefoil is not commutative.*

Indeed, the formulas $f(a) = (213)$, $f(b) = (132)$ define a homomorphism of the group $\pi_1(\mathbb{R}^3 - K)$ onto $S(3)$ [because (213) and (132) satisfy the above relation and generate the group $S(3)$].

Corollary. *The trefoil knot is not isotopic to an unknot.*

The general case is presented in Fig. 33. As before, we orient the gates according to the chosen orientation of the knot.

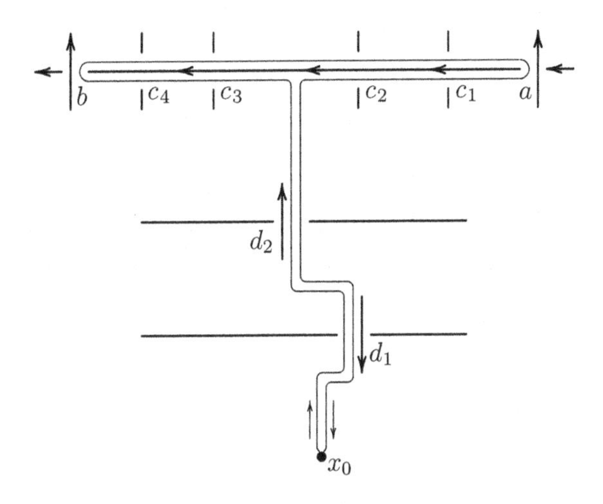

Fig. 33 Calculating the group of a knot: general case

The relation arising from this picture is

$$d_1^{-1}d_2c_3c_4bc_4^{-1}c_3^{-1}c_2^{-1}c_1^{-1}a^{-1}c_1c_2d_2^{-1}d_1 = 1.$$

First of all, a relation $ABA^{-1} = 1$ is always equivalent to the relation $B = 1$, so we can drop the ds, *and we can always ignore the gates that we pass on our way from* x_0 *to the curve*. The relation takes the form

$$c_3c_4bc_4^{-1}c_3^{-1}c_2^{-1}c_1^{-1}a^{-1}c_1c_2 = 1.$$

Next, we change this relation into

$$a^{-1}c_1c_2c_3c_4bc_4^{-1}c_3^{-1}c_2^{-1}c_1^{-1} = 1,$$

that is, $a^{-1}CbC^{-1}$, where $C = c_1c_2c_3c_4$, the product of generators, corresponding to the gates on our way from a to b. And this is what this relation always looks like.

We summarize our results in the following theorem.

Theorem 5. *Let K be an oriented knot in space presented by a knot diagram. This diagram has some number of components (oriented segments) and an equal number of gates. The group $\pi_1(\mathbb{R}^3 - K)$ has a system of generators corresponding to the gates with generating relations corresponding to the segments. Namely, if there is a segment beginning at the gate a, ending at the gate b, and passing through the gates $c_1, \ldots c_n$ (ordered according to the orientation of the segment), then the corresponding relation is*

$$a^{\varepsilon_1}c_1 \ldots c_nb^{\varepsilon_2}c_n^{-1} \ldots c_1^{-1} = 1,$$

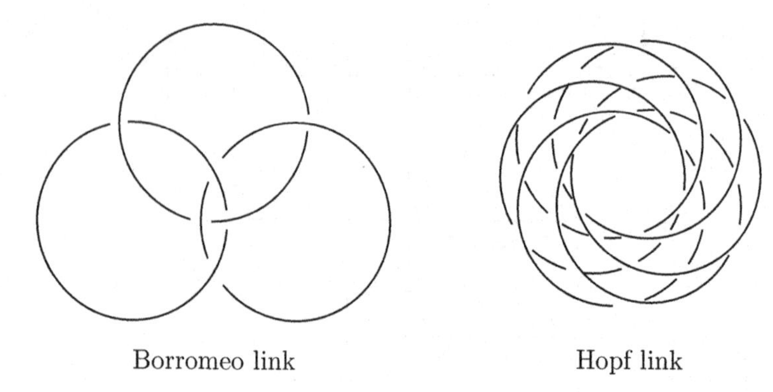

Fig. 34 Links for Exercises 12 and 13

where ε_1 *and* ε_2 *are determined by the following picture:*

a $\varepsilon_1 = -1$ a $\varepsilon_1 = 1$ b $\varepsilon_2 = 1$ b $\varepsilon_2 = -1$

EXERCISE 11. Find the fundamental group of the complement to the figure-eight knot (it should be presented as a group with two generators and one relation).

EXERCISE 12. Prove that the fundamental group of the complement of Borromeo link[4] (see Fig. 34, left) has three generators, a, b, and c, with three relations,

$$[a, [b, c^{-1}]] = 1, [b, [c, a^{-1}]] = 1, [c, [a, b^{-1}]] = 1.$$

EXERCISE 13. The Hopf link H_n is presented in Fig. 34, right (for $n = 6$; n is the number of components). Prove that the group $\pi_n(\mathbb{R}^3 - H_n)$ has n generators $a_1, a_2, \dots, a_n$ with $n - 1$ relations:

$$a_1a_2 \dots a_n = a_na_1a_2 \dots a_{n-1} = a_{n-1}a_na_1 \dots a_{n-2} = \dots = a_2a_3 \dots a_na_1.$$

Prove (algebraically) that the same group is isomorphic to a product of $\mathbb{Z}$ and a free group with $n - 1$ generators. (Actually, $S^3 - H_n$ is homeomorphic to the product of S^1 and S^2 minus n points; you can try to prove this.)

EXERCISE 14. Let $\Delta \in \mathbb{R}^2$ be a diagram of a knot $K \in \mathbb{R}^3$. An admissible 3-coloring of Δ is a coloring Δ into the colors #1, #2, and #3 such that at every crossing, either only one color is used or all three colors are used. Prove that the number of admissible colorings is an isotopy knot invariant; that is, Reidemeister

[4]"Borromeo" is not the name of a mathematician. It belongs to a family of Italian noblemen who had the picture of the link on their coat of arms.

moves (see Exercise 10) do not change this number. (Use this invariant to prove that the trefoil is not isotopic to an unknot.)

EXERCISE 15. (Sequel of Exercise 14) Prove that the number of admissible colorings from Exercise 14 is precisely 3 less than the number of homomorphisms $\pi_1(X, x_0) \to S_3$.

7.4 Another Presentation of the Fundamental Groups of Knots and Links

In Sect. 9.3, we derived a presentation of the fundamental group of a knot or a link, in which the generators correspond to gates and relations correspond to arcs. There exists a presentation, also coming from Van Kampen's theorem, in which generators correspond to arcs and relations correspond to gates. Many people find it more convenient than the presentation of Sect. 9.3. It is known by the name *Wirtinger presentation.*

To obtain his presentation, we cut, as in Sect. 7.3, the space with the knot deleted by a plane, but this time it is not the plane containing the knot or link diagram, but the parallel plane slightly below the plane of the diagram, so that the underpasses of the gates cut short curves in the intersection of the two pieces. The part U_1 below this dividing plane is contractible (the underpasses carve short pitches on its surface), while the part U_2 above the dividing plane has tunnels corresponding to the curves of the diagram. (As before, we assume that $U_1 \cap U_2$ has some small thickness, and the knot has small thickness exceeding, however, the thickness of $U_1 \cap U_2$.) The base point x_0 is chosen in $U_1 \cap U_2$. Thus, $\pi_1(U_1, x_0)$ is trivial, and each of $\pi_1(U_1 \cap U_2, x_0)$ and $\pi_1(U_2, x_0)$ is a free group with generators corresponding, respectively, to gates and curves of the diagram. To specify the relations, we need to orient the diagram. For every gate, there arises a relation between the three generators corresponding to the three curves involved; it is shown in Fig. 35.

(The orientations of the curves y and z are irrelevant.) We leave the details to the reader (see Exercise 16).

EXERCISE 16. Prove that the Wirtinger presentation is a valid presentation for a group of a knot or a link.

EXERCISE 17. Redo Exercises 11–15 using the Wirtinger presentation.

7.5 Fundamental Groups and Attaching Cells

Theorem. *Let X be a path connected topological space with a base point x_0, let $f: S^n \to X$ be a continuous map, and let s be a path in X joining, for the base point*

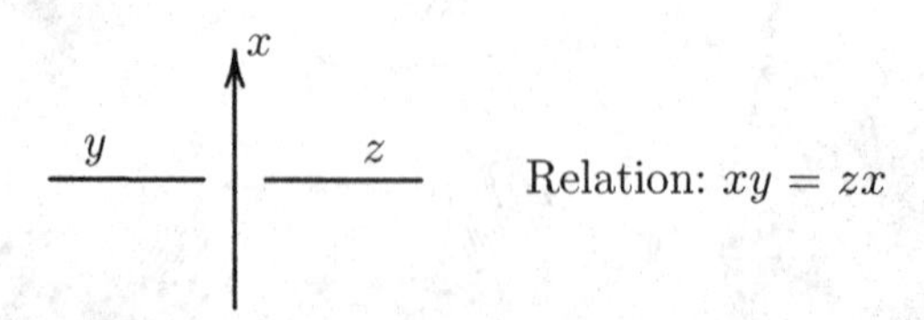

Fig. 35 Wirtinger's relations

z_0 of S^n, the point $f(z_0)$ with the point x_0. Let $Y = X \bigcup_f D^{n+1}$, and let $j: X \to Y$ be the inclusion map.

(1) *If $n > 1$, then $j_*: \pi_1(X, x_0) \to \pi_1(Y, x_0)$ is an isomorphism.*
(2) *If $n = 1$, then $j_*: \pi_1(X, x_0) \to \pi_1(Y, x_0)$ is onto and* $\operatorname{Ker} j_*$ *is the normal subgroup of $\pi_1(X, x_0)$ generated (as a normal subgroup) by $s_\#[f]$; in particular, this normal subgroup does not depend on the choice of s.*

Proof. Take two concentric balls in D^{n+1} and cover Y with two open sets: U_1 is the union of X and the complement to the smaller ball, and U_2 is the interior of the bigger ball (see Fig. 36).

Take a point $y_0 \in U_1 \cap U_2$ on the same radius as z_0. By Van Kampen's theorem,

$$\pi_1(Y, y_0) = \pi_1(U_1, y_0) *_{\pi_1(U_1 \cap U_2, y_0)} \pi_1(U_2, y_0).$$

Obviously, U_2 is contractible and $U_1 \cap U_2 \sim S^n$; also, $(U_1, y_0) \sim (X, f(z_0))$ (the latter is a deformation retract of the former). Hence, $\pi_1(U_2, y_0) = 1$ and if $n > 1$, then $\pi_1(U_1 \cap U_2, y_0) = 1$. Thus, if $n > 1$, then the inclusion map $U_1 \to Y$ induces an isomorphism between fundamental groups, and hence so does the inclusion map $X \to Y$; this proves (1).

If $n = 1$, then $\pi_1(Y, y_0) = \pi_1(U_1, y_0) *_{\pi_1(U_1 \cap U_2, y_0)} 1$. But this means that $\pi_1(Y, f(z_0)) = \pi_1(X, f(z_0)) *_{\mathbb{Z}} 1$, where $\mathbb{Z}$ is generated by $[f]$. In other words, $\pi_1(Y, f(z_0))$ is obtained from $\pi_1(X, f(z_0))$ by imposing an additional relation $[f] = 1$, or by factorizing by the normal subgroup generated by $[f]$. The same is true for $\pi_1(Y, x_0)$ and $\pi_1(X, x_0)$, only $[f]$ should be replaced by $s_\#[f]$. Since the conjugacy class of $s_\#[f]$ does not depend on s, the same is true for the normal subgroup generated by $s_\#[f]$.

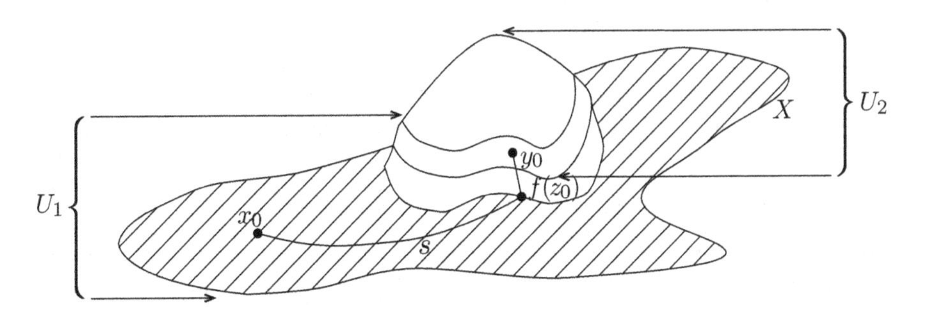

Fig. 36 Proof of the attaching cell theorem

7.6 Fundamental Groups of CW Complexes

Let X be a CW complex with precisely one zero-dimensional cell, x_0 (according to the theorem in Sect. 5.9, every connected CW complex is homotopy equivalent to such a CW complex). Obviously, the skeleton $\mathrm{sk}_1 X$ is a bouquet of circles. Thus, $\pi_1(\mathrm{sk}_1 X, x_0)$ is a free group with generators corresponding to the one-dimensional cells (see the corollary in Sect. 7.2; to specify the generators, we need to *orient* all one-dimensional cells).

The transition from $\mathrm{sk}_1 X$ to $\mathrm{sk}_2 X$ consists in attaching a certain number of two-dimensional cells. According to part (2) of the theorem in Sect. 7.5, attaching every cell imposes a relation on $\pi_1(\mathrm{sk}_1 X, x_0)$; this relation equates to 1 the class of the attaching map. Thus, $\pi_1(\mathrm{sk}_2 X, x_0)$ is a group with generators corresponding to one-dimensional cells and relations corresponding to two-dimensional cells; in particular, any group can be the fundamental group of a two-dimensional CW complex.

Finally, according to part (1) of the theorem in Sect. 7.5, attaching cells of dimensions 3 and more does not affect the fundamental group; thus, $\pi_1(X, x_0)$ is the same as $\pi_1(\mathrm{sk}_2 X, x_0)$.

Remarks. (1) If the number of cells is infinite, we need to make a reference to Axiom (W): Every spheroid, as well as every homotopy between spheroids, is contained in a finite CW subcomplex of X. (2) Some of our earlier statements follow directly from the cellular approximation theorem (Sect. 5.7). Moreover, in this way we can drop the assumption that X has only one zero-dimensional cell. Namely, let X be a connected CW complex, and let x_0 be a zero-dimensional cell. Then the inclusion maps $\mathrm{sk}_1 X \to X$ and $\mathrm{sk}_2 X \to X$ induce, respectively, an epimorphism (a homomorphism onto) and an isomorphism of the π_1 groups with the base point x_0. Indeed, any loop is a continuous map of the CW complex I to the CW complex X which is cellular on the CW subcomplex ∂I of I; hence, it is ∂I-homotopic to a cellular map, that is, to a loop in $\mathrm{sk}_1 X$. Similarly, a homotopy between two loops in $\mathrm{sk}_1 X$ is a continuous map $I \times I \to X$ which is cellular on $\partial(I \times I)$; hence, it is $\partial(I \times I)$-homotopic to a cellular map, that is, to a homotopy between the two loops in $\mathrm{sk}_2 X$.

We summarize everything just said in one proposition.

Theorem. *Let X be a connected CW complex, and let x_0 be a zero-dimensional cell. Then the inclusions of $\mathrm{sk}_1 X$ and $\mathrm{sk}_2 X$ into X induce an epimorphism $\pi_1(\mathrm{sk}_1 X, x_0) \to \pi_1(X, x_0)$ and an isomorphism $\pi_1(\mathrm{sk}_2 X, x_0) \to \pi_1(X.x_0)$. Moreover, if X has no zero-dimensional cells different from x_0, then $\pi_1(X, x_0)$ has a system of generators corresponding to one-dimensional cells (classes of characteristic maps $D^1 = I \to X$) with a system of relations corresponding to two-dimensional cells (classes of attaching maps $S^1 \to X$).*

Examples. We begin with classical surfaces (without holes). A CW structure of these surfaces is described in Sect. 2.4(F) (which is based, in turn, on the polygonal construction described in Exercise 14 of Lecture 1). We already know that $\pi_1(S^2) = 1$ and $\pi_1(\mathbb{R}P^2) \cong \mathbb{Z}_2$. The standard CW decomposition of the Klein bottle K has two one-dimensional cells and one two-dimensional cell. Thus, $\pi_1(K)$ has two generators, denoted by c and d, and one relation, which can be read in Fig. 4d in Lecture 1: $cdc^{-1}d = 1$. Every handle results in two additional generators to the system of generators and in modifying the only relation by the multiplication of its left-hand side by the commutators of these additional generators. With this in mind, we get the following description of the fundamental groups:

$$\pi_1(S^2 \text{ with } g \text{ handles}) = \langle a_1, b_1, \dots, a_g, b_g \mid [a_1, b_1] \dots [a_g, b_g] = 1 \rangle;$$
$$\pi_1(\mathbb{R}P^2 \text{ with } g \text{ handles}) = \langle c, a_1, b_1, \dots, a_g, b_g \mid c^2[a_1, b_1] \dots [a_g, b_g] = 1 \rangle;$$
$$\pi_1(K \text{ with } g \text{ handles}) = \langle c, d, a_1, b_1, \dots, a_g, b_g \mid cdc^{-1}d[a_1, b_1] \dots [a_g, b_g] = 1 \rangle.$$

As to the other classical spaces, we can easily deduce from the CW structure of projective spaces that $\pi_1(\mathbb{R}P^n) \cong \mathbb{Z}_2$ for all $n \geq 2$ (including $n = \infty$), which we already know (see Example 2 in Sect. 6.9). In addition to that, we can use the CW decompositions of the Grassmannian manifolds to compute the fundamental groups.

EXERCISE 18. Prove that $\pi_1(G(n,k)) \cong \mathbb{Z}_2$ for $1 \leq k \leq n-1$ and $n > 2$.

It is easy to see that complex and quaternion Grassmann and flag manifolds are all simply connected (as well as the Cayley projective plane).

Lecture 8 Homotopy Groups

8.1 Definition: Commutativity

Homotopy groups $\pi_n(X, x_0)$ $(n \geq 1)$ of a space X with a base point x_0 were defined in Lecture 4 as a particular case of a general group-valued homotopy functor. Recall that the set $\pi_n(X, x_0)$ was defined as the set of base point homotopy classes of continuous maps of the sphere S^n into X. These maps are called *spheroids*. In a different way, a spheroid can be defined as a continuous map of the cube I^n into X taking the boundary ∂I^n of the cube into x_0.

The *sum* of two spheroids, $f, g\colon S^n \to X$, is defined as the spheroid $f + g\colon S^n \to X$ which is constructed in the following way: First, the equator of the sphere S^n (containing the base point) is collapsed into a point, so the sphere becomes the bouquet of two spheres, and then these two spheres are mapped into X according to f and g (see Fig. 37).

Another description uses cubic language. If $f, g\colon I^n \to X$ are two "cubic" spheroids (each takes ∂I^n into x_0), then the spheroid $f + g$ can be defined as the continuous map of I^n into X, where on the left half of the cube, $\{x_1 < 1/2\}$ if the composition of f with the double compression of the cube in the direction of the x_1-axis, and on the right half is defined in a similar way, with g instead of f.

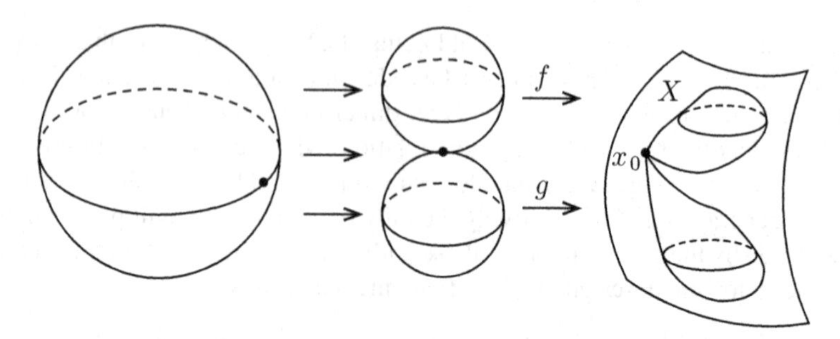

Fig. 37 The sum of spheroids

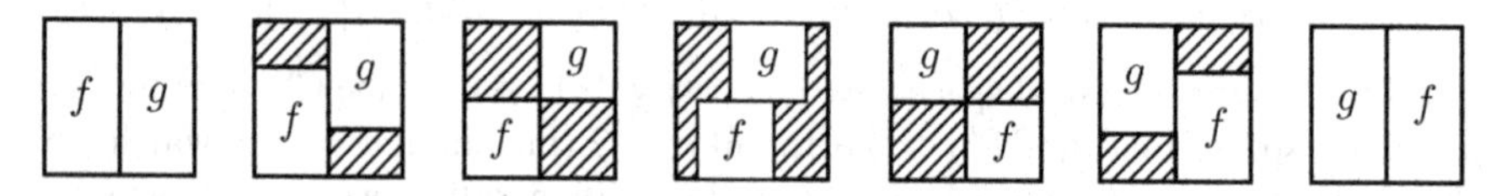

Fig. 38 A homotopy between $f + g$ and $g + f$

The operation of adding spheroids is not a group operation, but it becomes a group operation after the transition to homotopy classes. Thus, $\pi_n(X, x_0)$ becomes a group (the inverse to $[f]$ is $[f \circ r]$, where r is the reflection of the cube in the plane $x = \frac{1}{2}$; the identity element is the class of the constant spheroid). (We leave the details to the reader.)

EXERCISE 1. Prove that $\pi_n(X \times Y, (x_0, y_0)) = \pi_n(X, x_0) \times \pi_n(Y, y_0)$.

For $n = 1$ the homotopy group is the fundamental group. For $n > 1$ the homotopy group acquires a new feature: It is commutative.

Theorem. *If $n > 1$, then the group $\pi_n(X, x_0)$ is commutative for any (X, x_0).*

Proof. We need to prove that $f + g \sim g + f$. In the language of cubic spheroids, the homotopy is shown in Fig. 38. (The picture shows the homotopy for $n = 2$; if $n > 2$, then we need to take the direct product of this picture and the cube I^{n-2} in the plane perpendicular to the plane of the picture.)

There is a slightly different way to visualize the sum of spheroids, which makes the commutativity of π_n with $n \geq 2$ still more obvious. If n-dimensional spheroids f, g of some space with a base point are given, we choose two small balls on the sphere S^n and define a new spheroid which maps the complements of the balls into the base point and maps the balls according to f and g (see Fig. 39). It is clear that if $n \geq 2$, then the order of the balls is insignificant.

Notice in conclusion that a continuous map $\varphi: (X, x_0) \to (Y, y_0)$ can be applied to spheroids, $f \mapsto \varphi \circ f$, and, consequently, to a homomorphism $\varphi_*: \pi_n(X, x_0) \to \pi_n(Y, y_0)$. The latter depends only on the homotopy class of φ. It is clear also that $\mathrm{id}_* = \mathrm{id}$ and $(\varphi \circ \psi)_* = \varphi_* \circ \psi_*$. Hence, homotopy equivalent spaces with base points have isomorphic homotopy groups.

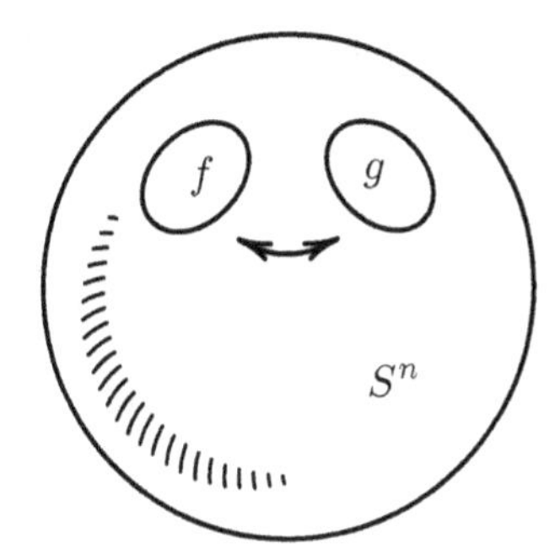

Fig. 39 Another proof of $(f+g) \sim (g+f)$

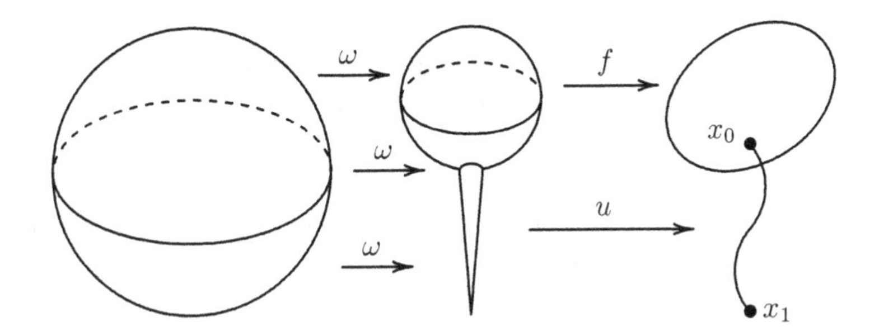

Fig. 40 Base point change

8.2 Dependence on the Base Point

A path $u: I \to X$ joining points $x_0, x_1 \in X$ gives rise to an isomorphism $u_\#: \pi_n(X, x_0) \to \pi_n(X, x_1)$. The construction of $u_\#$ is shown in Fig. 40: First, we construct a map ω of the sphere S^n onto a bouquet $S^n \vee I$ (taking the base point of S^n into the endpoint of I distant from S^n) and then assign to a spheroid $f: S^n \to X$ taking the base point of S^n into x_0 the spheroid

$$u_\#: S^n \xrightarrow{\omega} S^n \vee I \xrightarrow{f \vee u} X,$$

taking the base point of S^n into x_1.

It is easy to check that $u_\#(f+g) \sim u_\#(f) + u_\#(g)$ and that $(u^{-1})_\# = (u_\#)^{-1}$. It is clear also that for $n = 1$ the isomorphism $u_\#$ coincides with the isomorphism $u_\#$ constructed in Sect. 6.2.

As seen in the example of the fundamental group, the isomorphism $u_\#$ may be different for different paths u although it remains the same when the path u is replaced by a homotopic path. In particular, loops representing an element $\alpha \in \pi_1(X, x_0)$ determine the same automorphism of $\pi_n(X, x_0)$, which we can denote as $\alpha_\#$. In this way, we get a group action, or a representation, of $\pi_1(X, x_0)$ in $\pi_n(X, x_0)$.

If for some n the isomorphism $u_{\#}$ does not depend on the path u at all, the space X is called n-simple. It follows from results of Lecture 6 that X is 1-simple if and only if the group $\pi_1(X)$ is commutative. We should remark that for $n > 1$, the property of being n-simple has nothing to do with the commutativity of the fundamental group.

Spaces which are n-simple for all n are called simple. For example, simply connected spaces are simple.

EXERCISE 2. Prove that topological groups and H-spaces are simple (compare with Exercise 4 in Sect. 6.7).

8.3 Coverings and Homotopy Groups

Theorem 1. *Let $p: T \to X$ be a covering, let $\widetilde{x_0} \in T$, and let $x_0 = p(\widetilde{x_0}) \in X$. If $n \geq 2$, then $p_*: \pi_n(T, \widetilde{x_0}) \to \pi_n(X, x_0)$ is an isomorphism.*

This follows from results of Sect. 6.10 (see the corollary in this section) and from the simply connectedness of the sphere S^n for $n \geq 2$.

Theorem 1 may be immediately used for the computation of homotopy groups of some spaces. Here is an example.

Theorem 2.

$$\pi_n(S^1) = \begin{cases} \mathbb{Z}, & \text{if } n = 1, \\ 0, & \text{if } n \geq 2. \end{cases}$$

The first is already known (Sect. 6.3), the second follows from the fact that there is a covering $\mathbb{R} \to S^1$, and the line $\mathbb{R}$ is contractible.

EXERCISE 3. Prove that if X is a bouquet of circles, then $\pi_n(X) = 0$ for all $n \geq 2$. (Prove that the universal covering of X is contractible; see Example 5 in Sect. 6.9.)

EXERCISE 4. Prove that if X is a classical surface (Sect. 1.10) different from S^2 and $\mathbb{R}P^2$, then $\pi_n(X) = 0$ for all $n \geq 2$. (Classical surfaces with holes are homotopy equivalent to bouquets of circles; thus, the statement follows from Exercise 3. The universal covering of classical surfaces without holes different from S^2 and $\mathbb{R}P^2$ is homeomorphic to the plane, which is contractible. Another way of proving the statement in this case is to consider a nonuniversal covering, as described ahead.)

Let X be a surface S with a handle. We consider the infinite covering $p: \widetilde{X} \to X$, where X is a cylinder with infinitely many copies of S attached (see Fig. 41).

By Theorem 1, $\pi_n(\widetilde{X}) = \pi_n(X)$ (for $n \geq 2$). But every spheroid of $\widetilde{X}$ is contained in a "finite part of $\widetilde{X}$," like the one shown in Fig. 41. This finite part is a surface with (at least two) holes, and for this surface with holes, $\pi_n = 0$ by the first case considered. Thus, this spheroid is homotopic to a constant, and the group $\pi_n(\widetilde{X})$ is zero. (This proof does not cover the case of the Klein bottle, but the Klein bottle is doubly covered by a torus; thus, its π_n groups are zero for $n \geq 2$.)

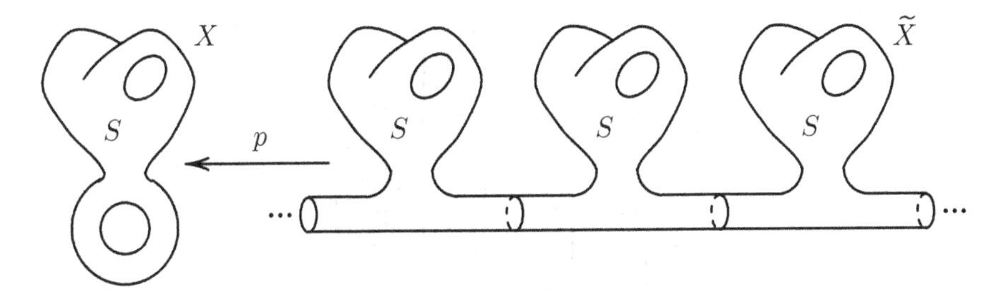

Fig. 41 A covering over a surface S with a handle

8.4 Relative Homotopy Groups

Let (X, A) be a pair with the base point $x_0 \in A$. Let $n \geq 2$. The *relative homotopy group* $\pi_n(X, A, x_0)$ is defined as the set of homotopy classes of *n-dimensional relative spheroids*. Relative spheroids (like the absolute[5] ones) can be defined in two ways: the ball one and the cubic one. A ball relative spheroid is a continuous map $f: D^n \to X$ such that $f(S^{n-1}) \subset A$ and $f(S_+^{n-1}) = x_0$. (S_+^{n-1} is the "upper hemisphere"; that is, the set $\{(x_1, \dots, x_n) \in \mathbb{R}^n \mid x_1^2 + \dots + x_n^2 = 1,\ x_n \geq 0\}$.) A cubic relative spheroid is a continuous map $f: I^n \to X$ such that $f(\partial I^n) \subset A$ and $f(\partial I^n - I^{n-1}) = x_0$. The sum of two cubic relative spheroids $f, g: I^n \to X$ is a cubic relative spheroid $f + g: I^n \to X$ defined by the formula (the same as in the absolute case)

$$(f+g)(x_1, x_2, \dots, x_n) = \begin{cases} f(2x_1, x_2, \dots, x_n), & \text{if } x_1 \leq \dfrac{1}{2}, \\ g(2x_1 - 1, x_2, \dots, x_n), & \text{if } x_1 \geq \dfrac{1}{2}. \end{cases}$$

It is clear that $f + g$ is a relative spheroid and that if $f \sim f'$ and $g \sim g'$, then $f + g \sim f' + g'$. The last property makes it possible to define the $+$ operation in the set $\pi_n(X, A, x_0)$.

EXERCISE 5. Check the group axioms for $\pi_n(X, A; x_0)$; in particular, the identity element is the class of the constant spheroid.

EXERCISE 6. Show that if $n \geq 3$, then the group $\pi_n(X, A; x_0)$ is commutative. (Like we did in the absolute case, it is convenient to use the construction of $f + g$ as shown in Fig. 42; the shadowed domain is mapped into x_0; if $n \geq 3$, then the order of domains marked as f and g is insignificant.)

[5]To distinguish relative homotopy groups and spheroids from homotopy groups and spheroids considered before, we will sometimes call the latter *absolute*.

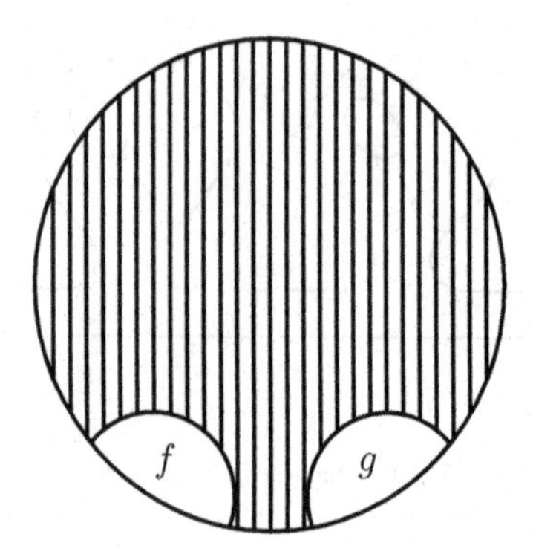

Fig. 42 Another description of $f + g$

EXERCISE 7. Find the relative analog of Exercise 1.

If $(X, A), (Y, B)$ are pairs with base points $x_0 \in A$, $y_0 \in B$, then a continuous map $f: X \to Y$ such that $f(A) \subset B$ and $f(x_0) = y_0$ [in writing, $f: (X, A, x_0) \to (Y, B, y_0)$] induces a homomorphism $f_*: \pi_n(X, A, x_0) \to \pi_n(Y, B, y_0)$. Homotopic maps $(X, A, x_0) \to (Y, B, y_0)$ induce the same homomorphism; in particular, homotopy equivalent pairs have isomorphic homotopy groups.

The dependence of relative homotopy groups on the base point is similar to that for absolute homotopy groups: A path $u: I \to A$ joining $x_0 \in A$ with $x_1 \in A$ gives rise to an isomorphism $u_\#: \pi_n(X, A, x_0) \to \pi_n(X, A, x_1)$.

8.5 "Homotopy Groups" $\pi_0(X, x_0)$ and $\pi_1(X, A, x_0)$

The definition of the sets $\pi_n(X, x_0)$ and $\pi_n(X, A; x_0)$ makes sense if we take $n = 0$ in the first case and $n = 1$ in the second case. In particular, $\pi_0(X, x_0)$ is the set of path components of X [the set $\pi_1(X, A, x_0)$ does not have such a transparent meaning]. However, there is no natural group structure in these sets [for $\pi_0(X, x_0)$, this follows from the results of Lecture 4; for $\pi_1(X, A, x_0)$, we leave the explanation to the reader]. Still these sets possess a distinguished element, "the unity": This is the class of the constant spheroid $S^0 \to x_0$ or $I^1 \to x_0$.

Although $\pi_0(X, x_0)$ and $\pi_1(X, A, x_0)$ are not groups, one should not totally ignore them. For example, the statement "the space X is n-connected if and only if $\pi_i(X) = 0$ for $i \leq n$" is valid for $n = 0$ as well as for $n > 0$ [the equality $\pi_0(X) = 0$ means that X is path connected]. Moreover, we will have the courage to say that although the notation $\pi_{-1}(X)$ makes no sense at all, the notion of (-1)-connectedness exists and means being nonempty (every map $\emptyset \to X$ can be extended to a map $\mathrm{pt} \to X$). One can say that $\pi_{-1}(X)$ is not a group and even is not a set, but that there are two possibilities: $\pi_{-1}(X) = 0$ (X is nonempty) and $\pi_{-1}(X) \neq 0$ (X is empty). All this gives the impression of idle talk, but it may clarify the similarity which exists sometimes between a proof of existence of, say, a map of some kind, or a solution of some equation (a computation of the π_{-1} group) and a description of the set of homotopy classes of such maps or of solutions of the equation (a computation of the π_0 group).

8.6 Relations Between Relative and Absolute Homotopy Groups

First, *absolute homotopy groups may be regarded as a particular case of relative homotopy groups.* Namely, $\pi_n(X, x_0) = \pi(X, x_0, x_0)$ for $n \geq 1$.

Second, *relative homotopy groups may be regarded as a particular case of absolute homotopy groups.* Namely, there exists a construction which assigns to a pair X, A with a base point x_0 a space Y with a base point y_0 such that $\pi_n(X, A, x_0) = \pi_{n-1}(Y, y_0)$ for $n \geq 1$. This explains why $\pi_n(X, A, x_0)$ is a group only for $n \geq 2$ and a commutative group only for $n \geq 3$. We postpone the construction of Y to Lecture 9 (see Sect. 9.10).

Third, there are natural homomorphisms $\pi_n(X, x_0) \rightarrow \pi_n(X, A, x_0)$. These homomorphisms arise from the observation that an absolute spheroid $(I^n, \partial I^n) \rightarrow (X, x_0)$ can be regarded as a relative spheroid

$$(I^n, \partial I^n, \partial I^n - I^{n-1}) \rightarrow (X, A, x_0);$$

differently, one can say that these homomorphisms coincide with j_*, where j is the identity map $X \rightarrow X$ regarded as a map $(X, x_0) \rightarrow (X, A)$.

EXERCISE 8. Prove that the image of the homomorphism

$$j_*: \pi_2(X, x_0) \rightarrow \pi_2(X, A, x_0)$$

is contained in the center of the group $\pi_2(X, A, x_0)$.

Fourth (and the most important!), there are *connecting homomorphisms*

$$\partial: \pi_n(X, A, x_0) \rightarrow \pi_{n-1}(A, x_0).$$

The homomorphism ∂ takes the class of a relative spheroid

$$f: (I^n, \partial I^n, \partial I^n - I^{n-1}) \rightarrow (X, A, x_0)$$

into the class of the absolute spheroid

$$f|_{I^{n-1}}: (I^{n-1}, \partial I^{n-1}) \rightarrow (A, x_0)$$

[or the class of a relative spheroid $f: (D^n, S^{n-1}, S_+^{n-1}) \rightarrow (X, A, x_0)$ into the class of the absolute spheroid $f|_{S^{n-1}}: S^{n-1} \rightarrow A$].

8.7 The Homotopy Sequence of a Pair

The "homotopy sequence of a pair" is the name given to the sequence

$$\dots \xrightarrow{\partial} \pi_n(A,x_0) \xrightarrow{i_*} \pi_n(X,x_0) \xrightarrow{j_*} \pi_n(X,A,x_0) \xrightarrow{\partial} \pi_{n-1}(A,x_0)$$
$$\dots \xrightarrow{i_*} \pi_1(X,x_0) \xrightarrow{j_*} \pi_1(X,A,x_0) \xrightarrow{\partial} \pi_0(A,x_0) \xrightarrow{i_*} \pi_0(X,x_0),$$

where j_* and ∂ are homomorphisms described earlier in this chapter and i_* is induced by the inclusion map $i: A \to X$.

The main property of this sequence is that it is *exact*; that is, the image of every map coincides with the kernel of the next map (for the last three arrows, the kernel is the inverse image of the "unity element"). We recommend that the reader proves this as an exercise; but for those who do not want to do this work, we present a proof now.

Proof of Exactness. (1) $\operatorname{Im} i_* \subset \operatorname{Ker} j_*$. We need to prove that for every spheroid $f: (I^n, \partial I^n) \to (A, x_0)$, $j_* \circ i_*[f] = 0$. The class $j_* \circ i_*[f]$ is represented by the same map f regarded as a map $(I^n, \partial I^n, \partial I^n - I^{n-1}) \to (X, A, x_0)$, and the spheroids $f_t: (I^n, \partial I^n, \partial I^n - I^{n-1}) \to (X, A, x_0), f_t(x_1, \dots, x_{n-1}, x_n) = f(x_1, \dots, x_{n-1}, t + (1 - t)x_n)$ form a homotopy connecting f with a constant spheroid.

(2) $\operatorname{Ker} j_* \subset \operatorname{Im} i_*$. We need to show that if a spheroid $f: (I^n, \partial I^n) \to (X, x_0)$ is homotopic to the constant within the class of relative spheroids, then it is homotopic (as an absolute spheroid) to a spheroid whose image is contained in A. Let $F: I^n \times I \to X$ be a homotopy between f and the constant spheroid in the class of constant spheroids within the class of relative spheroids of the pair (X, A). Then F is the map $I^{n+1} \to X$, which is f on the face $I^n = \{x_{n+1} = 0\}$ that maps the face $\{x_n = 0\}$ into A and maps the remaining part of ∂I^{n+1} into x_0. Let $I^n_t \subset I^{n+1}$ be the intersection of the cube I^{n+1} with the plane $tx_n + (1-t)x_{n-1} = 0$ (see Fig. 43). It is clear that $I^n_t \approx I^n$ and that $F|_{I^n_t}: I^n_t = I_n \to X$ is a homotopy joining (within the class of absolute spheroids of X) the spheroid f with a spheroid g whose image is contained in A.

(3) $\operatorname{Im} j_* \subset \operatorname{Ker} \partial$. Indeed, if $f: I^n \to X$ is an absolute spheroid, then $f|_{I^{n-1}}$ is a constant map.

(4) $\operatorname{Ker} \partial \subset \operatorname{Im} f_*$. Let $f: I^n \to X$ be a relative spheroid, and let $g_t: I^{n-1} \to A$ be a homotopy connecting the absolute spheroid $F|_{I^{n-1}}$ of A with the constant spheroid. Consider the homotopy $f_t: \partial I^n \to X$ coinciding with g_t on I^{n-1} and taking $\partial I^n - I^{n-1}$ into x_0 and extend it (using Borsuk's theorem) to a homotopy $h_t: I^n \to X$ of the spheroid f. It is clear that h^t is a homotopy connecting f with the constant spheroid within the class of relative spheroids.

(5) $\operatorname{Im} \partial \subset \operatorname{Ker} i_*$. If an absolute spheroid $f: I^{n-1} \to A$ is a restriction of an absolute spheroid $g: I^n \to X$, then $g_t = g|_{I^{n-1} \times t}: I^{n-1} \times t = I^{n-1} \to X$ is a homotopy connecting f with the constant spheroid.

(6) $\operatorname{Ker} i_* \subset \operatorname{Im} \partial$. If $g_t: I^{n-1}$ is a homotopy connecting (in X) a spheroid $f = g_0: I^{n-1} \to A$ with a constant spheroid, then $g: I^n \to X$, $g(x_1, \dots, x_n) = g_{x_n}(x_1, \dots, x_{n-1})$ is a relative spheroid of the pair (X, A) whose restriction to A is f.

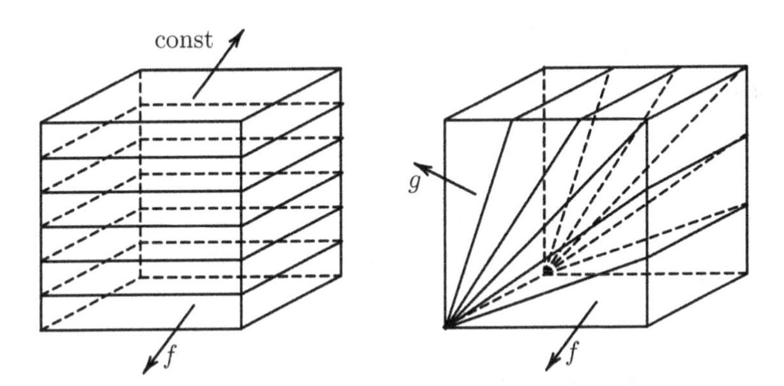

Fig. 43 Proof of $\operatorname{Ker} j_* \subset \operatorname{Im} i_*$

EXERCISE 9. Define a natural (right) action of the group $\pi_1(X, x_0)$ in the set $\pi_1(X, A, x_0)$ and prove that the orbits of this action coincide with the inverse images with respect to ∂ of elements of the set $\pi_0(A, x_0)$ [thus, the exactness in the term $\pi_1(X, A, x_0)$, which means the boundary between group and nongroup terms of the homotopy sequence of a pair has a group nature].

EXERCISE 10. Let (X, A, B) be a triple with a base point $x_0 \in B$. Prove the exactness of the "homotopy sequence of a triple"

$$\cdots \to \pi_n(A, B, x_0) \to \pi_n(X, B, x_0) \to \pi_n(X, A, x_0) \to \pi_{n-1}(A, B, x_0) \ldots$$

[in this sequence the dimension-preserving homomorphisms are induced by inclusion maps of the pairs $(A, B) \to (X, B) \to (X, A)$ and the "connecting homomorphism" $\partial\colon \pi_n(X, A, x_0) \to \pi_{n-1}(A, B, x_0)$ is the composition $\pi_n(X, A, x_0) \xrightarrow{\partial} \pi_{n-1}(A, x_0) \xrightarrow{j_*} \pi_n(A, B, x_0)$].

8.8 Properties of Exact Sequences and Corollaries of Exactness of the Homotopy Sequence of a Pair

In this section, we consider sequences of groups and homomorphisms. The trivial group (consisting of one element) is denoted by the symbol 1, but in the situation when all the groups considered are Abelian, we can use the symbol 0.

EXERCISE 11. The sequence $1 \to A \xrightarrow{\varphi} B$ is exact if and only if φ is a monomorphism (that is, $\operatorname{Ker} \varphi = 1$); a sequence $A \xrightarrow{\psi} B \to 1$ is exact if and only if ψ is an epimorphism $\operatorname{Im} \psi = B$. In particular, the sequence $1 \to A \xrightarrow{\varphi} B \to 1$ is exact if and only if φ is an isomorphism.

EXERCISE 12. A sequence $1 \to A \xrightarrow{\varphi} B \xrightarrow{\psi} C \to 1$ (such sequences are called *short*) is exact if and only if φ is a monomorphism, $C \cong B/\varphi(A)$ and ψ is the natural projection.

COROLLARIES. If A is contractible, then $\pi_n(X) \cong \pi_n(X,A)$ (more precisely, j_* is an isomorphism); if X is contractible, then $\pi_n(X,A) \cong \pi_{n-1}(A)$ (more precisely, ∂ is an isomorphism); if A is a deformation retract of X, then $\pi_n(X,A) = 0$ for $n \geq 1$.

EXERCISE 13. If A is a retract (not necessarily a deformation retract) of X, then, for all n,

- $i_*: \pi_n(A) \to \pi_n(X)$ is a monomorphism,
- $j_*: \pi_n(X) \to \pi_n(X,A)$ is an epimorphism,
- $\partial: \pi_n(X,A) \to \pi_{n-1}(A)$ is a zero homomorphism;

moreover, $\pi_n(X) \cong \pi_n(X,A) \oplus \pi_n(A)$.

EXERCISE 14. If A is contractible to a point within X, then

- $j_*: \pi_n(X) \to \pi_n(X,A)$ is a monomorphism,
- $\partial: \pi_n(X,A) \to \pi_{n-1}(A)$ is an epimorphism,
- $i_*: \pi_n(A) \to \pi_n(X)$ is a zero homomorphism;

moreover, $\pi_n(X,A) \cong \pi_n(X) \oplus \pi_{n-1}(A)$.

EXERCISE 15. If there exists a homotopy $f_t: X \to X$ driving X into A, that is, such that $f_0 = \text{id}$ and $f_1(X) \subset A$, then

- $\partial: \pi_n(X,A) \to \pi_{n-1}(A)$ is a monomorphism,
- $i_*: \pi_n(A) \to \pi_n(X)$ is an epimorphism,
- $j_*: \pi_n(X) \to \pi_n(X,A)$ is a zero homomorphism;

moreover, $\pi_n(A) = \pi_n(X) \oplus \pi_{n+1}(X,A)$.

Theorem ("Five-lemma"). *If*

$$\begin{array}{ccccccccc} A_1 & \xrightarrow{f_1} & A_2 & \xrightarrow{f_2} & A_3 & \xrightarrow{f_3} & A_4 & \xrightarrow{f_4} & A_5 \\ \downarrow{\scriptstyle\varphi_1} & & \downarrow{\scriptstyle\varphi_2} & & \downarrow{\scriptstyle\varphi_3} & & \downarrow{\scriptstyle\varphi_4} & & \downarrow{\scriptstyle\varphi_5} \\ B_1 & \xrightarrow{g_1} & B_2 & \xrightarrow{g_2} & B_3 & \xrightarrow{g_3} & B_4 & \xrightarrow{g_4} & B_5 \end{array}$$

is a commutative diagram with exact rows, and $\varphi_1, \varphi_2, \varphi_4, \varphi_5$ are isomorphisms, then φ_3 is also an isomorphism.

This theorem is covered by the following two propositions.

Proposition 1. *If*

$$\begin{array}{ccccccc} A_1 & \xrightarrow{f_1} & A_2 & \xrightarrow{f_2} & A_3 & \xrightarrow{f_3} & A_4 \\ \downarrow{\scriptstyle\varphi_1} & & \downarrow{\scriptstyle\varphi_2} & & \downarrow{\scriptstyle\varphi_3} & & \downarrow{\scriptstyle\varphi_4} \\ B_1 & \xrightarrow{g_1} & B_2 & \xrightarrow{g_2} & B_3 & \xrightarrow{g_3} & B_4 \end{array}$$

is a commutative diagram with exact rows, φ_1 is an epimorphism, and φ_2, φ_4 are monomorphisms, then φ_3 is also a monomorphism.

Proposition 2. *If*

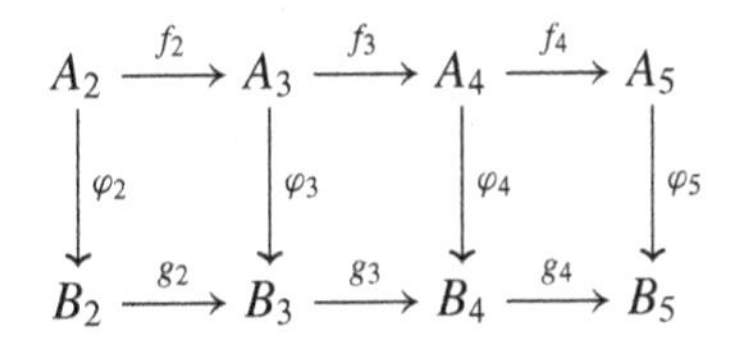

is a commutative diagram with exact rows, φ_5 is a monomorphism, and φ_2, φ_4 are epimorphisms, then φ_3 is also an epimorphism.

Remark. Thus, of eight assumptions of the theorem ($\varphi_1, \varphi_2, \varphi_3, \varphi_3$ are monomorphisms, $\varphi_1, \varphi_2, \varphi_3, \varphi_3$ are epimorphisms), three are needed to establish that φ_3 is a monomorphisms, three more are needed to establish that φ_3 is an epimorphism, and two are not needed at all. In the following, we will sometimes use these additional features of the five-lemma.

Proof of Proposition 1. Let $a_3 \in A_3$ and $\varphi_3(a_3) = 0$. Then $g_3 \circ \varphi_3(a_3) = 0 \Rightarrow \varphi_4 \circ f_3(a_3) = 0$ (commutativity of the third square) $\Rightarrow f_3(a_3) = 0$ (φ_4 is a monomorphism). Hence, there exists an $a_2 \in A_2$ such that $f_2(a_2) = a_3$ ($\operatorname{Ker} f_3 \subset \operatorname{Im} f_2$). Furthermore, $g_2 \circ \varphi_2(a_2) = \varphi_3 \circ f_2(a_2)$ (commutativity of the second square) $= \varphi_3(a_3) = 0$; hence, there exists a $b_1 \in B_1$ such that $g_1(b_1) = \varphi_2(a_2)$ ($\operatorname{Ker} g_2 \subset \operatorname{Im} g_1$). Choose an $a_1 \in A_1$ such that $\varphi_1(a_1) = b_1$ (φ_1 is an epimorphism). Then $\varphi_2 \circ f_1(a_1) = g_1 \circ \varphi_1(a_1)$ (commutativity of the first square $= g_1(b_1) = \varphi_2(a_2)$. Thus, $\varphi_2(f_1(a_1)) = \varphi_2(a_2) \Rightarrow f_1(a_1) = a_2$ (φ_2 is a monomorphism) and $a_3 = f_2(a_2) = f_2 \circ f_1(a_1) = 0$ ($\operatorname{Im} f_1 \subset \operatorname{Ker} f_2$).

Proof of Proposition 2. Let $b_3 \in B_3$. Choose an $a_4 \in A_4$ such that $\varphi_4(a_4) = g_3(b_3)$ (φ_4 is an epimorphism). Then $\varphi_5 \circ f_4(a_4) = g_4 \circ \varphi_4(a_4)$ (commutativity of the third square) $= g_4 \circ g_3(b_3) = 0$ ($\operatorname{Im} g_3 \subset \operatorname{Ker} g_4$). Hence, $f_4(a_4) = 0$ (φ_5 is a monomorphism), and hence there exists an $a_3 \in A_3$ such that $f_3(a_3) = a_4$. Then $g_3 \circ \varphi_3(a_3) = \varphi_4 \circ f_3(a_3)$ (commutativity of the second square) $= \varphi_4(a_4) = g_3(b_3)$; that is, $g_3(b_3 - \varphi_3(a_3) = 0$. Hence, there exists a $b_2 \in B_2$ such that $g_2(b_2) = b_3 - \varphi_3(a_3)$ ($\operatorname{Ker} g_3 \subset \operatorname{Im} g_2$). Choose an $a_2 \in A_2$ such that $\varphi_2(a_2) = b_2$ (φ_2 is an epimorphism). Then $\varphi_3 \circ f_2(a_2) = g_2 \circ \varphi_2(a_2)$ (commutativity of the first square) $= g_2(b_2) = b_3 - \varphi_3(a_3)$. Thus, $b_3 = \varphi_3(a_3 + f_2(a_2)) \in \operatorname{Im} \varphi_3$.

Remark. One can see from these proofs that the exactness of the rows is also used only partially. This may be less important than the previous remark, but we prefer to point this out.

EXERCISE 16. If one removes the arrow φ_3 from the diagram in the five-lemma, leaving all the other assumptions intact, will it be true that $A_3 \cong B_3$?

Corollary. *Let $(X,A),(Y,B)$ be pairs with base points $x_0 \in A, y_0 \in B$, and let $f{:}X \to Y$ be a continuous map such that $f(A) \subset B, f(x_0) = y_0$. Consider the following statements:*

$(1)_n$ *$f_*{:}\pi_m(X,x_0) \to \pi_m(Y,y_0)$ is an isomorphism for all $m < n$ and an epimorphism for $m = n$.*
$(2)_n$ *$f_*{:}\pi_m(X,A,x_0) \to \pi_m(Y,B,y_0)$ is an isomorphism for all $m < n$ and an epimorphism for $m = n$.*
$(3)_n$ *$(f|_A)_*{:}\pi_m(A,x_0) \to \pi_m(B,y_0)$ is an isomorphism for all $m < n$ and an epimorphism for $m = n$.*

Then $(2)_n \& (3)_n \Rightarrow (1)_n$; $(1)_n \& (2)_{n+1} \Rightarrow (3)_n$; $(3)_{n+1} \& (1)_{n+1} \Rightarrow (2)_n$. In particular, any two of the statements

(1) *$f_*{:}\pi_n(X,x_0) \to \pi_n(Y,y_0)$ is an isomorphism for all n,*
(2) *$f_*{:}\pi_n(X,A,x_0) \to \pi_n(Y,B,y_0)$ is an isomorphism for all n,*
(3) *$(f|_A)_*{:}\pi_n(A,x_0) \to \pi_n(B,y_0)$ is an isomorphism for all n*

imply the third.

EXERCISE 17. Let $1 \to A_0 \to \cdots \to A_n \to 1$ be an exact sequence. (1) Prove that if all the groups A_i are finite and $q_i = |A_i|$, then $\prod_{i=0}^n q_i^{(-1)^i} = 1$. (2) Prove that if all A_i are finitely generated Abelian groups and $r_i = \operatorname{rank} A_i$, then $\sum_{i=0}^n (-1)^i r_i = 0$.

Lecture 9 Fibrations

In Lecture 6, we considered coverings which locally look like products of more or less arbitrary topological spaces ("bases") and discrete spaces. Coverings turned out to be intimately related to fundamental groups. In this lecture (and many subsequent lectures) we will consider a more general notion of fibrations whose main difference from coverings is that the second factor is not assumed to be discrete any more. One can say that fibrations for homotopy groups are the same as coverings for fundamental groups; but it would be fair to say that the notion of a fibration by itself is at least not less important than the notion of a homotopy group.

Before proceeding to definitions, we will make a terminological remark. In topology, many different kinds of fibrations are considered, and the word "fibration" not accompanied by any explanatory adjectives may be ambiguous. What we call a fibration in this lecture (and in some subsequent lectures) is more usually called a locally trivial fibration. Some other kinds of fibrations (such as Serre fibrations or Hurewicz fibrations) will be introduced in this lecture.

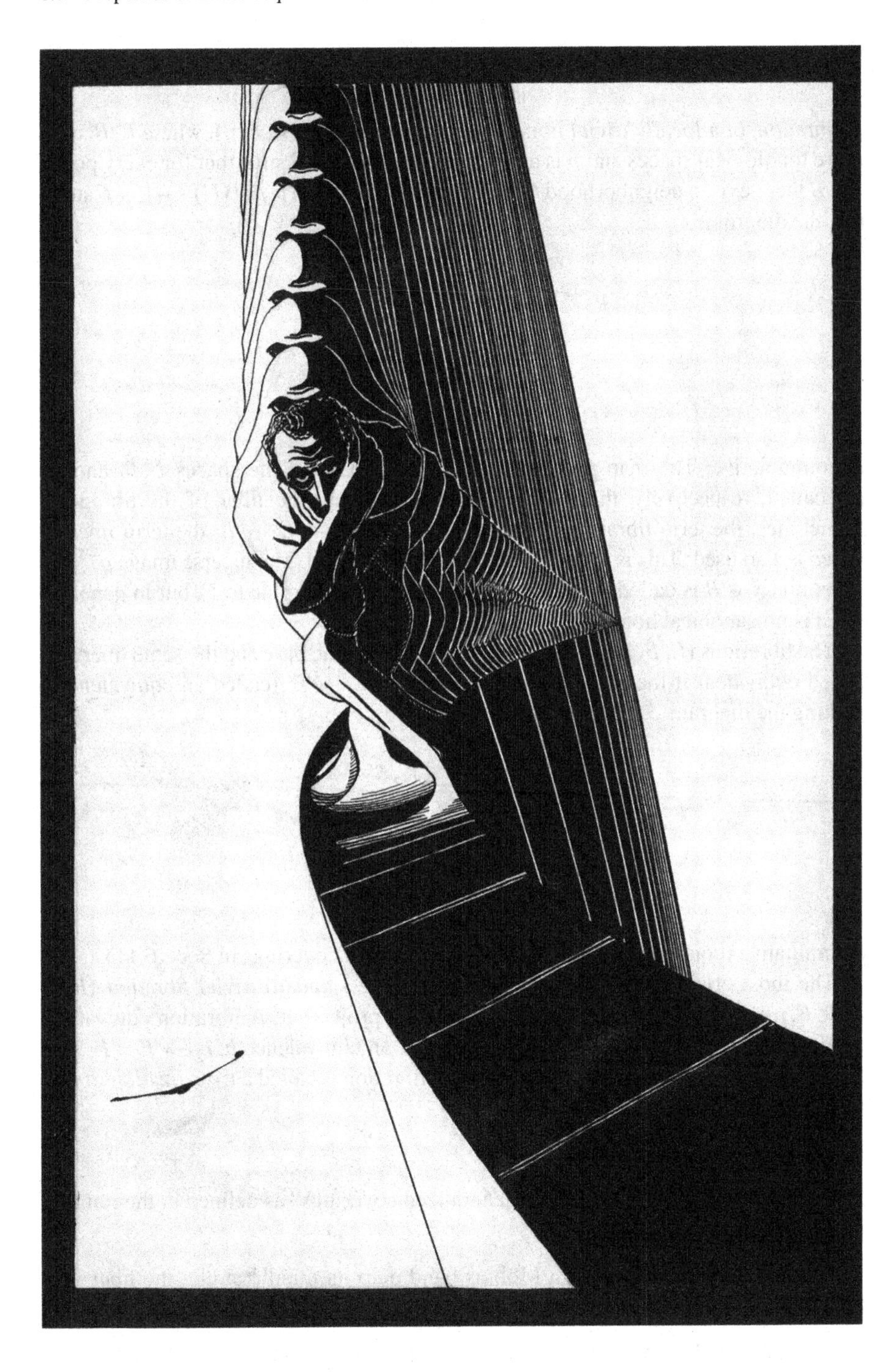

9.1 Definitions and Examples

A *fibration*, or a *locally trivial fibration*, is a quadruple (E, B, F, p), where E, B, and F are topological spaces and p is a continuous map $E \to B$ such that for every point $x \in B$ there exist a neighborhood U and a homeomorphism $f: p^{-1}(U) \to U \times F$ such that the diagram

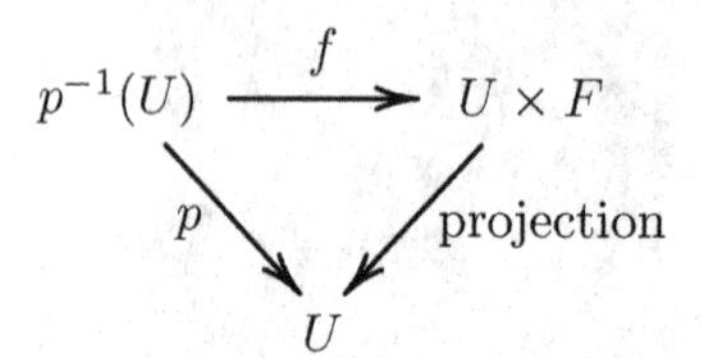

is commutative. The map p is called the *projection*, and the spaces E, B, and F are called, respectively, the *total space*, the *base*, and the *fiber* of the fibration. Sometimes, the term *fibration* is attributed to the map $p: E \to B$; the term *fibered space* is also used: This is what the space E may be called. The inverse image $p^{-1}(x)$ of a point $x \in B$ is called the *fiber over* x; it is homeomorphic to F, but in general, there is no canonical homeomorphism.

The fibrations (E, B, F, p), (E', B, F, p') with the same base and the same fiber are called *equivalent* if there is a homeomorphism $h: E \to E'$ (called an *equivalence*) making the diagram

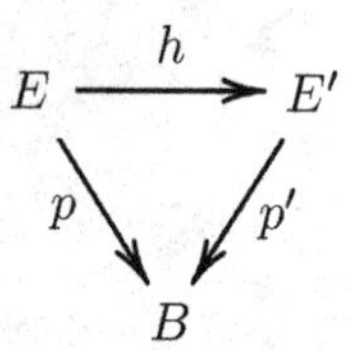

commutative (compare to the definition of equivalent coverings in Sect. 6.11).

The most obvious example of a fibration is the *standard trivial fibration* $(B \times F, B, F, p)$, where $p: B \times F \to B$ is the product projection. A fibration equivalent to the standard trivial fibration is called *trivial*; an equivalence $h: E \to B \times F$ of a fibration (E, B, F, p) with the standard trivial fibration is called a *trivialization* of the former.

Example -1. Trivial fibrations.

Example 0. Coverings (including "generalized coverings" as defined in the remark in Sect. 9.5).

Example 1. The projection of a Möbius band onto its middle circle; the fiber is I (this is probably the most popular example of a nontrivial fibration).

Example 2. Let $E = S^3 = \{(z_1, z_2) \in \mathbb{C}^2 \mid |z_1|^2 + |z_2|^2 = 1\}$, $B = S^2 = \mathbb{C}P^1$, $F = S^1 = \{z \in \mathbb{C} \mid |z| = 1\}$, $p(z_1, z_2) = (z_1 : z_2)$.

EXERCISE 1. Prove that (E, B, F, p) is a fibration. (This fibration is called the *Hopf fibration.*)

EXERCISE 2. In Lecture 1 (see Sect. 1.3) a map $S^{2n+1} \to \mathbb{C}P^n$ was introduced. Denote this map by p and prove that $(S^{2n+1}, \mathbb{C}P^n, S^1, p)$ is a fibration. (This fibration generalizes the fibration from Exercise 1 and is also called the Hopf fibration.)

Example 3. Let G be a Lie group and H be its compact subgroup. Let $p: G \to G/H$ be the natural projection.

EXERCISE 3. Prove that $(G, G/H, H, p)$ is a fibration. (The Hopf fibration from Exercise 1 is a particular case of this fibration.)

Example 4. Let a compact Lie group G act in a smooth manifold X.

EXERCISE 4. If the action is free, then $(X, X/G, G, p)$, where p is the projection $X \to X/G$, is a fibration. (All fibrations from Exercises 1–3 are particular cases of this fibration.)

Example 5. Let X, Y be compact smooth manifolds and $f: X \to Y$ be a submersion, that is, a smooth map whose differential at every point is an epimorphism. Let y_0 be a point of Y.

EXERCISE 5. Prove that if the space Y is (path) connected, then $(X, Y, f^{-1}(y_0), f)$ is a fibration.

9.2 Covering Homotopies

Fibrations, like coverings, possess a covering homotopy property (CHP). What is lost when we pass from coverings to fibrations is the uniqueness. Here is the precise statement.

Theorem. *Let (E, B, F, p) be a fibration, let X be a CW complex, let $\widetilde{\varphi}: X \to E$ be a continuous map, and let $\Phi: X \times I \to B$ be a homotopy such that $\widetilde{\Phi} |_{X \times 0} = p \circ \widetilde{\varphi}$. Then there exists a homotopy $\widetilde{\Phi}: X \times I \to E$ such that $\widetilde{\Phi} |_{X \times 0} = \widetilde{\varphi}$ and $p \circ \widetilde{\Phi} = \Phi$.*

We will prove this theorem in a stronger, relative version. Namely, *if for some CW subcomplex Y of X there is already given a homotopy $\widetilde{\Psi}: Y \times I \to E$ such that $\widetilde{\Psi} |_{Y \times 0} = \widetilde{\varphi} |_Y$ and $p \circ \widetilde{\Psi} = \Phi |_{Y \times I}$, then $\widetilde{\Phi}$ can be constructed with an additional property that $\widetilde{\Phi} |_{Y \times I} = \widetilde{G}$.*

To prove the theorem, we need two definitions and a lemma.

Definition 1. Let $\xi = (E, B, F, p)$ be a fibration, and let $B' \subset B, E' = p^{-1}(B')$. The locally trivial fibration $(E', B', F, p' = p |_{E'})$ is called the restriction of the fibration ξ to B' and is denoted as $\xi |_{B'}$.

Definition 2. Let $\xi = (E, B, F, p)$ be a fibration, and let $f: B' \to B$ be a continuous map. Denote by E' the subset of $E \times B'$ consisting of all points (e, b') such that

$f(b') = p(e)$. Then define a map $p': E' \to B'$ by the formula $p'(e, b') = b'$. The locally trivial fibration (E', B', F, p') (EXERCISE 6; check the local triviality of this fibration) is called the fibration *induced by ξ by means of f* and is denoted as $f^*\xi$.

Clarification of Definition 2. Obviously,

$$(p')^{-1}(b') = p^{-1}(f(b'))$$

[we mean the canonical homeomorphism established by the map $E' \to E$, $(e, b') \mapsto e$]. Thus, we can say that the fibered space E' is made out of fibers of the fibration ξ in such a way that the fiber over b is used as a fiber over b' whenever $f(b') = b$; if f is not one-to-one, the same fiber of ξ can be used many times as a fiber of $f^*\xi$.

Remark. The notions introduced by Definitions 1 and 2 are interrelated. First, $\xi|_{B'}$ is $i^*\xi$, where $i: B' \to B$ is the inclusion map. Second, $f^*\xi = (B' \times \xi)|_{\mathrm{graph}(f)}$, where $B' \times \xi = (B' \times E, B' \times B, F, \mathrm{id}_{B'} \times p)$ and $\mathrm{graph}(f) = \{(b', b) \in B' \times B \mid b = f(b')\}$; obviously, graph$(f)$ is canonically homeomorphic to B'.

EXERCISE 7. Let $\xi = (E, B, F, p)$ be a fibration, and let $f: B' \to B$ be a continuous map. Prove that if $\xi' = (E', B', F, p')$ is a fibration for which there exists a continuous map $h: E' \to E$ which maps every fiber $p^{-1}(b')$ of ξ' homeomorphically onto the fiber $(p')^{-1}(f(b'))$ of ξ, then the fibration ξ' is equivalent to $f^*\xi$.

Lemma (Feldbau's Theorem). *Every locally trivial fibration whose base is a cube (of any dimension) is trivial.*

Proof. Let $\xi = (E, I^n, F, p)$ be our fibration.

Step 1. Let

$$I_1^n = \left\{(x_1, \dots, x_n) \in I^n \mid x_n \leq \frac{1}{2}\right\},$$

$$I_2^n = \left\{(x_1, \dots, x_n) \in I^n \mid x_n \geq \frac{1}{2}\right\},$$

and let $\xi_1 = \xi|_{I_1^n}$, $\xi_2 = \xi|_{I_2^n}$. We will prove that *if ξ_1, ξ_2 are trivial, then ξ is trivial.* Let $h_1: p^{-1}(I_1^n) \to I_1^n \times F$, $h_2: p^{-1}(I_2^n) \to I_2^n \times F$ be trivializations of ξ_1, ξ_2. The maps h_1, h_2 do not form any map of $E = p^{-1}(I_1^n) \cup p^{-1}(I_2^n)$ into $I^n \times F$, because they are not compatible on $p^{-1}(I_1^n) \cap p^{-1}(I_2^n) = I^{n-1} \times \dfrac{1}{2}$. Actually, for $x \in I^{n-1}$, there arises a homeomorphism $\varphi_x: F = x \times F \xrightarrow{h_2^{-1}} p^{-1}(x) \xrightarrow{h_1} x \times F = F$. We define $h: E \to I^n \times F$ by the formula

$$h(e) = \begin{cases} h_1(e), & \text{if } p(e) \in I_1^n, \\ (\mathrm{id}_{I_2^n} \times \varphi_x) \circ h_2(e), & \text{if } p(e) \in x \times \left[\dfrac{1}{2}, 1\right] \subset I_2^n \end{cases}$$

(see Fig. 44). This is a trivialization of ξ.

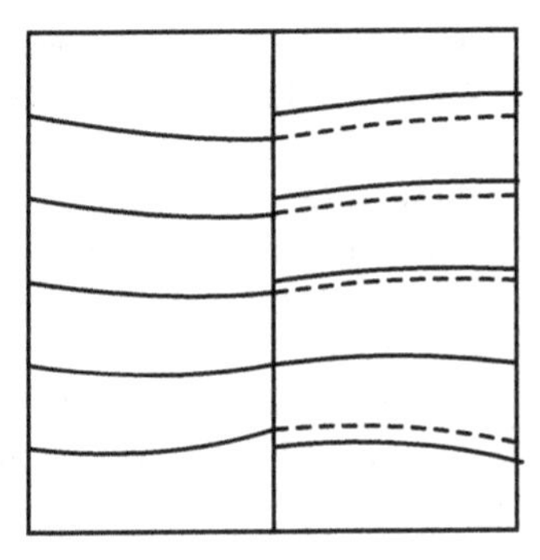

Fig. 44 Proof of Feldbau's theorem, step 1

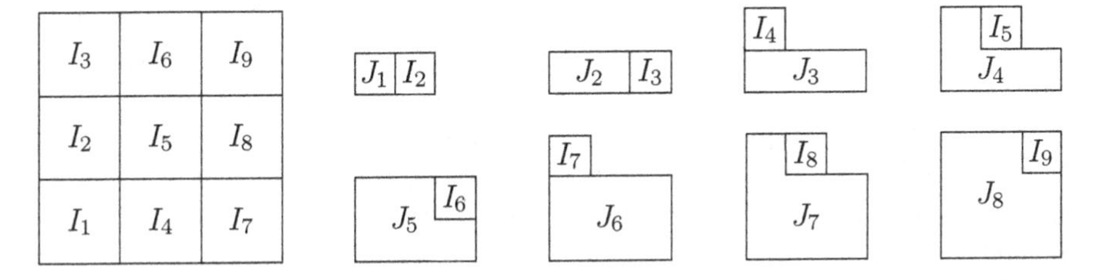

Fig. 45 Proof of Feldbau's theorem, step 2

Step 2. Cut the cube I^n into N^n small cubes with the side $\frac{1}{N}$, where N is so big that the fibration is trivial over every small cube, and numerate these small cubes as $I_1, I_2, \ldots, I_{N^n}$ in lexicographical order. For $1 \leq m \leq N^n$, let $J_m = I_1 \cup \cdots \cup I_m$. Then every J_m is homeomorphic to I^n, and for $m \geq 2$, the homeomorphism $q_m: J_m \to I^n$ can be chosen in such a way that $q_m(J_{m-1}) = I_1^n$ and $q_m(I_m) = I_2^n$. Then, according to step 1, if the fibration is trivial over J_{m-1}, it is also trivial over J_m.

Since the fibration is trivial over $J_1 = I_1$, the induction shows that it is trivial over $J_{N^n} = I^n$ (Fig. 45).

9.3 Proof of CHP

In this section we will prove (the relative version of) the theorem of Sect. 6.2. We will successively consider four cases.

Case 1: The given fibration is trivial. In this case we can assume that $E = B \times F$ and identify maps $X \to E$ with pairs of maps $X \to B$, $X \to F$. We are given a pair of maps $\varphi_1: X \to B$, $\varphi_2: X \to F$, a homotopy $\Phi_1: X \times I \to B$ of φ_1, and, in addition to $\Psi_1 = \Phi_1 |_{Y \times I}$, a homotopy $\Psi_2 \mid Y \times I$ of $\varphi_2 |_Y$. We need to extend the homotopy Ψ_2 to a homotopy $\Phi_2: X \times I \to F$ of φ_2. But this is precisely what Borsuk's theorem (Sect. 2.5) provides.

Case 2: The fibration is arbitrary, $(X, Y) = (D^n, S^{n-1})$. The induced fibration $\Phi^*(E, B, F, p) = (E', D^n \times I, F, p')$ is trivial by Feldbau's theorem ($D^n \times I$ is homeomorphic to I^{n+1}). Recall that $E' \subset (D^n \times I) \times E$. The map $\widetilde{\omega}: D^n \to E'$, $\widetilde{\omega}(x) = ((x, 0), \widetilde{\varphi}(x))$ and homotopies $\Omega = \text{id}: D^n \times I \to D^n \times I$ and $\widetilde{\Lambda}: S^{n-1} \times I \to E'$, $\widetilde{\Lambda}(x,t) = ((x,t), \widetilde{\Psi}(x,t))$ satisfy the requirements of the theorem, and, by case 1, there exists a homotopy $\widetilde{\Omega}: D^n \times I \to E'$ of $\widetilde{\omega}$ which covers Ω and extends $\widetilde{\Lambda}$. If $\widetilde{\Omega}(x,t) = ((x,t), \widetilde{\Phi}(x,t))$, then $\widetilde{\Phi}: D^n \times I \to E$ is a homotopy of $\widetilde{\varphi}$ which covers Φ and extends $\widetilde{\Psi}$.

Case 3: The fibration is arbitrary, and the CW complex X is finite. The obvious induction makes it possible to assume that $X - Y$ is one cell, e. Let $f: D^n \to X$ be a characteristic map of e [so $f(S^{n-1}) \subset Y$ and $X = Y \bigcup_{f|_{S^{n-1}}} D^n$]. The map $\widetilde{\sigma} = \widetilde{\varphi} \circ f: D^n \to X$ and homotopies $\Sigma = \Phi \circ (f \times I): D^n \times I \to B$ and $\widetilde{T} = \widetilde{\Psi} \circ (f |_{S^{n-1}} \times I): S^{n-1} \times I \to E$ satisfy the requirement of the theorem, and, by case 2, there exists a homotopy $\widetilde{\Sigma}: D^n \times I \to E$ of $\widetilde{\sigma}$ which covers Σ and extends $\widetilde{T}$. The homotopies $\widetilde{\Psi}: Y \times I \to E$ and $\widetilde{\Sigma}: D^n \times I \to E$ compose a homotopy $\widetilde{\Phi}: X \times I \to E$ that is required by the CHP. (We leave to the reader to check that $\widetilde{\Psi}$ and $\widetilde{\Sigma}$ are compatible with the attaching of $D^n \times I$ to $Y \times I$ by the map $f |_{S^{n-1}} \times I$.)

Case 4: General. If X has infinitely many cells of one dimension not contained in Y, then we need to apply the construction of case 3 to these cells simultaneously. If $X - Y$ contains cells of unlimited dimensions, then we have to apply this construction infinitely many times. In both cases, the continuity of the resulting homotopy follows from Axiom (W).

9.4 Serre Fibrations

A *Serre fibration* is a triple (E, B, p) where E, B are topological spaces and p is a continuous map $E \to B$ which satisfies the relative form of CHP (as stated in Sect. 9.2). A Serre fibration is not necessarily a locally trivial fibration (see Fig. 46) although the theorem in Sect. 9.2 states that a locally trivial fibration is a Serre fibration.

There are equivalent definitions of a Serre fibration.

Proposition 1. *The definition of a Serre fibration is equivalent to the definition which states CHP only for the case when $(X, Y) = (D^n, S^{n-1})$ for all n.*

Proof. Repeat cases 3 and 4 of the proof in Sect. 9.3.)

Proposition 2. *The definition of a Serre fibration is equivalent to the definition which states CHP only in the absolute form.*

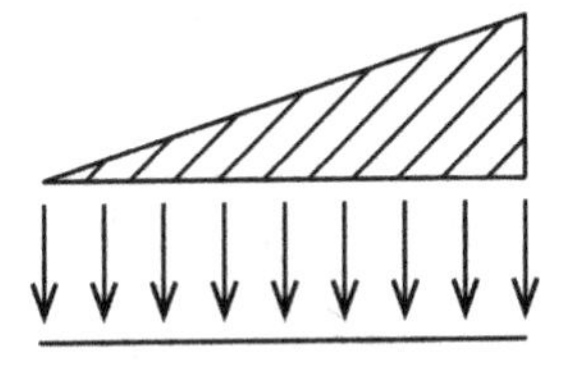

Fig. 46 A Serre fibration is not necessarily locally trivial

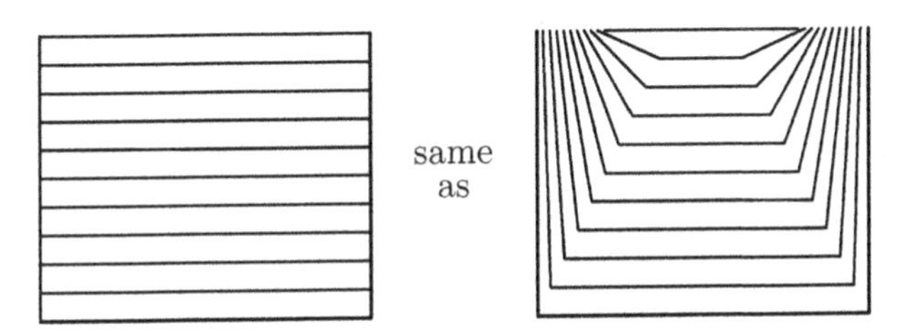

Fig. 47 The absolute CHP for D^n is the same as the relative CHP for (D^n, S^{n-1})

Proof. By Proposition 1, it is sufficient to deduce the relative CHP for $(X, Y) = (D^n, S^{n-1})$ from the absolute CHP for $X = D^n$. But the two statements are essentially the same, as Fig. 47 shows.

One more version of the definition of a Serre fibration (*not* equivalent to the initial one) may be obtained if we require the absolute CHP for X being an arbitrary topological space, not necessarily a CW complex. In this way, we arrive at a definition of a *strong Serre fibration*, or a *Hurewicz fibration*. Obviously, every Hurewicz fibration is a Serre fibration, but the converse is known to be not always true.

Notice also that definitions of a restriction of fibrations and induced fibrations as given in the local trivial case in Sect. 9.2 can be repeated for Serre fibrations and strong Serre fibrations with all accompanying remarks and clarifications.

Example 1. Locally trivial fibrations.

Example 2 (Path fibration). Let W be an arbitrary topological space with a base point w_0. Put $E = E(W, w_0)$ (the space of paths of W beginning at w_0), $B = W$, and define $p: E \to B$ by the formula $p(s) = s(1)$. Then (E, B, p) is a strong Serre fibration. Indeed, let $\widetilde{\varphi}: X \to E$ be a continuous map, and let $\Phi: X \times I \to B = W$ be a homotopy such that $\Phi(x, 0) = (\widetilde{\varphi}(x))(1)$ for every $x \in X$ (see Fig. 48). The covering homotopy $\widetilde{\Phi}: X \times I \to E$ may be defined by the formula

$$\left[\widetilde{\Phi}(x,t)\right](\tau) = \begin{cases} [\widetilde{\varphi}(x)]\,(\tau(1+t)), & \text{if } \tau(1+t) \leq 1, \\ \Phi(x, \tau(1+t)-1), & \text{if } \tau(1+t) \geq 1. \end{cases}$$

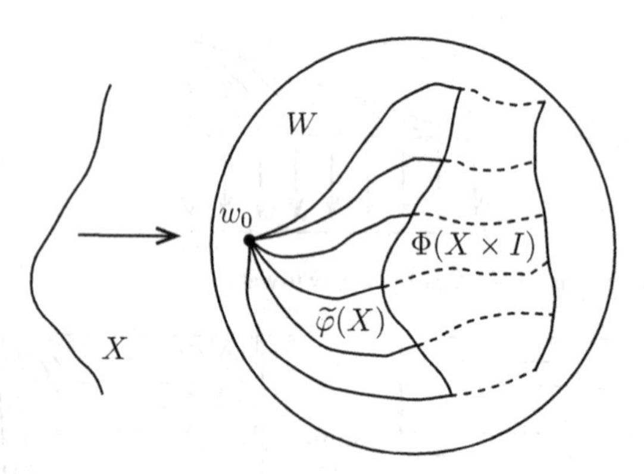

Fig. 48 The path fibration

Example 3 (A generalization of Example 2). Let (X, Y) be a Borsuk pair (for example, a CW pair), and let W be an arbitrary space. Put $E = W^X$, $B = W^Y$ (mapping spaces, see Sect. 2.5), and let $P: E \to B$ be a restriction map $(p(f) = f\mid_Y)$.

EXERCISE 8. Prove that (E, B, p) is a strong Serre fibration.

9.5 A Digression: Weak Homotopy Equivalences

The example of a Serre fibration in Fig. 46 shows that the fibers of a Serre fibration, that is, inverse images of points of the base, do not need to be homeomorphic to each other. Still these fibers turn out to have some resemblance to each other.

Definition. We will say that a topological space S is *weakly homotopy equivalent* to a topological space T if, for CW complexes X, there exist bijections $\pi(X, S) \leftrightarrow \pi(X, T)$, natural with respect to X. More precisely, for every CW complex X there is fixed a bijection $\varphi_X: \pi(X, S) \to \pi(Y, T)$ such that for every continuous (or cellular; it makes no difference in view of the cellular approximation theorem) map $f: X \to Y$, the diagram

$$\begin{array}{ccc} \pi(X,S) & \xrightarrow{\varphi_X} & \pi(X,T) \\ \uparrow{\scriptstyle f^*} & & \uparrow{\scriptstyle f^*} \\ \pi(Y,S) & \xrightarrow{\varphi_Y} & \pi(Y,T) \end{array}$$

is commutative (compare with Definition 3 of a homotopy equivalence in Sect. 3.3).

Remark. It is obvious that the relation of a weak homotopy equivalence is homotopy invariant: If $X \sim X'$, $y \sim Y'$, and X and Y are weakly homotopy equivalent, then so are X' and Y'. In particular, homotopy equivalent spaces are weakly homotopy equivalent.

According to Definition 1 in Sect. 3.3, usual homotopy equivalences are established by continuous maps. Weak homotopy equivalences are established by continuous maps sometimes, but not always. Namely, a continuous map $\varphi: S \to T$ is called a *weak homotopy equivalence* if $\varphi_*: \pi(X, S) \to \pi(X, T)$ is a bijection for every CW complex X. It is obvious that if there is a weak homotopy equivalence $\varphi: S \to T$, then S and T are weakly homotopy equivalent (just put $\varphi_X = \varphi_*$); but it is not true that weakly homotopy equivalent spaces can always be connected by a weak homotopy equivalence.

There are some important properties of weak homotopy equivalences which we can state but not prove now (they will be proved in Lecture 11; see Sects. 11.4 and 11.6). The main two statements are as follows. (1) *A continuous map between path connected spaces with base points $\varphi: (S, s_0) \to (T, t_0)$ is a weak homotopy equivalence if and only if $\varphi_*: \pi_n(S, s_0) \to \pi_n(T, t_0)$ is an isomorphism for all* $n \geq 1$. (2) *Every topological space is weakly homotopy equivalent to a CW complex, and this CW complex is unique up to a homotopy equivalence.* There is one more statement which we can prove now.

Proposition. *If CW complexes X and Y are weakly homotopy equivalent, then they are homotopy equivalent.*

The *proof* is the same as the proof of the equivalence of Definitions 1 and 2 in Sect. 3.3. The bijections

$$\varphi_X: \pi(X, X) \to \pi(X, Y) \text{ and } \varphi_Y: \pi(Y, X) \to \pi(Y, Y)$$

associate to the classes of the identity maps id_X and id_Y homotopy classes of continuous maps $f: X \to Y$ and $g: Y \to X$, and the commutativity of diagrams

$$\begin{array}{ccc} \pi(X,X) & \xrightarrow{\varphi_X} & \pi(X,Y) \\ \uparrow f^* & & \uparrow f^* \\ \pi(Y,X) & \xrightarrow{\varphi_Y} & \pi(Y,Y) \end{array}, \qquad \begin{array}{ccc} \pi(Y,X) & \xrightarrow{\varphi_Y} & \pi(Y,Y) \\ \uparrow g^* & & \uparrow g^* \\ \pi(X,X) & \xrightarrow{\varphi_X} & \pi(X,Y) \end{array}$$

show that

$$\begin{aligned} [f \circ g] &= f^*[g] = f^* \circ (\varphi_Y)^{-1}[\mathrm{id}_Y] \\ &\qquad = (\varphi_X)^{-1} \circ f^*[\mathrm{id}_Y] = (\varphi_X)^{-1}[f] = [\mathrm{id}_X] \\ [g \circ f] &= g^*[f] = g^* \circ \varphi_X[\mathrm{id}_X] = \varphi_Y \circ g^*[\mathrm{id}_X] = \varphi_Y[g] = [\mathrm{id}_Y]. \end{aligned}$$

9.6 Fibers of Serre Fibrations

Theorem. *If (E, B, p) is a Serre fibration, then for any points x_0, x_1 from the same path component of B, the fibers $p^{-1}(x_0), p^{-1}(x_1)$ are weakly homotopy equivalent. If (E, B, p) is a strong Serre fibration, then $p^{-1}(x_0), p^{-1}(x_1)$ are homotopy equivalent.*

Proof. We begin with the first statement. The proof is based on the following construction. Let X be a CW complex, let $f: X \to p^{-1}(x_0)$ be a continuous map, and let $s: I \to B$ be a path joining x_0 with x_1. Define $\widetilde{\varphi}: X \to E$ as the composition of f with the inclusion of $p^{-1}(x_0)$ into E and define $\Phi: X \times I \to B$ by the formula $\Phi(x, t) = s(t)$. CHP yields a homotopy $\widetilde{\Phi}: X \times I \to E$ which can be regarded as a family of maps $h_t: X \to p^{-1}(s(t))$; in particular, there arises a continuous map $g = h_1: X \to p^{-1}(x_1)$, and we want to use the correspondence $[f] \mapsto [g]$ to establish a weak homotopy equivalence between $p^{-1}(x_0)$ and $p^{-1}(x_1)$.

For this purpose, we will prove that the homotopy class of g does not depend on the choice of the covering homotopy $\widetilde{\Phi}$ and also on the path s within a homotopy class of paths. Let $s': I \to B$ be a path homotopic to s (and also joining x_0 with x_1), and let $\Phi', \widetilde{\Phi}', h'_t$, and g' be constructed with the use of s' in the same way as $\Phi, \widetilde{\Phi}, h_t$, and g are constructed with the use of s. Using the homotopy between s and s', we define a map $S: [-1, 1] \times I \to B$ that is a homotopy of the path $s^{-1}s'$ to the constant path $I \to x_1$ (see Fig. 49). Next, we define a map $\Psi: (X \times [-1, 1]) \times I \to B$ by the formula $\Psi((x, u), t) = S(u, t)$, and a map $\widetilde{\psi}: X \times [-1, 1] \to E$ by the formula $\widetilde{\psi}(x, u) = \begin{cases} \widetilde{\Phi}(x, -u), & \text{if } u \leq 0, \\ \widetilde{\Phi}'(x, u), & \text{if } u \geq 0. \end{cases}$ An application of CHP provides a homotopy $\widetilde{\Psi}: (X \times [-1, 1]) \times I \to E$, and the restriction of it to $X \times (((-1) \times I) \cup ([-1, 1] \times 0) \cup (1 \times I))$ is a homotopy between g and g'.

Thus, the correspondence $[f] \mapsto [g]$ provides a well-defined map $\psi_s: [X, p^{-1}(s(0))] \to [X, p^{-1}(s(1))]$ that depends only on the homotopy class of s, and it is obvious that $\psi_{s_1 s_2} = \psi_{s_2} \circ \psi_{s_2}$. In particular, $\psi_s \circ \psi_{s^{-1}} = \text{id}$, so ψ_s is a bijection. The bijections ψ_s for all CW complexes X compose the required weak homotopy equivalence.

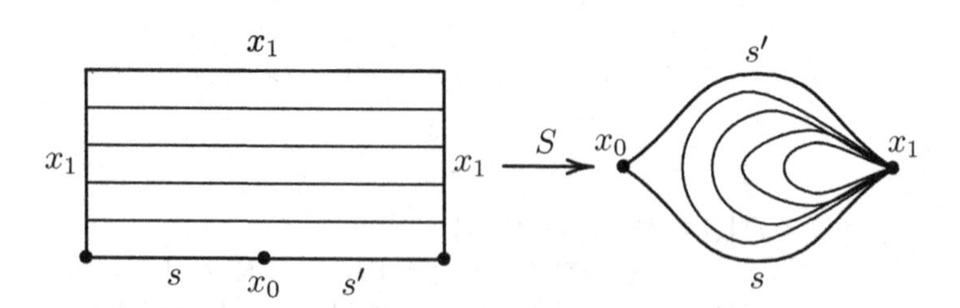

Fig. 49 To the proof of a weak homotopy equivalence of the fibers

The second statement can be proved in the same way with X being an arbitrary space, not a CW complex. An easier way to obtain a homotopy equivalence $p^{-1}(x_0) \to p^{-1}(x_1)$ is to apply the previous construction to $X = p^{-1}(x_0)$ and $f = \mathrm{id}$; then $g: p^{-1}(x_0) \to p^{-1}(x_1)$ will be the desired homotopy equivalence.

Example. A fiber of the path fibration $E(W, w_0), W, p$ (see Example 2 in Sect. 9.2) over a point $w_1 \in W$ is $E(W; w_0, w_1)$ (see Sect. 2.5). Thus, all path spaces $E(W; w_0, w_1)$ of a path connected space W, in particular, all loop spaces, are homotopy equivalent.

9.7 Every Continuous Map Is Homotopy Equivalent to a Serre Fibration

We say that continuous maps $f: X \to Y$ and $f': X' \to Y'$ are *homotopy equivalent* if there are homotopy equivalences $\varphi: X \to X'$ and $\psi: Y \to Y'$ which make the diagram

$$\begin{array}{ccc} X & \xrightarrow{f} & Y \\ \downarrow{\scriptstyle \varphi} & & \downarrow{\scriptstyle \psi} \\ X' & \xrightarrow{f'} & Y' \end{array}$$

homotopy commutative ($\psi \circ f \sim f' \circ \varphi$).

Theorem. *For every continuous map, there exists a strong Serre fibration homotopy equivalent to this map.*

ADDITIONAL PROPERTIES. First, a strong Serre fibration homotopy equivalent to a given continuous map $f: X \to Y$ is provided by a canonical construction. Second, this construction preserves Y; that is, a Serre fibration homotopy equivalent to f has the form $(\widetilde{X}, Y, p(f))$, and the homotopy equivalence $Y \to Y$ required by the definition of a homotopy equivalence between maps is just id_Y [and there is also a homotopy equivalence $\varphi(f): \widetilde{X} \to X$ such that $p(f) \sim f \circ \varphi(f)$]. Third, the construction is natural in the sense that for a homotopy commutative diagram

$$\begin{array}{ccc} X & \xrightarrow{f} & Y \\ \downarrow{\scriptstyle \alpha} & & \downarrow{\scriptstyle \beta} \\ X' & \xrightarrow{f'} & Y' \end{array}$$

there arises in a canonical way a continuous map $\widetilde{\alpha}: \widetilde{X} \to \widetilde{X}'$ such that the diagrams

$$\begin{array}{ccc} \widetilde{X} & \xrightarrow{\widetilde{\alpha}} & \widetilde{X}' \\ \downarrow{\scriptstyle p(f)} & & \downarrow{\scriptstyle p(f')} \\ Y & \xrightarrow{\beta} & Y' \end{array} \quad \text{and} \quad \begin{array}{ccc} \widetilde{X} & \xrightarrow{\widetilde{\alpha}} & \widetilde{X}' \\ \downarrow{\scriptstyle \varphi(f)} & & \downarrow{\scriptstyle \varphi(f')} \\ X & \xrightarrow{\alpha} & X' \end{array}$$

are homotopy commutative.

Proof of Theorem. For $\widetilde{X}$, we take the space of pairs (x, s), where $x \in X$ and s is a path if Y beginning at the point $f(x)$. The projection $p(f): \widetilde{X} \to Y$ and the homotopy equivalence $\varphi(f): \widetilde{X} \to X$ are defined by the formulas $[p(f)](x, s) = s(1)$ and $[\varphi(f)](x, s) = s$. The verification of all necessary properties is immediate.

Remark. This theorem is dual (in the sense of duality considered in Lecture 4) to the following simple statement: For every continuous map $f: X \to Y$ there exists an embedding $i(f): X \to \widetilde{Y}$ homotopy equivalent to f; moreover, we can request that $(\widetilde{Y}, [i(f)](X))$ be a Borsuk pair. Proof: We can take for $\widetilde{Y}$ the cylinder $\mathrm{Cyl}(f)$ of the map f (see Sect. 2.3) and for $i(f)$ the natural embedding. Because of this duality, the space $\widetilde{X}$ constructed in the proof is sometimes called the *cocylinder* of the map f.

9.8 The Homotopy Sequence of a Fibration

Lemma. *Let (E, B, p) be a Serre fibration, let $e_0 \in E$ be an arbitrary point, let $b_0 = p(e_0)$, and let $F = p^{-1}(b_0)$. Then the map*

$$p_*: \pi_n(E, F, e_0) \to \pi_n(B, b_0)$$

is an isomorphism for all n.

Proof. In this proof we use a slightly modified definition of relative spheroids: A relative spheroid of a pair (E, F) with the base point e_0 is defined as a map $(D^n, S^{n-1}, y_0) \to (E, F, e_0)$ (where y_0 is the base point in D^n and S^{n-1}) rather than a map $(D^n, S^{n-1}, S^{n-1}_+) \to (E, F, e_0)$; this does not make any difference, since $(D^n, S^{n-1}, y_0) \sim (D^n, S^{n-1}, S^{n-1}_+)$.

First, prove that p_* is a monomorphism. Let $\widetilde{f}: D^n \to E$ be a relative spheroid of (E, F), and let $f: S^n \to B$ be its projection into B. Let $H: S^n \times I \to B$ be a homotopy joining the spheroid f with the constant spheroid. It may be regarded as a homotopy of the map $p \circ \widetilde{f}: D^n \to B$. As such it is covered by a homotopy $\widetilde{H}: D^n \times I \to E$ of $\widetilde{f}$ which may be regarded as a homotopy of the relative spheroid $\widetilde{f}$ ($\widetilde{H}(S^{n-1} \times I) \subset F$, since H, regarded as a homotopy of $p \circ \widetilde{f}: D^n \to B$, maps $S^{n-1} \times I$ to b_0; also we can assume, using the relative version of CHP, that $\widetilde{H}(y_0 \times I) = e_0$). The homotopy $\widetilde{H}$ joins $\widetilde{f}$ with a relative spheroid whose image is contained in F, that is, with the relative spheroid of the zero class. Thus, $[\widetilde{f}] = 0$.

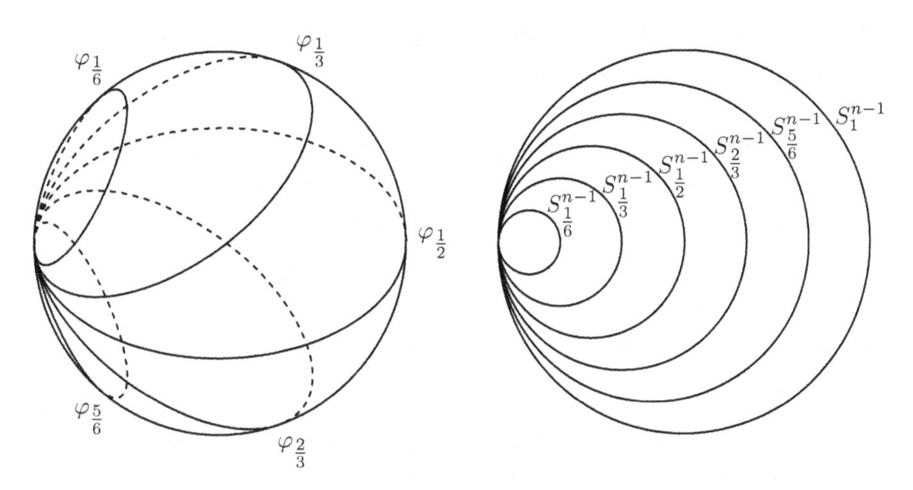

Fig. 50 Proof of the lemma: p_* is an epimorphism

Now prove that p_* is an epimorphism. Let $\varphi_t: S^{n-1} \to S^n$ be a homotopy sweeping the sphere S^n as shown in Fig. 50, left. Let $f: S^n \to B$ be a spheroid. Consider the homotopy $f \circ \varphi_t: S^n \to B$ and lift it, using the (relative) CHP to a homotopy $\widetilde{g}_t: S^n \to E$ such that $\widetilde{g}_0(S^n) = e_0$ and $\widetilde{g}_t(y_0) = e_0$ for all t. This homotopy may be considered as a map $\widetilde{G}: D^n \to E$ (the ball D^n is covered by spheres S^{n-1}_t, $0 < t \leq 1$, and $\widetilde{G}\,|_{S^{n-1}_t}$ is $\widetilde{g}_t$). This map $\widetilde{G}$ is a relative spheroid of the pair (E, F), and obviously $p_*[G] = [\widetilde{f}]$.

Now replace in the homotopy sequence of the pair (E, F) the groups $\pi_i(E, F)$ by the isomorphic groups $\pi_i(B)$. We get an exact sequence

$$\begin{aligned}\cdots \to \pi_n(F, e_0) \to \pi_n(E, e_0) \to \pi_n(B, b_0) \to \pi_{n-1}(F, e_0) \to \dots\\ \to \pi_1(B, b_0) \to \pi_0(F, e_0) \to \pi_0(E, e_0) \to \pi_0(B, b_0)\end{aligned}$$

consisting only of absolute homotopy groups (not all of them are groups, as we know). This sequence is called the *homotopy sequence of the fibration.*

Remark 1. The homomorphisms $\pi_n(F, e_0) \to \pi_n(E, e_0)$ and $\pi_n(E, e_0) \to \pi_n(B, b_0)$ of this sequence are induced by the inclusion map $F \to E$ and the projection $p: E \to B$. The construction in the second part of the proof of the lemma yields a direct construction, not involving relative homotopy groups, of the "connecting" homomorphism $\pi_n(B, b_0) \to \pi_{n-1}(F, e_0)$: Using a spheroid $f: S^n \to B$, we construct a spheroid homotopy $f \circ \varphi_t: S^{n-1} \to B$, lift it to a spheroid homotopy $g_t: S^{n-1} \to E$, and observe that g_1 is, actually, a spheroid of F.

Remark 2. The term $\pi_0(B)$ of the homotopy sequence of the fibration does not actually come from the homotopy sequence of the pair (E, F), so the exactness in the term $\pi_0(E, e_0)$ must be checked independently.

9.9 First Applications of the Exactness of the Homotopy Sequence of a Fibration

Let us begin with the Hopf fibration $p: S^3 \to S^2$ with fiber S^1 (see Example 2 in Sect. 9.1). The homotopy sequence of this fibration contains fragments

$$\pi_2(S^3) \to \pi_2(S^2) \to \pi_1(S^1) \to \pi_1(S^3),$$
$$\pi_n(S^1) \to \pi_n(S^3) \to \pi_n(S^2) \to \pi_{n-1}(S^1),$$

and, since $\pi_i(S^3) = 0$ for $i = 1, 2$ and $\pi_i(S^1) = 0$ for $i \geq 2$ (see Sects. 5.9 and 8.3), we obtain isomorphisms $\pi_2(S^2) \cong \pi_1(S^1)$ and $\pi_n(S^3) \cong \pi_n(S^2)$ for all $n \geq 3$. The first isomorphism shows that $\pi_2(S^2) \cong \mathbb{Z}$; in particular, we have a proof (at last!) of the fact that the sphere S^2 is not contractible. It is interesting that this result will not be covered by a more general theorem of Lecture 10, so it has its independent value. The second isomorphism looks unexpected, especially for $n = 3$. We expect that the group $\pi_3(S^3)$ is nontrivial (actually, we will prove soon that it is isomorphic to $\mathbb{Z}$); thus, $\pi_3(S^2)$ must also be nontrivial, a fact that is not readily offered by naive geometric intuition.

Some other applications of the exactness of the homotopy sequence of a fibration are contained in the following exercises.

EXERCISE 9. Analyze the homotopy sequence of a covering. Deduce from it the major results of Sects. 6.6 and 6.8.

EXERCISE 10. Deduce from the homotopy sequence of the Hopf fibration $p: S^{2n+1} \to \mathbb{C}P^n$ (Example 3 of Sect. 9.1) that $\pi_r(\mathbb{C}P^n)$ is $\mathbb{Z}$ for $r = 2$ and zero for $3 \leq r \leq 2n-1$; in particular, $\mathbb{C}P^\infty$ has only one nontrivial homotopy group: $\pi_2(\mathbb{C}P^\infty) \cong \mathbb{Z}$.

EXERCISE 11. Using the path fibration from Sect. 9.4, prove that $\pi_n(\Omega X) \cong \pi_{n+1}(X)$ for all X and $n \geq 0$. (This fact can be easily proved directly, by comparing spheroids of X and ΩX. By the way, it shows that the group $\pi_1(\Omega X)$ is always commutative, in accordance with Exercise 4 of Sect. 6.7).

EXERCISE 12. Prove that if the base of a Serre fibration is contractible, then the inclusion of (any) fiber in the total space induces an isomorphism of homotopy groups. Prove that if the base of a Serre fibration is connected and one of the fibers is contractible, then the projection induces an isomorphism of homotopy groups of the total space and the base. *Note.* According to a theorem promised in Sect. 9.5 (but not proven so far), these statements mean, respectively, that the inclusion map $F \to E$ and the projection $E \to B$ are weak homotopy equivalences.

EXERCISE 13. Prove that if all the homotopy groups of the base and the fiber are finite, then so are homotopy groups of the total space, and the orders of the homotopy groups of the total space do not exceed the product of orders of corresponding homotopy groups of the base and the fiber. Formulate and prove a similar statement concerning finitely generated groups and their ranks.

EXERCISE 14. Prove that if a Serre fibration (E, B, p) has a section (that is, a continuous map $s: B \to E$ such that $p \circ s = \mathrm{id}_B$) or if F is a retract of E, then $\pi_n(E) \cong \pi_n(B) \oplus \pi_n(F)$ for $n \geq 2$ (and for $n = 1$ if $\pi_1(E)$ is commutative).

EXERCISE 15. Prove that if the fiber of the Serre fibration E, B, p is contractible in E, then $\pi_n(B) \cong \pi_n(E) \oplus \pi_{n-1}(F)$ for all $n \geq 2$.

9.10 A Construction Promised in Lecture 8

In Sect. 8.6, we promised to construct for a topological pair (X, A) a space Y with an isomorphism $\pi_n(X, A) \cong \pi_{n-1}(Y)$. We present this construction here with an additional remark that both Y and the isomorphism will be natural in all possible senses.

Following Sect. 9.7, construct a (strong) Serre fibration $p': A' \to X$ homotopy equivalent to the inclusion map $A \to X$ and denote by Y a fiber of this fibration (over some point $x_0 \in A \subset X$). Let $f: A' \to A$ be a canonical homotopy equivalence. We assume that p' and f are constructed by the canonical construction, so Y is the space of paths in X beginning at x_0 and ending in A. This observation provides a canonical map $\pi_{n-1}(Y) \to \pi_n(X, A)$: For a spheroid $g: (S^{n-1}, y_0) \to (Y, \mathrm{const})$ we define a spheroid $G: D^n = CS^{n-1} \to X$ of the pair (X, A) by the formula $G(s, t) = [g(s)](t)$, $s \in S^{n-1}, t \in I$. It is obvious that these canonical maps are included in the commutative diagram

$$\begin{array}{ccccccccccc}
\cdots \to & \pi_n(A') & \to & \pi_n(X) & \to & \pi_{n-1}(Y) & \to & \pi_{n-1}(A') & \to & \pi_{n-1}(X) & \to \cdots \\
 & \downarrow & & \| & & \downarrow & & \downarrow & & \| & \\
\cdots \to & \pi_n(A) & \to & \pi_n(X) & \to & \pi_n(X, A) & \to & \pi_{n-1}(A) & \to & \pi_{n-1}(X) & \to \cdots
\end{array}$$

whose rows are homotopy sequences of the fibration (A', X, p') and the pair (X, A). It follows from the five-lemma that the maps $\pi_{n-1}(Y) \to \pi_n(X, A)$ are isomorphisms (it can be also proved directly: Our map is already bijective at the level of spheroids).

Thus, not only have we established the promised interpretation of relative homotopy groups as absolute homotopy groups, but we have also discovered one more relation between homotopy sequences of pairs and of fibrations.

Lecture 10 The Suspension Theorem and Homotopy Groups of Spheres

10.1 Main Theorem

Let $f: S^q \to X$ be a q-dimensional spheroid of a topological space X (with a base point). The map $\Sigma f: \Sigma S^q = S^{q+1} \to \Sigma X$, $[\Sigma f](y,t)_\Sigma = (f(y),t)_\Sigma$ is a $(q+1)$-dimensional spheroid of the space ΣX. It is clear also that if spheroids $f, g: S^q \to X$ are homotopic, then the spheroids $\Sigma f, \Sigma g: S^{q+1} \to \Sigma X$ are also homotopic, and the spheroid $\Sigma(f+g)$ is homotopic to the spheroid $\Sigma f + \Sigma g$. Thus, the correspondence $f \mapsto \Sigma f$ gives rise to a homomorphism $\pi_q(X) \to \pi_{q+1}(\Sigma X)$. This homomorphism is called the suspension homomorphism and is also denoted by Σ. In particular, for every q and n, there arises a homomorphism

$$\Sigma: \pi_q(S^n) \to \pi_{q+1}(S^{n+1}).$$

Theorem (Freudenthal). *This homomorphism is an isomorphism if $q < 2n-1$ and is an epimorphism if $q = 2n-1$.*

This statement is sometimes called the "easy part" of Freudenthal's theorem. We will discuss the "difficult part" later in this lecture. Let us mention one more a generalization of Freudenthal's theorem which we will be able to prove in Chap. 3: *If X is an n-connected CW complex, then $\Sigma: \pi_q(X) \to \pi_{q+1}(\Sigma X)$ is an isomorphism for $q < 2n+1$ and an epimorphism for $q = 2n+1$.* (This is a generalization of the easy part of Freudenthal's theorem; the difficult part has a similar generalization.)

Proof of the Epimorphism Part. Let $f: S^{q+1} \to S^{n+1}$ be a spheroid. We want to prove that (if $q \leq 2n-1$) there exists a spheroid $h: S^q \to S^n$ such that $f \sim \Sigma h$. We may assume that $n > 0$, in which case the sphere S^{n+1} is simply connected and we may forget the base points.

Let N and S be the poles of the sphere S^{n+1}. We will present the sphere S^{q+1} as $\mathbb{R}^{q+1} \cup \infty$ assuming that $f(\infty)$ is neither of the poles.

First, we apply a construction similar to that of Sect. 5.8 (where we used it to prove a free-point lemma): We assume that there are triangulations of neighborhoods U and V of N and S ("polar caps") and of a big ball $B \subset \mathbb{R}^{q+1}$ containing both $f^{-1}(U)$ and $f^{-1}(V)$ such that f is simplicial on the union of all simplices of B whose images are not disjoint from $U \cup V$. Also, we may assume that both N and S are interior points of $(n+1)$-dimensional simplices. Then $P = f^{-1}(N)$ and $Q = f^{-1}(S)$ are disjoint polyhedra of dimension $\leq q-n$ [that is, each of P and Q is compact and is contained in a finite union of $(q-n)$-dimensional planes].

Second, we want to construct a homotopy of f to such a map that the inverse images of N and S will be separated by a hyperplane in $\mathbb{R}^{q+1}$. Choose such a hyperplane Π such that P lies on one side of Π and on the other side of Π choose a point X such that the cone with the base Q and the vertex X is disjoint from P.

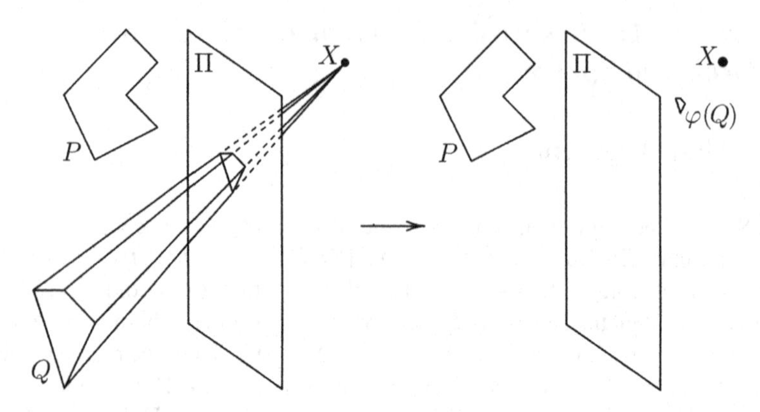

Fig. 51 Separating inverse images of poles

The existence of such an X follows from the inequality $q \leq 2n - 1$ (and this is the only place in the proof where we use this inequality): The point X should not belong to any line joining a point of P with a point of Q. If we replace P and Q by two planes of dimension $q - n$, then the union of such lines is (contained in) a plane of dimension $2(q-n)+1$, so for our P and Q the set of forbidden points X is contained in a finite union of such planes, and the inequality $q \leq 2n-1$ is the same as $2(q-n)+1 < q+1$, so we can choose a point not belonging to this union. Now we choose a positive continuous function F on the set of all lines through X such that $F(\ell)$ is some very big number K for lines hitting P and $F(\ell)$ is 1 for every line hitting Q. Consider a homeomorphism φ of $\mathbb{R}^{q+1}$ onto itself which compresses every line ℓ through X to X with the coefficient $F(\ell)$ (see Fig. 51). This homeomorphism does not move P and, if K is big enough, pulls Q into a polyhedron on the other (with respect to P) side of Π. (We can assume that our homeomorphism $\mathbb{R}^{q+1} \cup \infty \approx S^{q+1}$ takes the plane Π into the equator of S^{q+1}.) Moreover, this compression may be done gradually, so we get an isotopy (a homotopy consisting of homeomorphisms) of id to φ. So $f \circ \varphi^{-1}$ is a spheroid homotopic to F which possesses the desirable property: The inverse images of the poles are polyhedra separated by a hyperplane. For brevity's sake, we will denote a spheroid with this property again by f. (This argument, and actually the whole proof of Freudenthal's theorem, is based on the fact that polyhedra of dimensions p and q cannot be linked in a space of dimension $> p+q+1$. For example, two disjoint closed polygonal lines can be linked in $\mathbb{R}^3$, but not in $\mathbb{R}^4$; see Fig. 52.)

Thus, we now have a rather good spheroid: The inverse images of "polar caps" U and V of S^{n+1} are contained in, respectively, the northern and southern hemispheres of S^{q+1} and the image of the equator S^q of S^{q+1} does not touch the polar caps of S^{n+1} (see Fig. 53).

Next we make the spheroid f still better by combining it with a homotopy of S^{n+1} which stretches the polar caps U and V to the whole hemispheres and compresses the equatorial belt to the equator. The new spheroid, which we still denote by f, maps

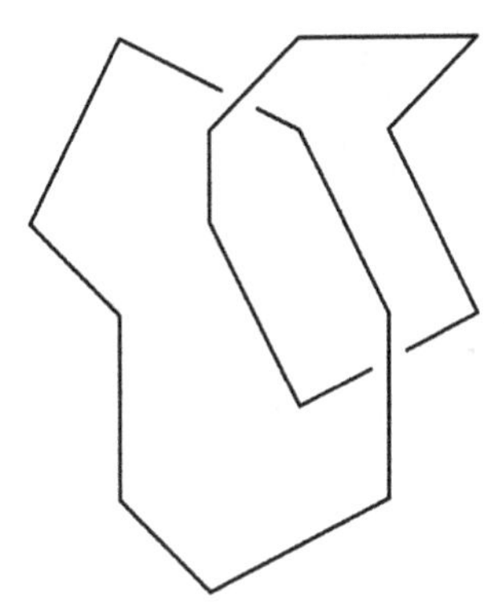

Fig. 52 Linked curves in space

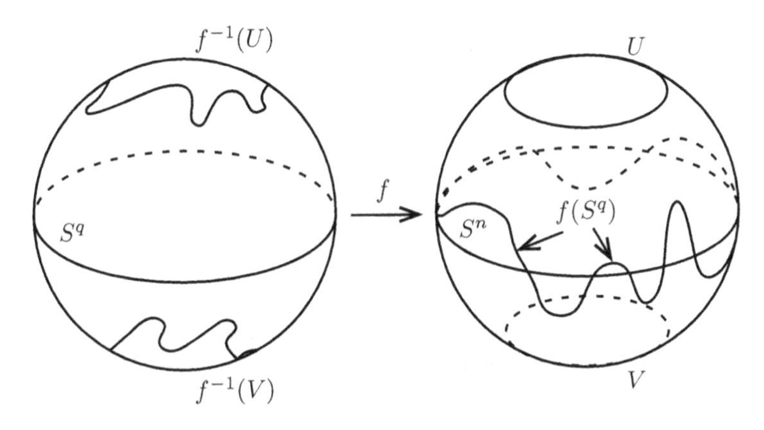

Fig. 53 The spheroid f with inverse images of polar caps separated

the equator S^q of S^{q+1} into the equator S^n of S^{n+1} (thus we get the map $S^q \to S^n$!) and maps the northern hemisphere into the northern hemisphere and the southern hemisphere into the southern hemisphere. If we look at the spheres from above, we will see the maps f and g as shown in Fig. 54.

The spheroid f is still different from Σg: It takes meridians into some arbitrary looking curves while Σg takes meridians into meridians (see Fig. 54). However, for no point of $y \in S^{q+1}$ are the points $f(y)$ and $\Sigma g(y)$ opposite, so there is a convenient big circle homotopy joining f and Σg. This completes the proof of Σ being onto.

Proof of the Monomorphism Part. Now we assume that $q < 2n-1$. Let $g_0, g_1 \colon S^q \to S^n$ be two spheroids, and let $f_t \colon S^{q+1} \to S^{n+1}$ be a homotopy between Σg_0 and Σg_1. We want to prove that there exists a homotopy $g_t \colon S^q \to S^n$ between g_0 and g_1. For this, we will deform the homotopy f_t to the homotopy of the form Σg_t. Let us apply the previous construction to every f_t. The only arbitrary choice in that construction was the choice of the point X. Now we have to require that this X depends continuously on t, which we will achieve by requesting that X not depend

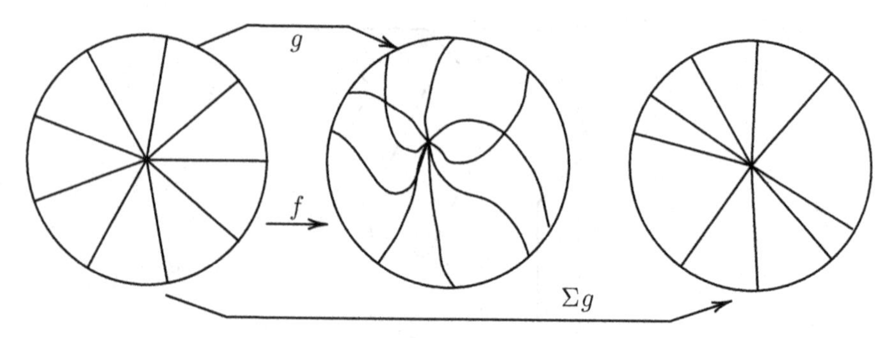

Fig. 54 The maps f, g, and Σg

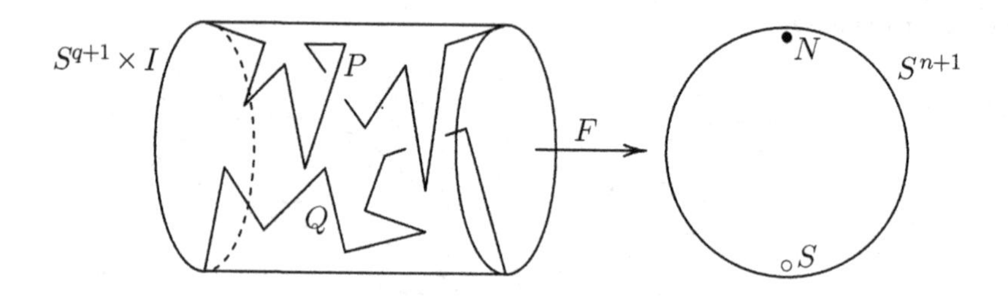

Fig. 55 Proof of the monomorphism part of Freudenthal's theorem

on t at all. In the previous construction there was a prohibited set for X, which was (the union of planes of) dimension $2(q-n)+1$. Now we have to deal with a one-parameter family of such sets, which creates one additional dimension. Thus, to apply the construction we now need the inequality $2(q-n)+2 < q+1$, which is the same as $q < 2n-1$.

In other words, the inverse images P, Q of the poles of S^{n+1} with respect to a (piecewise linear approximation of a) homotopy $\{f_t\} = F: S^{q+1} \times I \to S^{n+1}$ are polyhedra of dimensions $\leq q-n+1$ (see Fig. 55). These polyhedra live in $\mathbb{R}^{q+2}$, and they are not linked, if $2(q-n+1)+1 < q+2$, which again means that $q < 2n-1$.

This completes the proof of the easy part of Freudenthal's theorem.

10.2 First Applications

Theorem (Hopf). $\pi_n(S^n) \cong \mathbb{Z}$.

Proof. For $n = 1, 2$, we already know this (see Sect. 6.3 for $n = 1$ and Sect. 9.9 for $n = 2$). For $n \geq 3$, we have an isomorphism $\Sigma: \pi_{n-1}(S^{n-1}) \to \pi_n(S^n)$.

Additional remarks. 1. By Freudenthal's theorem, $\Sigma: \pi_1(S^1) \to \pi_2(S^2)$ is an epimorphism. But since $\pi_1(S^1) \cong \pi_2(S^2) \cong \mathbb{Z}$, it is actually an isomorphism.
2. The group $\pi_n(S^n)$ is generated by the class of the identity spheroid. For $n = 1$, it was proved in Sect. 6.3; for $n > 1$, this follows from the obvious fact that the suspension over the identity map is the identity map. From now on, we choose the class of the identity spheroid for a generator of $\pi_n(S^n)$ and thus establish a canonical isomorphism $\pi_n(S^n) = \mathbb{Z}$.

Corollary. *The sphere S^n is not contractible for any n.*

This statement, whose proof turned out to be unexpectedly long, is very important: If S^n were contractible, the group $\pi_n(X)$ would have been zero for any X.

One More Corollary. $\pi_3(S^2) \cong \mathbb{Z}$.

This follows from the isomorphism $\pi_3(S^2) \cong \pi_3(S^3)$ established in Sect. 9.9. Actually, it also follows from the results of Sect. 9.9 that the group $\pi_3(S^2)$ is generated by the class of the Hopf map $S^3 \to S^2$.

10.3 The Degree of a Map $S^n \to S^n$

A continuous map $f: S^n \to S^n$ regarded as an n-dimensional spheroid of S^n determines an element of $\pi_n(S^n) = \mathbb{Z}$, that is, an integer. This integer is called the *degree* of f and is denoted as $\deg f$. Let us observe the properties of the degree which are already known to us. There are maps $S^n \to S^n$ of an arbitrary degree. Two maps $S^n \to S^n$ are homotopic if and only if they have equal degrees. A continuous map $S^n \to S^n$ of degree d induces a homomorphism $\pi_n(S^n) \to \pi_n(S^n)$, that is, $\mathbb{Z} \to \mathbb{Z}$, which is a multiplication by d. Homeomorphisms have degrees ± 1. A suspension over a map $S^n \to S^n$ of degree d is a map $S^{n+1} \to S^{n+1}$ of the same degree d.

Now we will describe a way of computing the degree of a map $f: S^n \to S^n$. A point $y \in S^n$ is called a regular value of f if there is a neighborhood U in y homeomorphic to a ball D^n such that $f^{-1}(U)$ is a disjoint union of open sets U_α such that f maps every U_α homeomorphically onto U. [A ridiculous but important example: If $y \notin F(S^n)$, that is, if *y is not a value of f at all*, it is a regular value of f.] For example, smooth maps and piecewise linear maps have ample sets of regular values (in the piecewise linear case it is obvious; in the smooth case it is a standard theorem of analysis). If $y \in S^n$ is a regular value of f, then the inverse image $f^{-1}(y)$ is finite [otherwise, $f^{-1}(y)$ contains limit points, and no neighborhood of a limit point of $f^{-1}(y)$ can be homeomorphically mapped onto a neighborhood of y]. For every point $z \in f^{-1}(y)$, the map f either preserves or reverses the orientation [in the smooth case this is determined by the sign of the Jacobian of f at z; in the continuous case, we can say that z belongs to some U_α, and the map f determines a homeomorphism $S^n/(S^n - U_\alpha)$ onto $S^n/(S^n - U)$, which is a homeomorphism between two copies of S^n; its degree is ± 1 and we define the preserving or reversing orientation accordingly].

Theorem. *If y is a regular value of f, then*

$$\begin{aligned}\deg f = \#\{z \in f^{-1}(y) \mid f \text{ preserves the orientation at } z\} \\ -\#\{z \in f^{-1}(y) \mid f \text{ reverses the orientation at } z\} \\ = \sum_{z \in f^{-1}(y)} \varepsilon(z),\end{aligned}$$

where $\varepsilon(z)$ is 1 if f preserves the orientation in the neighborhood of z and is -1 otherwise.

Remark. As a rule, a map f has many regular values y, and the numbers of points in $f^{-1}(y)$ may be different for different ys. However, our theorem shows that the difference in our equality depends only on f.

Proof of Theorem. Let U be a neighborhood of y as in the definition of a regular value. We combine the spheroid f with a (homotopic to the identity) map $S^n \to S^n$ which collapses the complement of U into the base point and stretches U to the whole sphere. The new spheroid will be the sum of spheroids corresponding to points in f^{-1}. Each of these spheroids will be a homeomorphism, and the degree of a spheroid corresponding to $z \in f^{-1}(y)$ is $\varepsilon(z)$. Thus, $\deg f = \sum_{z \in f^{-1}(y)} \varepsilon(z)$.

10.4 Stable Homotopy Groups of Spheres and Other Spaces

Thus, the homotopy groups of spheres are arranged into stabilizing series of groups $\pi_{n+k}(S^n)$ with a fixed k:

$$\dots \xrightarrow{\Sigma} \pi_{n+k}(S^n) \xrightarrow{\Sigma} \pi_{n+k+1}(S^{n+1}) \xrightarrow{\Sigma} \pi_{n+k+2}(S^{n+2}) \xrightarrow{\Sigma} \dots$$

with the stabilization occurring in the term $\pi_{2k+2}(S^{k+2})$:

$$\dots \xrightarrow{\Sigma} \pi_{2k+1}(S^{k+1}) \xrightarrow{\text{epi}} \pi_{2k+2}(S^{k+2}) \xrightarrow{\text{iso}} \pi_{2k+3}(S^{k+3}) \xrightarrow{\text{iso}} \dots.$$

The groups $\pi_{n+k}(S^n)$ with $n \geq k+2$ do not depend on n. They are called *stable*, and for them the notation π_k^S is used; the group $\pi_{2k+1}(S^{k+1})$ is called *metastable*. So far, we have almost no information on the homotopy groups of spheres; what we know is contained in the following table.

$\pi_q(S^n)$	$q=1$	$q=2$	$q=3$	$q=4$	$q=5$	$q=6$
$n=1$	$\mathbb{Z}$ $\searrow^{i}$	0	0	0	0	0
$n=2$	0	$\mathbb{Z}$ $\searrow^{i}$	$\mathbb{Z}$ $\searrow^{e}$	? $\searrow$	? $\searrow$	?
			$\Vert$H	$\Vert$H	$\Vert$H	$\Vert$H
$n=3$	0	0	$\mathbb{Z}$ $\searrow^{i}$	? $\searrow^{i}$	? $\searrow^{e}$	?
$n=4$	0	0	0	$\mathbb{Z}$	?	?

In this table, slanted arrows denote Σ, the letter H means Hopf isomorphism, and the letters i and e mean, respectively, isomorphism and epimorphism. We can add that since $\Sigma\colon \pi_3(S^2) \to \pi_4(S^3)$ is an epimorphism and $\pi_3(S^2) = \mathbb{Z}$, the group $\pi_4(S^3)$ must be cyclic, and so must be the groups $\pi_4(S^2)$ and π_1^S.

The homotopy groups of spheres has not been calculated yet, but ample information about them exists, of both a general and a tabular nature. In this book, we will address these groups many times, and we will learn a lot about them.

In conclusion, we remark that stable homotopy groups do not exist only for spheres. For any topological space X, we can consider a sequence

$$\pi_k(X) \xrightarrow{\Sigma} \pi_{k+1}(\Sigma X) \xrightarrow{\Sigma} \pi_{k+2}(\Sigma^2 X) \xrightarrow{\Sigma} \pi_{k+3}(\Sigma^3 X) \xrightarrow{\Sigma} \dots.$$

This sequence has a "limit" (algebraists call it the *direct limit*), but we actually do not need it, since this sequence always stabilizes at the term $\pi_{2k+2}(\Sigma^{k+2}X)$. We will prove this later, but we have already mentioned the necessary result (see a remark after the statement of Freudenthal's theorem in Sect. 10.1).

For stable homotopy groups of X we use the notation $\pi_k^S(X)$; thus, $\pi_k^S = \pi_k^S(S^0)$.

10.5 Whitehead Product and the Difficult Part of Freudenthal's Theorem

The product $S^m \times S^n$ of two spheres has a CW decomposition into four cells, of dimensions $0, m, n$, and $m+n$. The union of the first three cells is the bouquet $S^m \vee S^n$. The attaching map of the fourth cell, $S^{m+n-1} \to S^m \vee S^n$, is called the *Whitehead map*.

This construction describes the Whitehead map up to a homotopy; there exists a canonical, completely concrete, description of the map $w\colon S^{m+n-1} \to S^m \vee S^n$. The sphere S^{m+n-1} is cut into the union of two closed domains,

$$U = \left\{(x_1, \dots, x_{m+n}) \in S^{m+n-1} \mid x_1^2 + \dots + x_m^2 \le \tfrac{1}{2}\right\},$$
$$V = \left\{(x_1, \dots, x_{m+n}) \in S^{m+n-1} \mid x_1^2 + \dots + x_m^2 \ge \tfrac{1}{2}\right\}.$$

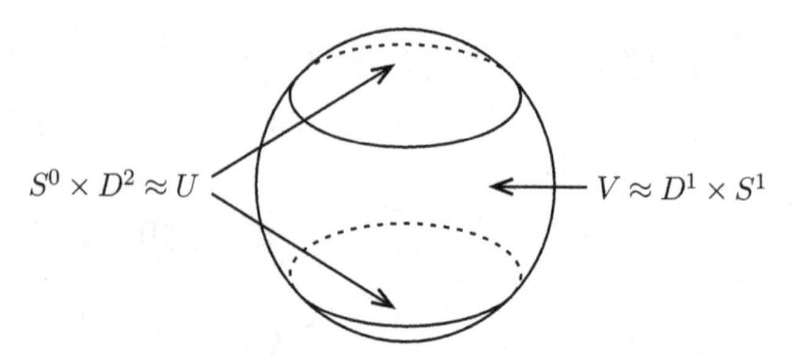

Fig. 56 $S^{m+n-1} = (D^m \times S^{n-1}) \cup (S^{m-1} \times D^n)$

Obviously, $U \approx D^m \times S^{n-1}$, $V \approx S^{m-1} \times D^n$, $U \cap V = S^{m-1} \times S^{n-1}$ (the case when $m = 2, n = 1$ is shown in Fig. 56; another important example is the cut of the sphere S^3 by a torus $S^1 \times S^1$ into the union of two solid tori; we mentioned this in a remark in Sect. 2.4).

The decomposition $S^{m+n-1} = U \cup V$ may also be constructed in the following way: $S^{m+n-1} = \partial D^{m+n} \approx \partial(D^m \times D^n) = (D^m \times \partial D^n) \cup (\partial D^m \times D^n) = (D^m \times S^{n-1}) \cup (S^{m-1} \times D^n)$.

Our map $w\colon S^{m+n-1} \to S^m \vee S^n$ consists of two projections,

$$U = D^m \times S^{n-1} \to D^m \to D^m/S^{m-1} = S^m \subset S^m \vee S^n,$$
$$V = S^{m-1} \times D^n \to D^n \to D^n/S^{n-1} = S^n \subset S^m \vee S^n,$$

and takes the "cutting surface" $S^{m-1} \times S^{n-1}$ into a point.

Now let $t\colon S^m \to X$, $g\colon S^n \to X$ be two spheroids of some space X with a base point x_0. Together, they form a map $S^m \vee S^n \to X$, and the composition of this map with w is a spheroid $h\colon S^{m+n-1} \to X$. It is clear that the homotopy class of h is determined by the homotopy classes of f and g. Thus, we get an operation which assigns to $\alpha \in \pi_m(X, x_0)$ and $\beta \in \pi_n(X, x_0)$ some element of $\pi_{m+n-1}(X, x_0)$; this element is called the *Whitehead product* of α and β and is denoted as $[\alpha, \beta]$.

In these exercises, $\alpha, \alpha_1, \alpha_2 \in \pi_m(X, x_0)$, $\beta, \beta_1, \beta_2 \in \pi_n(X, x_0)$, $\gamma \in \pi_p(X, x_0)$.

EXERCISE 1. Prove that if $m = n = 1$, then $[\alpha, \beta] = \alpha\beta\alpha^{-1}\beta^{-1}$.

EXERCISE 2. Prove that if $m = 1, n > 1$, then $[\alpha, \beta] = \alpha_\#(\beta) - \beta$.

EXERCISE 3. Prove that if $n > 1$, then $[\alpha, \beta_1 + \beta_2] = [\alpha, \beta_1] + [\alpha, \beta_2]$.

EXERCISE 4. Prove that if $m, n > 1$, then $[\beta, \alpha] = (-1)^{mn+m+n}[\alpha, \beta]$.

EXERCISE 5. Prove that if $m, n, k > 1$, then

$$(-1)^{mk+n}[\alpha, [\beta, \gamma]] + (-1)^{nm+k}[\beta, [\gamma, \alpha]] + (-1)^{kn+m}[\gamma, [\alpha, \beta]] = 0$$

("super-Jacobi identity").

COMMENT. Although the higher homotopy groups are commutative, the Whitehead product may be regarded as a substitute for a commutator in these groups. The properties above create for homotopy groups (more rigorously, for the direct sum $\bigoplus_{n=2}^{\infty} \pi_n$) a structure similar to that of a *Lie superalgebra*.

EXERCISE 6. Prove that the suspension over the Whitehead product $\Sigma[\alpha, \beta] \in \pi_{m+n}(\Sigma X)$ is 0. This implies (and, actually, is implied by) the fact that $\Sigma(S^m \times S^n)$ is homotopy equivalent to $S^{m+1} \vee S^{n+1} \vee S^{m+n+1}$; why?

EXERCISE 7. Let $\iota_n \in \pi_n(S^n)$ be the class of the identity spheroid and $\eta_2 \in \pi_3(S^2)$ be the class of the Hopf map. Prove that $[\iota_2, \iota_2] = 2\eta_2$.

EXERCISE 8. Prove that if X is a topological group, or an H-space, then $[\alpha, \beta] = 0$ for any $\alpha \in \pi_m(X), \beta \in \pi_n(X)$. (In view of Exercises 1 and 2, this is a generalization of Exercise 4 of Sect. 6.7 and of Exercise 2 of Sect. 8.2.)

Theorem (Difficult Part of Freudenthal's Theorem). *The kernel of the homomorphism* $\Sigma: \pi_{2n-1}(S^n) \to \pi_{2n}(S^{n+1})$ *is a cyclic group generated by the Whitehead square* $[\iota_n, \iota_n]$ *of the class* ι_n.

We will not prove this theorem here although it has a purely geometric proof based on the following argument. The inverse images of the poles with respect to a homotopy $S^{2n} \times I \to S^{n+1}$ cannot be deformed to polyhedra separated by $S^{2n-1} \times I$, but we can do it if we let these inverse images transversely cross each other finitely many times at isolated points. In this way, a homotopy $f_t: S^{2n} \to S^{n+1}$ between the suspensions of spheroids $g_0, g_1: S^{2n-1} \to S^n$ can be deformed to a piecewise continuous homotopy of the form Σg_t with finitely many discontinuity points t_i, and each such discontinuity can be compensated by adding or subtracting a spheroid of class $[\iota_n, \iota_n]$.

The reader who does not feel inclined to get involved in detailing this idea can return to this theorem after reading Chap. II (or, even better, Lecture 24) and use algebraic means developed there.

Notice that in combination with Exercise 6, this theorem shows that $\pi_4(S^3) \cong \mathbb{Z}_2$, and hence $\pi_1^S \cong \mathbb{Z}_2$. By the way, the alternative $\pi_4(S^3) = \mathbb{Z}_2$ or 0 follows directly from (rather easy) Exercises 6 and 7.

Notice also that if n is even, then the cyclic group generated by $[\iota_n, \iota_n]$ is infinite (we will prove this in Lecture 16). For n odd, this group is $\mathbb{Z}_2$ or 0 (this follows from Exercise 2).

Lecture 11 Homotopy Groups and CW Complexes

This lecture is devoted to different relations between homotopy groups and CW structures. It is rather heterogeneous. We will calculate the first nontrivial homotopy group of a CW complex, will clarify the role of homotopy groups in homotopy classification of CW complexes, and will construct CW complexes with homotopy groups prescribed.

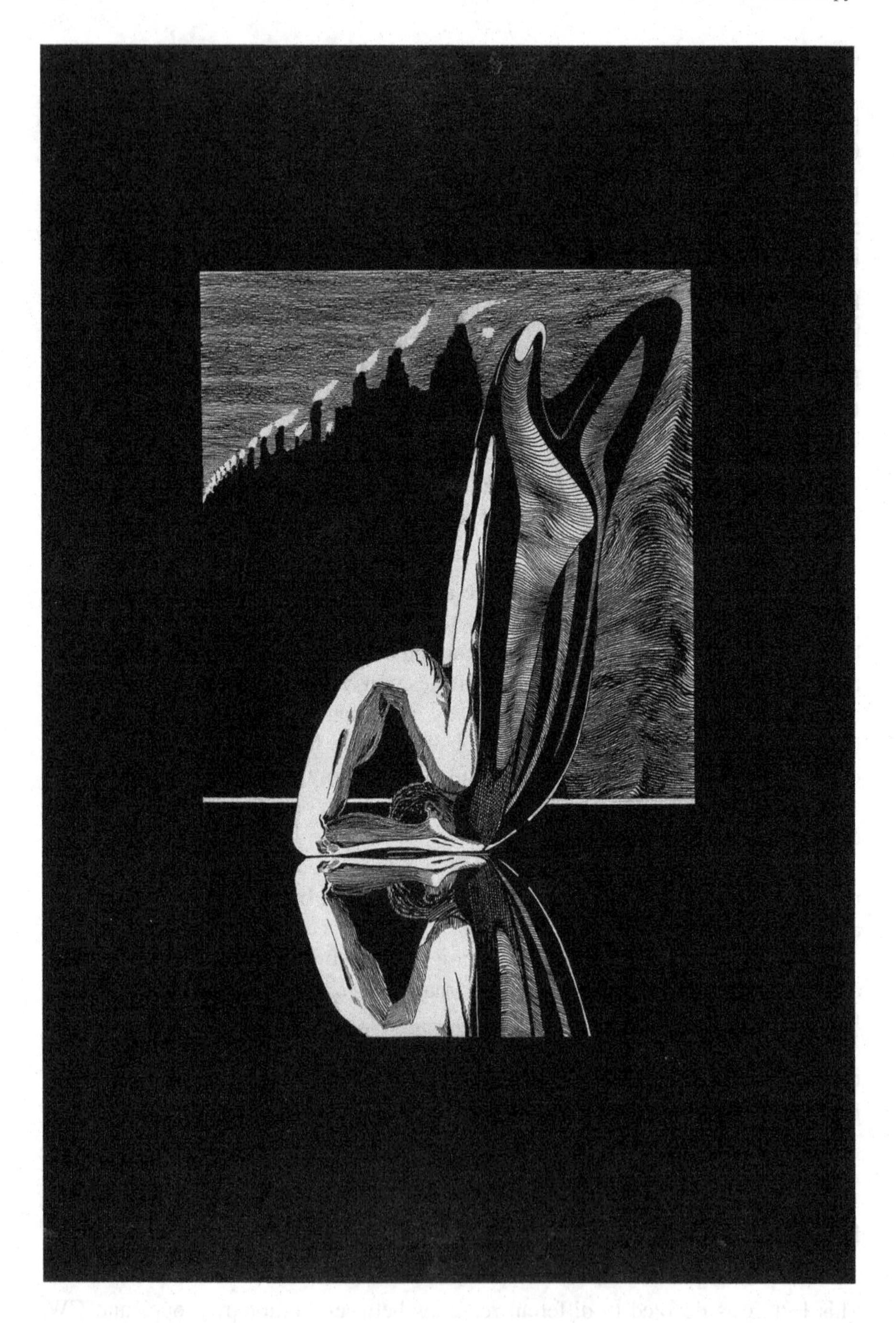

11.1 Homotopy Groups and Attaching Cells

Theorem. *Let X be a path connected topological space, and let $f: S^n \to X$ be a continuous map. Let $Y = X \cup_f D^{n+1}$. The homomorphism*

$$\pi_i(X, x_0) \to \pi_i(Y, x_0), \qquad (*)$$

where $x_0 \in X \subset Y$ is an arbitrarily chosen base point, induced by the inclusion map $X \to Y$ is an isomorphism if $i < n$; if $i = n$, then it is an epimorphism whose kernel is generated by all classes of the form $u_\#[f]$, where u is a path joining $f(y_0)$ with x_0 (here y_0 is a base point of S^n) and $[f] \in \pi_n(X, f(y_0))$ is the homotopy class of the spheroid f.

Remark 1. A similar theorem for fundamental groups was proved in Sect. 7.3. The proof, however, cannot be the same, because in Sect. 7.3 we could use Van Kampen's theorem, which has no satisfactory generalizations to higher-dimensional homotopy groups.

Remark 2. It is convenient to assume that $f(y_0) = x_0$. Then the kernel of the homomorphism $(*)$ is generated by the elements of the $\pi_1(X, x_0)$-orbit of $[f] \in \pi_n(X, x_0)$. The case when X is simply connected is especially important. In this case the theorem shows that attaching an $(n+1)$-dimensional cell does not affect homotopy groups of dimension less than n, and the n-dimensional homotopy group is factored by the cyclic group generated by the class of the attaching map.

Lemma. *Let E^m be the sphere S^m or the ball D^m where $m \leq n+1$, and let $T \subset E^m$ be the base point y_0 if $E^m = S^m$ and S^{m-1} if $E^m = D^m$. In both cases, we identify $E^m - T$ with $\mathbb{R}^m$. Then let $h: E^m \to Y$ be a continuous map such that $h(T) \subset X$. Then there exists a map $h_1: E^m \to Y$ which is T-homotopic to h and possesses the following properties:*

(1) *h_1 coincides with h on $h^{-1}(X)$.*
(2) *If $m \leq n$, then $h(E^m) \subset X$.*
(3) *If $m = n+1$, then there is a finite family of pairwise disjoint small balls $d_1, \dots, d_N$ in E^m such that $h_1(E^m - \bigcup_i d_i) \subset X$ and $h_1|_{\operatorname{Int} d_i}$ is, for every i, a linear (orientation preserving or reversing) homeomorphism of* $\operatorname{Int} d_i$ *onto* $\operatorname{Int} D^{n+1} \subset Y$.

This lemma is not much different from the free-point lemma proved in Sect. 5.8 and repeated several times after it (last time in Sect. 9.3). We leave further details to the reader.

Proof of Theorem. The facts that the map $(*)$ is an epimorphism if $i \leq n$ and a monomorphism if $i < n$ follow directly from the lemma. Suppose that the class of spheroid $g: S^n \to X$ belongs to the kernel of the map $(*)$; that is, g can be extended to a map $h: D^{n+1} \to X$. Using the lemma, we replace the map h with another extension h_1 of g which has the properties listed in part (3) of the lemma. On the boundary ∂d_i of each small ball d_1, the map h_1 coincides with a composition $\partial d_i \xrightarrow{\ell_i} S^n \xrightarrow{f} X$,

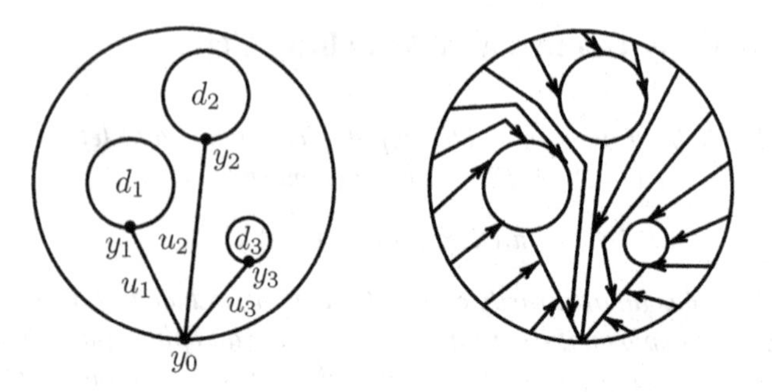

Fig. 57 The main construction in the proof of the theorem

where ℓ_i is a linear map. Set $y_i = \ell_i^{-1}(y_0)$ and for every i join the point y_i with y_0 by a path u_i such that the interiors of these paths are disjoint from the balls d_i and from each other. In $D^{n+1} - \bigcup_i \operatorname{Int} d_i$, there is a natural homotopy connecting the inclusion spheroid $S^n \to D^{n+1}$ with the sum of the spheroids $(u_i)_\#(\ell_i^{-1})$ (see Fig. 57).

Then we transfer to this homotopy to X by the map h_1. We obtain a homotopy connecting the spheroid g with the sum of spheroids $(h \circ u_i)_\#(f_i)$ where f_i is a spheroid obtained from f by a linear transformation of the sphere S^n. Since $[f_i] = \pm[f]$ (where the sign depends on this linear transformation preserving or reversing orientation), we see that $[g]$ is indeed a linear combination of generators listed in the theorem. This completes the proof of the theorem (the fact that $s_\#[f]$ always belongs to the kernel of the map $(*)$ is obvious).

Corollary. *If Y is a CW subcomplex of a CW complex X and the difference $X - Y$ does not contain cells of dimension $\leq n$, then the homomorphism $\pi_i(Y) \to \pi_i(X)$ induced by the inclusion map is an isomorphism for $i < n$ and an epimorphism for $i = n$. In particular, $\pi_n(X) = \pi_n(\operatorname{sk}_{n+1} X)$.*

11.2 Application of the Attaching Cell Theorem: The Homotopy Groups of Bouquets

Theorem. *Let X, Y be CW complexes.*

(1) *If X is p-connected and Y is q-connected where $p, q \geq 1$, then $\pi_n(X \vee Y) = \pi_n(X) \oplus \pi_n(Y)$ for $n \leq p + q$.*
(2) *For any n, $\pi_n(X \vee Y)$ contains a direct summand isomorphic to $\pi_n(X) \oplus \pi_n(Y)$.*

Proof. According to a theorem in Sect. 5.9, the spaces X and Y are homotopy equivalent to CW complexes with one vertex and without cells of dimensions $1, \dots, p$ and $1, \dots, q$, respectively. The bouquet $X \vee Y$ is a CW subcomplex of $X \times Y$, and all the cells in $(X \times Y) - (X \vee Y)$ have dimensions $\geq p + q + 2$. Thus, part (1) follows from the corollary in Sect. 11.1 and the fact that $\pi_n(X \times Y) = \pi_n(X) \oplus \pi_n(Y)$. To prove part (2), we note that the composition

$$\pi_n(X) \oplus \pi_n(Y) \to \pi_n(X \vee Y) \to \pi_n(X \times Y) = \pi_n(X) \oplus \pi_n(Y),$$

where the first arrow denotes the sum of homomorphisms induced by the inclusion maps $X \to X \vee Y$, $Y \to X \vee Y$ and the second arrow is induced by the inclusion map $X \vee Y \to X \times Y$, is the identity.

Corollary. *If $n \geq 2$, then $\pi_n(\underbrace{S^n \vee \cdots \vee S^n}_{q}) = \underbrace{\mathbb{Z} \oplus \cdots \oplus \mathbb{Z}}_{q}$; the system of free generators is composed by the classes of q natural embeddings $S^n \to S^n \vee \cdots \vee S^n$. (Here q may be ∞.)*

EXERCISE 1. For X, Y, p, and q as in the theorem, prove that $\pi_{p+q+1}(X \vee Y)$ is isomorphic to $\pi_{p+q+1}(X) \oplus \pi_{p+q+1}(Y) \oplus [\pi_{p+1}(X) \otimes \pi_{q+1}(Y)]$, where the last summand is embedded into $\pi_{p+q+1}(X \vee Y)$ be means of the map $\alpha \otimes \beta \mapsto [i_*\alpha, j_*\beta]$. In particular, $\pi_3(S^2 \vee S^2) \cong \mathbb{Z} \oplus \mathbb{Z} \oplus \mathbb{Z}$.

Remark. There is a result called the Hilton–Milnor theorem stating that the homotopy groups of an arbitrary bouquet of spheres, $S^{n_1} \vee \cdots \vee S^{n_r}$, are generated by elements of homotopy groups of spheres S^{m_i} and their Whitehead products.

11.3 The First Nontrivial Homotopy Group of a CW Complex

This is an extension of the result of Sect. 7.6 to higher-dimensional homotopy groups. Since the case of the fundamental group is settled by that result, we assume now that the CW complex considered is $(n-1)$-connected where $n > 1$. Then we can assume that X has only one vertex and has no cells of dimension $1, \dots, n-1$. We assume that n-dimensional and $(n+1)$-dimensional cells e_i^n, e_j^{n+1} of X are indexed by elements $i \in I, j \in J$ of some sets I and J. In this case, $\operatorname{sk}_n X$ is homeomorphic to the bouquet $\bigvee_i S_i^n$ of n-dimensional spheres corresponding to n-dimensional cells of X (the homeomorphism is established by characteristic maps of the n-dimensional cells). Thus, $\pi_n(\operatorname{sk}_n(X)) \cong \pi_n(\bigvee_i S_i^n) = \bigoplus_i \mathbb{Z}$ (the isomorphism depends on the choice of characteristic maps of the n-dimensional cells, but this dependence is limited to the multiplication of some generators of $\bigoplus_i \mathbb{Z}$ by -1). Let $f_j : S^n \to \operatorname{sk}_n X$ be an attaching map of a cell e_j^{n+1}.

Theorem. *Let X be a CW complex with one vertex and with no other cells of dimension $< n$. The group $\pi_n(X)$ has a system of generators corresponding to n-dimensional cells (the classes of characteristic maps of n-dimensional cells) and defining system of relations corresponding to $(n+1)$-dimensional cells [the classes of attaching maps of $(n+1)$-dimensional cells are equated to zero].*

This follows directly from the results of Sects. 11.1 and 11.2.

EXERCISE 2. Prove the following relative version of the theorem. Let (X, A) be a CW pair with connected A such that $X - A$ contains no cells of dimension $< n$, where $n \geq 3$. Then the first nontrivial group of the pair (X, A), that is, the group $\pi_n(X, A)$, is generated as a $\pi_1(A)$-module by n-dimensional cells in $X - A$ with relations corresponding to $(n + 1)$-dimensional cells in $X - A$.

EXERCISE 3. State and prove a similar statement in the case $n = 2$.

EXERCISE 4. Let (X, A) be a CW pair with simply connected A, and let all cells in $X - A$ have dimensions $\geq n \geq 2$. Prove that the natural map $\pi_n(X, A) \to \pi_n(X/A)$ is an isomorphism.

Remark. This proposition has a generalization: If A is k-connected and all cells in $X - A$ have dimensions $\geq n \geq 2$, then the natural map $\pi_q(X, A) \to \pi_q(X/A)$ is an isomorphism for $g \leq n + k - 1$ and an epimorphism for $g = n + k$. At the moment, we do not have the necessary technique to prove this theorem. The reader may return to it after reading Chap. II or, better, Lecture 24.

11.4 Weak Homotopy Equivalence Revisited

We are going to prove (a slightly enhanced version of) one of the propositions promised in Sect. 9.5 (another one will be proven in Sect. 11.6 ahead).

Theorem. *For a continuous map $f: X \to Y$ (X and Y are arbitrary topological spaces) the following properties are equivalent:*

(1) *f is a weak homotopy equivalence* (see Sect. 9.5).
(2) *$f_*: \pi_n(X, x_0) \to \pi_n(Y, f(x_0))$ is an isomorphism for every n and $x_0 \in X$.*
(3) *If (W, A) is a CW pair and $h: A \to X$, $g: W \to Y$ are such continuous maps that $f \circ h \sim g\,|_A$, then there exists a continuous map $\widetilde{h}: W \to X$ such that $\widetilde{h}\,|_A = h$ and $f \circ \widetilde{h} \sim g$.*

Proof. The implication (1) $\Rightarrow$ (2) is obvious. The implication (3) $\Rightarrow$ (1) is almost obvious: If we put $(W, A) = (Z, \emptyset)$, we arrive at the conclusion that the map $f_*: \pi(Z, X) \to \pi(Z, Y)$ is onto, and, taking $(W, A) = (Z \times I, (Z \times 0) \cup (Z \times 1))$, we see that f_* is also one-to-one.

It remains to prove the implication (2) $\Rightarrow$ (3). Let the map $f: X \to Y$ satisfy condition (2), and let (W, A) be a CW pair. We assume first that W is different from A by just one cell: $W = A \bigcup_\alpha D^{n+1}$, where $\alpha: S^n \to A$ is a continuous map. Since the composition $S^n \xrightarrow{\alpha} A \xrightarrow{\subset} W$ is homotopic to a constant (the natural map $D^{n+1} \to W$ is its extension to the ball), the composition $(g\,|_A) \circ \alpha: S^n \to Y$ is also homotopic to a constant. Hence, the spheroid $f \circ h \circ \alpha: S^n \to Y$ is homotopic to zero, and hence so is the spheroid $h \circ \alpha: S^n \to X$ [since $f_*: \pi_n(X) \to \pi_n(Y)$ is a monomorphism]. Hence, the map $h \circ \alpha: S^n \to X$ can be extended to a map $\beta: D^{n+1} \to X$, and we can combine the maps h and β into a continuous map $\widetilde{h}': W \to X$ (see Fig. 58).

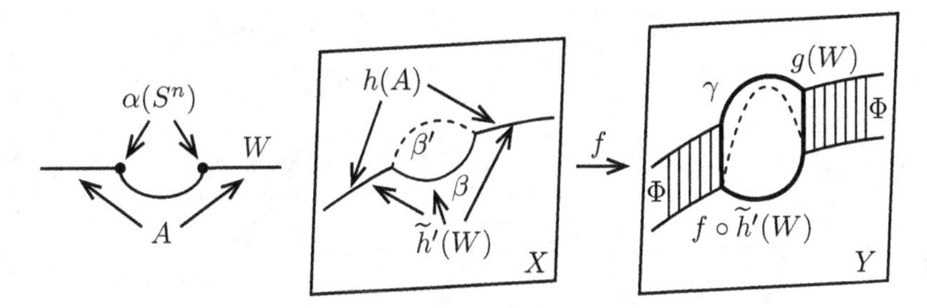

Fig. 58 The construction in the proof of the weak homotopy equivalence theorem

Here we use the prime notation because this is not the map we need. We want $f \circ \widetilde{h} \sim g$, but we have only $f \circ \widetilde{h} \mid_A \sim g \mid_a$. Choose a homotopy $\Phi: A \times I \to Y$ joining $f \circ \widetilde{h} \mid_A = f \circ h$ with $g \mid_A$ and consider the $(n+1)$-dimensional spheroid $\gamma: S^{n+1} = \partial(D^{n+1} \times I) \to Y$ composed of the maps

$$D^{n+1} \times 0 = D^{n+1} \xrightarrow{\subset} W \xrightarrow{g} Y,$$

$$S^n \times I \xrightarrow{\alpha \times I} A \times I \xrightarrow{\Phi} Y,$$

$$D^{n+1} \times 1 = D^{n+1} \xrightarrow{\beta} X \xrightarrow{f} Y.$$

If we want the homotopy Φ to be extendable to a homotopy between $f \circ \widetilde{h}$ and g, we need the spheroid γ to be homotopic to zero, but we cannot count on that, because we did nothing to achieve it. To amend the construction, we have only one thing at our disposition: the choice of the map β extending $h \circ \alpha: S^n \to X$. If we choose a different map, β', then γ will be replaced by $\gamma - (f \circ \delta)$, where $\delta: S^{n+1} \to X$ is the spheroid composed of the maps $\beta, \beta': D^{n+1} \to X$ (which are compatible on ∂D^{n+1}). It is clear that the spheroid δ can be made (homotopically) arbitrary, but then the class of the spheroid $f \circ \delta$ also can be made arbitrary [since $f_*: \pi_{n+1}(X) \to \pi_{n+1}(X)$ is an epimorphism]. This completes the proof in the case when $W - A$ is one cell. In the general case, we perform this construction simultaneously for all cells of the same dimension; if the total number of cells in $W - A$ is infinite, then the continuity of the final map h is secured by Axiom (W).

11.5 Whitehead's Theorem

Theorem. *Let X and Y be CW complexes, and let $f: X \to Y$ be s continuous map If*

$$f_*: \pi_n(X, x_0) \to \pi_n(Y, f(x_0))$$

is an isomorphism for all n and x_0, then f is a homotopy equivalence.

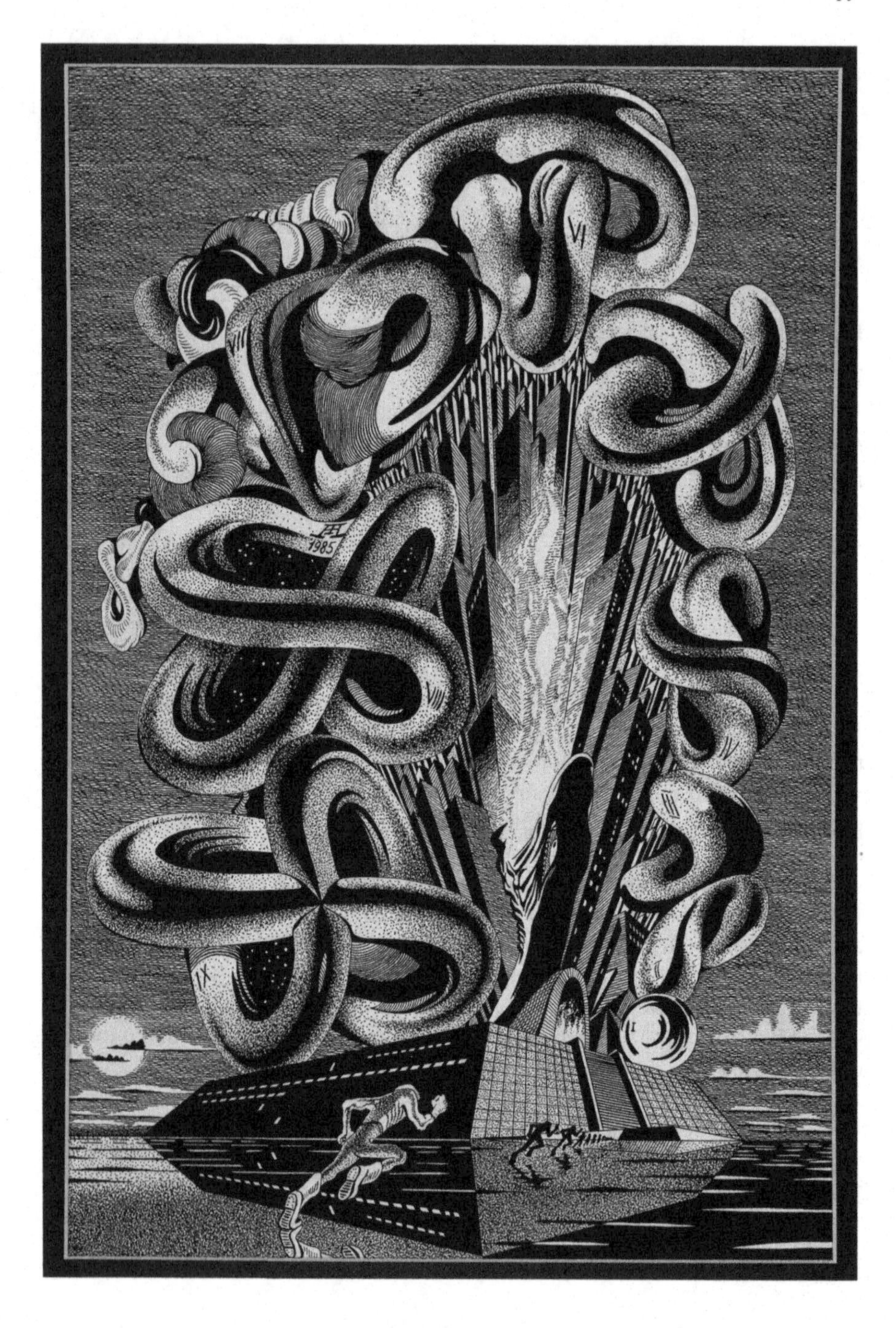
VI
VII
1985
IX

(If X and Y are connected, then it is sufficient to check the condition for one point x_0.)

This theorem is a direct corollary of the previous theorem and the fact that for CW complexes a weak homotopy equivalence is the same as the usual homotopy equivalence (see Sect. 9.5).

It follows from Whitehead's theorem that if all the homotopy groups of some (nonempty, connected) CW complex are trivial, then this CW complex is contractible (homotopy equivalent to a point). However, in a general case coincidence of homotopy groups is not sufficient for a homotopy equivalence; it is required additionally that the isomorphism between homotopy groups is established by some continuous map. (See, however, Sect. 11.8.)

EXERCISE 5. Show that the spaces S^2 and $S^3 \times \mathbb{C}P^\infty$ have equal homotopy groups, but are not homotopy equivalent.

EXERCISE 6. Show that the spaces $S^m \times \mathbb{R}P^n$ and $S^n \times \mathbb{R}P^m$ $(m \neq n)$ have equal homotopy groups, but are not homotopy equivalent.

11.6 Cellular Approximations of Topological Spaces

Theorem. *For every topological space X there exists a CW complex Y with a weak homotopy equivalence $f: Y \to X$.* [Such a pair (Y,f) is called a cellular approximation of X.]

Proof. We can restrict ourselves to the case when X is path connected. We are going to construct a chain of CW complexes (and cellular inclusions) $Y_0 \subset Y_1 \subset Y_2 \subset \dots$ and a chain of continuous maps $f_i: Y_i \to X$ that successively chain each other such that $(f_i)_*: \pi_q(Y_i) \to \pi_q(X)$ will be an isomorphism for $i \leq q$. For Y_0, we will take a point, while for f_0 we will take an arbitrary map. Assume that for some $n \geq 1$, the chain $Y_0 \subset Y_1 \subset \dots \subset Y_{n-1}$ and the maps $f_i: Y_i \to X$ with the required properties have already been constructed. Put $y_0 = Y_0, x_0 = f_0(y_0)$. Choose a family of generators $\{\varphi_\alpha\}$ in the group $\pi_n(X, x_0)$, fix for every generator φ_α a representing spheroid g_α, put $Y'_n = Y_{n-1} \vee \left(\bigvee_\alpha (S^n_\alpha = S^n)\right)$, and define the map $f'_n: Y'_n \to X$ as f_{n-1} on Y_{n-1} and as g_α on S^n_α. By the theorem in Sect. 11.2, the inclusion map $Y_{n-1} \to Y'_n$ induces an isomorphism for every group π_q with $q < n$, so $(f'_n)_*: \pi_q(Y'_n, y_0) \to \pi_q(X, x_0)$ is an isomorphism for $q < n$. For $q = n$ it is an epimorphism, since its image contains all generators of $\pi_n(X, x_0)$. It is not necessarily a monomorphism. Choose a system of generators $\{\psi_\beta\}$ in $\operatorname{Ker}((f'_n)_*: \pi_n(Y'_n, y_0) \to \pi_n(X, x_0))$, fix for every ψ_β a representing spheroid $h_\beta: S^n \to Y'_n$ (we can require that h_β be a cellular map, but our construction guarantees that $\dim Y'_n \leq n$, so it may be not necessary), and then attach to Y'_n an $(n+1)$-dimensional ball by every h_β. The CW complex arising is our Y_n. Since $(f'_n)_*(\psi_\beta) = 0$, the map f'_n can be continuously extended to every attached ball, and we get a continuous map of Y_n into X, and this is our f_n. Since attaching $(n+1)$-dimensional balls does not affect groups π_q with $q <$

n, the map $(f_n)_*$ is still an isomorphism for homotopy groups of dimensions less than n. In dimension n, it is still an epimorphism (since $Y_n \supset Y'_n$), and it is also a monomorphism, since $\pi_n(Y_n, y_0) = \pi_n(Y'_n, y_0)/\operatorname{Ker}((f'_n)_*)$ [here we used the fact that $\operatorname{Ker}((f'_n)_*)$ is a π_1-submodule of $\pi_n(Y_n, y_0)$].

As soon as we have the chain $Y_0 \subset Y_1 \subset Y_2 \subset \ldots$ constructed, we put $Y = \bigcup_n Y_n$ (with the weak topology) and get a CW complex Y with a continuous map $f: Y \to X$ which induces isomorphisms for all homotopy groups and hence is a weak homotopy equivalence.

EXERCISE 7. Prove that a cellular approximation of any topological space X is homotopy invariant. This means that if (Y, f) and (Z, g) are two cellular approximations of the same topological space X, then there exists a homotopically unique homotopy equivalence $h: Y \to Z$ such that $g \sim f \circ h$.

The main raison d'être of this theorem consists in a possibility to generalize some homotopy results from CW complexes to (more or less) arbitrary topological spaces. As an exercise, the reader can do this with the theorem in Sect. 11.2. Another application of this theorem is a new clarification of the notion of the weak homotopy equivalence, as stated in the following exercise.

EXERCISE 8. Prove that topological spaces X and Y are weakly homotopy equivalent if and only if there exist maps

$$X \xleftarrow{f} Z \xrightarrow{g} Y,$$

where Z is a CW complex and f, g are weak homotopy equivalences.

11.7 Eilenberg–MacLane Spaces ($K(\pi, n)$s)

Theorem. *Let n be a positive integer, and let π be a group which is supposed to be commutative if $n > 1$. Then there exists a CW complex X such that*

$$\pi_q(X) = \begin{cases} \pi, & \text{if } q = n, \\ 0, & \text{if } q \neq n. \end{cases}$$

(compare with the end of Lecture 4.)

Such spaces are called *Eilenberg–MacLane spaces* or *spaces of type $K(\pi, n)$*. People sometimes say that X is a $K(\pi, n)$.

Proof of Theorem. Nothing new for us. First, we choose a presentation of the group π by systems of generators and relations $\pi = \langle \varphi_\alpha, \alpha \in A \mid \psi_\beta, \beta \in B \rangle$. Then we take the bouquet of n-dimensional spheres labeled with elements of $A, X_n = \bigvee_\alpha \in A(S^n\alpha = S^n)$. Then $\pi_q(X_n) = 0$ for $q < n$, and $\pi_n(X_n)$ is a free group with generators labeled with elements of A. For every $\beta \in B$, we can regard ψ_β as an element of $\pi_n(X_n)$. Choose spheroids $h_\beta: S^n \to X_n$ representing ψ_β, and for every β attach to X_n an n-dimensional cell by the map h_β. We get a CW complex X_{n+1}, of dimension at most $n+1$, such that $\pi_q(X_{n+1}) = 0$ if $q < n$ and π if $q = n$. Then we take an arbitrary system of generators in $\pi_{n+1}(X_{n+1})$, represent these generators by spheroids, and attach $(n+2)$-dimensional cells by these spheroids. We get some $X_{n+2} \supset X_{n+1}$ with the same homotopy groups as X_{n+1} in dimensions $\leq n$ and with $\pi_{n+1}(X_{n+2}) = 0$. Then we do the same with $\pi_{n+2}(X_{n+2})$, get an X_{n+3}, and so on. In the end, we get a CW complex X with all the required properties.

Remark. The construction in the proof is very far from being explicit: We do not know the groups $\pi_{n+1}(X_{n+1}), \pi_{n+2}(X_{n+2}), \ldots$ and have no technical means to compute them. Certainly, we can extinguish all arbitrary choices in this construction attaching $(n+2)$-dimensional cells by *all* possible $(n+1)$-dimensional spheroids of X_{n+1}, then attach $(n+3)$-dimensional cells by *all* possible $(n+2)$-dimensional spheroids of X_{n+2}, and so on. This will make the construction natural (functorial), but it will become tremendously inconvenient. This makes especially interesting the relatively few known explicit constructions of $K(\pi, n)$s.

Explicit constructions. (1) The space $\mathbb{C}P^\infty$ is a space of type $K(\mathbb{Z}, 2)$. This is the only case when a $K(\pi, n)$ with $n > 2$ has a geometrically explicit construction. (2) S^1 is a $K(\mathbb{Z}, 1)$. (3) $\mathbb{R}P^\infty$ has the type $K(\mathbb{Z}_2, 1)$. (4) The infinite-dimensional lens space $L_m^\infty = S^\infty/\mathbb{Z}_m$, where the generator T of the group $\mathbb{Z}_m$ acts in $S^\infty \subset \mathbb{C}^\infty$ by the formula $T(z_1, z_2, \ldots) = (z_1 e^{2\pi i/m}, z_2 e^{2\pi i/m}, \ldots)$, and is a space of type $K(\mathbb{Z}_m, 1)$. (5) Since $K(\pi_1, n) \times K(\pi_2, n) = K(\pi_1 \times \pi_2, n)$, constructions (2)–(4) give us spaces of type $K(\pi, 1)$ for every finitely generated Abelian group π.

There are lots of known spaces of type $K(\pi, 1)$ with non-Abelian π, for example, all classical surfaces, except S^2 and $\mathbb{R}P^2$ (see Exercise 4 in Sect. 8.3), and also bouquets of circles (see Exercise 3 in Sect. 8.3).

EXERCISE 9. Prove that the space of all unordered sets of n points in $\mathbb{R}^\infty$ (or S^∞) is a $K(S_n, 1)$.

EXERCISE 10. Prove that the space of all unordered sets of n points in the plane is a $K(\pi, 1)$ for a certain group π. This group is called the *Artin n-thread braid group.*

EXERCISE 11. Do the same for the space of *ordered* n-point subsets of the plane. (The group arising is called the group of *pure braids.* It is better to do this exercise before the previous one.)

EXERCISE 12. Prove that a complete nonpositively curved Riemannian manifold is a space of type $K(\pi, 1)$ for some π. (The proof is based on the fact that in a simply connected complete negatively curved Riemannian manifold every two points are connected by a unique geodesic.)

Remark. A complement to a knot in S^3 is also a space of type $K(\pi, 1)$, but it is not likely that the reader is able to prove it with the technical means currently at hand.

EXERCISE 13. Prove that $\Omega K(\pi, n) = K(\pi, n-1)$. [This shows that every $K(\pi, n)$ with an Abelian π is an H-space, and even a homotopy commutative H-space. Actually, every $K(\pi, n)$ with an Abelian π can be constructed as an Abelian topological group.]

11.8 The Uniqueness of $K(\pi, n)$s

Theorem. *Any two spaces of type $K(\pi, n)$ are weakly homotopy equivalent. Hence, any two CW complexes of type $K(\pi, n)$ are homotopy equivalent.*

Proof. Let X be a $K(\pi, n)$, and let be a CW complex which is a $K(\pi, n)$ with one vertex and without other cells of dimension $< n$. Our theorem will be proved if we construct a weak homotopy equivalence $f: Y \to X$. The latter means only that $f_*: \pi_n(Y) \to \pi_n(X)$ is an isomorphism.

The nth skeleton of Y is a bouquet of n-dimensional spheres which represent generators of $\pi_n(Y) = \pi$. Since $\pi_n(X)$ is also π, we can choose the same generators for $\pi_n(X)$ and then map every sphere of the bouquet $\text{sk}_n Y$ into X according to some spheroid representing the corresponding element of $\pi_n(X)$. We get a continuous map $f_n: \text{sk}_n Y \to X$. Let e be an $(n+1)$-dimensional cell of Y, and let $g: S^n \to \text{sk}_n Y$ be the attaching map. The spheroid g represents a zero class in $\pi_n(X)$; hence, the composition $f \circ g$ is a homotopic to a zero spheroid of X, and we can extend the map f continuously to the cell e and, in this way, to the whole skeleton $\text{sk}_{n+1} Y$. The map $f_{n+1}: \text{sk}_{n+1} Y \to X$ induces an isomorphism $(f_{n+1})_*: \pi_n(\text{sk}_{n+1} Y) = \pi \to \pi_n(X)$. The extension of f_{n+1} to cells of dimensions $> n+1$ does not meet any obstructions: Attaching maps are maps $S^q \to \text{sk}_q Y$ with $q > n$, and their composition with f_q forms spheroids homotopic to zero. This completes the construction.

EXERCISE 14. Prove that if X, Y are $K(\pi, n)$s and Y is a CW complex, then a weak homotopy equivalence $Y \to X$ is homotopically unique. Moreover, if X is a $K(\pi, n)$ and Y is a $K(\rho, n)$ and also a CW complex, then for every homomorphism $\varphi: \rho \to \pi$ there exists a homotopically unique continuous map $f: Y \to X$ such that $f_*: \pi_n(Y) \to \pi_n(X)$ is φ.

Remark. This property of $K(\pi, n)$s that their (weak) homotopy type is determined by their homotopy groups is not generalizable to spaces with multiple homotopy groups. There exist CW complexes X, Y such that each has two nontrivial homotopy groups, and these groups for X and Y are the same, but, however, X and Y are not homotopy equivalent. In Chap. III we will be able not only to find such examples, but even to provide a sort of a classification for them.

11.9 Capping and Killing Homotopy Groups

In conclusion, we will discuss two constructions for CW complexes which affect their homotopy groups in a prescribed way. The first of them is known to us, and we have used it several times without naming it: It is *capping* homotopy groups. Namely, if X is a CW complex, then for any number n we can construct a CW complex X' which contains X and has homotopy groups

$$\pi_q(X') = \begin{cases} \pi_q(X), & \text{if } q \leq n, \\ 0, & \text{if } q > n; \end{cases}$$

moreover, the inclusion map $X \to X'$ induces isomorphisms for all homotopy groups of dimensions $\leq n$. This is achieved by multiple attaching cells of dimensions $> n+1$. This capping operation is homotopically unique, as the following exercise shows.

EXERCISE 15. Let a CW complex X and a number n be given. Let $X_1' \supset X$, $X_2' \supset X$ be two CW complexes with the properties listed above. Then there exists a (homotopically unique) homotopy equivalence $X_1' \to X_2'$ whose restriction to $X \subset X_1'$ is homotopic to the inclusion map $X \to X_2'$.

Let $\pi_n(X)$ be the first nontrivial homotopy group of a CW complex X. Then the capping operation gives rise to a homotopically unique map (embedding) $X \to K(\pi_n(X), n)$. We turn this map into a homotopy equivalent (strong) Serre fibration (see Sect. 9.7) and denote the fiber of this fibration as $X|_{n+1}$. This space which is defined canonically up to a homotopy equivalence is sometimes called the *first* and sometimes the $(n+1)$st killing space of X.

Theorem.

$$\pi_q(X|_{n+1}) = \begin{cases} \pi_q(X), & \text{if } q \neq n, \\ 0, & \text{if } q = n \end{cases}$$

moreover, the inclusion map $X|_{n+1}$ induces isomorphisms for all homotopy groups π_q with $q \neq n$.

Proof. The fragment

$$\pi_{q+1}(K(\pi, n)) \longrightarrow \pi_q(X|_{n+1}) \xrightarrow{(*)} \pi_q(X) \longrightarrow \pi_q(K(\pi, n))$$

[where $\pi = \pi_n(X)$] of the homotopy sequence of the fibration shows that the map $(*)$ is an isomorphism for $q \neq n, n-1$. The fragment

$$\pi_n(X|_{n+1}) \xrightarrow{0} \pi_n(X) \xrightarrow{\cong} \pi_n(K(\pi, n)) \xrightarrow{0} \pi_{n-1}(X|_{n+1}) \longrightarrow \pi_{n-1}(K(\pi, n))$$

(the homomorphisms of an exact sequence surrounding an isomorphism are both zeroes) shows all the rest.

One can iterate the construction of the killing space and get, for every m, the killing space $X|_m$ and a map $f: X|_m \to X$ such that

$$\pi_q(X|_m) = \begin{cases} \pi_q(X), & \text{if } q \geq m, \\ 0, & \text{if } q < m, \end{cases}$$

and the isomorphisms $\pi_q(X|_m) \to \pi_q(X)$ are established by f_*.

EXERCISE 16. Prove that for every connected CW complex X, the canonical map $X|_2 \to X$ is homotopy equivalent to the universal covering.

EXERCISE 17. Prove that $S^2|_3 \sim S^3$; generalization: $\mathbb{C}P^n|_3 = \mathbb{C}P^n|_{2n+1} \sim S^{2n+1}$.

EXERCISE 18. Prove that in the situation of the theorem above, the fiber of the fibration homotopy equivalent to a canonical map $X|_{n+1} \to X$ is $K(\pi, n-1)$.

Additionally, the cellular approximation of topological spaces (Sect. 11.6) makes possible the generalization of both capping and killing to arbitrary spaces, and only homotopy equivalences become weak homotopy equivalences.

Thus, we have two constructions of extinguishing homotopy groups, which are, actually, dual in the sense of Lecture 4. We can kill homotopy groups of a space X above a certain dimension, and X is canonically mapped into the new space. Or, we can kill homotopy groups of X below a certain dimension, and the new space is canonically mapped into X.

Chapter 2
Homology

Lecture 12 Main Definitions and Constructions

Besides the homotopy groups $\pi_n(X)$, there are other series of groups corresponding in a homotopy invariant way to a topological space X; the most notable are homology and cohomology groups, $H_n(X)$ and $H^n(X)$. Compared with homotopy groups, they have an important flaw—their accurate definition requires substantial algebraic work—and important advantages: Their computation is much easier, we will calculate them more or less immediately for the majority of topological spaces known to us, and also they are geometrically better visualizable [there are no counterintuitive phenomena like $\pi_3(S^2) \cong \mathbb{Z}$]. The information of a simply connected topological space contained in homology groups is comparable with that contained in homotopy groups.

The main geometric idea of homology is as follows. Spheroids are replaced by *cycles*; an n-dimensional cycle is, roughly, an n-dimensional surface, maybe a sphere, but it may be something different, say, a torus. The relation of being homotopic is replaced by a relation of being *homological*: Two n-dimensional cycles are homological if they cobound a piece of surface of dimension $n + 1$. How do we define cycles and those pieces of surfaces which they bound, the so-called chains? One can try to present them as continuous maps of some standard objects, spheres and something else (k-dimensional manifolds?). But this leads to severe difficulties, especially in dimensions > 2. It is easier to define cycles and chains as the union of standard "bricks." The role of these bricks is assumed by "singular simplices."

Notice that the construction of homology (and cohomology) groups does not require a fixation of a base point.

A. Fomenko, D. Fuchs, *Homotopical Topology*, Graduate Texts in Mathematics 273,
DOI 10.1007/978-3-319-23488-5_2

12.1 Singular Simplices, Chains, and Homology

Let $A_0, A_1, \ldots, A_q$ be points of the space $\mathbb{R}^n$, $n \geq q$, not contained in one $(q-1)$-dimensional plane. The convex hull of these points is called the *Euclidean simplex* with *vertices* $A_0, A_1, \ldots, A_q$ (this notion is known to us from Lecture 5; see Sect. 5.8). The convex hulls of (nonempty) subsets of the set of vertices are called *faces* of the simplex; they are also Euclidean simplices. Euclidean simplices of the same dimension are essentially the same, and this motivates us to choose one *standard* Euclidean simplex. The usual choice of the standard simplex is the simplex Δ^n in $\mathbb{R}^{n+1}$ with the ends of coordinate vectors taken for vertices. Thus,

$$\Delta^n = \left\{(t_0, t_1, \ldots, t_n) \in \mathbb{R}^{q+1} \mid t_0 \geq 0, t_1 \geq 0, \ldots, t_n \geq 0, \sum\nolimits_{i=0}^{n} t_i = 1\right\}.$$

Let X be an arbitrary topological space. We define an n-dimensional *singular simplex* of X simply as a continuous map of Δ^n into X. An n-dimensional *singular chain* of X is a formal finite linear combination of n-dimensional singular simplices with integral coefficients: $\sum_i k_i f_i$, $f_i\colon \Delta^n \to X$. The set of all n-dimensional singular chains of X is denoted as $C_n(X)$. The usual addition of linear combinations makes $C_n(X)$ an Abelian group; thus, $C_n(X)$ is the free Abelian group generated by the set of all n-dimensional singular simplices of X.

Next we describe the *boundary homomorphism* $\partial = \partial_n\colon C_n(X) \to C_{n-1}(X)$. Since the group $C_n(X)$ is free, it is sufficient to define ∂ for the generators, that is, for singular simplices. For a singular simplex f we put

$$\partial f = \sum_{i=0}^{n} (-1)^i \Gamma_i f,$$

where $\Gamma_i f$ is the ith face of f, which is defined as the restriction of f to the ith face $\Gamma_i \Delta^n$,

$$\Gamma_i \Delta^n = \{(t_0, t_1, \ldots, t_n) \in \Delta^n \mid t_i = 0\}$$

[we identify $\Gamma_i \Delta^n$ with Δ^{n-1} using the correspondence

$$(t_0, \ldots, t_{i-1}, 0, t_{i+1}, \ldots .t_n) \leftrightarrow (t_0, \ldots, t_{i-1}, t_{i+1}, \ldots .t_n)].$$

Theorem. *The composition*

$$C_{n+1}(X) \xrightarrow{\partial_{n+1}} C_n(X) \xrightarrow{\partial_n} C_{n-1}(X)$$

is trivial; in other words, $\operatorname{Im}(\partial_{n+1}) \subset \operatorname{Ker}(\partial_n)$.

Proof. A direct verification is based on the equality

$$\Gamma_i \Gamma_j f = \begin{cases} \Gamma_{j-1} \Gamma_i f, & \text{if } j > i, \\ \Gamma_j \Gamma_{i+1} f, & \text{if } j \leq i. \end{cases}$$

To make our upcoming life slightly easier, we assume that $C_n(X) = 0$ for $n < 0$ and extend the definition of ∂ accordingly. The theorem is not affected.

Main Definition. The quotient group

$$H_n(X) = \operatorname{Ker} \partial_n / \operatorname{Im} \partial_{n+1}$$

is called the *nth homology group* of X. In particular, $H_0(X) = C_0(X)/\operatorname{Im}\partial_1$ and $H_n(X) = 0$ for $n < 0$.

There are also common notations $\operatorname{Ker}\partial_n = Z_n(X)$ and $\operatorname{Im}\partial_{n+1} = B_n(X)$. Thus, $H_n(X) = Z_n(X)/B_n(X)$. Elements of the groups $Z_n(X)$ and $B_n(X)$ are called, respectively, *cycles* and *boundaries*. (Thus, every boundary is a cycle, but the converse is, generally, false.) If the difference of two cycles is a boundary, then these cycles are called *homologous*. Thus, the homology group is the group of classes of homologous cycles (which may be called *homology classes*).

If the group $H_n(X)$ is finitely generated, then its rank is called the nth *Betti number* of X.

12.2 Chain Complexes, Map, and Homotopies

A *chain complex*, or simply a complex, is an (infinite in both directions) sequence of groups and homomorphisms

$$\dots \xrightarrow{\partial_{n+2}} C_{n+1} \xrightarrow{\partial_{n+1}} C_n \xrightarrow{\partial_n} C_{n-1} \xrightarrow{\partial_{n-1}} \dots$$

such that $\partial_n \circ \partial_{n+1} = 0$ for all n.

The group $H_n = \operatorname{Ker}\partial_n / \operatorname{Im}\partial_{n+1}$ is called the nth homology group of the complex.

EXERCISE 1. Let

$$\dots \xrightarrow{\partial_{n+2}} C_{n+1} \xrightarrow{\partial_{n+1}} C_n \xrightarrow{\partial_n} C_{n-1} \xrightarrow{\partial_{n-1}} \dots$$

be a complex. Put $\widetilde{C}_n = C_n \oplus C_{n+1}$ and define $\widetilde{\partial}_n : \widetilde{C}_n \to \widetilde{C}_{n-1}$ by the formula $\widetilde{\partial}_n(c, c') = (\partial_n c, \partial_{n+1} c' + (-1)^n c)$, $c \in C_n, c' \in C_{n+1}$. Prove that

$$\dots \xrightarrow{\widetilde{\partial}_{n+2}} \widetilde{C}_{n+1} \xrightarrow{\widetilde{\partial}_{n+1}} \widetilde{C}_n \xrightarrow{\widetilde{\partial}_n} \widetilde{C}_{n-1} \xrightarrow{\widetilde{\partial}_{n-1}} \dots$$

is a complex and that the homology of this complex is trivial ($\widetilde{H}_n = 0$ for all n).

Our main example of a chain complex, so far, is the *singular complex* of a space X: $C_n = C_n(X)$. This complex is *positive*, which means that $C_n = 0$ for $n < 0$.

Mostly, we will consider positive complexes, but there will be exceptions, and the first exception appears immediately: The *augmented* or *reduced* singular complex of a space X,

$$\dots \xrightarrow{\partial_{n+2}} \widetilde{C}_{n+1}(X) \xrightarrow{\partial_{n+1}} \widetilde{C}_n(X) \xrightarrow{\partial_n} \widetilde{C}_{n-1}(X) \xrightarrow{\partial_{n-1}} \dots,$$

is defined by the formula

$$\widetilde{C}_n(X) = \begin{cases} C_n(X), & \text{if } n \neq -1, \\ \mathbb{Z}, & \text{if } n = -1, \end{cases}$$

and ∂_n are all as before, except $\partial_0 \colon C_0(X) \to \mathbb{Z}$, more commonly denoted as ϵ and called an *augmentation*, which takes every zero-dimensional singular simplex of X into $1 \in \mathbb{Z}$. Thus, the reduced complex of X looks like

$$\dots \xrightarrow{\partial_2} C_1(X) \xrightarrow{\partial_1} C_0(X) \xrightarrow{\epsilon} \mathbb{Z} \longrightarrow \dots.$$

[Thus, for a zero-dimensional chain $c = \sum k_i f_i$, $\epsilon(c) = \sum k_i$; the number $\epsilon(c)$ is sometimes called the *index* of the zero-dimensional chain c; it may be denoted as $\operatorname{ind}(c)$.] A natural question arises: Why is this complex called *reduced*? It looks bigger than the unreduced complex. The answer is in the following proposition.

Proposition 1. *The homology $\widetilde{H}_n(X)$ of the reduced singular complex* (called the *reduced homology* of X) *is related to the usual homology as follows. If X is not empty, then*

$$H_n(X) = \begin{cases} \widetilde{H}_n(X), & \text{if } n \neq 0, \\ \widetilde{H}_0(X) \oplus \mathbb{Z}, & \text{if } n = 0; \end{cases}$$

if X is empty, then the only nonzero reduced homology group of X is $\widetilde{H}_{-1}(X) = \mathbb{Z}$.

Proof. Obvious.

Back to algebra. If $\mathcal{C} = \{C_n, \partial_n\}$ and $\mathcal{C}' = \{C'_n, \partial'_n\}$ are two chain complexes, then a *chain map*, or a *homomorphism* $\varphi \colon \mathcal{C} \to \mathcal{C}'$, is defined as a sequence of group homomorphisms $\varphi_n \colon C_n \to C'_n$ which make the diagram

$$\begin{array}{ccccccccc} \dots & \xrightarrow{\partial_{n+2}} & C_{n+1} & \xrightarrow{\partial_{n+1}} & C_n & \xrightarrow{\partial_n} & C_{n-1} & \xrightarrow{\partial_{n-1}} & \dots \\ & & \downarrow{\scriptstyle \varphi_{n+1}} & & \downarrow{\scriptstyle \varphi_n} & & \downarrow{\scriptstyle \varphi_{n-1}} & & \\ \dots & \xrightarrow{\partial'_{n+2}} & C'_{n+1} & \xrightarrow{\partial'_{n+1}} & C'_n & \xrightarrow{\partial'_n} & C'_{n-1} & \xrightarrow{\partial'_{n-1}} & \dots \end{array}$$

commutative.

From this commutativity, $\varphi_n(\operatorname{Ker} \partial_n) \subset \operatorname{Ker}(\partial'_n)$ and $\varphi_n(\operatorname{Im} \partial_{n+1}) \subset \operatorname{Im}(\partial'_{n+1})$, so there arise homomorphisms $\varphi_* = \varphi_{*n} \colon H_n(\mathcal{C}) \to H_n(\mathcal{C}')$ with obvious properties, like $(\psi \circ \varphi)_* = \psi_* \circ \varphi_*$. For our main example, a continuous map $h \colon X \to Y$ naturally induces homomorphisms $h_\# = h_{\#n}\colon C_n(X) \to C_n(Y)$, $h_\#\left(\sum_i k_i f_i\right) = \sum_i k_i(h \circ f_i)$ and also $h_{\#n}\colon \widetilde{C}_n(X) \to \widetilde{C}_n(Y)$ (with $h_{\#,-1} = \mathrm{id}$) which comprise homomorphisms between both unreduced and reduced singular complexes. Thus, there arise maps $f_*\colon H_n(X) \to H_n(Y)$ and $\widetilde{H}_n(X) \to \widetilde{H}_n(Y)$ (with the same obvious properties).

Again back to algebra. Let $\mathcal{C} = \{C_n, \partial_n\}$ and $\mathcal{C}' = \{C'_n, \partial'_n\}$ be two chain complexes and $\varphi = \{\varphi_n\}, \psi = \{\psi_n\}\colon \mathcal{C} \to \mathcal{C}'$ be two chain maps. A *chain homotopy* between φ and ψ is a sequence $D = \{D_n\colon C_n \to C'_{n+1}\}$ satisfying the identities

$$D_{n-1} \circ \partial_n + \partial'_{n+1} \circ D_n = \psi_n - \varphi_n.$$

For the reader's convenience (or inconvenience?) we show all the maps involved in this definition in one diagram (which, certainly, is not commutative):

$$\begin{array}{ccccccccc}
\cdots & \longrightarrow & C_{n+1} & \xrightarrow{\partial_{n+1}} & C_n & \xrightarrow{\partial_n} & C_{n-1} & \longrightarrow & \cdots \\
 & & \psi_{n+1}\Big\downarrow\Big\downarrow\varphi_{n+1} & \swarrow D_n & \psi_n\Big\downarrow\Big\downarrow\varphi_n & \swarrow D_{n-1} & \psi_{n+1}\Big\downarrow\Big\downarrow\varphi_{n+1} & & \\
\cdots & \longrightarrow & C'_{n+1} & \xrightarrow[\partial'_{n+1}]{} & C'_n & \xrightarrow[\partial'_n]{} & C'_{n-1} & \longrightarrow & \cdots
\end{array}$$

If chain maps φ, ψ can be connected by a chain homotopy, they are called (*chain*) *homotopic*.

Proposition 2. *If chain maps $\varphi, \psi\colon \mathcal{C} \to \mathcal{C}'$ are homotopic, then the induced homology maps $\varphi_*, \psi_*\colon H_n(\mathcal{C}) \to H_n(\mathcal{C}')$ are equal.*

Proof. Let $D = \{D_n\}$ be a homotopy between φ and ψ. If $c \in \operatorname{Ker} \partial_n \subset C_n$, then

$$\psi_n(c) - \varphi_n(c) = D_{n-1} \circ \partial_n(c) + \partial'_{n+1} \circ D_n(c) = \partial'_{n+1}(D_n(c)) \in \operatorname{Im} \partial'_{n+1};$$

that is, $\varphi_n(c)$ and $\psi_n(c)$ are homologous for every cycle $c \in C_n$. Thus, $\varphi_{*n} = \psi_{*n}$.

EXERCISE 2. A complex $(\mathcal{C})$ is called *contractible* if the identity map $\mathrm{id}\colon \mathcal{C} \to \mathcal{C}$ is homotopic to the zero map $0\colon \mathcal{C} \to \mathcal{C}$. A complex $(\mathcal{C})$ is called acyclic if $H_n(\mathcal{C}) = 0$ for all n.

(Warmup) Prove that a contractible complex is acyclic.

(a) Prove that the complex $\{\widetilde{C}_n, \widetilde{\partial}_n\}$ from Exercise 1 is not only acyclic but also contractible.

(b) Prove that the complex

$$\cdots \leftarrow 0 \leftarrow 0 \leftarrow \mathbb{Z}_2 \xleftarrow{\text{onto}} \mathbb{Z} \xleftarrow{\cdot 2} \mathbb{Z} \leftarrow 0 \leftarrow 0 \ldots$$

is acyclic but not contractible.

(c) Let $(\mathcal{C}) = \{C_n, \partial_n\}$ be a positive ($C_n = 0$ for $n < 0$) free (all C_n are free Abelian groups) complex. Prove that if $(\mathcal{C})$ is acyclic, then it is contractible.

Finally, we will establish a connection between chain homotopies considered here with homotopies between continuous maps. (This connection is actually a justification for the term "chain homotopy.") Namely, we will show how a homotopy between continuous maps $f, g: X \to Y$ determines a chain homotopy between the maps $f_\#, g_\#$ of singular complexes.

We begin with a geometric construction which presents a covering of a cylinder $\Delta^n \times I$ by $n+1$ Euclidean simplices (in the language of Sect. 5.8, it is a *triangulation* of $\Delta^n \times I$). Recall that $\Delta^n = \{(t_0, \dots, t_n) \in \mathbb{R}^{n+1} \mid t_i \geq 0, \sum t_i = 1\}$. The vertices of Δ^n are $v_i = (0, \dots, 0, 1, 0, \dots, 0)$ with $1 = t_i$. For $0 \leq i \leq n$, put

$$A_i = \{((t_0, \dots, t_n), t) \in \Delta^n \times I \mid t_0 + \dots + t_{i-1} \leq t \leq t_0 + \dots t_i\}$$

(where the empty sum is regarded as 0). It is easy to see that A_i is the convex hull of $(v_0, 1), \dots, (v_i, 1), (v_i, 0), \dots, (v_n, 0)$, that is, the Euclidean simplex with the vertices $(v_0, 1), \dots, (v_i, 1), (v_i, 0), \dots, (v_n, 0)$. Indeed, all these points belong to A_i, and if $y = ((t_0, \dots, t_n), t)$, then $y = t_0(v_0, 1) + \dots + t_{i-1}(v_{i-1}, 1) + t_i'(v_i, 1) + t_i''(v_i, 0) + t_{i+1}(v_{i+1}, 0) + \dots + t_n(v_n, 0)$, where $t_i' = t - (t_0 + \dots + t_{i-1})$ and $t_i'' = t_i - t_i' = (t_0 + \dots + t_i) - t$, so if $y \in A_i$, then the sum of the coefficients is 1 and all of them are between 0 and 1.

For $n = 1$ and 2, this triangulation is shown in Fig. 59 (familiar to the reader from elementary geometry textbooks).

Let $\alpha_i = \alpha_i(\Delta^n): \Delta^{n+1} \to \Delta^n \times I$ be the affine homeomorphism of Δ^{n+1} onto A_i preserving the order of vertices. These α_is are singular simplices of $\Delta^n \times I$. Consider the faces $\Gamma_j\alpha_i$ $(0 \leq i \leq n,\ 0 \leq j \leq n+1)$. First, $\Gamma_i\alpha_i = \Gamma_i\alpha_{i-1}$ $(1 \leq i \leq n)$; in addition to that, $\Gamma_0\alpha_0 = \mathrm{id}_{\Delta^n} \times 0$, $\Gamma_{n+1}\alpha_n = \mathrm{id}_{\Delta^n} \times 1$. Second,

$$\Gamma_j\alpha_i(\Delta^n) = \begin{cases} \alpha_{i-1}(\Gamma_j\Delta^n), & \text{if } j < i, \\ \alpha_i(\Gamma_{j-1}\Delta_n), & \text{if } j > i+1. \end{cases}$$

Next, let us calculate the boundary of $\alpha(\Delta^n) = \sum_i (-1)^i \alpha_i(\Delta^n)$.

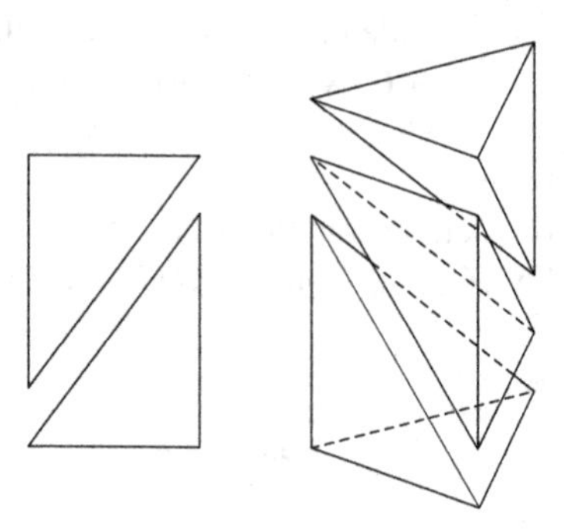

Fig. 59 Triangulations of cylinders over simplices

$$\partial\alpha(\Delta^n) = \sum_{j=0}^{n+1}\sum_{i=0}^{n}(-1)^{i+j}\Gamma_j\alpha_i(\Delta^n) = \mathrm{id}_{\Delta^n}\times 0 + \left[\sum_{j=2}^{n+1}\sum_{i=0}^{j-2} + \sum_{j=0}^{n-1}\sum_{i=j+1}^{n}\right](-1)^{i+j}\Gamma_j\alpha_i(\Delta^n) - \mathrm{id}_{\Delta^n}\times 1$$
$$= \mathrm{id}_{\Delta^n}\times 0 + \sum_{i=0}^{n-1}\sum_{j=0}^{n}(-1)^{i+j+1}\alpha_i(\Gamma_j\Delta^n) - \mathrm{id}_{\Delta^n}\times 1$$
$$= \mathrm{id}_{\Delta^n}\times 0 - \mathrm{id}_{\Delta^n}\times 1 - \alpha(\partial\Delta^n).$$

Now let $f, g: X \to Y$ be two continuous maps and let $H: X\times I \to Y$ be a homotopy connecting f with g. For an n-dimensional singular simplex $b: \Delta^n \to X$, define an $(n+1)$-dimensional singular chain B of Y as $(H\circ(b\times I))_\#\alpha(\Delta^n)$; the correspondence $b \mapsto B$ is extended to a homomorphism $C_n(X) \to C_{n+1}(Y)$, which we take for D_n. The previous computations show that for any chain $c \in C_n(X)$,

$$\partial D_n(c) = f_\#(c) - g_\#(c) - D_{n-1}(\partial c),$$

which means that $\{D_n\}$ is a chain homotopy between $f_\#$ and $g_\#$ (see Fig. 60).

We arrive at the following result.

Theorem. *If continuous maps $f, g: X \to Y$ are homotopic, then the chain maps $f_\#, g_\#$ are chain homotopic.*

Corollary 1. *If continuous maps $f, g: X \to Y$ are homotopic, then for all n the induced homology homomorphisms $f_*, g_*: H_n(X) \to H_n(Y)$ coincide.*

Corollary 2. *A homotopy equivalence $f: X \to Y$ induces for all n isomorphisms $f_*: H_n(x) \xrightarrow{\cong} H_n(Y)$. In particular, homotopy equivalent spaces have isomorphic homology groups.*

(Question: And what about weak homotopy equivalence? The answer is in Lecture 14.)

EXERCISE 3. Prove the last three statements for reduced homology.

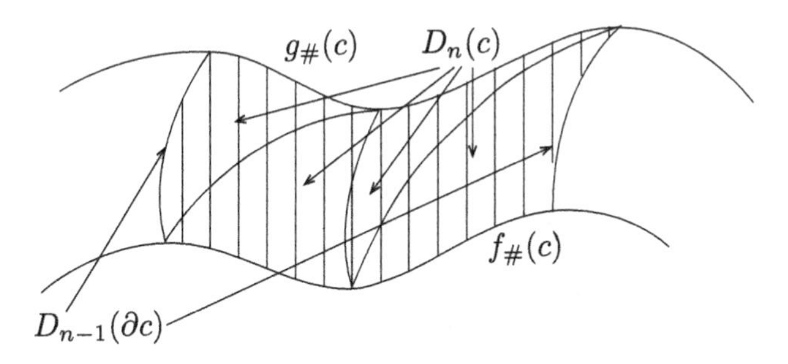

Fig. 60 From a homotopy to a chain homotopy

12.3 First Calculations

The groups of singular chains are usually huge and difficult to deal with; they are not fit for systematic calculations of homology groups. There are some efficient indirect methods of homology calculations which will be presented in the nearest future. Still, some direct calculations are possible and, actually, necessary for developing those indirect methods.

A: Homology of the One-Point Space

Let pt denote the one-point space. Then in every dimension $n \geq 0$ there is only one singular simplex $f_n\colon \Delta^n \to \text{pt}$. In particular, $\Gamma_i f_n = f_{n-1}$ for all i, and $\partial f_n = f_{n-1} - f_{n-1} + f_{n-1} - \cdots + (-1)^n f_{n-1}$, which is 0 if n is odd and f_{n-1} if n is even and positive. Thus, the (unreduced) singular complex of pt has the form

$$\dots \xrightarrow{\text{id}} \mathbb{Z} \xrightarrow{0} \mathbb{Z} \xrightarrow{\text{id}} \mathbb{Z} \xrightarrow{0} \mathbb{Z} \to 0 \to 0 \to \dots,$$

and

$$H_n(\text{pt}) = \begin{cases} \mathbb{Z}, \text{ if } n = 0, \\ 0, \text{ if } n \neq 0. \end{cases}$$

Add to this that $\widetilde{H}_0(\text{pt}) = 0$; this shows that $\widetilde{H}_n(\text{pt}) = 0$ for all n.

A space whose homology is the same as that of pt is called *acyclic*.

Corollary (of homotopy invariance of homology). *Contractible spaces are acyclic.*

The converse is not true; fans of the function $\sin \frac{1}{x}$ will appreciate an example in Fig. 61. There are more interesting examples, say, the Poincaré sphere with one point deleted.

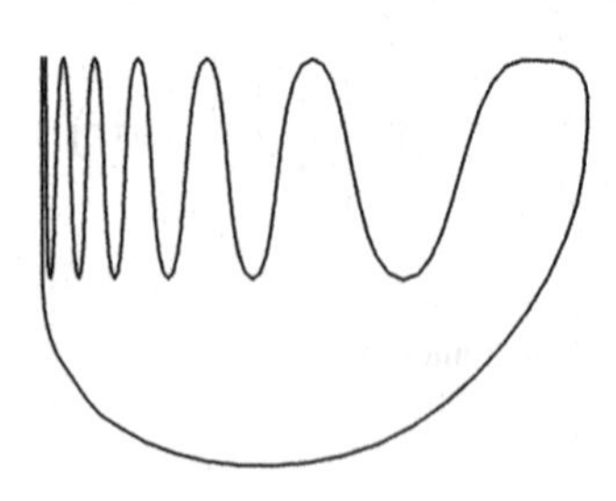

Fig. 61 A noncontractible acyclic space

B: Zero-Dimensional Homology

Theorem. *If X is path connected, then* $H_0(X) = \mathbb{Z}$.

Proof. Zero-dimensional singular simplices of X are just points of X; one-dimensional simplices are paths, and the boundary of a path joining x_0 with x_1 is $x_1 - x_0$. If X is connected, then every zero-dimensional chain $\sum_i k_i f_i$ (which is always a cycle) is homological to $\left(\sum_i k_i\right) f_0$, where f_0 is an arbitrarily fixed zero-dimensional singular simplex; indeed, if s_i is a path joining f_0 with f_i, then $\partial \sum_i k_i s_i = \sum_i k_i(f_i - f_0) = \sum_i k_i f_i - \left(\sum_i k_i\right) f_0$. We see that if $\sum_i k_i = 0$, then the chain is homological to zero. The converse is also true: The sum of the coefficients of the boundary of a one-dimensional singular simplex, and hence of the boundary of every zero-dimensional singular chain, is zero. We see that the map $\epsilon : C_0(X) = Z_0(x) \to \mathbb{Z}$ establishes an isomorphism $H_0(X) \to \mathbb{Z}$.

Equivalent statement (for a path connected X): $\widetilde{H}_0(X) = 0$.

EXERCISE 4. Prove that if $f: X \to Y$ is a continuous map between two path connected spaces, then $f_*: H_0(X) \to H_0(Y)$ is an isomorphism.

C: Homology and Components

Standard simplices are connected. Hence, every singular simplex of a space belongs to one of the path components of this space. This shows that $C_n(X) = \bigoplus_\alpha C_n(X_\alpha)$, where the X_αs are path components of X, and also $Z_n(X) = \bigoplus_\alpha Z_n(X_\alpha)$, $B_n(X) = \bigoplus_\alpha B_n(X_\alpha)$, $H_n(X) = \bigoplus_\alpha H_n(X_\alpha)$. In particular, the two previous computations imply the following. (1) For an arbitrary X, $H_0(X)$ is a free Abelian group generated by the path components of X; (2) If the space X is discrete, then $H_n(X) = 0$ for any $n \neq 0$.

12.4 Relative Homology

Let (X, A) be a topological pair; that is, A is a subset of a space X. Then $C_n(A) \subset C_n(X)$. The group $C_n(X, A) = C_n(X)/C_n(A)$ is called the groups of (*relative*) *singular chains of the pair* (X, A) or *of X modulo A*. Obviously, $C_n(X, A)$ is a free Abelian group generated by singular simplices $f: \Delta^n \to X$ such that $f(\Delta^n) \not\subset A$. Since $\partial(C_n(A)) \subset C_{n-1}(A)$, there arise a quotient homomorphism $\partial: C_n(X, A) \to C_{n-1}(X, A)$ and a complex

$$\dots \xrightarrow{\partial} C_{n+1}(X, A) \xrightarrow{\partial} C_n(X, A) \xrightarrow{\partial} C_{n-1}(X, A) \xrightarrow{\partial} \dots .$$

The homology groups of this complex are denoted $H_n(X, A)$ and are called *relative homology groups*. One can say that $H_n(X, A)$ is the quotient $Z_n(X, A)/B_n(X, A)$ of the group of *relative cycles* over the group of *relative boundaries*. Here a relative cycle is a singular chain of X whose boundary lies in A, and a relative boundary is a chain of X which becomes a boundary after adding a chain from A. (Obviously, relative boundaries are relative cycles.)

EXERCISE 5. Compute $H_0(X, A)$ in the case when X and A are both connected and in the general case.

EXERCISE 6. Construct for an arbitrary space X and an arbitrary point $x_0 \in X$ a natural isomorphism $\widetilde{H}_n(X) = H_n(X, x_0)$.

The boundary of a relative cycle is an absolute (that is, usual) cycle in A; the correspondence $c \mapsto \partial c$ determines (for every n) a *boundary homomorphism*

$$\partial_*: H_n(X, A) \to H_{n-1}(A)$$

(indeed, if $c - c'$ is a relative boundary, then $\partial c - \partial c'$ is an absolute boundary in A). The homomorphism ∂_* is included in a *homology sequence of a pair* (similar to a homotopy sequence of a pair; see Sect. 8.7; but it looks simpler than the homotopy sequence, since it involves only Abelian groups):

$$\dots \xrightarrow{\partial_*} H_n(A) \xrightarrow{i_*} H_n(X) \xrightarrow{j_*} H_n(X, A) \xrightarrow{\partial_*} H_{n-1}(A) \xrightarrow{i_*} \dots,$$

where i_* is induced by the inclusion map $i: A \to X$ and j_* is induced by the projection $C_n(X) \to C_n(X)/C_n(A) = C_n(X, A)$.

Theorem. *The homology sequence of a pair is exact.*

We prefer to have this theorem in a "more general" algebraic form. Let $\mathcal{C} = \{C_n, \partial_n\}$ be a complex and let $\mathcal{C}' = \{C'_n, \partial'_n\}$ be a subcomplex which means $C'_n \subset C_n$, $\partial_n(C'_n) \subset C'_{n-1}$ for all n and $\partial'_n(c) = \partial_n(c)$ for all $c \in C'_n$. There arise a quotient complex $\mathcal{C}'' = \mathcal{C}/\mathcal{C}' = \{C''_n = C_n/C'_n, \partial''_n\}$ with a naturally defined ∂''_n, and also inclusion and projection homomorphisms $\iota: \mathcal{C}' \to \mathcal{C}$ and $\pi: \mathcal{C} \to \mathcal{C}''$. There also arise "connecting homomorphisms"

$$\partial_*: H_n(\mathcal{C}'') \to H_{n-1}(\mathcal{C}').$$

Namely, let $\gamma'' \in H_n(\mathcal{C}'')$ be an arbitrary homology class and let $c'' \in \operatorname{Ker} \partial''_n \subset C''_n = C_n/C'_n$ be a representative of γ''. Let $c \in C_n$ be a representative of (the coset) c''. The equality $\partial''_n c'' = 0$ means precisely that $c' = \partial_n c \in C'_{n-1}$. Moreover, $\partial'_{n-1} c' = \partial_{n-1} c' = \partial_{n-1} \circ \partial_n c = 0$. Thus, $c' \in \operatorname{Ker} \partial'_{n-1}$ and hence belongs to the homology class in $\gamma' \in H_{n-1}(\mathcal{C}')$; we take this class for $\partial_*(\gamma'')$.

EXERCISE 7. Prove that the correspondence $\gamma'' \mapsto \gamma'$ provides a well-defined homomorphism $\partial_*: H_n(\mathcal{C}'') \to H_{n-1}(\mathcal{C}')$. In particular, γ' does not depend on the choice of c'' in γ'' and of c in c''. Moreover, one needs to check that ∂_* is a homomorphism.

Algebraic Theorem. *The sequence*

$$\dots \xrightarrow{\partial_*} H_n(C') \xrightarrow{\iota_*} H_n(C) \xrightarrow{\pi_*} H_n(C'') \xrightarrow{\partial_*} H_{n-1}(C') \xrightarrow{\iota_*} \dots$$

is exact.

EXERCISE 8. Prove the algebraic theorem. (The proof has some resemblance to the proof of exactness of the homotopy sequence of a pair in Sect. 8.7.)

The algebraic theorem implies the theorem above; it will be used many more times in this book, including exercises later in this section.

Notice that a map $f:(X,A) \to (Y,B)$ between topological pairs (that is, a map $f:X \to Y$ such that $f(A) \subset B$) induces homomorphisms $f_*:H_n(X,A) \to H_n(Y,B)$ and a homomorphism of the homology sequence of the pair (X,A) into the homology sequence of the pair (Y,B), that is, a "commutative ladder"

$$\begin{array}{ccccccccc}
\dots \to & H_n(A) & \to & H_n(X) & \to & H_n(X,A) & \to & H_{n-1}(A) & \to \dots \\
& \downarrow{\scriptstyle (f|_A)_*} & & \downarrow{\scriptstyle f_*} & & \downarrow{\scriptstyle f_*} & & \downarrow{\scriptstyle (f|_A)_*} & \\
\dots \to & H_n(B) & \to & H_n(Y) & \to & H_n(Y,B) & \to & H_{n-1}(B) & \to \dots
\end{array}$$

with exact rows. Add to that $H_n(X) = H_n(X,\emptyset)$ (in this sense relative homology is a generalization of absolute homology) and that the mysterious homomorphism $j_*:H_n(X) \to H_n(X,A)$ is actually induced by the map $j = \text{id}:(X,\emptyset) \to (X,A)$.

EXERCISE 9. Construct the *homology sequence of a triple*,

$$\dots \to H_n(A,B) \to H_n(X,B) \to H_n(X,A) \to H_{n-1}(A,B) \to \dots$$

$(B \subset A \subset X)$ and prove its properties, including the exactness. (Compare to Exercise 10 in Sect. 8.7.) (In the case when A is not empty, a combination of this exercise with Exercise 5 gives rise to a *reduced* homology sequence of a pair, with the absolute groups H replaced by $\widetilde{H}$).

The exactness of homology sequences of pairs and triples (combined with the five-lemma; see Sect. 8.8) has a standard set of corollaries. Among them, there is a homotopy invariance of relative homology: If $f:X \to Y$ is a homotopy equivalence, $f(A) \subset B$, and the map $A \to B$ arising is also a homotopy equivalence, then $f_*:H_n(X,A) \to H_n(Y,B)$ is an isomorphism for all n.

(We have to disappoint a reader who expects an exact "homology sequence of a fibration" relating homology groups of the total space, the base, and the fiber of a fibration. The relations between homology and fibrations are more complicated, and we will thoroughly study them in the subsequent chapters of this book.)

12.5 Relative Homology as Absolute

The results here provide the main technical tool to effectively compute homology.

Theorem. *Let (X, A) be a topological pair.*

(1) *The inclusion $X \to X \cup CA$, where $X \cup CA$ is obtained from X by attaching the cone over A, induces for every n an isomorphism*

$$H_n(X, A) \cong H_n(X \cup CA, CA) = H_n(X \cup CA, v) = \widetilde{H}_n(X \cup CA),$$

where v is the vertex of the cone CA.

(2) *If (X, A) is a Borsuk pair* (see Sect. 5.6), *for example, a CW pair* (see again Sect. 5.6), *then*

$$p_*: H_n(X, A) \to H_n(X/A, a) = \widetilde{H}_n(X/A)$$

[where $p: X \to X/A$ is the projection and $a = p(A)$] is an isomorphism for all n.

COMMENTS. 1. Part (2) follows from part (1) because of the homotopy equivalence $X \cup CA \sim X/A$ for Borsuk pairs (see Sect. 5.6 again). Thus, we need to prove only part (1).

2. In Sect. 9.10, we showed how relative homotopy groups can be presented as absolute homotopy groups of a certain space. Here we do the same for homology groups, and it is obvious that for homology the construction is much simpler than for homotopy. This may be regarded as a first illustration of a reason why homology groups are way easier to compute than homotopy groups.

The proof of the theorem is based on the so-called refinement lemma, whose proof is based on the so-called transformator lemma. Both lemmas (especially, the first) have considerable independent value. We arrange the proof in the following order. First, we state the refinement lemma. Then we state and prove the transformator lemma. Then we prove the refinement lemma. And after that we prove our theorem.

Let X be a topological space and let $\mathcal{U} = \{U_\alpha\}$ be an open covering of X. We say that a singular simplex $f: \Delta^n \to X$ is subordinated to the covering $\mathcal{U}$ if $f(\Delta^n)$ is contained in U_α for some α. Let $C_n^{\mathcal{U}}(X)$ be a subgroup of $C_n(X)$ generated by singular simplices subordinated to $\mathcal{U}$. It is obvious that $\partial(C_n^{\mathcal{U}}(X)) \subset C_{n-1}^{\mathcal{U}}(X)$: If a singular simplex is subordinated to $\mathcal{U}$, then all its faces are subordinated to $\mathcal{U}$. Thus, the groups $C_n^{\mathcal{U}}(X)$ form a subcomplex of the singular complex of X.

Refinement Lemma. *The inclusion of the complex $\{C_n^{\mathcal{U}}(X)\}$ into the complex $\{C_n(X)\}$ induces a homology isomorphism. In other words, (1) every singular cycle of X is homologous to a cycle composed of singular simplices subordinated to $\mathcal{U}$ and (2) if two such cycles are homologous in X, then their difference equals a boundary of a chain composed of singular simplices subordinated to $\mathcal{U}$.*

To prove this lemma, we need "transformators."

Definition. A *transformator* τ is a rule which assigns to every topological space X and every integer n a homomorphism $\tau_n^X\colon C_n(X) \to C_n(X)$ such that

(1) $\tau_0^X = \mathrm{id}$ for every X.
(2) $\partial_n \circ \tau_n^X = \tau_{n-1}^X \circ \partial_n$ for every X and every n.
(3) If $h\colon X \to Y$ is a continuous map, then $h_\# \circ \tau_n^X = \tau_n^Y \circ h_\#$ for every n.

Example 1 (Barycentric Transformator). The barycentric subdivision of the standard simplex Δ^n (see Fig. 21 in Sect. 5.8) consists of $(n+1)!$ n-dimensional Euclidean simplices corresponding to chains $\delta^0 \subset \delta^1 \subset \cdots \subset \delta^n$ of faces of dimensions $0, 1, \dots, n$; the vertices of the simplex corresponding to this chain are centers of $\delta^0, \delta^1, \dots, \delta^n$. In other words, simplices of the subdivision correspond to permutations $\sigma \in S_{n+1}$: The simplex $\beta_\sigma \Delta^n$ corresponding to a permutation σ of $0, 1, \dots, n$ has vertices

$$u_k^\sigma = \frac{v_{\sigma(0)} + v_{\sigma(1)} + \cdots + v_{\sigma(k)}}{k+1}, \ k = 0, 1, \dots, n,$$

where $v_0, v_1, \dots, v_n$ are the vertices of Δ^n in their natural order. The correspondence $v_i \mapsto u_i^\sigma$ is extended to an affine map $\beta_\sigma\colon \Delta^n \to \Delta^n$, which may be regarded as an n-dimensional singular simplex of Δ^n. Put $\beta(\Delta^n) = \sum_{\sigma \in S_{n+1}} \mathrm{sgn}(\sigma)\beta_\sigma$. A direct computation shows that $\partial(\beta(\Delta^n)) = \sum_{i=0}^n (-1)^i \beta(\Gamma_i \Delta^n)$ (the faces inside Δ^n are cancelled; there remain only simplices of barycentric subdivisions of faces of Δ^n, and they appear in $\partial(\beta(\Delta^n))$ with proper signs).

EXERCISE 10. Reconstruct the details of this direct computation.

Now to the transformator. For a chain $c = \sum_i k_i f_i \in C_n(X)$, we put $\beta_n^X(c) = \sum_i k_i (f_i)_\#(\beta(\Delta^n))$. This is a transformator: Properties (1) and (3) are immediately clear, and property (2) follows from the formula for $\partial(\beta(\Delta^n))$.

Example 2 (Backward Transformator). Let $\omega\colon \Delta^n \to \Delta^n$ be the affine homeomorphism reversing the order of vertices $(\omega(v_i) = v_{n-1})$. For $c = \sum_i k_i f_i \in C_n(X)$, put $\omega_n^X(c) = \sum_i k_i (-1)^{\frac{n(n+1)}{2}} (f_i \circ \omega)$. It is immediately clear that $\{\omega_n^X\}$ satisfies conditions (1) and (3) from the definition of a transformator, and a direct computation shows that condition (2) is also satisfied.

EXERCISE 11. Reconstruct the details of this direct computation.

We will use the backward transformator later, in Lecture 16.

Transformator Lemma. *Let $\tau = \{\tau_n^X\}$ be a transformator. Then for every X the chain map $\tau^X = \{\tau_n^X\colon C_n(X) \to C_n(X)\}$ is homotopic to the identity. Thus, $(\tau^X)_{*n}\colon H_n(X) \to H_n(X)$ is $\mathrm{id}_{H_n(X)}$.*

Moreover, a homotopy $D_n^X\colon C_n(X) \to C_{n+1}(X)$ between τ^X and id *can be defined in such a way that $f_{\#,n+1} \circ D_n^X = D_n^Y \circ f_{\#n}$ for every continuous map $f\colon X \to Y$.*

Proof of Transformator Lemma. We put $D_0^X = 0$ for all X. Let $n > 0$. Assume that for all X and $m < n$ we have already defined homomorphisms $D_m^X\colon C_m(X) \to$

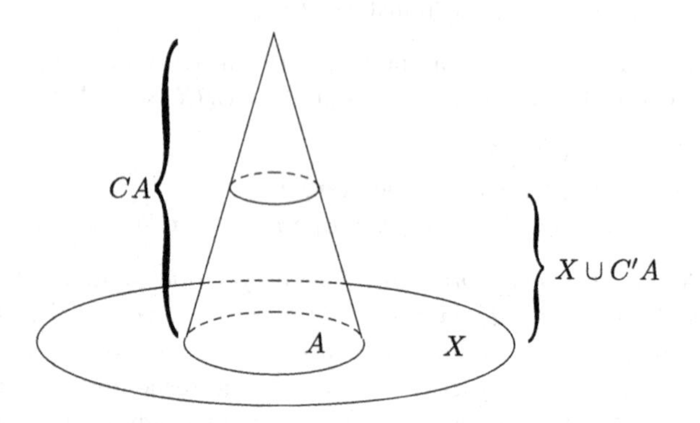

Fig. 62 The two-set covering of $X \cup CA$

$C_{m+1}(X)$ which satisfy all the conditions required (including the condition $\partial_{m+1} \circ D_m^X + D_{m-1}^X \circ \partial_m = \tau_m^X - \mathrm{id}$). The construction of D_n^X we begin with is $D_n^{\Delta^n}(\mathrm{id})$. The desired property is

$$\partial D_n^{\Delta^n}(\mathrm{id}) = \tau_n^{\Delta^n}(\mathrm{id}) - \mathrm{id} - D_{n-1}^{\Delta^n}(\partial\, \mathrm{id}).$$

But $\partial \circ D_{n-1}^{\Delta^n}(\partial\, \mathrm{id}) = \tau_{n-1}^{\Delta^n}(\partial\, \mathrm{id}) - \partial\, \mathrm{id} - D_{n-2}^{\Delta^n}(\partial\partial\, \mathrm{id}) = \partial(\tau_n^{\Delta^n}(\mathrm{id}) - \mathrm{id})$, which shows that $\partial(\tau_n^{\Delta^n}(\mathrm{id}) - \mathrm{id} - D_{n-1}^{\Delta^n}(\partial\, \mathrm{id})) = 0$. Since $H_n(\Delta^n) = 0$ (Δ^n is connected), the cycle $\tau_n^{\Delta^n}(\mathrm{id}) - \mathrm{id} - D_{n-1}^{\Delta^n}(\partial\, \mathrm{id}) \in C_n(\Delta^n)$ is a boundary of some chain in $C_{n+1}(\Delta^n)$; we choose such a chain and take it for $D_n^{\Delta^n}(\mathrm{id})$. After that, for an arbitrary X and arbitrary $c = \sum_i k_i f_i \in C_n(X)$, we put $D_n^X(c) = \sum_i k_i (f_i)_\#(D_n^{\Delta^n}(\mathrm{id}))$. This D_n^X obviously satisfies the conditions in the "moreover" part of the lemma.

Proof of the Refinement Lemma. We use the barycentric transformator β. We need to prove that (1) every cycle from $C_n(X)$ is homologous to a cycle in $C_n^{\mathcal{U}}(X)$ and (2) if a cycle from $C_n^{\mathcal{U}}(X)$ is a boundary of some chain from $C_{n+1}(X)$, then it is a boundary of some chain from $C_{n+1}^{\mathcal{U}}(X)$. This follows from the following three facts. (A) For every chain $c \in C_n(X)$ the chain $(\beta_n^X)^N(c)$ with a sufficiently big N is contained in $C_n^{\mathcal{U}}(X)$ (it is obvious). (B) A cycle c is homologous to $\beta(c)$, and hence to $\beta^N(c)$ (the transformator lemma). (C) If a cycle c belongs to $C_n^{\mathcal{U}}(X)$, then the difference $c - \beta(c)$, and hence the difference $c - \beta^N(c)$, is a boundary of a chain from $C_{n+1}^{\mathcal{U}}(X)$ (the "moreover" part of the transformator lemma).

Proof of Theorem. We need to prove only part (1). Consider the covering $\mathcal{U}$ of $C \cup CA$ by two open sets: CA (without the base) and $X \cup C'A$, where $C'A$ is the lower half of the cone (without the upper base): See Fig. 62.

It follows from the relative version of the transformator lemma (which, on one side, can be proved precisely as the absolute version, and, on the other side, follows from the absolute version and the five-lemma) that the homology of the pair $(X \cup CA, CA)$ can be computed with the chain groups

$$C_n^{\mathcal{U}}(X \cup CA, CA) = C_n^{\mathcal{U}}(X \cup CA)/C_n^{\mathcal{U}}(CA);$$

the covering of the cone CA, induced by the covering $\mathcal{U}$, we denote again by $\mathcal{U}$. But obviously

$$C_n^{\mathcal{U}}(X \cup CA)/C_n^{\mathcal{U}}(CA) = C_n(X \cup C'A)/C_n(C'A) = C_n(X \cup C'A, C'A).$$

Thus,

$$\begin{aligned}\widetilde{H}_n(X \cup CA) = H_n(X \cup CA, \mathrm{pt}) &= H_n(X \cup CA, CA) \\ &= H_n(X \cup C'A, C'A) = H_n(X, A)\end{aligned}$$

(the last equality follows from the homotopy invariance of homology).

12.6 Generalizations of the Refinement Lemma: Sufficient Sets of Singular Simplices

The refinement lemma says that for computing homology groups of spaces and pairs it is possible to consider only singular simplices satisfying some additional condition. This additional condition (for the refinement lemma this is the condition of being subordinated to an open covering) may be different.

Definition. A set $\mathcal{S}$ of singular simplices is called sufficient if all faces of a singular simplex from $\mathcal{S}$ also belong to $\mathcal{S}$, so the groups $C_n^{\mathcal{S}}(X) \subset C_n(X)$ form a subcomplex of the singular complex of X, and if the inclusion map of this subcomplex induces a homology isomorphism. In other words, for every n, every cycle from $C_n(X)$ is homologous to some cycle belonging to $C_n^{\mathcal{S}}(X)$, and if a cycle belonging to $C_n^{\mathcal{S}}(X)$ equals the boundary of some chain from $C_{n+1}(X)$, then it is also a boundary of a chain in $C_{n+1}^{\mathcal{S}}(X)$. The usual procedure of proving sufficiency of some set $\mathcal{S}$ of singular simplices is to find some way of "approximating" singular simplices with all faces in $\mathcal{S}$ by chains in $C_n^{\mathcal{S}}(X)$ with the same boundary. We will not prove any general result of this kind but will list several sufficient sets in the form of exercises (the statement in the last of these exercises will actually be proved quite soon).

EXERCISE 12. If X is a smooth manifold (say, a smooth surface of some dimension in some Euclidean space), then smooth singular simplices form a sufficient set.

EXERCISE 13. If X is a domain in an Euclidean space, then affine singular simplices form a sufficient set.

EXERCISE 14. If X is a triangulated space, then affine isomorphisms of standard simplices onto the simplices of the triangulation form a sufficient set.

12.7 More Applications of the Refinement Lemma

We will give here in the form of exercises two additional properties of homology groups. In the next lecture we will prove similar statements in the CW context.

EXERCISE 15. Let (X, A) be a topological pair, and let $B \subset A$. The inclusion map $(X - B, A - B) \to (X, A)$ induces a homomorphism

$$H_n(X - B, A - B) \to H_n(X, A)$$

called an *excision homomorphism*. Prove that if $\overline{B} \subset \operatorname{Int} A$, then the excision homomorphism is an isomorphism. (This statement is called the excision theorem, or, within a certain axiomatic approach to homology theory, the excision axiom. The conditions on X, A, B which imply the excision isomorphism may be different.)

EXERCISE 16. Let $X = A \cup B$, $A \cap B = C$. We suppose that the excision homomorphisms $H_n(B, C) \to H_n(X, A)$ and $H_n(A, C) \to H_n(X, B)$ are isomorphisms. Then the homomorphisms

$$H_n(X) \xrightarrow{j_*} H_n(X, A) \xrightarrow{\text{exc.}^{-1}} H_n(B, C) \xrightarrow{\partial_*} H_{n-1}(C)$$
$$H_n(X) \xrightarrow{j_*} H_n(X, B) \xrightarrow{\text{exc.}^{-1}} H_n(A, C) \xrightarrow{\partial_*} H_{n-1}(C)$$

are the same, and we denote them as γ_n. The sequence

$$\cdots \to H_n(C) \xrightarrow{\alpha_n} H_n(A) \oplus H_n(B) \xrightarrow{\beta_n} H_n(X) \xrightarrow{\gamma_n} H_{n-1}(C) \to \dots,$$

where α_n is the *difference* of the homomorphisms induced by the inclusions $C \to A$ and $C \to B$ and β_n is the *sum* of the homomorphisms induced by the inclusions $A \to X$ and $B \to X$, is called the *Mayer–Vietoris homology sequence* or the *homology sequence of the triad* $(X; A, B)$. Prove that this sequence is exact.

Lecture 13 Homology of CW Complexes

In this lecture, we will see that it is possible to compute the homology groups of CW complexes via a complex way narrower than the singular complex. We have to begin with the homology of spheres and bouquets of spheres.

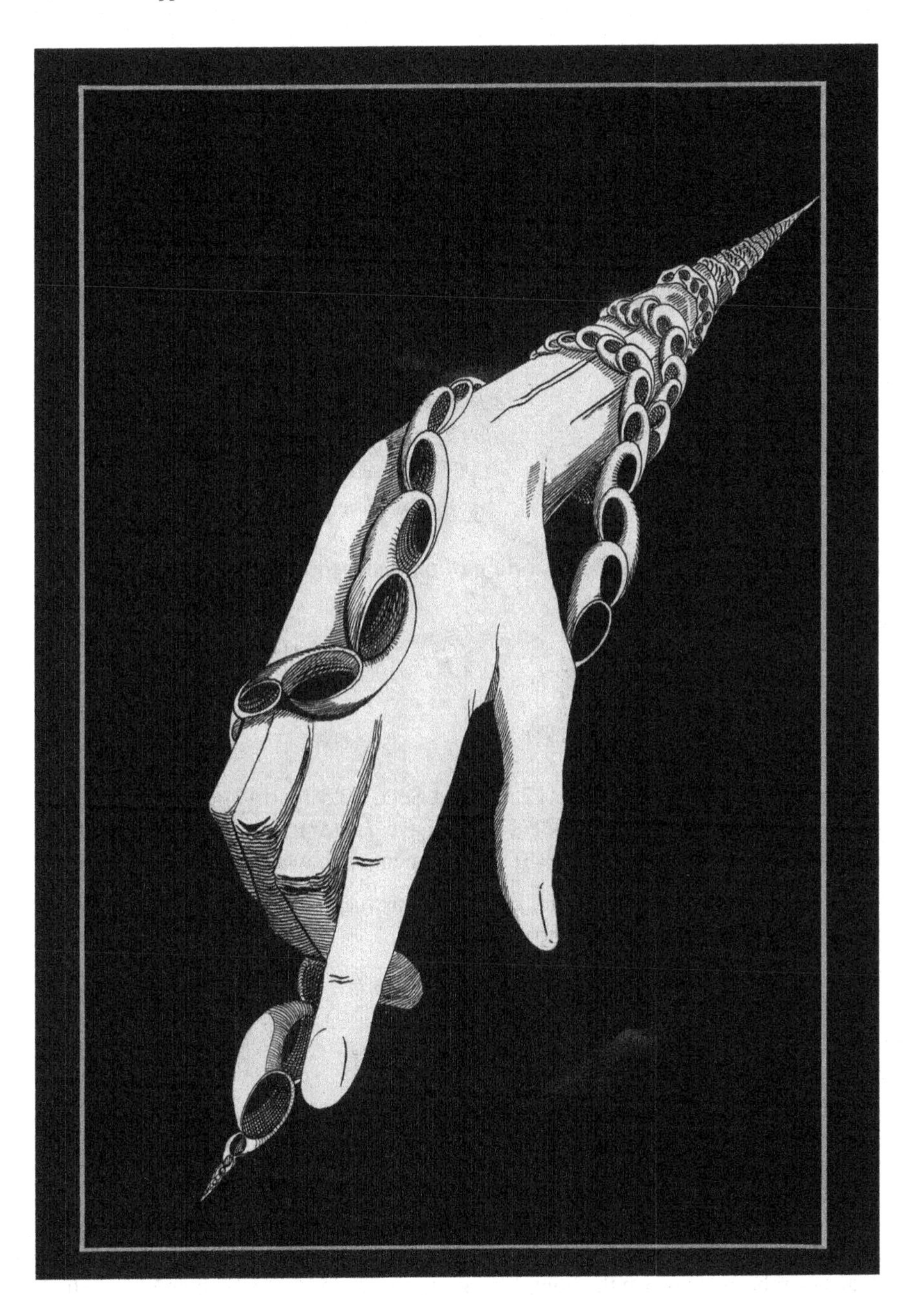

13.1 Homology of Spheres: Suspension Isomorphism

Theorem 1. *If $n > 0$, then*

$$H_m(S^n) = \begin{cases} \mathbb{Z}, \text{ if } m = 0, n, \\ 0, \text{ if } m \neq 0, n. \end{cases}$$

The homology of the (two-point) sphere S^0 looks different: $H_0(S^0) = \mathbb{Z} \oplus \mathbb{Z}$, $H_m(S^0) = 0$, if $m \neq 0$. To make the statement better looking, we may consider the reduced homology.

Theorem $\widetilde{1}$. *For all n,*

$$\widetilde{H}_m(S^n) = \begin{cases} \mathbb{Z}, \text{ if } m = n, \\ 0, \text{ if } m \neq n. \end{cases}$$

Proof of Theorem 1 Consider a portion of the reduced homology sequence of the pair (D^n, S^{n-1}):

$$\begin{array}{ccccccc} \widetilde{H}_m(D^n) & \to & H_m(D^n, S^{n-1}) & \to & \widetilde{H}_{m-1}(S^{n-1}) & \to & \widetilde{H}_{m-1}(S^{n-1}) \\ \| & & \| & & & & \| \\ 0 & & \widetilde{H}_m(S^n) & & & & 0 \end{array}$$

[the equalities come from Sect. 12.3.A and Sect. 12.5 (part (2) of the theorem)]. From the exactness of the sequence, we have $\widetilde{H}_m(S^n) = \widetilde{H}_{m-1}(S^{n-1})$, which completes the proof, since for $n = 0$ the statement is known to us.

The isomorphism $\widetilde{H}_m(S^n) = \widetilde{H}_{m-1}(S^{n-1})$ constructed in the proof is generalized as the following suspension isomorphism.

Theorem 2. *For any topological space X and any n,*

$$\widetilde{H}_n(\Sigma X) = \widetilde{H}_{n-1}(X).$$

Proof. It follows from the reduced homology sequence of the pair (CX, X), the contractibility of CX, the equality $\Sigma X = CX/X$, and the (obvious) fact that (CX, X) is a Borsuk pair.

Remark. From the point of view of the Eckmann–Hilton duality (Lecture 4), this isomorphism is dual to $\pi_n(X) = \pi_{n-1}(\Omega X)$. Freudenthal's theorem (Lecture 10) is dual to a relation between the homology groups of X and ΩX which will be studied in Chap. 3.

EXERCISE 1 (A more precise version of Theorem 2). Let $f\colon \Delta^{n-1} \to X$ be a singular simplex of X. The composition

$$\Delta^n = C\Delta^{n-1} \xrightarrow{Cf} CX \xrightarrow{\text{proj.}} \Sigma X$$

is a singular simplex of ΣX, which we denote as Σf. Prove that the maps

$$\Sigma\colon C_{n-1}(X) \to C_n(\Sigma X),\ \sum_i k_i f_i \mapsto \sum_i k_i(\Sigma f_i)$$

commute with ∂ and induce the isomorphism $\widetilde{H}_{n-1}(X) \xrightarrow{\Sigma} \widetilde{H}_n(\Sigma X)$.

EXERCISE 2. Using Exercise 1, construct singular cycles representing the homology of spheres.

EXERCISE 3. Prove that a generator of a group $H_n(D^n, S^{n-1}) \cong \mathbb{Z}$ is represented by a one-simplex relative cycle $f\colon \Delta^n \to D^n$, where f is a homeomorphism.

EXERCISE 4. Construct a relative version of the isomorphism Σ of Exercise 1 and prove that it commutes with maps f_* and ∂_*.

13.2 Homology of Bouquets of Spheres and Other Bouquets

Theorem 1. *Let A be an arbitrary set and let S^n_α, $\alpha \in A$, be copies of the standard n-dimensional sphere. Then*

$$\widetilde{H}_n\left(\bigvee_{\alpha\in A} S^n_\alpha\right) = \begin{cases} \bigoplus_{\alpha\in A} \mathbb{Z}\alpha, & \text{if } m = n,\\ 0, & \text{if } m \neq n. \end{cases}$$

Here $\bigoplus_{\alpha\in A} \mathbb{Z}\alpha$ is the free Abelian group generated by the set A, that is, the sum of groups $\mathbb{Z}$ corresponding to the spheres of the bouquet.

Proof. This follows from Theorem 2 of Sect. 13.1, since $\bigvee_{\alpha\in A} S^n_\alpha$ is homotopy equivalent to the suspension of $\bigvee_{\alpha\in A} S^{n-1}_\alpha$ (and even is homeomorphic to this suspension if the latter is understood in the base point version), and for the bouquet of the zero-dimensional spheres the statement is true. Also, this follows from the next theorem.

Theorem 2. *If (X_α, x_α) are base point spaces which are Borsuk pairs, then for any m,*

$$\widetilde{H}_m\left(\bigvee_{\alpha\in A} X_\alpha\right) = \bigoplus_{\alpha\in A} \widetilde{H}_m(X_\alpha).$$

Proof. A bouquet is the quotient space of a disjoint union under the union of the base points.

EXERCISE 5. Construct the previous isomorphism at the level of cycles, establish its relative version, and prove the compatibility with f_* and ∂_*.

13.3 Maps of Spheres into Spheres and of Bouquets of Spheres into Bouquets of Spheres

Recall that a continuous map of S^n into S^n has a *degree*, an integer which characterizes its homotopy class (Sect. 10.3). A continuous map

$$g\colon \bigvee_{\alpha\in A} S^n_\alpha \to \bigvee_{\beta\in B} S^n_\beta$$

(where S^n_α, S^n_β are copies of the sphere S^n) has a whole *matrix of degrees* $\{d_{\alpha\beta} \mid \alpha \in A, \beta \in B\}$, where $d_{\alpha\beta}$ is the degree of the map

$$S^n \xrightarrow{i_\alpha} \bigvee S^n_\alpha \xrightarrow{g} \bigvee S^n_\beta \xrightarrow{p_\beta} S^n,$$

where i_α is the identity map of S^n onto S^n_α and p_β is the identity map of S^n_β of S^n and the constant map on the other spheres of the bouquet.

EXERCISE 6. Do the degrees $d_{\alpha\beta}$ determine the homotopy class of the map g?

Theorem. *The matrix of the map*

$$\begin{array}{ccc} H_n\left(\bigvee_{\alpha\in A} S^n_\alpha\right) & \xrightarrow{g_*} & H_n\left(\bigvee_{\beta\in B} S^n_\beta\right) \\ \| & & \| \\ \bigoplus_{\alpha\in A} \mathbb{Z}\alpha & & \bigoplus_{\beta\in B} \mathbb{Z}\beta \end{array}$$

coincides with $\{d_{\alpha\beta}\}$. *In particular, the map*

$$\begin{array}{ccc} H_n(S^n) & \xrightarrow{f_*} & H_n(S^n) \\ \| & & \| \\ \mathbb{Z} & & \mathbb{Z} \end{array}$$

induced by the map $f\colon S^n \to S^n$ *of degree* d *is the multiplication by* d.

Proof. Since Σ preserves the degrees, both for maps $S^n \to S^n$ and homomorphisms $H_n(S^n) \to H_n(S^n)$, our statement for some dimension n and some matrix $\{d_{\alpha\beta}\}$ implies our statement for dimension $n+1$ and the same matrix. On the other side, in dimension 0 everything is known (obvious). However, this does not resolve our problem: The trouble is that a base point–preserving map $S^0 \to S^0$ can have only degree 0 or 1. Thus, this suspension argumentation proves our theorem only for maps $g\colon \bigvee_\alpha S^n_\alpha \to \bigvee_\beta S^n_\beta$ which are n-fold suspensions of maps $\bigvee_\alpha S^0_\alpha \to \bigvee_\beta S^0_\beta$. Still, there are such maps, in particular, i_α and p_β. Thus, $(i_\alpha)_*\colon \mathbb{Z} \to \bigoplus_\alpha \mathbb{Z}\alpha$ takes a $c \in \mathbb{Z}$ into $c\alpha$ and $(p_\beta)_*\colon \bigoplus_\beta \mathbb{Z}\beta \to \mathbb{Z}$ takes $\sum c_\beta\beta$ into c_β. We want to prove that g_* takes $\sum_\alpha c_\alpha\alpha$ into $\sum_{\alpha,\beta} d_{\alpha\beta}c_\alpha\beta$, which (because of the computation of $(i_\alpha)_*$ and $(p_\beta)_*$ above) is the same as proving that $(p_\beta \circ g \circ i_\alpha)_*\colon \mathbb{Z} \to \mathbb{Z}$ is the multiplication by $d_{\alpha\beta}$. In other words, all we need is to prove that a map $S^n \to S^n$ of degree d

induces a homomorphism $H_n(S^n) \to H_n(S^n)$ which is the multiplication by d. Let us prove this (for $d = 1$, it is obvious).

Let $B = S_1^n \vee \cdots \vee S_d^n$, and let $r: S^n \to B$ be a map whose composition with each $p_k: B \to S^n (k = 1, \ldots, d)$ has degree 1 (obviously, such a map exists). Let $s: B \to S^n$ map every sphere of the bouquet onto S^n by the identity map. Then $s \circ r$ is a map of degree d. Since $\deg(p_k \circ r) = 1$, the homomorphism $r_*: \mathbb{Z} \to \mathbb{Z} \oplus \cdots \oplus \mathbb{Z}$ (d summands) takes a $c \in \mathbb{Z}$ into $(c, \ldots, c)$. Since $\deg(s \circ i_k) = 1$, the homomorphism $s_*: \mathbb{Z} \oplus \cdots \oplus \mathbb{Z} \to \mathbb{Z}$ takes $(c_1, \ldots, c_d)$ into $c_1 + \cdots + c_d$. Hence, $(s \circ r)_*(c) = c + \cdots + c = dc$, which is what we needed to prove.

13.4 Cellular Complex

Let X be a CW complex and let $X^n = \mathrm{sk}_n X$ $(n = 0, 1, 2, \ldots)$ be its skeletons. Let $\{e_\alpha^n \mid \alpha \in A_n\}$ be the set of all n-dimensional cells of X.

Pre-lemma. *The space X^n/X^{n-1} is homeomorphic to the bouquet $\bigvee_{\alpha \in A_n} S_\alpha^n$; if characteristic maps $f_\alpha: (D^n, S^{n-1}) \to (X^n, X^{n-1})$ are fixed, then there arises a canonical homeomorphism between X^n/X^{n-1} and $\bigvee_{\alpha \in A_n} S_\alpha^n$.*

Indeed, the maps f_α compose a continuous map $\coprod_\alpha (D_\alpha^n, S_\alpha^{n-1}) \to (X^n, X^{n-1})$, and it is obvious [follows from the properties of characteristic maps and Axiom (W)] that the map $(\coprod_\alpha D_\alpha^n)/(\coprod_\alpha S_\alpha^{n-1}) = \bigvee_\alpha S_\alpha^n \to X^n/X^{n-1}$ is a homeomorphism.

Lemma.

$$H_m(X^n, X^{n-1}) \cong \begin{cases} \text{free Abelian group generated by} \\ \qquad\qquad n\text{-dimensional cells of } X, \text{ if } m = n, \\ 0, \qquad\qquad\qquad\qquad\qquad\qquad \text{if } m \neq n. \end{cases}$$

Proof. $H_m(X^n, X^{n-1}) = \widetilde{H}_m(X^n/X^{n-1}) = \widetilde{H}_m(\bigvee_{\alpha \in A_n} S_\alpha^n)$.

The group $\mathcal{C}_n(X) = H_n(X^n, X^{n-1})$ is called the *groups of cellular chains* of X. The *cellular differential* or *cellular boundary operator* $\partial = \partial_n: \mathcal{C}_n(X) \to \mathcal{C}_{n-1}(X)$ is defined as the connecting homomorphism

$$\begin{array}{ccc} H_n(X^n, X^{n-1}) & \xrightarrow{\partial_*} & H_{n-1}(X^{n-1}, X^{n-2}) \\ \| & & \| \\ \mathcal{C}_n(X) & & \mathcal{C}_{n-1}(X) \end{array}$$

from the homology sequence of the triple (X^n, X^{n-1}, X^{n-2}) (see Exercise 7 from Sect. 12.4).

AN OBVIOUS FACT: $\partial_{n-1} \circ \partial_n: \mathcal{C}_n(X) \to \mathcal{C}_{n-1}(X)$ is zero (follows from the equality $\partial \circ \partial = 0$ in the singular complex).

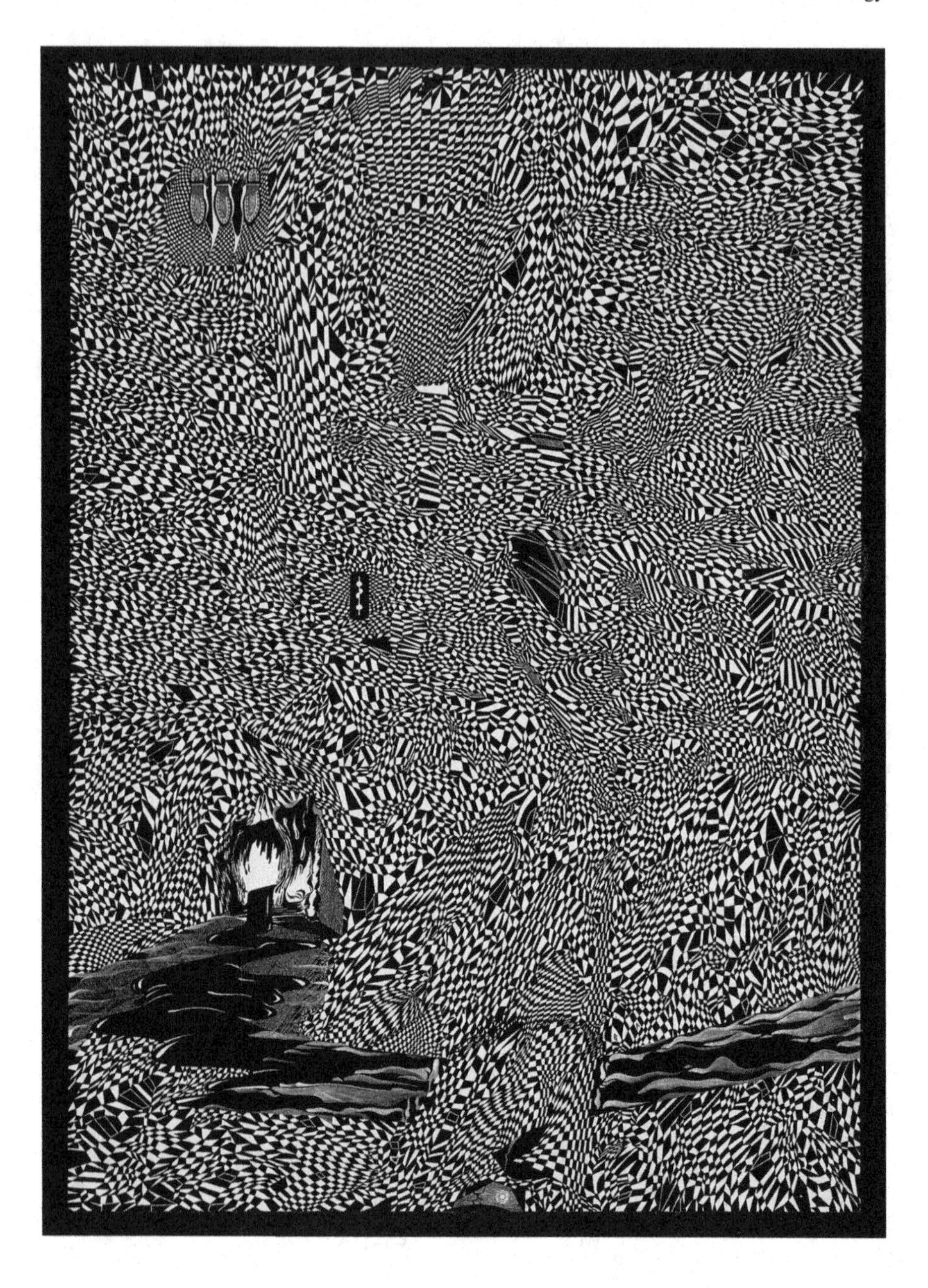

We obtain a complex

$$\dots \xrightarrow{\partial} \mathcal{C}_n(X) \xrightarrow{\partial} \dots \xrightarrow{\partial} \mathcal{C}_2(X) \xrightarrow{\partial} \mathcal{C}_1(X) \xrightarrow{\partial} \mathcal{C}_0(X) \xrightarrow{\partial} 0 \dots,$$

which is called the *cellular complex of* X. If we add the term $\mathcal{C}^{-1}(X) = \mathbb{Z}$ and augmentation $\partial_0 = \epsilon : \mathcal{C}_0(X) = H_0(X^0) \to \mathbb{Z}$, and then replace the notation $\mathcal{C}$ by $\widetilde{\mathcal{C}}$, we will get a definition of a reduced or augmented cellular complex.

There are two important things concerning cellular complexes. First, it is far from being as big as the singular complex; for example, for finite CW complexes the cellular chain groups are finitely generated. Moreover, not only the cellular chain groups, but also the cellular boundary operators have an explicit description that is easy to deal with. Second, we will prove that the homology of the cellular complex is the same as the homology of the singular complex. We will show how these results can be applied to calculating the homology of many classical CW complexes.

We will begin with the second part of this program.

13.5 Cellular Homology

Theorem. *For an arbitrary CW complex X, the homology of the cellular complex* $\{\mathcal{C}_n(X), \partial\}$ *coincides with the singular homology* $H_n(X)$.

Proof The proof consists of three steps.

Step 1. $H_n(X) = H_n(X^{n+1})$. Let $m > n$. From the exactness of homology sequence of the pair (X^{m+1}, X^m),

$$\begin{array}{ccccccc} H_{n+1}(X^{m+1}, X^m) & \to & H_n(X^m) & \to & H_n(X^{m+1}) & \to & H_n(X^{m+1}, X^m) \\ \| & & & & & & \| \\ 0 & & & & & & 0 \end{array}$$

we see that all homomorphisms

$$H_n(X^{n+1}) \to H_n(X^{n+2}) \to H_n(X^{n+3}) \to \dots$$

induced by the inclusion maps are isomorphisms. If X is finite dimensional, this settles our statement. In the general case, consider the map $H_n(X^{n+1}) \to H_n(X)$. Every $\alpha \in H_n(X)$ is represented by a finite sum of singular simplices, and every singular simplex is covered by a finite number of cells. This implies that α is represented by a cycle contained in some X^N, that is, belongs to the image of the map $H_n(X^N) \to H_n(X)$ (and we can assume that $N > n$). Since $H_n(X^{n+1}) \to H_n(X^N)$ is an isomorphism, α also belongs to the image of the map $H_n(X^{n+1}) \to H_n(X)$, so the latter is onto. Now let $\beta \in H_n(X^{n+1})$ be annihilated by the map $H_n(X^{n+1}) \to H_n(X)$. Then a cycle representing β is the boundary of some singular chain of X. But, as before, this chain must be contained in some X^N. Hence, β is also annihilated by some map $H_n(X^{n+1}) \to H_n(X^N)$, which is an isomorphism. Thus, $\beta = 0$ and our map $H_n(X^{n+1}) \to H_n(X)$ is one-to-one.

Step 2. $H_n(X^{n+1}) = H_n(X^{n+1}, X^{n-2})$. Let $m < n-1$. From the exactness of the homology sequence of the triple (X^{n+1}, X^m, X^{m-1}),

$$\begin{array}{ccccccc} H_n(X^m, X^{m-1}) & \to & H_n(X^n, X^{m-1}) & \to & H_n(X^n, X^m) & \to & H_{n-1}(X^m, X^{m-1}) \\ \| & & & & & & \| \\ 0 & & & & & & 0 \end{array}$$

we see that all homomorphisms

$$\begin{array}{ccccccc} H_n(X^{n+1}, X^{n-2}) & \leftarrow & H_n(X^{n+1}, X^{n-3}) & \leftarrow \cdots \leftarrow & H_n(X^{n+1}, X^{-1}) \\ & & & & \| \\ & & & & H_n(X^{n-1}) \end{array}$$

are isomorphisms. This proves our statement.

Step 3. $H_n(X^{n+1}, X^{n-2}) = \dfrac{\operatorname{Ker}(\partial_n : \mathcal{C}_n(X) \to \mathcal{C}_{n-1}(X))}{\operatorname{Im}(\partial_{n+1} : \mathcal{C}_{n+1}(X) \to \mathcal{C}_n(X))}$. Consider the diagram

$$\begin{array}{ccccccc} & & H_n(X^{n-1}, X^{n-2}) = 0 & & & & \\ \mathcal{C}_{n+1}(X) & & \downarrow & & & & \\ \| & & & & & & \\ H_{n+1}(X^{n+1}, X^n) & \xrightarrow{\partial_*} & H_n(X^n, X^{n-2}) & \xrightarrow{\alpha} & H_n(X^{n+1}, X^{n-2}) & \longrightarrow & H_n(X^{n+1}, X^n) \\ & \searrow^{\partial_{n+1}} & \downarrow \beta & & & & \| \\ & & H_n(X^n, X^{n-1}) = \mathcal{C}_n(X) & & & & 0 \\ & & \downarrow \partial_n & & & & \\ & & H_{n-1}(X^{n-1}, X^{n-2}) = \mathcal{C}_{n-1}(X) & & & & \end{array}$$

where the row is a fragment of the homology sequence of the triple (X^{n+1}, X^n, X^{n-2}) and the column is a fragment of the homology sequence of the triple (X^n, X^{n-1}, X^{n-2}); in particular, both are exact. There are two zeroes in the diagram, and they show that α is an epimorphism, and β is a monomorphism. From this (and again the exactness of the sequences) we obtain

$$\begin{aligned} H_n(X^{n+1}, X^{n-2}) &= H_n(X^n, X^{n-2})/Ker\alpha = H_n(X^n, X^{n-2})/\operatorname{Im}\partial_* \\ &= \beta(H_n(X^n, X^{n-2}))/\beta(\operatorname{Im}\partial_*) = \operatorname{Im}\beta/\operatorname{Im}(\beta \circ \partial_*) \\ &= \operatorname{Ker}\partial_n/\operatorname{Im}\partial_{n+1}. \end{aligned}$$

This completes step 3, and the combination of the three steps gives the isomorphism we need.

13.6 A Closer Look at the Cellular Complex

We already know that for a CW complex X, the group $\mathcal{C}_n(X)$ is isomorphic to a free Abelian group generated by n-dimensional cells of X. But the isomorphism is not genuinely canonical: It depends on a choice of characteristic maps of cells, which is not convenient because usually characteristic maps are not fixed—we know only that they exist. Actually, what we need to fix for every cell is not a characteristic map, but an *orientation*. A characteristic map of an n-dimensional cell establishes an isomorphism between two groups isomorphic to $\mathbb{Z}$: $H_n(D^n, S^n - 1) = \widetilde{H}_n(S^n)$ and $H_n(X^{n-1} \cup e, X^{n-1}) = \widetilde{H}_n((X^{n-1} \cup e)/X^{n-1})$ or $\widetilde{H}_n(X^n/(X^n - e))$ (which is the same group). One can say that the orientation of e is a choice of a generator in $\widetilde{H}_n((X^{n-1} \cup e)/X^{n-1}) \cong \mathbb{Z}$. Geometrically this indeed is an orientation: Say, if $n = 1$, then a choice of orientation is a choice of a direction of an arrow on e. In other words, characteristic maps f and $f \circ r$ always determine opposite orientations. (Zero-dimensional cells have canonical orientations.)

Thus, chains in $\mathcal{C}_n(X)$ can be presented as finite integral linear combinations of oriented n-dimensional cells, $\sum k_i e_i$. An orientation change for e_i results in a sign change for k_i.

There also exists a good description of the boundary homomorphism $\partial_{n+1}: \mathcal{C}_{n+1}(X) \to \mathcal{C}_n(X)$. Let e and f be cells of X of dimensions $n + 1$ and n. In the homology sequence of the triple $(X^n \cup e, X^n, X^n - f)$, there is a homomorphism

$$\mathbb{Z} \cong H_{n+1}(X^n \cup e, X^n) \xrightarrow{\partial_*} H_n(X^n, X^n - f) \cong \mathbb{Z}.$$

The choice of the isomorphisms with $\mathbb{Z}$ corresponds to the orientations of the cells e, f. Every homomorphism $\mathbb{Z} \to \mathbb{Z}$ is a multiplication by some integer. This integer is called the *incidence number* of the *oriented cells* e and f and is denoted as $[e : f]$ (certainly, if $\overline{e}$ and f are disjoint, then $[e : f] = 0$). The orientation change for any of the cells e and f results in the sign change for $[e : f]$.

Theorem. *Let e be an oriented $(n+1)$-dimensional cell of X regarded as an element of $\mathcal{C}_{n+1}(X)$. Then*

$$\partial_{n+1}(e) = \sum_f [e : f] f,$$

where the sum is taken over all n-dimensional cells of X with fixed orientations. [This sum is always finite: The intersection $\overline{e} \cap f$ may be nonempty for only finitely many n-dimensional cells f—this is Axiom (C).]

EXERCISE 7. Prove this. *Recommendation:* It may be useful to consider the commutative diagram

$$\begin{array}{ccc} H_{n+1}(X^n \cup e, X^n) & \xrightarrow{\partial_*} & H_n(X^n, X^n - f) \\ \downarrow & & \uparrow \\ H_{n+1}(X^{n+1}, X^n) & \xrightarrow{\partial_{n+1}} & H_n(X^n, X^{n-1}), \end{array}$$

where the vertical maps are induced by the inclusion maps between pairs.

(A clarification is needed and possible in the case when $n = 0$. An oriented one-dimensional cell e is a path joining two zero-dimensional cells, f_0 and f_1. Then $\partial e = f_1 - f_0$; in particular, if $f_0 = f_1$, then $\partial e = 0$.)

The description of the boundary map in the preceding theorem motivates a better understanding of the incidence numbers. They can be described as degrees of maps $S^n \to S^n$. Namely, if $\varphi: S^n \to X^n$ is an attaching map for e (determined by a certain characteristic map for e) and $\psi: X^n/(X^n - f) = \overline{f}/\dot{f} \to S^n$ is a homeomorphism determined by a certain characteristic map for F, then $[e : f]$ is nothing but the degree of the map

$$S^n \xrightarrow{\varphi} X^n \xrightarrow{\text{proj.}} X^n/(X^n - f) \xrightarrow{\psi} S^n.$$

The description of the degree of a map $S^n \to S^n$ given in Sect. 10.3 may be used as a geometric description of incidence numbers. Namely, take a regular value $x \in f$ of the attaching map $\varphi: S^n \to X^n$ [rather of the map $\varphi: \varphi^{-1}(f) \to f$] and compute the "algebraic number" of inverse images of x (that is, the number of inverse images where φ preserves the orientation minus the number of inverse images where φ reverses the orientation); this is $[e : f]$.

Having this in mind, we can give our theorem an aggressively tautological form: The boundary of a cell is the sum of cells which appear in the boundary of this cell with coefficients equal to the multiplicity of their appearance in this boundary.

13.7 First Applications

Theorem 1. *If the number of n-dimensional cells of a CW complex X is N, then the group $H_n(X)$ is generated by at most N generators; in particular, the nth Betti number $B_n(X)$ does not exceed N. For example, if X does not have n-dimensional cells at all, then $H_n(X) = 0$; in particular, if X is finite dimensional, then $H_n(X) = 0$ for all $n >$* $\dim X$. (Compare with homotopy groups!)

It follows directly from previous results.

Algebraic Lemma (Euler–Poincaré). *Let*

$$\dots \xrightarrow{\partial_{n+2}} C_{n+1} \xrightarrow{\partial_{n+1}} C_n \xrightarrow{\partial_n} C_{n-1} \xrightarrow{\partial_{n-1}} \dots$$

be a complex with the "total group" $\bigoplus_n C_n$ finitely generated. Let c_n be the rank of the group C_n and h_n be the rank of the homology group H_n. Then

$$\sum_n (-1)^n c_n = \sum_n (-1)^n h_n.$$

EXERCISE 8. Prove this.

Corollary. *Let X be a finite CW complex, and let c_n be the number of n-dimensional cells of X. Then*

$$\sum_n (-1)^n c_n = \sum_n (-1)^n B_n(X).$$

Thus, the number $\sum_n (-1)^n c_n$ does not depend on the CW structure; it is determined by the topology (actually, be the homotopy type) of X. This number is called the *Euler characteristic* of X and is traditionally denoted by $\chi(X)$.

Historical Remark. This number is attributed to Euler because of the Euler polyhedron theorem, which states that for every convex polyhedron in space, the numbers V, E, and F of vertices, edges, and faces are connected by the relation $V - E + F = 2$. Certainly, this is a computation of the Euler characteristic of the surface of the polyhedron, that is, of the sphere. It is worth mentioning that Euler was not the first to prove this theorem: It was proved, a century before Euler, by Descartes.

Now let us revisit the excision theorem and the Mayer–Vietoris sequence (Exercises 13 and 14 of Lecture 12).

Theorem 2 (Excision Theorem). *Let X be a CW complex and let A, B be CW subcomplexes of X such that $A \cup B = X$. Then (for every n)*

$$H_n(X, A) = H_n(B, A \cap B).$$

Indeed, X/A and $B/(A \cap B)$ are the same as CW complexes.

Theorem 3 (Mayer–Vietoris Sequence). *Let X be a CW complex and let A, B be CW subcomplexes of X such that $A \cup B = X$. Then there exists an exact sequence*

$$\dots \to H_n(A \cap B) \to H_n(A) \oplus H_n(B) \to H_n(X) \to H_{n-1}(A \cap B) \to \dots$$

(see the description of maps in Exercise 14 of Lecture 12).

Proof. Let $Y = (A \times 0) \cup ((A \cap B) \times I) \cup (B \times 1) \subset X \times I$ and let $C \subset Y$ be $(A \cap B) \times I$. Then Y/C and $\Sigma(A \cap B)$ (actually with the vertices merged; this slightly affects the case of dimension 0) are the same CW complexes (see schematic picture in Fig. 63).

Notice, in addition, that $C = A \coprod B$ and $Y \sim X$. The last homotopy equivalence is established by the obvious map $f: Y \to X$ (the restriction of the projection $X \times I \to X$) and a map $g: X \to Y$ which is defined in the following way. The homotopy $h_t: A \cap B \to Y$, $h_t(x) = (x, 1 - t)$ is extended, by Borsuk's theorem, to a homotopy $H_t: A \to Y$ of the map $A \to Y$, $x \to (x, 1)$. Then maps $H_1: A \to Y$ and $B \to Y$, $x \mapsto (x, 0)$ agree on $A \cap B$ and hence compose a map $X \to Y$; this is g; the relations $f \circ g \sim \text{id}$, $g \circ f \sim \text{id}$ are obvious. Thus, $H_n(Y) = H_n(X)$, $H_n(C) = H_n(A) \oplus H_n(B)$, and $H_n(Y, C) = H_{n-1}(A \cap B)$ (with small corrections in dimension 0), and the homology sequence of the pair (Y, C) is the Mayer–Vietoris sequence of the triad $(X; A, B)$.

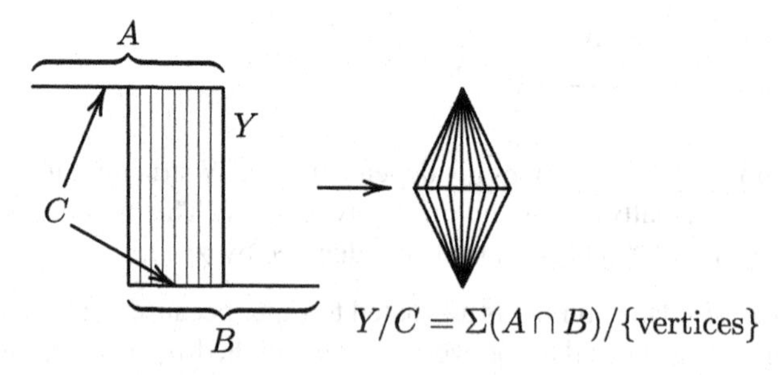

Fig. 63 To the proof of the Mayer–Vietoris theorem

13.8 Some Calculations

A: Spheres

We already know the homology of spheres, but let us calculate them again for practice in the technique based on cellular complexes. The sphere S^n has a CW structure with two cells, of dimensions 0 and n. Thus (if $n > 0$), $\mathcal{C}_0(S^n) = \mathcal{C}_n(S^n) = \mathbb{Z}$, and all other cellular chain groups are trivial. The differential ∂ has to be 0 (if $n > 1$, then this follows from the "dimension argumentations"; for $n = 1$, we use the remark after Exercise 7); hence,

$$H_i(S^n) = \mathcal{C}_i(S^n) = \begin{cases} \mathbb{Z}, \text{ if } i = 0, n, \\ 0, \text{ if } i \neq 0, n. \end{cases}$$

EXERCISE 9. Prove this using another CW decomposition of S^n described in Sect. 5.4.

B: Projective Spaces

The cases of complex, quaternionic, and Cayley projective spaces are not more difficult than the cases of spheres: For the CW structures described in Sect. 5.4, there are no cells of adjacent dimensions, the differential ∂ is trivial, and the homology groups coincide with the cellular chain groups. Thus,

$$H_i(\mathbb{C}P^n) = \begin{cases} \mathbb{Z}, \text{ if } i = 0, 2, 4, \dots, [2n, \text{if } n \text{ is finite}], \\ 0 \quad \text{for all other } i; \end{cases}$$

$$H_i(\mathbb{H}P^n) = \begin{cases} \mathbb{Z}, \text{ if } i = 0, 4, 8, \dots [, 4n, \text{if } n \text{ is finite}], \\ 0 \quad \text{for all other } i; \end{cases}$$

$$H_i(\mathbb{C}\mathbf{a}P^2) = \begin{cases} \mathbb{Z}, \text{ if } i = 0, 8, 16, \\ 0 \quad \text{for all other } i. \end{cases}$$

The real case is more complicated, since $\mathbb{R}P^n$ has cells $e^0, e^1, e^2, \dots$ [, e^n if n is finite].

Lemma. $[e^{i+1} : e^i] = \begin{cases} \pm 2, \text{ if } n \text{ is odd}, \\ 0, \quad \text{if } n \text{ is even}. \end{cases}$

Proof. The attaching map $f: S^i \to \mathbb{R}P^i$ is the standard twofold covering. The inverse image of (actually, any) point of $\mathbb{R}P^i$ consists of two points, and the restrictions of f to neighborhoods of these points are related by the antipodal map $S^i \to S^i$.

This antipodal map preserves the orientation if i is odd and reverses the orientation if i is even. Thus, the contributions of these two points in $[e^{i+1}, e^i]$ have the same sign if i is odd and have different signs if i is even. This implies the formula of the lemma.

Thus, the cellular complex of $\mathbb{R}P^n$ is as shown below.

$$\begin{array}{lccccccccccccc}
\text{(if } n \text{ is odd)} & 0 & & 2 & & 0 & & & & & & & & \\
& & & & & & & 0 & & 2 & & 0 & & \\
\text{(if } n \text{ is even)} & 2 & & 0 & & 2 & & & & & & & & \\
& \mathbb{Z} & \longrightarrow & \mathbb{Z} & \longrightarrow & \mathbb{Z} & \longrightarrow & \cdots & \longrightarrow & \mathbb{Z} & \longrightarrow & \mathbb{Z} & \longrightarrow & \mathbb{Z} \\
& C_n & & C_{n-1} & & C_{n-2} & & & & C_2 & & C_1 & & C_0
\end{array}$$

Since $\mathrm{Im}(\xrightarrow{0}) = 0$, $\mathrm{Ker}(\xrightarrow{0}) = \mathbb{Z}$, $\mathrm{Im}(\xrightarrow{2}) = 2\mathbb{Z}$, and $\mathrm{Ker}(\xrightarrow{2})$ $= 0$, the factorization yields

$$H_i(\mathbb{R}P^n) = \begin{cases} \mathbb{Z}, & \text{if } i = 0 \text{ or } i = n \text{ and } n \text{ is odd,} \\ \mathbb{Z}_2, & \text{if } i \text{ is odd and } i < n, \\ 0 & \text{in all other cases.} \end{cases}$$

EXERCISE 10. Find the Euler characteristics of all finite-dimensional projective spaces.

C: Grassmann Manifolds

Again, in the complex and quaternion cases, there are no cells of adjacent dimensions, so the ith homology group is a free Abelian group of rank (= Betti number) equal to the number of i-dimensional cells. The Betti numbers are as follows. For i odd, $B_i(G(n,k)) = 0$; for i even, this is the number of Young diagrams of $\frac{i}{2}$ cells contained in the $k \times (n-k)$ rectangle. For quaternionic Grassmann manifolds everything is doubled: $B_i(\mathbb{H}G(n,k)) = B_{i/2}(\mathbb{C}G(n,k))$; in particular, $B_i(\mathbb{H}G(n,k)) = 0$ if i is not divisible by 4.

In the real case the situation is more complicated.

EXERCISE 11. Let Δ and Δ' be two Young diagrams with i and $i-1$ cells contained in the $k \times (n-k)$ rectangle. Prove that if $\Delta' \not\subset \Delta$, then $[e(\Delta) : e(\Delta')] = 0$. If $\Delta' \subset \Delta$ and the difference $\Delta - \Delta'$ consists of one cell with the coordinates (s,t), then

$$[e(\Delta) : e(\Delta')] = \begin{cases} \pm 2, & \text{if } s+t \text{ is even,} \\ 0, & \text{if } s+t \text{ iodd.} \end{cases}$$

Use this for computation of the homology of $G(n,k)$ with reasonably small n, k. Also, compute $H_{k(n-k)}(G(n,k))$.

EXERCISE 12. Find incidence numbers for the case of the manifold $G_+(n,k)$. In particular, find $H_{k(n-k)}(G_+(n,k))$.

D: Flag Manifolds

Again, the complex and quaternionic cases are relatively easy. The reader can try to investigate the real case.

E: Classical Surfaces

Classical surfaces with holes are homotopy equivalent to bouquets of circles, so we will consider classical surfaces without holes. The cellular complex for such a surface has the form

$$\begin{array}{ccccc} \mathbb{Z} & \xrightarrow{\partial_2} & \mathbb{Z} \oplus \cdots \oplus \mathbb{Z} & \xrightarrow{\partial_1} & \mathbb{Z} \\ \mathcal{C}_2 & & \mathcal{C}_1 & & \mathcal{C}_0, \end{array}$$

where the number of the summands $\mathbb{Z}$ in $\mathcal{C}_1$ is $2g$, $2g+1$, or $2g+2$ if our surface is a sphere with g handles, a projective plane with g handles, or a Klein bottle with g handles, respectively. The differential ∂_1 is zero (every one-dimensional cell has equal endpoints). To find ∂_2, we consider the construction of the classical surface from a polygon (Sect. 1.10). Each of the $2g$ one-dimensional cells arising from the handles is obtained by attaching differently oriented sides of the polygon, so the incidence numbers of the two-dimensional cell with each of these 1-cells is 0. On the other hand, the other one-dimensional cells (if there are any) are obtained by attaching coherently oriented sides, and the incidence number with these cells is 2. Thus,

$$\partial_2(1) = \begin{cases} (0,\dots,0) & \text{for a sphere with } g \text{ handles,} \\ (0,\dots,0,2) & \text{for a projective plane with } g \text{ handles,} \\ (0,\dots,0,2,2) & \text{for the Klein bottle with } g \text{ handles.} \end{cases}$$

This leads to the results for homology:

$$H_0(X) = \mathbb{Z} \text{ always,}$$

$$H_1(X) = \begin{cases} \underbrace{\mathbb{Z} \oplus \cdots \oplus \mathbb{Z}}_{2g}, & \text{if X is a sphere with } g \text{ handles,} \\ \underbrace{\mathbb{Z} \oplus \cdots \oplus \mathbb{Z}}_{2g} \oplus \mathbb{Z}_2, & \text{if } X \text{ is a projective plane with } g \text{ handles,} \\ \underbrace{\mathbb{Z} \oplus \cdots \oplus \mathbb{Z}}_{2g+1} \oplus \mathbb{Z}_2, & \text{if } X \text{ is a Klein bottle with } g \text{ handles,} \end{cases}$$

$$H_2(X) = \begin{cases} \mathbb{Z}, & \text{if } X \text{ is a sphere with handles,} \\ 0 & \text{in all other cases} \end{cases}$$

EXERCISE 13. Find the Euler characteristics of classical surfaces.

13.9 Chain Maps of Cellular Complexes

Let $h: X \to Y$ be a cellular map of a CW complex into a CW complex. Then $h(X^n) \subset Y^n$ for all n, and hence h induces a map $H_n(X^n, X^{n-1}) \to H_n(Y^n, Y^{n-1})$, that is, $C_n(X) \to C_n(Y)$, which we denote as $h_\#$ or $h_{\#n}$. Such maps induce a homomorphism between cellular complexes of X and Y, and the induced homology map is just $h_*: H_n(X) \to H_n(Y)$. To prove this, we need to consider every step of the proof of the theorem in Sect. 13.5, and to consider maps between the diagrams in these steps for X into similar diagram for Y. The commutativity of (three-dimensional) diagrams arising will imply our statement.

We can add that if $c = \sum_i k_i e_i \in C_n(X)$, where e_i are n-dimensional cells of X, then $h_\#(c) = \sum_i k_i \big(\sum_j d_h(e_i, f_j) f_j\big)$, where the f_j are n-dimensional cells of Y and the number $d_h(e,f)$ is defined with the help of characteristic maps φ and ψ of e and f as the degree of the map

$$S^n = D^n/S^{n-1} \xrightarrow{\varphi} X^n/X^{n-1} \xrightarrow{h} Y^n/Y^{n-1} \xrightarrow{\text{proj.}} Y^n/(Y^n - f) \xrightarrow{\psi^{-1}} D^n/S^{n-1} = S^n.$$

Using the description of the degree of a map $S^n \to S^n$ in Sect. 10.3, we can say that $d_h(e,f)$ is the algebraic number of inverse images of a regular value $x \in f$ of the map $h: e \cap h^{-1}(f) \to f$.

Certainly, this construction works only for cellular maps, but it is not a big deal, since every continuous map is homotopic to a cellular map. (Not a big deal? We will cast a doubt on this statement in Lecture 16.) Thus, one can say that the cellular theory can be used as a substitute for the singular theory. But without the singular theory (which is topologically invariant from the very beginning) we would have had to prove that homeomorphic CW complexes have isomorphic homology groups.

13.10 Classical Complex

A cellular complex appears especially attractive when a CW structure is actually a triangulation (see Sect. 5.8). We consider a triangulated space X with an additional structure (a substitute for fixing characteristic maps): We suppose that the set of vertices of X is *ordered*, or, at least, vertices of every simplex are ordered in such a way that the ordering of vertices of a face of a simplex is always compatible with the ordering of vertices of this simplex. We refer to such triangulations as ordered triangulations. (For example, the barycentric subdivision of any triangulation is naturally ordered: Vertices of simplices of a barycentric subdivisions are centers of faces of simplices of the given triangulation and these are ordered by the dimensions of the faces.)

For a simplex with the vertices ordered, there is a canonical affine homeomorphism of the standard simplex onto this simplex; this homeomorphism can be regarded as a singular simplex of an ordered triangulated space X. We obtain a set of special singular simplices of X, and it is clear that faces of "special singular simplices" are also special. By this reason, linear combinations of special singular simplices form a subcomplex of the singular complex, and it is also clear that this subcomplex is precisely the cellular complex of the triangulation.

Historically, the complex described above is the first chain complex of a (orderly triangulated) topological space ever considered. It can be described very directly: Chains are integral linear combinations of simplices (remember the ordering!), and the boundary is given by the very familiar formula $\partial\left(\sum_i k_i s_i\right) = \sum_i k_i\left(\sum_j (-1)^j \Gamma_j s_i\right)$, where the s_i are simplices of our triangulation and the $\Gamma_j s_i$ are their faces. Obviously, the inclusion of the classical complex into the singular complex induces the isomorphism of the homology groups [to show this, the only thing we need to add to what we already know is that n-dimensional simplices regarded as singular simplices are relative cycles of (X^n, X^{n-1}), and their homology classes form the usual basis in $\mathcal{C}_n(X) = H_n(X^n, X^{n-1})$].

For the classical chain groups, the notation $\mathcal{C}_n^{\text{class}}(X)$ is often used.

Historical Remark. The classical definition of homology created the necessity of proving a topological invariance theorem: Homeomorphic triangulated spaces have isomorphic homology groups. The initial proof, due to J. Alexander, was long and complicated (hundreds of pages in old topology textbooks). There was an attempt to deduce the topological invariance of classical homology from the so-called *Hauptvermutung* (German for *main conjecture*) of combinatorial topology: Any two triangulations of a topological space have simplicially equivalent subdivisions. But the *Hauptvermutung* turns out to be false: The first counterexample was found by J. Milnor in 1961, and many other counterexamples were constructed later, in particular for simply connected smooth manifolds. The whole problem of topological invariance disappeared mysteriously when singular homology was defined. The first definition of singular homology was given by O. Veblen in the late 1920s but became broadly known some 10 years later.

EXERCISE 14. Using the classical complex, find the Betti numbers of the skeletons of the standard simplex. (Make your computations as explicit as possible.)

EXERCISE 15. (An algebraic lemma) Let $\{C_n, \partial_n\}, \{C'_n, \partial'_n\}$ be two positive $C_n = C'_n = 0$ (for $n < 0$) complexes of *free* Abelian groups, and let f be a homomorphism of the first complex into the second one. Prove that if f_{*n} is an isomorphism for all, then f is a homotopy equivalence. Deduce that the classical complex is homotopy equivalent to the singular complex.

EXERCISE 16. Prove that the cellular complex of a CW complex is homotopy equivalent to its singular complex. (There are several different ways of proving that, so we refrain from giving any hint.)

13.11 The Singular Complex as a CW Complex

We finish this lecture with a construction which may seem amusing to some readers but actually is quite useful (we will use it in the beginning of the next lecture). Let X be a topological space, and let $\mathrm{Sing}_n(X)$ be the set of all n-dimensional singular simplices of X. Consider a (monstrous, we agree) topological space $Y = \coprod_{n=0}^{\infty} \coprod_{\alpha\in\mathrm{Sing}_n(X)} \Delta^n_\alpha$ (where Δ^n_α is a copy of the standard simplex Δ^n) and make, for every n and every α, the identification $\Gamma_i\Delta^n_\alpha = \Delta^{n-1}_{\Gamma_i\alpha}$ (both are copies of Δ^{n-1} contained in Y). We denote the resulting space as $\mathcal{S}ing(X)$. This space has a natural CW structure (images of $\mathrm{Int}\,\Delta^n_\alpha \subset Y$ are cells of $\mathcal{S}ing(X)$ and the maps $\Delta^n \xrightarrow{=} \Delta^n_\alpha \xrightarrow{\subset} Y \xrightarrow{\mathrm{proj.}} \mathcal{S}ing(X)$ can be taken for characteristic maps. [Notice that although the cells of $\mathcal{S}ing(X)$ look like simplices, its CW structure is not a triangulation: The intersection of closed simplices is not a face.] There is also a natural map $\mathcal{S}ing(X) \to X$, which induces the identity homomorphism in homology [just take $\alpha\colon \Delta^n \to X$ on $\Delta^n_\alpha \subset \mathcal{S}ing(X)$].

It is immediately obvious that the cellular complex of $\mathcal{S}ing(X)$ is the same as the singular complex of X; in particular, $H_n(\mathcal{S}ing(X)) = H_n(X)$ for all X. Actually, the spaces $\mathcal{S}ing(X)$ and X are weakly homotopy equivalent (and homotopy equivalent if X is a CW complex). We will see that later.

Let us add that the $\mathcal{S}ing$ construction is natural in the sense that a continuous map $X \to Y$ gives rise to a cellular map $\mathcal{S}ing(X) \to \mathcal{S}ing(Y)$ with the same induced map in homology. Also, if $A \subset X$, then $\mathcal{S}ing(A) \subset \mathcal{S}ing(X)$ and there arises a continuous map

$$(\mathcal{S}ing(X), \mathcal{S}ing(A)) \to (X, A)$$

which induces isomorphisms

$$H_n(\mathcal{S}ing(X), \mathcal{S}ing(A)) \to H_n(X, A).$$

Lecture 14 Homology and Homotopy Groups

The connection between homology and homotopy groups is seen always from the preliminary description of homology in the beginning of Lecture 12: Spheroids are cycles and homotopical spheroids are homological cycles. This suggests that there must be a natural map from homotopy groups into homology groups. This map, called the *Hurewicz homomorphism*, is the main subject of this lecture. We will see that the connection between homotopy and homology groups is deeper than it may seem at the beginning, but we also will show examples which should serve as a warning to a reader who expects too much of this connection.

14.1 Homology and Weak Homotopy Equivalences

Theorem. *If $f: X \to Y$ is a weak homotopy equivalence, then $f_*: H_n(X) \to H_n(Y)$ is an isomorphism for all n.*

Proof. Since both weak homotopy equivalences and homology homomorphisms are homotopy equivalent, we can replace the map f by the inclusion map $X \to Cyl(f)$ of X into the mapping cylinder of f (see Sects. 2.3 and 3.3). Because of this, we can assume that the given map f is an inclusion, so we have a pair (Y, X). Also, we have a pair $(Sing(Y), (Sing(X))$ and a continuous map $h: (Sing(Y), Sing(X)) \to (Y, X)$ which induces isomorphisms

$$h_*: H_n(Sing(Y), Sing(X)) \to H_n(Y, X)$$

(see Sect. 13.11).

On the other hand, since f is a weak homotopy equivalence, the map $f_*: \pi(Sing(Y), X) \to \pi(Sing(Y), Y)$ is a bijection, which means that the map $h: Sing(Y) \to Y$ is homotopic to a map whose image is contained in X. Hence, the map $h_*: H_n(Sing(Y), Sing(X)) \to H_n(Y, X)$ is zero, which shows that $H_n(Y, X) = 0$ for all n. By exactness of the homology sequence of the pair (Y, X), this shows that all the homomorphisms $f_*: H_n(X) \to H_n(Y)$ are isomorphisms.

Recall that according to another result from Sect. 11.4, a map is a weak homotopy equivalence if and only if it induces an isomorphism in homotopy groups. Because of this, our theorem assumes the following memorable form.

Corollary. *If a continuous map induces an isomorphism between homotopy groups, then it also induces an isomorphism between homology groups.*

This will be further developed in the last section of this lecture.

To finish this section, we will formulate some exercises which will show that some statements looking similar to the preceding theorem and corollary are actually false.

EXERCISE 1. Prove that the spaces S^2 and $S^3 \times \mathbb{C}P^\infty$ have isomorphic homotopy groups but nonisomorphic homology groups. Same for the spaces $S^m \times \mathbb{R}P^n$ and $S^n \times \mathbb{R}P^m$ with $m \neq n, m \neq 1, n \neq 1$. (Compare with Exercises 5 and 6 in Lecture 11.)

EXERCISE 2. Prove that the spaces $S^1 \times S^1$ and $S^1 \vee S^1 \vee S^2$ have isomorphic homology groups but nonisomorphic homotopy groups.

EXERCISE 3. Prove that the Hopf map $S^3 \to S^2$ induces a trivial homomorphism in reduced homology groups but a nontrivial homomorphism in homotopy groups.

EXERCISE 4. Prove that the projection map $S^1 \times S^1 \to (S^1 \times S^1)/(S^1 \vee S^1) = S^2$ induces a trivial homomorphism in homotopy groups but a nontrivial homomorphism in reduced homology groups.

14.2 The Hurewicz Homomorphism

Let X be a topological space with a base point X_0. Let s_n be the canonical generator of the group $H_n(S^n) = \mathbb{Z}, n = 1, 2, \ldots$. For a $\varphi \in \pi_n(X, x_0)$ put

$$h(\varphi) = f_*(s_n) \in H_n(X),$$

where $f: S^n \to X$ is a spheroid of the class φ [obviously, $h(\varphi)$ does not depend on the choice of the spheroid f]. The function $\varphi \mapsto h(\varphi)$ is a homomorphism

$$h: \pi_n(X, x_0) \to H_n(X).$$

Indeed, let the spheroid f be the sum of spheroids $f', f'': S^n \to X$, that is, f is the composition

$$S^n \xrightarrow{\mu} S^n \vee S^n \xrightarrow{f' \vee f''} X$$

(see Fig. 37). Then $\mu_*(s) = s' + s''$ where $s', s'' \in H_n(S^n \vee S^n)$ are generators corresponding to the two spheres of the bouquet, and $f_*(s) = (f' \vee f'')_*(s' + s'') = f'_*(s) + f''_*(s)$.

This homomorphism is called the Hurewicz homomorphism; it is natural with respect to continuous maps (taking a base point into a base point).

EXERCISE 5. Prove that the diagram

$$\begin{array}{ccc} \pi_n(X, x_0) & \xrightarrow{u_\#} & \pi_n(X, x_1) \\ & {\scriptstyle h}\searrow \quad \swarrow{\scriptstyle h} & \\ & H_n(X) & \end{array}$$

is commutative for any path u joining x_0 with x_1.

Theorem (Hurewicz). *Let $\pi_0(X, x_0) = \cdots = \pi_{n-1}(X, x_0) = 0$, where $n \geq 2$. Then $H_1(X) = \cdots = H_{n-1}(X, x_0) = 0$ and $h: \pi_n(X, x_0) \to H_n(X, x_0)$ is an isomorphism.*

Proof. By the theorem of Sect. 11.6, there exists a CW complex weakly homotopy equivalent to X. Since a weak homotopy equivalence induces isomorphisms both in homotopy groups and in homology groups (the first by Sect. 13.11, the second by Sect. 14.1), we can assume that X itself is a CW complex. Then Sect. 5.9 allows us to make an additional assumption that X has one vertex and no cells of dimensions $1, \ldots, n-1$. This already shows that $H_1(X) = \cdots = H_{n-1}(X) = 0$ (Theorem 1 in Sect. 13.7), and $H_n(X) = C_n(X)/\operatorname{Im}\partial_{n-1}$ is not different from $\pi_n(X)$ according to the theorem in Sect. 11.3.

Corollary (The Inverse Hurewicz Theorem). *If X is simply connected and $H_2(X) = \cdots = H_{n-1}(X) = 0$ ($n \geq 2$), then $\pi_2(X) = \cdots = \pi_{n-1}(X) = 0$ and $h: \pi_n(X) \to H_n(X)$ is an isomorphism.*

Together these theorems mean that the first nontrivial homotopy and homology groups of a simply connected space occur in the same dimension and are isomorphic.

EXERCISE 6. Prove that a simply connected CW complex with the same homology groups as S^n is homotopy equivalent to S^n. [*Hint:* Apply Whitehead's theorem to a spheroid $S^n \to X$ representing a generator of the group $\pi_n(X) \cong \mathbb{Z}$.] Do the same for the bouquet of spheres of the same dimensions.

Remark. Thus, we see that the triviality of the homotopy groups, as well as the triviality of the homology groups, implies the homotopy triviality (contractibility) of a simply connected CW complex. At the same time, we have the examples which show that neither the triviality of induced homotopy groups homomorphisms nor the triviality of induced homology homomorphisms secures homotopy triviality of a continuous map. It turns out that even these two trivialities together do not imply the homotopy triviality of a continuous map.

EXERCISE 7. Prove that the composition

$$S^1 \times S^1 \times S^1 \xrightarrow{\text{proj.}} (S^1 \times S^1 \times S^1)/\operatorname{sk}_2(S^1 \times S^1 \times S^1) = S^3 \xrightarrow{\text{Hopf}} S^2$$

induces a trivial map of both homotopy and homology groups but is not homotopic to a constant map.

EXERCISE 8. Do the same for the map

$$S^{2n-2} \times S^3 \xrightarrow{\text{proj.}} (S^{2n-2} \times S^3)/(S^{2n-2} \vee S^3) = S^{2n+1} \xrightarrow{\text{Hopf}} \mathbb{C}P^n.$$

14.3 The Case $n = 1$

Theorem (Poincaré). *For an arbitrary path connected space X, the Hurewicz homomorphism $h: \pi_1(X) \to H_1(X)$ is an epimorphism whose kernel is the commutator subgroup $[\pi_1(X), \pi_1(X)]$ of the group $\pi_1(X)$. Thus,*

$$H_1(X) \cong \pi_1(X)/[\pi_1(X), \pi_1(X)].$$

(Recall that the commutator subgroup $[G, G]$ of a group G is its subgroup generated by commutators $[g_1, g_2] = g_1 g_2 g_1^{-1} g_2^{-1}$ for all $g_1, g_2 \in G$. The commutator subgroup is always normal. The group $G/[G, G]$ is obtained from g by *Abelianization*, that is, by imposing additional relations: Any two generators commute with each other.)

Proof of Theorem is a copy of the proof of the theorem in Sect. 14.2: We can assume that X is a CW complex with only one vertex, and for such spaces, it is sufficient to compare the procedures of computing the groups π_1 and H_1; see Sects. 7.6 and 13.5.

EXERCISE 9. Show that a loop $f: S^1 \to X$ determines an element of the kernel of the map $h: \pi_1(X) \to H_1(X)$ ("homologous to zero") if and only if it can be extended to the map into X of the disk (with the boundary S^1) with handles. Moreover, the minimal number of these handles is equal to the minimal number of commutators in $\pi_1(X)$ whose product is $[f]$.

EXERCISE 10. The space X_{Ab} is called an *Abelianization*, or *Quillenization*, of a path connected space X if the fundamental group of X_{Ab} is Abelian and there exists a continuous map $X \to X_{\text{Ab}}$ inducing an isomorphism $H_n(X) \to H_n(X_{\text{Ab}})$ for every n. Prove that X possesses an Abelianization if and only if

$$[\pi_1(X), \pi_1(X)] = [\pi_1(X), [\pi_1(X), \pi_1(X)]],$$

that is, if every element of $[\pi_1(X), \pi_1(X)]$ can be presented as a product of commutators of elements of $\pi_1(X)$ with elements of $[\pi_1(X), \pi_1(X)]$.

Remark. Our definition of an Abelianization is a simplified version of a more common definition in which the space X_{Ab} is assumed simple (see Sect. 8.2), or even an H-space (see Exercise 2 in Sect. 8.2) or even a loop space (see Lecture 4). This enhanced definition of an Abelianization plays an important technical role in one of the versions of constructing an algebraic K-functor. The problem of the existence of an Abelianization in this sense is much more complicated, and there are no general theorems about it. But there are several remarkable examples of the Abelianization, two of which we will mention. The first was discovered in 1971 by M. Barratt, D. Kahn, and S. Priddy: The Abelianization of the space $X = K(S_\infty, 1)$, where $S_\infty = \cup_n S_n$ is the group of finite permutations of the set $\mathbb{Z}_{>0}$, is $X_{\text{Ab}} = (\Omega^\infty S^\infty)_0 = \cup_n (\Omega^n S^n)_0$ (the subscript 0 indicates that we consider only one component of the set). Another example belongs to G. Segal (1973) and states that if $X = K(B(\infty), 1)$ where $B(\infty)$ is the infinite braid group and hence X is the set of (unordered) countable subsets of the plane consisting, for some N (depending on the subset), of points $(n+1, 0), (n+2, 0), \ldots$ and n more points different from each other and from the points listed above, then X_{Ab} is $\Omega^2 S^3$. In both cases, the space X has a complicated fundamental group and trivial higher homotopy groups, and the space X_{Ab} has a simple fundamental group ($\mathbb{Z}_2$ in the first case and $\mathbb{Z}$ in the second case) and complicated, so far unknown, homotopy groups. For further details, see Barratt and Priddy [20], Segal [74], and Fuchs [37].

EXERCISE 11. Prove that any two-dimensional homology class of an arbitrary space X can be represented by a sphere with handles; that is, for every $\alpha \in H_2(X)$, there exist a sphere with handles S and a continuous map $f: S \to X$ such that the map $f_*: H_2(S) \to H_2(X)$ takes the canonical generator of $H_2(S) = \mathbb{Z}$ into α.

14.4 The Relative Hurewicz Theorem

The relative Hurewicz homomorphism $h: \pi_n(X, A) \to H_n(X, A)$ is defined similarly to the absolute one. If $f: (D^n, S^{n-1}) \to (X, A)$ is a relative spheroid representing the class $\varphi \in \pi_n(X, A)$, then $h(\varphi)$ is the image of the canonical generator if the group $H_n(D^n, S^{n-1}) = \mathbb{Z}$ with respect to the homomorphism $f_*: H_n(D^n, S^{n-1}) \to H_n(X, A)$.

Theorem. *Let (X, A) be a topological pair such that the space X is path connected and A is simply connected. Let $n \geq 3$.*

(1) *Suppose that $\pi_2(X, A) = \cdots = \pi_{n-1}(X, A) = 0$. Then $H_1(X, A) = H_2(X, A) = \cdots = H_{n-1}(X, A) = 0$ and $h: \pi_n(X, A) \to H_n(X, A)$ is an isomorphism.*
(2) *Suppose that $H_2(X, A) = \cdots = H_{n-1}(X, A) = 0$. Then $\pi_2(X, A) = \cdots = \pi_{n-1}(X, A) = 0$ and $h: \pi_n(X, A) \to H_n(X, A)$ is an isomorphism.*

Proof The proof can be obtained from the proof of the theorem in Sect. 14.2 by modifications characteristic for a transition from the absolute case to a relative case.

We begin by constructing a cellular approximation of the pair (X, A). For this purpose, we first find a cellular approximation (B, g) of A (see Sect. 11.6). Then we attach additional cells to B and successively expand the map $i \circ g: B \to X$ (where i is the inclusion map of A into X) to the new cells in such a way that B is expanded to a CW complex Y and $i \circ g$ is expanded to a weak homotopy equivalence $f: Y \to X$ (this is a replica of the construction in the proof of the theorem in Sect. 11.6). Since $f|_B = g$, the maps f and g compose a map $(Y, B) \to (X, A)$. We already know that f and g induce isomorphisms in both homotopy and homology groups, and the five-lemma implies that the map between the pairs induces isomorphisms for relative homotopy and homology groups. After this, we can assume that the pair (X, A) in the theorem is actually a CW pair.

According to Exercise 22 in Sect. 5.9, there exists a CW pair (X', A') homotopy equivalent to (X, A) and such that A' contains all cells of X' of dimension less than n. We can assume that the pair (X, A) itself has these properties. Then the relative version of the theorem in Sect. 11.3 (see Exercise 2, or, even better, Exercise 4 in Sect. 11.3) describes the group $\pi_n(X, A)$, and this description is not different from the description of $H_n(X, A)$.

EXERCISE 12. If A is not simply connected, then part (1) of the theorem remains true with the following modification: $H_n(X, A)$ is isomorphic to $\pi_n(X, A)$ *factorized over the natural action of* $\pi_1(A)$.

14.5 Whitehead's Theorem

(Not to be confused with a different theorem of the same Whitehead, in Sect. 11.5.)

Theorem. *Let X and Y be simply connected spaces, and let $f: X \to Y$ be a continuous map such that $f_*: \pi_2(X) \to \pi_2(Y)$ is an epimorphism.*

(1) *If the homomorphism $f_*: \pi_m(X) \to \pi_m(Y)$ is an isomorphism for $m < n$ and an epimorphism for $m = n$, then the same is true for $f_*: H_m(X) \to H_m(Y)$.*
(2) *The same with π and H swapped.*

Proof. We may assume that f is an embedding, so (Y, X) is a topological pair. The exactness of homotopy and homology sequences of this pair yields a translation of conditions and claims of the theorem into the language of relative homotopy and homology groups. Namely, the condition "$f_*: \pi_2(X) \to \pi_2(Y)$ is an epimorphism" means precisely that $\pi_2(Y, X) = 0$; the condition "$f_*: \pi_m(X) \to \pi_m(Y)$ is an isomorphism for $m < n$ and an epimorphism for $m = n$" means that $\pi_m(Y, X) = 0$ for $m \leq n$; the same for homology groups. Thus, the theorem is equivalent to the relative Hurewicz theorem in Sect. 14.4.

Corollary. *If a continuous map $f: X \to Y$ between simply connected topological spaces induces an epimorphism $f_*: \pi_2(X) \to \pi_2(Y)$ and isomorphisms $f_*: H_m(X) \to H_m(Y)$ for all m, then f is a weak homotopy equivalence (a homotopy equivalence, if X and Y are CW complexes).*

Lecture 15 Homology with Coefficients and Cohomology

One can apply to the singular or cellular complex of a topological space the standard algebraic operations $- \otimes G$ and $\mathrm{Hom}(-, G)$. In this way, we obtain new complexes which also have homologies; these homologies are called *homology* and *cohomology* of the space *with coefficients (values) in G*. Certainly, the transition to these homology and cohomology may be regarded as a purely algebraic operation, but the experience shows that a too frankly algebraic presentation of this subject may scare a geometrically oriented reader off. To avoid hurting the feelings of such a reader, we will refer to tensor products, Homs, and other such things only when it is absolutely necessary. Still, we will have numerous such necessities.

15.1 Definitions

Let G be an Abelian group. A singular n-dimensional chain of a space X with coefficients in G is a formal linear combination of the form $\sum_i g_i f_i$ where $g_i \in G$ and $f_i \colon \Delta^n \to X$ are singular simplices. The group of n-dimensional singular chains of X with coefficients in G is denoted as $C_n(X;G)$; obviously, $C_n(X;G) = C_n(X) \otimes G$. Our previous group of chains, $C_n(X)$, is, in this notation, $C_n(X;\mathbb{Z})$. A singular n-dimensional *cochain* of X with coefficients (values) in G is defined as a function on the set of all n-dimensional singular simplices of X with values in G (no conditions like continuity are imposed). The group of n-dimensional cochains of X with coefficients in G is denoted as $C^n(X;G)$; obviously, $C^n(X;G) = \mathrm{Hom}(C_n(X),G)$. The value of a cochain c on a chain a is denoted as $\langle c,a\rangle$; thus, $\left\langle c, \sum_i g_i f_i\right\rangle = \sum_i c(f_i) g_i$. A generalization: if a bilinear multiplication (pairing) $G_1 \times G_2 \to G_3$ is given, then for $c \in C^n(X;G_1)$ and $a \in C_n(X;G_2)$ there arises the "value" $\langle c,a\rangle \in G_3$.

Boundary and coboundary operators

$$\begin{aligned} \partial = \partial_n &\colon C_n(X;G) \to C_{n-1}(X;G),\\ \delta = \delta^n &\colon C^n(X;G) \to C^{n+1}(X;G) \end{aligned}$$

are defined by the formulas

$$\partial \sum_i g_i f_i = \sum_i g_i \sum_{j=0}^{n} (-1)^j \Gamma_j f_i, \quad (\delta c)(f) = \sum_{j=0}^{n} (-1)^j c(\Gamma_j f).$$

Obviously, for every $c \in C^n(X;G)$ and $a \in C_{n+1}(X;G)$,

$$\langle c, \partial a\rangle = \langle \delta c, a\rangle.$$

A simple computation shows that $\partial\partial = 0$ and $\delta\delta = 0$ (the second follows from the first and the formula for $\langle -,-\rangle$ above), and we set

$$H_n(X;G) = \frac{\mathrm{Ker}[\partial_n \colon C_n(X;G) \to C_{n-1}(X;G)]}{\mathrm{Im}[\partial_{n+1} \colon C_{n+1}(X;G) \to C_n(X;G)]},$$

$$H^n(X;G) = \frac{\mathrm{Ker}[\delta^n \colon C^n(X;G) \to C^{n+1}(X;G)]}{\mathrm{Im}[\delta^{n-1} \colon C^{n-1}(X;G) \to C^n(X;G)]}.$$

The related terminology is *homology, cohomology, cycles, cocycles, boundaries. coboundaries, homological cycles, cohomological cocycles.*

Chain and cochain complexes may be augmented by maps

$$\begin{gathered} \epsilon \colon C_0(X;G) \to G,\ \epsilon^* \colon G \to C^0(X;G)\\ \epsilon \sum_i g_i f_i = \sum_i g_i \text{ and } [\epsilon^*(g)](f) = g. \end{gathered}$$

The *reduced* homology and cohomology, $\widetilde{H}_n(X;G), \widetilde{H}^n(X;G)$, are the same as unreduced ones with obvious exceptions: $H_0(X;G) = \widetilde{H}_0(X;G) \oplus G$, $H^0(X;G) = \widetilde{H}^0(X;G) \oplus G$, if X is nonempty, and $\widetilde{H}_{-1}(X;G) = G = \widetilde{H}^{-1}(X;G)$ if X is empty.

15.2 Transfer of the Known Results

All major results of Lectures 12 and 13 and some results of Lecture 14 can be transferred to the new context without serious changes, either in statements or in proofs (for the proofs, we have an option to deduce new results from the old results using simple algebraic means; we will not do this, at least now).

A continuous map $h: X \to Y$ induces homology and cohomology homomorphisms, the latter of which acts in the "opposite direction":

$$h_*: H_n(X;G) \to H_n(Y;G),\ h^*: H^n(Y;G) \to H^n(X;G)$$

[the cochain map $h^\#: C^n(Y,G) \to C^n(X;G)$ is defined by the formula $[h^\#(c)](f) = c(h \circ f)$, where f is a singular simplex of X].

Homology with coefficients and cohomology are homotopy invariant: If $g \sim h$, then $g_* = h_*$ and $g^* = h^*$; in particular, homology with coefficients and cohomology of homotopy equivalent spaces are the same.

For a disjoint union $X = X_1 \sqcup \cdots \sqcup X_N$,

$$H_n(X;G) = \bigoplus_i H_n(X_i;G),\ H^n(X;G) = \bigoplus_i H^n(X;G).$$

For *infinite* disjoint unions, a difference appears between homology and cohomology: $H_n(X;G)$ is the *direct sum* of the groups $H_n(X_i;G)$, while $H^n(X;G)$ is the *direct product* of the groups $H^n(X_i;G)$.

For the one-point space pt,

$$\begin{aligned} H_0(\text{pt};G) &= G = H^0(\text{pt};G), \\ H_n(\text{pt};G) &= 0 = H^n(\text{pt};G) \text{ for } n \neq 0, \\ \widetilde{H}_n(\text{pt};G) &= 0 = \widetilde{H}^n(\text{pt};G) \text{ for all } n. \end{aligned}$$

Relative homology with coefficients is defined precisely as usual (integral) relative homology, while in the definition of relative cohomology there arises a small (and expectable) new feature: The group $C^n(X,A;G)$ is a subgroup, not a quotient group, of $C^n(X;G)$; it consists of cochains from $C^n(X;G)$ which have zero restriction to $C_n(A) \subset C_n(X)$ (or, equivalently, assume zero value at every singular simplex in A).

The homology sequence of a pair (X,A) with coefficients in G looks the same as in the integral case (just insert "; G" where necessary). The cohomology sequence has all the arrows reversed:

$$\cdots \to H^{n-1}(A;G) \xrightarrow{\delta^*} H^n(X,A;G) \to H^n(X;G) \to H^n(A;G) \to \ldots.$$

The homomorphism $\delta^*: H^{n-1}(A;G) \to H^n(X;A)$ is defined in the following (expectable) way. For a class $\gamma \in H^(A;G)$, choose a representing cocycle $c \in C^{n-1}(A;G)$. Then expand the function c [on $(n-1)$-dimensional singular simplices of A] to all $(n-1)$-dimensional singular simplices of X (for example, set it equal to 0 on simplices not contained in A) and take the coboundary of the chain $c' \in C^{n-1}(X;G)$ arising. Then $\delta c'$ is zero on $C^n(A;G)$ (since c is a cocycle). Thus, $\delta c' \in C^n(X,A)$. It is a (relative) cocycle (since $\delta\delta = 0$), and its cohomology class $\beta \in H^n(X,A;G)$ does not depend on the arbitrary choices of the construction (c in γ and the extension c' of c; it is similar to Exercise 7 in Lecture 12). The function $\gamma \mapsto \beta$ is δ^*.

Both homology with coefficients and cohomology sequences of a pair are exact. There are also exact reduced homology with coefficients and cohomology sequences of pairs (no reducing for relative homology and cohomology groups) and exact homology with coefficients and cohomology sequences of triples.

For a Borsuk pair (X,A), there are isomorphisms

$$H_n(X,A;G) = \widetilde{H}_n(X/A;G), \ H^n(X,A;G) = \widetilde{H}^n(X/A;G)$$

established by the projection $X \to X/A$. For an arbitrary pair there are similar isomorphisms with X/A replaced by $X \cup CA$. Under the same assumptions as in Sect. 12.7, there are excision isomorphisms $H_n(X-B, A-B;G) = H_n(X,A;G)$ and $H^n(X-B, A-B;G) = H^n(X,A;G)$ and exact Mayer–Vietoris sequences; the cohomology Mayer–Vietoris sequences assume the form

$$\begin{aligned} \cdots \to H^{n-1}(A \cap B;G) &\to H^n(X;G) \\ &\to H^n(A;G) \oplus H^n(B;G) \to H^n(A \cap B;G) \to \ldots. \end{aligned}$$

For a CW complex, homology with coefficients and cohomology can be calculated through the cellular complex. Namely, for a CW complex X, $\mathcal{C}_n(X;G)$ is the group of linear combinations $\sum_i g_i e_i$, where e_i are oriented n-dimensional cells (an orientation change for a cell e_i results in a replacement of g_i by $-g_i$). Furthermore, $\mathcal{C}^n(X;G)$ is the group of G-valued functions on the set of oriented n-dimensional cells of X, where the orientation change for e_i leads to a sign change for the value at e_i. The boundary and coboundary operations act by the formulas

$$\partial\left(\sum_i g_i e_i\right) = \sum_i g_i \sum_f [e_i : f] f, \ [\delta c](e) = \sum_f [e : f] c(f),$$

where the inner summation on the right-hand side of the first formula is spread to all $(n-1)$-dimensional cells f of X and the summation in the second formula is spread to all n-dimensional cells of X.

Let us now show the results of calculating homology with coefficients and cohomology for the most important CW complexes. For spheres,

$$\widetilde{H}_m(S^n;G) = \widetilde{H}^m(S^n;G) = \begin{cases} G, \text{ if } m = n, \\ 0, \text{ if } m \neq n \end{cases}$$

(this fact certainly can be obtained with the cellular complexes, but the reader who wants to reconstruct all the proofs will have to do it at an earlier stage, as in Sect. 13.1). For complex, quaternion, and Cayley projective spaces, as well as for complex and quaternion Grassmann manifolds and flag manifolds, the homology with coefficients and cohomology are not different from the corresponding cellular chains and cochains. For example,

$$H_m(\mathbb{C}P^n;G) = H^m(\mathbb{C}P^n;G) = \begin{cases} G, \text{ if } m = 0, 2, 4, \dots [, 2n, \\ \qquad\qquad\qquad\qquad \text{if } n \text{ is finite}], \\ 0 \quad \text{for all other } m. \end{cases}$$

In the real case, the computation may be more complicated (compare Sect. 13.8), but it becomes much simpler if $G = \mathbb{Z}_2$, since in this case all the boundary and coboundary operators (in cellular complexes) are zero and homology with coefficients and cohomology again do not differ from the corresponding cellular chain and cochain groups. For example,

$$H_m(\mathbb{R}P^n;\mathbb{Z}_2) = H^m(\mathbb{R}P^n;\mathbb{Z}_2) = \begin{cases} \mathbb{Z}_2, \text{ for } 0 \leq m \leq n, \\ 0 \quad \text{for all other } m. \end{cases}$$

Notice in addition that for a classical surface X (without holes),

$$H_0(X;\mathbb{Z}_2) = H^0(X;\mathbb{Z}_2) = H_2(X;\mathbb{Z}_2) = H^2(X;\mathbb{Z}_2) = \mathbb{Z}_2,$$
$$H_1(X;\mathbb{Z}_2) = H^1(X;\mathbb{Z}_2) = \underbrace{\mathbb{Z}_2 \oplus \dots \oplus \mathbb{Z}_2}_{r},$$

where

$$r = \begin{cases} 2g, \qquad \text{if } X \text{is a sphere with } g \text{ handles}, \\ 2g+1, \text{ if } X \text{ is a projective plane with } g \text{ handles}, \\ 2g+2, \text{ if } X \text{ is a Klein bottle with } g \text{ handles}. \end{cases}$$

EXERCISE 1. Find the homology and cohomology of real projective spaces and real Grassmann manifolds with coefficients in $\mathbb{Z}_m$ where m is odd.

To finish the section, let us notice that if $f: X \to Y$ is a weak homotopy equivalence, then

$$f_*: H_n(X;G) \to H_n(Y;G) \text{ and } f^*: H^n(Y;G) \to H^n(X;G)$$

are isomorphisms for all G and n.

15.3 Coefficient Sequences

We begin studying relations between homologies and cohomologies with different coefficients. There is an obvious fact that any homomorphism $\varphi\colon G_1 \to G_2$ between Abelian groups induces, for every X and n, homomorphisms

$$\varphi_*\colon H_n(X;G_1) \to H_n(X;G_2) \text{ and } \varphi_*\colon H_n(X;G_1) \to H_n(X;G_2)$$

(in the same direction). However, as many examples (including some known to us) show, the homomorphism φ being a monomorphism, or an epimorphism, or just nontrivial, does not imply similar properties for any of the φ_*s. For a deeper understanding of the subject, let us consider the following situation. Let G be an Abelian group, H be a subgroup of G, and F be the quotient group G/H. Usually, all of this is presented as a short exact sequence,

$$0 \to H \to G \to F \to 0.$$

Besides the homomorphisms $H_n(X;H) \to H_n(X;G) \to H_n(X;F)$ and $H^n(X;H) \to H^n(X;G) \to H^n(X;F)$, there arise "connecting homomorphisms"

$$\delta_*\colon H_n(X;F) \to H_{n-1}(X;H) \text{ and } \delta^*\colon H^n(X;F) \to H^{n+1}(X;H).$$

Here is the construction of the first of them. For an $\alpha \in H_n(X;F)$, choose a representative $a \in C_n(X;F)$. Since $G \to F$ is an epimorphism, a possesses an inverse image $\widetilde{a} \in C_n(X;G)$. The projection $C_{n-1}(X;G) \to C_{n-1}(X;F)$ takes $\partial\widetilde{a}$ into $\partial a = 0$; hence, $\partial\widetilde{a}$ actually belongs to $C_{n-1}(X;H)$. This is a cycle, and its homology class in $H_{n-1}(X;H)$ is taken for $\partial_*(\alpha)$. The construction of the homomorphism δ^* is similar $[(\gamma \in H^n(X;F)) \mapsto (c \in C^n(X;F)) \mapsto (\widetilde{c} \in C^n(X;G)) \mapsto (\delta\widetilde{c} \in C^{n+1}(X;H)) \mapsto (\delta^*(\gamma) \in H^{n+1}(X;H))]$.

EXERCISE 2. Check that the preceding constructions provide well-defined homomorphisms ∂_* and δ^*.

EXERCISE 3. Prove that the *coefficient sequences*

$$\cdots \to H_n(X;H) \to H_n(X;G) \to H_n(X;F) \to H_{n-1}(X;H) \to \dots,$$
$$\cdots \to H^n(X;H) \to H^n(X;G) \to H^n(X;F) \to H^{n+1}(X;H) \to \dots$$

are exact.

HISTORICAL AND TERMINOLOGICAL REFERENCE. The homomorphisms ∂_* and δ^* were discovered, in a particular case, by M. Bockstein long before exact sequences became commonplace in algebraic topology. Here is how the Bockstein homomorphism was first described. Let $\alpha \in H_n(X;\mathbb{Z}_m)$. Take a representative a of α. All the coefficients involved in a are residues modulo m; we can regard them as integers $0, 1, \dots, m-1$. Then the cycle a becomes an integral chain $\widetilde{a}$. The boundary $\partial\widetilde{a}$ is divisible by m; let us divide. The result, $\frac{1}{m}\partial\widetilde{a}$, is an integral cycle. It represents some class $B_m(\alpha) \in H_{n-1}(X;\mathbb{Z})$ [by the way, $mB_m(\alpha) = 0$]; after reducing modulo m, it becomes a class $b_m(\alpha) \in H_{n-1}(X;\mathbb{Z}_m)$. We have constructed "Bockstein homomorphisms"

$$B_m: H_n(X;\mathbb{Z}_m) \to H_{n-1}(X;\mathbb{Z}) \text{ and } b_m: H_n(X;\mathbb{Z}_m) \to H_{n-1}(X;\mathbb{Z}_m).$$

In a very similar way, cohomological Bockstein homomorphisms

$$B^m: H^n(X;\mathbb{Z}_m) \to H^{n+1}(X;\mathbb{Z}) \text{ and } b^m: H^n(X;\mathbb{Z}_m) \to H^{n+1}(X;\mathbb{Z}_m)$$

are defined.

Actually, all of these Bockstein homomorphisms are connecting homomorphisms ∂_* and δ^* of coefficient sequences induced by the short exact sequences

$$0 \to \mathbb{Z} \xrightarrow{\cdot m} \mathbb{Z} \to \mathbb{Z}_m \to 0 \text{ and } 0 \to \mathbb{Z}_m \to \mathbb{Z}_{m^2} \to \mathbb{Z}_m \to 0.$$

From the exactness of the coefficient sequences, it follows then that (1) an element of $H_n(X;\mathbb{Z}_m)$ belongs to the kernel of B_m if and only if it is "integral," that is, belongs to the image of the reducing homomorphism $H_n(X;\mathbb{Z}) \to H_n(X;\mathbb{Z}_m)$; an element of $H_n(X;\mathbb{Z})$ belongs to the image of B_m if and only if it is annihilated by the multiplication by m; similarly for the cohomological Bockstein homomorphisms.

15.4 Algebraic Preparation to Universal Coefficients Formulas

Let A and B be Abelian groups. Then let $B = F_1/F_2$, where F_1 is a free Abelian group and F_2 is a subgroup of F_1 which must also be free (such a presentation exists for any Abelian group). What are the interrelations between $A \otimes F_1$, $A \otimes F_2$, and $A \otimes B$? To answer this question, we need a lemma which can be regarded as the most fundamental property of tensor products.

Lemma 1. *The tensor product operation is right exact. This means that if the sequence*

$$A \xrightarrow{\alpha} B \xrightarrow{\beta} C \longrightarrow 0$$

is exact, then the sequence

$$G \otimes A \xrightarrow{G\otimes\alpha} G \otimes B \xrightarrow{G\otimes\beta} G \otimes C \longrightarrow 0$$

is exact.

Proof. Recall that, by definition, the tensor product $K \otimes L$ is $F(K \times L)/R(K, L)$, where $F(K \times L)$ is the free Abelian group generated by the set $K \times L$ and $R(K, L)$ is the subgroup of $F(K \times L)$ generated by elements of the form $(k, \ell)+(k', \ell)-(k+k', \ell)$ and $(k, \ell) + (k, \ell') - (k, \ell + \ell')$. The image of (k, ℓ) in $K \otimes L$ is denoted as $k \otimes \ell$.

It is obvious that $G \otimes \beta$ is onto: $\sum_i (g_i \otimes c_i) = [G \otimes \beta] \left(\sum_i (g_i \otimes b_i)\right)$, where the b_i are chosen to satisfy the condition $\beta(b_i) = c_i$. It is also obvious that $(G \otimes \beta) \circ (G \otimes \alpha) = 0$. It remains to prove that $\mathrm{Ker}(G \otimes \beta) \subset \mathrm{Im}(G \otimes \alpha)$.

Let $[G \otimes \beta] \left(\sum_i (g_i \otimes b_i)\right) = 0$. This means that $\sum_i (g_i, \beta(b_i)) \in R(G, C)$; that is, $\sum_i (g_i, \beta(b_i))$ is a linear combination of elements of $F(G \times C)$ of the form $(g, c) + (g', c) - (g + g', c)$ and $(g, c) + (g, c') - (g, c + c')$. For all c, c' involved, find $b, b' \in B$ whose β-images are c, c', and the subtract from $\sum_i (g_i, \beta(b_i))$ the same linear combination with c, c' replaced by the corresponding b, b'. We get an element of $F(G \times B)$ which also represents $\sum_i (g_i \otimes b_i)$ but also belongs to the kernel of the map $F(G \times \beta): F(G \times B) \to F(G \times C)$. This kernel is generated by differences $(g, b') - (g, b'')$ with $\beta(b' - b'') = 0$, that is, $b' - b'' \in\in \alpha$. Thus, $\sum_i (g_i \otimes b_i) = \sum_j (g')j \otimes (b'_j - b''_j))$ and hence $\sum_i (g_i \otimes b_i) = [G \otimes \alpha] \left(\sum_j (g'_j \otimes a_j)\right)$, where $\alpha(a_j) = b'_j - b''_j$.

Lemma 1 shows that the sequence

$$A \otimes F_2 \to A \otimes F_1 \to A \otimes B \to 0$$

is exact; that is, $A \otimes B$ is a quotient of $A \otimes F_1$ over the image of the natural map $A \otimes F_2 \to A \otimes F_1$, but this map is not necessarily a monomorphism.

Lemma 2. *The kernel* $\mathrm{Ker}(A \otimes F_2 \to A \otimes F_1)$ *does not depend on the choice of presentation* $B = F_2/F_1$.

Proof The proof consists in constructing a canonical isomorphism

$$\mathrm{Ker}(A \otimes F'_2 \to A \otimes F'_1) \cong \mathrm{Ker}(A \otimes F_2 \to A \otimes F_1)$$

for an arbitrary other presentation $B = F'_1/F'_2$. First, we construct homomorphisms $\alpha_1: F'_1 \to F_1$, $\alpha_2: F'_2 \to F_2$, making the diagram

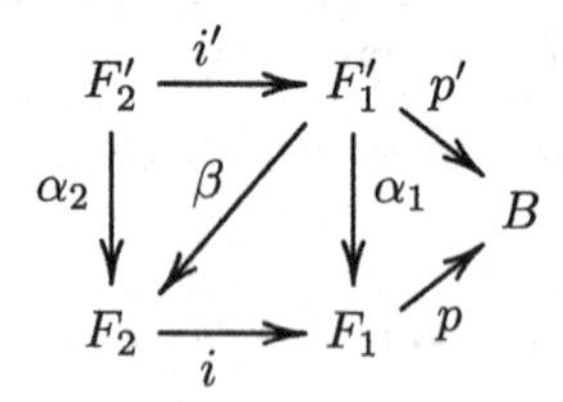

(where the i, i' are inclusion maps and the p, p' are projections) commutative. Here α_1 takes a generator x of F'_1 into $y \in F_1$ such that $p(y) = p'(x)$ (which exists, since p is an epimorphism). This α_1 takes $\mathrm{Ker}\, p' = F'_2$ into $\mathrm{Ker}\, p = F_2$, thus giving rise to an $\alpha_2: F'_2 \to F_2$. Since y in the previous construction is determined (by x) up to an element of $\mathrm{Ker}\, p = F_2$, any other choice of α_1 has the form $\alpha_1 + i' \circ \beta$, where

β is a homomorphism $F'_1 \to F_2$, and then the new α_2 is $\alpha_2 + \beta \circ i$. Take the tensor product of (the square part of) this diagram with A:

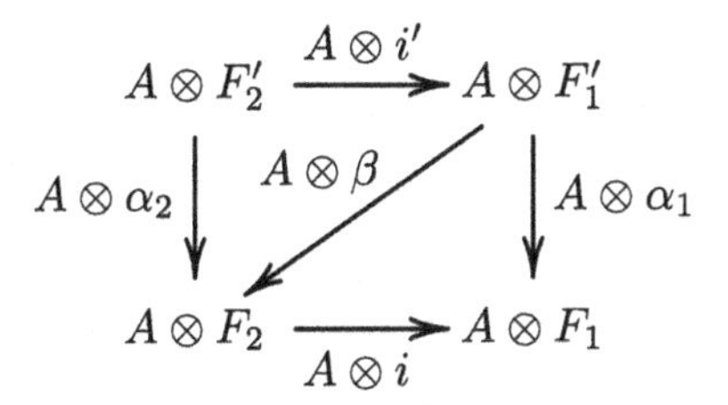

The map $A \otimes \alpha_2$ takes $\mathrm{Ker}(A \otimes i'$ into $\mathrm{Ker}(A \otimes i)$. This map does not depend on the choice of α_1 and α_2, since $A \otimes (\beta \circ i') = (A \otimes \beta) \circ (A \otimes i')$ is zero on $\mathrm{Ker}(A \otimes i')$. The map $\mathrm{Ker}(A \otimes i') \to \mathrm{Ker}(A \otimes i)$ is constructed in the same way, and the composition of these maps in any order is the identity, because of the same uniqueness (this time, applied to $F'_1 = F_1$, $F'_2 = F_2$).

Definition. The kernel $\mathrm{Ker}(A \otimes F_2 \to A \otimes F_1)$ is called the *periodic product* of A and B and is denoted as $\mathrm{Tor}(A, B)$.

EXERCISE 4. Show that the operation Tor is natural with respect to both arguments; that is, homomorphisms $A \to A', B \to B'$ induce a homomorphism $\mathrm{Tor}(A, B) \to \mathrm{Tor}(A', B')$ with all expectable properties (for A it is obvious, while for B this requires a construction like the one in the beginning of the proof of the lemma).

EXERCISE 5. Prove a natural isomorphism $\mathrm{Tor}(A, B) \to \mathrm{Tor}(B, A)$. (This might be harder than one can expect. The most common idea of proving that is the following. Consider two presentations $A = F_1/F_2$, $B = G_1/G_2$ with free Abelian F_1, F_2, G_1, G_2, form the complex

$$0 \to F_2 \otimes G_2 \to [(F_1 \otimes G_2) \oplus (F_2 \otimes G_1)] \to F_1 \otimes G_1 \to 0,$$

and prove that the homology groups H_2, H_1, and H_0 of this complex are 0, $\mathrm{Tor}(A, B)$, and $\mathrm{Hom}(A, B)$. This provides a definition of Tor symmetric in A, B.)

EXERCISE 6. Prove that if A (or B) is a free Abelian group, then $\mathrm{Tor}(A, B) = 0$.

EXERCISE 7. Prove that $\mathrm{Tor}(\mathbb{Z}_m, \mathbb{Z}_n) \cong \mathbb{Z}_m \otimes \mathbb{Z}_n$ $[= \mathbb{Z}_{\gcd(m,n)}]$ [this isomorphism is *not* canonical; it depends on the choice of generators in $\mathbb{Z}_m$ and $\mathbb{Z}_n$]. Thus, for finitely generated Abelian groups A, B,

$$\mathrm{Tor}(A, B) \cong \mathrm{Tors}\, A \otimes \mathrm{Tors}\, B$$

($\mathrm{Tors}\, A =$ torsion of A, the group of elements of finite order).

EXERCISE 8. For infinitely generated A, B, the last isomorphism, in general, does not hold: Construct an example.

EXERCISE 9. Prove that is $A = \mathbb{Q}, \mathbb{R}$, or $\mathbb{C}$, then $\mathrm{Tor}(A, B) = 0$ for any B.

The "dual" operation Ext is defined in a similar way. First, we dualize Lemma 1:

Lemma 3. *If the sequence*

$$A \xrightarrow{\alpha} B \xrightarrow{\beta} C \longrightarrow 0$$

is exact, then the sequence

$$\mathrm{Hom}(A,G) \xleftarrow{\mathrm{Hom}(\alpha,G)} \mathrm{Hom}(B,G) \xleftarrow{\mathrm{Hom}(\beta,G)} \mathrm{Hom}(C,G) \longleftarrow 0$$

is exact.

Proof The proof is left to the reader; it is easier than the proof of the Lemma 1.

EXERCISE 10. Prove that the operation $\mathrm{Hom}(G,-)$ is *left exact*. This means that if the sequence

$$0 \longrightarrow A \xrightarrow{\alpha} B \xrightarrow{\beta} C$$

is exact, then the sequence

$$0 \longrightarrow \mathrm{Hom}(G,A) \xrightarrow{\mathrm{Hom}(G,\alpha)} \mathrm{Hom}(G,B) \xrightarrow{\mathrm{Hom}(G,\beta)} \mathrm{Hom}(G,C)$$

is exact.

Let A, B be Abelian groups, and let $A = F_1/F_2$, where F_1 and F_2 are free Abelian groups. Lemma 3 says that the kernel of the map $\mathrm{Hom}(F_1,B) \to \mathrm{Hom}(F_2,B)$, $f \mapsto f\,|_{F_2}$ is $\mathrm{Hom}(A,B)$, but this map is not onto. The cokernel of this map, which is the quotient of $\mathrm{Hom}(F_2,B)$ over the image of this map, is taken for $\mathrm{Ext}(A,B)$.

EXERCISE 11. Prove that Ext is well defined (this is a dualization of Lemma 2).

EXERCISE 12. Show that the operation Ext is natural with respect to both arguments; that is, homomorphisms $A \to A', B \to B'$ induce a homomorphism $\mathrm{Ext}(A',B) \to \mathrm{Ext}(A,B')$ with all expectable properties. (Notice the reversion of the arrow $A \to A'$.)

EXERCISE 13. Prove that $\mathrm{Ext}(\mathbb{Z},B) = 0$ for any B; prove also that $\mathrm{Ext}(\mathbb{Z}_m,\mathbb{Z}_n) \cong \mathrm{Hom}(\mathbb{Z}_m,\mathbb{Z}_n) \cong \mathbb{Z}_m \otimes \mathbb{Z}_n \cong \mathbb{Z}_{(m,n)}$ (not canonically!), and $\mathrm{Ext}(\mathbb{Z}_m,\mathbb{Z}) \cong \mathbb{Z}_m$ (unlike $\mathrm{Tor}(\mathbb{Z}_m,\mathbb{Z}) = 0$).

EXERCISE 14. The set $\mathrm{Ext}(A,B)$ has another definition (due to Yoneda) as the set of equivalence classes of "extensions" of A by B, that is, short exact sequences

$$0 \to B \to C \to A \to 0$$

where C is an Abelian group. Prove the equivalence of the two definitions of Ext and make up a direct definition of a group structure in the set $\mathrm{Ext}(A,B)$ described as the set of extensions.

EXERCISE 15. Prove that if one of the groups A, B is $\mathbb{Q}, \mathbb{R}$, or $\mathbb{C}$, then $\mathrm{Ext}(A, B) = 0$.

15.5 The Universal Coefficients Formula

Now we will show that the usual (integral) homology of X (actually, of any complex consisting of free Abelian groups) determine homology and cohomology of X with arbitrary coefficients.

Theorem. *For any X, n, and G,*

$$H_n(X;G) \cong (H_n(X) \otimes G) \oplus \mathrm{Tor}(H_{n-1}(X), G)$$

$$H^n(X;G) \cong (H^n(X) \otimes G) \oplus \mathrm{Tor}(H^{n+1}(X;\mathbb{Z}), G)$$

$$H^n(X;G) \cong \mathrm{Hom}(H_n(X), G) \oplus \mathrm{Ext}(H_{n-1}(X), G).$$

IMPORTANT ADDITION. *The isomorphisms of the theorem are not canonical. What is canonical are the following three exact sequences:*

$$0 \to H_n(X) \otimes G \to H_n(X;G) \to \mathrm{Tor}(H_{n-1}(X), G) \to 0,$$

$$0 \to H^n(X;\mathbb{Z}) \otimes G \to H^n(X;G) \to \mathrm{Tor}(H^{n+1}(X;\mathbb{Z}), G) \to 0,$$

$$0 \leftarrow \mathrm{Hom}(H_n(X), G) \leftarrow H^n(X;G) \leftarrow \mathrm{Ext}(H_{n-1}(X), G) \leftarrow 0.$$

Proof. The first two exact sequences are easily obtained from coefficient sequences. The first sequence is obtained in the following way. Let $G = F_1/F_2$, where F_1 and F_2 are free Abelian groups. Then $F_1 = \mathbb{Z} \oplus \mathbb{Z} \oplus \dots,$, and hence

$$H_n(X;F_1) = H_n(X;\mathbb{Z} \oplus \mathbb{Z} \oplus \dots) = H_n(X) \oplus H_n(X) \oplus \dots = H_n(X) \otimes F_1,$$

and, similarly, $H_n(X;F_2) = H_n(X) \otimes F_2$. Hence, the fragment

$$H_n(X;F_2) \to H_n(X;F_1) \to H_n(X;G) \to H_{n-1}(X;F_2) \to H_{n-1}(X;F_2)$$

of the coefficient sequence takes the form

$$\begin{aligned} H_n(X) \otimes F_2 \to H_n(X) \otimes F_1 &\to H_n(X;G) \\ &\to H_{n-1}(X) \otimes F_2 \to H_{n-1}(X) \otimes F_2. \end{aligned}$$

A five-term exact sequence $A \xrightarrow{\varphi} B \to C \to D \xrightarrow{\psi} E$ can be transformed into a short exact sequence $0 \to \mathrm{Coker}\,\varphi \to C \to \mathrm{Ker}\,\psi \to 0$ (where Coker is the quotient over the image, $\mathrm{Coker}\,\varphi = B/\mathrm{Im}\,\varphi$). This transformation converts the last sequence into the first of the three exact sequences in the theorem. The second sequence is obtained in the way from the cohomological coefficient sequence (and the isomorphisms $H^n(X; F_i) = H^n(X; \mathbb{Z}) \otimes F_i$). The last sequence can hardly be obtained in a similar way, because it contains both homology and cohomology. But there exists a different approach which yields isomorphisms from the theorem rather than the exact sequences.

Since for every n, $B_n(X) = \operatorname{Im}[\partial_{n+1}: C_{n+1}(X) \to C_n(X)]$ is a free Abelian group, there exists a (nonunique) homomorphism $s_n: B_n(X) \to C_{n+1}(X)$ such that $\partial_{n+1} \circ s_n = \mathrm{id}$. Thus,

$$C_{n+1}(X) = \operatorname{Ker} \partial_{n+1} \oplus \operatorname{Im} s_n = Z_{n+1}(X) \oplus B_n(X).$$

The boundary operator looks like this:

$$\begin{array}{ccc} C_{n+1}(X) = Z_{n+1}(X) & \oplus & B_n(X) \\ \Big\downarrow \partial_{n+1} & & \Big\downarrow \text{inclusion} \\ C_n(X) \quad = & & Z_n(X) \oplus B_{n-1}(X)\,. \end{array}$$

This shows that the whole singular complex $\mathcal{C} = \{C_n(X), \partial_n\}$ is isomorphic (not canonically) to the direct sum of very short complexes $\mathcal{C}(n)$,

$$\begin{array}{cccc} \ldots 0 \to 0 \to & B_n(X) & \xrightarrow{\text{incl.}} & Z_n(X) \to 0 \to 0 \ldots \\ & (n+1) & & (n) \end{array}$$

[for this complex, the n-dimensional homology is $H_n(X)$; all the other homology groups are zero]. Since the tensor product has the distributivity property, the complex $\mathcal{C} \otimes G = \{C_n(X; G) = C_n(X) \otimes G, \partial_n \otimes G\}$ is the sum of complexes $\mathcal{C}(n) \otimes G$,

$$\ldots 0 \to 0 \to B_{n-1} \otimes G \to Z_{n-1} \otimes G \to 0 \to 0 \ldots .$$

Since $B_n(X)$ and $Z_n(X)$ are free Abelian groups and $Z_n(X)/B_n(X) = H_n(X)$, the homology groups of the complex $\mathcal{C} \otimes G$ are

$$\text{dimension } n+1\text{: } \operatorname{Tor}(H_n(X), G);$$
$$\text{dimension } n\text{: } H_n(X) \otimes G.$$

The summation over n gives the first formula of the theorem: $H_n(X; G) \cong (H_n(X) \otimes G) \oplus \operatorname{Tor}(H_{n-1}(X), G)$. The second formula is obtained in the same way; we leave this job to the reader.

To prove the last part of the theorem, consider again the decomposition of the singular complex $\mathcal{C}$ of X into the sum of "very short complexes" $\mathcal{C}(n)$:

$$\begin{array}{ccccccccc}
\cdots & C_{n+2}(X) & \xrightarrow{\partial} & C_{n+1}(X) & \xrightarrow{\partial} & C_n(X) & \xrightarrow{\partial} & C_{n-1}(X) & \cdots \quad (\mathcal{C}) \\
 & & & \downarrow & & \uparrow & & & \\
\cdots & 0 & \longrightarrow & B_n(X) & \xrightarrow{\text{incl.}} & Z_n(X) & \longrightarrow & 0 & \cdots \quad (\mathcal{C}(n))
\end{array}$$

We see that although the decomposition $\mathcal{C} = \bigoplus \mathcal{C}(n)$ is not canonical, and hence there is neither a canonical projection $\mathcal{C} \to \mathcal{C}(n)$ or a canonical embedding $\mathcal{C}(n) \to \mathcal{C}$, there are still the canonical projection $C_{n+1}(X) \to B_n(X)$ and the canonical embedding $Z_n(X) \to C_n(X)$, as shown in the diagram. Now apply to this diagram the operation $\mathrm{Hom}(-, G)$. We obtain the diagram

$$\begin{array}{ccc}
\mathrm{Hom}(C_{n+2}(X), G) & \longrightarrow & 0 \\
\uparrow & & \uparrow \\
\mathrm{Hom}(C_{n+1}(X), G) & \longrightarrow & \mathrm{Hom}(B_n(X), G) \\
\uparrow & & \uparrow \\
\mathrm{Hom}(C_n(X), G) & \longrightarrow & \mathrm{Hom}(Z_n(X), G) \\
\uparrow & & \uparrow \\
\mathrm{Hom}(C_{n-1}(X), G) & \longrightarrow & 0
\end{array}$$

For the (co)homology $H^m(\mathcal{C}; G)$ of the complex $\mathrm{Hom}(\mathcal{C}, G)$, we have

$$\begin{aligned} H^n(\mathcal{C}(n); G) &= \mathrm{Ker}[\mathrm{Hom}(Z_n(X), G) \to \mathrm{Hom}(B_n(X), G)] \\ &= \mathrm{Hom}(H_n(X), G), \end{aligned}$$

$$\begin{aligned} H^{n+1}(\mathcal{C}(n); G) &= \mathrm{Coker}[\mathrm{Hom}(Z_n(X), G) \to \mathrm{Hom}(B_n(X), G)] \\ &= \mathrm{Ext}(H_n(X), G) \end{aligned}$$

and $H^m(\mathcal{C}(n)) = 0$ for $m \neq n, n+1$. From this,

$$H^n(X; G) \cong \bigoplus_k H^n(\mathcal{C}; G) = \mathrm{Hom}(H_n(X), G) \oplus \mathrm{Ext}(H_{n-1}(X), G),$$

as stated. Moreover, as we have seen, there are canonical homomorphisms

$$H^n(X; G) \to \mathrm{Hom}(H_n(X), G), \ \mathrm{Ext}(H_{n-1}(X), G) \to H^n(X; G),$$

which form the exact sequence

$$0 \leftarrow \mathrm{Hom}(H_n(X), G) \leftarrow H^n(X; G) \leftarrow \mathrm{Ext}(H_{n-1}(X), G) \leftarrow 0.$$

This completes the proof of the theorem.

We can add that the map

$$C^n(X; G) = \mathrm{Hom}(C_n(X), G) \to \mathrm{Hom}(Z_n(X), G)$$

considered above is simply the restriction to $Z_n(X)$; moreover, if $c \in C^n(X; G)$ is a cocycle, then the restriction of c to $B_n(X)$ is zero, which provides an element of $\mathrm{Hom}(H_n(X), G)$ depending only on the cohomology class of c; this is how our homomorphism $H^n(X; G) \to \mathrm{Hom}(H_n(X), G)$ acts. In other words, this homomorphism sends a cohomology class $\gamma \in H^n(X; G)$ to a homomorphism $\alpha \mapsto \langle \gamma, \alpha \rangle$ of $H_n(X)$ into G. The fact that this homomorphism is onto yields the following important proposition.

Corollary 1. *For every homomorphism $f: H_n(X) \to G$, there exists a cohomology class $\gamma \in H^n(X; G)$ such that $f(\alpha) = \langle \gamma, \alpha \rangle$ for every $\alpha \in H_n(G)$.*

Remark also that this γ is defined up to an element of $\mathrm{Ext}(H_n(X), G)$; in particular, if $H_n(X)$ and G are finitely generated, then this Ext group is finite, so γ is defined by f up to adding an element of finite order.

Before the final exercises of this section, we will mention one more interesting corollary.

Corollary 2. *If the groups $H_n(X)$ are finitely generated, then*

$$H^n(X; \mathbb{Z}) \cong \text{Free part of } H_n(X) \oplus \text{Torsion part of } H_{n-1}(X).$$

In particular, $H^1(X; \mathbb{Z})$ is a free Abelian group.

EXERCISE 16. If $\mathbb{K} = \mathbb{Q}, \mathbb{R}$, or $\mathbb{C}$, then

$$H_n(X; \mathbb{K}) = H_n(X) \otimes \mathbb{K} \text{ and } H^n(X; \mathbb{K}) = \mathrm{Hom}(H_n(X), \mathbb{K}).$$

Thus, the transition from the integral coefficients to the rational, real, or complex coefficients kills the torsion. On the other hand, the Betti numbers of X become the dimension of homology or cohomology with coefficients in $\mathbb{Q}, \mathbb{R}$ or $\mathbb{C}$. (Actually, the same is true for any field of characteristic zero.)

EXERCISE 17. If $\mathbb{K}$ is a field, then homology and cohomology with coefficients in $\mathbb{K}$ possess a natural structure of vector spaces over $\mathbb{K}$. Prove that

$$H^n(X; \mathbb{K}) = \mathrm{Hom}_{\mathbb{K}}(H_n(X; \mathbb{K}), \mathbb{K}).$$

[It is better not to deduce this formula from the universal coefficients formula, but rather to prove it directly using the equality $C^n(X; \mathbb{K}) = \mathrm{Hom}_{\mathbb{K}}(C_n(X; \mathbb{K}), \mathbb{K})$.]

EXERCISE 18. Prove that if X is a finite CW complex and $\mathbb{K}$ is a field, then

$$\sum (-1)^m \dim_{\mathbb{K}} H_m(X; \mathbb{K})$$

does not depend on $\mathbb{K}$ and is equal to the Euler characteristic of X (see Sect. 13.7).

15.6 Künneth's Formula

By its contents, Künneth's formula is closer to the next lecture than to the current one. But by sight, this formula has so strong resemblance to the universal coefficients formulas (actually, these formulas can be deduced from the same general algebraic result; thus, they have a common ancestor) that it would be unfair to try to separate them.

Theorem 1. *Let X_1, X_2 be topological spaces. Then for any n,*

(1) *There is a (noncanonical) isomorphism*

$$H_n(X_1 \times X_2) \cong \bigoplus_{i+j=n} (H_i(X_1) \otimes H_j(X_2)) \quad \oplus \bigoplus_{i+j=n-1} \operatorname{Tor}(H_i(X_1), H_j(X_2)).$$

(2) *There is a canonically defined exact sequence*

$$0 \to \bigoplus_{i+j=n} (H_i(X_1) \otimes H_j(X_2)) \to H_n(X_1 \times X_2) \to \bigoplus_{i+j=n-1} \operatorname{Tor}(H_i(X_1), H_j(X_2)) \to 0.$$

We will deduce Theorem 1 from an algebraic result related to the tensor product of complexes.

Definition. Let

$$\begin{array}{ll} (\mathcal{C}) & \dots \xrightarrow{\partial_{n+1}} C_n \xrightarrow{\partial_n} C_{n-1} \xrightarrow{\partial_{n-1}} \dots, \\ (\mathcal{C}') & \dots \xrightarrow{\partial'_{n+1}} C'_n \xrightarrow{\partial'_n} C'_{n-1} \xrightarrow{\partial'_{n-1}} \dots \end{array}$$

be two *positive* complexes. Let

$$T_n = \bigoplus_{i+j=n} (C_i \otimes C'_j)$$

and let $\tau_n \colon T_n \to T_{n-1}$ take $c \otimes c' \in C_i \otimes C_j \subset T_n$ into

$$\tau_n(c \otimes c') = (\partial_i c \otimes c') + (-1)^i (c \otimes \partial'_j c') \in (C_{i-1} \otimes C_j) \oplus (C_i \otimes C_{j-1}) \subset T_{n-1}.$$

A direct verification (see below) shows that $\tau_{n-1} \circ \tau_n = 0$. The complex arising,

$$\dots \xrightarrow{\tau_{n+1}} T_n \xrightarrow{\tau_n} T_{n-1} \xrightarrow{\tau_{n-1}} \dots,$$

is called the *tensor product* of the complexes $\mathcal{C}$ and $\mathcal{C}'$ and is denoted as $\mathcal{C} \otimes \mathcal{C}'$.

VERIFICATION OF $\tau_{n-1} \circ \tau_n = 0$. Let $c \in C_i$, $c' \in C'_j$. Then

$$\begin{aligned}\tau_{n-1}\circ\tau_n(c\otimes c') &= \tau_{n-1}(\partial_i c\otimes c')+(-1)^i\tau_{n-1}(c\otimes\partial'_j c')\\ &= (\partial_{i-1}\circ\partial_i(c)\otimes c')+(-1)^{i-1}(\partial_i c\otimes\partial'_j c')\\ &+(-1)^i((\partial_i c\otimes\partial'_j c')+(-1)^{j-1}(c\otimes\partial'_{j-1}\circ\partial'_j(c')))\\ &= (-1)^{i-1}(\partial_i c\otimes\partial'_j c')+(-1)^i(\partial_i c\otimes\partial'_j c')=0.\end{aligned}$$

Our next goal is to express the homology of the tensor product of two complexes in terms of homologies of these complexes.

Theorem 2. *If the complexes $\mathcal{C}$, and $\mathcal{C}'$ are free (that is, all C_n, C'_n are free Abelian groups), then, for every n,*

(1) *There is a (noncanonical) isomorphism*

$$H_n(\mathcal{C}\otimes\mathcal{C}')\cong \bigoplus_{i+j=n}(H_i(\mathcal{C})\otimes H_j(\mathcal{C}')\quad\bigoplus\bigoplus_{i+j=n-1}\operatorname{Tor}(H_i(\mathcal{C}),H_j(\mathcal{C}')).$$

(2) *There is a canonically defined exact sequence*

$$0\to\bigoplus\nolimits_{i+j=n}(H_i(\mathcal{C})\otimes H_j(\mathcal{C}'))\to H_n(\mathcal{C}\otimes\mathcal{C}')\to\bigoplus\nolimits_{i+j=n-1}\operatorname{Tor}(H_i(\mathcal{C}),H_j(\mathcal{C}'))\to 0.$$

Proof. Begin with part (2). Let $Z_n=\operatorname{Ker}\partial_n, B_{n-1}=\operatorname{Im}\partial_n$. Consider the diagram

$$\begin{array}{ccccccccc}
 & & 0 & & 0 & & 0 & & \\
 & & \downarrow & & \downarrow & & \downarrow & & \\
\dots & \xrightarrow{0} & Z_{n+1} & \xrightarrow{0} & Z_n & \xrightarrow{0} & Z_{n-1} & \xrightarrow{0} & \dots\\
 & & \downarrow\subset & & \downarrow\subset & & \downarrow\subset & & \\
\dots & \xrightarrow{\partial} & C_{n+1} & \xrightarrow{\partial} & C_n & \xrightarrow{\partial} & C_{n-1} & \xrightarrow{\partial} & \dots\\
 & & \downarrow\partial & & \downarrow\partial & & \downarrow\partial & & \\
\dots & \xrightarrow{0} & B_n & \xrightarrow{0} & B_{n-1} & \xrightarrow{0} & B_{n-2} & \xrightarrow{0} & \dots\\
 & & \downarrow & & \downarrow & & \downarrow & & \\
 & & 0 & & 0 & & 0. & &
\end{array}$$

The rows of this diagram are complexes, the columns are exact sequences, and the diagram is commutative. Thus, this diagram can be regarded as a short exact sequence of complexes:

$$0\to\mathcal{Z}\to\mathcal{C}\to\mathcal{B}\to 0,$$

where $\mathcal{Z}$ and $\mathcal{B}$ are complexes with trivial differential composed of groups Z_n and B_n [but the nth group of the complex $\mathcal{B}$ is B_{n-1}]. Since the complex $\mathcal{C}'$ is free, the sequence remains exact after tensoring with $\mathcal{C}'$:

$$0 \to \mathcal{Z} \otimes \mathcal{C}' \to \mathcal{C} \otimes \mathcal{C}' \to \mathcal{B} \otimes \mathcal{C}' \to 0.$$

Since $\mathcal{Z}$ and $\mathcal{B}$ have trivial differentials and consist of free Abelian groups,

$$H_n(\mathcal{Z} \otimes \mathcal{C}') = \bigoplus_{i+j=n} (Z_i \otimes H_j(\mathcal{C}')), \; H_n(\mathcal{B} \otimes \mathcal{C}') = \bigoplus_{i+j=n-1} (B_i \otimes H_j(\mathcal{C}')).$$

Thus, the homology sequence corresponding to the last short exact sequence of complexes takes the form

$$\bigoplus_{i+j=n} (B_i \otimes H_j(\mathcal{C}')) \xrightarrow{\varphi} \bigoplus_{i+j=n} (Z_i \otimes H_j(\mathcal{C}')) \to H_n(\mathcal{C} \otimes \mathcal{C}')$$
$$\to \bigoplus_{i+j=n-1} (B_i \otimes H_j(\mathcal{C}')) \xrightarrow{\psi} \bigoplus_{i+j=n-1} (Z_i \otimes H_j(\mathcal{C}')).$$

It is easy to see also that the connecting homomorphisms φ and ψ are induced by the inclusion maps $B_i \to Z_i$ [before tensoring with $\mathcal{C}'$, they consist first in applying ∂^{-1} and then ∂; tensoring with $\mathcal{C}'$ does not change anything]. Since the Abelian groups B_i and Z_i are free and $H_i(\mathcal{C}) = Z_i/B_i$, the exact sequence $0 \to \operatorname{Coker} \varphi \to H_n(\mathcal{C} \otimes \mathcal{C}') \to \operatorname{Ker} \psi \to 0$ is precisely the exact sequence from part (2) of Theorem 2.

To prove part (1), first notice that if $H_n(\mathcal{C}) = 0$ for $n \neq i$ and $H_n(\mathcal{C}') = 0$ for $n \neq j$, then part (2) shows that the homology of $\mathcal{C} \otimes \mathcal{C}'$ is zero, except

$$H_{i+j}(\mathcal{C} \otimes \mathcal{C}') = H_i(\mathcal{C}) \otimes H_j(\mathcal{C}'),$$
$$H_{i+j-1}(\mathcal{C} \otimes \mathcal{C}') = \operatorname{Tor}(H_i(\mathcal{C}), H_j(\mathcal{C}')),$$

so the isomorphism of part (1) holds. In general,

$$\mathcal{C} \cong \bigoplus \mathcal{C}(i), \text{ where } \mathcal{C}(i) \text{ is } \ldots 0 \to 0 \to \underset{(i+1)}{B_i} \xrightarrow{\text{incl.}} \underset{(i)}{Z_i} \to 0 \to 0 \ldots,$$

$$\mathcal{C}' \cong \bigoplus \mathcal{C}'(j), \text{ where } \mathcal{C}'(j) \text{ is } \ldots 0 \to 0 \to \underset{(j+1)}{B'_j} \xrightarrow{\text{incl.}} \underset{(j)}{Z'_j} \to 0 \to 0 \ldots$$

(noncanonical isomorphisms; compare with Sect. 15.5), and all the homology groups of $\mathcal{C}(i)$ and $\mathcal{C}'(j)$ are zero besides $H_i(\mathcal{C}(i)) = H_i(\mathcal{C})$ and $H_j(\mathcal{C}'(j)) = H_j(\mathcal{C}')$. This implies part (1) in full generality.

Proof of Theorem 1. In the case when X_1 and X_2 are CW complexes, it is sufficient to remark that the cellular chain complex of $X_1 \times X_2$ is the tensor product of the cellular complexes of X_1 and X_2 ($e \times e' \leftrightarrow e \otimes e'$). To extend the result to arbitrary topological spaces, we use two previous results: (1) Every topological space is weakly homotopy equivalent to a CW complex (Sect. 11.6); and (2) homology is weakly homotopy invariant (Sect. 14.1).

Remarks. (1) It is not true, in general, that the singular complex of the product $X_1 \times X_2$ of two topological spaces is isomorphic the tensor product of the singular complexes of X_1 and X_2. But these complexes are homotopy equivalent (there exists a homotopy equivalence canonically defined up to a homotopy between them). This fact, known as the Eilenberg–Zilber theorem, is proved in many textbooks in topology.

(2) A comparison of the universal coefficients formula with Künneth's formula gives the following result (which may be useful in Chap. 3):

$$H_n(X_1 \times X_2) = \bigoplus_{i+j=n} H_i(X_1; H_j(X_2)).$$

EXERCISE 19. The last equality can be modified to the case of homology and cohomology with coefficients:

$$H_n(X_1 \times X_2; G) = \bigoplus\nolimits_{i+j=n} H_i(X_1; H_j(X_2; G))$$
$$H^n(X_1 \times X_2; G) = \bigoplus\nolimits_{i+j=n} H^i(X_1; H^j(X_2; G)).$$

(These equalities, as well as the equality in the preceding remark, can be proven without any references to the universal coefficients and Künneth's formulas: They hold, actually, at the level of cellular chains. This provides a direct way to deduce the noncanonical part of Künneth's formula from the similar part of the universal coefficients formulas.)

Here is a small but significant application of Künneth's formula.

EXERCISE 20. Find the homology of $\mathbb{R}P^2 \times \mathbb{R}P^2$. (If the result seems unexpected to you, check it using a direct cellular computation.)

Like the universal coefficients formula, Künneth's formula simplifies a lot in the case of coefficients in a field.

EXERCISE 21. Prove that if $\mathbb{K}$ is a field, then

$$H_n(X_1 \times X_2; \mathbb{K}) = \bigoplus\nolimits_{i+j=n} H_i(X_1; \mathbb{K}) \otimes_{\mathbb{K}} H_j(X_2; \mathbb{K}),$$
$$H^n(X_1 \times X_2; \mathbb{K}) = \bigoplus\nolimits_{i+j=n} H^i(X_1; \mathbb{K}) \otimes_{\mathbb{K}} H^j(X_2; \mathbb{K}).$$

In conclusion, here are two more formulas.

EXERCISE 22. $B_n(X_1 \times B_2) = \sum_{i+j=n} B_i(X_1)B_j(X_2)$.

EXERCISE 23. $\chi(X_1 \times X_2) = \chi(X_1)\chi(X_2)$. (In both exercises, we assume that the right-hand sides of the formulas are defined.)

Lecture 16 Multiplications

16.1 Introduction

Although homology is geometrically much more transparent than cohomology, cohomology is immensely more useful because it possesses many naturally defined additional structures. The first of these structures is a multiplication: If G is a ring, then for $\alpha \in H^{n_1}(X;G)$ and $\beta \in H^{n_2}(X;G)$ there exists a naturally defined "product" $\alpha\beta \in H^{n_1+n_2}(X;G)$ which has good algebraic properties. Nothing like this is possible for homology (see Exercise 14 ahead). We will discuss these products (and some other products) in this lecture and will describe many other structures in later chapters (starting with Chap. 4).

The simplest way to introduce the cohomological multiplication is as follows. Let G be a commutative ring, and let X_1, X_2 be two CW complexes. For cellular cochains $c_1 \in \mathcal{C}^{n_1}(X_1;G), c_2 \in \mathcal{C}^{n_2}(X_2;G)$, we define a cellular cochain $c_1 \times c_2 \in \mathcal{C}^{n_1+n_2}(X_1 \times X_2;G)$ in the most natural way: For the oriented cells $e_1 \subset X_1, e_2 \subset X_2$ of dimensions n_1, n_2, the value of $c_1 \times c_2$ on $e_1 \times e_2$ is $c_1(e_1)c_2(e_2)$ (product in G). It is easy to check that $\delta(c_1 \times c_2) = (\delta c_1) \times c_2 + (-1)^{n_1} c_1 \times \delta c_2$; thus, if c_1, c_2 are cocycles, then $c_1 \times c_2$ is also a cocycle. The same formula shows that the cohomology class of the cocycle $c_1 \times c_2$ depends only on the cohomology classes of cocycles c_1, c_2, so we get a valid (bilinear, associative) multiplication

$$[\gamma_1 \in H^{n_1}(X_1;G), \gamma_2 \in H^{n_2}(X_2;G)] \mapsto \gamma_1 \times \gamma_2 \in H^{n_1+n_2}(X_1 \times X_2;G).$$

A similar construction exists for homology. Namely, if $a_1 = \sum_i g_i e_{1i} \in \mathcal{C}_{n_1}(X_1;G), a_2 = \sum_j g_j e_{2j} \in \mathcal{C}_{n_2}(X_2;G)$, then we put

$$a_1 \times a_2 = \sum_{i,j} (g_i g_j)(e_{1i} \times e_{2j}) \in \mathcal{C}_{n_1+n_2}(X_1 \times X_2;G).$$

A check shows that $\partial(a_1 \times a_2) = (\partial a_1) \times a_2 + (-1)^{n_1} a_1 \times \partial a_2$, which gives rise to a homological multiplication

$$[\alpha_1 \in H_{n_1}(X_1;G), \alpha_2 \in H_{n_2}(X_2;G)] \mapsto \alpha_1 \times \alpha_2 \in H_{n_1+n_2}(X_1 \times X_2;G).$$

The two $\times$-products (usually called *cross-products*) are connected by the formula

$$\langle \gamma_1 \times \gamma_2, \alpha_1 \times \alpha_2 \rangle = (-1)^{n_1 n_2} \langle \gamma_1, \alpha_1 \rangle \langle \gamma_2, \alpha_2 \rangle.$$

EXERCISE 1. Another definition of the homological $\times$-product can be obtained from Künneth's formula: This formula yields a canonical map $H_{n_1}(X_1) \otimes H_{n_2}(X_2) \to H_{n_1+n_2}(X_1 \times X_2)$, and the image of $\alpha_1 \otimes \alpha_2$ with respect to this map is taken for $\alpha_1 \times \alpha_2$. Prove the equivalence of the two definitions.

At this moment, however, the difference between homology and cohomology becomes important. For any topological space X, there exists the *diagonal map* $\Delta: X \to X \times X$, $\Delta(x) = (x, x)$. This maps induces homomorphisms

$$\Delta_*: H_n(X; G) \to H_n(X \times X; G),$$
$$\Delta^*: H^n(X \times X; G) \to H^n(X; G);$$

of these homomorphisms; the first one is useless for us now, but the second one provides cohomological multiplication: For $\gamma_1 \in H^{n_1}(X; G), \gamma_2 \in H^{n_2}(X; G)$, we put

$$\gamma_1 \smile \gamma_2 = \Delta^*(\gamma_1 \otimes \gamma_2) \in H^{n_1+n_2}(X; G).$$

(The classical notation $\smile$, "cup," is not very convenient, so often instead of $\gamma_1 \smile \gamma_2$ we will simply write $\gamma_1\gamma_2$.)

However, this way of defining the cohomological product has two important disadvantages. First, we must still prove the independence of the CW structure. Second, the diagonal map is not cellular, and to apply it to a cellular cochain we need to choose a cellular approximation, which cannot be done in a canonical way, at least, in the context of arbitrary CW complexes. To avoid these difficulties we will use the opposite order of the definition. First, we will define a $\smile$-product (usually called the *cup-product*) by a singular, topologically invariant, construction, and then we will use it to define the cross-product.

Terminological Remark. The cup-product was initially called the *Kolmogorov–Alexander product*, after the two remarkable mathematicians who (independently of each other) conceived of this operation in the mid-1930s. Unfortunately, the next generation of topologists found this term too long.

16.2 The Cup-Product: A Direct Construction

In the standard simplex Δ^n, $n = n_1 + n_2$ with the vertices $v_0, \dots, v_n$, consider two faces of dimensions n_1 and n_2: $\Gamma_-^{n_1}\Delta^n$ with the vertices $v_0, \dots, v_{n_1}$ and $\Gamma_+^{n_2}\Delta^n$ with vertices $v_{n_1}, \dots, v_n$. These faces have dimensions n_1 and n_2 and have one common vertex, v_{n_1}. Accordingly, for an n-dimensional singular simplex $f: \Delta^n \to X$, we will consider faces $\Gamma_-^{n_1}f = f\mid_{\Gamma_-^{n_1}\Delta^n}$ and $\Gamma_+^{n_2}f = f\mid_{\Gamma_+^{n_2}\Delta^n}$, which are singular simplices of dimensions n_1 and n_2.

Let X be an arbitrary topological space and let G be a commutative ring. Then let $c_1 \in C^{n_1}(X; G)$ and $c_2 \in C^{n_2}(X; G)$. We define a cochain $c_1 \smile c_2 \in C^{n_1+n_2}(X; G)$ by the formula

$$[c_1 \smile c_2](f) = c_1(\Gamma_-^{n_1}f)c_2(\Gamma_+^{n_2}f),$$

where f is $(n_1 + n_2)$-dimensional singular simplex of X.

Proposition (Properties of the Cochain Cup-Product). *Let $c_1 \in C^{n_1}(X; G)$, $c_2 \in C^{n_2}(X; G)$. Then*

(0) $\delta(c_1 \smile c_2) = (\delta c_1) \smile c_2 + (-1)^{n_1} c_1 \smile \delta c_2$.
(1) $c_1 \smile (c_2 \smile c_3) = (c_1 \smile c_2) \smile c_3$ $[c_3 \in C^{n_3}(X; G)]$.
(2) *Let ω be the backward transformator (Example 2 in Sect. 12.5). Then for any $(n_1 + n_2)$-dimensional singular chain a,*

$$[c_1 \smile c_2](a) = (-1)^{n_1 n_2}[c_2 \smile c_1](\omega^X_{n_1+n_2} a).$$

(3) *For a continuous map $g: X \to Y$,*

$$g^\#(c_1 \smile c_2) = (g^\# c_1) \smile (g^\# c_2).$$

(4) *For a ring homomorphism $h: G \to H$,*

$$h_*(c_1 \smile c_2) = (h_* c_1) \smile (h_* c_2).$$

Proof The proof is obvious [only property (0) requires a simple calculation] and is left to the reader.

Remark. The noncommutativity (even the non-plus-minus-commutativity) of the chain cup-product is an unavoidable property which has important consequences (which will show themselves in Chap. 4).

Property (0) shows that the cup-product of two cocycles is a cocycle whose cohomology class depends only on the cohomology classes of the factors. This gives rise to the cohomological cup-product

$$[\gamma_1 \in H^{n_1}(X_1; G), \gamma_2 \in H^{n_2}(X_2; G)] \mapsto \gamma_1 \times \gamma_2 \in H^{n_1+n_2}(X_1 \times X_2; G).$$

Theorem (Properties of the Cohomology Cup-Product). *Let $\gamma_1 \in H^{n_1}(X; G)$, $\gamma_2 \in H^{n_2}(X; G)$. Then*

(1) $\gamma_1 \smile (\gamma_2 \smile \gamma_3) = (\gamma_1 \smile \gamma_2) \smile \gamma_3$ $[\gamma_3 \in H^{n_3}(X; G)]$.
(2) $\gamma_1 \smile \gamma_2 = (-1)^{n_1 n_2} \gamma_2 \smile \gamma_1$.
(3) *For a continuous map $g: X \to Y$,*

$$g^*(\gamma_1 \smile \gamma_2) = (g^* \gamma_1) \smile (g^* \gamma_2).$$

(4) *For a ring homomorphism $h: G \to H$,*

$$h_*(\gamma_1 \smile \gamma_2) = (h_* \gamma_1) \smile (h_* \gamma_2).$$

This follows from the proposition [the proof of property (2) uses the transformator lemma; see Sect. 12.5].

Notice that there is an obvious generalization of the previous construction: If $\gamma_1 \in H^{n_1}(X; G_1)$, $\gamma_2 \in H^{n_2}(X; G_2)$ and there is a pairing $\mu: G_1 \times G_2 \in G$, then there arises a cup-product $\gamma_1 \smile_\mu \gamma_2 = \gamma_1 \smile \gamma_2 \in H^{n_1+n_2}(X; G)$. For example, if $\gamma_1 \in H^{n_1}(X; G)$ (where G is just an Abelian group) and $\gamma_2 \in H^{n_2}(X; \mathbb{Z})$, then there is a cup-product $\gamma_1 \smile \gamma_2 \in H^{n_1+n_2}(X; G)$.

EXERCISE 2. Prove that if X is connected and $\gamma \in H^0(X; G) = G$, then $\gamma \smile \gamma_1 = \gamma\gamma_1$ for any $\gamma_1 \in H^n(X; G)$. In particular, if $1 \in G$ is the unity of the ring G, then $1 \in G = H^0(X; G)$ is the unity of the cohomological multiplication.

EXERCISE 3. Construct a relative version of cup-product: If $\gamma_1 \in H^{n_1}(X, A; G)$ and $\gamma \in H^{n_2}(X, B; G)$, then $\gamma_1 \smile \gamma_2 \in H^{n_1+n_2}(X, A \cup B; G)$. [To prove this, it is convenient to regard $H_n(X, A \cup B)$ not as the homology of the complex consisting of the groups $C_n(X)/C_n(A \cup B)$, but rather as the complex of groups $C_n(X)/(C_n(A) \oplus C_n(B))$; the homology remains the same (for sufficiently good A and B) by the refinement lemma.]

16.3 The Cross-Product: A Construction via the Cup-Product

As before, let X_1, X_2 be topological spaces, let G be a commutative ring, and let $\gamma_1 \in H^{n_1}(X_1; G)$, $\gamma_2 \in H^{n_2}(X_2; G)$ be cohomology classes. Put

$$\gamma_1 \times \gamma_2 = (p_1^*\gamma_1) \smile (p_2^*\gamma_2) \in H^{n_1+n_2}(X_1 \times X_2; G),$$

where p_1 and p_2 are projections of $X_1 \times X_2$ onto X_1 and X_2.

EXERCISE 4. Make up a definition of the relative cross-product,

$$[\gamma_1 \in H^{n_1}(X_1, A_1; G), \gamma_2 \in H^{n_2}(X_2, A_2; G)]$$
$$\mapsto \gamma_1 \times \gamma_2 \in H^{n_1+n_2}(X_1 \times X_2, (A_1 \times X_2) \cup (X_1 \times A_2); G).$$

EXERCISE 5. Check all kinds of naturalness for the cross-product.

Theorem. *This definition of the cross-product is equivalent to that in Sect. 16.1.*

Proof. It turns out to be sufficient to compute explicitly the cross-product in one particular case. Since standard simplices and their products are homeomorphic to balls,

$$H^{n_1}(\Delta^{n_1}, \partial\Delta^{n_1}; \mathbb{Z}) = \mathbb{Z}, \ H^{n_2}(\Delta^{n_2}, \partial\Delta^{n_2}; \mathbb{Z}) = \mathbb{Z};$$
$$H^{n_1+n_2}(\Delta^{n_1} \times \Delta^{n_2}, \partial(\Delta^{n_1} \times \Delta^{n_2}); \mathbb{Z}) = \mathbb{Z}.$$

Similar formulas hold for homology.

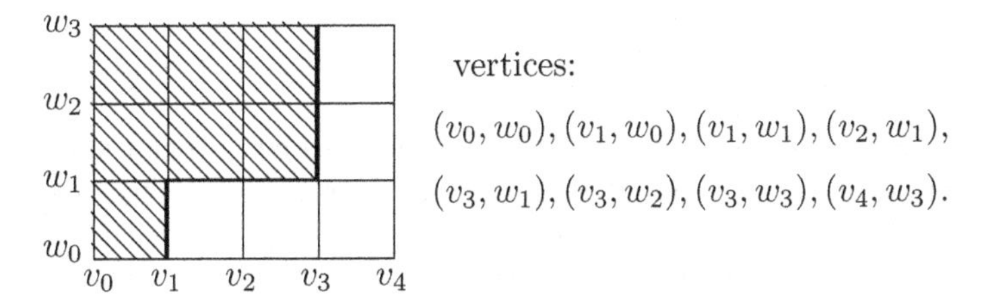

Fig. 64 Triangulation of a product of simplices

What we want to check is that the cross-product of the generators of the groups $H^{n_1}(\Delta^{n_1}, \partial\Delta^{n_1}; \mathbb{Z}) = \mathbb{Z}$, $H^{n_2}(\Delta^{n_2}, \partial\Delta^{n_2}; \mathbb{Z}) = \mathbb{Z}$ is, up to a sign, the generator of $H^{n_1+n_2}(\Delta^{n_1} \times \Delta^{n_2}, \partial(\Delta^{n_1} \times \Delta^{n_2}); \mathbb{Z}) = \mathbb{Z}$.

Obviously, the singular simplex id: $\Delta^{n_1} \to \Delta^{n_1}$ is a relative cycle representing the generator of $\mathbb{Z} = H_{n_1}(\Delta^{n_1}, \partial\Delta^{n_1})$, and similarly for Δ^{n_2}. As to $\mathbb{Z} = H^{n_1+n_2}(\Delta^{n_1} \times \Delta^{n_2}, \partial(\Delta^{n_1} \times \Delta^{n_2}); \mathbb{Z})$, to describe the generator, we will construct a triangulation (actually, quite standard) of the product $\Delta^{n_1} \times \Delta^{n_2}$, generalizing the triangulation of the product $\Delta^n \times I$ constructed in Sect. 12.2; see Fig. 59.

Let $v_0, v_1, \dots, v_{n_1}$ be the vertices of Δ^{n_1}, and let $w_0, w_1, \dots, w_{n_2}$ be the vertices of Δ^{n_2}. In $\Delta^{n_1} \times \Delta^{n_2}$, take $(n_1 + n_2)$-dimensional affine simplices whose vertices make a sequence of the form

$$(v_{i_0}, w_{j_0}), (v_{i_1}, w_{j_1}), (v_{i_2}, w_{j_2}), \dots, (v_{i_{n_1+n_2}}, w_{j_{n_1+n_2}}),$$

where

$$\begin{gathered} 0 = i_0 \le i_1 \le i_2 \le \dots \le i_{n_1+n_2} = n_1; \\ 0 = j_0 \le j_1 \le j_2 \le \dots \le j_{n_1+n_2} = n_2; \\ i_s + j_s = s. \end{gathered}$$

In other words, in an $(n_1 + 1) \times (n_2 + 1)$ grid with horizontal bars labeled by $w_0, \dots, w_{n_2}$ and vertical bars labeled by $v_0, \dots, v_{n_1}$, we choose a path from (v_0, w_0) to (v_{n_1}, w_{n_2}) and take the sequence of crossings of the bars on this path (see an example in Fig. 64).

There are $\begin{pmatrix} n_1 + n_2 + 2 \\ n_1 + 1 \end{pmatrix}$ such paths, and accordingly $\Delta^{n_1} \times \Delta^{n_2}$ falls into the union of this amount of $(n_1 + n_2)$-dimensional simplices. These simplices can be described in terms of barycentric coordinates: to which of them the point $((t_0, \dots, t_{n_1}), (u_0, \dots, u_{n_2}) \in \Delta^{n_1} \times \Delta^{n_2}$ belongs depends on the ordering of numbers

$$t_0, t_0 + t_1, \dots, t_0 + t_1 + \dots + t_{n_1-1}; u_0, u_0 + u_1, \dots, u_0 + u_1 + \dots + u_{n_2-1}.$$

For example, the seven-dimensional simplex corresponding to the path in Fig. 64 is described in $\Delta^3 \times \Delta^4$ by the inequalities

$$0 \le t_0 \le u_0 \le t_0 + t_1 \le t_0 + t_1 + t_2 \le u_0 + u_1$$
$$\le u_0 + u_1 + u_2 \le t_0 + t_1 + t_2 + t_3 \le 1$$

(the rule is as follows: We move along the path and after a horizontal edge we place the sum of ts, and after a vertical edge we place the sum of us). Since the vertices of each simplex of the subdivision are ordered, there arise canonical maps of the standard simplex onto the simplices of the subdivision, that is, singular simplices of $\Delta^{n_1} \times \Delta^{n_2}$. Let $c(n_1, n_2) \in C_{n_1+n_2}(\Delta^{n_1} \times \Delta^{n_2})$ be the sum of these singular simplices with the coefficients ± 1 where the sign is determined by the parity of the number of squares of grid below the chosen path (left unshadowed in Fig. 64; for the path shown there this number is 5 and the sign is minus). It is obvious that $c(n_1, n_2)$ is a relative cycle modulo $\partial(\Delta^{n_1} \times \Delta^{n_2})$: Two of our simplices have a common $(n_1 + n_2 - 1)$-dimensional face in the interior of $\Delta^{n_1} \times \Delta^{n_2}$ if and only if the two paths have precisely one square between them; then they appear in $c(n_1, n_2)$ with opposite signs, and the faces have the same number in them; so the faces cancel. To prove that $\alpha_1 \times \alpha_2$ is plus–minus the standard generator of $H^{n_1+n_2}(\Delta^{n_1} \times \Delta^{n_2}, \partial(\Delta^{n_1} \times \Delta^{n_2}); \mathbb{Z}) = \mathbb{Z}$, it is sufficient to check that $\langle \alpha_1 \times \alpha_2, c(n_1, n_2) \rangle = \pm 1$. For an $(n_1 + n_2 - 1)$-dimensional singular simplex f of $\Delta^{n_1} \times \Delta^{n_2}$, the value of $\alpha_1 \times \alpha_2$ of f (here by α_1, α_2 we mean rather cochains than cohomology classes) is $\alpha_1(p_1 \circ \Gamma_-^{n_1} f)\alpha_2(p_2 \circ \Gamma_+^{n_2} f)$. But for a simplex f with vertices

$$(v_{i_0}, w_{j_0}), (v_{i_1}, w_{j_1}), (v_{i_2}, w_{j_2}), \dots, (v_{i_{n_1+n_2}}, w_{j_{n_1+n_2}}),$$

the simplex $p_1(\Gamma_-^{n_1} f)$ has the vertices $v_{i_0}, \dots, v_{i_{n_1}}$ and the simplex $p_2(\Gamma_+^{n_2} f)$ has the vertices $w_{j_{n_1}}, \dots, w_{j_{n_1+n_2}}$. The only case when these two simplices are not contained in $\partial\Delta^{n_1}$ and $\partial\Delta^{n_1}$ is when

$$i_0 = 0, \dots, i_{n_1-1} = n_1 - 1, i_{n_1} = i_{n_1+1} = \dots = i_{n_1+n_2} = n_1;$$
$$j_0 = j_1 = \dots = j_{n_1} = 0, j_{n_1+1} = 1, \dots, j_{n_1+n_2} = n_2.$$

Thus, only one summand in $c(n_1, n_2)$ makes a contribution into $\langle \alpha_1 \times \alpha_2, c(n_1, n_2) \rangle$, and this contribution is ± 1.

The rest of the proof uses only the naturalness of the cross-product. It consists of six steps.

Step 1. The cross-product

$$H^{n_1}(S^{n_1}, \mathrm{pt}; \mathbb{Z}) \times H^{n_2}(S^{n_2}, \mathrm{pt}; \mathbb{Z}) \to H^{n_1+n_2}(S^{n_1} \times S^{n_2}, S^{n_1} \vee S^{n_2}; \mathbb{Z})$$

is, up to a sign, the standard multiplication $\mathbb{Z} \times \mathbb{Z} \to \mathbb{Z}$. Indeed, the projections $(\Delta^{n_1}, \partial\Delta^{n_1}) \to (S^{n_1}, \mathrm{pt}), (\Delta^{n_2}, \partial\Delta^{n_2}) \to (S^{n_2}, \mathrm{pt}), (\Delta^{n_1} \times \Delta^{n_2}, \partial(\Delta^{n_1} \times \Delta^{n_2})) \to (S^{n_1} \times S^{n_2}, S^{n_1} \vee S^{n_2})$ induce isomorphisms in the cohomology of dimensions $n_1, n_2, n_1 + n_2$.

Step 2. The cross-product

$$H^{n_1}(S^{n_1}; \mathbb{Z}) \times H^{n_2}(S^{n_2}; \mathbb{Z}) \to H^{n_1+n_2}(S^{n_1} \times S^{n_2}; \mathbb{Z})$$

is, up to a sign, the standard multiplication $\mathbb{Z}\times\mathbb{Z} \to \mathbb{Z}$. Indeed, the maps $(S^{n_1}, \text{pt}) \to (S^{n_1}, \emptyset), \dots$ induce isomorphisms in the cohomology of appropriate dimensions.

Step 3. Similar statements for the bouquets of spheres (we leave precise statements to the reader).

Step 4. The ring $\mathbb{Z}$ can be replaced by an arbitrary ring G. This follows from the naturalness of the cross-product with respect to ring homomorphisms $\mathbb{Z} \to G$.

Step 5. X_1, X_2 are CW complexes of the respective dimensions n_1, n_2, and cohomology classes $\gamma_1 \in H^{n_1}(X_1; G), \gamma_2 \in H^{n_2}(X_2; G)$ are represented by cellular cocycles c_1, c_2; then $\gamma_1 \times \gamma_2 \in H^{n_1+n_2}(X_1 \times X_2; G)$ is represented by the cellular cocycle

$$[c_1 \times c_2](e_1 \times e_2) = \pm c_1(e_1)c_2(e_2).$$

For the proof we can consider the projections $X_1 \to X_1/\operatorname{sk}_{n_1-1} X_1, X_2 \to X_2/\operatorname{sk}_{n_2-1} X_2$; the induced cohomology homomorphisms are epimorphisms.

Step 6. The general case. For the transition to this case we consider the inclusion maps $\operatorname{sk}_{n_1} X_1 \to X_1, \operatorname{sk}_{n_2} X_2 \to X_2, \operatorname{sk}_{n_1} X_1 \times \operatorname{sk}_{n_2} X_2 \to X_1 \times X_2$; the induced cohomology homomorphisms in the appropriate dimensions are monomorphisms.

This completes the proof.

16.4 Cup-Product and Diagonal Map

Now let us briefly investigate the connection between the definition of the cup-product in Sect. 16.2 and the preliminary definition from the introduction (Sect. 16.1). The first statement is almost obvious.

Theorem. *For any X, G, and $\gamma_1 \in H^{n_1}(X; G), \gamma_2 \in H^{n_2}(X; G)$,*

$$\gamma_1 \smile \gamma_2 = \Delta^*(\gamma_1 \times \gamma_2),$$

where $\Delta: X \to X \times X$ is the diagonal map.

Proof. Obviously, $p_1 \circ \Delta = p_2 \circ \Delta = \text{id}$. Hence,

$$\Delta^*(\gamma_1 \times \gamma_2) = \Delta^*(p_1^*\gamma_1 \smile p_2^*\gamma_2) = (p_1 \circ \Delta)^*\gamma_1 \smile (p_2 \circ \Delta)^*\gamma_2 = \gamma_1 \smile \gamma_2.$$

In addition to that, we remark that actually the definition of cup-product in Sect. 16.2 can be regarded as a combination of the definition in Sect. 16.1 and a particular choice of a cellular approximation of the diagonal map. Let us describe the latter, first in the case when X is a triangulated space. First, in the product $\Delta^n \times \Delta^n$, let us consider the CW subcomplex $\bigcup_{p+q=n}(\Gamma_-^p \Delta^n \times \Gamma_+^q \Delta^n)$; for $n_1 = n_2 = 2$, it is shown in Fig. 65 (surely, a picture of a four-dimensional figure on a two-dimensional paper sheet cannot be awfully clear). The dashed triangle is the diagonal image of Δ^2; it is not a cellular subspace of $\Delta^2 \times \Delta^2$. The cellular

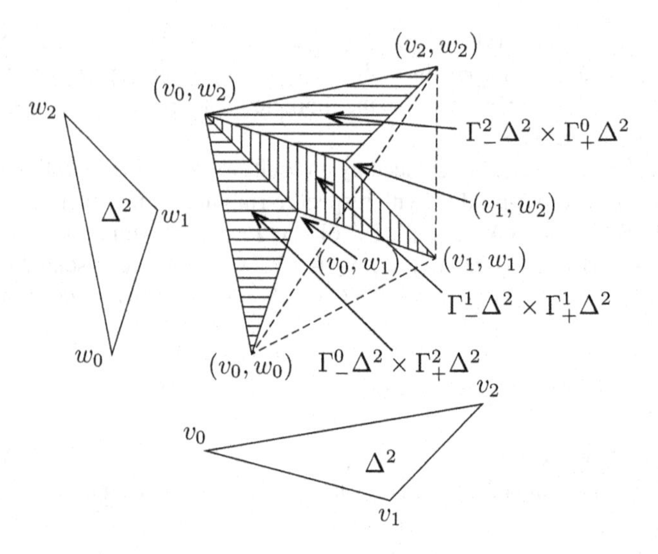

Fig. 65 A cellular approximation of the diagonal in $\Delta^2 \times \Delta^2$

approximations of the diagonal edges $[(v_0, w_0), (v_1, w_1)]$, $[(v_1, w_1), (v_2, w_2)]$, and $[(v_2, w_2), (v_0, w_0)]$ are broken lines $[(v_0, w_0), (v_0, w_1), (v_1, w_1)]$, $[(v_1, w_1), (v_1, w_2), (v_2, w_2)]$, and $[(v_2, w_2), (v_2, w_0), (v_0, w_0)]$; the diagonal triangle is approximated by the union of three pieces: two triangles and one parallelogram, as shown in Fig. 65.

In general, the approximation $\Delta_0: \Delta^n \xrightarrow{\cong} \bigcup_{p+q=n}(\Gamma^p_-\Delta^n \times \Gamma^q_+\Delta^n) \subset \Delta^n \times \Delta^n$ is defined by the formula

$$\begin{aligned}(t_0, \dots, t_n) \mapsto ((2t_0, \dots, 2t_{p-1}, 2(t_p + \dots + t_n) - 1, 0 \dots, 0),\\ (0, \dots, 0, 2(t_0 + \dots + t_p) - 1, 2t_{p+1}, \dots, 2t_n))),\\ \text{if } t_0 + \dots + t_p \geq \tfrac{1}{2},\ t_p + \dots + t_n \geq \tfrac{1}{2}.\end{aligned}$$

It is clear that the restriction of Δ_0 to any face of Δ^n (of any dimension) is a similar map for this face.

If X is an *ordered* triangulated space (see Sect. 13.10), then this construction can be applied to each simplex of the triangulation, and we obtain a canonical cellular approximation $\Delta_0: X \to X \times X$ of the diagonal map (here we mean the CW structure of $X \times X$ which is obtained as the product of two copies of the triangulation of X regarded as a CW structure; thus, the cells of $X \times X$ are products of simplices). Now it is clear that for the two cochains $c_1 \in C^{n_1}(X; G), c_2 \in C^{n_2}(X; G)$, the cochain $c_1 \smile c_2 \in C^{n_1+n_2}(X \times X; G)$ is nothing but $(\Delta_0)_\#(c_1 \times c_2)$; this sheds light on the connection between the definitions of cup-product given in Sects. 16.1 and 16.2. We can add that the construction above can be applied not only to triangulated spaces; for example, it works perfectly well for the cellular realization $Sing(X)$ of the singular complex of an arbitrary topological space, and hence gives an explanation for the construction of the $\smile$-product of singular cochains.

16.5 First Application: The Hopf Invariant

To demonstrate at once the power of the cohomological multiplication, we will immediately, before any serious computations of this multiplication, prove a highly nontrivial statement concerning the homotopy groups of spheres.

Theorem. *The group $\pi_{4n-1}(S^{2n})$ is infinite for any $n \geq 1$. Moreover, the Whitehead square $[\iota_{2n}, \iota_{2n}]$ of the generator of $\pi_{2n}(S^{2n})$ has an infinite order in $\pi_{4n-1}(S^{2n})$.* (Compare this theorem with the results of Sects. 9.9 and 10.5.)

The proof of this theorem is based on the *Hopf invariant*, which is an integer assigned to every element of $\varphi \in \pi_{4n-1}(S^{2n})$. Its definition is as follows. Consider a spheroid $f: S^{4n-1} \to S^{2n}$ and form the space $X_\varphi = S^{2n} \cup_f D^{4n}$ (aka the cone of f). The space X_φ depends, up to a homotopy equivalence, only on φ (which justifies the notation). It has a natural CW structure with three cells of dimensions $0, 2n$, and $4n$. Thus,

$$H^q(X_\varphi; \mathbb{Z}) = \begin{cases} \mathbb{Z} \text{ for } q = 0, 2n, 4n, \\ 0 \text{ for } q \neq 0, 2n, 4n. \end{cases}$$

The groups $H^{2n}(X_\varphi; \mathbb{Z}), H^{4n}(X_\varphi; \mathbb{Z})$ (isomorphic to $\mathbb{Z}$) have natural generators (determined by the canonical orientations of S^{2n} and D^{4n}), and we denote these generators by a and b. Since the cup-square $a^2 = a \smile a$ has dimension $4n$, we have $a^2 = hb$, where $h \in \mathbb{Z}$. The number $h = h(\varphi)$ is, by definition, the Hopf invariant of φ. [1] Our theorem is covered by the following two lemmas.

Lemma 1. *The Hopf invariant is additive: $h(\varphi + \psi) = h(\varphi) + h(\psi)$.*

Lemma 2. *The Hopf invariant is nontrivial; in particular,*

$$h([\iota_{2n}, \iota_{2n}] = 2.$$

Proof of Lemma 1. In addition to the spaces $X_\varphi, X_\psi, X_{\varphi+\psi}$ (constructed using the spheroids $f, g, f + g: S^{4n-1} \to S^{2n}$), we will consider the space

$$Y_{\varphi,\psi} = (S^{2n} \cup_f D^{4n}) \cup_g D^{4n} = S^{2n} \cup_{f \vee g} (D^{4n} \vee D^{4n}).$$

This space has a CW structure with four cells of dimensions $0, 2n, 4n, 4n$ and has the following cohomology:

$$H^q(Y_{\varphi,\psi}; \mathbb{Z}) = \begin{cases} \mathbb{Z} \oplus \mathbb{Z} & \text{for } q = 4n, \\ \mathbb{Z} & \text{for } q = 0, 2n, \\ 0 & \text{for } q \neq 0, 2n, 4n. \end{cases}$$

[1] In the homotopy theory, there are interesting generalizations of the Hopf invariant; see Whitehead [88] and Hilton [44].

Denote the canonical generators of the cohomology groups $H^{2n}(Y_{\varphi,\psi};\mathbb{Z})$ and $H^{4n}(Y_{\varphi,\psi};\mathbb{Z})$ by a' and b'_1, b'_2. There are natural CW embeddings $X_\varphi \to Y_{\varphi,\psi}$ and $X_\psi \to Y_{\varphi,\psi}$. There is also a natural map $X_{\varphi+\psi} \to Y_{\varphi,\psi}$; it consists of the identity map $S^{2n} \to S^{2n}$ and the map $D^{4n} \to D^{4n} \vee D^{4n}$ which collapses the equatorial plane to a point (these maps compose a continuous map $X_{\varphi+\psi} \to Y_{\varphi,\psi}$ because the diagram

$$\begin{array}{ccc} S^{4n-1} & \xrightarrow{\text{proj.}} & S^{4n-1}/\text{equator} = S^{4n-1} \vee S^{4n-1} \\ & {\searrow}^{f+g} \quad {\swarrow}_{f \vee g} & \\ & S^{2n} & \end{array}$$

is commutative by the definition of the sum of spheroids). The induced cellular chain maps for all three maps described above are obvious; the cohomology maps act like this:

$$\begin{aligned} X_\varphi \to Y_{\varphi,\psi}: &\quad a' \mapsto a,\ b'_1 \mapsto b,\ b'_2 \mapsto 0 \\ X_\psi \to Y_{\varphi,\psi}: &\quad a' \mapsto a,\ b'_1 \mapsto 0,\ b'_2 \mapsto b \\ X_{\varphi+\psi} \to Y_{\varphi,\psi}: &\quad a' \mapsto a,\ b'_1 \mapsto b,\ b'_2 \mapsto b. \end{aligned}$$

We must have $(a_1)^2 = h_1 b'_1 + h_2 b'_2$, where $h_1, h_2 \in \mathbb{Z}$. By the naturalness of the cup-product,

$$a^2 = h_1 b \text{ in } X_\varphi,\ a^2 = h_2 b \text{ in } X_\psi,\ a^2 = (h_1 + h_2) b \text{ in } X_{\varphi+\psi}.$$

On the other hand,

$$a^2 = h(\varphi) b \text{ in } X_\varphi,\ a^2 = h(\psi) b \text{ in } X_\psi,\ a^2 = h(\varphi + \psi) b \text{ in } X_{\varphi+\psi}.$$

Hence, $h_1 = h(\varphi), h_2 = h(\psi), h_1 + h_2 = h(\varphi + \psi)$, from which $h(\varphi + \psi) = h(\varphi) + h(\psi)$.

Proof of Lemma 2. Consider the product $S^{2n} \times S^{2n}$. Its cohomology is $H^{2n}(S^{2n} \times S^{2n};\mathbb{Z}) = \mathbb{Z} \oplus \mathbb{Z}$ (the generators c_1, c_2) and $H^{4n}(S^{2n} \times S^{2n};\mathbb{Z}) = \mathbb{Z}$ (the generator d). The multiplication: $c_1^2 = c_2^2 = 0$ (proof: Consider the projections $S^{2n} \times S^{2n} \to S^{2n}$) and $c_1 c_2 = d$ (follows from step 2 of the proof in Sect. 16.3 plus the definition of the cup-product in Sect. 16.2).

Make a factorization of $S^{2n} \times S^{2n}$ using the relation $(x_0, x) \sim (x, x_0)$ for all $x \in S^{2n}$, where x_0 is the zero-dimensional cell of S^{2n}. That is, we glue to each other the two two-dimensional cells of $S^{2n} \times S^{2n}$. The resulting space X has three cells, of dimensions $0, 2n$, and $4n$; that is, it has the form $S^{2n} \cup_f D^{4n}$, where f is a certain map $S^{4n-1} \to S^{2n}$. Moreover, if we compare this construction with the definition of the Whitehead product in Sect. 10.5, we notice that this f is nothing but the canonical spheroid representing the Whitehead product $[\iota_{2n}, \iota_{2n}]$. Thus, $X = X_{[\iota_{2n}, \iota_{2n}]}$. The cohomology of X is $H^{2n}(X;\mathbb{Z}) = H^{4n}(X;\mathbb{Z}) = \mathbb{Z}$, and if a, b are canonical generators of these cohomology groups, then $a^2 = h([\iota_{2n}, \iota_{2n}])b$. But the cohomology homomorphism induced by the projection $S^{2n} \times S^{2n} \to X$ takes a and b into $c_1 + c_2$ and d. Thus, in the cohomology of $S^{2n} \times S^{2n}$, $(c_1 + c_2)^2 = h([\iota_{2n}, \iota_{2n}])d$, and, since $(c_1 + c_2)^2 = c_1^2 + 2c_1c_2 + c_2^2 = 2d$, we have $h([\iota_{2n}, \iota_{2n}]) = 2$.

Remark 4. As we will see in Chap. 3, $\pi_{4n-1}(S^{2n}) = \mathbb{Z}\oplus$ a finite group. In particular, $\pi_3(S^2) = \mathbb{Z}$ (we already know this), $\pi_7(S^4) = \mathbb{Z} \oplus \mathbb{Z}_{12}, \pi_{11}(S^6) = \mathbb{Z}, \pi_{15}(S^8) = \mathbb{Z} \oplus \mathbb{Z}_{120}$. It is also true that all the homotopy groups of spheres are finite besides $\pi_n(S^n) = \mathbb{Z}$ and $\pi_{4n-1}(S^{2n})$.

Remark 5. Lemma 2 shows that the image of the Hopf homomorphism $h\colon \pi_{4n-1}(S^{2n}) \to \mathbb{Z}$ is either the whole group $\mathbb{Z}$ or the group of even integers. The choice between these two options is reduced to the question: Does $\pi_{4n-1}(S^{2n})$ contain an element with the Hopf invariant one? This question has several remarkable equivalent statements. For example, it is possible to show that S^m possesses an H-space structure if and only if m is odd, that is, $m = 2n - 1$, and $\pi_{4n-1}(S^{2n})$ contains an element with the Hopf invariant one. The same condition is necessary and sufficient for the existence in $\mathbb{R}^{m+1}$ of a bilinear multiplication with a unique division. The combination of Lemma 2 and Exercise 7 in Lecture 10 shows that the Hopf invariant of the Hopf class $\eta_2 \in \pi_3(S^2)$ equals 1 (this corresponds to the complex number multiplication in $\mathbb{R}^2$ or to the natural group structure in S^1). In 1960, J. Adams showed that elements with the Hopf invariant one are contained only in $\pi_3(S^2), \pi_7(S^4)$, and $\pi_{15}(S^8)$ (we mentioned his results in Sect. 1.4; we will discuss two proofs of it: in Chaps. 5 and 6).

16.6 An Addendum: Other Multiplications

A: Homological ×-Product

We already mentioned this in the introduction. Its definition corresponds to the general spirit of this lecture: Singular simplices $f_1\colon \Delta^{n_1} \to X_1, f_2\colon \Delta^{n_2} \to X_2$ give rise to a map $f_1 \times f_2\colon \Delta^{n_1} \times \Delta^{n_2} \to X_1 \times X_2$; then we triangulate the product $\Delta^{n_1} \times \Delta^{n_2}$ as in the proof of the theorem in Sect. 16.3. Then we define the product of the singular simplices f_1 and f_2 the singular chain of $X_1 \times X_2$, which is the sum with the coefficients ± 1 (the same as in Sect. 16.3) of the singular simplices which are restrictions of the map $f_1 \times f_2$ to the $(n_1 + n_2)$-dimensional simplices of the triangulation. This chain is also denoted as $f_1 \times f_2$. By bilinearity, this $\times$-product is extended to singular chains: $\left(\sum_i g_{1i}f_{1i}\right) \times \left(\sum_j g_{2j}f_{2j}\right) = \sum_{i,j} g_{1i}g_{2j}\left(f_{1i} \times f_{2j}\right)$ (where g_{1i}, g_{2j} are elements of the coefficient ring G). A verification shows that $\partial(c_1 \times c_2) = (\partial c_1) \times c_2 + (-1)^{n_1} c_1 \times \partial c_2$ (where $n_1 = \dim c_1$). Thus, there arises a homology multiplication: For $\alpha_1 \in H_{n_1}(X_1; G), \alpha_2 \in H_{n_2}(X_2; G)$, there is the product $\alpha_1 \times \alpha_2 \in H_{n_1+n_2}(X_1 \times X_2; G)$. The proof of coincidence of this product with the homological cross-product described in Sect. 16.1 is a replica of the proof of the similar cohomological result in Sect. 16.3.

EXERCISE 6. Prove that for $\alpha_1 \in H_{n_1}(X_1; G), \alpha_2 \in H_{n_2}(X_2; G), \gamma_1 \in H^{n_1}(X_1; G)$, $\gamma_2 \in H^{n_2}(X_2; G)$,

$$\langle \gamma_1 \times \gamma_2, \alpha_1 \times \alpha_2 \rangle = (-1)^{n_1 n_2} \langle \gamma_1, \alpha_1 \rangle \langle \gamma_2, \alpha_2 \rangle.$$

B: Cap-Product

This is a mixed operation involving both homology and cohomology. Let $a = \sum_i g_i f_i \in C_{n_1}(X; G), c \in C^{n_2}(X; G)$, where $n_1 \geq n_2$. Put

$$a \frown c = \sum_i g_i c\left(\Gamma_-^{n_2}\right) \Gamma_+^{n_1 - n_2} \in C_{n_1 - n_2}(X; G)$$

(we use the notation introduced in Sect. 16.2).

EXERCISE 7. Prove the formula

$$(\partial a) \frown c = a \frown \delta c + (-1)^{n_2} \partial (a \frown c).$$

EXERCISE 8. Deduce from this that if a is a cycle representing a homology class $\alpha \in H_{n_1}(X; G)$ and c is a cocycle representing a cohomology class $\gamma \in H^{n_2}(X; G)$, then $a \frown c$ is a cycle whose homology class is fully determined by α and γ.

In the notation of Exercise 9, the homology class of $a \frown c$ is denoted as $\alpha \frown \gamma$. Thus, we get the *cap-product*

$$[\alpha \in H_{n_1}(X; G), \gamma \in H^{n_2}(X; G)] \mapsto \alpha \frown \gamma \in H_{n_1 - n_2}(X; G).$$

EXERCISE 9. Prove that if $n_1 = n_2$ and X is connected, then $\alpha \frown \gamma = \langle \gamma, \alpha \rangle \in G = H_0(X; G)$.

EXERCISE 10. Prove the "mixed associativity": $\alpha \frown (\gamma_1 \smile \gamma_2) = (\alpha \frown \gamma_1) \frown \gamma_2$.

EXERCISE 11. Prove the naturalness of the cap-product: If $\alpha \in H_{n_1}(X; G), \gamma \in H^{n_2}(Y; G)$, and $f: X \to Y$ is a continuous map, then $(f_* \alpha) \frown \gamma = f_*(\alpha \frown f^* \gamma)$.

C: Pontryagin–Samelson Multiplication

EXERCISE 12. Prove that if n_1, n_2 are positive integers, then there is no way to introduce for all X a nonzero bilinear multiplication

$$H_{n_1}(X; G) \times H_{n_2}(X; G) \to H_{n_1 + n_2}(X; G)$$

natural with respect to continuous maps.

However, it is possible to define a multiplication in homology groups of X if X itself possesses a multiplication making it a topological group or, at least, an H-space. The definition is obvious: If $\mu: X \times X \to X$ is the multiplication in X and $\alpha_1 \in H_{n_1}(X; G), \alpha_2 \in H_{n_2}(X; G)$ where G is a ring, then $\alpha_1\alpha_2 = \mu_*(\alpha_1 \times \alpha_2)$. This product is called the Pontryagin–Samelson product. We have no opportunity to discuss this product in detail, but we recommend to the reader, after reading Chap. 3, to return to this product and to calculate it for the homology groups of major topological groups and H-spaces.

Final Remark. All multiplications considered in this lecture can be generalized, in an obvious way, from the case of ring coefficients to the case when there is a pairing $G_1 \times G_2 \to G$, the factors lie in the homology/cohomology with coefficients in G_1 and G_2, and the product belongs to the homology/cohomology with coefficients in G.

Lecture 17 Homology and Manifolds

Among the natural computational tools used by homology theory, the most efficient ones are delivered by the topology of smooth manifolds, and we cannot help considering this subject. However, the foundations of the theory of manifolds, rooted in geometry and analysis, require a thick volume by themselves. The most common way to overcome this difficulty is to replace the notion of a smooth manifold by various combinatorial substitutes like homology manifolds or pseudomanifolds (see Sects. 17.2 and 17.3 ahead). By doing this, we can achieve a rigor of the proofs at the expense of geometric visuality. To compensate for the latter, we will sometimes provide geometric explanations based on statements which are easy to believe, but not always easy to prove.

We begin with a short sightseeing tour in the theory of smooth manifolds.

17.1 Smooth Manifolds

A Hausdorff topological space with a countable base of open sets (these topological assumptions are not in the spirit of this book, but we have to impose them, since without them many statements that follow would be plainly wrong) is called an n-dimensional (topological) *manifold* if every point of it possesses a neighborhood homeomorphic to the space $\mathbb{R}^n$ or the half-space $\mathbb{R}^n_- = \{(x_1, \dots, x_n) \in \mathbb{R}^n \mid x_n \leq 0\}$. A point of an n-dimensional manifold X which has no neighborhood homeomorphic to $\mathbb{R}^n$ is called a *boundary point*. Boundary points of X form an $(n-1)$-dimensional manifold ∂X called the *boundary* of X. Obviously, ∂X is a manifold without boundary: $\partial\partial X = \emptyset$.

Examples of manifolds: Euclidean spaces, spheres, balls, classical surfaces, projective spaces, Grassmann manifolds, flag manifolds, Lie groups, Stiefel manifolds, products of the spaces listed above, open sets in these spaces, closed domains with smooth boundaries in these spaces, and so on.

A homeomorphism between $\mathbb{R}^n$ or $\mathbb{R}^n_-$ (or an open set in one of these spaces) and an open set U in a manifold X determines coordinates in U which are called *local coordinates*. If the domains U, V of local coordinate systems $f: U \to \mathbb{R}^n_{(-)}, g: V \to \mathbb{R}^n_{(-)}$ (also called *charts*) overlap, then there arises a *transition map*

$$\begin{array}{ccccc} f(U \cap V) & \xrightarrow{f^{-1}} & U \cap V & \xrightarrow{g} & g(U \cap V) \\ \cap & & & & \cap \\ \mathbb{R}^n & & & & \mathbb{R}^n, \end{array}$$

which is described by usual functions of n variables. These functions can be smooth (as usual in topology, we understand the word *smooth* as belonging to the class C^∞), analytic, algebraic, etc. A set of charts which cover the manifold is called an *atlas*. An atlas is called smooth (analytic) if such functions are all transition functions between charts of this atlas. Two smooth (analytic) atlases are called smoothly (analytically) equivalent if their union is smooth (analytic) atlas. A class of equivalent smooth (analytic) atlases is called a *smooth (analytic) structure* on a manifold. A manifold with a smooth (analytic) structure is called a smooth (analytic) manifold. The boundary of a smooth (analytic) manifold is, in a natural way, a smooth (analytic) manifold. In the following, we will not consider analytic manifolds any seriously.

All manifolds listed above possess a natural smooth structure. Add one more example: Smooth surfaces in a Euclidean space, that is, closed subsets of $\mathbb{R}^m$ locally determined by systems of equations

$$f_i(x_1, \ldots, x_m) = 0, \; i = 1, \ldots, k$$

and, possibly, one inequality

$$f_{k+1}(x_1, \ldots, x_m) \geq 0,$$

where $f_1, \ldots, f_k$ $(, f_{k+1})$ are smooth functions whose gradients in their common domain are linearly independent.

There are two fundamental theorems in the theory of smooth manifolds (also called *differential topology*).

Theorem 1. *Every smooth manifold is diffeomorphic (that is, homeomorphic with preserving the smooth structure) to a smooth surface in an Euclidean space.*

Theorem 2. *Every compact smooth manifold is homeomorphic to a triangulated subset of an Euclidean space, and the homeomorphism can be made smooth on every simplex of the triangulation.*

Remarks. (1) In both theorems, the dimension of the Euclidean space can be as small as twice the dimension of the manifold.

(2) Theorem 2 also holds for noncompact manifolds, but the triangulation in this case has to be infinite.

We do not prove these theorems. Theorem 1 is proved in many textbooks in differential topology. Its proof is not hard. The situation with Theorem 2 is worse. Since the 1920s, the topologist regarded this fact as obvious. There are many geometric approaches to this result which look promising. For example, take a compact smooth surface in an Euclidean space and decompose this space into a union of small cubes. If the decomposition satisfies some general position condition with respect to the surface, we can expect that the intersections of the surface with the cubes will be close to convex polyhedra and we can easily triangulate these polyhedra. Or, choose a random finite subset of the smooth surface which is sufficiently dense, and take the Dirichlet domain; again we should get a subdivision of the surface into smooth polyhedra. However, numerous attempts to make this proof rigorous turned out to be unsuccessful. The first flawless proof of this theorem (actually, of a stronger relative result) was given in the 1930s by H. Whitney. This proof was based on entirely different ideas and did not look easy. We know two textbook presentations of this proof, in the books Whitney [89] and Munkres [64].

EXERCISE 1. Construct a realization as smooth surfaces in Euclidean spaces of projective spaces, Grassmann manifolds, flag manifolds, and Stiefel manifolds.

EXERCISE 2. Prove that all classical surfaces can be presented as smooth surfaces in $\mathbb{R}^n$ with $n \leq 4$.

EXERCISE 3. Construct smooth triangulations of classical surfaces; try to minimize the number of simplices needed.

EXERCISE 4. Prove that the number of n-dimensional simplices adjacent to an $(n-1)$-dimensional simplex of a smooth triangulation of an n-dimensional smooth manifold is 2 if this $(n-1)$-dimensional simplex is not contained in the boundary, and is 1 otherwise.

EXERCISE 5 (a generalization of Exercise 4). Let s be a k-dimensional simplex of a smooth triangulation of an n-dimensional smooth manifold. Consider the simplices of the triangulation which contain s, and in each of these simplices take the face opposite s (that is, spanned by the vertices not belonging to s). Prove that the union of these faces (which is called the *link* of the simplex s) is homeomorphic to S^{n-k-1} if s is not contained in the boundary and is homeomorphic to D^{n-k-1} otherwise. (For a warmup, begin with the case when $n = 3$ and $k = 1$.)

Remark. The notion of a link will be used later, so the reader who is not interested in this exercise still has to understand the definition of a link.

An atlas of a smooth manifold is called *oriented* if for every two overlapping charts the transition map has a positive determinant at every point. Two oriented atlases determine (belong to) the same *orientation* if their union is an oriented atlas.

A manifold is called *orientable* (*oriented*) if it possesses (is furnished by) an oriented atlas, that is, an orientation.

EXERCISE 6. Which projective spaces and Grassmann manifolds are orientable? (*Answer*: Only real projective spaces and Grassmann manifolds can be nonorientable. Namely, $\mathbb{R}P^n$ is orientable if and only if n is odd, and $G(n,k)$ is orientable if and only if n is even.)

EXERCISE 7. Prove that spheres with handles are orientable and projective planes and Klein bottles are nonorientable; drilling holes does not affect the orientability.

EXERCISE 8. Prove that a connected orientable manifold of positive dimension has precisely two orientations.

EXERCISE 9. Prove that every connected chart of an orientable manifold can be included in an oriented atlas; thus, if an orientable manifold is connected, then every connected chart determines an orientation.

EXERCISE 10. Prove that a manifold is orientable if and only if a neighborhood of every closed curve on this manifold is orientable.

EXERCISE 11. Prove that every simply connected manifold is orientable.

EXERCISE 12. Prove that every connected nonorientable manifold possesses an orientable twofold covering.

EXERCISE 13. Prove that the boundary of an orientable manifold is orientable.

It is also possible to define orientations using the language of triangulations. An orientation of an n-dimensional simplex is the order of its vertices given up to an even permutation. An orientation of an n-dimensional simplex induces orientations of its $(n-1)$-dimensional faces (using an even permutation of the order of vertices, we make the number of the vertex complementary to the face to be n, after which we orient the face by the order of remaining vertices). (Some modification is needed in the cases of $n = 0, 1$: An orientation of a zero-dimensional simplex is just $+$ or $-$, the orientation of faces v_0 and v_1 of a one-dimensional simplex $[v_0, v_1]$ are $-$ and $+$.) If two n-dimensional simplices share an $(n-1)$-dimensional face, then their orientations are *coherent* if they induce opposite orientations on this face. A triangulated n-dimensional manifold is orientable if all its n-dimensional simplices can be coherently oriented.

EXERCISE 14. An orientation of a connected orientable n-dimensional manifold is determined by an orientation of any of its n-dimensional simplices. [It may be reasonable to do this exercise after reading (the beginning of) the next section.]

17.2 Pseudomanifolds and Fundamental Classes

Definition. A triangulated space X is called an *n-dimensional pseudomanifold* if it satisfies the following three axioms.

1 (Dimensional homogeneity). X is the union of its n-dimensional simplices.
2 (Strong connectedness). For any two n-dimensional simplices s, s' of X, there exists a finite chain of n-dimensional simplices, $s_0, s_1, \dots, s_k$, such that $s_0 = s, s_k = s'$, and for every $i = 1, \dots, k$, the simplices s_{i-1}, s_i share an $(n-1)$-dimensional face.
3 (Nonbranching property). Every $(n-1)$-dimensional simplex of X is a face of precisely two n-dimensional simplices of X.

If X is a connected smooth n-dimensional manifold without boundary furnished with a smooth triangulation, then the triangulation obviously satisfies Axiom 1, satisfies Axiom 3 as stated in Exercise 4, and satisfies Axiom 2 as stated in Exercise below.

EXERCISE 15. Prove that a smoothly triangulated smooth connected manifold without boundary is strongly connected (see Axiom 2). [All we need to establish is that two interior points of n-dimensional simplices can be joined by a path avoiding an $(n-2)$-dimensional skeleton.]

Thus, a *smoothly triangulated connected smooth manifold without boundary is a pseudomanifold.* The converse is wrong: A pseudomanifold is not always a manifold. See the simplest example in Fig. 66.

There are fewer artificial examples of pseudomanifolds topologically different from manifolds: complex algebraic varieties, and Thom spaces of vector bundles (these will be extensively studied later, in Lecture 31 and further lectures).

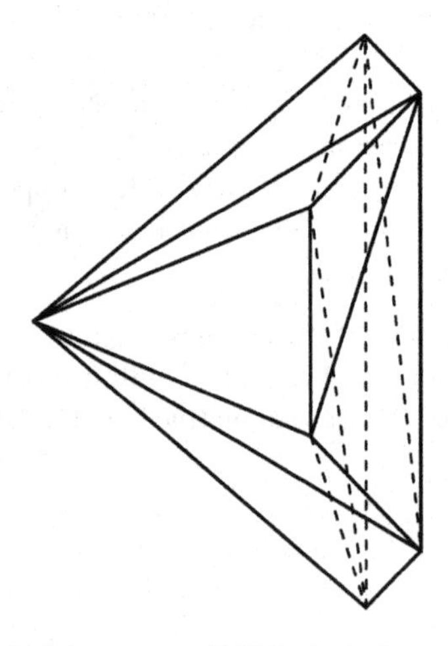

Fig. 66 A pseudomanifold which is not a manifold (a pinched torus)

An orientation of a pseudomanifold is defined as in the end of the previous section (Exercise 14 is also applied to this case). If a pseudomanifold is a manifold, then an orientation of this pseudomanifold is the same as an orientation of the manifold (in the sense of Sect. 17.1).

Theorem. *Let X be an n-dimensional pseudomanifold. Then*

$$H_n(X) = \begin{cases} \mathbb{Z}, & \text{if } X \text{ is compact and orientable,} \\ 0 & \text{otherwise;} \end{cases}$$

$$H_n(X;\mathbb{Z}_2) = \begin{cases} \mathbb{Z}_2, & \text{if } X \text{ is compact,} \\ 0 & \text{otherwise.} \end{cases}$$

Proof. We consider the classical complex $\{C_n(X), \partial_n\}$, corresponding to an arbitrary ordering of vertices (see Sect. 13.10). Since $C_{n+1}(X) = 0$, $H_n(X) = Z_n(X)$, the group of n-dimensional cycles of the classical complex. Let $c = \sum_i k_i s_i$ be such a cycle (k_i are integers, s_i are n-dimensional simplices). If the simplices s_i and s_j share an $(n-1)$-dimensional face, then this face does not belong to any other simplex, and $\partial c = 0$ implies $k_i = \pm k_j$ (the sign depends on the orientations). Since X is strongly connected, this shows that c involves all n-dimensional simplices of X, with all the coefficients of the form $\pm k$, where k is a nonnegative integer, the same for all the simplices. From this we immediately see that if the number of simplices is infinite, then there are no nonzero cycles, and $H_n(X) = 0$. If the number of simplices is finite, then let us reverse the orientations of simplices with a negative value of the coefficient. Since c is a cycle, these new orientations induce opposite orientations on every $(n-1)$-dimensional face; that is, they are coherent. We see that a nonzero cycle exists if and only if X is orientable. This proves our result for $H_n(X)$. The case of $\mathbb{Z}_2$-coefficients is similar, but it does not involve signs, and hence does not involve orientations.

This proof provides a canonical generator for the group $H_n(X)$ for a compact oriented pseudomanifold X: This is the homology class of the cycle, which is the sum of all n-dimensional simplices of X with orientations compatible with the orientation of X and with the coefficients all equal to 1. This homology class is called the *fundamental class* of X (and the cycle is called the *fundamental cycle*). In the orientation-free case, we have fundamental classes and fundamental cycles with coefficients in $\mathbb{Z}_2$ (certainly, only for compact pseudomanifolds). Notation: $[X] \in H_n(X)$ or $H_n(X;\mathbb{Z}_2)$.

Since connected smooth manifolds without boundary are pseudomanifolds, the preceding theorem holds for them. In particular, for compact connected smooth manifolds without boundary there are fundamental classes. (It is time to mention a broadly used term: A compact manifold without boundary is called *closed*.) This has an obvious generalization to the disconnected case: For a closed oriented n-dimensional manifold X, $H_n(X) = \bigoplus_\alpha H_n(X_\alpha)$, where X_α are components of X, and $[X]$ is simply $\{[X_\alpha]\}$.

EXERCISE 16. Prove that if X is a connected n-dimensional manifold with nonempty boundary, then $H_n(X) = H_n(X; \mathbb{Z}_2) = 0$.

EXERCISE 17. Prove that if X has a boundary, then the same construction as above gives a class $[X, \partial X] \in H_n(X, \partial X)$ or $H_n(X, \partial X; \mathbb{Z}_2)$ and $\partial_*[X, \partial X] = [\partial X]$.

EXERCISE 18. Prove the relation $[X_1 \times X_2] = [X_1] \times [X_2]$ in all possible versions (including the boundary one).

EXERCISE 19. Prove that for any homology class $\alpha \in H_n(Y)$ of an arbitrary topological space Y there exists a compact oriented (not necessarily connected) pseudomanifold X and a continuous map $f: X \to Y$ such that $f_*[X] = \alpha$. Prove a similar statement for an $\alpha \in H_n(Y; \mathbb{Z}_2)$ and nonoriented pseudomanifolds. (Actually, the $\mathbb{Z}_2$-case is easier, and so it may be advisable to begin with it; a construction in Sect. 13.11 may serve as a pattern for both the oriented and nonoriented cases.)

There arises a natural question regarding the possibility to present a homology class of a topological space as an image of the fundamental class of a manifold. The answer is negative, for homology classes with coefficients in $\mathbb{Z}$ as well as for those with coefficients in $\mathbb{Z}_2$. We will return to the discussion of this in the last lecture of this book.

A more popular question arises in the topology of manifolds: If Y is a manifold and $\alpha \in H_n(Y)$, then when is it possible to find a closed oriented n-dimensional submanifold X of Y (we assume that the reader understands what it is) such that the homomorphism induced by the inclusion map sends $[X]$ into α (as people say, *X realizes α*)? Again, a similar question exists for the $\mathbb{Z}_2$ homology classes and nonoriented submanifolds. There are many remarkable results regarding submanifold realizations; for example, for any homology class α of a manifold, there exists a number N such that $N\alpha$ can be realized by a submanifold. (For this result and other results, see the classical paper by Thom [84].)

EXERCISE 20. Prove that the generators of groups

$$H_m(\mathbb{R}P^n; \mathbb{Z}_2),\ H_m(\mathbb{R}P^n),\ H_{2m}(\mathbb{C}P^n),\ H_{4m}(\mathbb{H}P^n)$$

are realized by projective subspaces of $\mathbb{R}P^n$, $\mathbb{C}P^n$, $\mathbb{H}P^n$. (Compare also to Exercise 11 in Lecture 14.)

Mention in conclusion that if X, Y are oriented pseudomanifolds of the same dimension, and $f: X \to Y$ is a continuous map, then $f_*[X] = k \cdot [Y]$, where k is an integer. This k is called the degree of f and is denoted as $\deg f$; it is a homotopy invariant. In the nonoriented case, the degree $\deg f$ may be defined as an element of $\mathbb{Z}_2$. We have already had this notion in the particular case $X = Y = S^n$ (see Sects. 10.3 and 13.3). In the manifold case, there exists a description of the degree similar to the description given in Sect. 10.3 for spheres; we formulate the result in the form of an exercise.

EXERCISE 21. Let $f{:}X \to Y$ be a (piecewise) smooth map between two closed oriented n-dimensional manifolds, and let $y{:}\,Y$ be a regular value of this map. Then there is a neighborhood U of y such that $f^{-1}(U)$ is a disjoint union of a finite collection of sets U_i with all restrictions $f|_{U_i}$ being homeomorphisms $U_i \to U$. Prove that $\deg f$ is the number of i for which this homeomorphism preserves the orientation minus the number of i for which it reverses the orientation.

17.3 Homology Manifolds

The most general definition of a homology manifold is formulated in terms of *local homology*: For a topological space X, its mth local homology at the point $x_0 \in X$ is defined as $H^{\text{loc}}_{m,x_0}(X) = H_m(X, X - x_0)$.

Definition. A space X is called an n-dimensional homology manifold if, for any m, $H^{\text{loc}}_{m,x_0}(X) = \widetilde{H}_m(S^n)$, that is,

$$H^{\text{loc}}_{m,x_0}(X) = \begin{cases} \mathbb{Z}, & \text{if } m = n, \\ 0, & \text{if } m \neq n. \end{cases}.$$

For us, the most important will be the case when X is triangulated. Recall that the star $\text{St}(s)$ of a simplex s of triangulation is the union of simplices that contain s. The link $\text{Lk}(s)$ is the union of faces of simplices that contain s opposite to s. Figure 67 shows examples of stars and links of a vertex and a one-dimensional simplex of the standard triangulation of the plane.

Proposition 1. (1) *A triangulated space X is an n-dimensional homology manifold if and only if for every vertex v of X, the link* $\text{Lk}(v)$ *is a homological $(n-1)$-dimensional sphere* (that is, has the same homology groups as S^{n-1}).

(2) *A triangulated space X is an n-dimensional homology manifold if and only if for every simplex s of X, the link* $\text{Lk}(s)$ *is a homological $(n-k-1)$-dimensional sphere where $k = \dim s$.*

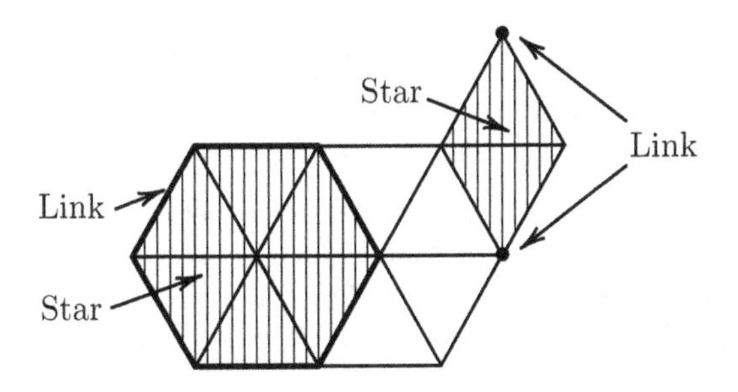

Fig. 67 Stars and links

Proof. Open stars of vertices, $\mathrm{st}(v) = \mathrm{St}(v) - \mathrm{Lk}(v)$, for an open cover of X. Also, $\mathrm{St}(v)$ is a cone over $\mathrm{Lk}(v)$ with the vertex v. Thus, if $x_0 \in \mathrm{st}(v)$, then

$$\begin{aligned} H^{\mathrm{loc}}_{m,x_0}(X) = H_m(X, X - x_0) &= H_m(X, X - \mathrm{st}(v)) \\ &= H_m(\mathrm{St}(v), \mathrm{Lk}(v)) = \widetilde{H}_{m-1}(\mathrm{Lk}(v)) \end{aligned}$$

[the four equalities follow from the definition of local homology, homotopy invariance of homology, excision theorem, and reduced homology sequence of the pair $(\mathrm{St}(v), \mathrm{Lk}(v))$]. This proves (1).

To prove (2), notice that for a simplex s, $\mathrm{St}(s) = s * \mathrm{Lk}(s)$. Hence, for every interior point x_0 of s,

$$\begin{aligned} H^{\mathrm{loc}}_{m,x_0}(X) &= H_m(X, X - x_0) = H_m(s * \mathrm{Lk}(s), (\partial s) * \mathrm{Lk}(s)) \\ &= \widetilde{H}_{m-1}((\partial s) * \mathrm{Lk}(s)) = \widetilde{H}_{m-1}(\Sigma^k \mathrm{Lk}(s)) = \widetilde{H}_{m-k-1}(\mathrm{Lk}(s)), \end{aligned}$$

where $k = \dim s$. This proves (2).

Proposition 2. *Every connected n-dimensional homology manifold is an n-dimensional pseudomanifold.*

Proof. Let X be an n-dimensional homology manifold. Since the link of every vertex of X is an $(n-1)$-dimensional homological sphere, this link contains simplices of dimension $\geq n-1$; hence, every vertex is a vertex of an n-dimensional simplex. There cannot be simplices of dimension $> n$, because the link of every n-dimensional simplex must be empty (homological S^{-1}). Every simplex of dimension $< n$ must have a nonempty link, so it must be a face of a simplex of a bigger dimension. Hence, X must be the union of n-dimensional simplices (dimensional homogeneity axiom holds). The link of an $(n-1)$-dimensional simplex s consists of isolated points, one for every n-dimensional simplex containing s; since the link is a homological S^0, this number is 2 (unbranching axiom holds). A path connecting two points of X can be made straight within every simplex; since the links of simplices of dimension $\leq n-2$ are connected, the path can be pushed from every point of a simplex of dimension $\leq n-2$ to simplices of bigger dimensions. Hence, there is a path disjoint from the $(n-2)$nd skeleton of X (the strong connectedness holds).

Remark 1. Proposition 2 shows that everything said in Sect. 17.2 about pseudomanifolds can be applied to homological manifolds. In particular, homological manifolds can be orientable or nonorientable, there are fundamental cycles and classes, and the theorem of Sect. 17.2 holds for a connected homology manifold.

Remark 2. This argumentation shows a difference between pseudomanifolds and homology manifolds. While in homology manifolds all links are homological spheres of appropriate dimensions, in n-dimensional pseudomanifolds this holds for links of simplices of dimensions n and $n-1$. Add to that that a pseudomanifold in Fig. 66 is not a homology manifold.

Remark 3. A smooth manifold without boundary is a homology manifold (and in the smooth case, links are homeomorphic to spheres, not just are homological spheres).

Remark 4. A homology manifold is not always a topological manifold. For example, there are manifolds with the same homology as a sphere, but not simply connected (the best known example is the *Poincaré sphere* defined in $S^5 = \{(z_1, z_2, z_3) \in \mathbb{C}^3 \mid |z_1|^2 + |z_2|^2 + |z_3|^2 = 1\}$ by the equation $z_1^5 + z_2^3 + z_3^2 = 0$). The suspension over such a manifold is a homology manifold, but no neighborhoods of vertices are homeomorphic to a Euclidean space.

17.4 Poincaré Isomorphism

The main result of the homological theory of manifolds is the following:

Theorem. *Let X be a compact n-dimensional homology manifold, and let $0 \leq m \leq n$. If X is orientable, then for any G,*

$$H_m(X; G) \cong H^{n-m}(X; G).$$

In the general case,

$$H_m(X; \mathbb{Z}_2) \cong H^{n-m}(X; \mathbf{Z}_2).$$

In both cases, there are canonical isomorphisms

$$D: H^{n-m}(X; G) \to H_m(X; G)$$

which act by the formula $D(\alpha) = [X] \frown \alpha$, where $[X]$ is the fundamental class (see Sect. 17.2) and $\frown$ denotes the cap-product (see Sect. 16.6).

Remarks. (1) The isomorphism D is usually referred to as the *Poincaré isomorphism.*

(2) By Remark (3) in Sect. 17.3, the theorem holds for closed (compact and boundary-less) smooth manifold.

The proof of the theorem will consist of two parts: First we will give (the most classical) construction of Poincaré isomorphism, and then we will prove the formula involving the cap-product. This formula will show, in particular, that the isomorphism provided by the classical construction does not depend on the triangulation.

For a simplex s of the triangulation of X, denote as $\text{Bast}(s)$ the union of all simplices of the barycentric triangulation whose intersection with s is the center of s. Using the fact that the simplices of the barycentric triangulation correspond to the increasing chains $s_0 \subset \cdots \subset s_j$ of the initial triangulation, we can describe

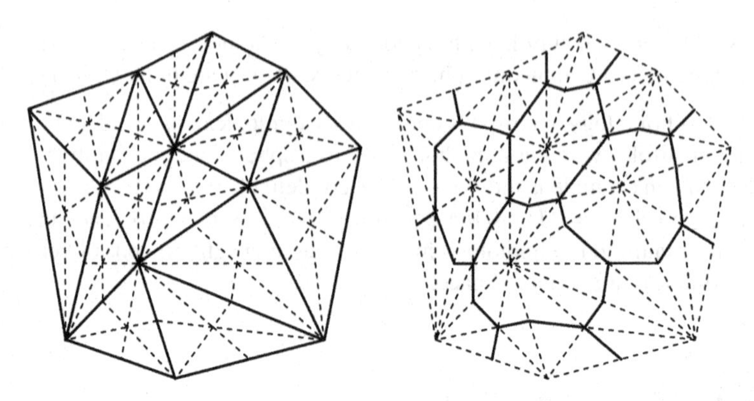

Fig. 68 Barycentric stars

Bast(s) as the union of simplices of barycentric triangulation corresponding to chains as above with $s_0 = s$. Obviously, Bast(s) is the union of its simplices of the maximal dimension, $n-k$ (where $k = \dim s$), that is, simplices corresponding to chains $s = s_0 \subset s_1 \cdots \subset s_{n-k}$ with $\dim s_i = k+i$. This important that $\dim \mathrm{Bast}(s) = n - \dim(s)$.

The reader may see in Fig. 68 (where $n = 2$) what barycentric stars look like. Barycentric stars of vertices are polyhedra of dimension 2 ("centered" at these vertices), barycentric stars of one-dimensional simplices have dimension 1, and barycentric stars of two-dimensional simplices are centers of these simplices (this is true for any dimension n: The barycentric star of an n-dimensional simplex is its center).

Besides barycentric stars, there are barycentric links: For a simplex s, Balk(s) is the union of faces of barycentric simplices in Bast(s) opposite the center of s. Obviously, Bast(s) is the cone over Balk(s) and Balk(s) is homeomorphic to Lk(s) (the reader who has any doubt can observe all this in Fig. 68). Also, there are *open barycentric stars*, bast(s) = Bast(s) − Balk(s). Obviously, X is a disjoint union of open barycentric stars of all its simplices.

If X is a homology manifold, then

$$\begin{aligned} H_m(\mathrm{Bast}(s), \mathrm{Balk}(s)) &= H_m(C(\mathrm{Balk}(s)), \mathrm{Balk}(s)) \\ &= \widetilde{H}_{m-1}(\mathrm{Balk}(s)) = \widetilde{H}_{m-1}(\mathrm{Lk}(s)) \\ &= \begin{cases} \mathbb{Z}, \text{ if } m = n - \dim(s), \\ 0 \quad \text{otherwise} \end{cases} \end{aligned}$$

In other words, although the decomposition of X into open barycentric cells is not necessarily a CW structure, still it can be used for computing homology in the same way. We can define "skeletons" $\mathrm{sk}^m_{\mathrm{bast}}(X)$ as unions of barycentric stars of dimensions $\leq m$ (that is, barycentric stars of simplices of dimensions $\geq n-m$), and the complex $\{C^{\mathrm{bast}}_m(X), \partial_m\}$, where

$$C_m^{\text{bast}}(X) = H_m(\text{sk}_{\text{bast}}^m(X), \text{sk}_{\text{bast}}^{m-1}(X)),$$

$$\partial_m = \partial_*: H_m(\text{sk}_{\text{bast}}^m(X), \text{sk}_{\text{bast}}^{m-1}(X)) \to H_{m-1}(\text{sk}_{\text{bast}}^{m-1}(X), \text{sk}_{\text{bast}}^{m-2}(X)),$$

has homology equal to that of X.

Our next remark is that if the homology manifold X is oriented, then there exists a natural way to establish a correspondence between orientations of a simplex s and of the barycentric star $\text{Bast}(s)$. Namely, let the orientation of s be determined by an order of its vertices, $v_0, v_1, \ldots, v_k$. Consider an $(n-k)$-dimensional (barycentric) simplex u belonging to $\text{Bast}(s)$; it corresponds to a sequence $s = s_0 \subset \cdots \subset s_{n-k}$ with $\dim s_i = k + i$. For $i = 1, \ldots, n-k$, let v_{k+i} be the vertex of s_i not belonging to s_{i-1}. Then $v_0, \ldots, v_k, v_{k+1}, \ldots, v_n$ is the full set of vertices of the n-dimensional simplex s_{n-k}, and we assign to u the orientation determined by the order $v_k, \ldots, v_n$ of its vertices if the order $v_0, \ldots, v_n$ of vertices of the simplex v_{n-k} determines the orientation of v_{n-k} compatible with the orientation of X, and we assign the opposite orientation otherwise. If the simplex u shares an $(n-k-1)$-dimensional face with another simplex $u' \subset \text{Bast}(s)$, then u' corresponds to a sequence $s = s_0 \subset \ldots s_{j-1} \subset s'_j \subset s_{j+1} \cdots \subset s_{n-k}$ with $s'_j \neq s_j$. If $j < n-k$, then the simplex s_{n-k} stays the same, but the vertices v_j, v_{j+1} are swapped; thus, the orientation of u' is determined by the order of vertices $v_0, \ldots, v_{j+1}, v_j, \ldots, v_{n-k}$ only if the orientation of u is *not* determined by the order of vertices $v_0, \ldots, v_j, v_{j+1}, \ldots, v_{n-k}$; their common $(n-k-1)$-dimensional face has the vertices $v_0, \ldots, v_{j-1}, v_{j+1}, \ldots, v_{n-k}$, and it obtains opposite orientations from u and u'. The case $j = n-k$ is similar: In this case $s'_{n-k} \neq s_{n-k}$, the simplices s'_{n-k} and s_{n-k} have a common $(n-1)$-dimensional face, let it be t, and t obtains opposite orientations from s_{n-k} and s'_{n-k}. The orientations of the common face of u and u' are determined by the orientations of s and t (precisely as the orientation of u is determined by the orientations of s and s_{n-k}) and thus they are also opposite each other.

$\mathcal{C}_{n-k}^{\text{bast}}(X; G)$ is the group of linear combinations $\sum_i g_i \,\text{Bast}(s_i)$ where the summation is taken over oriented k-dimensional simplices s_i and $g_i \in G$. If X is a compact oriented n-dimensional homology manifold, consider an isomorphism

$$D: C_{\text{class}}^k(X; G) \to \mathcal{C}_{n-k}^{\text{bast}}(X; G), \quad D(s^*) = \text{Bast}(s),$$

where s^* is a k-dimensional cochain of the classical complex of X which takes value 1 on s and value 0 on every other k-dimensional simplex, and the orientations of s and $\text{Bast}(s)$ are compatible as above. Fact: For a cochain $c \in C_{\text{class}}^k(X; G)$,

$$D(\delta c) = (-1)^k \partial D(c) \qquad (*)$$

(see ahead). This shows that D established a dimension-reversing isomorphism between cohomology and homology of X; this is Poincaré isomorphism (also denoted by D).

It remains to establish two facts: the relation $(*)$ and the relation $D(\alpha) = [X] \frown \alpha$. Begin with the first. The boundary of $\mathrm{Bast}(s) \in C^{\mathrm{bast}}_{n-k}(X)$ consists of barycentric simplices lying in $\mathrm{Balk}(s)$ [faces inside $\mathrm{Bast}(s)$ are cancelled as follows from the preceding argumentations regarding the orientations]. The face of the barycentric simplex corresponding to the sequence $s = s_0, s_1, \dots, s_{n-k}$ lying in $\mathrm{Balk}(s)$ corresponds to the sequence $s_1, \dots, s_{n-k}$ and thus is contained in $\mathrm{Bast}(s_1)$. In this way, we see that $\mathrm{Bast}(s_1)$ is contained in the boundary of $\mathrm{Bast}(s)$ if and only if s is a face of s_1. The coefficient is $(-1)^k$ (this requires comparing the orientations, which we leave to the reader). Now, go to the second relation. Let bX be the barycentric subdivision of X with the ordering of vertices described in Sect. 13.10, and let $c \in C^k_{\mathrm{class}}(bX; G)$ and $[X]$ be the fundamental cycle of bX. The cellular map $\mathrm{id}\colon X \to bX$ induces a map

$$\mathrm{id}^{\#}\colon C^k_{\mathrm{class}}(bX; G) \to C^k_{\mathrm{class}}(X; G),$$

and the cochain $\mathrm{id}^{\#} c$ takes on a k-dimensional simplex s of X on s, the value equal to the sum of the values, with appropriate signs, of c of k-dimensional simplices of bX contained in s. On the other hand, the chain $[bX] \frown c$ is the sum of faces of n-dimensional simplices of bX spanned by the last vertices (see the definition of $\frown$ in Sect. 16.6). These are simplices in barycentric stars of k-dimensional simplices of X; each barycentric star of s appears in $[X] \frown c$ with the coefficient equal to the sum of values of c on the barycentric parts of s, that is, to $\mathrm{id}^{\#} c(s)$. Thus, $\mathrm{id}_{\#}(D(\mathrm{id}^{\#} c)) = [bX] \frown c$, where the last $\mathrm{id}_{\#}$ is

$$\mathrm{id}_{\#}\colon C^{\mathrm{bast}}_{n-k}(bX; G) \to C^{\mathrm{class}}_{n-k}(X; G).$$

This finishes the proof in the oriented case. In the nonoriented case everything is the same with the usual simplification—we do not need to care about orientations and signs (since the coefficient group is $\mathbb{Z}_2$).

Corollary. *The Euler characteristic of a closed homology manifold of odd dimension equals 0.*

For the proof, it is more convenient to use Poincaré isomorphism with coefficients in $\mathbb{Z}_2$, since it also holds in the nonorientable case. If $n = \dim X$, then

$$\begin{aligned}\chi(X) &= \textstyle\sum_m (-1)^m \dim_{\mathbb{Z}_2} H_m(X; \mathbb{Z}_2) = \sum_m (-1)^m \dim_{\mathbb{Z}_2} H^{n-m}(X; \mathbb{Z}_2)\\ &= \textstyle\sum_m (-1)^m \dim_{\mathbb{Z}_2} H_{n-m}(X; \mathbb{Z}_2) = \sum_m (-1)^{n-m} \dim_{\mathbb{Z}_2} H_m(X; \mathbb{Z}_2)\\ &= \textstyle(-1)^n \sum_m (-1)^m \dim_{\mathbb{Z}_2} H_m(X; \mathbb{Z}_2) = -\chi(X).\end{aligned}$$

17.5 Intersection Numbers and Poincaré Duality

The results of Sect. 15.5 give the possibility to restate Poincaré isomorphisms between homology and cohomology as (noncanonical) isomorphisms between homology and homology. Namely,

$$H_m(X;\mathbb{Z}_2) \cong H_{n-m}(X;\mathbb{Z}_2)$$

for an arbitrary n-dimensional homology manifold X and

$$\begin{gathered}\text{Free Part of } H_m(X) \cong \text{Free Part of } H_{n-m}(X)\\ \text{Torsion Part of } H_m(X) \cong \text{Torsion Part of } H_{n-m-1}(X)\end{gathered}$$

in the oriented case. It turns out that these noncanonical isomorphisms reflect a very canonical duality called *Poincaré duality* which is much more classical than Poincaré isomorphisms. We will postpone (until Sect. 17.7) a discussion of torsion parts and concentrate our attention on the free parts of homology groups.

Poincaré duality is based on the notion of the *intersection number*. Let $c_1 = \sum_i k_i \operatorname{Bast}(s_i)$ be some m-dimensional chain of the barycentric star complex of some compact triangulated oriented n-dimensional homology manifold X, and let $c_2 = \sum_j \ell_j s_j$ be some $(n-m)$-dimensional chain of the classical complex of X. Thus, both summations are taken over the set of $(n-m)$-dimensional simplices of X. The integer

$$\phi(c_1, c_2) = \sum_i k_i \ell_i = \langle D^{-1} c_1, c_2 \rangle$$

is called the intersection number of c_1 and c_2. It follows from the last formula and the properties of Poincaré isomorphism that the intersection number of two cycles depends only on the homology classes of these cycles, and we can speak of intersection numbers of homology classes: If $\alpha_1 \in H_m(X)$ and $\alpha_2 \in H_{n-m}(X)$, then $\phi(\alpha_1, \alpha_2) = \langle D^{-1}\alpha_1, \alpha_2 \rangle$, or $\phi(\alpha_1, \alpha_2) = \alpha_2 \frown D^{-1}\alpha_1 \in H_0(X) = \mathbb{Z}$ (see Exercise 10 in Sect. 16.6). Differently, the homology invariance of intersection numbers can be deduced from the formula $\phi(\partial c_1, c_2) = \phi(c_1, \partial c_2)$, which follows, in turn, from relation $(*)$ in Sect. 17.4:

$$\phi(\partial c_1, c_2) = \langle D^{-1}\partial c_1, c_2 \rangle = \langle \delta D^{-1} c_1, c_2 \rangle = \langle D^{-1} c_1, \partial c_2 \rangle = \phi(c_1, \partial c_2).$$

Another interesting relation arises from the "mixed associativity" of cup- and cap-products (see Exercise 11 in Sect. 16.6):

$$\begin{aligned}\phi(\alpha_1, \alpha_2) &= \alpha_2 \frown D^{-1}\alpha_1 = ([X] \frown D^{-1}\alpha_2) \frown D^{-1}\alpha_1\\ &= [X] \frown (D^{-1}\alpha_2 \smile D^{-1}\alpha_1) = D(D^{-1}\alpha_2 \smile D^{-1}\alpha_1).\end{aligned}$$

This provides a more symmetric definition of the intersection number, which implies, in particular [in view of commutativity relation for the cup-product; see the theorem in Sect. 16.2, part (2)], the commutativity relation

$$\phi(\alpha_1, \alpha_2) = (-1)^{m(n-m)}\phi(\alpha_2, \alpha_1) \quad (\alpha_1 \in H_m(X), \alpha_2 \in H_{m-k}(X)).$$

In the nonoriented case, the intersection number can be defined for cycles and homology classes modulo 2; they take values in $\mathbb{Z}_2$. It is also possible to define "intersection numbers" corresponding to an arbitrary pairing $G_1 \times G_2 \to G$.

A remarkable property of the intersection numbers is their geometric visualizability. A simplex and its barycentric star transversely intersect each other at one point, so the intersection number of two cycles may be regarded as the number of their intersection points taken with the signs determined by their orientations. This statement has a convenient differential statement.

Theorem 1. *Let X be a smooth closed oriented n-dimensional manifold, and let $\alpha_1 \in H_m(X), \alpha_2 \in H_{n-m}(X)$. Let Y_1 and Y_2 be closed oriented submanifolds of X of dimensions m and $n - m$ which realize α_1 and α_2 in the sense that $\alpha_1 = (i_1)_*[Y_1]$ and $\alpha_2 = (i_2)_*[Y_2]$ where i_1, i_2 are inclusion maps. We assume also that Y_1, Y_2 are in general position (which means that they intersect in finitely many points and transverse to each other at each of these points). We assign a sign to every intersection point: plus if the orientations of Y_1 and Y_2 (in this order) compose the orientation of X at this point, and minus otherwise. Then the intersection number $\phi(\alpha_1, \alpha_2)$ equals to the number of the intersection points of Y_1 and Y_2 counted with the signs described above.*

Similar statements hold for homology classes modulo 2 (in which case no orientation is needed) and for manifolds with pseudomanifold-like singularities (away from the intersection points).

As usual (see the warning in the beginning of this lecture), we do not give a rigorous proof of these statements; but from the point of view of common sense they are obvious. We can make the simplices of a triangulation of X much smaller than the distances between the intersection points of Y_1 and Y_2 and then approximate Y_1 and Y_2 by cycles of, respectively, classical and barycentric star complexes. Then the statements become obvious.

Notice that the general position condition is not really harmful: We can make the position of Y_1 and Y_2 general by a small perturbation of one of those.

Example. Natural generators y_r, y_{n-r} of the groups $H_{2r}(\mathbb{C}P^n)$, $H_{2(n-r)}(\mathbb{C}P^n)$ have the intersection number 1. Indeed, they are realized by projective subspaces $\mathbb{C}P^r, \mathbb{C}P^{n-r}$ of $\mathbb{C}P^n$ which (in the general position) intersect in one point. Regarding the sign, we will make an *important remark*. If X is a *complex manifold*, that is, its charts are maps into $\mathbb{C}^n$ and the transition maps are holomorphic, then X possesses a natural, "complex," orientation. The matter is that the Jacobian of a holomorphic map $\mathbb{C}^n \to \mathbb{C}^n$ regarded as a smooth map $\mathbb{R}^{2n} \to \mathbb{R}^{2n}$ is equal to the square of the absolute value of the complex Jacobian and, hence,

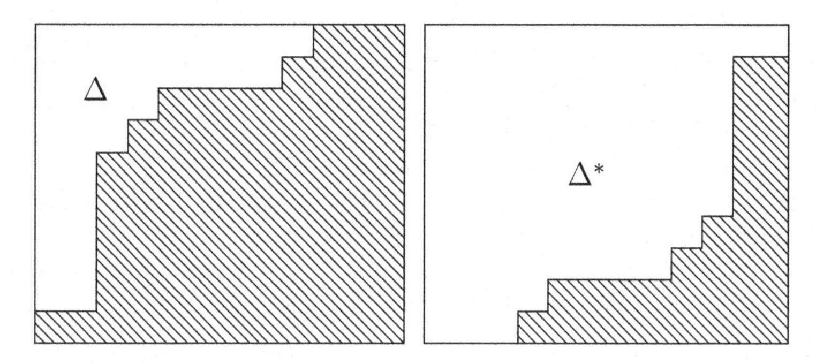

Fig. 69 Dual Young diagrams

is always positive. Moreover, if Y_1, Y_2 are complex (that is, locally determined by holomorphic equations) submanifolds of X of complementary dimensions in a general position, then every point in Y_1, Y_2 contributes $+1$ into the intersection number of the homology classes. Thus, $\phi(y_r, y_{n-r}) = 1$, not -1.

EXERCISE 22. Let Δ be a Young diagram inscribed into a rectangle $k \times (n-k)$, and let Δ^* be the "dual" Young diagram obtained from the complement of Δ in the rectangle by the reflection in the center of the rectangle (see Fig. 69). Then the intersection number of the homology classes of $\mathbb{C}G(n, k)$ corresponding to the Young diagrams Δ, Δ' (see Sects. 5.4.C and 13.8.C) is 1 if $\Delta' = \Delta^*$ and is 0 otherwise. (The same is true for modulo 2 intersection numbers for real Grassmann manifolds; the proof is the same).

The fact that the intersection number of two cycles depends only on the homology classes of these cycles is often used in solving geometric problems. Of a huge set of problems of this kind we give two.

EXERCISE 23. Prove that on any smooth closed orientable surface in $\mathbb{R}^4 = \mathbb{C}^2$, there exist at least two different points for which the tangent planes are complex lines. (*Hint*: The orientation takes care of the existence of more than one such point.)

EXERCISE 24. Prove that if X_1, X_2 are two closed orientable surfaces in $\mathbb{R}^4$, then there are at least four pairs of points ($x_1 \in X_1$, $x_2 \in X_2$) such that the tangent planes to $X_1 m, x_2$ at x_1, x_2 are parallel.

Return to our definition of the intersection number. Together with Corollary 1 in Sect. 15.5, it implies the following statement.

Theorem 2. *Let X be compact oriented homology manifold.* (1) *For every homomorphism $f: H_m(X) \to \mathbb{Z}$, there exists a homology class $\alpha \in H_{n-m}(X)$ such that $f(\alpha) = \phi(\alpha, \beta)$ for every $\beta \in H_m(X)$.* (2) *The class β is determined by f uniquely, up to adding an element of finite order.*

A similar result holds in the nonoriented case for homology and intersection numbers modulo 2; moreover, in this case β, for a given f, is genuinely unique.

Thus, the intersection numbers determine a nondegenerate duality between the free parts of the groups $H_m(X)$ and $H_{n-m}(X)$ in the oriented case and between the vector spaces $H_m(X;\mathbb{Z}_2)$ and $H_{n-m}(X;\mathbb{Z}_2)$ in general. This duality is called *Poincaré duality*. (One can notice that in the topological literature confusion exists between the terms "Poincaré isomorphism" and "Poincaré duality." It is especially surprising, since in other cases mathematicians have a tendency to be supersensitive to the difference between a vector space and a dual vector space.)

Notice that in the middle-dimensional homology of an even-dimensional manifold, Theorem 2 has the following, more algebraic restatement.

Theorem 3. *Let X be a connected closed orientable manifold of even dimension $2k$, and let $H_k^0(X)$ be the free part of $H_k(X)$. Then the integral bilinear form ϕ (the intersection index) on $H_k^0(X)$ is unimodular [that is, the matrix $\|\phi(\alpha_i,\alpha_j)\|$ where $\alpha_1,\alpha_2,\dots$ is a system of generators in $H_k^0(X)$ has determinant ± 1].*

This matrix is symmetric if k is even and is skew-symmetric if k is odd. Since any skew-symmetric matrix of odd order is degenerate, we have the following:

Corollary. *The middle Betti number of any closed orientable manifold of dimension* $\equiv 2 \bmod 4$ *is even; hence, the Euler characteristic of such a manifold is even.*

For nonorientable manifolds neither is true; examples: the first Betti number of the Klein bottle is 1, and the Euler characteristic of the real projective plane is 1.

Proof of Theorem 3. Consider the homomorphism $\omega_i\colon H_k^0(X) \to \mathbb{Z}$, $\omega_i(\alpha_j) = \delta_{ij}$. By part (2) of Theorem 2, there exists a $\beta_i \in H^k(X;\mathbb{Z})$ such that $\langle\beta_i,\alpha\rangle = \omega_i(\alpha)$, in particular, $\langle\beta_i,\alpha_j\rangle = \phi(D\beta_i,\alpha_j) = \delta_{ij}$. Let $D\beta_i = \sum_k b_{ik}\alpha_k +$ a finite order element (where b_{ki} are integers). Then

$$\phi(D\beta_i,\alpha_j) = \sum_k b_{ik}\phi(\alpha_k,\alpha_j) = \delta_{ij}.$$

That is, the product of integer matrices $\|b_{ij}\|$ and $\|\phi(\alpha_i,\alpha_j)\|$ is the identity matrix; hence, each of them has the determinant ± 1.

Theorem 3 demonstrates the importance of the theory of integral unimodular ($\det = \pm 1$) forms in topology of manifolds, especially of dimensions divisible by 4: For an oriented closed manifold of such dimension, there arises a unimodular integral quadratic form as the intersection form in the middle dimension. For example, the famous Pontryagin theorem states that a homotopy type of a simply connected closed four-dimensional manifold is fully determined by this form. A lot is known about the classification of such forms (the best source is Milnor and Husemoller [58]), but the question of which forms can be intersection forms for smooth closed four-dimensional simply connected manifolds is very far from being resolved.

In conclusion, let us prove a useful statement on Poincaré duality in products of manifolds.

Theorem 4. *Let X_1, X_2 be a compact oriented homology manifold of dimensions n_1, n_2, and let $\gamma_1 \in H^{q_1}(X_1; G)$, $\gamma_2 \in H^{q_2}(X_2; G)$. Then*

$$D_{X_1 \times X_2}(\gamma_1 \times \gamma_2) = (-1)^{(n_1-q_1)q_2} D_{X_1}\gamma_1 \times D_{X_2}\gamma_2.$$

(Here D_X denotes Poincaré isomorphism in X.)

Proof. We use the obvious relation $(\alpha_1 \times \alpha_2) \frown p_1^*\gamma = (a \frown \gamma) \times \beta)$, where $\alpha_1 \in H_{q_1}(X_1), \alpha_2 \in H_{q_2}(X_2), \gamma \in H^r(X_1; G), p_i: X_1 \times X_2 \to X_i$ is the projection (this relation holds at the chain–cochain level), and the relation $(\alpha_1 \times \alpha_2) \frown p_2^*\gamma = (-1)^{q_1 r}\alpha_1 \times (\alpha_2 \frown \gamma)$, which is obtained from the previous relation by applying the swapping homeomorphism $X_1 \times X_2 \leftrightarrow X_2 \times X_1$.

Back to the theorem:

$$\begin{aligned}
[X_1 \times X_2] \frown (\gamma_1 \times \gamma_2) &= [X_1 \times X_2] \frown (p_1^*\gamma_1 \smile p_2^*\gamma_2) \\
&= ([X_1 \times X_2] \frown p_1^*\gamma_1) \frown p_2^*\gamma_2 \\
&= (([X_1] \times [X_2]) \frown p_1^*\gamma_1) \frown p_2^*\gamma_2 \\
&= (([X_1] \frown \gamma_1)) \times [X_2]) \frown p_2^*\gamma_2 \\
&= (-1)^{(n_1-q_1)q_2}([X_1] \frown \gamma_1) \times ([X_2] \frown \gamma_2).
\end{aligned}$$

17.6 Application: The Lefschetz Formula

Let X be a compact topological space with finitely generated homology $\bigoplus_n H_n(X)$, and let $f: X \to X$ be a continuous map. The number

$$\mathcal{L}(f) = \sum_n (-1)^n \operatorname{Tr} f_{*n}$$

is called the *Lefschetz number* of f [here $\operatorname{Tr} f_{*n}$ denotes the trace of the lattice homomorphism

$$f_{*n}: H_n(X)/\operatorname{Tors} H_n(X) \to H_n(X)/\operatorname{Tors} H_n(X)].$$

Obviously, $\mathcal{L}(f)$ is a homotopy invariant of f. The goal of this section is to establish a relation between the Lefschetz number of f and the behavior of fixed points of f.

Algebraic Lemma. *Let*

$$(\mathcal{C}) \qquad \dots \longrightarrow C_{n+1} \xrightarrow{\partial_{n+1}} C_n \xrightarrow{\partial_n} C_{n-1} \longrightarrow \dots$$

be a complex with finitely generated $\bigoplus_n C_n$, *and let* $f = \{f_n: C_n \to C_n\}$ *be an endomorphism of* $\mathcal{C}$. *Let* $f_{*n}: H_n(\mathcal{C}) \to H_n(\mathcal{C})$ *be the induced homology endomorphism. Then*

$$\sum_n (-1)^n \operatorname{Tr} f_n = \sum_n (-1)^n \operatorname{Tr} f_{*n}.$$

EXERCISE 25. Prove the algebraic lemma.

For example, if X is a finite CW complex, then the Lefschetz number of a continuous map $f: X \to X$ can be calculated as the alternated sum of traces of homomorphisms $g_{\#}: \mathcal{C}_n(X) \to \mathcal{C}_n(X)$ induced by a cellular approximation g of f. This observation alone yields the first, and maybe the most important, application of Lefschetz numbers (not related to manifolds, the more so to Poincaré duality).

Theorem 1. *Let* X *be a finitely triangulated space, and let* $f: X \to X$ *be a continuous map. If* f *has no fixed points, then* $\mathcal{L}(f) = 0$.

Proof. We assume that X is furnished with a metric in which every simplex is isometric to the standard simplex. Then there is a positive δ such that $\operatorname{dist}(x, f(x)) > \delta$ for every $x \in X$. By applying to X the barycentric subdivision sufficiently many times, we can make the diameters of the simplices much less than δ. After this, a simplicial approximation g of f will be such that $g(s) \cap s = \emptyset$ for every simplex s of X. In this case, the simplicial chain $g_{\#}(s)$ will not involve s, so all the diagonal entries of the matrix of $g_{\#n}$ will be zero. Hence, all the traces are zero, and the Lefschetz number is 0.

Let us return to manifolds (but, for now, not to Poincaré duality).

Theorem 2. *Let* X *be a compact smooth manifold (not necessarily orientable, and maybe with a nonempty boundary), and let* ξ *be a vector field on* X. *Suppose that* ξ *has no zeroes and that on the boundary* ∂X *it is directed inside* X. *Then* $\chi(X) = 0$.

This result implied the immensely popular "hairy ball theorem": There is no nowhere vanishing vector field on S^2 (one cannot comb a hairy ball).

Proof of Theorem 2. A vector field ξ on X (with or without zeroes) determines a "flow" $f_t: X \to X$, and for a sufficiently small positive ε the fixed points of f_ε are zeroes of ξ. Since f_ε is homotopic to the identity, $\mathcal{L}(f_\varepsilon) = \mathcal{L}(\mathrm{id}) = \chi(X)$, and if ξ has no zeroes, then $\chi(X) = 0$.

(We will see in Lecture 18 that the converse is also true: If a closed manifold, orientable or not, has zero Euler characteristic, then it possesses a nowhere vanishing vector field.)

So far, regarding Lefschetz numbers, we were interested only in their being zero or not zero. But in reality, in the case of manifolds, the Lefschetz number gives some count of fixed points. This can be expressed by the following proposition.

Theorem 3. *Let* X *be a triangulated compact orientable* n*-dimensional homology manifold* (we will discuss later how much the orientability is really needed) *and let*

$f: X \to X$ be a continuous map. Let $F: X \to X \times X$, $F(x) = (x, f(x))$ be the graph of f, and let $\Delta: X \to X \times X$ be the diagonal map, $\Delta(x) = (x, x)$. Then

$$\phi(F_*[X], \Delta_*[X]) = \mathcal{L}(f).$$

Before proving this theorem, let us briefly discuss its meaning. The intersection points of $F(X)$ and $\Delta(X)$ correspond precisely to fixed points of f. In the smooth case, the intersection number is described in Theorem 1 of Sect. 17.5. First, we need to assume that all the intersections of the graph and the diagonal are transverse. This condition may be formulated in the language of calculus. If x_0 is a fixed point of a smooth map $f: X \to X$, then there arises the differential, $d_{x_0}f: T_{x_0}X \to T_{x_0}X$. The graph and the diagonal are transverse at x_0 if the matrix of $d_{x_0}f - \mathrm{id}$ is nondegenerate, that is, if $f_{x_0}f$ has no eigenvalues equal to 1. If this condition holds, then every intersection point acquires some sign, and the intersection number, equal to the Lefschetz number by Theorem 3, is the "algebraic number of fixed points." The sign can be described as the parity of the number of real eigenvalues of $d_{x_0}f$ less than 1.

A very similar thing can be said about the vector fields. A nondegenerate zero of a vector field can be assign a sign, and then the algebraic number of zeroes of a vector field must be equal to the Euler characteristic of the manifold.

Now, let us turn to proving Theorem 3. We will need a couple of lemmas.

Lemma 1. $\phi(f_*\alpha_1, \alpha_2) = (-1)^{\dim \alpha_1} \phi(F_*[X], \alpha_1 \times \alpha_2)$.

(On the left-hand side the intersection number is taken in X, while on the right-hand side it is taken in $X \times X$.)

Proof of Lemma 1. Let $\alpha_1 = D\gamma_1, \alpha_2 = D\gamma_2$. Then

$$\begin{aligned}
\phi(F_*[X], \alpha_1 \times \alpha_2) &= \phi((\mathrm{id} \times f)_* \circ \Delta_*[X], \alpha_1 \times \alpha_2) \\
&= \langle D^{-1}(\alpha_1 \times \alpha_2), (\mathrm{id} \times f)_* \circ \Delta_*[X] \rangle \\
&= \pm \langle \gamma_1 \times \gamma_2, (\mathrm{id} \times f)_* \circ \Delta_*[X] \rangle = \pm \langle \Delta^*(\gamma_1 \times f^*\gamma_2), [X] \rangle \\
&= \pm \langle \gamma_1 \smile f^*\gamma_2, [X] \rangle = \pm [X] \frown (\gamma_1 \smile f^*\gamma_2) \\
&= \pm([X] \frown \gamma_1) \frown f^*\gamma_2 = \pm \alpha_1 \frown f^*\gamma_2 = \pm \langle f^*\gamma_2, \alpha_1 \rangle \\
&\qquad\qquad = \pm \langle \gamma_2, f_*\alpha_1 \rangle = \pm \phi(f_*\alpha_1, \alpha_2)
\end{aligned}$$

(the signs are determined in Theorem 3 of Sect. 17.5).

Lemma 2. *Let $\alpha_1, \ldots, \alpha_N$ be a basis in the free part of the full homology group of a compact oriented homology manifold X [first, the basis in $H_0(X)$, then $H_1(X)$, and so on], and let $\alpha_1^*, \ldots, \alpha_N^*$ be the dual basis [that is, $\phi(\alpha_i^*, \alpha_j) = \delta_{ij}$]. Then, up to a summand of finite order, $\Delta_*[X] = \sum_i (\alpha_i^* \times \alpha_i)$.*

Proof. By part (2) of Theorem 2 in Sect. 17.5, it is sufficient to prove that

$$\phi(\Delta_*[X], \alpha_p \times \alpha_q) = \phi\left(\sum\nolimits_i (\alpha_i^* \times \alpha_i), \alpha_p \times \alpha_q\right)$$

for every p, q. But

$$\phi(\Delta_*[X], \alpha_p \times \alpha_q) = (-1)^{\dim \alpha_p}\phi(\alpha_p, \alpha_q)$$

by the lemma, and

$$\begin{aligned}\phi\left(\sum_i(\alpha_i^* \times \alpha_i), \alpha_p \times \alpha_q\right) &= \sum_i \phi((\alpha_i^* \times \alpha_i), (\alpha_p \times \alpha_q))\\ &= \sum_i(-1)^{\dim \alpha_i \dim \alpha_p}\phi((\alpha_i^*, \alpha_p)\phi(\alpha_i, \alpha_q))\\ &= (-1)^{(\dim \alpha_p)^2}\phi(\alpha_p, \alpha_q)\end{aligned}$$

by Exercise 7 in Sect. 16.6. This proves Lemma 2.

Proof of Theorem 3. Since the intersection numbers are not sensitive to terms of finite order, we can replace in Theorem 3 $\Delta_*[X]$ by $\sum_i \alpha_i^* \times \alpha_i$ and $F_*[X] = (\mathrm{id} \times f)* \circ \Delta_*[X]$ by $\sum_j \alpha_j^* \times f_*\alpha_j$. Also, since the diagonal Δ is invariant with respect to the coordinate swapping map $X \times X \to X \times X$, we have $\sum_i \alpha_i^* \times \alpha_i = \sum_i (-1)^{d_i(n-d_i)}\alpha_i \times \alpha_i^*$ where $d_i = \dim \alpha_i$. Put $f_*\alpha_j = \sum_k a_{jk}\alpha_k$ and perform the calculations:

$$\begin{aligned}\phi(F_*[X],\, \Delta_*[X]) &= \phi\left(\sum_{j,k} \alpha_j^* \times a_{jk}\alpha_k, \sum_i (-1)^{d_i(n-d_i)}\alpha_i \times \alpha_i^*\right)\\ &= \sum_{i,j,k}(-1)^{d_i(n-d_i)}(-1)^{d_i d_k} a_{jk}\phi(a_j^*, a_i)(-1)^{(n-d_i)d_k}\phi(\alpha_i^*, \alpha_k)\\ &= \sum_{i,j,k}(-1)^{d_i(n-d_i)+d_i d_k+(n-d_i)d_k} a_{jk}\delta_{ji}\delta_{ik} = \sum_i (-1)^{d_i^2} a_{ii} = \mathcal{L}(f).\end{aligned}$$

Let us now briefly discuss the applicability of the Lefschetz theory to the nonorientable and boundary cases. We begin with vector fields. For a nonoriented (even nonorientable) closed manifold the equality between the algebraic number of zeroes of a vector field and the Euler characteristic obviously holds modulo 2. But in reality, mod 2 reduction is not needed. First, the definition of signs attributed to zeroes of vector fields does not require orientation. Second, a connected nonorientable manifold X has an orientable twofold covering, $\widehat{X}$, and a vector field ξ on X can be lifted to a vector field $\widehat{\xi}$ on $\widehat{X}$. It is clear also that $\chi(\widehat{X}) = 2\chi(X)$ (follows from Corollary in Sect. 13.7) and the (algebraic) number of zeroes of $\widehat{\xi}$ is twice the same number for ξ. This implies the statement.

EXERCISE 26. Let X be a connected closed nonorientable manifold, and let $f: X \to X$ be a smooth map which takes orientation preserving loops into orientation preserving loops and orientation reversing loops into orientation reversing loops. Prove that if all fixed points of f are nondegenerate, then the algebraic number of these points is $\mathcal{L}(f)$.

Another extension of the Lefschetz theory may be obtained by admitting, for a manifold considered, a nonempty boundary. Namely, if X is a compact manifold

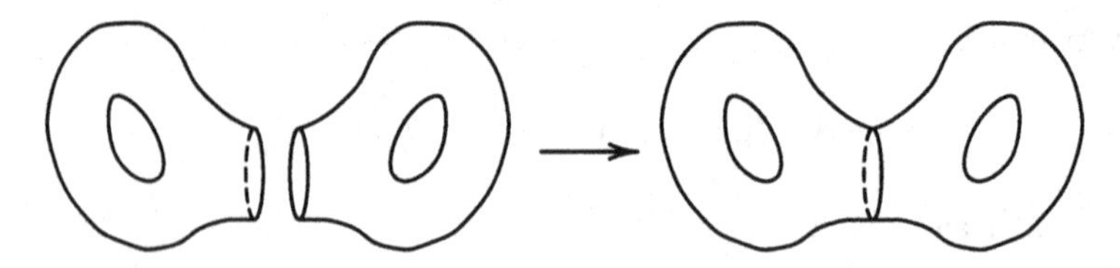

Fig. 70 Doubling a manifold with boundary

with the boundary ∂X, then we can *double* X by attaching to it a second copy of X to the common boundary of the two copies (see Fig. 70).

Let $f: X \to X$ be a continuous map without fixed points on ∂X, and let XX be the double of X. We can extend f to a map $ff: XX \to X \subset XX$ defining this map on the second half to be the same as on the first half [thus $ff(XX)$ is contained in the first half of XX]. It is obvious that ff has the same fixed points as f and $\mathcal{L}(f) = \mathcal{L}(ff)$; hence, the statement of the relation of Lefschetz numbers with fixed points holds for compact manifolds with boundary (orientable or not). Also, we can state that the algebraic number of zeroes of a vector field ξ on a manifold X with boundary such that ξ has no zeroes and directed inside X on ∂X is equal to $\chi(X)$.

EXERCISE 27. There exists a different approach to the Lefschetz theory. First we prove Theorem 1: The Lefschetz number of a fixed-point–free map is zero. Then we consider a map $f: X \to X$ with a nondegenerate fixed point, and, at a neighborhood of this point, we modify both X and f in such a way that the fixed point disappears and the Lefschetz number is changed in a controllable way. Try to recover the details.

In conclusion, let us give one of countless applications of the Lefschetz theory.

EXERCISE 28. The n-dimensional torus T^n can be regarded as $\mathbb{R}^n/\mathbb{Z}^n$. Hence, a linear map $\mathbb{R}^n \to \mathbb{R}^n$ determined by an integral matrix A can be factorized to some continuous map $T^n \to T^n$; denote it as f_A. (Certainly, every continuous map $T^n \to T^n$ is homotopic to a unique map of the form f_A; you may try to prove this.) Calculate the Lefschetz number for f_A (the best possible answer expresses this Lefschetz number in terms of the eigenvalues of A).

EXERCISE 29. Denote the Lefschetz number from Exercise 28 as $\mathcal{L}_A$. Prove that a map homotopic to f_A has at least $|\mathcal{L}_A|$ different fixed points.

EXERCISE 30. Prove that a map $f: T^n \to T^n$ homotopic to f_A with $A = \begin{pmatrix} 2 & 1 \\ 1 & 1 \end{pmatrix}$ has infinitely many periodic points. [A point $y \in Y$ is called a periodic point of a map $g: Y \to Y$ if $g^n(y) = y$ for some n.]

[The last two statements are taken from the note by Ginzburg [43] (Russian).]

17.7 Secondary Intersection Numbers and Secondary Poincaré Duality

Let us return to Poincaré duality. The duality between

$$\operatorname{Tors} H_m(X) \text{ and } \operatorname{Tors} H_{n-m-1}(X)$$

is based on *secondary intersection numbers*, which are defined ahead. (We need to warn the reader that the main results of this section will be given in the form of exercises.)

Let X be a compact oriented n-dimensional homology manifold, and let $\alpha \in H_m(X)$ and $\beta \in H_{n-m-1}(X)$ be homology classes of finite order. Let a and b be cycles representing α and β in the barycentric star and classical complexes of X, and assume that $Na = \partial c$. We define $\omega(\alpha, \beta)$ to be the rational number $\frac{1}{N}\phi(c, b)$ reduced modulo 1 [thus $\omega(\alpha, \beta) \in \mathbb{Q}/\mathbb{Z}$].

EXERCISE 31. Check that $\omega(\alpha, \beta)$ is well defined. (It is this statement that requires the assumption that β has a finite order.)

EXERCISE 32. Prove that if $N\alpha = 0$ and $M\beta = 0$, then $K\omega(\alpha, \beta) = 0$, where $K = \gcd(M, N)$.

EXERCISE 33. Prove that $\omega(\beta, \alpha) = \pm\omega(\alpha, \beta)$ (what is the sign?).

The main property of secondary intersection numbers is the following *secondary Poincaré duality*.

Theorem. *The correspondence* $\alpha \mapsto \{\beta \mapsto \omega(\alpha, \beta)\}$ *yields an isomorphism*

$$\operatorname{Tors} H_m(X) \xrightarrow{\cong} \operatorname{Hom}(\operatorname{Tors} H_{n-m-1}(X), \mathbb{Q}/\mathbb{Z}).$$

EXERCISE 34. Prove this theorem.

17.8 Inverse Homomorphisms

Let X and Y be compact oriented homology manifolds of, possibly, different dimensions m and n, and let $f: X \to Y$ be a continuous map. Poincaré isomorphism allows us to construct "wrong direction" homology and cohomology homomorphisms

$$f^!: H_q(Y; G) \xrightarrow{D^{-1}} H^{n-q}(Y; G) \xrightarrow{f^*} H^{n-q}(X; G) \xrightarrow{D} H_{m-n+q}(X; G),$$
$$f_!: H^q(X; G) \xrightarrow{D} H_{m-q}(X; G) \xrightarrow{f_*} H_{m-q}(Y; G) \xrightarrow{D^{-1}} H^{n-m+q}(Y; G).$$

Both homomorphisms change dimensions by $m - n$: The homomorphism $f^!$ "increases" the dimension by $m-n$ (we use quotation marks because $m-n$ may be negative or zero), and the homomorphism $f_!$ "decreases" the dimension by $m-n$. We will not say much about the cohomology homomorphism $f_!$. It can be regarded as the simplest case of a general construction called "direct image." Its analytic sense (and it belongs rather to analysis than to topology), at least in the case when f is the projection of a smooth fibration, can be best described by the words "fiberwise integration" (people familiar with the de Rham theory can easily understand them). As to the homology homomorphism $f^!$ (called the *inverse Hopf homomorphism*), it has a transparent geometric sense which is described, in the smooth case, by the following proposition.

Theorem. *Let a homology class $\alpha \in H_q(Y)$ be represented by a q-dimensional submanifold Z of Y* (that is, $\alpha = i_*[Z]$, where $i: Z \to Y$ is the inclusion map), *and let f be transversely regular with respect to Z* (that is, the composition

$$T_yY \xrightarrow{d_yf} T_{f(y)}X \xrightarrow{\text{proj}} T_{f(y)}X/T_{f(y)}Z$$

is onto for every point $y \in f^{-1}(Z)$). *Then $f^{-1}(Z)$ is a $(q+m-n)$-dimensional submanifold of X which represents the homology class $f^!(\alpha) \in H_{q+m-n}(X)$.*

We will not prove this theorem but will restate it in a form in which it can be easily translated into an easy-to-prove statement concerning homology manifolds. Let W be an oriented $(q+m-n)$-dimensional submanifold of X transverse to $f^{-1}(Z)$ which may have pseudomanifold-like singularities not in a neighborhood of $f^{-1}(Z)$. Then, at least in a neighborhood of Z, $f(W)$ is an $(n-q)$-dimensional manifold of Y, and f establishes a (sign-preserving) bijection between $W \cap f^{-1}Z$ and $f(W) \cap Z$. Now let us turn to the homology manifold case.

Proposition 1. *Let X, Y, and f be as above, and let $\alpha \in H_q(Y)$, $\beta \in H_{m-q}(X)$. Then*

$$\phi_X(f^!\alpha, \beta) = \phi_Y(\alpha, f_*\beta)$$

(ϕ_X and ϕ_Y denote the intersection number in X and Y).

Proof.

$$\begin{aligned}\phi_X(f^!\alpha, \beta) = \phi_X(Df^*D^{-1}\alpha, \beta) &= \langle f^*D^{-1}\alpha, \beta\rangle \\ &= \langle D^{-1}\alpha, f_*\beta\rangle = \phi_Y(\alpha, f_*\beta).\end{aligned}$$

By part (2) of Theorem 2 in Sect. 17.5, this relation determines $f^!\alpha$ up to a summand of finite order.

Here is one more illustration of the fact that geometrically $f^!$ may be regarded as a preimage.

Proposition 2. *Let X, Y be compact oriented homological manifolds, and let $p: X \times Y \to Y$ be the projection. Then, for any $\alpha \in H_m(Y)$,*

$$p^!\alpha = [X] \times \alpha.$$

Proof. Let $\alpha = D\gamma, \gamma \in H^{n-m}(Y; \mathbb{Z})$. Then

$$p^!\alpha = D_{X\times Y}p^*\gamma = D_{X\times Y}(1 \times \gamma) = D_X 1 \times D_Y\gamma = [X] \times \alpha.$$

EXERCISE 34. Prove the formula $\langle \alpha, f^!\beta \rangle = \langle f_!\alpha, \beta \rangle$.

Let us now turn to the case when $\dim X = \dim Y$.

Proposition 3. *Let X, Y be connected compact oriented manifolds of the same dimension n, and let $f: X \to Y$ be a continuous map of degree d. Then the compositions*

$$H_m(Y) \xrightarrow{f^!} H_m(X) \xrightarrow{f_*} H_m(Y),$$
$$H^m(Y; \mathbb{Z}) \xrightarrow{f^*} H^m(X; \mathbb{Z}) \xrightarrow{f_!} H^m(Y; \mathbb{Z})$$

are both multiplication by d.

Here is a proof of the first statement. Let $\alpha \in H_m(Y)$, $\alpha = D_Y\gamma, \gamma \in H^{n-m}(Y; \mathbb{Z})$. Then $f_* f^! \alpha = f_* D_X f^*\gamma = f_*([X] \frown f^*\gamma) = f_*[X] \frown \gamma = d[Y] \frown \gamma = dD_Y\gamma = d\alpha$ (we used Exercise 12 of Sect. 16.6).

EXERCISE 35. Prove the second statement of Proposition 3.

Corollary. *If $d = \pm 1$, then f_* is an epimorphism, and f^* is a monomorphism.*

GENERALIZATION. *If $d \neq 0$, then every homology class of Y multiplied by d belongs to the image of f_*, and every cohomology class of Y belonging to* Ker f^* *is annihilated by the multiplication by d.*

For example, there is no map $S^2 \to S^1 \times S^1$ of a nonzero degree, but there is a map $S^1 \times S^1 \to S^2$ of degree 1: factorization over $S^1 \vee S^1$.

Everything said in this section has an obvious nonorientable $\mathbb{Z}_2$-analog.

17.9 Poincaré Duality and the Cup-Product

Again, we begin with a statement for the smooth case.

Theorem 1. *Let Y_1, Y_2 be closed oriented submanifolds of a smooth closed oriented manifold X transverse to each other; the latter means that the inclusion map i_1 of Y_1 in X is transversely regular to Y_2. Then the intersection $Z = Y_1 \cap Y_2 = i_1^{-1}(Y_2)$ is a submanifold of X whose dimension k is related to the dimensions n, m_1, m_2 of X, Y_1, Y_2 by the formula $k = m_1 + m_2 - n$. Let $\alpha_1 \in H^{n-m_1}(X; \mathbb{Z}), \alpha_2 \in H^{n-m_2}(X; \mathbb{Z})$,* *and* $\beta \in H^{2n-m_1-m_2}(X;\mathbb{Z})$ *be cohomology classes such that homology classes* $D\alpha_1, D\alpha_2$*, and* $D\beta$ *are represented by* Y_1, Y_2*, and* Z*. Then*

$$\alpha_1 \smile \alpha_2 = \beta.$$

There is a similar $\mathbb{Z}_2$-statement for the nonorientable case.

Proof of Theorem 1.

$$\begin{aligned} D(\alpha_1 \smile \alpha_2) &= [X] \frown (\alpha_1 \smile \alpha_2) = ([X] \frown \alpha_1) \frown \alpha_2 = (D\alpha_1) \frown \alpha_2 \\ &= i_{1*}[Y_1] \frown \alpha_2 = i_{1*}([Y_1] \frown i_1^* \alpha_2) = i_{1*}(Di_1^*\alpha_2) \\ &= i_{1*}(Di_1^* D^{-1}(i_{2*}[Y_2]) = i_{1*}(i_1^!(i_{2*}[Y_2])) \\ &= i_{1*}[i_1^{-1}(Y_2)] = i_*[Z] = D\beta. \end{aligned}$$

(Here i_2 and i are inclusion maps of Y_2 and Z in X; we used in this the proof of Theorem 1 from Sect. 17.9, which was not proven there; if we use instead Theorem 2, then the equality $\alpha_1 \smile \alpha_2 = \beta$ will be proven in a broader context of homology manifolds, but only modulo summand of a finite order.)

This theorem provides a very powerful tool for determining multiplicative structure in cohomology, mainly for manifolds, but actually for all spaces, because of the naturality of the multiplicative structure.

Example. If $q + r \leq n$, then the product of canonical generators of the groups $H^{2q}(\mathbb{C}P^n;\mathbb{Z})$ and $H^{2r}(\mathbb{C}P^n;\mathbb{Z})$ is the canonical generator of $H^{2(q+r)}(\mathbb{C}P^n;\mathbb{Z})$; indeed, Poincaré isomorphism takes the three generators into the homology classes of projective subspaces of dimensions $n - q, n - r$, and $n - q - r$, and, in general position, the intersection of the first two is the third. Thus, the ring $H^*(\mathbb{C}P^n;\mathbb{Z}) = \bigoplus_i H^i(\mathbb{C}P^n;\mathbb{Z})$ has the following structure: There is $1 \in H^0(\mathbb{C}P^n;\mathbb{Z})$ and the generator $x \in H^2(\mathbb{C}P^n;\mathbb{Z})$; the group $H^{2q}(\mathbb{C}P^n;\mathbb{Z})$ with $1 \leq q \leq n$ is generated by x^q. If n is finite, then $x^{n+1} = 0$. In more algebraic terms, $H^*(\mathbb{C}P^n;\mathbb{Z})$ is the ring of polynomials of one variable x factorized by the ideal generated by x^{n+1},

$$H^*(\mathbb{C}P^n;\mathbb{Z}) = \mathbb{Z}[x]/(x^{n+1}), \ \dim x = 2;$$

similarly,

$$\begin{aligned} H^*(\mathbb{H}P^n;\mathbb{Z}) &= \mathbb{Z}[x]/(x^{n+1}), \ \dim x = 4; \\ H^*(\mathbb{R}P^n;\mathbb{Z}_2) &= \mathbb{Z}_2[x]/(x^{n+1}), \ \dim x = 1; \\ H^*(\mathrm{Ca}P^2;\mathbb{Z}) &= \mathbb{Z}[x]/(x^3), \ \dim x = 8. \end{aligned}$$

In all cases, excluding $\mathbb{R}P^n$, the ring $\mathbb{Z}$ may be replaced by any commutative ring.

EXERCISE 36. Prove that the integral cohomology ring of the sphere S_g^2 with g handles is as follows: there are generators $a_1, \dots, a_g, b_1, \dots, b_g$ of $H^1(S_g^2;\mathbb{Z})$ such that $a_1 b_1 = a_2 b_2 = \dots = a_g b_g$ is the generator of $H^2(S_g^2;\mathbb{Z})$ and all other products of generators of $H^1(S^2_g;\mathbb{Z})$ are zeroes. Describe the multiplicative structure in $\mathbb{Z}_2$-cohomology of the projective plane with handles and the Klein bottle with handles.

EXERCISE 37. Prove that any continuous map $\mathbb{C}P^n \to \mathbb{C}P^m$ with $n > m$ induces a trivial map in cohomology of any positive dimension (with any coefficients). Prove a similar statement for real projective spaces.

EXERCISE 38. Prove that if $g < h$, then there are no continuous maps $S^2_g \to S^2_h$ of a nonzero degree.

Theorem 1 shows that the multiplicative structure in cohomology of a closed orientable manifold is rich (many nonzero products). Actually, we already have a strong statement of this kind: Theorems 2 and 3 of Sect. 17.5 show that if X is a compact oriented n-dimensional homology manifold, then for every infinite order class $\alpha \in H^m(X;\mathbb{Z})$ there exists a $\beta \in H^{n-m}(X;\mathbb{Z})$ such that $\langle \alpha \smile \beta, [X]\rangle = 1$. If $\dim X = 2k$ and $\alpha_1, \alpha_2, \dots$ is a basis in the free part of $H^k(X;\mathbb{Z})$, then the matrix $\| \langle \alpha_i \smile \alpha_j, [X]\rangle \|$ is unimodular (that is, its determinant is ± 1).

The remaining part of this lecture is devoted to several modifications (generalizations) of Poincaré duality.

17.10 The Noncompact, Relative, and Boundary Cases of Poincaré Isomorphism

Suppose that a connected triangulated space X is an oriented n-dimensional homology manifold which, however, is not assumed to be compact; that is, the triangulation may be not finite. In this case we still have a correspondence between (oriented) simplices and barycentric stars of complementary dimensions, but no isomorphism between chains and cochains, since chains are supposed to be finite linear combinations of simplices (or barycentric stars), and cochains are allowed to take nonzero values on infinitely many simplices. To construct Poincaré isomorphism, we need to modify the definition either of chains or of cochains. Both modifications are well known in topology; moreover, they exist on the singular level. Here, we restrict ourselves to a brief description of these modifications.

Let X be a locally compact topological space. An n-dimensional *open singular chain* of X is a possibly infinite, linear combination of n-dimensional singular simplices of X with integer coefficients, $\sum_i k_i f_i$, $f_i\colon \Delta^n \to X$, such that for any compact subset $K \subset X$ the coefficients k_i may be nonzero only for finitely many singular simplices f_i such that $f_i(\Delta^n) \cap K \neq \emptyset$. Open chains form a group $C_n^{\text{open}}(X)$, and the usual definition of the boundary operator gives homomorphisms $\partial\colon C_n^{\text{open}}(X) \to C_{n-1}^{\text{open}}(X)$ with $\partial\partial = 0$ and, finally, *open homology groups* $H_n^{\text{open}}(X)$. *Proper* (preimages of compact sets are compact) continuous maps $f\colon X \to Y$ induce chain and homology homomorphisms $f_\#\colon C_n^{\text{open}}(X) \to C_n^{\text{open}}(Y)$ and $f_*\colon H_n^{\text{open}}(X) \to H_n^{\text{open}}(Y)$ with all usual properties (including *proper* homotopy invariance for open homology). In particular, if X is a locally finite CW complex, then $H_n^{\text{open}}(X)$ can be calculated by means of cellular chains which are not assumed to be finite.

There is also a similar (dual) definition of *compact* or *compactly supported* cohomology of a locally compact topological space X. Namely, a cochain $c \in C^n(X;G)$ is called compactly supported if there exists a compact set $K \subset X$ such that $c(f) = 0$ for any singular simplex $f\colon \Delta^n \to X$ such that $f(\Delta^n) \cap K = \emptyset$. There arise groups of compactly supported cochains, $C^n_{\text{comp}}(X;G)$, coboundary operators, $\delta\colon C^n_{\text{comp}}(X;G) \to C^{n+1}_{\text{comp}}(X;G)$, and compact(ly supported) cohomology $H^n_{\text{comp}}(X;G)$. For compactly supported cochains and cohomologies, homomorphisms $f^\#$ and f^* are induced by proper continuous maps. For locally finite CW complexes, compact cohomology can be calculated by means of complexes of finite cochains. Remark also that the usual definition of multiplications gives (in the presence of a pairing $G_1 \times G_2 \to G$) the following binary operations:

$$\left[\gamma_1 \in H^{q_1}_{\text{comp}}(X;G_1), \gamma_2 \in H^{q_2}(X;G_2)\right] \mapsto \gamma_1 \smile \gamma_2 \in H^{q_1+q_2}_{\text{comp}}(X;G);$$
$$\left[\alpha \in H^{\text{open}}_{q_1}(X;G_1), \gamma \in H^{q_2}(X;G_2)\right] \mapsto \alpha \frown \gamma \in H^{\text{open}}_{q_1-q_2}(X;G);$$
$$\left[\alpha \in H^{\text{open}}_{q_1}(X;G_1), \gamma \in H^{q_2}_{\text{comp}}(X;G_2)\right] \mapsto \alpha \frown \gamma \in H_{q_1-q_2}(X;G).$$

All these operations are defined in the usual way on the chain/cochain level.

Consider again a connected triangulated oriented n-dimensional homology manifold X. The barycentric star construction of Sect. 17.4 provides Poincaré isomorphisms

$$D\colon H^m(X;G) \to H^{\text{open}}_{n-m}(X;G) \text{ and } D\colon H^m_{\text{comp}}(X;G) \to H_{n-m}(X;G);$$

both can be expressed by the formula $D\gamma = [X] \frown \gamma$, where the fundamental class $[X]$ is an element of $H_n^{\text{open}}(X)$. These isomorphisms may not look appealing because they involve exotic homology and cohomology groups. However, in many important cases this may be avoided. This possibility is provided by the following general proposition.

Proposition 1. *Let X be a compact topological space and let $A \subset X$ be a closed subset. Then there are natural* (make the statement precise: in what sense natural?) *isomorphisms*

$$H_n^{\text{open}}(X - A;G) \cong H_n(X,A;G) \text{ and } H^n_{\text{comp}}(X - A;G) \cong H^n(X,A;G).$$

In particular, if X is locally compact and $X^\bullet$ is the one-point compactification of X, then

$$H_n^{\text{open}}(X;G) \cong \widetilde{H}_n(X^\bullet;G) \text{ and } H^n_{\text{comp}}(X;G) \cong \widetilde{H}^n(X^\bullet;G).$$

Proposition 1 shows that the preceding Poincaré isomorphisms, in the case when the given homology manifold is a complement to a CW subcomplex A of a compact CW complex X, take the form

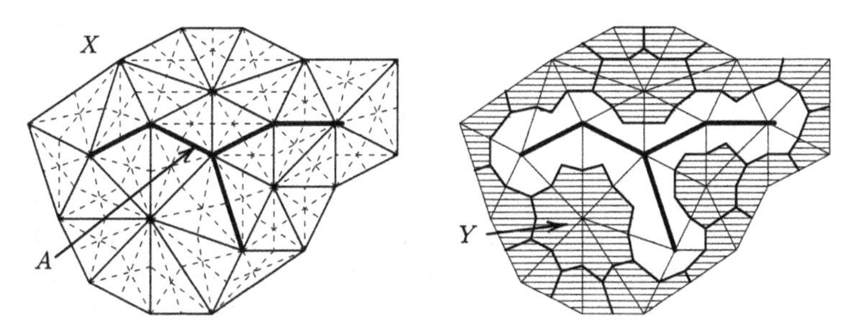

Fig. 71 A barycentric star complex approximation of $X - A$

$$D: H^m(X - A; G) \to H_{n-m}(X, A; G)$$
$$\text{and } H^m(X, A; G) \to H_{n-m}(X - A; G).$$

(Moreover, both isomorphisms can be described as cap-products with the "fundamental class" $[X, A] \in H_n(X, A)$.) We do not prove this proposition, and we do not even offer it as an exercise. Instead, we will give a direct construction of the last isomorphisms, at least in the triangulated case.

Let X be a compact triangulated space, and let A be a triangulated subspace of X such that $X - A$ is a homology manifold. We assume that A satisfies the "regularity condition": If all vertices of some simplex s of X belong to A, then s is contained in A. Let Y be the union of barycentric stars of simplices of X not contained in A (see Fig. 71). Then Y is a closed subset of X, even a triangulated subspace of the barycentric subdivision of X; moreover, Y is homotopy equivalent to $X - A$ (we do not give a formal proof of this homotopy equivalence, but we hope that Fig. 71 may serve as a convincing confirmation of that). The correspondence between simplices and their barycentric stars provides isomorphisms between free Abelian groups generated by simplices in X not contained in A and barycentric stars in Y. These isomorphisms may be considered as either $C^m_{\text{bast}}(Y; \mathbb{Z}) \cong C^{\text{class}}_{n-m}(X, A)$ or $C^m_{\text{class}}(X, A; \mathbb{Z}) \cong C^{\text{bast}}_{n-m}(Y)$; in both cases, the commutativity with ∂ and δ [similar to $(*)$ in Sect. 17.4] holds, so there arise homology/cohomology isomorphisms

$$D: H^m(X - A; \mathbb{Z}) \to H_{n-m}(X, A) \text{ and } D: H^m(X, A; \mathbb{Z}) \to H_{n-m}(X - A)$$

as stated above (it is easy to extend them to an arbitrary coefficient group G).

EXERCISE 39. Prove that both isomorphisms can be expressed as $[X, A] \frown$. (For one of them, we will have to reverse the ordering of vertices in the barycentric subdivision.)

EXERCISE 40. For homology classes $\alpha \in H_m(X - A)$, $\beta \in H_{n-m}(X, A)$, define the intersection number $\phi(\alpha, \beta)$ which has the usual geometric sense. (This must be a replica of Sect. 17.5.) Prove the relative Poincaré duality: The homomorphism

$$\text{Free}\, H_m(X - A) \to \text{Hom}(\text{Free}\, H_{n-m}(X, A), \mathbb{Z}),\ \alpha \mapsto \{\beta \mapsto \phi(\alpha, \beta)\}$$

is an isomorphism. Do similar work with the torsion subgroup and the secondary intersection numbers.

There are two especially important cases of the relative Poincaré duality: the case when X is a sphere and the case when X is a manifold with boundary and $A = \partial X$. We postpone the first case to the next section and will consider the second case now.

Although there exists a theory of homology manifolds with boundary (see, for example, Mitchel [62]), we will not discuss it here; instead of this, we will restrict ourselves to the smooth case. Let X be a connected oriented compact $(n + 1)$-dimensional smooth manifold with a boundary ∂X; we suppose that X possesses a smooth triangulation such that simplices contained in ∂X form a smooth triangulation of ∂X. Since, obviously, $X - \partial X$ is a homology manifold, the previous construction yields (for an arbitrary coefficient group G) Poincaré isomorphisms

$$D: H^m(X; G) \to H_{n+1-m}(X, \partial X; G),$$
$$D: H^m(X, \partial X; G) \to H_{n+1-m}(X; G)$$

(we use the obvious fact that $X - \partial X$ is homotopy equivalent to X). Both isomorphisms have the form $\gamma \mapsto [X, \partial X] \frown \gamma$, where $[X, \partial X] \in H_{n+1}(X, \partial X)$ is the fundamental class of X [represented in the classical complex by the sum of all $(n + 1)$-dimensional simplices of X oriented in accordance to the orientation of X].

Proposition 2. *Poincaré isomorphisms described above, together with Poincaré's isomorphisms for the manifold ∂X, form an isomorphism between homology and cohomology sequences of the pair $(X, \partial X)$; more precisely, there arises a plus–minus commutative diagram*

$$\begin{array}{ccccccccc}
\cdots & H_m(\partial X; G) & \to & H_m(X; G) & \to & H_m(X, \partial X; G) & \to & H_{m-1}(\partial X; G) & \cdots \\
& \uparrow D & & \uparrow D & & \uparrow D & & \uparrow D & \\
\cdots & H^{n-m}(\partial X; G) & \to & H^{n+1-m}(X, \partial X; G) & \to & H^{n+1-m}(X; G) & \to & H^{n+1-m}(\partial X; G) & \cdots
\end{array}$$

Proof. We will prove the plus–minus commutativity of the first square; for the third square the proof is more or less the same, while the commutativity of the second square is obvious.

Take a $c \in C^{n-m}(\partial X; G)$ and extend it to $\widetilde{c} \in C^{n-m}(X; G)$. Here we use the notations $[X, \partial X]$ and $[\partial X]$ for chains; thus, $[X, \partial X] \in C_{n+1}(X)$ and $\partial[X, \partial X] = [\partial X] \in C_n(\partial X) \subset C_n(X)$. As we know from Sect. 16.6 (Exercise 8),

$$\partial([X, \partial X] \frown \widetilde{c}) = \pm(\partial[X, \partial X] \frown \widetilde{c}) \pm ([X, \partial X] \frown \delta\widetilde{c}). \qquad (*)$$

Since $\partial[X, \partial X] = [\partial X] \in C_n(\partial X) \subset C_n(X)$, the cap-product $\partial[X, \partial X] \frown \widetilde{c} \in C_m(X; G)$ belongs to $C_m(\partial X; G)$ and, in this capacity, is $[\partial X] \frown (\widetilde{c}\big|_{\partial X}) = [\partial X] \frown c$. If c is a cocycle representing a class $\gamma \in H^{n-m}(\partial X; G)$, then $\partial[X, \partial X] \frown \widetilde{c}$ and $[X, \partial X] \frown \delta\widetilde{c}$ are cycles (in $C_m(X; G)$) representing $i_*([\partial X] \frown \gamma) = i_* \circ D\gamma$ and

$[X, \partial X] \frown \delta^*\gamma = D \circ \delta^*\gamma$. Since the sum or difference of these cycles is a boundary [formula $(*)$], this proves the plus–minus commutativity of the first square.

We will reformulate the last proposition by passing from Poincaré isomorphisms to Poincaré duality. To avoid separately considering free parts and torsion, we will assume that the coefficient domain is $\mathbb{Q}$, and, for brevity's sake, we will omit the indication of the coefficient domain. We will replace the bottom line of the diagram in Proposition 2 by the dual (with respect to $\langle\ ,\ \rangle$) homology sequence. We get the following "duality diagram."

$$\begin{array}{ccccccccc} \cdots & H_m(\partial X) & \xrightarrow{i_*} & H_m(X) & \xrightarrow{j_*} & H_m(X, \partial X) & \xrightarrow{\partial_*} & H_{m-1}(\partial X) & \cdots \\ & \text{dual} & & \text{dual} & & \text{dual} & & \text{dual} & \\ \cdots & H_{n-m}(\partial X) & \xleftarrow{\partial_*} & H_{n+1-m}(X, \partial X) & \xleftarrow{j_*} & H_{n+1-m}(X) & \xleftarrow{i_*} & H_{n+1-m}(\partial X) & \cdots \end{array}$$

The spaces of each vertical are dual to each other with respect to the intersection number, while the arrows of each vertical are plus–minus dual to each other. The last fact (equivalent to Proposition 2) means the following:

$$\begin{aligned} \phi(i_*\alpha, \beta) &= \pm\phi(\alpha, \partial_*\beta) \text{ for every } \alpha \in H_m(\partial X),\ \beta \in H_{n-m+1}(X, \partial X), \\ \phi(j_*\alpha, \beta) &= \pm\phi(\alpha, j_*\beta) \text{ for every } \alpha \in H_m(X),\ \beta \in H_{n-m+1}(X), \\ \phi(\partial_*\alpha, \beta) &= \pm\phi(\alpha, i_*\beta) \text{ for every } \alpha \in H_m(X, \partial X),\ \beta \in H_{n-m+1}(\partial X). \end{aligned}$$

These results appear the most interesting when n is even: $n = 2k$. Consider the fragment

$$H_{k+1}(X, \partial X) \xrightarrow{\partial_*} H_k(\partial X) \xrightarrow{i_*} H_k(X)$$

of the homology sequence of the pair $(X, \partial X)$ (with the coefficient in $\mathbb{Q}$). The middle space is self-dual, the left and right groups are dual to each other, as well as the homomorphisms i_* and ∂_* (all the dualities are with respect to the intersection number ϕ). The exactness of the sequence implies the equality $\dim H_k(\partial X) = \operatorname{rank} \partial_* + \operatorname{rank} i_*$, and the duality shows that $\operatorname{rank} \partial_* = \operatorname{rank} i_*$. Together, these equalities show that $B_k(\partial X) = \dim H_k(\partial X) = 2 \operatorname{rank} \partial_*$. In other words, the space $H_k(\partial X)$ is even-dimensional (we already know this in the case when k is odd; see Theorem 3 of Sect. 17.5), and the dimension of $\operatorname{Ker} i_* = \operatorname{Im} \partial_* \subset H_k(\partial X)$ is half of $\dim H_k(\partial X)$. For example, the torus T can be presented as a boundary of an orientable compact three-dimensional manifold in many different ways (for example, the torus is the boundary of the solid torus). But if $T = \partial X$ (where X is a compact orientable three-dimensional manifold), then the inclusion homomorphism $i_*: H_1(T) \to H_1(X)$ must have a one-dimensional kernel, not less and not more (if X is a solid torus, then i_* annihilates the homology class of the meridian, but not the homology class of the parallel).

Furthermore, if $\alpha, \beta \in H_{k+1}(X, \partial X)$, then, since ∂_* and i_* are ϕ-dual to each other,

$$\phi(\partial_*\alpha, \partial_*\beta) = \phi(\alpha, i_*\partial_*\beta) = \phi(\alpha, 0) = 0,$$

which shows that the restriction of the form ϕ to this subspace is zero. In the case when k is odd, the form ϕ determines a symplectic structure in $H_k(\partial X)$, and the last statement means that $\operatorname{Ker} i_* = \operatorname{Im} \partial_*$ is a *Lagrangian subspace* of $H_k(\partial X)$. This, however, does not impose any condition on the manifold ∂X. The case when k is even, however, is very much different. A real vector space V with a nondegenerate symmetric bilinear form ω can have a subspace W of dimension one half of $\dim V$ with a zero restriction $\omega|_W$ if and only if the *signature* of ω (the difference between the positive and negative inertia indices) is zero. For a compact oriented 4ℓ-dimensional manifold Y, the signature of the form ϕ in $H_{2\ell}(Y)$ is called *the signature of Y* and is denoted as $\tau(Y)$.

EXERCISE 41. Prove that τ is multiplicative: If Y_1 and Y_2 are two closed oriented manifolds of dimensions divisible by 4, then $\tau(Y_1 \times Y_2) = \tau(Y_1)\tau(Y_2)$.

EXERCISE 42. Prove that if Y_1 and Y_2 are two closed orientable manifolds whose dimensions are not divisible by 4, but sum up to a number divisible by 4, then $\tau(Y_1 \times Y_2) = 0$.

EXERCISE 43. Prove that the reversion of the orientation leads to the negation of the signature.

EXERCISE 44. Let Y_1 and Y_2 be two connected orientable closed manifolds of the dimension 4ℓ, and let $Y = Y_1 \# Y_2$ be the connected sum of Y_1, Y_2 (that is, Y is obtained from Y_1, Y_2 by drilling holes in both of them and then attaching to the boundaries of the holes the tube $S^{4\ell-1} \times I$). Prove that $\tau(Y) = \tau(Y_1) + \tau(Y_2)$.

Theorem. *If a closed oriented 4ℓ-dimensional manifold Y is a boundary of a compact oriented manifold X, then $\tau(Y) = 0$ [in particular, $B_{2\ell}(Y)$ is even].*

Proof. We showed that $B_{2\ell}(\partial X)$ must be even and that $H_{2\ell}(\partial X)$ contains a subspace of dimension $\frac{1}{2}B_{2\ell}(\partial X)$ with zero restriction of ϕ. Hence, $\tau(\partial X) = 0$.

Example. The manifold $\mathbb{C}P^{2\ell}$ cannot be a boundary of a compact orientable $(4\ell + 1)$-dimensional manifold, because $B_{2\ell}(\mathbb{C}P^{2\ell}) = 1$ is odd. But the connected sum $\mathbb{C}P^{2\ell} \# \mathbb{C}P^{2\ell}$ (see Exercise 44), which has even middle Betti number, is also not a boundary since its signature is not zero (it is 2). The same is true for a connected sum of a number of copies of $\mathbb{C}P^{2\ell}$. But the connected sum $\mathbb{C}P^{2\ell} \# (-\mathbb{C}P^{2\ell})$ (where the minus sign stands for the orientation reversion) has zero signature and may be a boundary. Actually, it is a boundary (see Exercise 45 ahead).

EXERCISE 45. Let Y be a connected closed oriented manifold. Prove that the manifold $Y \# (-Y)$ is a boundary of some compact manifold. (*Hint*: Drill a hole in Y and then multiply by I.)

17.11 Alexander Duality

Let $A \subset S^n$ be a simplicial subset of S^n, that is, a union of some simplices of some triangulation of S^n. The goal of this section is to construct *Alexander isomorphisms*,

$$L\colon \widetilde{H}^m(A;G) \xrightarrow{\cong} \widetilde{H}_{n-1-m}(S^n - A;G)$$
$$\text{and } L\colon \widetilde{H}^m(S^n - A;G) \xrightarrow{\cong} \widetilde{H}_{n-1-m}(A;G),$$

and then to reformulate them as a duality between homology groups of A and $S^n - A$. We begin with an obvious remark: If A is empty of is equal to S^n, then the existence of the isomorphisms follows from the definition of groups $\widetilde{H}_{-1}$ and $\widetilde{H}^{-1}$ (which demonstrates one more time that these definitions are right). From now on, we assume that neither A, nor $S^n - A$, is empty. For brevity's sake, we will always omit the indication to the coefficient group (which may be arbitrary).

Remember that, according to Sect. 17.10, the cap-product $[S^n, A]\frown$ yields isomorphisms

$$D\colon H^m(S^n - A) \to H_{n-m}(S^n, A)$$
$$\text{and } D\colon H^{m+1}(S^n, A) \to H_{n-1-m}(S^n - A).$$

Consider the reduced homology sequence of the pair (S^n, A):

$$\dots \widetilde{H}_{n-m}(S^n) \to H_{n-m}(S^n, A) \to \widetilde{H}_{n-1-m}(A) \to \widetilde{H}_{n-1-m}(S^n) \dots . \qquad (*)$$

If $m \neq 0, 1$, then the first and last groups in this exact sequence are zeroes, and we obtain an isomorphism $\partial_*\colon H_{n-m}(S^n, A) \to \widetilde{H}_{n-1-m}(A)$ and the composition

$$L = \partial_* \circ D\colon H^m(S^n - A) \xrightarrow{\cong} \widetilde{H}_{n-1-m}(A)$$

as was promised [for these m, $H^m(S^n - A) = \widetilde{H}^m(S^n - A)$]. It remains to settle the cases $m = 0, 1$.

Lemma. *If $A \neq S^n$, then the inclusion homomorphism $H_n(A) \to H_n(S^n)$ is zero.*

Proof. If $x_0 \notin A$, then this homomorphism factorizes as $H_n(A) \to H_n(S^n - x_0) \to H_n(S^n)$, and $H_n(S^n - x_0) = 0$, since $S^n - x_0$ is homeomorphic to $\mathbb{R}^n$.

[Actually, $H_n(A) = 0$, since $H_{n+1}(S^n, A) = 0$; but we do not need this.]

If $m = 1$, then the last homomorphism of the sequence $(*)$ is zero, and ∂_* remains an isomorphism. If $m = 0$, we get the exact sequence

$$\xrightarrow{0} \widetilde{H}_n(S^n)\,(= \mathbb{Z}) \to H_n(S^n, A) \to \widetilde{H}_{n-1}(A) \to 0,$$

which provides an isomorphism $H_n(S^n, A)/\mathbb{Z} \to \widetilde{H}_{n-1}(A)$ which gives, in combination with D, the promised isomorphism

$$L\colon \widetilde{H}^0(S^n - A) = H^0(S^n - A)/\mathbb{Z} \xrightarrow{D} H_n(S^n, A)/\mathbb{Z} \to \widetilde{H}_{n-1}(A)$$

(the reader is granted the right to replace $\mathbb{Z}$ everywhere with G).

The isomorphism $L\colon \widetilde{H}^m(A) \to \widetilde{H}_{n-1-m}(S^n - A; G)$ is obtained from the isomorphism $D\colon H^{m+1}(S^n, A) \to H_{n-1-m}(S^n - A)$ precisely in the same way, with use of the reduced *co*homology sequence of the pair (S^n, A).

Like Poincaré isomorphism, Alexander isomorphism may be turned into a homology–homology duality, with the role of intersection numbers played by so-called *linking numbers*. From the point of view of Alexander isomorphism, the definition of linking numbers is immediately clear. Let $A \subset S^n$ be as above, and let $\alpha \in H_p(S^n - A)$, $\beta \in H_q(A)$ be two homology classes with $p + q = n - 1$. Then

$$\lambda(\alpha, \beta) = \langle L^{-1}\alpha, \beta\rangle$$

is called the linking number of α and β, and the isomorphism L (rather L^{-1}) becomes a duality

$$\operatorname{Free} H_q(A) \xrightarrow{\cong} \operatorname{Hom}(\operatorname{Free} H_p(S^n - A), \mathbb{Z}),\ \beta \mapsto \{\alpha \mapsto \lambda(\alpha, \beta)\}.$$

But, like intersection numbers, linking numbers have a clear geometric sense, which we will describe now.

Let a, b be two cycles of a compact oriented n-dimensional homology manifold X whose dimensions p, q sum up to $n - 1$. [It is convenient to assume that $a \in C_p^{\text{class}}(X)$ and $b \in C_q^{\text{bast}}(X)$.] Suppose also that both a, b are homological to zero. Choose a c with $\partial c = b$ and put

$$\lambda(a, b) = \phi(a, c)$$

(see Fig. 72).

EXERCISE 46. Prove that $\lambda(a, b)$ does not depend on the choice of c.

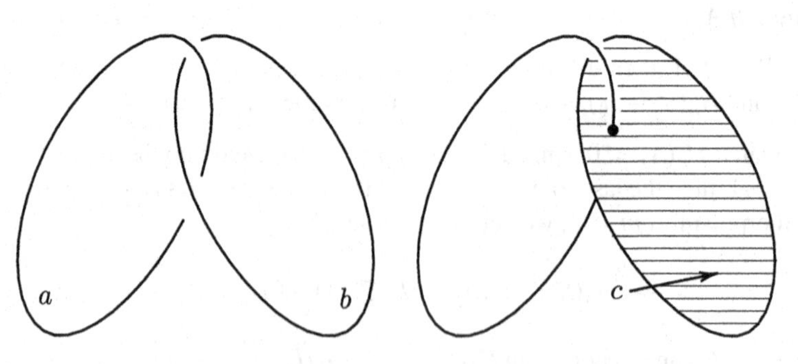

Fig. 72 Definition of the linking number $\lambda(a, b)$

EXERCISE 47. Prove that $\lambda(a, b) = (-1)^{pq+1}\lambda(b, a)$. (For example, the linking number of two disjoint oriented closed curves in $\mathbb{R}^3$ is symmetric with respect to these curves.)

Let us now transfer the definition of a linking number into a context closer to the Alexander duality. Let A, B be disjoint closed subsets of a compact oriented n-dimensional homology manifold X (we can conveniently assume that both are union of simplices of X), and let $\alpha \in H_p(A)$, $\beta \in H_q(B)$ be homology classes which are annihilated by homology homomorphisms induced by the inclusions $A \to X$, $B \to X$. Then $\beta = \partial_*\gamma$ for some $\gamma \in H_{q+1}(X, B)$, and we put $\lambda(\alpha, \beta) = \phi(\alpha, \gamma)$ (in the last formula, we can think of α on the right-hand side as of the image of α in the homology of $X - B$).

EXERCISE 46′. Prove that $\lambda(\alpha, \beta)$ does not depend on the choice of γ.

EXERCISE 47′. Prove that $\lambda(\alpha, \beta) = (-1)^{pq+1}\lambda(\beta, \alpha)$.

In particular, we can take S^n for X, and the complement to a thin neighborhood of A (which is as above) for B (that is, B may look like Y in Fig. 71). Then linking numbers are defined for any $\alpha \in \widetilde{H}_p(A)$, $\beta \in \widetilde{H}_q(B)$ with $p + q = n - 1$.

Theorem. *The equality*

$$\lambda(\alpha, \beta) = \langle L^{-1}\alpha, \beta\rangle$$

holds.

This follows from the definition of L: $L = \partial_* \circ D$.

Thus, linking numbers provide Alexander duality similar to the Poincaré duality.

EXERCISE 48. Make up the definition of "secondary linking numbers" $\mu(\alpha, \beta) \in \mathbb{Q}/\mathbb{Z}$ for $\alpha \in \operatorname{Tors} H_p(A)$, $\beta \in \operatorname{Tors} H_q(S^n - A)$ with $p + q = n - 2$ and prove that

$$\operatorname{Tors} H_p(A) \to \operatorname{Hom}(\operatorname{Tors} H_q(S^n - A), \mathbb{Q}/\mathbb{Z}),\ \alpha \mapsto \{\beta \mapsto \mu(\alpha, \beta)\}$$

(where $p + q = n - 2$) is an isomorphism.

In conclusion, several exercises.

EXERCISE 49. (The Alexander isomorphism in $\mathbb{R}^n$) Let A be a compact polyhedron in $\mathbb{R}^n$. Prove that $H_p(A) \cong \widetilde{H}_q(\mathbb{R}^n - A)$ for $p + q = n - 1$.

EXERCISE 50. Let A be a k-component link (= the union of k disjoint non-self intersecting closed curves in S^3). Find the homology of $S^3 - A$.

EXERCISE 51. (A continuation of Exercise 50) Assume that the linking numbers of the components of A are known. Find the multiplicative structure in the integral cohomology of $S^3 - A$.

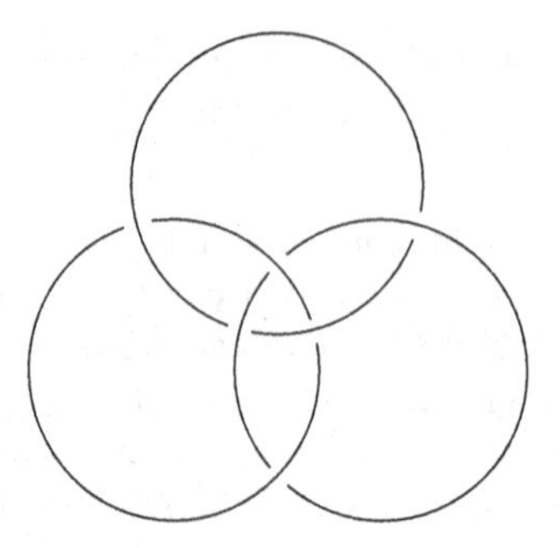

Fig. 73 Borromeo rings

EXERCISE 52. The following is a description of a "secondary multiplicative structure in cohomology" provided by "Massey products." Let $\alpha \in H^p(X;G)$, $\beta \in H^q(X;G)$, $\gamma \in H^r(X;G)$ be cohomology classes of some topological space with coefficients in a ring. Assume that $\alpha \smile \beta = 0$ and $\beta \smile \gamma = 0$. Let $a \in C^p(X;G)$, $b \in C^q \in C^q(X;G)$, $c \in C^r(X;G)$ be (singular) cocycles representing α, β, γ, and let $a \smile b = \delta u$, $b \smile c = \delta v$. Then $h = u \smile c - (-1)^p a \smile v \in C^{p+q+r-1}(X;G)$ is a cocycle, and its cohomology class is determined by α, β, and γ up to a summand of the form $\alpha \smile \sigma + \tau \smile \gamma$ with $\sigma \in H^{q+r-1}(X;G)$, $\tau \in H^{p+q-1}(X;G)$. This (not always and not uniquely) defined cohomology class is called the (triple) Massey product of α, β, γ and is denoted as $\langle \alpha, \beta, \gamma \rangle$. Check all this and compute the cohomology, with cup-products and Massey products, of the complement of the "Borromeo rings" (see Fig. 73).

There exists an extensive theory of "triple linking numbers" and their relations to Massey products (with further generalizations); see Milnor [54] and Turaev [87].

17.12 Integral Poincaré Isomorphism for Nonorientable Manifolds

These isomorphisms have the form

$$H^m(X;\mathbb{Z}) \cong H^{n-m}(X;\mathbb{Z}_T),\ H^m(X;\mathbb{Z}_T) \cong H_{n-m}(X;\mathbb{Z}).$$

Here X is a connected compact n-dimensional nonorientable homology manifold, and homology and cohomology with coefficients in $\mathbb{Z}_T$ ("twisted" integers) are defined in the following way. Let $\widetilde{X}$ be the oriented twofold covering of X. Then there is a canonical orientation reversing involution $t : \widetilde{X} \to \widetilde{X}$. There arise a transformation $t_\#\colon C_q(\widetilde{X}) \to C_q(\widetilde{X})$ with the square 1, and a decomposition

$$C_q(\widetilde{X}) = C_q^+(\widetilde{X}) \oplus C_q^-(\widetilde{X}),$$

where $C_q^{\pm}(\widetilde{X}) = \{c \in C_q(\widetilde{X}) \mid t_\#(c) = \pm c\}$. Obviously, $C_q^+(\widetilde{X})$ is the same as $C_q(X)$; we take the other summand, $C_q^-(\widetilde{X})$, for $C_q(X; \mathbb{Z}_T)$. The groups $C_q(X; \mathbb{Z}_T)$ form, in the obvious way, a complex. The homology of this complex is denoted as $H_q(X; \mathbb{Z}_T)$, and the corresponding cohomology is taken for $H^q(X, \mathbb{Z}_T)$. We will not discuss in any detail these homology and cohomology with "twisted coefficients"; moreover, we will have to do it in a much bigger generality in Chap. 3. Now we restrict ourselves to a recommendation to the reader to reconstruct Poincaré isomorphism given above [they are cap-products with a "fundamental class" $[X] \in H_n(X; \mathbb{Z}_T)$], and Poincaré duality with appropriately defined intersection numbers and secondary intersection numbers.

Lecture 18 The Obstruction Theory

18.1 Obstructions to Extending a Continuous Map

Most problems in homotopy topology consist in a homotopy classification of continuous maps between two topological spaces. A natural intermediate problem is the question of whether a given continuous map $A \to Y$ can be extended to a continuous map $X \to Y$ for some $X \supset A$ (with a subsequent classification of such extensions). This is what the obstruction theory was designed for. We will begin with a technically important particular case.

Let X be a CW complex, and let Y be a connected topological space which is assumed homotopically simple (that is, the action of the fundamental group in all homotopy groups is trivial; later, we will discuss several possibilities of removing or, at least, weakening this condition). Consider the problem of extending a continuous map $f: X^n \to Y$ to a continuous map $X^{n+1} \to Y$ (where X^n, X^{n+1} are skeletons). Let $e \subset X$ be a cell of dimension $n + 1$, and let $h: D^{n+1} \to X$ be a corresponding characteristic map. There arises a continuous map $f_e = f \circ h|_{S^n}: S^n \to Y$. It is obvious that f can be continuously extended to $X^n \cup e$ if and only if f_e is homotopic to a constant, that is, if f_e represents the class $0 \in \pi_n(Y)$ (since Y is homotopically simple, we do not need to fix a base point in Y).

Furthermore, the possibility of extension of f to X^{n+1} is the same as the possibility of its extension to every $(n + 1)$-dimensional cell of X. If we construct, as above, a map $f_e: S^n \to Y$ for every e and denote by φ_e the class of f_e in $\pi_n(Y)$, we arrive at the following, essentially tautological, statement: A continuous map $f: X^n \to Y$ can be extended to a continuous map $X^{n+1} \to Y$ if and only if every φ_e is equal to 0.

The function $e \mapsto \varphi_e$ can be regarded as an $(n + 1)$-dimensional cellular cochain c_f of X with coefficients in $\pi_n(Y)$. (This cochain does not depend on the choice of characteristic maps. Indeed, from the homotopy point of view there are only two characteristic maps corresponding to the two orientations of e; the replacement of h

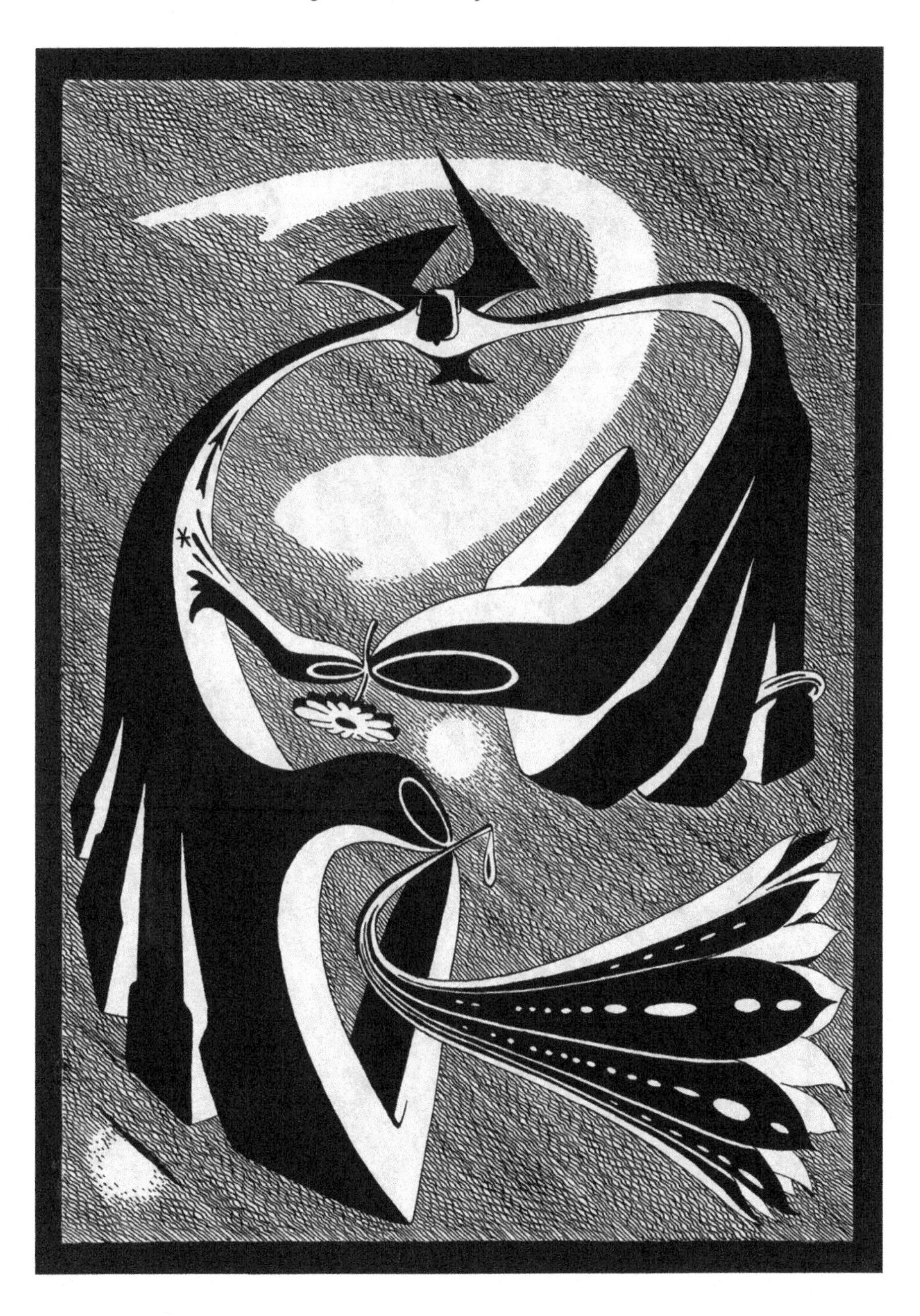

by a characteristic map of the opposite orientation changes the sign at φ_e, but also reverses the orientation of e, so the cochain c_f stays unchanged.) Thus,

$$c_f \in C^{n+1}(X; \pi_n(Y)),$$

and f can be extended to X^{n+1} if and only if $c_f = 0$. The cochain c_f is called the *obstruction cochain* to the extension of f to X^{n+1}.

Notice that the obstruction cochains have a naturality property: If $\varphi: X' \to X$ is a cellular map and $\psi: Y \to Y'$ is a continuous map, then $c_{\psi \circ f \circ \varphi} = \varphi^{\#}\psi_{\#}c_f$.

Up to now, everything said was a sheer triviality. Here is the first nontrivial statement.

Theorem 1. *The obstruction cochain is a cocycle:* $\delta c_f = 0$.

Proof. The statement may be regarded as a variation on the theme of $\partial\partial = 0$ [we need to prove that $c_f(\partial a) = 0$, but the cochain c_f itself is defined by means of boundaries], but the accurate proof requires some work. For example, it can be deduced from the relative Hurewicz theorem (Sect. 14.4). According to this theorem, if X satisfies some conditions (we will discuss them later), then the Hurewicz homomorphism $h: \pi_q(X^q, X^{q-1}) \to H_q(X^q, X^{q-1})$ is an isomorphism. Consider the diagram

$$\begin{array}{ccc}
\mathcal{C}_{n+2}(X) = H_{n+2}(X^{n+2}, X^{n+1}) & \xrightarrow{h^{-1}} & \pi_{n+2}(X^{n+2}, X^{n+1}) \\
 & & \downarrow \partial \\
\downarrow \partial & & \pi_{n+1}(X^{n+1}) \\
 & & \downarrow \\
\mathcal{C}_{n+1}(X) = H_{n+1}(X^{n+1}, X^{n}) & \xrightarrow{h^{-1}} & \pi_{n+1}(X^{n+1}, X^{n}) \\
 & & \downarrow \partial \\
 & \searrow c_f & \pi_n(X^n) \\
 & & \downarrow f_* \\
 & & \pi_n(Y)
\end{array}
\qquad \left.\begin{array}{c} \pi_{n+1}(X^{n+1}) \to \pi_{n+1}(X^{n+1}, X^n) \to \pi_n(X^n) \end{array}\right\} 0$$

This diagram is commutative by the definition of the cochain c_f and the homomorphism $\partial: \mathcal{C}_{n+2}(X) \to \mathcal{C}_{n+1}(X)$. Also, the part of the vertical column marked by a brace is a fragment of the homotopy sequence of the pair (X^{n+1}, X^n), and hence the composition os homomorphism within this part is 0. Thus, $c_f \circ \partial = \delta c_f = 0$.

However, the reference to the relative Hurewicz theorem forces us to respect its assumptions, that is, to assume that X is simply connected and that $n + 1 > 1$.

We will ignore the second assumption (it is easy to see that our arguments are valid when $n = 0$), and we can get rid of the simply connectedness assumption in the following way. Let $p:\widetilde{X} \to X$ be the universal covering of X. The CW decomposition of X induces a CW decomposition of $\widetilde{X}$, and the map $p^{\#}: C^q(X) \to C^q(\widetilde{X})$ is a monomorphism. For a map $f: X^n \to Y$, the obstruction cochain $c_{f\circ p} \in C^{n+1}(\widetilde{X}; \pi_n(Y))$ is $p^{\#}c_f$, $p^{\#}\delta c_f = \delta p^{\#} c_f = \delta c_{f\circ p} = 0$, and hence $\delta c_f = 0$.

The cohomology class $C_f \in H^{n+1}(X; \pi_n(Y))$ of the cocycle c_f is called the *cohomology obstruction*, or simply the *obstruction* to extension of f to X^{n+1}.

Theorem 2. *The condition $C_f = 0$ is necessary and sufficient to the existence of extending $f|_{X^{n-1}}$ to X^{n+1}. In other words, C_f=0 if and only if it is possible to extend f to X^{n+1} after, possibly, a changing f on $X^n - X^{n-1}$.*

[One can apply this theorem to successive extensions of f from a skeleton to a skeleton. Say, let us have a continuous map $f: X^n \to Y$. There arises an obstruction $C_f \in H^{n+1}(X; \pi_n(Y)$. If it is 0, we can extend f to X^{n+1} at the price of some modification of f on X^n not touching f on X^{n-1}. In this case (that is, if $C_f = 0$), we get a new obstruction in $H^{n+2}(X; \pi_{n+1})$. If it is zero, we extend f to X^{n+2} (maybe, after changing the previous extension), and get the next obstruction in $H^{n+3}(X; \pi_{n+2}(Y))$, and so on. One should remember, however, that every new obstruction depends from the previous extension, and hence these obstructions are defined with a growing indeterminacy.]

Before proving Theorem 2, we will give a new definition which will be useful in the proof but will also have a considerable independent value. Let $f, g: X^n \to Y$ be two continuous maps which agree on X^{n-1}. Consider an arbitrary n-dimensional cell e with a characteristic map $h: D^n \to X$. The maps $f \circ h, g \circ h: D^n \to Y$ agree on S^{n-1} [since $h(S^{n-1}) \subset X^{n-1}$, and f and g agree on X^{n-1}] and together compose a map $k_e: S^n \to Y$ (which is $f \circ h$ on the lower hemisphere and $g \circ h$ on the upper hemisphere). We define the *difference cochain*

$$d_{f,g} \in C^n(X; \pi_n(Y)),$$

whose value on e is the class of k_e in $\pi_n(Y)$. It is clear that the condition $d_{f,g} = 0$ is necessary and sufficient for the existence of a homotopy between f and g which is fixed on X^{n-1} (in the terminology of Chap. 1, an X^{n-1}-homotopy; see Sect. 5.7). In the important case when f and g are defined on the whole X and agree on X^{n-1}, the condition $d_{f,g} = 0$ is necessary and sufficient for the existence of an X^{n-1}-homotopy of f making f agree with g on X^n (for this statement, we need to use Borsuk's theorem, Sect. 5.5). Notice also that the difference cochains have a naturality property similar to that of the obstruction cochains: $d_{\psi\circ f\circ\varphi, \psi\circ f\circ\varphi} = \varphi^{\#}\psi_{\#}d_{f,g}$.

Lemma 1. *For any continuous map $f: X^n \to Y$ and any cochain $d \in C^n(X; \pi_n(Y))$, there exists a continuous map $g: X^n \to Y$ which agrees with f on X^{n-1} and is such that $d_{f,g} = d$.*

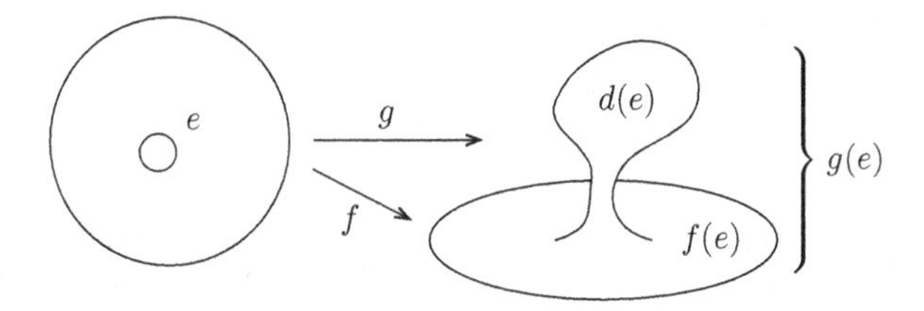

Fig. 74 Proof of Lemma 1

Proof. Consider an n-dimensional cell e of X and distinguish a small ball in e. Than change the map f on this ball in such a way that the two maps of the ball, the old one and the new one, compose a spheroid of the class $d(e)$ (see Fig. 74). Having such a change made on each n-dimensional cell, we get the map g with the required properties.

Lemma 2. $\delta d_{f,g} = c_g - c_f$.

Proof. Consider, for simplicity's sake, the case when f and g are different on only one n-dimensional cell $e \subset X$ (the general case, essentially, is not different from this case). Let σ be an $(n+1)$-dimensional cell of X; we want to show that

$$c_g(\sigma) - c_f(\sigma) = [\sigma : e] d_{f,g}(e).$$

Let $h: D^{n+1} \to X$ be a characteristic map for σ. We can assume that $h^{-1}(e)$ consists of several open balls, of which every one is mapped by h homeomorphically onto e, with preserving or reversing the orientation, and $[\sigma : e]$ is the difference of the number of balls where the orientation is preserved and the number of balls where it is reversed (compare the description of the incidence numbers in Sect. 13.6). This makes the desired equality obvious: A spheroid representing $c_g(\sigma)$ is obtained from a spheroid representing $c_f(\sigma)$ by adding spheroids of the class $\pm d_{f,g}(e)$, and the algebraic number of these spheroids is $[\sigma : e]$.

Proof of Theorem 2. If $C_f = 0$, then $c_f = \delta d$ and, by Lemma 1, there exists a map $g: X^n \to Y$ such that $g\,|_{X^{n-1}} = f\,|_{X^{n-1}}$ and $d_{f,g} = -d$. But then, by Lemma 2, $c_g = c_f + \delta d_{f,g} = \delta d - \delta d = 0$; thus, g can be extended to X^{n+1}. Conversely, if there exists a map $g: X^n \to Y$ which agrees with f on X^{n-1} and can be extended to X^{n+1}, then $c_g = 0$ and $c_f = c_f - c_g = \delta d_{f,g}$, and hence $C_f = 0$.

Remark. The two lemmas of this proof are not less important than the theorem; we will use them later.

EXERCISE 1. Prove that $d_{g,f} = -d_{f,g}$ and $d_{f,h} = d_{f,g} + d_{g,h}$.

18.2 The Relative Case

Let A be a CW subcomplex of a CW complex X, and let the continuous map f be defined on $A \cup X^n$. The obstruction cochain c_f to an extension of this map to $A \cup X^{n+1}$ is contained in $C^{n+1}(X, A; \pi_n(Y))$, it is a cocycle, and its cohomology class $C_f \in H^{n+1}(X, A; \pi_n(Y))$ is called an obstruction. The theory of these relative obstructions is absolutely parallel to its absolute prototype; in particular, it contains the notion of difference cochains, and there are precise analogies (for both the statements and the proofs) of all theorems and lemmas of the previous section. We will point out the following important consequence of the relative theory in the absolute theory.

Let $f, g: X \to Y$ (or $X^{n+1} \to Y$) be two maps with $f|_{X^{n-1}} = g|_{X^{n-1}}$ [or with a fixed homotopy connecting $f|_{X^{n-1}}$ and $g|_{X^{n-1}}$]. We consider the problem of constructing a homotopy between f and g fixed (or coinciding with the given homotopy) on X^{n-1}. This problem is equivalent to extending to $X \times I$ (or to $X^{n+1} \times I$) the map which is given on $(X \times 0) \cup (X^{n-1} \times I) \cup (X \times 1)$ by the formula

$$(x, t) \mapsto \begin{cases} f(x), \text{ if } t = 0 \text{ or } x \in X^{n-1}, \\ g(x), \text{ if } t = 1 (\text{or } x \in X^{n-1}) \end{cases}$$

(this formula is for the case when f and g agree on X^{n-1}; if a homotopy between $f|_{X^{n-1}}$ and $g|_{X^{n-1}}$ is given, the formula will be slightly different; we leave the details to the reader). The obstruction to an extension of this map to $(X \times 0) \cup (X^n \times I) \cup (X \times 1)$ lies in $C^{n+1}(X \times I, (X \times 0) \cup (X \times 1); \pi_n(Y)) = C^n(X; \pi_n(Y))$, and it is easy to see that it is nothing but $d_{f,g}$. By the way, $\delta d_{f,g} = c_g - c_f = 0$, since f and g are both defined on the whole X (or, at least, on X^{n+1}). If we apply to this situation the relative version of Theorem 2 of Sect. 18.1, we will get the following result.

Theorem. *If $f, g: X \to Y$ are two continuous maps which agree on X^{n-1}, then the difference cochain $d_{f,g}$ is a cocycle whose cohomology class $D_{f,g} \in H^n(X; \pi_n(Y))$ is equal to 0 if and only if $f|_{X^n}$ and $g|_{X^n}$ are X^{n-2}-homotopic.*

18.3 The First Application: Cohomology and Maps into $K(\pi, n)$s

The main result of this section was promised in Lecture 4. Let π be an Abelian group.

Recall that the construction of a $K(\pi, n)$ space begins with taking a bouquet of n-dimensional spheres set into a correspondence with some system of generators of π (see Sect. 11.7); then we attach to this bouquet cells of dimensions $> n$. If we assign to every n-dimensional cell of $K(\pi, n)$ the corresponding element of π, we get a cochain $c \in C^n(K(\pi, n); \pi)$ [we admit here a certain abuse of notation, using the symbol $K(\pi, n)$ for a CW complex obtained by some concrete construction].

Lemma. *c is a cocycle.*

First Proof (Direct). The cells of dimension $n + 1$ correspond to the defining relations between the chosen generators. If the cell σ corresponds to the relation $\sum k_i g_i = 0$ between the generators g_i, then for the n-dimensional cell e_i corresponding to the generator g_i, the incidence number $[\sigma : e_i]$ is k_i. Then

$$\delta c(\sigma) = \sum_i [\sigma : e_i] c(e_i) = \sum_i k_i g_i = 0.$$

Second Proof (Indirect). Actually, $c = d_{\text{const,id}}$; thus, $\delta c = 0$ by Lemma 2 of Sect. 18.1.

The cohomology class $F_\pi \in H^n(K(\pi, n); \pi)$ of the cocycle c is called the *fundamental cohomology class* of $K(\pi, n)$. Another description of this class: According to the universal coefficients formula,

$$H^n(K(\pi, n); \pi) = \text{Hom}(H_n(K(\pi, n)), \pi),$$

and, by Hurewicz's theorem, $H_n(K(\pi, n)) = \pi_n(K(\pi, n)) = \pi$. The class F_π corresponds to the identity homomorphism

$$\text{id}_\pi \in \text{Hom}(H_n(K(\pi, n)), \pi).$$

EXERCISE 2. Prove the equivalence of the two definitions of the fundamental homology class.

Notice that the second definition of the fundamental class can be applied to an arbitrary $(n-1)$-connected space X. In this case, it yields a cohomology class $F_X \in H^n(X; \pi_n(X))$. We will return to this class later.

Now we turn to the main result of this section.

Theorem. *Let X be a CW complex. For any Abelian group π and for any $n > 0$, the map*

$$\pi(X, K(\pi, n)) \to H^n(X; \pi), \ [f] \to f^*(F_\pi), \tag{$*$}$$

is a bijection.

Proof. First, let $\gamma \in H^n(X; \pi)$, and let $c \in C^n(X; \pi)$ be a cocycle of the class γ. We want to construct a continuous map $f: X \to K(\pi, n)$ which takes the cocycle of class F_π (constructed above) into c. By Lemma 1 of Sect. 18.1, there exists a map $f: X^n \to K(\pi, n)$ such that $f(X^{n-1})$ is the (only) vertex of $K(\pi, n)$ and $d_{\text{const},f} = c$. Then, obviously, $f^\#: C^n(K(\pi, n); \pi) \to C^n(X; \pi)$ takes $d_{\text{const,id}}$ into $d_{\text{const},f} = c$ (by the naturality property of the difference cochains; see Sect. 18.1). Then we extend this map f to $X^{n+1}, X^{n+2}, \ldots$, and it is possible, since $\pi_{n+1}(K(\pi, n)), \pi_{n+2}(K(\pi, n)), \ldots$ are all zeroes. We obtain a map $f: X \to K(\pi, n)$. By construction, f^* takes F_π into γ. Thus, the map $(*)$ is onto.

Now let $f, g: X \to K(\pi, n)$ be two continuous maps with $f^*F_\pi = g^*F_\pi$. We want to prove that $f \sim g$; we can assume that f and g are cellular maps (in particular, they are constant on X^{n-1}). Then f^*F_π and g^*F_π are represented by $f^\# d_{\text{const,id}} = d_{\text{const},f}$ and $g^\# d_{\text{const,id}} = d_{\text{const},g}$. Hence, the cocycles $d_{\text{const},f}$ and $d_{\text{const},g}$ are cohomological, so the difference $d_{\text{const},g} - d_{\text{const},f} = d_{f,g}$ is cohomological to 0, or $D_{f,g} = 0$. According to the theorem in Sect. 18.2, this shows that f and g are X^{n-2}-homotopic (the homotopy being fixed on X^{n-2} is not important to us) on X^n. They are also homotopic on further skeletons, since the further difference cochains belong to the cochain groups with trivial coefficients. Thus, the map $(*)$ is one-to-one.

Corollary 1. *A CW complex of the type $K(\pi, n)$ is homotopically unique. Hence, a topological space of the type $K(\pi, n)$ is weakly homotopically unique.*

Proof. Let X, X' be CW complexes of the type $K(\pi, n)$, and let $F_\pi \in H^n(X; \pi), F'_\pi \in H^n(X'; \pi)$ be the fundamental classes. According to the theorem, there exist continuous maps $f: X \to X'$, $g: X' \to X$ such that $f^*(F'_\pi) = F_\pi$ and $g^*(F_\pi) = F'_\pi$. Since $(g \circ f)^*(F_\pi) = f^* \circ g^*(F_\pi) = F_\pi = (\text{id}_X)^*(F_\pi)$, we have $g \circ f \sim \text{id}_X$ and, similarly, $f \circ g \sim \text{id}_{X'}$.

EXERCISE 3. Since $K(\pi, n) \sim \Omega K(\pi, n+1)$ is an H-space, the set $\pi(X, K(\pi, n))$ is a group (see Lecture 4), and the bijection $H^n(X; \pi) \leftrightarrow \pi(X, K(\pi, n))$ is a bijection between two groups. Prove that it is a group isomorphism.

Actually, for every Abelian group π and every n, there exists an Abelian topological group of the type $K(\pi, n)$. The reader may try to prove it by an appropriate enhancing of the construction of the (second) loop space.

Corollary 2. *For a CW complex X, there is a group isomorphism $H^1(X; \mathbb{Z}) \cong \pi(X, S^1)$ (where S^1 is regarded as an Abelian topological group).*

EXERCISE 4. Prove that every continuous map $\underbrace{S^1 \times \dots \times S^1}_{n} \to \underbrace{S^1 \times \dots \times S^1}_{m}$ is homotopic to a linear map (that is, to a map obtained by a factorization from a linear map $\mathbb{R}^n \to \mathbb{R}^m$ determined by an integral matrix).

18.4 The Second Application: Hopf's Theorems

Theorem 1 (Hopf). *For every n-dimensional CW complex X, there is a bijection*

$$H^n(X; \mathbb{Z}) \leftrightarrow \pi(X, S^n),\ [f] \mapsto f^*(s),$$

where $s = 1 \in \mathbb{Z} = H^n(S^n; \mathbb{Z})$.

Proof. This classical theorem (proved, actually, before the appearance of not only the obstruction theory, but also cohomology) is, from a modern point of view, a corollary of the theorem in Sect. 18.3. Indeed, the construction of the space

$K(\pi, n)$, as given in Sect. 11.7, begins with a bouquet of n-dimensional spheres corresponding to generators of π; if $\pi = \mathbb{Z}$, we can take one sphere. On the next step, we attach $(n+1)$-dimensional cells corresponding to relations between the chosen generators; but in the case $\pi = \mathbb{Z}$ there are no relations, and no $(n+1)$-dimensional cells are needed. Then we attach cells of dimensions $\geq n+2$. We see that the $(n+1)$st skeleton of (such constructed) $K(\mathbb{Z}, n)$ is S^n. Hence, by the cellular approximation theorem, if X is n-dimensional, every map $X \to K(\mathbb{Z}, n)$ is homotopic to a map $X \to S^n \subset K(\mathbb{Z}, n)$ and every two maps $X \to S^n \subset K(\mathbb{Z}, n)$ homotopic in $K(\mathbb{Z}, n)$ are homotopic in S^n.

[There is a more direct proof which is a replica of the proof of the theorem in Sect. 18.3. The main difference is that the higher obstruction and difference cochains are equal to zero not because the higher homotopy groups of S^n are zeroes (which is not true), but because X has no cells of higher dimensions.]

Theorem 2 (Hopf). *Let an n-dimensional CW complex X contain as a CW subcomplex a sphere S^{n-1}. This sphere is a retract of X if and only if the inclusion homomorphism $H^{n-1}(X; \mathbb{Z}) \to H^{n-1}(S^{n-1}; \mathbb{Z})$ is an epimorphism.*

Proof. The only if part is obvious: If $r: X \to S^{n-1}$ is a retraction, then the composition

$$H^{n-1}(S^{n-1}; \mathbb{Z}) \xrightarrow{r^*} H^{n-1}(X; \mathbb{Z}) \xrightarrow{j^*} H^{n-1}(S^{n-1}; \mathbb{Z}),$$

where j is the inclusion map, is the identity, and hence j^* is an epimorphism. Assume now that j^* is an epimorphism and fix a class $\alpha \in H^{n-1}(X; \mathbb{Z})$ such that $j^*(\alpha) = 1 \in \mathbb{Z} = H^{n-1}(S^{n-1}; \mathbb{Z})$. Let $a \in C^{n-1}(X; \mathbb{Z})$ be a cocycle of the class α. Construct a map $q: X \to S^{n-1}$ in the following way. All the cells of dimensions $\leq n-2$ we map into a point. On every $(n-1)$-dimensional cell e define the map as the spheroid of the class $a(e)$. This requirement means precisely that the map $q^\#$ takes $1 \in \mathbb{Z} = C^{n-1}(S^{n-1}; \mathbb{Z})$ into a. On the other side, it means that the cochain a is the difference cochain between the already constructed part of the map q and the map $\mathrm{const}: X^{n-1} \to S^{n-1}$. Hence,

$$0 = \delta a = \delta d_{q,\mathrm{const}} = c_q - c_{\mathrm{const}} = c_q,$$

so the map q can be extended to $X^n = X$. The composition

$$S^{n-1} \xrightarrow{j} X \xrightarrow{q} S^{n-1}$$

induces the identity map in cohomology: $(q \circ j)^*(1) = j^*(q^*(1)) = j^*(\alpha) = 1$, and hence homotopic to id. We can extend the homotopy between this map and id to the homotopy of the map q. As a result, we will get a map $r: X \to S^{n-1}$ which is the identity on S^{n-1}, that is, a retraction.

18.5 Obstructions to Extensions of Sections

Let $\xi = (E, B, F, p)$ be a locally trivial fibration. We assume that the fiber F is homotopically simple (for example, simply connected), and the base B is simply connected. (The last assumption can be weakened to the assumption of the *homotopical simplicity of the fibration.* The latter means that for every continuous map $S^1 \to B$, the induced fibration over S^1 is trivial. In the next lecture, we will encounter important examples of this situation.)

Assume that the base B is a CW complex and that there given a section $s: B^n \to E$ [which means that $p \circ s = \mathrm{id}$] over the nth skeleton of the base. We are going to describe an obstruction to extending this section to B^{n+1}. Let e be an $(n+1)$-dimensional cell over B. The fibration $h^*\xi$ over D^{n+1}, induced by means of a characteristic map $h: D^{n+1} \to B$ for the cell e, is trivial. The section s induces a section $S^n \to D^{n+1} \times F$ of the restriction of the last fibration to $S^n \subset D^{n+1}$, and hence an element of $\pi_n(D^{n+1} \times F) = \pi_n(F)$ (rather of the fiber $p^{-1}(x)$ over some point $x \in e$, but the simply connectedness of the base, or the homotopical simplicity of the fibration ξ, provides a canonical homomorphism between homotopy groups of all fibers—the reader will reconstruct a detailed explanation of this). We get a cochain $c_s \in C^{n+1}(B; \pi_n(F))$. This is the *obstruction cochain to extending s to B^{n+1}*. The properties of this obstruction cochain are the same as those of the obstruction cochains considered in Sect. 18.1. Namely:

(1) The section s can be extended to a section over the $(n+1)$st skeleton of B if and only if $c_s = 0$.
(2) $\delta c_s = 0$.
(3) The cohomology class $C_s \in H^{n+1}(B; \pi_n(F))$ of c_s (which is called the *obstruction*) is equal to 0 if and only if the section s can be extended to a section over B^{n+1}.

There are also *difference cochains* $d_{s,s'}$ whose definition and properties are the same as before.

Obstructions to extending maps may be regarded as particular cases of obstructions to extending sections. Namely, a continuous map $f: X \to Y$ can be represented by the graph $F: X \to X \times Y$, $F(x) = (x, f(x))$, which, in turn, is a section of the trivial fibration $(X \times Y, X, Y, p)$, where $p: X \times Y \to X$ is the projection of the product onto a factor. Obstructions to extending a map are the same as obstructions to extending its graph. On the other hand, the theory of obstructions to sections cannot be reduced to the theory of obstructions to maps. In particular, the latter does not have any analogy of the next construction.

Suppose that $\pi_0(F) = \pi_1(F) = \cdots = \pi_{n-1}(F) = 0$, and $\pi_n(F) \neq 0$. Then there are no obstructions to extending a section from B^0 (where it obviously exists) to $B^1, \ldots, B^{n-1}$ and the first obstruction emerges in $H^{n+1}(B; \pi_n(F))$: It is the obstruction to extending the section from B^{n-1} to B^n. This obstruction could depend, however, on the sections on the previous skeletons; however, the next proposition states that it is not the case.

Proposition 1. *Let* $\pi_0(F) = \pi_1(F) = \cdots = \pi_{n-1}(F) = 0$, *and let* $s, s'\colon B^n \to E$ *be two sections. Then* $C_s = C_{s'} \in H^{n+1}(B; \pi_n(F))$.

To prove this, we need a slightly modified version of the homotopy extension property (Borsuk's theorem; see Sect. 5.5).

Lemma (Borsuk's Theorem for Sections). *Let* $\xi = (E, B, F, p)$ *be a locally trivial fibration with a CW base, let* $S\colon B \to E$ *be a section of* ξ, *let* A *be a CW subspace of* B, *and let* $s_t\colon A \to E$ *be a homotopy consisting of sections of* $\xi \mid_A$ *such that* $s_0 = S \mid_A$. *Then there exists a homotopy* $S_t\colon B \to E$ *consisting of sections of* ξ *and such that* $S_0 = S$, $S_t \mid_A = s_t$.

Proof of Lemma. This lemma is not different from Borsuk's theorem in the case when the fibration is (standard) trivial: $E = B \times F$, p is the projection of the product onto a factor. Indeed, in this case, a section is the same as a continuous map $B \to F$. Passing to the general case, we can restrict ourselves to the situation when A and B differ by one cell: $B = A \cup e$, where e is a cell of B. Take a characteristic map $h\colon D^n \to B$ (where $n = \dim e$). Then the sections S, s_t of ξ and $\xi \mid_A$ give rise to sections S', s'_t of the fibrations $h^*\xi, h^*\xi \mid_{S^{n-1}}$ [such that $s'_0 = S' \mid_{S^{n-1}}$]. Since the fibration $h^*\xi$ is trivial (Feldbau's theorem, Sect. 9.2), the lemma has already been proved for this fibration, which provides a homotopy S'_t consisting of sections of this fibration such that $S'_0 = S'$ and $S'_t \mid_{S^{n-1}} = s'_t$. The homotopies s_t and S'_t together form a homotopy $S_t\colon B \to E$ with the required properties.

Proof of Proposition 1. It is clear that a homotopy of a section $s\colon B^k \to E$ will not affect either c_s or C_s. Suppose that the given sections s, s' are homotopic over B^k for some k, $0 \leq k < n - 1$ (since the fiber F is connected, this is obviously true for $k = 0$). A homotopy of s' to s on B^k can be extended, by the lemma, to a homotopy of s' on B^n, without any changes for $c_{s'}$ and $C_{s'}$ so we can assume that $s' = s$ on B^k. The difference cochain $d_{s,s'} \in C^{k+1}(B; \pi_{k+1}(F))$ is zero, because $\pi_{k+1}(F) = 0$; thus, $s' \sim s$ on B^{k+1}. In this way, we can reduce the general case of the proposition to the case when $s' = s$ on B^{n-1}. Then we have a difference cochain $d_{s,s'} \in C^n(B, \pi_n(X))$, and $\delta d_{s,s'} = c_{s'} - c_s$. Thus, the cocycles c_s and $c_{s'}$ are cohomological and hence $C_s = C_{s'}$.

Proposition 1 shows that the *first obstruction* to extending a section to the nth skeleton of the base is determined by the fibration, so we obtain a well-defined class $C(\xi) \in H^{n+1}(B; \pi_n(F))$ (recall that n is the number of the first nontrivial homotopy group of F); this class is called the *characteristic class* of ξ; we will also use the term *primary characteristic class* to distinguish it from numerous characteristic classes of vector bundles, which will be studied in Lecture 19.

One can say that a fibration as above has a section over the nth skeleton of the base if and only if its characteristic class is zero.

EXERCISE 5 (The main property of characteristic classes). Let ξ be a fibration as above, and let $f\colon B' \to B$ be a continuous map of some CW complex into B. Then

$$C(f^*\xi) = f^*(C(\xi)).$$

EXERCISE 6. Prove that a characteristic class is homotopy invariant (we leave to the reader not only the proof, but also a precise statement of this fact). In particular, the characteristic class does not depend on the CW structure of the base.

EXERCISE 7. Using previous exercises, make up a definition of a characteristic class in the case when the base is not a CW complex.

Example. (Since this example concerns smooth manifolds, the definitions and statements will not be genuinely rigorous.) Let X be a connected closed oriented n-dimensional manifold and let T be the manifold of all nonzero tangent vectors of M. The projection $p: T \to X$ (which assigns to a tangent vector the tangency point) gives rise to a locally trivial fibration $\tau_X = (T, X, \mathbb{R}^n - 0, p)$. Since the fiber is homotopy equivalent to S^{n-1}, there arises a characteristic class $C(\tau_X) \in H^n(X;\mathbb{Z})$. (It is easy to understand that the fibration τ_X is simple if and only if the manifold X is orientable.)

Proposition 2. $\langle C(\tau_X), [X]\rangle = \chi(X)$.

Proof. A section of the fibration τ_X is the same as a nowhere vanishing vector field on X. It is easy to understand that a generic vector field on X has only isolated zeroes. Take a local coordinate system with the origin at the isolated zero x_0 of a vector field ξ, take a small sphere $S \approx S^{n-1}$ centered at x_0, and consider the map $S = S^{n-1} \to S^{n-1}$ which takes $x \in S$ into $\xi(x)/\|\xi(x)\|$. Denote by $d_\xi(x_0)$ the degree of this map. We can assume (although it is actually not necessary) that all the zeroes of ξ are *nondegenerate*, that is, $d(x_0) = \pm 1$. Now consider a smooth triangulation of X such that all zeroes of ξ lie inside n-dimensional simplices, at most one in every simplex. Then ξ is a section of the fibration τ_X over the $(n-1)$st skeleton of X, and the obstruction c_ξ to extending this section to an n-dimensional simplex s is zero if s does not contain zeroes of ξ and is $d(x_0)$ if s contains a zero x_0 of ξ. Since the fundamental cycle of $[X]$ is the sum of all (oriented) n-dimensional simplices of the triangulation, $\langle c_\xi, [X]\rangle = \sum_{x_0 \in \{\text{zeroes of } \xi\}} d(x_0)$. The left-hand side of this equality is $\langle C(\tau_X), [X]\rangle$, the right-hand side, as explained in Sect. 17.6 (see Theorem 3 and the discussion after it), is $\chi(X)$. This completes the proof of Proposition 2.

Corollary. *A connected closed orientable manifold possesses a nowhere vanishing vector field if and only if $\chi(X) = 0$.*

The only if part of this statement has been proved before: See Theorem 2 in Sect. 17.6. The if part was promised there. The orientability condition is not needed; it also was explained in Sect. 17.6.

In conclusion, a couple of additional exercises.

EXERCISE 8. Make up a theory of obstructions to extending sections in the context of Serre fibrations (see Sect. 9.4).

EXERCISE 9. Let X be a CW complex with $\pi_0(X) = \pi_1(X) = \cdots = \pi_{n-1}(X) = 0$, $\pi_n(X) \neq 0$. Prove that the characteristic class of the Serre fibration $EX \to X$ with the fiber ΩX which belongs to $H^n(X; \pi_{n+1}(\Omega X)) = H^n(X; \pi_n(X))$ is just the fundamental class of X.

Lecture 19 Vector Bundles and Their Characteristic Classes

19.1 Vector Bundles and Operations over Them

A: Definitions

We consider three types of vector bundles: *real, oriented,* and *complex*. A real n-dimensional vector bundle with the base B is a locally trivial fibration with the base B and the fiber homeomorphic to $\mathbb{R}^n$ with an additional structure: Each fiber is furnished by a structure of an n-dimensional vector space, in such a way that the vector space operations $(\lambda, x) \mapsto \lambda x$ and $(x, y) \mapsto x + y$ depend continuously on the fiber, in the sense that the arising maps $\mathbb{R} \times E \to E$ and $\{(x, y) \in E \times E \mid p(x) = p(y)\} \to E$ (where E is the total space and p is the projection of the fibration) are continuous. Complex vector bundles are defined precisely in the same way, only the field $\mathbb{R}$ is replaced by the field $\mathbb{C}$; oriented vector bundles are real vector bundles whose fibers are furnished with orientation depending continuously on the fiber. The last property can be formalized in the following way. For simplicity's sake, assume that B is connected. Let $\widetilde{E}$ be the set of all bases in all fibers of the fibration; there is a natural topology in $\widetilde{E}$. The fibration is orientable if and only if $\widetilde{E}$ has two (not one) components; a choice of one of these components is an *orientation* of the fibration.

For vector bundles of all three kinds there are natural definitions of *equivalences, restrictions* (over subspaces of the base) and *induced bundles* (by a continuous map of some space into the base). A *trivial bundle* is a bundle equivalent (in its class) to the projection bundle $B \times \mathbb{R}^n \to B$ or $B \times \mathbb{C}^n \to B$ furnished by the obvious structure.

Important Example. The *Hopf* or *tautological* vector bundle over $\mathbb{R}P^n$ is the one-dimensional vector bundle whose total space is the set of pairs (ℓ, x), where $\ell \in \mathbb{R}P^n$ is a line in $\mathbb{R}^{n+1}$ and $x \in \ell$ is a point on this line [topology in this set is defined by the inclusion into $\mathbb{R}P^n \times \mathbb{R}^{n+1}$]. Precisely in the same way, the Hopf, or tautological, one-dimensional complex vector bundle over $\mathbb{C}P^n$ is defined. An obvious generalization of this construction provides tautological vector bundles over the Grassmannians $G(m, n), G_+(m, n)$, and $\mathbb{C}G(m, n)$, which are n-dimensional, respectively, real, oriented, and complex vector bundles.

B: Realification and Complexification

One can make a complex vector bundle real by removing a part of its structure, namely the multiplication by nonreal scalars. If ξ is an n-dimensional complex vector bundle, then the realification provides a $2n$-dimensional real vector bundle which is denoted as $\mathbb{R}\xi$. The bundle $\mathbb{R}\xi$ possesses a canonical orientation: If $x_1, \ldots, x_n$ is a complex basis in a fiber of ξ, then $x_1, ix_1, \ldots, x_n, ix_n$ is a real basis in the same space, and the orientation of this basis does not depend on the choice of the complex basis $x_1, \ldots, x_n$ [this follows from the fact that the image of the natural embedding $c: GL(n, \mathbb{C}) \to GL(2n, \mathbb{R})$ consists of matrices with positive determinant; the last statement follows from the fact that $GL(n, \mathbb{C})$ is connected, or, more convincingly, from the formula $\det(cA) = |\det A|^2$; compare with the "important remark" in the example after Theorem 1 in Sect. 17.5]. The definition of the complexification $\mathbb{C}\xi$ of a real vector bundle $(E, B, \mathbb{R}^n, p)$ is a bit more complicated. In the product $\mathbb{C} \times E$, make an identification $(rx, \lambda) = (x, r\lambda)$ for every $x \in E, r \in \mathbb{R}, \lambda \in \mathbb{C}$. The resulting space $\mathbb{C}E$ is the space of our fibration; the projection $\mathbb{C}E \to B$ is defined by the formula $(x, \lambda) = p(x)$, and the vector operations act as $(x, \lambda) + (x, \lambda') = (x, \lambda + \lambda')$ and $\mu(x, \lambda) = (x, \mu\lambda)$ (it is obvious that these formulas are compatible with the preceding factorization). It is clear also that $\mathbb{C}\xi$ is an n-dimensional complex vector bundle.

There is one more operation related to the two previous ones. Let ξ be a complex vector bundle. Denote by $\overline{\xi}$ a complex vector bundle (of the same dimension as ξ) which differs from ξ only by the operation of multiplication by scalars: λx with respect to the structure of $\overline{\xi}$ is the same as $\overline{\lambda} x$ in ξ.

EXERCISE 1. Let ξ be a complex vector bundle. Prove that the following two statements are equivalent:

(i) The vector bundles ξ and $\overline{\xi}$ are equivalent to each other.
(ii) There exists a real vector bundle η such that ξ is equivalent to $\mathbb{C}\eta$.

C: Direct Sums and Tensor Products

If ξ_1, ξ_2 are two vector bundles of the same type (real, complex, oriented) and with the same base, then the (direct or Whitney) sum $\xi_1 \oplus \xi_2$ and the tensor product $\xi_1 \otimes \xi_2$ are defined as vector bundles with the same base whose fibers are, respectively, direct sums or tensor products of the fibers of the bundles ξ_1 and ξ_2. Here is a more formal definition of the sum (here and below, $\mathbb{K}$ denotes $\mathbb{R}$ or $\mathbb{C}$). Let $\xi_1 = (E_1, B_1, \mathbb{K}^{n_1}, p_1), \xi_2 = (E_2, B_2, \mathbb{K}^{n_2}, p_2)$ be two vector bundles (the bases may not be the same). Put $\xi_1 \times \xi_2 = (E_1 \times E_2, B_1 \times B_2, \mathbb{K}^{n_1+n_2}, p_1 \times p_2)$; this is a vector bundle over $B_1 \times B_2$ of dimension $n_1 + n_2$. If $B_1 = B_2 = B$, then we define $\xi_1 \oplus \xi_2$ as the restriction of $\xi_1 \times \xi_2$ to the diagonal $B \subset B \times B$. Another formal definition: Let $B_1 = B_2 = B$ and let $p_2^*\xi_1 = (\widetilde{E}, E_2, \mathbb{K}^{n_1}, \widetilde{p})$ be the bundle over E_2 induced by ξ_1. Then $\xi_1 \oplus \xi_2 = (\widetilde{E}, B, \mathbb{K}^{n_1+n_2}, p_2 \circ \widetilde{p})$.

There exists a different approach to the definition of $\oplus$ and $\otimes$ (see Sect. 19.4). At the moment, we speak of tensor products of vector bundles not specifying any formal definition; we hope that the reader will be able to create this definition without our help (Exercise 5).

EXERCISE 2. Prove the equivalence of the two definitions of $\xi_1 \oplus \xi_2$. (This will show, in particular, that the second definition is actually symmetric with respect to ξ_1 and ξ_2.)

EXERCISE 3. Introduce an orientation into the sum of two oriented bundles.

EXERCISE 4. Make up a formal definition of a tensor product of two (real or complex) vector bundles.

EXERCISE 5. For real or complex vector bundles ξ_1, ξ_2 with the same base, make up a definition of a vector bundle $\mathrm{Hom}(\xi_1, \xi_2)$.

Two vector bundles of the same type, but, possibly, of different dimensions, are called *stably equivalent* if they become equivalent after adding trivial bundles. To make up a more formal definition, notice that a standard trivial n-dimensional bundle $B \times \mathbb{K}^n \to B$ is usually denoted simply as n. With this notation,

$$\xi \sim_{\text{stab}} \eta \Leftrightarrow \exists m, n \colon \xi \oplus n \sim \eta \oplus m.$$

In conclusion, let us point out a connection of the sum construction with previous constructions.

EXERCISE 6. Make up a canonical real vector bundle equivalence $\mathbb{R}\mathbb{C}\xi \sim \xi \oplus \xi$ (where ξ is a real vector bundle).

EXERCISE 7. Make up a canonical complex vector bundle equivalence $\mathbb{C}\mathbb{R}\xi \sim \xi \oplus \overline{\xi}$ (where ξ is a complex vector bundle).

D: Linear Maps Between Vector Bundles, Subbundles, and Quotient Bundles

A linear map of a vector bundle $\xi_1 = (E_1, B_1, \mathbb{K}^{n_1}, p_1)$ into a vector bundle $\xi_2 = (E_2, B_2, \mathbb{K}^{n_2}, p_2)$ (as before, $\mathbb{K}$ denotes $\mathbb{R}$ or $\mathbb{C}$) is a pair of continuous maps $F \colon E_1 \to E_2$, $f \colon B_1 \to B_2$ such that $f \circ p_1 = p_2 \circ F$ and for every $x \in B$, the appropriate restriction of F is a linear map $p_1^{-1}(x) \to p_2^{-1}(f(x))$. The subbundle of a vector bundle $\xi = (E, B, \mathbb{K}^n, p)$ is a vector bundle $\xi' = \left(E', B, \mathbb{K}^{n'}, p|_{E'}\right)$ with $E' \subset E$ whose fibers are subspaces of the fibers of ξ. The inclusion map $E' \to E$ and the identity map $B \to B$ compose a linear map (inclusion) $\xi' \to \xi$. If ξ' is a subbundle of ξ, then a fiberwise factorization creates a quotient bundle ξ/ξ'. More formally, the total space of ξ/ξ' is obtained from E by a factorization over the equivalence relation: $x_1 \sim x_2$ if $p(x_1) = p(x_2)$ and $x_2 - x_1 \in E'$. There is an obvious linear map (projection) $\xi \to \xi/\xi'$.

Let us mention two important subbundles: $S^k\xi \subset \underbrace{\xi \otimes \dots \otimes \xi}_{k}$ and $\Lambda^k\xi \subset \underbrace{\xi \otimes \dots \otimes \xi}_{k}$.

E: Coordinate Presentation of a Vector Bundle

Let ξ be an n-dimensional vector bundle (of one of our three types). Fix an open covering $\{U_i\}$ of the base B such that the restrictions $\xi|_{U_i}$ are all trivial vector bundles; let $\varphi_i: p^{-1}(U_i) \to \mathbb{K}^n$ be a trivialization, that is, a map which is a vector space isomorphism on every $p^{-1}(x), x \in U_i$. For every $y \in U_i \cap U_j$, there arises a composition

$$\mathbb{K}^n \xrightarrow{\varphi_j^{-1}} p^{-1}(y) \xrightarrow{\varphi_i} \mathbb{K}^n;$$

the function which assigns this composition to y is a continuous map $\varphi_{ij}: U_i \cap U_j \to G$ where $G = GL(n, \mathbb{K})$ $[GL_+(n, \mathbb{R})$ in the case of an oriented bundle$]$. Moreover, (i) $\varphi_{ii}(y) = I$ for $y \in U_i$, (ii) $\varphi_{ji}(y) = \left(\varphi_{ij}(y)\right)^{-1}$ for $y \in U_i \cap U_j$, and (iii) $\varphi_{ik}(y)\varphi_{kj}(y) = \varphi_{ij}(y)$ for $y = U_i \cap U_j \cap U_k$. It is easy to understand that a set of maps $\varphi_{ij}: U_i \cap U_j \to G$ with properties (i)–(iii) gives rise to a vector bundle. This presentation of a vector bundle is called the *coordinate presentation*.

An obvious generalization of the so presented vector bundles consists in specifying a topological group G and a G-space F. Suppose that there are an open covering $\{U_i\}$ of a space B and a set of continuous functions $\varphi_{ij}: U_i \cap U_j \to G$ with properties (i)–(iii) just listed. In the disjoint union $\coprod_i (U_i \times F)$, make, for every $i, j, y \in U_i \cap U_j$, an identification $\left[(y, f) \in U_j \times F\right] \sim \left[(y, \varphi_{ij}(y)f) \in U_i \times F\right]$; the space arising we take for E. The projections $U_i \times F \to U_i \subset B$ form a projection $p: E \to F$, and there arises a locally trivial fibration (E, B, F, p) with a certain additional structure similar to a structure of a vector bundle. Such fibrations are called fiber bundles (or Steenrod fibrations); according to this terminology, G is the *structure group*, and F is the *standard fiber*. The reader can find details in the classical book by Steenrod [80], or in a variety of more modern books, for example, Husemoller [49]; here we only mention some examples.

There are many obvious examples. Take a coordinate presentation of a real, complex, or oriented vector bundle and assume that the functions φ_{ij} take values not in the group $GL(n, \mathbb{R}), GL(n, \mathbb{C})$ or $GL_+(n; \mathbb{R})$, but in some subgroup of one of these groups, say, in $O(n), SO(n)$, or $U(n)$. It is clear that the fiber bundles arising have an adequate description as real, complex, or oriented vector bundles with an additional structure, for the examples above, with an Euclidean or Hermitian structure, in every fiber. If the subgroup is the group of block diagonal matrices, $GL(p, \mathbb{K}) \times GL(q, \mathbb{K}) \subset GL(n; \mathbb{K}), n = p + q$, then the fiber bundle arising is the usual n-dimensional vector bundle presented as the sum of two vector bundles, of dimensions p and q. In a similar way, we can present vector bundles with a fixed nonvanishing section, or with a fixed subbundle, and so on. An example of a different nature: Take an arbitrary G and put $F = G$ with the left translation action; the fibrations arising are called *principal*. Some other examples will appear in the next sections.

19.2 Tangent and Normal Bundles

The notion of a *tangent vector* to a smooth manifold is very important, and for this reason it has many equivalent definitions. The most natural definition is based on local coordinates. Let x be a point of an n-dimensional manifold X, and let $\varphi\colon U \to \mathbb{R}^n$ be a chart such that $x \in U$; then a tangent vector to X at x is defined as a vector v of the space $\mathbb{R}^n$ at the point $\varphi(x)$. If there is another chart, $\psi\colon V \to \mathbb{R}^n$, also covering x, then the tangent vector corresponding to the chart φ and the vector v is identified with the tangent vector corresponding to the chart ψ and the vector $w = d_{\varphi,\psi}(v)$, where $d_{\varphi,\psi}$ is the differential of the map $\varphi(U \cap V) \to \psi(U \cap V)$, $y \mapsto \psi(\varphi^{-1}(y))$. Another possibility, which does not require a fixation of a chart, is to define a tangent vector at x as a class of parametrized smooth curves $\gamma\colon (-\varepsilon, \varepsilon) \to X$ such that $\gamma(0) = x$, where the curves γ, γ' are equivalent if $\operatorname{dist}(\gamma(t), \gamma'(t)) = o(t)$ (the distance is calculated with respect to any local coordinate system). An algebraically more convenient approach consists in defining a tangent vector of X at x as a linear map $v\colon C^\infty(X) \to \mathbb{R}$ ($C^\infty(X)$ is the space of real C^∞-functions) such that $v(fg) = v(f)g(x) + f(x)v(g)$ (in other words, tangent vectors are identified with directional derivatives). Finally, if X is presented as a smooth surface in an Euclidean space, then a tangent vector to X is simply a tangent vector to this surface. To make this definition compatible with previous definitions, we can say that a tangent vector at some point to the Euclidean space regarded as a smooth manifold is simply a vector of this space at this point, and tangent vectors to a submanifold are tangent vectors to the manifold tangent to the submanifold.

The set of tangent vectors to an n-dimensional manifold X at a point x is an n-dimensional vector space which is denoted as T_xX. The union of all spaces T_xX possesses a natural topology and, moreover, a structure of a $2n$-dimensional smooth manifold; this manifold is denoted as TX. The natural projection $TX \to X$ makes TX a total space of a vector bundle over X; this vector bundle is called the *tangent bundle* of X and is denoted as $\tau(X)$. A section of a tangent bundle is a vector field on the manifold. A manifold whose tangent bundle is trivial is called *parallelizable*; a manifold is parallelizable if it is possible to choose bases in all tangent spaces depending continuously of a point or, equivalently, if there exist $n = \dim X$ vector fields on X which are linearly independent at every point. For example, the circle is parallelizable, the torus is parallelizable, while the two-dimensional sphere is not parallelizable. The three-dimensional sphere is parallelizable: If it is presented as the space of unit quaternions, then the basis at the space T_xS^3 is formed by quaternions ix, jx, kx where i, j, k are quaternion units. If you replace quaternions by octonions, you will prove that the sphere S^7 is parallelizable. There is a remarkable fact that no spheres besides S^1, S^3, S^7 are parallelizable: This is one of the versions of the Frobenius conjecture proven by Adams (two different proofs, both belonging to Adams, will be presented in Chaps. V and VI later). Notice that the problem of parallelization of spheres is equivalent to the problem of existence of spheroids with the invariant Hopf equal to one (see Remark 5 in Sect. 16.5).

EXERCISE 8. Prove that the orientability of a manifold X (in the sense of Sect. 17.1) is equivalent to the orientability of the tangent bundle $\tau(X)$.

If Y is a submanifold of a manifold X, then there arise two vector bundles with the base Y: $\tau(Y)$ and $\tau(X)|_Y$, and $\tau(Y) \subset \tau(X)|_Y$ (a tangent vector to a submanifold is also a tangent vector to the manifold). The quotient bundle $\tau(X)|_Y/\tau(Y)$ is called the *normal bundle* of Y in X and is denoted as $\nu_X(Y)$ or $\nu(Y)$. The word "normal" is an indication of the fact that if X is a submanifold of an Euclidean space, then the total space of $\nu(Y)$ may be regarded as consisting of vectors at points of Y which are tangent to X and normal to Y.

Mark an isomorphism $\tau(Y) \oplus \nu(Y) = \tau(X)|_Y$. In particular, if $X = \mathbb{R}^n$, then $\tau(Y) \oplus \nu(Y) = n$.

Notice that the construction of normal bundles with all properties listed can be applied not only to submanifolds, that is, to embeddings of a manifold Y to a manifold X, but also to *immersions* $\iota\colon Y \to X$; the only significant change is that the restriction bundle $\tau(X)|_Y$ should be replaced by the induced bundle $\iota^*\tau(X)$.

EXERCISE 9. Deduce from the last equality that normal bundles of a manifold corresponding to different embeddings or immersions of this manifold to Euclidean spaces (possibly, of different dimensions) are stably equivalent.

EXERCISE 10. Prove that the normal bundle to an n-dimensional oriented surface embedded (or immersed) into the $(n+1)$-dimensional Euclidean space is trivial. Deduce from this that the tangent bundle to such a surface (for example, to an arbitrary sphere with handles) is stably trivial (that is, stably equivalent to a trivial bundle). A manifold whose tangent bundle is stably trivial is called *stably parallelizable*. Obviously, a manifold is stably parallelizable if and only if its normal bundle is stably trivial.

FYI (this is not an exercise). A closed connected manifold is stably parallelizable if and only if it is parallelizable in the complement to a point. A noncompact connected manifold if stably parallelizable if and only if it is parallelizable. A manifold is stably parallelizable if and only if it is orientable and admits an immersion in the Euclidean space of the dimension bigger by 1.

EXERCISE 11. Let ζ be the Hopf bundle over $\mathbb{R}P^n$. Prove that

$$\tau(\mathbb{R}P^n) \oplus 1 \sim \underbrace{\zeta \oplus \cdots \oplus \zeta}_{n+1} = (n+1)\zeta.$$

Prove a similar statement for $\mathbb{C}P^n$ [notice that the bundle $\tau(\mathbb{C}P^n)$ possesses a natural structure of a complex vector bundle].

19.3 Associated Fibrations and Characteristic Classes

A: An Introduction

Choose one of the three types of vector bundles, and choose integers n and q and an Abelian group G. A characteristic class c of n-dimensional vector bundles on the chosen type with values in q-dimensional cohomology with the coefficients in G is a function which assigns to every n-dimensional vector bundle ξ of the chosen type with a CW base B a cohomology class $c(\xi) \in H^q(B;G)$ such that if $f\colon B' \to B$ is a continuous map of another CW complex into B, then $c(f^*\xi) = f^*c(\xi)$. Here f^* on the left-hand side of the formula means the inducing operation for vector bundles, and on the right-hand side it means the induced cohomology homomorphism.

The term "characteristic class" is not new for us: In Sect. 18.5, we called the first obstruction to extending a section of a locally trivial fibration a characteristic class (or a primary characteristic class) of this fibration, and the equality $c(f^*\xi) = f^*c(\xi)$ held for that characteristic classes. However, that construction cannot be applied to vector bundles directly, because their fiber is contractible. (Recall that the coefficient domain for the characteristic classes of Sect. 18.5 is the first nontrivial homotopy group of the fiber.) What we still can do is to apply the construction to some fibration which can be constructed from the given vector bundle. An ample variety of such fibrations is delivered by the construction of an *associated fibration*.

B: A Construction of Associated Fibrations

This construction was actually described in Sect. 19.1.E. We take a coordinate presentation $\{\{U_i\}, \{\varphi_{ij}\colon U_i \cap U_j \to G\}\}$ of a vector bundle with the base B [where $G = GL(n,\mathbb{R}), GL_+(n;\mathbb{R})$ or $GL(n,\mathbb{C})$] and choose an arbitrary space F with an action of the group G. After this, we construct the total space E of a new fibration as

$$\coprod_i (U_i \times F) \mid [(y,f) \in U_j \times F] \sim [(y, \varphi_{ij}(y)f) \in U_i \times F] \quad \text{for all } y \in U_i \cap U_j, f \in F.$$

The fibration (E,B,F,p) [where $p\colon E \to B$ is the projection $(y,f) \mapsto y$] is the associated (by the given vector bundle) fibration with the standard fiber F. However, usually we will not need this general construction: Almost always, we will restrict ourselves to one particular case of it, which is described ahead. Let $\xi = (E,B,\mathbb{R}^n,p)$, or $(E,B,\mathbb{C}^n,p)$, be a given vector bundle, and let $1 \leq k \leq n$. Put

$$E_k = \{(x_1,\dots,x_k) \in E \times \dots \times E \mid p(x_1) = \dots = p(x_k); \; x_1,\dots,x_k \text{ are linearly independent}\}.$$

There is an obvious projection $p_k: E_k \to B$, and there arises a locally trivial fibration $\xi_k = (E_k, B, R_k, p_k)$ where R_k is the space of all linearly independent k-frames in $\mathbb{R}^n$ or $\mathbb{C}^n$. (This is the fibration associated with ξ with the standard fiber R_k.) The case $k = 1$ is especially simple: E_1 is $E - B$, where B is embedded into E as the zero section, and R_1 is $\mathbb{R}^n - 0$ or $\mathbb{C}^n - 0$.

Point out a small defect of this construction (rather more aesthetic than mathematical). The fibers are noncompact spaces which would have better been replaced by homotopy equivalent classical manifolds: Stiefel manifolds and spheres. This can be done with the help of the following simple lemma.

Lemma. *If a vector bundle has a CW base, then it is possible to introduce in all fibers an Euclidean or Hermitian structure which depends continuously on the point of the base; moreover, this can be done in a homotopically unique way.*

Proof. The set of all Euclidean (Hermitian) structures in fibers of a vector bundle is a total space of a fibration whose fiber is the space of all Euclidean (Hermitian) structures in a given vector space (this is also a fibration associated with the vector bundle). Obviously, the fiber of this fibration is contractible (it is a convex subset of the space of all symmetric bilinear (Hermitian) forms in this vector space. This fibration has a section (all the obstructions are zeroes) and this section is homotopically unique (all difference cochains are zeroes). This is precisely the statement of the lemma.

Using these Euclidean or Hermitian structures in the fibers, we can replace the fibration ξ_k into the fibration ξ_k^0 whose total space is the space of all orthonormal (unitary) frames in the fibers of ξ. The fiber of ξ_k^0 is the Stiefel manifold $V(n,k)$ or $\mathbb{C}V(n,k)$; in particular, ξ_1^0 is the fibration whose fiber is the sphere S^{n-1} (S^{2n-1} in the complex case); this fibration is called *spherical.*

C: Classical Characteristic Classes of Vector Bundles

Let ξ be an n-dimensional oriented (real) vector bundle with the CW base B. Consider the corresponding spherical fibration ξ_1^0. It is easy to see that the orientability of the bundle ξ implies the orientability of the fibration ξ_1^0; that is, the fibration ξ_1^0 is homologically simple. (The reader may prove that a Steenrod bundle whose structure group is connected is always simple.) Thus, there arises the first obstruction to extending a section of ξ_1^0, and this first obstruction is an element of $H^n(B;\mathbb{Z})$. Regarded as a characteristic class of the bundle ξ, this element is called the *Euler class* of ξ; the notation: $e(\xi)$.

Pass to the fibrations ξ_k^0.

Lemma. *Let* $1 \leq k < n$. *Then*

(i) $\pi_i(V(n,k)) = 0$ *for* $i < n-k$.

(ii) $\pi_{n-k}(V(n,k)) \cong \begin{cases} \mathbb{Z}, \text{ if } k = 1 \text{ or } n-k \text{ is even;} \\ \mathbb{Z}_2 \text{ in all other cases.} \end{cases}$

Proof. The case $k = 1$ is trivial: $\pi_i(V(n,1)) = \pi_i(S^{n-1})$ is zero for $i < n-1$ and $\mathbb{Z}$ for $i = n-1$. Let $k \geq 2$, and consider the fibration

$$V(n,k) \xrightarrow{V(n-1,k-1)} S^{n-1}$$

[the projection assigns to $\{v_1, \ldots, v_k\} \in V(n,k)$ the last vector v_k]. Consider the fragment

$$\pi_{i+1}(S^{n-1}) \to \pi_i(V(n-1,k-1)) \to \pi_i(V(n,k)) \to \pi_i(S^{n-1})$$

of the homotopy sequence of this fibration. If $i < n-2$, then the first and last terms are zeroes, and we get an isomorphism $\pi_i(V(n-1,k-1)) \cong \pi_i(V(n,k))$. Thus, if $i < n-k$ and $k > 1$, then

$$\pi_i(V(n,k)) \cong \pi_i(V(n-1,k-1)) \cong \cdots \cong \pi_i(V(n-k+1,1)) = \pi_i(S^{n-k}) = 0.$$

For $i = n-k$, this chain of isomorphisms becomes shorter:

$$\pi_{n-k}(V(n,k)) \cong \pi_{n-k}(V(n-1,k-1)) \cong \cdots \cong \pi_{n-k}(V(n-k+2,2)),$$

and the general case of the lemma is reduced to the case of $V(n,2)$. We need to prove that $\pi_{n-2}(V(n,2)) = \mathbb{Z}$ for n even and $\mathbb{Z}_2$ for n odd. For $k = 2$ and $i = n-2$, our homotopy sequence becomes

$$\begin{array}{ccccccc} \pi_{n-1}(S^{n-1}) & \to & \pi_{n-2}(S^{n-2}) & \to & \pi_{n-2}(V(n,2)) & \xrightarrow{\partial_*} & \pi_{n-2}(S^{n-1}) \\ \| & & \| & & & & \| \\ \mathbb{Z} & & \mathbb{Z} & & & & 0. \end{array}$$

Thus, $\pi_{n-2}(V(n,2)) = \operatorname{Coker}[\partial_*\colon \pi_{n-1}(S^{n-1}) \to \pi_{n-2}(S^{n-2})]$. The space $V(n,2)$ is the space T_1S^{n-1} of unit tangent vectors to the sphere S^{n-1}, the fibration $V(n,2) \xrightarrow{S^{n-2}} S^{n-1}$ is the natural fibration of the space of unit tangent vectors. The construction of the homomorphism ∂_* is the following. We take a homotopy of an $(n-2)$-dimensional spheroid of S^{n-1} sweeping an $(n-1)$-dimensional spheroid, lift this homotopy to T_1S^{n-1}, and obtain a spheroid of the fiber. If we apply this construction to the identity spheroid $S^{n-1} \to S^{n-1}$, the lifting provides a vector field on S^{n-1}, and the resulting element of $\pi_{n-2}(S^{n-2})$ is the value of the obstruction to extending a vector field on S^{n-1}. As proved in Sect. 18.5 (see Proposition 2), this value is the Euler characteristic of S^{n-1}, that is, 2 for n odd and 0 for n even. Thus, the homomorphism $\partial_*\colon \pi_{n-1}(S^{n-1}) \to \pi_{n-2}(S^{n-2})$ is trivial if n is even and is a multiplication by 2 if n is odd. This completes the proof of the lemma.

The lemma shows that the first obstruction to extending a section of fibration ξ_k^0 (or ξ_k) takes value in $H^{n-k+1}(B; \mathbb{Z} \text{ or } \mathbb{Z}_2)$. Reduced modulo 2, this obstruction is a characteristic class of ξ with the values in $H^j(B; \mathbb{Z}_2)$, $j = n - k + 1$. This class is called the jth *Stiefel–Whitney class* of ξ and is denoted as $w_j(\xi)$. We also put $w_i(\xi) = 0$ for $i > \dim \xi$ and $w_0(\xi) = 1 \in H^0(B; \mathbb{Z}_2)$.

Notice that the orientability of the vector bundle ξ which was needed for the simplicity of the fibration ξ_k becomes unnecessary after reducing modulo 2; thus, the Stiefel–Whitney classes are defined for arbitrary real vector bundles.

For an n-dimensional oriented vector bundle ξ, $w_n(\xi) = \rho_2 e(\xi)$, where ρ_2 is the reduction modulo 2.

The complex version of the previous construction is a simplified version of it.

Lemma. *Let* $1 \leq k < n$. *Then*

$$\pi_i(\mathbb{C}V(n,k)) \cong \begin{cases} 0 \text{ for } i < 2(n-k)+1, \\ \mathbb{Z} \text{ for } i = 2(n-k)+1. \end{cases}$$

Proof This repeats the first, easier, part of the proof of the previous lemma and is based on the equality $\mathbb{C}V(n,1) = S^{2n-1}$ and the homotopy sequence

$$\pi_{i+1}(S^{2n-1}) \to \pi_i(\mathbb{C}V(n,k)) \to \pi_i(\mathbb{C}V(n-1,k-1) \to \pi_i(S^{2n-1})$$

of the fibration $\mathbb{C}V(n,k) \xrightarrow{\mathbb{C}V(n-1,k-1)} S^{2n-1}$.

Let ξ be an n-dimensional complex vector bundle with a CW base B. The lemma shows that the first obstruction to extending a section in the fibration ξ_k^0 (or ξ_k) is a class $c_j(\xi) \in H^{2j}(B; \mathbb{Z})$ where $j = n-k+1$. We get a characteristic class of complex vector bundles which is called the jth *Chern class*. Precisely as in the real case, we put $c_i(\xi) = 0$ for $i > \dim_{\mathbb{C}} \xi$ and $c_0(\xi) = 1$.

Finally, if ξ is again an n-dimensional vector bundle, then we put $p_j(\xi) = (-1)^j c_{2j}(\mathbb{C}\xi) \in H^{4j}(B; \mathbb{Z})$ and call the classes $p_j(\xi)$ *Pontryagin classes* of the bundle ξ. [The sign $(-1)^j$ has a historic origin. The reason why we restrict ourselves to even-numbered Chern classes is that the odd-numbered Chern classes of a complexification of a real vector bundle have order at most 2; see Exercise 15 in Sect. 19.5 later.] It is possible to define Pontryagin classes directly: We can associate with an n-dimensional vector bundle a fibration whose standard bundle is the space of all systems of $n-2j+2$ vectors of rank $> n-2j$; the first obstruction to extending sections in this fibration is $p_j(\xi)$ (the reader can try to prove this although it is not awfully interesting).

EXERCISE 12. Prove that $w_1(\xi) = 0$ if and only if the bundle ξ is orientable.

EXERCISE 13. Prove that if ξ is an n-dimensional complex vector bundle, then

$$e(\mathbb{R}\xi) = c_n(\xi),\ w_{2j}(\mathbb{R}\xi) = \rho_2 c_j(\xi),\ w_{2j+1}(\mathbb{R}\xi) = 0.$$

D: Geometric Construction of Euler, Stiefel–Whitney, and Chern Classes

Let ξ be an n-dimensional oriented vector bundle with a CW base B. Then there exists a nowhere vanishing section of ξ over the $(n-1)$st skeleton B^{n-1} of B. We can extend this section to B^n, but it may have zeroes over n-dimensional cells. If we assume these zeroes to be transverse intersections with the zero section, then we can count the "algebraic number" of these zeroes (that is, we assign a $+$ or $-$ sign to every zero), and a function which assigns this number to every cell is an n-dimensional integral cellular cocycle. Its cohomology class is the Euler class $e(\xi)$ (this is the construction of the first obstruction).

If ξ is not assumed oriented, then the previous construction gives a cohomology class modulo 2, and this is $w_n(\xi)$. We can construct in this way the other Stiefel–Whitney classes. Namely, let us assume that ξ has an Euclidean structure (in the fibers), and consider again a nowhere vanishing section of ξ over B^{n-1}. Let us try to construct a second nowhere vanishing section of ξ orthogonal to the first section. This can be done over B^{n-2}, but if we want to extend the second section to B^{n-1}, we have to admit that it will have zeroes over $(n-1)$-dimensional cells. Assuming these zeroes transverse, we can count their number modulo 2 in every $(n-1)$-dimensional cell, and in this way we get an $(n-1)$-dimensional cellular cocycle with coefficients in $\mathbb{Z}_2$, and the cohomology class of this cocycle is $w_{n-1}(\xi)$. Then we construct a third section orthogonal to the first two, it can be made nowhere vanishing over B^{n-3}, but to extend this third section to B^{n-2}, we have to admit transverse zeroes over $(n-2)$-dimensional cells, and in this way we obtain a cocycle representing $w_{n-2}(\xi)$. And so on.

The Chern classes of complex vector bundles may be constructed in a similar way; we leave the details to the reader.

19.4 Characteristic Classes and Classifying Spaces

A: The Classification Theorem

In Sect. 19.1.A, we mentioned tautological bundles over Grassmannians. They will be of primary importance now.

The theory here has three absolutely parallel versions for the three types of vector bundles. We will consider in detail the real case; the transition to the two other cases does not require any efforts: One should just replace the Grassmannians $G(\infty, n)$ by $G_+(\infty, n)$ and $\mathbb{C}G(\infty, n)$.

Recall that the total space of the tautological bundle (which we denote as η or η_n) over the Grassmannian $G(\infty, n)$ is the space of pairs (π, x) where $\pi \in G(\infty, n)$ is an n-dimensional subspace of $\mathbb{R}^N$ and $x \in \pi \subset \mathbb{R}^\infty$; the projection acts as $(\pi, x) \mapsto \pi$.

Theorem. *Let X be a finite CW complex. Then*

(i) *For every n-dimensional vector bundle ξ over X, there exists a continuous map $f: X \to G(\infty, n)$ such that $f^*\eta = \xi$.*
(ii) *This map f is unique up to a homotopy; that is, if $f_1^*\eta \sim f_2^*\eta$, then $f_1 \sim f_2$ (the second $\sim$ means a homotopy).*
(iii) *Conversely, if $f_1 \sim f_2$, then $f_1^*\eta \sim f_2^*\eta$.*

Corollary. *The correspondence $f \mapsto f^*\eta$ establishes a bijection between the set $\pi(X, G(\infty, n))$ of homotopy classes of continuous maps $X \to G(\infty, n)$ and equivalence classes of n-dimensional vector bundles with the base X.*

Proof of Theorem. First, notice that since X is compact and $G(\infty, n) = \varinjlim G(N, n)$, a continuous map $X \to G(\infty, n)$ is the same as a continuous map $X \to G(N, n)$ (with sufficiently large N) composed with the inclusion map $G(N, n) \to G(\infty, n)$. Same for homotopies: Maps $X \to G(N_1, n) \to G(\infty, n)$ and $X \to G(N_2, n) \to G(\infty, n)$ are homotopic if and only if maps $X \to G(N_1, n) \to G(M, n)$ and $X \to G(N_2, n) \to G(M, n)$ are homotopic for sufficiently large M.

Second, notice that statements (i) and (ii) are covered by the following relative version of statement (i):

(i′) *Let X be a finite CW complex, and ξ be an n-dimensional vector bundle over X. Then let A be a CW subcomplex of X, and let $g: A \to G(\infty, n)$ be a continuous map such that $g^*\eta \sim \xi|_A$. Then there exists a continuous map $f: X \to G(\infty, n)$ such that $f^*\eta \sim \xi$ and $f|_A = g$.*

We begin with proving statement (i), that is, (i′) with $A = \emptyset$, and then we will explain what we need to add to handle the case $A \neq \emptyset$. A *linear functional* on the total space E of ξ is a continuous function $E \to \mathbb{R}$ which is linear on every fiber of the bundle ξ. To construct a linear functional on E, it is sufficient to take some linear function $\varphi: p^{-1}(x) \to \mathbb{R}$ (where $x \in X$), then extend it to a linear functional $p^{-1}(U) \to \mathbb{R}$ where U is a neighborhood of x such that the restriction $\xi|_U$ is trivial [there is a retraction $\rho: p^{-1}(U) \approx U \times p^{-1}(x) \to \pi^{-1}(x)$ which is linear on every fiber, and the composition $\varphi \circ \rho$ is a required functional], and then we multiply the last functional by a continuous function $X \to \mathbb{R}$, which is 1 in a neighborhood V of x such that $\overline{V} \subset U$ and is 0 in the complement of U. We apply this construction to some linearly independent functionals $\varphi_i: p^{-1}(x) \to \mathbb{R}$, $i = 1, \dots, n$, and we get linear functionals $\varphi_{x,i}: E \to \mathbb{R}$ whose restrictions to $p^{-1}(x)$ are linearly independent; hence, for some neighborhood U_x of x the restrictions of these functionals to $p^{-1}(y)$ are linearly independent for all $y \in U_x$. Since X is compact, there exist some $x_1, \dots, x_m$ such that the sets U_{x_j}, $j = 1, \dots, m$ cover X. Then the functionals $\varphi_{x_j,i}$ have the following property: *For every $z \in X$ there are n of these functionals which are linearly independent on $p^{-1}(z)$.*

Together, the $N = mn$ functionals $\varphi_{x_j,i}: E \to \mathbb{R}$ form a map $\Phi: E \to \mathbb{R}^N$, and for every $z \in X$, the restriction $\Phi|_{p^{-1}(z)}$ is a linear monomorphism. The image $\Phi(p^{-1}(z))$ is an n-dimensional subspace of $\mathbb{R}^N$, and we define the map $f: X \to G(N, n)$ by the formula $f(z) = \Phi(p^{-1}(z)$. Since Φ maps isomorphically the fiber of ξ over $z \in X$ onto the fiber of η over $f(z)$, we have $f^*\eta = \xi$, as required.

Now, let us adjust this proof to the relative version. We assume that there are some $A \subset X$, $g: A \to G(M,n) \subset G(\infty,n)$, and an equivalence between $\xi|_A$ and $g^*\eta$. The last equivalence is the same as a continuous map $\psi: p^{-1}(A) \to \mathbb{R}^M$ which maps isomorphically $p^{-1}(z)$ $(z \in A)$ onto the subspace $g(z)$ of $\mathbb{R}^M$. First, we extend this map to a continuous map $\Psi: E \to \mathbb{R}^M$ which is linear on each fiber of ξ [to do this, we need to extend each of the M coordinate functions of ψ to a function $E \to \mathbb{R}$ linear on fibers; this is the same as extending from A to X a section of a certain vector bundle (composed of dual spaces $(p^{-1}(x))^*$) which does not meet any obstruction, since the fibers of a vector bundle are contractible]. The linear maps $\Psi|_{p^{-1}(y)}$ are isomorphisms for $y \in A$, and hence they are isomorphisms for $y \in W$, where W is some open neighborhood of A. To finish the construction, we take a $\Phi: X \to \mathbb{R}^N$ as constructed above and multiply this Φ by a continuous function $h: X \to \mathbb{R}$ which is 0 on A and 1 in the complement of W, $\Phi' = h\Phi$. The functions Ψ and Φ' together form a map $\Omega: E \to \mathbb{R}^{M+N}$, which is a linear monomorphism on every fiber of ξ such that $\Omega|_{p^{-1}(A)}$ is the composition $p^{-1}(A) \xrightarrow{\psi} \mathbb{R}^M \xrightarrow{\subset} \mathbb{R}^{M+N}$. This Ω gives rise to a continuous map $f: X \to G(M+N,n)$ such that $f^*\eta = \xi$ and $f|_A$ is the composition $A \xrightarrow{g} G(M,n) \xrightarrow{\subset} G(M+N,n)$. This completes a proof of (i′).

It remains to prove (iii). Our proof is based on the following simple observation. We say that n-dimensional subspaces π_1, π_2 of $\mathbb{R}^n$ are *close* to each other if no nonzero vector of π_1 is orthogonal to π_2 (this condition is symmetric in π_1, π_2); equivalently: π_1 is close to π_2 if the orthogonal projection $\pi_1 \to \pi_2$ is an isomorphism. Obviously, every $\pi \in G(N,n)$ has a neighborhood U in $G(N,n)$ such that every $\pi \in U$ is close to π.

Lemma. *Let $f_1, f_2: X \to G(N,n)$ (no restrictions on X) be two continuous maps such that, for every $x \in X$, the subspaces $f_1(x), f_2(x)$ of $\mathbb{R}^n$ are close to each other. Then $f_1^*\eta \sim f_2^*\eta$.*

Proof of Lemma. Let $p_1: E_1 \to X, p_2: E_2 \to X$ be the bundles $f_1^*\eta, f_2^*\eta$. For every $x \in X$, the definition of the inducing operation provides isomorphisms $\eta_1: p_1^{-1}(x) \to f_1(x), \eta_2: p_2^{-1}(x) \to f_2(x)$; also, there is the orthogonal projection $\pi: f_1(x) \to f_2(x)$. The composition $\eta_2^{-1} \circ \pi \circ \eta_1: p_1^{-1}(x) \to p_2^{-1}(x)$ is an isomorphism depending continuously on x; and these isomorphisms form an equivalence $f_1^*\eta \sim f_2^*\eta$.

Proof of (iii). If X is compact (otherwise, arbitrary), and $\{f_t: X \to G(N,n)\}$ is a homotopy, then there exists an m such that, for every $i, 0 \le i < m$, the maps $f_{i/m}, f_{(i+1)/m}$ satisfy the condition of the lemma. Hence, for every $i, f_{i/m}^*\eta \sim f_{(i+1)/m}^*\eta$. Hence, $f_0^*\eta \sim f_1^*\eta$, which is the statement of (iii).

B: More General Constructions

The space $G(\infty,n)$ is called a *classifying space* for real n-dimensional vector bundles, and η is called a *universal bundle*; a similar terminology is applied to $G_+(\infty,n)$ and $\mathbb{C}G(\infty,n)$. There exists a far-reaching generalization of the preceding construction. For a topological group G, there exists a principal fibration (see Sect. 19.1.E) (EG, BG, G, p_G) with a cellular base and contractible space EG; for a given G, a principal fibration with these properties is unique up to a homotopy equivalence. The space BG is called the *classifying space* for G; in particular, $BGL(n,\mathbb{R}) = BO(n) = G(\infty,n)$, $BGL_+(n,\mathbb{R}) = BSO(n) = G_+(\infty,n)$, $BGL(n,\mathbb{C}) = BU(n) = \mathbb{C}G(\infty,n)$. If F is a space with a faithful action of G, then, for a finite CW complex X, there is a bijection between the set of equivalence classes of Steenrod bundles over X with the structure group G and the standard fiber F and the set $\pi(X, BG)$ of homotopy classes of continuous maps $X \to BG$. This construction belongs to J. Milnor [55]. It has further generalizations to the cases when G is not a topological group, but an H-space or a topological groupoid.

C: Immediate Applications of the Classification Theorem

Some definitions and theorems of the previous sections can be clarified with the help of the classification theorem of Sect. 19.4.A. For example, the lemma of Sect. 19.3.B, which states that every vector bundle whose base is a finite CW complex can be furnished by an Euclidean or Hermitian structure in the fibers, follows immediately from the theorem of Sect. 19.4.A. Namely, if we fix an Euclidean structure in $\mathbb{R}^N$ [or a Hermitian structure in $\mathbb{C}^N$], then all n-dimensional subspaces inherit this structure. This provides an Euclidean or Hermitian structure in the fibers of η, and hence in the fibers of all vector bundles induced by η, that is, of all vector bundles whose bases are finite CW complexes.

The definition of the sum of vector bundles can be done in the following way: If $f: X \to G(N,n)$ and $g: X \to G(M,m)$ are two continuous maps, then there arises a map $f \oplus g: X \to G(M+N, m+n)$, $(f \oplus g)(x) = f(x) \oplus g(x) \subset \mathbb{R}^N \oplus \mathbb{R}^M$, and $f^*\eta \oplus g^*\eta = (f \oplus g)^*\eta$, which gives an alternative construction of the sum of vector bundles. The same for tensor products: We consider a map $f \otimes g: X \to \mathbb{R}^{NM}$, $f \otimes g(x) = f(x) \otimes g(x) \subset \mathbb{R}^N \otimes \mathbb{R}^M = \mathbb{R}^{NM}$, and $f^*\eta \otimes g^*\eta = (f \otimes g)^*\eta$, which can be regarded as a definition of a tensor product of vector bundles (same with complex vector bundles). In a similar way, for a vector bundle ξ, we can define $S^r\xi, \Lambda^r\xi, \xi^*$, etc.

D: Characteristic Classes and Cohomology of Classifying Spaces

Theorem. *The group of q-dimensional characteristic classes of n-dimensional real [resp. n-dimensional oriented, resp. n-dimensional complex] vector bundles with coefficients in G is isomorphic to the group* $H^q(G(\infty,n); G)$ *[resp.* $H^q(G_+(\infty,n); G)$*, resp.* $H^q(\mathbb{C}G(\infty,n); G)$*].*

Proof. We restrict ourselves to the real case; the proof in the other two cases is the same. If c is a characteristic class of the type considered, we can compute it for the bundle η over $G(\infty, n)$ [or over $G(N, n)$ with $N \gg n, q$]. We get a $c(\eta) \in H^q(G(\infty, n); G)$. We need to check two things: (i) If $c(\eta) = 0$, then $c = 0$; (ii) for every $\gamma \in H^q(G(\infty, n); G)$, there exists a characteristic class c such that $c(\eta) = \gamma$.

(i) Let $c(\eta) = 0$. If ξ is an n-dimensional vector bundle whose base X is a finite CW complex, then $\xi = f^*\eta$ for some $f: X \to G(\infty, n)$, and therefore $c(\xi) = c(f^*\eta) = f^*c(\eta) = f^*(0) = 0$. If the base X of ξ is an arbitrary CW complex, and $0 \neq c(\xi) = \alpha \in H^q(X; G)$, then there exists a finite CW subcomplex Y of X such that $\alpha|_Y \neq 0$; then $0 = c(\xi|_Y) = c(\xi)|_Y = \alpha|_Y \neq 0$, which is a contradiction.

(ii) Let $\gamma \in H^q(G(\infty, n); G)$; we want to define a characteristic class c with $c(\eta) = \gamma$. Let ξ be an n-dimensional real vector bundle with a CW base X. Then, for every finite CW subcomplex Y of X, we can define $c(\xi|_Y) \in H^q(Y; G)$ as $f^*\gamma$, where $f: Y \to G(\infty, n)$ is a continuous map with $f^*\eta = \xi|_Y$. Then, obviously, there exists a unique $\alpha \in H^q(X; G)$ such that $\alpha|_Y = c(\xi|_Y)$ for every finite $Y \subset X$. We set $c(\xi) = \alpha$.

(Both parts of this proof implicitly use the following property of cohomology. Let X be a CW complex, and let $\mathcal{F}$ be the category of finite CW subcomplexes of X and inclusions. Then $H^q(X; G) = \overleftarrow{\lim}_{\mathcal{F}} H^q(Y; G)$. This follows, for example, from a similar property for homology and the universal coefficients formula for cohomology. We leave the details to the reader.)

GENERALIZATION. *Characteristic classes of Steenrod fibrations with the structure group G taking values in the q-dimensional cohomology of the base with coefficients in A correspond bijectively to elements of* $H^q(BG; A)$.

E: Completeness of Systems of Euler, Stiefel–Whitney, Chern, and Pontryagin Characteristic Classes

Theorem. (i) *Every characteristic class of n-dimensional real vector bundles with coefficients in* $\mathbb{Z}_2$ *is a polynomial of the Stiefel–Whitney classes* $w_1, \dots, w_n$, *and different polynomials are different characteristic classes.*

(ii) *Every characteristic class of n-dimensional complex vector bundles with coefficients in* $\mathbb{Z}$ *is a polynomial of the Chern classes* $c_1, \dots, c_n$, *and different polynomials are different characteristic classes.*

(iii) *Every characteristic class of n-dimensional real vector bundles with coefficients in* $\mathbb{Q}$, *or* $\mathbb{R}$, *or* $\mathbb{C}$ *is a polynomial of the (images with respect to the inclusion of* $\mathbb{Z}$ *into the coefficient domain) of the Pontryagin classes* $p_1, \dots, p_{[n/2]}$, *and different polynomials are different characteristic classes.*

(iv) *Every characteristic class of n-dimensional orientable vector bundles with coefficients in* $\mathbb{Q}$, *or* $\mathbb{R}$, *or* $\mathbb{C}$ *is a polynomial of the (images with respect* *to the inclusion of $\mathbb{Z}$ into the coefficient domain) of the Pontryagin classes $p_1, \dots, p_{[n/2]}$ and, if n is even, the Euler class e, and different polynomials are different characteristic classes.*

We postpone the details of the proof to the next section. Here we only notice that the proof of every part consists of two parts. First, we need to show that the corresponding group

$$H^q(G(\infty, n), G_+(\infty, n), \text{ or } \mathbb{C}(\infty, n); \ \mathbb{Z}_2, \mathbb{Z}, \mathbb{Q}, \mathbb{R}, \text{ or } \mathbb{C})$$

has precisely the same size as the group of polynomials of the form indicated. This can be easily deduced from the computation of the cohomology of Grassmannians in Sect. 13.8.C. Next, we need to check that no one of these polynomials is zero as a characteristic class [in the nonfield case (ii) we will need slightly more]. For this purpose, we need a sufficient supply of explicit computations of characteristic classes. At the moment, we do not have such a supply, but it will appear in the next section.

19.5 The Most Important Properties of the Euler, Stiefel–Whitney, Chern, and Pontryagin Classes

A: The Properties of the Stiefel–Whitney Classes

Theorem. *The Stiefel–Whitney classes possess the following properties.*

(1) *For the Hopf (tautological) bundle ζ over $\mathbb{R}P^n$ $(n \geq 2)$, $0 \neq w_1(\zeta) \in H^1(\mathbb{R}P^n; \mathbb{Z}_2) = \mathbb{Z}_2$ and $w_i(\zeta) = 0$ for $i > 1$.*
(2) *For arbitrary real vector bundles ξ, η with (the same) CW base,*

$$w_i(\xi \oplus \eta) = \sum_{p+q=i} w_p(\xi) w_q(\eta).$$

Remark. Statements (1) and (2) are often considered as axioms for Stiefel–Whitney classes: Together with the property that Stiefel–Whitney classes are characteristic classes, these axioms uniquely determine them. We will not return to this axiomatic definition of Stiefel–Whitney classes, but the reader will be able to deduce all necessary statements from the results of the current section. In details, this axiomatic approach to all classical characteristic classes is developed in the book *Characteristic Classes* by Milnor, Stasheff [60].

Proof of Part (1) is immediate. The restriction of ζ to $\mathbb{R}P^1 = S^1$ is the Möbius bundle, and obviously it has no nowhere vanishing section. Thus, ξ has no section over the first skeleton, which means that the first obstruction $w_1(\zeta) \in H^1(\mathbb{R}P^n; \mathbb{Z}_2)$ is not zero.

Part (2) is equivalent to the statement (2′): *for arbitrary real vector bundles* ξ, η *with, possibly different, CW bases,*

$$w_i(\xi \times \eta) = \sum_{p+q=i} w_p(\xi) \times w_q(\eta).$$

The proof of (2′) consists of three steps.

Step 1. Stiefel–Whitney classes invariant with respect to stable equivalence, which is the same as the statement $w_i(\xi \oplus 1) = w_i(\xi)$. This follows from the inductive construction of Stiefel–Whitney classes outlined in Sect. 19.3.D. For the first section of $\xi \oplus 1$ we can take the natural nonzero section of the summand 1. Then the second section of $\xi \oplus 1$ is the first section of ξ, the third section of $\xi \oplus 1$ is the second section of ξ, and so on. We see that if $\dim \xi = n$, then, for every k, $w_{(n+1)-(k+1)+1}(\xi \oplus 1) = w_{n-k+1}(\xi)$, which is our statement.

Step 2. Let $\dim X = \dim \xi = p$, $\dim Y = \dim \eta = q$ (where X and Y are the bases of ξ and η); statement (2′) in this case means $w_{p+q}(\xi \times \eta) = w_p(\xi) \times w_q(\eta)$. Fix a section of ξ which has no zeroes on X^{p-1} and has transverse zeroes on p-dimensional cells; for a p-dimensional cell e of X, let $n_e \in \mathbb{Z}_2$ be the number of zeroes of the section on e reduced modulo 2. Similarly, fix a section of η, without zeroes on Y^{q-1} and with transverse zeroes on q-dimensional cells, and let $m_f \in \mathbb{Z}_2$ be the number of zeroes of η on a cell f reduced modulo 2. Then $w_p(\xi)$ is represented by the cocycle $e \mapsto n_e$, and $w_q(\eta)$ is represented by the cocycle $f \mapsto m_f$. The two sections together form a section of $\xi \times \eta$ with no zeroes on $(X \times Y)^{p+q-1}$ and with transverse zeroes on cells $e \times f$, the number of which modulo 2 is $n_e m_f$. Thus, $(e \times f \mapsto n_e m_f)$ is a cocycle of the class $w_{p+q}(\xi \times \eta)$ which shows that $w_{p+q}(\xi \times \eta) = w_p(\xi) \times w_q(\eta)$.

Step 3. The general case. Fix p, q with $p + q = i, p \leq \dim \xi, q \leq \dim \eta$, and consider the restrictions $\xi|_{X^p}, \eta|_{Y^q}$. We know that ξ has $\dim \xi - p$ linearly independent sections over X^p and η has $\dim \eta - q$ linearly independent sections over Y^q. From this, we conclude that

$$\xi|_{X^p} = \xi_p \oplus (\dim \xi - p), \ \ \eta|_{Y^q} = \eta_q \oplus (\dim \eta - q)$$

where ξ_p and η_q are bundles over X^p and Y^q of dimensions p and q. Certainly, it is also true that

$$(\xi \times \eta)|_{X^p \times Y^q} = (\xi_p \times \eta_q) + (\dim \xi + \dim \eta - p - q).$$

Let

$$u = w_i(\xi \times \eta) - \sum_{p'+q'=i} w_{p'}(\xi) \times w_{q'}(\eta) \in H^i(X \times Y; \mathbb{Z}_2).$$

Then

$$\begin{aligned} u|_{X^p\times Y^q} &= w_i((\xi\times\eta)|_{X^p\times Y^q}) - \textstyle\sum_{p'+q'=i} w_{p'}(\xi|_{X^p})\times w_{q'}(\eta|_{Y^q}) \\ &= w_i(\xi_p\times\eta_q) - \textstyle\sum_{p'+q'=i} w_{p'}(\xi_p)\times w_{q'}(\eta_q) \\ &= w_i(\xi_p\times\eta_q) - w_p(\xi_p)\times w_q(\eta_q) = 0 \end{aligned}$$

(here the first equality is obvious, the second equality follows from the result of step 1, the third equality follows from triviality of Stiefel–Whitney classes in dimensions exceeding the dimension of the bundle, and the last equality is the result of step 2). We see that $u|_{X^p\times Y^q} = 0$ for any p, q with $p+q=i$. Consider the homomorphism

$$\begin{aligned} H^i(X\times Y;\mathbb{Z}_2) &= \textstyle\bigoplus_{p+q=i} H^p(X;\mathbb{Z}_2)\otimes H^q(Y;\mathbb{Z}_2) \\ &\to \textstyle\bigoplus_{p+q=i} H^p(X^p;\mathbb{Z}_2)\otimes H^q(Y^q;\mathbb{Z}_2) \\ &= \textstyle\bigoplus_{p+q=i} H^i(X^p\times Y^q;\mathbb{Z}_2); \end{aligned}$$

here the two equalities follow from Künneth's formula, and the arrow denotes the sum of homomorphisms induced by the inclusion maps $X^p \to X$, $Y^q \to Y$. On the one hand, every homomorphism $H^p(X;\mathbb{Z}_2) \to H^p(X^p;\mathbb{Z}_2)$ is a monomorphism (since $H^p(X,X^p;\mathbb{Z}_2) = 0$), and similarly for Y; thus, the preceding homomorphism is a monomorphism. On the other hand, this homomorphism acts as

$$\gamma \mapsto (\gamma|_{X^i\times Y^0}, \gamma|_{X^{i-1}\times Y^1}, \dots, \gamma|_{X^1\times Y^{i-1}}, \gamma|_{X^0\times Y^i}).$$

Hence, it takes u to 0, and hence $u=0$. This completes the proof.

It is convenient to write the formulas from (2) and (2′) as

$$w(\xi\oplus\eta) = w(\xi)w(\eta),\ \ w(\xi\times\eta) = w(\xi)\times w(\eta)$$

where w is the formal sum $1 + w_1 + w_2 + \dots$.

B: The Splitting Principle for the Stiefel–Whitney Classes

We begin with a computation of the Stiefel–Whitney classes for a very important example.

Proposition. *Consider the vector bundle* $\underbrace{\zeta\times\dots\times\zeta}_{n}$ *over the space* $\underbrace{\mathbb{R}P^\infty\times\dots\times\mathbb{R}P^\infty}_{n}$. *Let* $x_1,\dots,x_n \in H^1(\mathbb{R}P^\infty\times\dots\times\mathbb{R}P^\infty;\mathbb{Z}_2)$ *be the generators of* $H^*(\mathbb{R}P^\infty\times\dots\times\mathbb{R}P^\infty;\mathbb{Z}_2)$. *Then*

$$w_i(\zeta\times\dots\times\zeta) = e_i(x_1,\dots,x_n),$$

where e_i is ith elementary symmetric polynomial.

Proof. Since $w(\zeta) = 1 + x$, the preceding formula (2′) shows that

$$w(\zeta \times \cdots \times \zeta) = (1+x) \times \cdots \times (1+x) = (1+x_1)\dots(1+x_n)$$
$$= 1 + \sum_{i=1}^{n} e_i(x_1, \dots, x_n),$$

as required.

Now we can prove a result announced in the previous section.

Proof of the Theorem in Sect. 19.4.E, Part (i). It is well known in algebra that every symmetric polynomial in n variables with coefficients in an arbitrary integral domain R has a unique presentation as a polynomial in the elementary symmetric polynomial; the uniqueness statement means that no nonzero polynomial in $e_1, \dots, e_n$ is equal to zero. If $W = P(w_1, \dots, w_n)$ is a nonzero polynomial of the Stiefel–Whitney classes, then $W(\zeta \times \cdots \times \zeta) = P(e_1(x_1, \dots, x_n), \dots, e_n(x_1, \dots, x_n)) \neq 0$, which shows that W is not zero as a characteristic class. Hence, the dimension (over $\mathbb{Z}_2$) of the full space of q-dimensional characteristic classes with coefficients in $\mathbb{Z}_2$ of n-dimensional real vector bundles is at least the number of partitions $q = 1 \cdot r_1 + 2 \cdot r_2 + \cdots + n \cdot r_n$ with nonnegative r_is. But this number is precisely the number of q-dimensional cells in the standard (Schubert) CW decomposition of $G(\infty, n)$, which, in turn, does not exceed $\dim_{\mathbb{Z}_2} H^q(G(\infty, n); \mathbb{Z}_2)$, that is, the dimension of the space of characteristic classes. Thus, all these numbers and dimensions are the same. This proves that all the characteristic classes of n-dimensional vector bundles with coefficients in $\mathbb{Z}_2$ are polynomials in $w_1, \dots, w_n$, as stated in part (i) of the theorem in Sect. 19.4.E.

Remark 1. This proof shows that $\dim_{\mathbb{Z}_2} H^q(G(\infty, n); \mathbb{Z}_2)$ is actually equal to the number of q-dimensional Schubert cells, which means, in turn, that all the incidence numbers in the cellular complex corresponding to the Schubert cell decomposition of the Grassmannian are even. This fact is not new for us; it was offered as Exercise 11 in Sect. 13.8.C. Now we have a proof of this fact, thus a (rather indirect) solution of that exercise.

Remark 2. We see also that a nonzero characteristic class with coefficients in $\mathbb{Z}_2$ of n-dimensional vector bundles takes a nonzero value on the bundle $\zeta \oplus \cdots \oplus \zeta$. This provides a method of finding relations between characteristic classes: A relation holds if it holds for $\zeta \oplus \cdots \oplus \zeta$. Usually, this statement is formulated in a seemingly weaker, but actually equivalent form: *To establish a relation between characteristic classes it is sufficient to check it for splitting bundles, that is, for bundles isomorphic to sums of one-dimensional bundles.* This proposition is known under the name of the *splitting principle* (later, we will deal with different versions of this principle).

EXERCISE 14. Prove the following version of the splitting principle (and explain why it is equivalent to the splitting principle). The $\mathbb{Z}_2$-cohomology homomorphism induced by the map

$$\underbrace{\mathbb{R}P^\infty \times \cdots \times \mathbb{R}P^\infty}_{n} \to G(\infty, n),$$

$$\big(\{\ell_1 \subset \mathbb{R}^\infty\}, \dots, \{\ell_n \subset \mathbb{R}^\infty\}\big) \mapsto \ell_1 \times \cdots \times \ell_n \subset \mathbb{R}^\infty \times \cdots \times \mathbb{R}^\infty = \mathbb{R}^\infty$$

is a monomorphism; moreover, its image in $H^*(\mathbb{R}P^\infty \times \cdots \times \mathbb{R}P^\infty; \mathbb{Z}_2) = \mathbb{Z}_2[x_1, \dots, x_n]$ is precisely the space of symmetric polynomials.

C: Stiefel–Whitney Classes and Operations over Vector Bundles

Formulas expressing the Stiefel–Whitney classes of the bundles $\xi \otimes \eta$, $\Lambda^k \xi$, $S^k \xi$, and so on via the Stiefel–Whitney classes of ξ and η (and the dimensions of ξ and η) exist, but more complicated and less convenient, than the formulas for the Stiefel–Whitney classes of the sum (or direct product). We will give a brief overview of this subject.

Lemma. *Let ξ, η be one-dimensional real vector bundle over the same CW base. Then*

$$w_1(\xi \otimes \eta) = w_1(\xi) + w_1(\eta).$$

Proof. Fix sections s, t of ξ and η over the 1-skeleton X^1 of the base X of ξ η. We may assume that these sections have no zeroes over X^0 and have transverse zeroes over one-dimensional cells, and the zeroes of s are different from the zeroes of t. Then $s \otimes t$ is a section of $\xi \otimes \eta$, and the set of zeroes of $s \otimes t$ is the union of the set of zeroes of s and the set of zeroes of t. Let m_e, n_e be residues modulo 2 of the numbers of zeroes of the sections s and t within a one-dimensional cell of X. Then the functions $e \mapsto m_e$, $e \mapsto n_e$, $e \mapsto m_e + n_e$ are cocycles representing $w_1(\xi), w_1(\eta), w_1(\xi \otimes \eta)$, whence our result.

For our next statement, we will need some notations from algebra of symmetric polynomials. Consider the ith symmetric polynomial of mn variables $y_j + z_k$, $1 \le j \le m$, $1 \le k \le n$, and express it as a polynomial in elementary symmetric polynomials separately in $y_1, \dots y_m$ and $z_1, \dots, z_n$ (we assume that $i \le m$ and $i \le n$):

$$e_i(y_j + z_k) = E_{m,n;i}(e_1(y), e_2(y), \dots; e_1(z), e_2(z), \dots);$$

for example,

$$E_{m,n;1} = \sum_{j=1}^{m}\sum_{k=1}^{n}(y_j + z_k) = n\sum_{j=1}^{m} y_j + m\sum_{k=1}^{n} z_k = ne_1(y) + me_1(z);$$

$$\begin{aligned} E_{m,n;2} &= \sum_{(j',k')\neq(j,k)} (y_j + z_k)(y_{j'} + z_{k'}) \\ &= n(n-1)\sum_{j=1}^{m} y_j^2 + n^2\sum_{j'\neq j} y_j y_{j'} + 2(mn-1)\sum_{j=1}^{m}\sum_{k=1}^{n} y_j z_k \\ &\qquad\qquad +m(m-1)\sum_{k=1}^{n} z_k^2 + m^2\sum_{k'\neq k} z_k z_{k'} \\ &= n(n-1)(e_1(y)^2 - 2e_2(y)) + 2n^2 e_2(y) + 2(mn-1)e_1(y)e_1(z) \\ &\quad +m(m-1)(e_1(z)^2 - 2e_2(z)) + 2m^2 e_2(z), \end{aligned}$$

that is,

$$\begin{aligned} E_{m,n;2} = &\frac{n(n-1)}{2} e_1(y)^2 + ne_2(y) + (mn-1)e_1(y)e_1(z) \\ &+\frac{m(m-1)}{2} e_1(z)^2 + me_2(z). \end{aligned}$$

These examples show that it is possible to find explicit expressions for the polynomials $E_{m,n;i}$, but the formula may be complicated.

In addition, consider the elementary symmetric polynomials of $\binom{n}{r}$ variables $x_{j_1} + \cdots + x_{j_r}$ $1 \le j_1 < \cdots < j_r \le n$. Obviously, they are symmetric polynomials in $x_1, \ldots, x_n$, and we can write

$$e_i(x_{j_1} + \cdots + x_{j_r}\ 1 \le j_1 < \cdots < j_r \le n) = F_{n;r;i}(e_1(x), e_2(x), \ldots),$$

where $F_{n;r;i}$ is a polynomial. For example,

$$F_{n;r;1} = \binom{n-1}{r-1} e_1(x),\ F_{n;2;2} = \frac{(n-1)(n-2)}{2} e_1(x)^2 + (n-2)e_2(x).$$

The polynomials F are related to the polynomials E. Namely, if we put $x_1 = y_1, \ldots, x_m = y_m, x_{m+1} = z_1, \ldots, x_{m+n} = z_n$, then, obviously, $\{x_j + x_k | 1 \le j < k \le m+n\} = \{y_j + y_k | 1 \le j < k \le m\} \bigcup \{y_j + z_k | 1 \le j \le m,\ 1 \le k \le n\} \bigcup \{z_j + z_k | 1 \le j < k \le n\}$, which shows that

$$\begin{aligned} F_{m+n;2;i}(e_1(x), e_2(x), \ldots) = &\sum_{p+q+r=i} F_{m;2;p}(e_1(y), e_2(y), \ldots) \\ \cdot E_{m,n;q}(e_1(y), e_2(y), \ldots, &e_1(z), e_2(z), \ldots) \cdot F_{n;2;r}(e_1(z), e_2(z), \ldots). \end{aligned} \tag{$*$}$$

And one more family of polynomials:

$$G_{n;r;i}(e_1(x), e_2(x), \dots) = e_i(x_{j_1} + \dots + x_{j_r} | 1 \le j_1 \le \dots \le j_r \le n);$$

a computation shows that

$$\begin{aligned} G_{n;r;1} &= \binom{n+r-1}{r-1} e_1(x), \\ G_{n;2} &= \frac{(n-1)(n+2)}{2} e_1^2 + (n+2)e_2, \ \text{if } n > 1. \end{aligned}$$

The formula $(*)$ with the polynomials F replaced by polynomials G is also true.

Now, let us formulate the main results of this section.

Theorem 1. *Let ξ and η be real vector bundles of dimensions m and n over the same CW base. Then*

$$w_i(\xi \otimes \eta) = E_{m,n;i}(w_1(\xi), w_2(\xi), \dots; w_1(\eta), w_2(\eta), \dots);$$

thus, in particular, $w_1(\xi \otimes \eta)$ and $w_2(\xi \otimes \eta)$ are, respectively,

$$nw_1(\xi) + mw_1(\eta)$$

and

$$\frac{n(n-1)}{2} w_1(\xi)^2 + nw_2(\xi) + (mn-1)w_1(\xi)w_1(\eta) + \frac{m(m-1)}{2} w_1(\eta)^2 + mw_2(\eta).$$

Theorem 2. *Let ξ be an n-dimensional real vector bundle with a CW base. Then*

$$w_i(\Lambda^r \xi) = F_{n;r;i}(w_1(\xi), w_2(\xi), \dots);$$

in particular,

$$\begin{aligned} w_1(\Lambda^r \xi) &= \binom{n-1}{r-1} w_1(\xi), \\ w_2(\Lambda^2 \xi) &= \frac{(n-1)(n-2)}{2} w_1(\xi) + (n-2)w_2(\xi). \end{aligned}$$

Theorem 3. *Let ξ be an n-dimensional real vector bundle with a CW base. Then*

$$w_i(S^r \xi) = G_{n;r;i}(w_1(\xi), w_2(\xi), \dots);$$

in particular,

$$w_1(S^r\xi) = \binom{n+r-1}{r-1} w_1(\xi),$$
$$w_2(S^2\xi) = \frac{(n-1)(n+2)}{2} w_1(\xi) + (n+2)w_2(\xi), \text{ if } n > 1.$$

Proofs. We begin with Theorem 1 in the case when $\xi = \eta$. The class $w_i(\xi \otimes \xi)$ is a characteristic class of a real vector bundle. Hence, by part (i) of the theorem in Sect. 19.4.E, it must be a polynomial in Stiefel–Whitney classes. To identify this polynomial, we need to compute the class for the bundle $\xi = \underbrace{\zeta \times \cdots \times \zeta}_{n}$ over $\underbrace{\mathbb{R}P^\infty \times \cdots \times \mathbb{R}P^\infty}_{n}$; this bundle is the same as $\zeta_1 \oplus \cdots \oplus \zeta_n$, where ζ_j is the bundle induced by ζ with respect to the projection of $\mathbb{R}P^\infty \times \cdots \times \mathbb{R}P^\infty$ onto the jth factor. Then $\xi \otimes \xi = \bigoplus_{j,k}(\zeta_j \otimes \zeta_k)$ and

$$\begin{aligned} w(\xi \otimes \xi) &= \prod_{j,k} w(\zeta_j \otimes \zeta_k) = \text{ (by Lemma) } \prod_{j,k}(1 + x_j + x_k) \\ &= 1 + \sum_{i \geq 1} e_i(x_j + x_k | 1 \leq j \leq n, 1 \leq k \leq n) \\ &= 1 + \sum_{i \geq 1} E_{n,n;i}(e_1(x), e_2(x), \ldots, e_1(x), e_2(x), \ldots) \\ &= 1 + \sum_{i \geq 1} E_{n,n;i}(w_1(\xi), w_2(\xi), \ldots, w_1(\xi), w_2(\xi), \ldots), \end{aligned}$$

which is the statement of the theorem (for $\xi = \eta$).

Next, we prove Theorem 2. The proof is the same as the previous proof, and it is based on the relation, for $\xi = \zeta_1 \oplus \cdots \oplus \zeta_n$,

$$\Lambda^r \xi = \bigoplus_{1 \leq j_1 < \cdots < j_r \leq n} (\zeta_{j_1} \otimes \cdots \otimes \zeta_{j_r}),$$

which gives, by the lemma,

$$\begin{aligned} w(\Lambda^r \xi) &= \prod_{1 \leq j_1 < \cdots < j_r \leq n} (1 + (j_1 + \cdots + j_r)) \\ &= 1 + \sum_{1 \leq j_1 < \cdots < j_r \leq n} e_1(x_{j_1} + \cdots + x_{j_r} | 1 \leq j_1 < \cdots < j_r \leq n). \end{aligned}$$

The rest of the proof repeats the previous proof.

The proof of Theorem 3 is so close to the proof of Theorem 2 that we do not feel any necessity in detailing it [just mention that it is based on the relation $S^r\xi = \bigoplus_{1\le j_1\le\dots\le j_r\le n} (\zeta_{j_1}\otimes\dots\otimes\zeta_{j_r})$].

Finally, let us prove Theorem 1 in the general case. Notice that for any vector bundles ξ and η,

$$\Lambda^2(\xi\oplus\eta) = \Lambda^2\xi\oplus(\xi\otimes\eta)\oplus\Lambda^2\eta,$$

and hence

$$w(\Lambda^2(\xi\oplus\eta)) = w(\Lambda^2\xi)w(\xi\otimes\eta)w(\Lambda^2\eta).$$

Since $w = 1 + w_1 + w_2 + \dots$ is an invertible element of the ring $H^*(B;\mathbb{Z}_2)$, this formula determines $w(\xi\otimes\eta)$ if $w(\Lambda^2\xi)$, $w(\Lambda^2\eta)$ and $\Lambda^2(\xi\oplus\eta)$ are known. The formula from Theorem 1 follows from the formula of Theorem 2 and the relation $(*)$.

D: Properties of the Euler, Chern, and Pontryagin Classes

For the Euler classes, a multiplication formula $e(\xi\otimes\eta) = e(\xi)e(\eta)$ holds.

All the major properties of the Stiefel–Whitney classes can be repeated with appropriate changes for the Chern classes. In particular, the class c_1 of the Hopf bundle $\zeta_{\mathbb{C}}$ is the standard generator of the group $H^2(\mathbb{C}P^\infty;\mathbb{Z})$. There are the multiplication formula

$$c_i(\xi\oplus\eta) = \sum_{p+q=i} c_p(\xi)c_q(\eta)$$

and the splitting principle. Like Stiefel–Whitney classes, the Chern classes are invariant with respect to stable equivalence. The computation of the Chern classes of tensor product, exterior powers, and symmetric powers of complex vector bundles repeats the computations in Sect. 19.5.C.

EXERCISE 15. Prove that $c_i(\overline{\xi}) = (-1)^i c_i(\xi)$. Deduce from this that for every *real* vector bundle ξ and every *odd* i the equality $2c_i(\mathbb{C}\xi) = 0$ (compare the comment to the definition of the Pontryagin classes in Sect. 19.3.C).

EXERCISE 16. Define a polynomial Q_r of r variables by the formula

$$N_r = Q_r(e_1,\dots,e_r),$$

where the e_i are elementary symmetric polynomials and N_i are sums of ith powers of variables (so $Q_1 = e_1$, $Q_2 = e_1^2 - 2e_2$, $Q_3 = e_1^3 - 3e_1e_2 + 3e_3, \dots$). For a complex vector bundle ξ with the base X, put

$$\mathrm{ch}_r(\xi) = \frac{1}{r!} Q_r(c_1(\xi), \dots, c_r(\xi)) \in H^{2r}(X; \mathbb{Q}).$$

The (nonhomogeneous) characteristic class ch with coefficients in $\mathbb{Q}$ defined by the formula

$$\mathrm{ch} = \mathrm{ch}_0 + \mathrm{ch}_1 + \mathrm{ch}_2 + \dots \in H^{\mathrm{even}}(X; \mathbb{Q})$$

is called the *Chern character.* Notice that $\mathrm{ch}_0(\xi) \in H^0(X; \mathbb{Q})$ is just $\dim \xi$.

Prove that

$$\mathrm{ch}(\xi \oplus \eta) = \mathrm{ch}(\xi) + \mathrm{ch}(\eta) \text{ and } \mathrm{ch}(\xi \otimes \eta) = \mathrm{ch}(\xi)\,\mathrm{ch}(\eta).$$

For the Pontryagin classes, the multiplication formulas and all the other formulas are deduced from the corresponding formulas for the Chern classes and hold "modulo 2-torsion"; for example,

$$2\left(p_i(\xi \oplus \eta) - \sum_{p+q=i} p_p(\xi) p_q(\eta)\right) = 0.$$

EXERCISE 17. Prove that stably equivalent bundles have equal Pontryagin classes.

E: More Relations Between Stiefel–Whitney, Chern, and Euler Classes

In conclusion, we give two more formulas expressing the Stiefel–Whitney and Chern classes via the Euler class. Let ξ be an n-dimensional real vector bundle with a CW base X and ζ be the Hopf bundle over $\mathbb{R}P^\infty$. Consider the bundle $\xi \otimes \zeta$ over $X \times \mathbb{R}P^\infty$ (more precisely, it is the tensor product of bundles induced by ξ and ζ with respect to the projections of the product $X \times \mathbb{R}P^\infty$ onto the factors). Then

$$\rho_2 e(\xi \otimes \zeta) = w_n(\xi \otimes \zeta) = \sum_{i=0}^{n} (w_i(\xi) \times x^{n-i}) \in H^n(X \times \mathbb{R}P^\infty; \mathbb{Z}_2),$$

where $x \in H^1(\mathbb{R}P^\infty; \mathbb{Z}_2)$ is the generator. Similarly, if ξ is an n-dimensional complex vector bundle with a CW base X and $\zeta_{\mathbb{C}}$ is the (complex) Hopf bundle over $\mathbb{C}P^\infty$, then

$$e(\xi \otimes \zeta_{\mathbb{C}}) = c_n(\xi \otimes \zeta) = \sum_{i=0}^{n} (c_i(\xi) \times x^{n-i}) \in H^{2n}(X \times \mathbb{C}P^\infty; \mathbb{Z}),$$

where $x \in H^2(\mathbb{C}P^\infty; \mathbb{Z})$ is the generator.

These formulas may be regarded as definitions of the Stiefel–Whitney and Chern classes.

EXERCISE 18. Prove these formulas.

19.6 Characteristic Classes in Differential Topology

We can only touch on this vast subject.

A: Geometric Interpretation of the First Obstruction

Let (E, B, F, p) be a homotopically simple locally trivial fibration where E and B are smooth manifolds and the manifold B is closed, n-dimensional and oriented, and p is a submersion, that is, a smooth map whose differential at every point has rank equal to n. Assume also that $\pi_0(F) = \cdots = \pi_{k-2}(F) = 0$ and $\pi_{k-1}(F) = \pi$. Then the first obstruction to extending a section of our fibration lies in $H^k(B; \pi)$. Suppose also that we were able to construct a section over $B - X$ where X is a submanifold of B (possibly, with singularities of codimension ≥ 2) of dimension $n - k$ or a union of a finite number of such submanifolds which are connected and transversally intersect each other, $X = \bigcup X_i$ (simple general position argumentations show that it is always possible to do this). For every i, choose a nonsingular point x_i of X_i and construct a small $(k-1)$-dimensional sphere s_i centered at x_i in a k-dimensional surface transversally intersecting X_i at x_i. Since there is a section over s_i, and the fibration is trivial in a proximity of x_i, we obtain a continuous map $S^{k-1} \to p^{-1}(x_i)$ which determines, since the fibration and the fiber are homotopically simple, an element $\alpha_i \in \pi_{k-1}(F) = \pi$.

Claim: The homology class $\sum_i \alpha_i[X_i] \in H_{n-k}(B; \pi)$ is the Poincaré dual of the first obstruction to extending a section in our fibration. The proof is left to the reader. (*Hint*: Triangulate the manifold B in such a way that X is disjoint from the simplices of dimension less than k and intersects each k-dimensional simplex transversally in at most one point.)

B: Differential Topology Interpretations of the Euler Class

For a closed oriented manifold X, the value of the Euler class of the tangent bundle $e(X) = e(\tau_X)$ on the fundamental class $[X]$ is equal to the Euler characteristic $\chi(X)$ of X (this is Proposition 2 of Sect. 18.5). This implies that a closed manifold possesses a nonvanishing vector field if and only if its Euler characteristic is zero (corollary in Sect. 18.5).

Some other properties of the Euler class are given here as exercises.

EXERCISE 19. Prove that a closed manifold X (orientable or not) possesses a continuous family of tangent lines (equivalently: The tangent bundle $\tau(X)$ possesses a one-dimensional subbundle) if and only if $\chi(X) = 0$.

EXERCISE 20. Let $(E, B, \mathbb{R}^n, p)$ be a smooth vector bundle (that is, a vector bundle such that E and B are smooth manifolds, p is a submersion, and the vector space operations in E are smooth). Suppose that B is closed and oriented. Let $s: B \to E$ be a section of ξ in a general position with the zero section. Show that the intersection $B \cap s(B)$ (we assume that B is embedded into E as the zero section) represents the homology class of B which is the Poincaré dual of the Euler class $e(\xi)$ of ξ.

EXERCISE 21. Let Y be a closed oriented submanifold of a closed oriented manifold X, and let $\nu_X(Y) = (\tau(X))|_Y/\tau(Y)$ be the corresponding normal bundle. Prove the formula

$$D(e(\nu_X(Y))] = i^![Y],$$

where D is Poincaré isomorphism (in Y), $i: Y \to X$ is the inclusion map, and $[Y]$ is a homology class of X represented by Y. Corollary: If $[Y] = 0$, then $e(\nu_X(Y)) = 0$; in particular, the Euler class of the normal bundle of a manifold embedded into an Euclidean space or a sphere is zero.

EXERCISE 22. The last statement does not hold for immersions. Show, in particular, that if f is an immersion of a closed oriented manifold of even dimension n into $\mathbb{R}^{2n}$ with transverse self-intersections, then the algebraic number of the self-intersection points (the reader will have to make up the definition of a sign corresponding to a transverse self-intersection) is equal to one half of the "normal Euler number," that is, of the value of the Euler class of the normal bundle on the fundamental class of the manifold. Example: Construct an immersion of S^2 into $\mathbb{R}^4$ with one transverse self-intersection (such a two-dimensional figure-eight) and find the Euler class of the corresponding normal vector bundle.

C: Differential Topology Interpretations of the Stiefel–Whitney Classes

In this section, we deal only with cohomology and homology with coefficients in $\mathbb{Z}_2$ and understand accordingly Poincaré isomorphism D.

The Stiefel–Whitney classes of the tangent bundle of a smooth manifold X are called the Stiefel–Whitney classes of X and are denoted as $w_i(X)$. [In a similar way, people consider the Pontryagin classes $p_i(X)$ of a smooth manifold X and the Chern classes $c_i(X)$ of a complex manifold X.] Since the normal bundle of a smooth manifold embedded into a Euclidean space does not depend, up to a stable equivalence, on the embedding, we can speak of the "normal Stiefel–Whitney classes," $\overline{w}_i(X)$, of a smooth manifold X. It follows from the multiplication formula and the fact that the sum of the tangent and normal bundles is trivial that

$$\sum_{p+q=i} w_p(X)\overline{w}_q(Y) = 0 \text{ for } i > 0,$$

or $\overline{w} = w^{-1}$ (we already remarked in the end of Sect. 19.5.C that w is invertible in the cohomology ring). Thus, the normal Stiefel–Whitney classes are expressed via the usual (tangent) Stiefel–Whitney classes.

EXERCISE 23. Consider a generic smooth map (the reader is supposed to clarify the meaning of the word *generic*) of a closed n-dimensional manifold X into $\mathbb{R}^q$, $q \leq n$; let $Y \subset X$ be the set of points where this map is not a submersion (the rank of the differential is less than q). Prove that Y is a $(q-1)$-dimensional submanifold of X [maybe, with singularities, but the class $[Y] \in H_{q-1}(X; \mathbb{Z}_2)$ is defined] and that

$$D^{-1}[Y] = w_{n-q+1}(X).$$

EXERCISE 24. Consider a generic smooth map of a closed n-dimensional manifold X into $\mathbb{R}^q$, $q \geq n$; let $Y \subset X$ be the set of points where this map is not an immersion (the rank of the differential is less than n). Prove that Y is a $(2n-q-1)$-dimensional submanifold of X (maybe, with singularities) and that

$$D^{-1}[Y] = \overline{w}_{q+1-n}(X).$$

EXERCISE 25. If an n-dimensional manifold X possesses an immersion into $\mathbb{R}^{n+q}$, then $\overline{w}_i(X) = 0$ for $i > q$. (For closed manifolds, this follows from Exercise 24, but actually this fact is much easier than Exercise 24, and it is more natural to prove it directly.)

EXERCISE 26. If an n-dimensional manifold X possesses an embedding into $\mathbb{R}^{n+q}$, then $\overline{w}_i(X) = 0$ for $i \geq q$. (To prove this, one needs to use, in addition to Exercise 25, the corollary part of Exercise 21.)

EXERCISE 27. Prove that if $2^k \leq n < 2^{k+1}$, then $\mathbb{R}P^n$ has no immersion in $\mathbb{R}^{2^{k+1}-2}$ and no embedding in $\mathbb{R}^{2^{k+1}-1}$. (To prove this, one needs to use, besides Exercises 25 and 26, Exercise 12 (Sect. 19.2) and the theorem in Sect. 19.5.A.

Remark 1. Thus, if $n = 2^k$, the n-dimensional manifold $\mathbb{R}P^n$ cannot be embedded into $\mathbb{R}^{2n-1}$. This is a very rare phenomenon. The classical Whitney theorem asserts that an n-dimensional manifold (with a positive n) can always be embedded into $\mathbb{R}^{2n}$ (this result should not be confused with an earlier theorem of Whitney stating that any smooth map of an n-dimensional manifold into any manifold of dimension $\geq 2n+1$ can be smoothly approximated by smooth embeddings); embeddings into

$\mathbb{R}^{2n-1}$ are almost always possible: For a nonexistence of such an embedding, it is necessary and sufficient that n is a power of 2, and there exists a one-dimensional cohomology class with coefficients in $\mathbb{Z}_2$ whose nth power is not zero (these conditions imply the nonorientability).

Remark 2. Further information concerning embeddability of (real and complex) projective spaces into Euclidean spaces can be obtained with the help of K-theory (see Sect. 42.6 in Chap. 6).

EXERCISE 28. Let X be a triangulated smooth manifold. Denote by C_i the i-dimensional classical chain of the barycentric subdivision of the triangulation of X equal to the sum of all i-dimensional simplices of this subdivision. Prove that C_i is a cycle and that

$$D^{-1}[C_i] = w_i(X)$$

($[C_i]$ is the homology class of C_i).

The values of the cohomology classes of the form $w_{i_1}(X) \dots w_{i_r}(X)$ with $i_1 \dots i_r = n$ on the fundamental class of closed n-dimensional manifold (they are residues modulo 2) are called *Stiefel–Whitney numbers* of the manifold X; notation: $w_{i_1 \dots i_r}[X]$. For example, two-dimensional manifolds have two Stiefel–Whitney numbers: $w_{11}[X]$ and $w_2[X]$.

EXERCISE 29. Find Stiefel–Whitney numbers of classical surfaces.

Remark. The reader will see that for any classical surface X, $w_{11}[X] = w_2[X]$. A classical theorem in the topology of a manifold asserts any connected closed two-dimensional manifold is a classical surface. Hence, the two Stiefel–Whitney numbers, $w_{11}[X]$ and $w_2[X]$, are always the same. Later in this section, we will see that there are more relations between Stiefel–Whitney numbers.

Theorem. *If a closed manifold is a boundary of a compact manifold, then all its Stiefel–Whitney numbers are zeroes.*

Proof. If $X = \partial Y$ and $i: X \to Y$ is the inclusion map, then $\tau(X) = \tau(Y)|_X \oplus 1$ (the normal bundle of the boundary is always trivial!). Hence, $w_j(X) = i^* w_j(Y)$ for every j, and

$$\begin{aligned} \langle w_{j_1}(X) \dots w_{j_r}(X), [X] \rangle &= \langle i^*(w_{j_1}(Y) \dots w_{j_r}(Y)), [X] \rangle \\ &= \langle w_{j_1}(Y) \dots w_{j_r}(Y), i_*[X] \rangle = 0 \end{aligned}$$

since $i_*[X] = 0$ (the fundamental cycle of the boundary of a compact manifold is the boundary of the fundamental cycle of this manifold).

This theorem provides a powerful necessary condition for a closed manifold to be a boundary of a compact manifold.

EXERCISE 30. Prove that if $n + 1$ is not a power of 2, then neither $\mathbb{R}P^n$ nor $\mathbb{C}P^n$ is a boundary of a compact manifold.

The most striking fact, however, is that this necessary condition is also sufficient for a closed manifold to be a boundary of a compact manifold (R. Thom, Fields Medal of 1952). We will discuss the proof of this result briefly in Chaps. 5 and 6. As we mentioned before, the Stiefel–Whitney numbers are not linearly independent ($w_1[X] = 0$ for any one-dimensional X, $w_{11}[X] = w_2[X]$ for any two-dimensional X). The fact is that a maximal linear independent system of Stiefel–Whitney numbers of a closed n-dimensional manifold is formed by the numbers $w_{j_1 \dots j_r}[X]$ such that $j_1 + \dots + j_r = n, j_1 \leq \dots \leq j_r$ and no one of the numbers $j_s + 1$ is a power of 2. (Corollary: Every closed three-dimensional manifold is the boundary of some compact four-dimensional manifold; this is a classical theorem of Rokhlin.)

D: Differential Topology Interpretations of the Pontryagin Classes

The following statement is similar to Exercise 23. Let X be a closed oriented n-dimensional manifold and $f: X \to \mathbb{R}^{n-2q+2}$ be a generic smooth map. Let $Y \subset X$ be the set of points where the rank of the differential of f does not exceed $n-2q$ (that is, is at least 2 less than its maximal possible value). Then Y is an oriented $(n-4q)$-dimensional manifold (maybe, with singularities), and the class $[Y] \in H_{n-4q}(X)$ is the Poincaré dual to the Pontryagin class $p_q(X) \in H^{4q}(X; \mathbb{Z})$ of (the tangent bundle of) the manifold X. A similar statement holds for the normal Pontryagin classes (compare to Exercise 24.)

(The orientedness, and even orientability, of manifold X is actually not needed, but, in general, we will need the version of Poincaré isomorphism developed in Sect. 17.12.)

If X is a closed oriented manifold of dimension $4m$, then the value of the class $p_{j_1}(X) \dots p_{j_r}(X), j_1 + \dots + j_r = m$ on the fundamental homology class of X is called a *Pontryagin number* and is denoted as $p_{j_1 \dots j_r}[X]$. (It is convenient to assume that X is not necessarily connected; the fundamental class of a disconnected X is defined as the sum of the fundamental classes of the components.) If X is a boundary of a compact oriented manifold, then all the Pontryagin numbers of X are zeroes (this fact is proved precisely as the similar fact for the Stiefel–Whitney numbers). There also is a *Thom theorem* which asserts that if all the Pontryagin numbers of a closed orientable manifold are zeroes (for example, if its dimension is not divisible by 4), then a union of several copies of X (taken all with the same orientation) is a boundary of some compact manifold. Moreover, every set of integers $\{p_{j_1 \dots j_r} \mid j_1 + \dots + j_r = m\}$ becomes, after a multiplication of all the numbers in the set by the same positive integer, the set of Pontryagin numbers of some closed oriented manifold of dimension $4m$. (Actually, this theorem is way easier than the similar theorem for the Stiefel–Whitney numbers; we will see this in Chap. 6.)

A useful corollary of the Thom theorem (and the fact that if $Y = X_1 \bigsqcup X_2$ is the disjoint union of two closed oriented $4m$-dimensional manifolds, then

$$p_{j_1 \dots j_r}[Y] = p_{j_1 \dots j_r}[X_1] + p_{j_1 \dots j_r}[X_2]$$

for every $j_1, \dots, j_r$ with $j_1 + \dots + j_r = m$) is the following statement.

EXERCISE 31. Suppose that for every closed oriented n-dimensional manifold there is assigned an integer $\sigma(X)$ with the following properties: (1) If X is a boundary of a compact oriented manifold, then $\sigma(X) = 0$; (2) $\sigma(X_1 \bigsqcup X_2) = \sigma(X_1) + \sigma(X_2)$. Prove that

$$\sigma(X) = \sum_{j_1 + \dots + j_r = n/4} a_{j_1 \dots j_r} p_{j_1 \dots j_r}[X],$$

where $a_{j_1 \dots j_r}$ are some *rational* numbers not depending on X. In particular, $\sigma(X) = 0$ if n is not divisible by 4.

This statement has only one broadly known application, but what an application it is! Denote by $\sigma(X)$ the signature of the intersection index form in the $2m$-dimensional homology of a $4m$-dimensional closed oriented manifold X. The theorem in Sect. 17.10 shows that σ satisfies condition (1); condition (2) for the signature is obvious. Hence, the signature is a rational linear combination of Pontryagin numbers. In particular, $\sigma(X) = ap_1[X]$ if $\dim X = 4$, $\sigma(X) = bp_2[X] + cp_{11}[X]$ if $\dim X = 8$, and so on. To find $a, b, c, \dots$, we need to have a sufficient supply of computations in concrete examples. For example, $H_{2m}(\mathbb{C}P^{2m}) = \mathbb{Z}$. The matrix of the intersection form is just (1); hence, $\sigma(\mathbb{C}P^{2m}) = 1$. Furthermore,

$$\tau(\mathbb{C}P^{2m}) \oplus 1 = (2m+1)\zeta_{\mathbb{C}}$$

(this is the complex version of Exercise 12), and hence

$$\mathbb{C}\tau(\mathbb{C}P^{2m}) \oplus 1_{\mathbb{C}} = (2m+1)(\zeta_{\mathbb{C}} \oplus \overline{\zeta}_{\mathbb{C}})$$

(see Exercise 8), and

$$\begin{aligned}(p_0 - p_1 + p_2 - \dots + (-1)^m p_m)(\mathbb{C}P^{2m}) &= [(1+x)(1-x)]^{2m+1} \\ &= (1-x^2)^{2m+1}\end{aligned}$$

where $x \in H^2(\mathbb{C}P^{2m}) = \mathbb{Z}$ is the canonical generator (see Sect. 19.5.D, including Exercise 16). Hence,

$$p_i(\mathbb{C}P^{2m}) = \begin{cases} \binom{2m+1}{i} x^{2i}, & \text{if } i \le m, \\ 0, & \text{if } i > m. \end{cases}$$

In particular, $p_1(\mathbb{C}P^2) = 3x^2$, $p_1[\mathbb{C}P^2] = 3$, and, since $\sigma(\mathbb{C}P^2) = 1$, then for every (closed, orientable) four-dimensional manifold X,

$$\sigma(X) = \frac{1}{3} p_1[X]. \tag{$*$}$$

(In particular, the Pontryagin number $p_1[X]$ of every closed orientable four-dimensional manifold X is divisible by 3.) Furthermore, $p_{11}[\mathbb{C}P^4] = 25, p_2[\mathbb{C}P^4] = 10, \sigma(\mathbb{C}P^4) = 1$. In addition,

$$(p_0 + p_1 + p_2)(\mathbb{C}P^2 \times \mathbb{C}P^2) = (p_0 + p_1)(\mathbb{C}P^2) \times (p_0 + p_1)(\mathbb{C}P^2)$$

(the multiplication formula for the Pontryagin classes holds only modulo 2-torsion, but there is no torsion in the cohomology of complex projective spaces), and hence

$$\begin{gathered} p_1(\mathbb{C}P^2 \times \mathbb{C}P^2) = (1 \times 3x^2) + (3x^2 \times 1), \\ p_1^2(\mathbb{C}P^2 \times \mathbb{C}P^2) = 18(x^2 \times x^2), \\ p_2(\mathbb{C}P^2 \times \mathbb{C}P^2) = p_1(\mathbb{C}P^2) \times p_1(\mathbb{C}P^2) = 3x^2 \times 3x^2, \\ p_{11}[\mathbb{C}P^2 \times \mathbb{C}P^2] = 18, \; p_2[\mathbb{C}P^2 \times \mathbb{C}P^2] = 9, \\ \text{and also } \sigma(\mathbb{C}P^2 \times \mathbb{C}P^2) = 1. \end{gathered}$$

EXERCISE 32. Prove that the signature is multiplicative: $\sigma(X \times Y) = \sigma(X)\sigma(Y)$.)

Hence, $1 = 10b + 25c, 1 = 9b + 18c$, whence $b = \frac{7}{45}, c = -\frac{1}{45}$. Thus, for $\dim X = 8$,

$$\sigma(X) = \frac{7p_2[X] - p_{11}[X]}{45}. \tag{$**$}$$

(Hence, $7p_2[X] - p_{11}[X]$ is divisible by 45, and if the first Pontryagin class of a closed orientable eight-dimensional manifold is zero, then its signature is divisible by 7.) The formulas $(*)$, $(**)$ form the beginning of an infinite chain of formulas relating the signature to the Pontryagin numbers. The work of explicitly writing these formulas was done in the 1950s by F. Hirzebruch. He calculated the Pontryagin numbers of manifolds of the form $\mathbb{C}P^{2m_1} \times \cdots \times \mathbb{C}P^{2m_k}$ (which, essentially, we have done) and, using the fact that the signatures of all these manifolds are equal to 1, he found the coefficients of the Pontryagin numbers in the formulas for signatures. The resulting formulas are presented in his book [46].

E: Invariance Problems for Characteristic Classes of Manifolds

As we know, the Euler class of a manifold can be expressed through the Betti numbers of this manifold. It turns out that although the Stiefel–Whitney classes are not determined by either homology groups or even a cohomology ring of this manifold, still they are homotopy invariant; that is, a map of one closed manifold into another one which is a homotopy equivalence takes the Stiefel–Whitney classes into the Stiefel–Whitney classes. This fact will be established (or, at least, discussed) in Chap. 4 (Sect. 31.2). For Pontryagin classes, however, the homotopy invariance fails (the only homotopy invariant nonzero polynomial in Pontryagin classes is the signature). In the 1960s, S. Novikov proved the difficult theorem of topological invariance of rational Pontryagin classes (a homeomorphism between two smooth closed orientable manifolds takes Pontryagin classes into Pontryagin classes modulo elements of finite order; these elements of finite order may be nonzero—there are examples). A decade before that, V. Rokhlin, A. Schwarz, and R. Thom proved this statement for homeomorphisms, establishing a correspondence between some smooth triangulations of two smooth manifolds (see Rokhlin and Schwarz [72], Thom [85]). This result leads naturally to the problem of "combinatorial calculation of Pontryagin classes," that is, their calculation via triangulation (compare to Exercise 29). At present, this problem has been solved only for the first Pontryagin class (see the article by Gabrielov, Gelfand, and Losik [42]).

Chapter 3
Spectral Sequences of Fibrations

Lecture 20 An Algebraic Introduction

20.1 Preliminary Definitions

Let C be an Abelian group. We will consider three kinds of structures in C.

Definition 1. A *differential* in C is a homomorphism $d: C \to C$ such that $d^2 = 0$. An Abelian group with a differential is called a *differential group*. We will use notations $Z = \operatorname{Ker} d$, $B = \operatorname{Im} d$, $H = Z/B$ (the condition $d^2 = 0$ is equivalent to the inclusion $B \subset Z$). The group H [also denoted as $H(C, d)$] is called the homology group of the differential group (C, d). Sometimes we will call elements of the groups Z, B, and H *cycles, boundaries*, and *homology classes*. A homomorphism $f: (C, d) \to (C', d')$ between differential groups is a homomorphism $f: C \to C'$ commuting with the differentials: $d' \circ f = f \circ d$. Such a homomorphism induces homomorphisms $Z \to Z'$, $B \to B'$, and $H \to H'$ [where Z', B', and H' mean for (C', d') the same as Z, B, and H mean for (C, d)]. The homomorphism $H \to H'$ may be denoted by f_*.

Definition 2. A *filtration* of C is a family of subgroups $F_pC \subset C$, $p \in \mathbb{Z}$ such that if $p < q$, then $F_pC \subset F_qC$. We will also assume that $\bigcup F_pC = C$ and $\bigcap F_pC = 0$. A filtration $\{F_pC\}$ of C is called *finite* if for some m and $n \geq m$, $F_pC = 0$ when $p < m$ and $F_pC = C$ when $p \geq n$. A filtration $\{F_pC\}$ is called *positive* if $F_pC = 0$ for all $p < 0$. Usually, we will assume that the filtration is positive and finite, in which case it is essentially a chain

$$0 = F_{-1}C \subset F_0C \subset F_1C \subset \ldots \subset F_{n-1}C \subset F_nC = C$$

(but even in this case we have the right to use the notation F_pC for $p < -1$ when it is 0, and for $p > n$ when it is C).

A. Fomenko, D. Fuchs, *Homotopical Topology*, Graduate Texts in Mathematics 273,
DOI 10.1007/978-3-319-23488-5_3

Definition 3. A *grading* of C is a family of subgroups $C_r \subset C$, $r \in \mathbb{Z}$ such that $C = \bigoplus_r C_r$. Usually (not always), we will assume that the grading is positive and finite, meaning that actually $C = \bigoplus_{r=0}^n C_r$ and $C_r = 0$ for $r < 0$ and $r > n$. For a filtered group C (as in Definition 2) we define the *adjoint graded group* $\operatorname{Gr} C = \bigoplus_r (F_r C / F_{r-1} C)$. The groups C and $\operatorname{Gr} C$ are not always isomorphic as Abelian groups (for example, the adjoint group to the group $C = \mathbb{Z}$ with a filtration $0 \subset 2\mathbb{Z} \subset \mathbb{Z}$ is $\mathbb{Z}_2 \oplus \mathbb{Z} \not\cong \mathbb{Z}$), but they may be regarded as closely related. For example, if a filtered group C is finite, then the group $\operatorname{Gr} C$ is also finite and has the same order; if C has a finite rank, then $\operatorname{Gr} C$ also has a finite rank, the same as C, if the filtration is finite; if C is a vector space (over some field), filtered by subspaces, then $\operatorname{Gr} C$ is also a vector space over the same field, of the same dimension as C, if C is finite dimensional.

Structures of these kinds may co-exist in the same Abelian group; then we usually assume that they satisfy some compatibility conditions; actually, we will not even explicitly state that these conditions are met; rather, we will state the opposite in the rare cases when the structures considered are not supposed to be compatible.

If an Abelian group C possesses a differential d and a filtration $\{F_p C\}$, then we assume that for all p, $d(F_p C) \subset F_p C$. In this case, we have differential groups $(F_p C, d\,|_{F_p C})$, and the inclusion map $F_p C \to C$ induces a homology homomorphism $H(F_p C, d\,|_{F_p C}) \to H = H(C, d)$, and its image is denoted as $F_p H$. [Thus, $F_p H = (F_p C \cap \operatorname{Ker} d)/(F_p C \cap d(C))$.] In this way, we obtain a filtration $\{F_p H\}$ of H.

If C has a differential d and a grading $C = \bigoplus_{r \in \mathbb{Z}} C_r$, then we usually assume that d is homogeneous of some degree $u \in \mathbb{Z}$, which means that for all r, $d(C_r) \subset C_{r+u}$. We are best familiar with the case $u = -1$. Then C is the same as a (chain) complex

$$\dots \xleftarrow{d_{r-1}} C_{r-1} \xleftarrow{d_r} C_r \xleftarrow{d_{r+1}} C_{r+1} \xleftarrow{d_{r+2}} \dots$$

in the sense of Sect. 12.2. The case $u = 1$ is represented by cochain complexes (Sect. 15.1). Ahead, we will deal with differential graded groups with differentials of all possible degrees. Notice that the homology group of a differential graded group with homogeneous differential (of some degree u) has a natural grading:

$$H_r = \frac{\operatorname{Ker}(d\colon C_r \to C_{r+u})}{\operatorname{Im}(d\colon C_{r-u} \to C_r)}.$$

EXERCISE 1. Prove that if the differential in C is homogeneous with respect to the grading, then (whatever the degree of d is) $H = \bigoplus_r H_r$.

If C has a filtration, $\{F_p C\}$, and a grading, $C = \bigoplus_r C_r$, then the two structures are called compatible if for every p, $F_p C = \bigoplus_r (F_p C \cap C_r)$. This condition is quite restrictive. It is stronger than it may seem at the first glance: A randomly chosen filtration and grading do not satisfy it. Here is the simplest (?) example. Let C be a free Abelian group with two generators: a and b (so $C = \mathbb{Z}a \oplus \mathbb{Z}b$). Consider the filtration $0 = F_{-1}C \subset F_0 C \subset F_1 C = C$ with $F_0 C = \mathbb{Z}(a+b)$ and the grading $C = C_0 \oplus C_1$ with $C_0 = \mathbb{Z}a$, $C_1 = \mathbb{Z}b$. Then

$$\mathbb{Z} \cong F_0 C \neq (F_0 C \cap C_0) \oplus (F_) C \cap C_1) = 0 \oplus 0 = 0.$$

For a better understanding, we can notice that the filtration and the grading are compatible if every F_pC is generated by "homogeneous elements," that is, by elements belonging to the groups C_r.

EXERCISE 2. Prove that if the differential, the filtration, and the grading are mutually compatible, then (whatever the degree of the differential is) the filtration and the grading in the homology are compatible.

And the last common situation is when C has two (or more, but we will never encounter this case) gradings, $C = \bigoplus_r C_r$ and $C = \bigoplus_s C'_s$. These two gradings are compatible (or form a *bigrading*) if $C = \bigoplus_{r,s} C_{r,s}$, where $C_{r,s} = C_r \cap C'_s$ (or, equivalently, if $C_r = \bigoplus_s C_{r,s}$; or, equivalently, if $C'_s = \bigoplus_r C_{r,s}$). A typical example is contained in Exercise 3.

EXERCISE 3. Prove that if C possesses a compatible filtration $\{F_p\}$ and grading $C = \bigoplus_s C_s$, then $\operatorname{Gr} C$ acquires a natural bigrading: $\operatorname{Gr} C = \bigoplus_{p,r}(F_pC \cap C_r)/(F_{p-1}C \cap C_r)$.

20.2 The Spectral Sequence of a Filtered Differential Group

Let C be a differential group with a differential d and a filtration $\{F_pC\}$ compatible with d. We will assume that the filtration is finite and positive, and we will briefly consider the case of the infinite filtration at the end of the section. In the next section, we will adjust our construction to the case when C also possesses a grading.

Begin with a simple observation. Since $d(F_pC) \subset F_pC$, the differential d induces a differential $d_p^0: F_pC/F_{p-1}C \to F_pC/F_{p-1}C$ [obviously, $(d_p^0)^2 = 0$] and the direct sum of all d_p^0 becomes a homogeneous differential of degree 0, $d^0: \operatorname{Gr} C \to \operatorname{Gr} C$. Question: Are $H(\operatorname{Gr} C, d^0)$ and $\operatorname{Gr} H(C, d)$ the same? Answer: not, in general. Indeed, when we compute $H(C)$, we first restrict ourselves to $\operatorname{Ker} d = \{c \in C \mid dc = 0\}$. But when we compute $H(\operatorname{Gr} C)$, we take those $c \in F_pC$ for which $dc \in F_{p-1}C$; that is, the group of "cycles" is bigger in the second computation. On the other hand, when we compute $H(C)$, we factorize over $d(C)$, while in the computation of $H(\operatorname{Gr} C)$ we factorize over $d(F_pC)$, which is not as big. This shows that the group $H(\operatorname{Gr} C)$ should be bigger than $\operatorname{Gr} H(C)$.

This is what the spectral sequence exists for: a gradual, "monotonic" transition from $H(\operatorname{Gr} C)$ to $\operatorname{Gr} H(C)$.

Now we pass to main definitions. For $p, r \geq 0$, put

$$E_p^r = \frac{F_pC \cap d^{-1}(F_{p-r}C)}{[F_{p-1}C \cap d^{-1}(F_{p-r}C)] + [F_pC \cap d(F_{p+r-1}C)]}, \quad E^r = \bigoplus_p E_p^r.$$

In words: We take elements of F_pC whose differentials lie in a smaller group, $F_{p-r}C$; then we factorize over those chosen elements which happen to be in $F_{p-1}C$ and also for those which are differentials, not of arbitrary elements of C, but only of elements of $F_{p+r-1}C$. Consider three particular cases.

$$E_p^0 = \frac{F_pC \cap d^{-1}(F_pC)}{[F_{p-1}C \cap d^{-1}(F_pC)] + [F_pC \cap d(F_{p-1}C)]}$$
$$= \frac{F_pC}{F_{p-1}C + d(F_{p-1}C)} = \frac{F_pC}{F_{p-1}C};$$

thus, $E^0 = \mathrm{Gr}\, C$;

$$E_p^1 = \frac{F_pC \cap d^{-1}(F_{p-1}C)}{[F_{p-1}C \cap d^{-1}(F_{p-1}C)] + [F_pC \cap d(F_pC)]}$$
$$= \frac{F_pC \cap d^{-1}(F_{p-1}C)}{F_{p-1}C + d(F_pC)} = \frac{\mathrm{Ker}\, d_p^0}{\mathrm{Im}\, d_p^0};$$

thus, $E^1 = H(\mathrm{Gr}\, C, d^0)$. It is also clear that if r is big enough, then $F_{p-r}C = 0$ and $F_{p+r-1}C = C$. In this case, E_p^r does not depend on r, and we will use the notations E_p^∞, E^∞. We have

$$E_p^\infty = \frac{F_pC \cap \mathrm{Ker}\, d}{[F_{p-1}C \cap \mathrm{Ker}\, d] + [F_pC \cap \mathrm{Im}\, d]} = \frac{F_pH(C)}{F_{p-1}H(C)}, \quad E^\infty = \mathrm{Gr}\, H(C).$$

To understand the relations between the (graded) groups $E^r, r = 0, 1, \ldots, \infty$, we introduce "differentials" $d_p^r : E_p^r \to E_{p-r}^r$.

Let $\alpha \in E_p^r$, and let $a \in F_pC \cap d^{-1}(F_{p-r}C)$ be a representative of α. Then $da \in F_{p-r}C$ and $dda = 0$, and hence $da \in d^{-1}(0) \subset d^{-1}(F_{p-2r}C)$. Thus, $da \in F_{p-r}C \cap d^{-1}(F_{p-2r}C)$. But E_{p-r}^r is the quotient of $F_{p-r}C \cap d^{-1}(F_{p-2r}C)$ over some subgroup; the class of da in E_{p-r}^r is taken for $d_p^r\alpha$. We need to check the following properties of this construction.

Proposition (J. Leray).

(1) *d_p^r is well defined; that is, $d_p^r\alpha$ does not depend on the choice of a in α.*
(2) $d_{p-r}^r \circ d_p^r = 0$.
(3) $\mathrm{Ker}\, d_p^r / \mathrm{Im}\, d_{p+r}^r \cong E_p^{r+1}$. (The proof will contain a construction of a canonical isomorphism.)

Proof. (1) Let a' be a different representative of α in $F_pC \cap d^{-1}(F_{p-r})$, that is, $a' = a + b + c$, where $b \in F_{p-1}C \cap d^{-1}(F_{p-r}C)$ and $c \in F_pC \cap d(F_{p+r-1}C)$. Then $da' = da + db + dc$. But $dc = 0$ and $db \in F_{p-r}C \cap d(F_{p-1}C) = F_{p-r}C \cap d(F_{(p-r)+r-1}C)$, which is a part of the denominator in the definition of E_{p-r}^r. This shows that da' belongs to the same class in E_{p-r}^r as da.

(2) A representative of $d_{p-r}^r \circ d_p^r\alpha$ in E_{p-2r}^r is dda, where a is a representative of α. But $dda = 0$.

Our proof of part (3) consists of three steps.

Step 1. We prove that if for an $\alpha \in E_p^r$, $d_p^r\alpha = 0$, then there exists a representative $a \in F_pC \cap d^{-1}(F_{p-r}C)$ of α such that $da \in F_{p-r-1}C$; that is, a belongs actually to $F_pC \cap d^{-1}(F_{p-r-1}C)$. We begin with an arbitrary representative $a' \in F_pC \cap d^{-1}(F_{p-r}C)$ of α. The condition $d_p^r\alpha = 0$ means precisely that da' belongs to the "denominator" of E_{p-r}^r, that is, to $[F_{p-r-1}C \cap d^{-1}(F_{p-2r}C)] + [F_{p-r}C \cap d(F_{p-1}C)]$; in other words, $da' = b + dc$, where $c \in F_{p-1}C, dc \in F_{p-r}C, b \in F_{p-r-1}C$, and $db \in F_{p-2r}C$ (we will not need the last inclusion). Put $a = a' - c$. Since $c \in F_{p-1}C \cap d^{-1}(F_{p-r}C)$, a is just another representative of α. And $da = da' - dc = b \in F_{p-r-1}C$, as required.

Step 2. For an $\alpha \in \operatorname{Ker} d_p^r$, choose a representative $a \in F_pC$ with $da \in F_{p-r-1}C$, as in step 1. Then $a \in F_pC \cap d^{-1}(F_{p-r-1}C)$; that is, a represents a certain element $\beta \in E_p^{r+1}$. Let us prove that this β is determined by α, just providing a well-defined homomorphism $\operatorname{Ker} d_p^r \to E_p^{r+1}$. Let $\widetilde{a} \in F_pC$ be another representative of α with $d\widetilde{a} \in F_{p-r-1}C$. We want to prove that $\widetilde{a}$ represents the same element of E_p^{r+1} as a, that is, $\widetilde{a} - a \in [F_{p-1}C \cap d^{-1}(F_{p-r-1}C)] + [F_pC \cap d(F_{p+r}C)]$. But we already know that $\widetilde{a} - a \in [F_{p-1}C \cap d^{-1}(F_{p-r}C)] + [F_pC \cap d(F_{p+r-1}C)]$, and also $\widetilde{a} - a \in d^{-1}(F_{p-r-1}C)$ and $F_{p+r-1}C \subset F_{p+r}C$. This gives us the required inclusion.

Step 3. It remains to check that $\operatorname{Ker}[\operatorname{Ker} d_p^r \to E_p^{r+1}] = \operatorname{Im} d_{p+r}^r$. If $\alpha = d_{p+r}^r\beta$, then α is represented by $a = db$, where $b \in F_{p+r}C$ represents β. But $db \in d(F_{p+r}C)$ and $d(F_{p+r}C)$ is a part of the denominator of E_p^{r+1}. Thus, a represents zero in E_p^{r+1}; that is, $\alpha \in \operatorname{Ker}[\operatorname{Ker} d_p^r \to E_p^{r+1}]$—that is, $\operatorname{Im} d_{p+r}^r \subset \operatorname{Ker}[\operatorname{Ker} d_p^r \to E_p^{r+1}]$. To prove the opposite inclusion, take an $\alpha \in \operatorname{Ker}[\operatorname{Ker} d_p^r \to E_p^{r+1}]$. Then α is represented by some $a \in [F_{p-1}C \cap d^{-1}(F_{p-r-1}C)] + [F_pC \cap d(F_{p+r}C)]$, that is, $a = b + dc$, where $b \in F_{p-1}C \cap d^{-1}(F_{p-r-1}C)$, $c \in F_{p+r}C \cap d^{-1}(F_pC)$. This c represents some element γ of E_{p+r}^r, and $d_{p+r}^r\gamma$ is represented by $dc = a - b$. Since $b \in F_{p-1}C \cap d^{-1}(F_{p-r-1}C) \subset F_{p-1}C \cap d^{-1}(F_{p-r}C)$, b represents 0 in E_p^r, and dc represents the same element of E_p^r as a. Thus, $d_{p+r}^r\gamma = \alpha$ and $\operatorname{Ker}[\operatorname{Ker} d_p^r \to E_p^{r+1}] \subset \operatorname{Im} d_{p+r}^r$.

This completes the proof of Leray's proposition.

EXERCISE 4. Prove that the notations d^0, d_p^0 match the notations introduced in the "simple observation" in the beginning of the section.

EXERCISE 5 (concerning Step 1 of the previous proof). Prove that if $\alpha \in E_p^1$ and $d_p^1\alpha = 0$, then the inclusion $da \in F_{p-2}C$ holds for an arbitrary, not specially chosen, representative a of α in $F_pC \cap d^{-1}(F_{p-1}C)$.

We have completed the main construction of this chapter: that of a *spectral sequence*. This construction will be enriched in subsequent sections and lectures, but now let us observe what we have already achieved.

Input: a differential group (C, d) with a positive finite filtration $\{F_pC\}$ compatible with the differential.

Output: a sequence of differential graded groups

$$\left\{E^r = \bigoplus\nolimits_p E_p^r, d^r = \sum\nolimits_p [d_p^r \colon E_p^r \to E_{p-r}^r]\right\},$$

which is called the *spectral sequence* associated with the filtered differential groups of the Input and which possesses the following properties:

(1) $E^0 = \mathrm{Gr}(C,d)$, $E^1 = H(\mathrm{Gr}(C,d))$.
(2) For every r, the graded group E^{r+1} is the homology group of the differential graded group E^r, d^r (with a homogeneous differential of degree $-r$).
(3) For big r, the groups E^r do not depend on r (that is, $d^r = 0$), which justifies the notation E^∞; the claim is that $E^\infty = \mathrm{Gr}\,H(C,d)$.

Informally speaking, the spectral sequence begins "almost at C," that is, at $\mathrm{Gr}\,C$, and "converges" "almost to $H(C)$," that is, to $\mathrm{Gr}\,H(C)$, in this way breaking the computation of the homology of C into a sequence of elementary steps.

The construction of a spectral sequence is natural in the obvious sense: If there is another differential group, C', d', with a filtration compatible with a differential, then there arises a spectral sequence $\{'E^r, 'd^r\}$ associated with this group, and any homomorphism $C \to C'$ compatible with differentials and filtrations induces homomorphisms $E^r \to {}'E^r, 0 \le r \le \infty$ compatible with differentials and gradings and coinciding with the induced homomorphisms $\mathrm{Gr}\,H(C,d) \to \mathrm{Gr}\,H(C',d')$ for $r = \infty$.

We conclude the section with two remarks, one methodical and one historical.

Remark 1. The construction of the spectral sequence can be easily adapted to the case when the filtration is infinite although still positive. The definitions and stated properties of E_p^r and d_p^r (with finite r) remain the same. The difference is that the sequence fails to stabilize. However, for every p, $d_p^r = 0$ for $r > p$. Thus, in the sequence of groups E_p^r with p fixed and $r > p$, every group is a pure quotient of the previous group, which gives us the right to speak of the "limit group" E_p^∞. The graded group E^∞ is still $\mathrm{Gr}\,H(C)$.

Remark 2. Spectral sequences were first introduced in 1945 by J. Leray in the context of the sheaf theory (see Leray [53]). The significance of spectral sequences for algebraic topology was demonstrated in 1951 by J.-P. Serre in his doctoral dissertation [75]. With the appearance of homological algebra (the term was used as the title of the famous book by Cartan and Eilenberg [29]), spectral sequences became the main technical tool in this area. Modern mathematics contains dozens of spectral sequences named after remarkable mathematicians. We will have to restrict ourselves in this book to topological applications of spectral sequences.

20.3 The Case of a Graded Filtered Differential Group

Suppose that besides filtration and differential, the group C has a grading, $C = \bigoplus_m C_m$, compatible with filtration $F_pC = \bigoplus_p F_pC_m$ where $F_pC_m = F_pC \cap C_m$, and the differential; we suppose that the differential has the degree -1: $d(C_m) \subset C_{m-1}$;

we suppose also that the grading is finite (there are only finitely many nonzero C_m). In this case, all terms of the spectral sequence acquire an additional structure: the second grading. There will be no additional proposition to prove; all we need is a modification of all notations.

For $p, r \geq 0$, $q \in \mathbb{Z}$, put

$$E^r_{pq} = \frac{F_pC_{p+q} \cap d^{-1}(F_{p-r}C_{p+q-1})}{[F_{p-1}C_{p+q} \cap d^{-1}(F_{p-r}C_{p+q-1})] + [F_pC_{p+q} \cap d(F_{p+r-1}C_{p+q+1})]},$$

$E^r = \bigoplus_{p,q} E^r_{pq}$. In particular,

$$\begin{aligned} E^0_{pq} &= \frac{F_pC_{p+q}}{F_{p-1}C_{p+q}}, \\ E^1_{pq} &= \frac{\operatorname{Ker}\left(\dfrac{F_pC_{p+q}}{F_{p-1}C_{p+q}} \xrightarrow{d} \dfrac{F_pC_{p+q-1}}{F_{p-1}C_{p+q-1}}\right)}{\operatorname{Im}\left(\dfrac{F_pC_{p+q+1}}{F_{p-1}C_{p+q+1}} \xrightarrow{d} \dfrac{F_pC_{p+q}}{F_{p-1}C_{p+q}}\right)}, \\ E^\infty_{pq} &= \frac{F_pH_{p+q}(C)}{F_{p-1}H_{p+q}(C)}. \end{aligned}$$

(The reader may wonder why instead of C_{p+q}, C_{p+q-1}, etc. we do not take C_q, C_{q-1}, etc. However, it is impossible to explain the reason for this before we consider examples. We will see that for the most important examples, E^r_{pq} will be zero for $p < 0$ and for $q < 0$. At the moment, we can only say that for E^r_{pq}, p is called the *filtration degree*, q is called the *complementary degree*, and $p + q$ is called the *full degree*.)

The differential d^r is defined in the same way as before (without an additional grading): An $\alpha \in E^r_{pq} \subset E^r$ has a representative in $a \in F_pC_{p+q}$ with $da \in F_{p-r}C_{p+q-1}$, and the class of this da is taken for $d^r\alpha$. This differential reduces the filtration degree by r and reduces the full degree by 1; hence, the component d^r_{pq} of this differential on E^r_{pq} is

$$d^r_{pq}: E^r_{pq} \to E^r_{p-r,q+r-1}.$$

Part (3) of Leray's proposition of Sect. 20.2 holds with an obvious enriching of notations:

$$E^{r+1}_{pq} = \frac{\operatorname{Ker}[d^r_{pq}: E^r_{pq} \to E^r_{p-r,q+r-1}]}{\operatorname{Im}[d^r_{p+r,q-r+1}: E^r_{p+r,q-r+1} \to E^r_{pq}]}.$$

There is a common graphic presentation of terms of the spectral sequence. For every r, the groups $E^r_{p,q}$ are placed into the cell of a graph paper and the differentials d^r_{pq} are shown by arrows. Fragments of such diagrams (around a randomly chosen

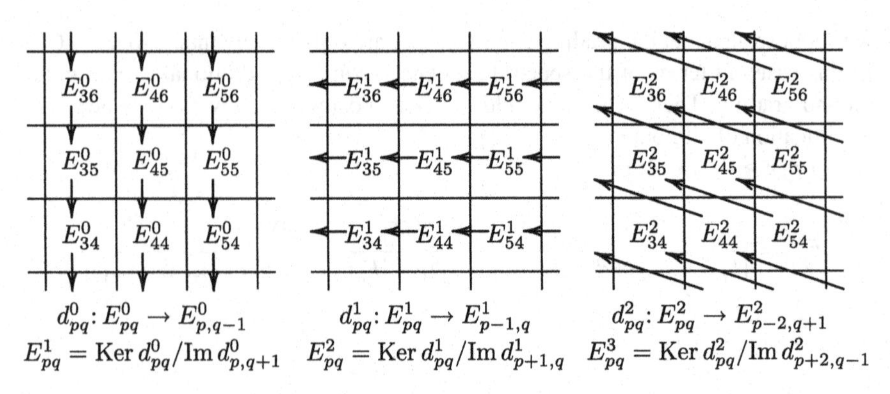

Fig. 75 Diagrams for the terms E^0, E^1, E^2 of a spectral sequence

cell with $p = 4, q = 5$) are shown in Fig. 75. Differentials d^0 act downward, differentials d^1 act from the right to the left, and differentials d^2 act by what has been cherished by generations of topologists—the "knight's move." Further differentials act by a "mad knight's move": The arrow corresponding to the differential d^r points r cells left and $r - 1$ cells up.

The relations between the term E^∞ and $H(C)$ are shown in Fig. 76. On the graph paper diagram for E^∞, consider the line $p + q = n$ and mark on this line all nonzero groups as $J_1, J_2, \ldots, J_m$, numbered in the order of increasing p and decreasing q. Then

$$J_1 \subset H_n(C),\ J_2 \subset H_n(C)/J_1,\ J_3 \subset (H_n(C)/J_1)/J_2, \ldots,$$
$$J_m = (\ldots(H_n(C)/J_1)/J_2 \ldots)/J_{m-1}.$$

In particular, if $E^\infty_{pq} = 0$ for all pairs p, q with $p + q = n$, then $H_n(C) = 0$; if, for some n, there is only one pair p, q, $p + q = n$ with $E^\infty_{pq} \neq 0$, then $H_n(C)$ equals this E^∞_{pq}.

20.4 A Cohomological Version

No wonder that, both in topology and in algebra, spectral sequences are more often applied to cohomology than to homology: Cohomological spectral sequences possess natural multiplicative structures (Lecture 24) as well as many other structures, which we will study later (Chap. 4). From a purely algebraic point of view, the main difference between chain and cochain complexes lies in the degree of the differential: It is $+1$ rather than -1, and this does not affect the general theory in any significant way. However, precisely how in general homology theory, the transition from homology to cohomology results in replacing the notations C_n, H_n, etc. by C^n, H^n, etc., the cohomological version of the previous theory requires some changes in notation. Let us observe these changes.

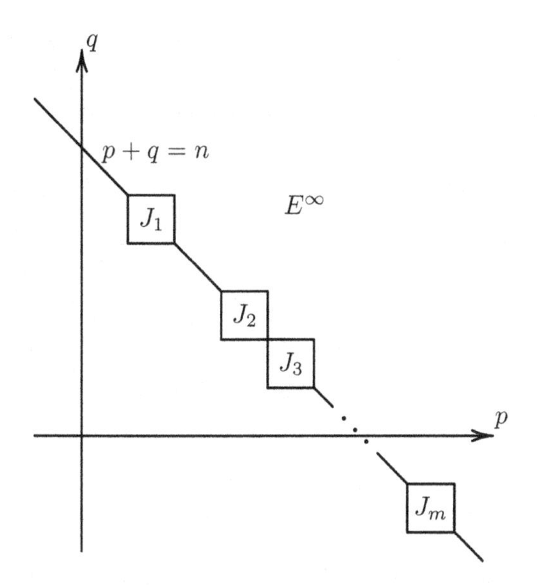

Fig. 76 Diagrams for the E^{∞}-term

For a filtration in an Abelian group C we use the notation $\{F^pC\}$; we assume that $F^pC \supset F^{p+1}C$. A filtration is called positive if $F^pC = C$ for $p < 0$ and is called finite if there are only finitely many terms F^pC different from both 0 and C. Thus, a finite positive filtration has the form

$$C = F^0C \supset F^1C \supset F^2C \supset \ldots \supset F^nC \supset F^{n+1}C = 0,$$

with the understanding that $F_pC = C$ if $p < 0$ and $F^pC = 0$ if $p > n$. We assume also that C possesses a grading $C = \bigoplus C^r$ (usually, with finitely many nonzero C^r) and a differential $d: C \to C$ such that

$$F^pC = \bigoplus_r F^pC^r \text{ where } F^pC^r = F^pC \cap C^r,\ d(F^pC) \subset F^pC, \\ \text{and } d(C^r) \subset C^{r+1}.$$

Then we put

$$E_r^{pq} = \frac{F^pC^{p+q} \cap d^{-1}(F^{p+r}C^{p+q+1})}{[F^{p+1}C^{p+q} \cap d^{-1}(F^{p+r}C^{p+q+1})] + [F^pC^{p+q} \cap d(F^{p-r+1}C^{p+q-1})]},$$

$E_r = \bigoplus_{p,q} E_r^{pq}$, and define a differential

$$d_r^{pq}: E_r^{pq} \to E_r^{p+r,q-r+1}$$

in the usual way: For $\alpha \in E_r^{pq}$, we choose a representative $a \in F^pC^{p+q} \cap d^{-1}(F^{p+r}C^{p+q+1})$ and take for $d_r^{pq}\alpha$ the class of $da \in F^{p+r}C^{p+q+1}$ in $E_r^{p+r,q-r-1}$.

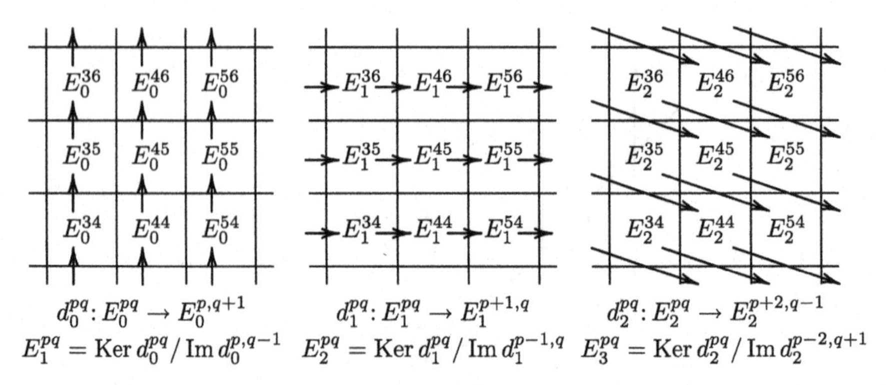

Fig. 77 Diagrams for the terms E^0, E^1, E^2 in the cohomological case

Leray's proposition (and its proof), including the graded version of Sect. 20.3, can be repeated with an appropriate notation change. In particular, $E_{r+1}^{pq} = \operatorname{Ker} d_r^{pq} / \operatorname{Im} d_r^{p-1,q+r-1}$. The diagrams of Fig. 75 assume the form shown in Fig. 77.

The relations between E_∞ and $H(C)$ are still described by Fig. 76. The difference is that now we need to order the groups in the line $p+q=n$ from the bottom to the top:

$$\begin{aligned} J_m \subset H^n(C), \dots, J_2 &\subset (\dots (H^n(C)/J_m) \dots)/J_3, \\ J_1 &= ((\dots (H^n(C)/J_m) \dots)/J_3)/J_2. \end{aligned}$$

20.5 Some Famous Spectral Sequences in Algebra (Exercises)

In this section we will consider some very mighty applications of spectral sequences in algebraic topology. However, many famous spectral sequences arise in other parts of mathematics. In this section, we will describe several of them (the reader who wants to become familiar with more should use any comprehensive text in homological algebra). But since this is not our subject, we will not prove much, so this section will look like a sequence of exercises.

A: Double Complexes and Their Spectral Sequences

This is a general construction which provides a very rich source of more specific spectral sequences. We will use here what can be called "homological notation." The reader can reverse all the arrows, switch the upper and lower indices, and obtain a cohomological version of the construction.

Let $C = \bigoplus_{p\geq 0, q\geq 0} C_{pq}$ be a doubly graded group endowed with two differentials: $d_{pq}^{\mathrm{I}}\colon C_{pq} \to C_{p-1,q}$ and $d_{pq}^{\mathrm{II}}\colon C_{pq} \to C_{p,q-1}$. In addition to the expectable assumptions that $d^{\mathrm{I}}_{p-1,q} \circ d^{\mathrm{I}}_{pq} = 0$ and $d^{\mathrm{II}}_{p,q-1} \circ d^{\mathrm{II}}_{pq} = 0$, we assume that $d^{\mathrm{II}}_{p-1,q} \circ d^{\mathrm{I}}_{pq} + d^{\mathrm{I}}_{p,q-1} \circ d^{\mathrm{II}}_{pq} = 0$. A doubly graded group C with differentials $d^{\mathrm{I}}, d^{\mathrm{II}}$ satisfying the conditions above is called a double complex.

For a double complex as above, we put $\mathbf{C}_n = \bigoplus_{p+q=n} C_{pq}$. The differentials d^{I}_{pq} and d^{II}_{pq} with $p+q=n$ sum up to a homomorphism $\mathbf{d}_n\colon \mathbf{C}_n \to \mathbf{C}_{n-1}$. The conditions formulated above mean precisely that $\mathbf{d}_{n-1} \circ \mathbf{d}_n = 0$. That is, $\mathbf{C} = \bigoplus \mathbf{C}_n$ becomes a graded differential group with the differential of degree -1; it is called the *total complex* of the double complex $C = \{C_{pq}; d^{\mathrm{I}}_{pq}, d^{\mathrm{II}}_{pq}\}$.

There are several homology theories related to a double complex. First, there are homologies

$$H^{\mathrm{I}}_{pq}(C) = \operatorname{Ker} d^{\mathrm{I}}_{pq} / \operatorname{Im} d^{\mathrm{I}}_{p+1,q} \text{ and } H^{\mathrm{II}}_{pq}(C) = \operatorname{Ker} d^{\mathrm{II}}_{pq} / \operatorname{Im} d^{\mathrm{I}}_{p,q+1}.$$

Next, since the differentials d^{I} and d^{II} commute (up to a sign), they give rise to "differentials" $d^{\mathrm{II}}_{pq}\colon H^{\mathrm{I}}_{pq}(C) \to H^{\mathrm{I}}_{p,q-1}(C)$ and $d^{\mathrm{I}}_{pq}(C)\colon H^{\mathrm{II}}_{pq}(C) \to H^{\mathrm{II}}_{p-1,q}(C)$, which gives rise to "double homologies"

$$H^{\mathrm{II}} H^{\mathrm{I}}_{pq}(C) \text{ and } H^{\mathrm{I}} H^{\mathrm{II}}_{pq}(C)$$

(the reader should not think that they are the same). Also, there is the homology of the total complex: $\mathbf{H}_n = \operatorname{Ker} \mathbf{d}_n / \operatorname{Im} \mathbf{d}_{n+1}$.

For the total complex $\mathbf{C}$, introduce two filtrations (sometimes they are called *stupid*, in contrast to the fact that they form a foundation for numerous important applications):

$$F^{\mathrm{I}}_p \mathbf{C} = \bigoplus_{s \le p} C_{sq}, \; F^{\mathrm{II}}_q \mathbf{C} = \bigoplus_{s \le q} C_{ps}.$$

There arise two spectral sequences, $\{{}^{\mathrm{I}}E^r_{pq}, {}^{\mathrm{I}}d^r_{pq}\}$ and $\{{}^{\mathrm{II}}E^r_{pq}, {}^{\mathrm{II}}d^r_{pq}\}$, whose properties are described in Exercise 6. We should warn the reader that in the second of these spectral sequences the roles p and q are swapped: q is the filtering degree and p is the complementary degree; this leads the differentials to act unusually: ${}^{\mathrm{II}}d^r_{pq}\colon {}^{\mathrm{II}}E^r_{pq} \to {}^{\mathrm{II}}E^r_{p+r-1,q-r}$.

EXERCISE 6. Prove that

(1) ${}^{\mathrm{I}}E^0_{pq} = {}^{\mathrm{II}}E^0_{pq} = C_{pq}$.
(2) ${}^{\mathrm{I}}E^1_{pq} = H^{\mathrm{II}}_{pq}(C)$, ${}^{\mathrm{II}}E^1_{pq} = H^{\mathrm{I}}_{pq}(C)$.
(3) ${}^{\mathrm{I}}E^2_{pq} = H^{\mathrm{I}} H^{\mathrm{II}}_{pq}(C)$, ${}^{\mathrm{II}}E^2_{pq} = H^{\mathrm{II}} H^{\mathrm{I}}_{pq}(C)$.
(4) ${}^{\mathrm{I}}E^\infty_{pq} = \dfrac{F^{\mathrm{I}}_p \mathbf{H}_{p+q}}{F^{\mathrm{I}}_{p-1} \mathbf{H}_{p+q}}$, ${}^{\mathrm{II}}E^\infty_{pq} = \dfrac{F^{\mathrm{II}}_q \mathbf{H}_{p+q}}{F^{\mathrm{II}}_{q-1} \mathbf{H}_{p+q}}$.

B: Application of Spectral Sequences of Double Complexes: Coefficient Spectral Sequences

The majority of applications of double complex spectral sequences use the following scheme. To solve some homology problem, we make up a double complex. There arise two spectral sequences. Usually, one of these sequences turns out to be severely degenerate; for example, its E^1-term has only one nontrivial row or column. Then all the higher differentials are zeroes, $E^\infty = E^1$, and the homology of the total complex is equal to the E^1-term, which gives us an expression for it. After it, we know the limit term of the other spectral sequence, and we also have information on its initial term, E^1 or E^2. Thus, this spectral sequence relates something we know with something we want to find.

The spectral sequence we are going to discuss may be regarded, as we will explain later, as a generalization of the (exact) coefficient sequence. Let X be a topological space, and let

$$0 \longleftarrow G \xleftarrow{\varepsilon} G_0 \xleftarrow{\alpha_1} G_1 \xleftarrow{\alpha_2} G_2 \xleftarrow{\alpha_3} \dots$$

be an exact sequence of Abelian groups (we assume it infinite to the right, but it is possible –and desirable—that $G_n = 0$ for n big enough). Consider the double complex $C = \bigoplus_{p\geq 0, q\geq 0} C_q(X; G_p)$ (where C_q denotes a chain group, maybe singular, maybe cellular—it does not matter) with the differentials $\partial_q\colon C_q(X; G_p) \to C_{q-1}(X; G_p)$ and $(-1)^p(\alpha_p)_*\colon C_q(X; G_p) \to C_q(X; G_{p-1})$ [here we assume G_{-1} to be zero, not G, and, accordingly $(\alpha_0)_*$ to be zero, not ε; $(-1)^p$ at $(\alpha_p)_*$ is necessary to ensure that the condition relating ${}^{\mathrm{I}}d$ to ${}^{\mathrm{II}}d$ is met].

EXERCISE 7. (1) Prove that

$${}^{\mathrm{I}}E^1_{pq} = \begin{cases} H_q(X : G), \text{ if } p = 0, \\ 0, \qquad\qquad \text{if} p \neq 0. \end{cases}$$

Deduce that $\mathbf{H}_n = H_n(X; G)$.

(2) Prove that

$${}^{\mathrm{II}}E^1_{pq} = H_q(X; G_p) \text{ and } {}^{\mathrm{II}}d^1_{pq} = [(-1)^p(\alpha_p)_*\colon H_q(X; G_p) \to H_{q-1}(X; G_{p-1})].$$

Exercise 7 shows that the spectral sequence ${}^{\mathrm{II}}E$ connects homology groups with coefficients in G with homology with coefficients in G_p. In this sense, it is similar to the coefficient homology sequence. Actually, if $G_p = 0$ for all $p \geq 2$, this spectral sequence is algebraically equivalent to the coefficient homology sequence (see Sect. 15.3). We prefer to state this fact in a more precise way later (see Exercise 2 in Sect. 5.3).

Cohomological versions of all these constructions exist.

C: The Hochschild–Serre Spectral Sequence

There are several spectral sequences named after G. Hochschild and J.-P. Serre (see Hochschild and Serre [47, 48]); each exists in homological and cohomological versions. We will consider here the spectral sequence in the theory of Lie algebra cohomology.

Let $\mathfrak{g}$ be a Lie algebra over the field $\mathbb{C}$, and let M be a $\mathfrak{g}$-module (that is a complex vector space endowed by a linear map $\rho\colon \mathfrak{g} \to \operatorname{End} M$ such that $\rho[g,h] = \rho(g)\circ\rho(h) - \rho(h)\circ\rho(g)$; we will abbreviate the formula $[\rho(g)](x)$ to gx). Define the "cochain space" $C^n(\mathfrak{g};M)$ as $\operatorname{Hom}_{\mathbb{C}}(\Lambda^n\mathfrak{g}, M)$ and the differential $d\colon C^n(\mathfrak{g};M) \to C^{n+1}(\mathfrak{g};M)$ by the formula

$$\begin{aligned} dc(g_1\wedge\ldots\wedge g_{n+1}) &= \\ \sum_{1\le s<t\le n+1} &(-1)^{s+t-1}c([g_s,g_t]\wedge g_1\wedge\ldots\widehat{g_s}\ldots\widehat{g_t}\ldots\wedge g_{n+1}) \\ &- \sum_{1\le u\le n+1}(-1)^u g_u c(g_1\wedge\ldots\widehat{g_u}\ldots\wedge g_{n+1}). \end{aligned}$$

EXERCISE 8. Prove that $d^2 = 0$.

The (co)homology of the complex $\{C^n(\mathfrak{g};M), d\}$ is called the cohomology of the Lie algebra $\mathfrak{g}$ with the coefficients in M and is denoted as $H^n(\mathfrak{g};M)$. The space $C^n(\mathfrak{g};M)$ possesses a natural structure of a $\mathfrak{g}$-module:

$$(gc)(g_1\wedge\ldots\wedge g_n) = \sum_{r=1}^{n} c(g_1\wedge\ldots\wedge[g,g_r]\wedge\ldots\wedge g_n) - g(c(g_1\wedge\ldots\wedge g_n)),$$

and the differentials are $\mathfrak{g}$-homomorphisms. This implies a structure of a $\mathfrak{g}$-module in $H^n(\mathfrak{g};M)$, but

EXERCISE 9. Prove that the structure of a $\mathfrak{g}$-module in $H^n(\mathfrak{g};M)$ is trivial: $g\alpha = 0$ for any $g\in\mathfrak{g}$, $\alpha\in H^n(\mathfrak{g};M)$.

Let $\mathfrak{h}\subset\mathfrak{g}$ be a Lie subalgebra. Put

$$\begin{aligned} F^pC^{p+q}(\mathfrak{g};M) = \{c\in C^{p+q}(\mathfrak{g};M) \mid c(g_1\wedge\ldots\wedge g_{p+q}) &= 0, \\ \text{if } g_1,\ldots,g_{q+1} &\in \mathfrak{h}\}. \end{aligned}$$

EXERCISE 10. (1) Prove that the spaces $F^pC_{p+q}(\mathfrak{g};M)$ form a filtration compatible with the differential; thus, there arises a spectral sequence with the limit term $\operatorname{Gr} H^n(\mathfrak{g};M)$; this is the Hochschild–Serre spectral sequence.

(2) Prove that in this spectral sequence $E_1^{pq} = H^q(\mathfrak{h};\operatorname{Hom}(\Lambda^q(\mathfrak{g}/\mathfrak{h}), M))$ [we expect that the reader will reconstruct the structure of an $\mathfrak{h}$-module in $\operatorname{Hom}(\Lambda^q(\mathfrak{g}/\mathfrak{h}), M)$].

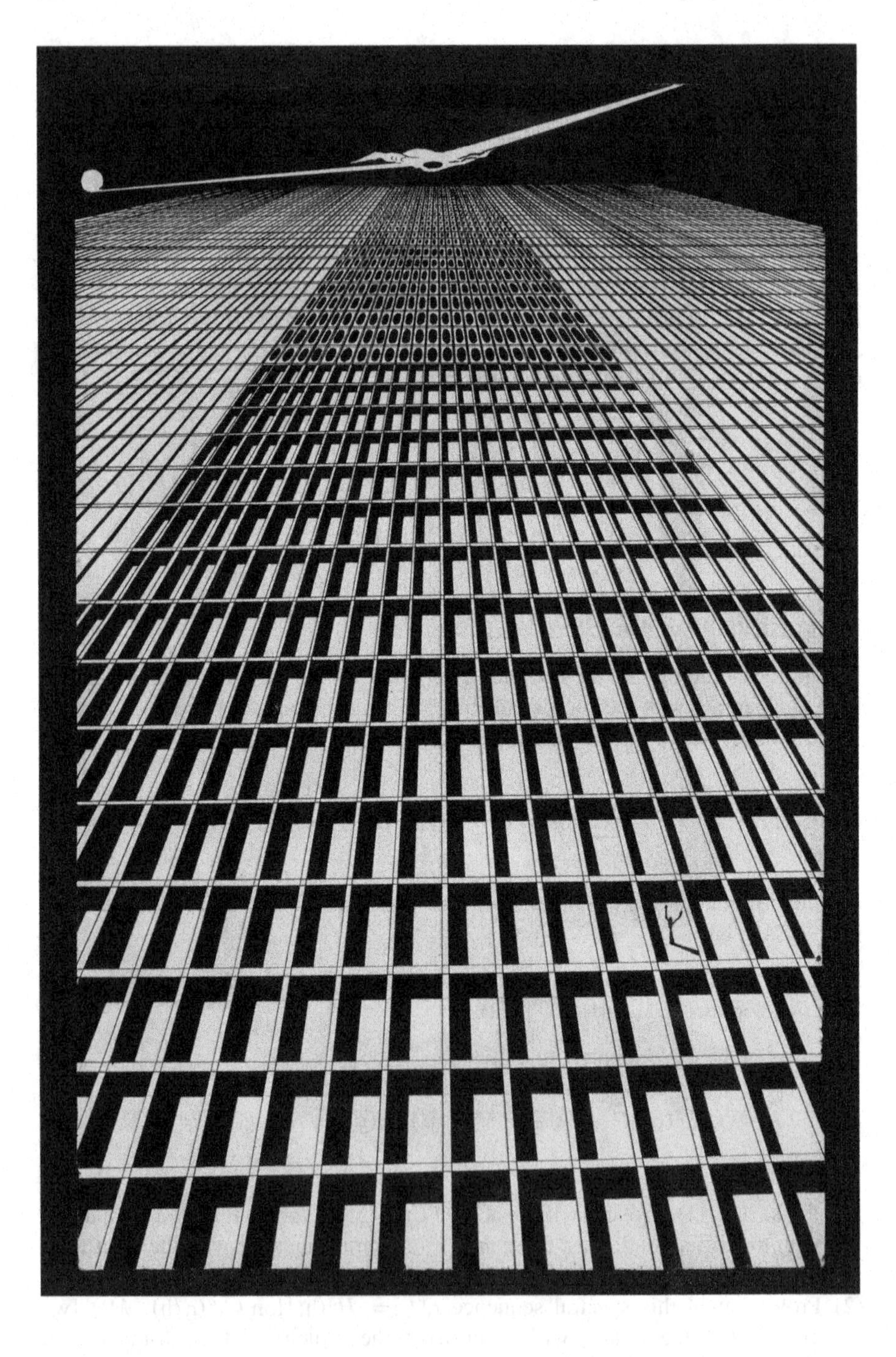

(3) Prove that if $\mathfrak{h}$ is an ideal in $\mathfrak{g}$, then $E_2^{pq} = H^p(\mathfrak{g}/\mathfrak{h}; H^q(\mathfrak{h}; M))$. [If $\mathfrak{h}$ is an ideal, then the spaces $C^n(\mathfrak{h}; M)$, $H^n(\mathfrak{h}; M)$ have a natural $\mathfrak{g}$-module structure; but, according to Exercise 9, the $\mathfrak{h}$-module structure in $H^n(\mathfrak{h}; M)$ is trivial; hence, the $\mathfrak{g}$-module structure in $H^n(\mathfrak{g}; M)$ is factorized to a $\mathfrak{g}/\mathfrak{h}$-module structure.]

Lecture 21 Spectral Sequences of a Filtered Topological Space

21.1 General Constructions

A (positive, finite) filtration in a topological space X is a chain of subspaces,

$$\emptyset = X_{-1} \subset X_0 \subset X_1 \subset \ldots \subset X_n = X$$

(when necessary, we will use the notation $X_p = \emptyset$ for $p < -1$ and $X_p = X$ for $p > n$). (The case of infinite filtration will be briefly discussed later.) The full chain group, $C = C_*(X; G) = \bigoplus_r C_r(X; G)$, has a grading (as shown), the differential $\partial\colon C_r(X; G) \to C_{r-1}(X; G)$ (of degree -1), and the filtration $F_pC = C_*(X_p; G) \subset C_*(X; G)$, and these three structures are compatible with each other as described in Sect. 20.1. Then the constructions of Sects. 20.2 and 20.3 can be applied, and they lead to a spectral sequence $\{E^r_{pq}, d^r_{pq}\colon E^r_{pq} \to E^r_{p-r,q+r-1}\}$. In our current notation, $E^r_{pq} =$

$$\frac{C_{p+q}(X_p; G) \cap \partial^{-1}(C_{p+q-1}(X_{p-r}; G))}{[C_{p+q}(X_{p-1}; G) \cap \partial^{-1}(C_{p+q-1}(X_{p-r}; G))] + [C_{p+q}(X_p; G) \cap \partial(C_{p+q+1}(X_{p+r-1}; G))]}$$

and

$$E^\infty_{pq} = \frac{\operatorname{Im}[H_{p+q}(X_p; G) \to H_{p+q}(X; G)]}{\operatorname{Im}[H_{p+q}(X_{p-1}; G) \to H_{p+q}(X; G)]}.$$

In addition, we can state that

$$\begin{aligned} E^0_{pq} &= C_{p+q}(X_p, X_{p-1}; G), \\ d^0_{pq} &= \left[\partial\colon C_{p+q}(X_p, X_{p-1}; G) \to C_{p+q-1}(X_p, X_{p-1}; G)\right], \end{aligned}$$

$$\begin{aligned} E^1_{pq} &= H_{p+q}(X_p, X_{p-1}; G), \\ d^1_{pq} &= \left[\partial_*\colon H_{p+q}(X_p, X_{p-1}; G) \to H_{p+q-1}(X_{p-1}, X_{p-2}; G)\right], \end{aligned}$$

where the homomorphism ∂_* in the second line belongs to the homology sequence of the triple (X_p, X_{p-1}, X_{p-2}).

EXERCISE 1. Check the previous statements concerning E^∞_{pq}, E^0_{pq}, and E^1_{pq}.

There is also a cohomology version of this spectral sequence. We consider $C = C^*(X;G) = \bigoplus_r C^r(X;G)$ with the differential $\delta\colon C^r(X;G) \to C^{r+1}(X;G)$ and the (decreasing) filtration $F^pC = C^*(X, X_{p-1};G) \subset C^*(X;G)$. The compatibility conditions hold, and the construction of Sect. 20.4 yields a spectral sequence $\{E_r^{pq}, d_r^{pq}\colon E_r^{pq} \to E_r^{p+r,q-r+1}\}$ with

$$E_r^{pq} = \frac{C^{p+q}(X,X_{p-1};G) \cap \delta^{-1}(C^{p+q+1}(X,X_{p+r-1};G))}{[C^{p+q}(X,X_p;G) \cap \delta^{-1}(C^{p+q+1}(X,X_{p+r-1};G))] + [C^{p+q}(X,X_{p-1};G) \cap \delta(C^{p+q-1}(X,X_{p-r};G))]},$$

$$\begin{aligned} E_0^{pq} &= C^{p+q}(X_p, X_{p-1};G),\\ d_0^{pq} &= [\delta\colon C^{p+q}(X_p,X_{p-1};G) \to C^{p+q+1}(X_p,X_{p-1};G)],\\ E_1^{pq} &= H^{p+q}(X_p,X_{p-1};G),\\ d_1^{pq} &= [\delta^*\colon H^{p+q}(X_p,X_{p-1};G) \to H^{p+q+1}(X_{p+1},X_p;G)] \end{aligned}$$

[the last δ^* belongs to the cohomology sequence of the triple (X_{p+1}, X_p, X_{p-1})],

$$E_\infty^{pq} = \frac{\operatorname{Ker}[H^{p+q}(X;G) \to H^{p+q}(X_{p-1};G)]}{\operatorname{Ker}[H^{p+q}(X;G) \to H^{p+q}(X_p;G)]}.$$

Notice that the explicit formulas for E^1 and d^1 (for E_1 and d_1) allow us, as a rule, to completely ignore the zeroth terms of spectral sequences; certainly, the higher differentials d^r and d_r are described at the chain/cochain level, but we will see that these direct descriptions are not really useful.

Notice also that if there is a different space, Y, with a (positive) filtration $\{Y_p\}$ and a continuous map $f\colon X \to Y$ such that $f(X_p) \subset Y_p$ for all p, then there arise homomorphisms between homological and cohomological spectral sequences of the filtered spaces X and Y, and these homomorphisms are compatible with the descriptions of the zeroth, first, and ∞-th terms given above.

Recall in conclusion that all the constructions of this section can be applied to a positive infinite filtration, $\emptyset = X_{-1} \subset X_0 \subset X_1 \subset X_2 \subset \dots \subset X$, on the condition that $X = \bigcup_p X_p$ and that X is furnished with a "weak topology": A set $F \subset X$ is closed in X if and only if every intersection $F \cap X_p$ is closed in X_p.

21.2 A New Understanding of the Cellular Computation of Homology and Cohomology

Here we restrict ourselves to the case of usual (integral) homology; the cases of homology and cohomology with coefficients are not significantly different.

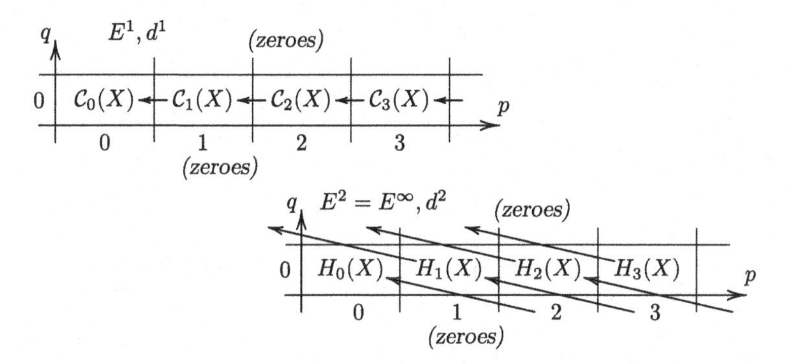

Fig. 78 The spectral sequence of a CW complex filtered by the skeletons

Let X be a CW complex; let us filter it by skeletons: $X_p = \text{sk}_p X = X^p$. In the corresponding homological spectral sequence,

$$E^1_{pq} = H_{p+q}(X^p, X^{p-1}) = \begin{cases} C_p(X), & \text{if } q = 0, \\ 0, & \text{if } q \neq 0, \end{cases}$$

so the diagrams for the terms of this spectral sequence are as shown in Fig. 78.

Thus, the term E^1 (as well as all the subsequent terms) contains only one nonzero row, E^1_{p0}; with the differential d^1_{p0}, this row is nothing but the cellular chain complex of X. Hence, in the term E^2, the same row contains the cellular homology of X. All the subsequent differentials are zero, since no one of them may connect two nonzero groups. Thus, $E^2 = E^\infty$, and since for every n there is at most one nonzero group E^∞_{pq} with $p + q = n$, that is, the group E^∞_{n0}, then (according to a remark in the end of Sect. 20.3), $H_n(X) = E^\infty_{n0} = E^2_{n0} =$ the nth cellular homology group.

We also see that the property of the skeleton filtration that $H_r(X^p,\ X^{p-1}) = 0$ for $r \neq p$ is crucial for this calculation of homology: The spectral sequence exists independently of this property, but if the property does not hold, higher differentials may appear, and the calculation becomes much less automatic.

21.3 A New Understanding of the Homology Sequence of a Pair

Let $(X.A)$ be a topological pair. It can be regarded as a "two-term filtration,"

$$(\emptyset = X_{-1}) \subset (A = X_0) \subset (X = X_1).$$

The corresponding (homological) spectral sequence has the term E^1 as shown in Fig. 79, left, with the differential $d^1 = \partial_*$. Hence, the E^2-term looks like Fig. 79, right. The differentials $d^r, r \geq 2$, are all zero, and hence $E^\infty = E^2$.

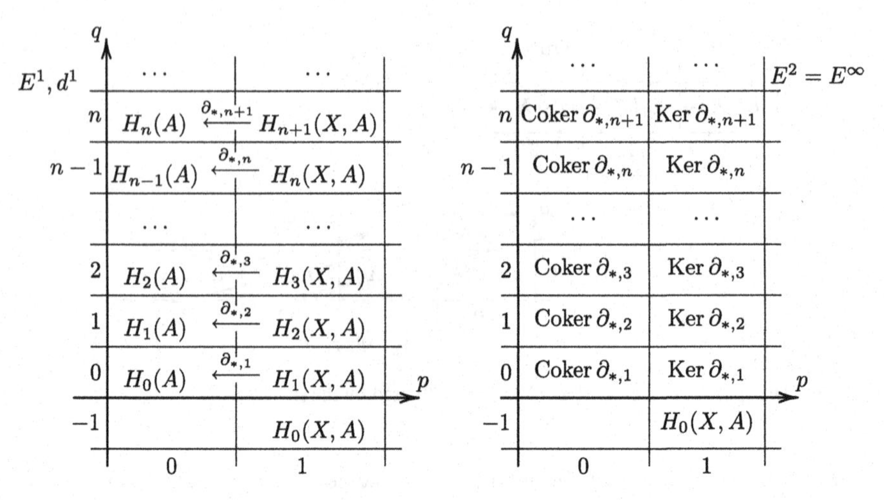

Fig. 79 The spectral sequence of a two-term filtration

According to Sect. 20.3, $H_n(X)/E^\infty_{0,n} = E^\infty_{1,n-1}$. This yields an exact sequence

$$\text{Coker}\ \partial_{*,n+1} \to H_n(X) \to \text{Ker}\ \partial_{*,n},$$

that is, an exact sequence

$$H_{n+1}(X,A) \xrightarrow{\partial_{*,n+1}} H_n(A) \longrightarrow H_n(X) \longrightarrow H_n(X,A) \xrightarrow{\partial_{*,n}} H_{n-1}(A),$$

which is the same as the homological sequence of the pair (X,A). In a similar way, homological and cohomological sequences with arbitrary coefficients can be presented as spectral sequences of two-term filtrations.

EXERCISE 2. Find an interpretation in terms of spectral sequences for the homological and cohomological sequences related to short exact sequences of coefficient groups (Example C in Sect. 20.5 may be useful).

The reader may expect that spectral sequences of three-term filtrations, $\emptyset \subset B \subset A \subset X$, must be related to homological and cohomological sequences of triples. In reality, a relation exists, but it is not as direct as in the case of pairs.

Lecture 22 Spectral Sequences of Fibrations: Definitions and Basic Properties

22.1 The Main Construction

Let $\xi = (E, B, F, p)$ be a locally trivial fibration with a CW base B with skeletons B^p. Consider a filtration $\{F_pE\}$ of the space E with $F_pE = p^{-1}(B^p)$ (we apologize for these two ps, but these notations are so common!). The spectral sequence (homology or cohomology, with arbitrary coefficients) corresponding to this filtration is called the *spectral sequence of the fibration* ξ. Let us begin with a calculation of the initial terms, E^1 and E^2 (we begin with the homological case). For this computation, we need an assumption which may provide some inconvenience later. In the next section, we will show how to get rid of this assumption by modifying some fundamental definitions.

Let $s: I \to B$ be a path joining points $b_0, b_1 \in B$. As explained in Sect. 9.6 (for Serre fibrations, but locally trivial fibrations are Serre fibrations), this path determines a (homotopically uniquely defined) homotopy equivalence $p^{-1}(b_0) \to p^{-1}(b_1)$, and hence isomorphisms $H_n(p^{-1}(b_0); G) \xleftarrow{\cong} H_n(p^{-1}(b_1); G)$. This isomorphism may depend on the path s (although it stays the same if the path s is replaced by a homotopic path). Actually, the fibration ξ is called *homologically simple* if this isomorphism does not depend on the path for any n and G. (We can weaken this condition to a "homological simplicity with the coefficients in G," keeping the group G fixed in the definition.) For example, if the base B is simply connected, then the fibration is homologically simple.

From now on, in this section, we assume the fibration ξ homologically simple.

Theorem. *In the spectral sequence of a homologically simple (with coefficients in G) fibration,*

(1) $E^1_{pq} = \mathcal{C}_p(B; H_q(F; G))$.
(2) $d^1_{pq} = [\partial_p : \mathcal{C}_p(B; H_q(F; G)) \to \mathcal{C}_{p-1}(B; H_q(F; G))]$.
(3) $E^2_{pq} = H_p(B; H_q(F; G))$.

Proof. As we know from Sect. 21.1,

$$E^1_{pq} = H_{p+q}(p^{-1}(B^p), p^{-1}(B^{p-1}); G).$$

Let $\{e_i \mid i \in I_p\}$ be the set of all p-dimensional cells of B, and let $c_i \in e_i, d_i \subset e_i$ be the center and a small ball around the center of the cell e_i (with respect to some characteristic map). Obviously, the pair $(p^{-1}(B^p), p^{-1}(B^{p-1}))$ is homotopy equivalent to the pair $(p^{-1}(B^p), p^{-1}(B^p) - \bigcup_i p^{-1}(c_i))$, and, by the excision theorem,

$$\begin{aligned} H_{p+q}(p^{-1}(B^p), p^{-1}(B^p) &- \textstyle\bigcup_i p^{-1}(c_i); G) \\ &= H_{p+q}(\textstyle\bigcup_i p^{-1}(d_i), \bigcup_i p^{-1}(d_i - c_i); G) \\ &= \textstyle\bigoplus_i H_{p+q}(p^{-1}(d_i), p^{-1}(d_i - c_i); G) \\ &= \textstyle\bigoplus_i H_{p+q}(p^{-1}(d_i), p^{-1}(d_i - \operatorname{Int} d_i); G). \end{aligned}$$

On the other hand, the fibration over the disk d_i is trivial with the fiber $F_i = p^{-1}(c_i) \approx F$, and the disk d_i may be regarded as a copy of the standard disk D^p (provided that a characteristic map for the cell e_i is fixed). Because of this,

$$\begin{aligned} H_{p+q}(p^{-1}(d_i), p^{-1}(d_i - \operatorname{Int} d_i); G) &= H_{p+q}(D^p \times F_i, S^{p-1} \times F_i; G) \\ &= H_q(F_i; G) \end{aligned}$$

[the last equality follows from Künneth's formula because $H_i(D^p, S^{p-1}) = \mathbb{Z}$ if $i = p$ and is 0 if $i \neq p$]. The final result is

$$E^1_{pq} = \bigoplus_i H_q(F_i; G),$$

which means that elements of E^1_{pq} can be written as $\sum_i a_i e_i$, where $a_i \in H_q(F_i; G)$. However, if the fibration ξ is homologically simple, we can consider all a_i as belonging to one group, $H_q(F; G)$. Namely, if B is connected, then there are paths joining every pair of points of B, each path establishes an isomorphism between homology groups of fibers over its endpoints, and for a homologically simple fibration, these isomorphisms do not depend on paths. In an unconnected case, we obtain in this way only isomorphisms between homologies of fibers within every component of the base, but we can also arbitrarily choose these isomorphisms between fibers over points of different components. This completes the construction of the isomorphism of part (1). To prove part (2), it is sufficient to compare the description of d^1_{pq} in Sect. 21.1 and the definition of the cellular boundary operator in Sect. 13.4. Part (3) directly follows from part (2).

EXERCISE 1. Generalize the theorem to Serre fibrations.

Thus, for a homologically simple fibration, $E^2_{pq} = H_p(B; H_q(F; G))$. In particular, $E^2_{pq} = 0$ if $p < 0$ or $q < 0$. If the base B and the fiber F are connected, then

$$E^2_{p0} = H_p(B; H_0(F; G)) = H_p(B; G),$$
$$E^2_{0q} = H_0(B; H_q(F; G)) = H_q(F; G).$$

Also,

$$E^2_{pq} = [H_p(B) \otimes H_q(F; G)] \oplus \mathrm{Tor}(H_{p-1}(B), H_q(G; F)).$$

If $G = \mathbb{Z}$, then the last equality takes the form

$$E^2_{pq} = [E^2_{p0} \otimes E^2_{0q}] \oplus \mathrm{Tor}(E^2_{p-1,q}, E^2_{0,q}),$$

and the second summand disappears if the homology of the base or of the fiber has no torsion. Also, if the coefficient domain is a field $\mathbb{K}$, then $E^2_{pq} = E^2_{p0} \otimes_{\mathbb{K}} E^2_{0q}$. All this can be presented on the diagram of the E^2-term, which we display in Fig. 80 in the case when $G = \mathbb{Z}$ and the homology of B or F is torsion-free.

Corollary. $\chi(E) = \chi(B)\chi(F)$.

Proof. Consider the homological spectral sequence with coefficients in $\mathbb{Z}$ and put

$$\chi(E^r) = \sum_m (-1)^m \operatorname{rank}\left(\bigoplus_{p+q=m} E^r_{pq}\right).$$

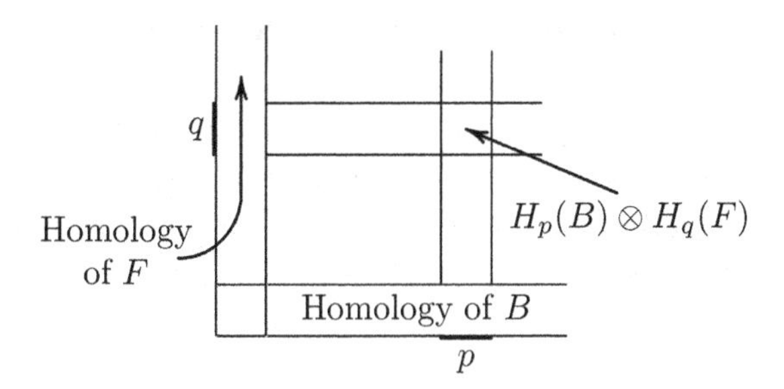

Fig. 80 The E^2-term of the spectral sequence of a fibration

From the theorem we have

$$\begin{aligned}\chi(E^2) &= \sum_m (-1)^m \sum_{p+q=m} \operatorname{rank} H_p(B) \operatorname{rank} H_q(F) \\ &= \sum_{p,q} (-1)^p \operatorname{rank} H_p(B) (-1)^q \operatorname{rank} H_q(F) = \chi(B)\chi(F).\end{aligned}$$

It follows from the Euler–Poincaré lemma (see Sect. 13.7) that $\chi(E^2) = \chi(E^3) = \dots = \chi(E^\infty)$. Finally, $\chi(E^\infty) = \chi(E)$, since $\operatorname{rank} H_m(E) = \sum_{p+q=m} \operatorname{rank} E^\infty_{pq}$.

EXERCISE 2. Prove that if the fibration ξ is trivial, that all the differentials in the spectral sequence of this fibration beginning from d^2 are zero (a spectral sequence with this property is called *degenerate*).

Remarks Concerning Exercise 2. (1) A usual way to prove that a spectral sequence of a fibration (and, actually, any spectral sequence) is degenerate is to show that every element of E^2_{pq} is represented by a genuine cycle of E, not just by a chain whose boundary has a filtration not exceeding $p - 2$. But the result is actually quite expectable. Indeed, according to Exercise 17 in Sect. 15.6, $\bigoplus_{p+q=m} H_p(B; H_q(F)) \cong H_m(B \times F)$, and a similar result holds for homology with arbitrary coefficients. This shows that if $E = B \times F$, then E^2 and E^∞ have "the same size," which makes any nontrivial differential impossible. This informal argument becomes absolutely rigorous if we consider the case when the coefficient domain is a field and the homologies of B and F are finite dimensional. The same can be said in the case when the coefficient domain is $\mathbb{Z}$ and the homologies of B and F have no torsion.

(2) Exercise 2 shows that of fibered spaces with a given base and fiber, the direct product has "the biggest homology" (because nontrivial differentials make a spectral sequence "decreasing").

To finish this section, we remark that all definitions and statements can be repeated with the obvious modification for cohomology. In particular, in the cohomological spectral sequence of a (co)homologically simple fibration with a cellular base,

$$E_2^{pq} = H^p(B; H^q(F; G)).$$

22.2 The Case When the Fibration Is Not Simple

To make this section shorter, we will give many results in the form of exercises. This does not mean that the proofs are difficult; we are sure that a reader who finds the subject interesting (and it is interesting indeed) will be able to reconstruct all missing proofs without much effort.

A: Local Systems

Let X be a topological space. A local system of groups (= an ensemble of groups = a locally trivial sheaf) over X is a function which assigns to every point $x \in X$ a group G_x and to every path $s: I \to X$ joining x_0 with x_1 an isomorphism $\tau_s: G_{x_0} \to G_{x_1}$ which depends only on the homotopy class of the path s and possesses the property $\tau_{ss'} = \tau_{s'} \circ \tau_s$. Examples: (1) $G_x = \pi_n(X, x)$ $(n \geq 1)$, $\tau_s = s_\#$ (see Sects. 6.2, 8.2); (2) X is the base of a (locally trivial or Serre) fibration $p: Y \to X$, $G_x = H_n(p^{-1}(x))$, and τ_x is an isomorphism constructed in Sect. 22.1 (similar for homology and cohomology with coefficients).

From now on, we assume the space X path connected; if it is not, a local system over X is just a collection of independently chosen local systems over path components of X.

Choose a base point x_0 in the base X of a local system $\{G_x, \tau_s\}$. Then every loop with the beginning at x_0 determines an automorphism of the group G_{x_0}, and in this way there arises a group action of the group $\pi_1(X, x_0)$ in G_{x_0}.

EXERCISE 3. Let $\{G_x, \tau_s\}$, $\{G'_x, \tau'_s\}$ be two local systems over X. Prove that if there exists an isomorphism $G_{x_0} \cong G'_{x_0}$ compatible with the actions of the group $\pi_1(X, x_0)$ described above, then the systems $\{G_x, \tau_s\}$ and $\{G'_x, \tau'_s\}$ are isomorphic (in the obvious sense).

EXERCISE 4. Show that an arbitrary group G with an arbitrary group action of the group $\pi_1(X, x_0)$ is a group G_{x_0} for some local system $\{G_x, \tau_x\}$ with the base X.

These statements create a way to construct a large number of new local systems. For example, let X be a (connected) homology manifold, $G = \mathbb{Z}$, and an element α of $\pi_1(X, x_0)$ determines the multiplication by 1 or -1 if α preserves or reverses orientation. The resulting system is denoted as $\mathbb{Z}_T$; it was considered, implicitly, in Sect. 17.12.

B: Homology and Cohomology with Coefficients in a Local System

Let $\mathcal{G} = \{G_x, \tau_x\}$ be a local system of Abelian groups over X. Denote by c_n the center of the standard n-dimensional simplex Δ^n and by $s_{n,i}$ the straight path in Δ^n from

c_n to the center of the ith $(n-1)$-dimensional face of Δ^n. A singular n-dimensional chain of X with coefficients in $\mathcal{G}$ is defined as a (finite) linear combination $\sum_i g_i f_i$, where $f_i\colon \Delta^n \to X$ is a singular simplex and $g_i \in G_{f_i(c_n)}$. The group of all such chains is denoted as $C_n(X;\mathcal{G})$, and the boundary operator $\partial = \partial_n\colon C_n(X;\mathcal{G}) \to C_{n-1}(X;\mathcal{G})$ is defined by the formula

$$\partial(gf) = \sum_{i=0}^{n} (-1)^i \tau_{f\circ s_{n,i}}(g)\Gamma_i f.$$

The homology arising is denoted as $H_n(X;\mathcal{G})$; the cohomology $H^n(X;\mathcal{G})$ is defined in a similar way.

A different approach to homology and cohomology with coefficients in local systems exists, similar to that in Sect. 17.12. It works when the space X is "good enough" in the sense that it has a universal covering $\pi\colon \widehat{X} \to X$ with the deck transformation group equal to $\pi_1(X, x_0)$ (see Sect. 6.8). Let $G = G_{x_0}$. Consider the subgroup of the group $C_n(\widehat{X}; G)$ consisting of those $c = \sum_i g_i f_i$ (where $g_i \in G$ and $f_i\colon \Delta^n \to \widehat{X}$) such that for every $\alpha \in \pi_1(X, x_0)$, $\alpha_\# c = \sum_i \alpha(g_i) f_i$ (on the left-hand side α is regarded as a transformation $\widehat{X} \to \widehat{X}$; on the right-hand side it is regarded as an automorphism of G). We denote this subgroup as $C_n(X;\mathcal{G})$ and notice that $\partial(C_n(X;\mathcal{G})) \subset C_{n-1}(X;\mathcal{G})$ (here ∂ is the boundary operator in the singular chains of $\widehat{X}$).

EXERCISE 5. Prove that this description of the singular chain complex of X with coefficients in $\mathcal{G}$ is equivalent to the previous definition.

EXERCISE 6. Prove that (for a path connected space X) $H^0(X;\mathcal{G}) = \{g \in G_{x_0} \mid \alpha g = g$ for all $\alpha \in \pi_1(X, x_0)$ and $H_0(X;\mathcal{G}) = G_{x_0}/G^*_{x_0}$, where $G^*_{x_0}$ is the subgroup of G_{x_0} generated by all differences $\alpha g - g$, $\alpha \in \pi_1(X, x_0)$, $g \in G_{x_0}$.

Homology and cohomology of a CW complex X with the coefficients in a local system $\{G_x\}$ can be calculated by means of a cellular complex. Namely, an n-dimensional cellular chain is a finite linear combination $\sum_i g_i e_i$, where e_i is an oriented n-dimensional cell of X, and $g_i \in G_{x_i}$, where $x_i \in e_i$ (the groups G_{x_i} for all points $x_i \in e_i$ are canonically isomorphic; we will denote G_{x_i} simply by G_{e_i}); the group of such cellular chains is denoted by $\mathcal{C}_n(X;\mathcal{G})$. The boundary operator $\partial = \partial_n\colon \mathcal{C}_n(X;\mathcal{G}) \to \mathcal{C}_{n-1}(X;\mathcal{G})$ acts via the formula

$$\partial(ge) = \sum_{\substack{\dim f = n-1 \\ f \cap \bar{e} \neq \emptyset}} \eta_{e,f}(g) f,$$

where the operator $\eta_{e,f}\colon G_e \to G_f$ is defined in the following way. Choose a characteristic map $h\colon D^n \to X$ for the cell e (representing the chosen orientation of e) with the following property. In f, there exists a small ball d (with the center $y \in f$) whose inverse image $h^{-1}(d)$ is a finite union of balls $d_i \in S^{n-1}$ such that each d_i is mapped onto d homeomorphically; we put $\varepsilon_i = \pm 1$ depending on whether $d_i \xrightarrow{h} d$ preserves or reverses orientation. Let c be the center of the ball D^n, $c_i = h^{-1}(y) \cap d_i$, and let s_i be the straight path in D^n joining c with c_i. We define the homomorphism $\eta_{e,f}: G_{h(c)} \to G_y$ by the formula $\eta_{e,f}(g) = \sum_i \varepsilon_i \tau_{h \circ s_i}(g)$.

EXERCISE 7. Prove that the homologies of the complexes

$$\{C_n(X; \mathcal{G}), \partial_n\} \text{ and } \{\mathcal{C}_n(X; \mathcal{G}), \partial_n\}$$

are the same. Make up and prove a similar statement for cohomology.

EXERCISE 8. Let $X = \mathbb{R}P^n$, G be an Abelian group and let $T: G \to G$ be an automorphism with $T^2 = \text{id}$. Let the generator of the group $\pi_1(\mathbb{R}P^n)$ act in G as T. Denote by $\mathcal{G}$ the local system arising. Prove that the homomorphism

$$\begin{array}{ccc} \partial: \mathcal{C}_r(\mathbb{R}P^n; \mathcal{G}) & \to & \mathcal{C}_{r-1}(\mathbb{R}P^n; \mathcal{G}), \ 0 < r \leq n \\ \| & & \| \\ G & & G \end{array}$$

acts like $\text{id} + T$ for r even and like $\text{id} - T$ for r odd. Compute the homology $H_r(\mathbb{R}P^n; \mathcal{G})$ in the general case and in the case $G = \mathbb{Z}, T = -\text{id}$.

Remark. If n is even, then the last $\mathcal{G}$ is nothing but $\mathbb{Z}_T$ from Sect. 17.12. Observe Poincaré isomorphism for the nonorientable manifold $\mathbb{R}P^n$ described in Sect. 17.12.

C: Main Theorem for Nonsimple Fibrations

Theorem.

$$E^2_{pq} = H_p(B; \{H_q(p^{-1}(x))\}),$$

where $\{H_p(p^{-1}))\}$ *is the local system described in Example* (2) *following the definition of a local system (see the beginning of Sect. 22.2.A). A similar thing holds for the homology and cohomology spectral sequences with arbitrary coefficients.*

EXERCISE 9. Prove the theorem (a proof is basically the same as that of the theorem in Sect. 22.1).

EXERCISE 10. Prove that the equality $\chi(E) = \chi(B)\chi(F)$ (see the corollary in Sect. 22.1) also holds for nonsimple fibrations.

D: Obstruction Theory for Nonsimple Spaces and Fibrations

EXERCISE 11. Using cohomology with coefficients in local systems, extend the obstruction theory (for both continuous maps and sections) to the cases of nonsimple target space and nonsimple fibrations. Make up a definition of the (integral) Euler class of nonorientable vector bundles.

22.3 First Applications

A reader who has the impression (which we do not share) that the material of the previous section is auxiliary and unnecessary is encouraged to switch on his or her full attention. We will demonstrate now how even the preliminary information we have can be used for quite nontrivial computations (although more serious applications of spectral sequences are still ahead).

A: Homology of the Special Unitary Group $SU(n)$

As we know, the coset space $SU(n)/SU(n-1)$ is nothing but the sphere S^{2n-1}. Thus, there arises a fibration

$$SU(n) \xrightarrow{SU(n-1)} S^{2n-1} \ (n \geq 2)$$

(we follow the tradition of writing the notation for a fiber of a fibration over the arrow which denotes the projection of this fibration). If $n = 2$, then the fiber of this fibration is one point. Hence, $SU(2) = S^3$ (we know this from Sect. 1.7). Thus, for $n = 3$ we obtain a fibration

$$SU(3) \xrightarrow{S^3} S^5.$$

Since we know the (integral) homology of the base and the fiber, we can display a full diagram of the E^2-term of this spectral sequence (Fig. 81).

It is clear from this diagram that no one of the differentials $d^2, d^3, d^4, \ldots$ (some of them are shown in the diagram) connects two nonzero groups. Thus, $E^\infty = E^2$. Moreover, for every n, there is at most one nonzero group E^∞_{pq} with $p + q = n$. This implies a full result for the homology of $SU(3)$:

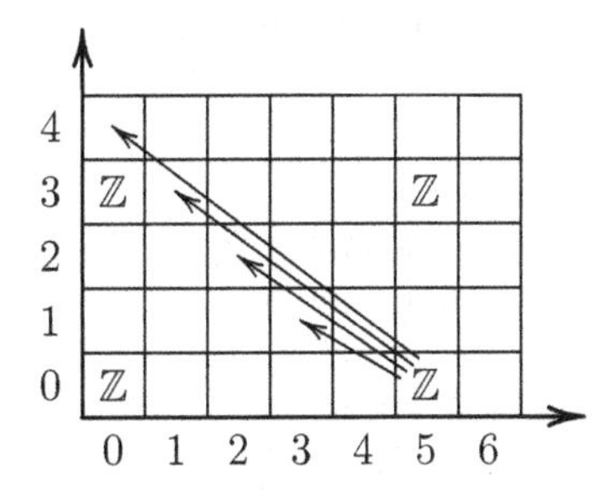

Fig. 81 The E^2-term for the fibration $SU(3) \xrightarrow{S^3} S^5$

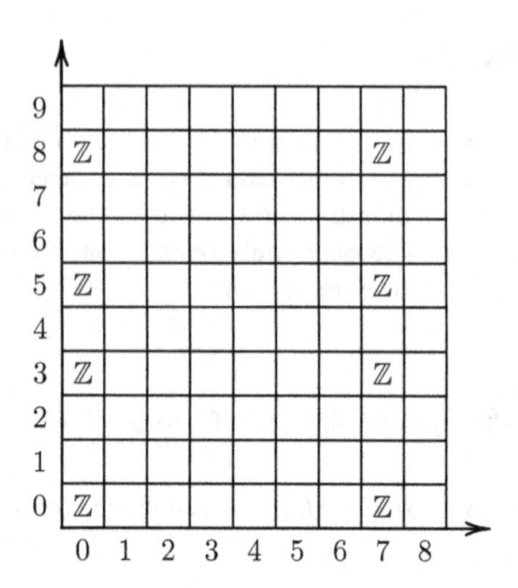

Fig. 82 The E^2-term for the fibration $SU(4) \xrightarrow{SU(3)} S^7$

$$H_n(SU(3)) = \begin{cases} \mathbb{Z} \text{ for } n = 0, 3, 5, 8, \\ 0 \text{ for all other } n. \end{cases}$$

Consider now the spectral sequence of the fibration $SU(4) \xrightarrow{SU(3)} S^7$. The diagram of the E^2-term is shown in Fig. 82.

The same arguments as before (they are usually called "dimension arguments") work, the differentials $d^2, d^3, d^4, \ldots$ are all trivial, $E^\infty = E^2$, and we find that

$$H_n(SU(4)) = \begin{cases} \mathbb{Z} \text{ for } n = 0, 3, 5, 7, 8, 10, 12, 15, \\ 0 \text{ for all other } n. \end{cases}$$

At this point the hope that we can proceed with the dimension arguments and show that $E^\infty = E^2$ for spectral sequences of all our fibrations arises. However, it turns out that the case $n = 4$ is the last case when the dimension arguments (and at the moment we have no other arguments) are sufficient for computing the homology of $SU(n)$. Indeed, let us consider the initial term of the spectral sequence of the next fibration, $SU(5) \xrightarrow{SU(4)} S^9$, shown in Fig. 83.

The dimension arguments show that in this spectral sequence $E^2 = \ldots = E^9$ and $E^{10} = \ldots = E^\infty$. However, the transition from E^9 to E^{10} involves two potentially nonzero differentials, $d^9_{9,0}\colon \mathbb{Z} \to \mathbb{Z}$ and $d^9_{9,7}\colon \mathbb{Z} \to \mathbb{Z}$ (shown by arrows in our diagram). Actually, these differentials are trivial, and so are all the differentials of all the spectral sequences of the fibrations considered. But at the moment we have no means to prove this; we will do so in Lecture 24, where we will show that

$$H_r(SU(n)) \cong H_r(S^3 \times S^5 \times \ldots \times S^{2n-1}).$$

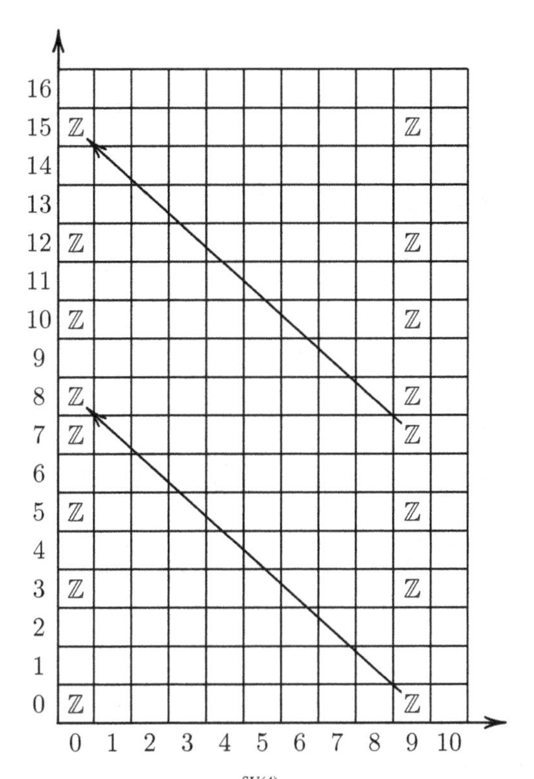

Fig. 83 The E^2-term for the fibration $SU(5) \xrightarrow{SU(4)} S^9$

[Notice that if $n \geq 3$, then the spaces $SU(n)$ and $S^3 \times S^5 \times \ldots \times S^{2n-1}$ are not homeomorphic: They have different groups π_4. The reader can try to prove this, but it is better to postpone this until the next chapter.]

B: Homology of Loop Spaces

Theorem. *Let X be a topological space (with a base point), and let the space X be (n − 1)-connected, that is,*

$$\pi_0(X) = \pi_1(X) = \ldots = \pi_{n-1}(X) = 0.$$

Then

$$H_r(X) \cong H_{r-1}(\Omega X) \text{ for } r \leq 2n - 2,$$

and a similar isomorphism holds for homology and cohomology with arbitrary coefficients.

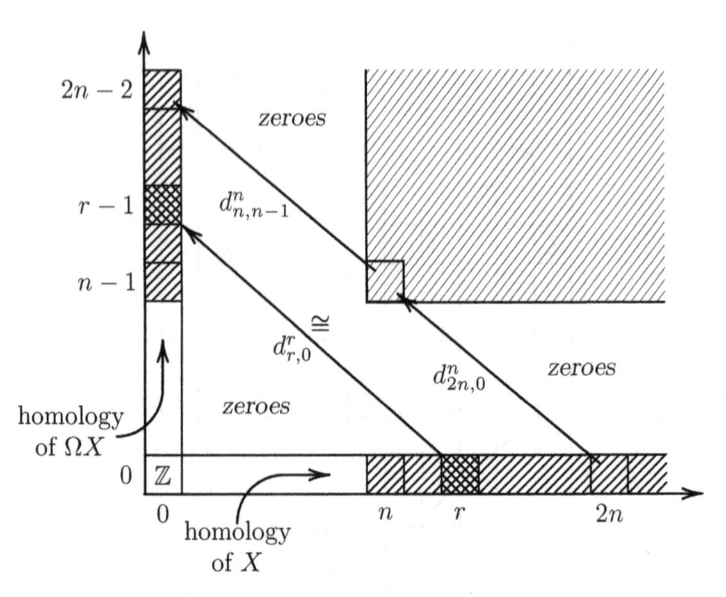

Fig. 84 The spectral sequence of the fibration $EX \xrightarrow{\Omega X} X$

Proof. We can restrict ourselves to the case when X is a CW complex; the transition to the general case (which actually is not very important) may be based on the results of Sects. 11.6 and 14.1.

Consider the spectral sequence of the fibration $EX \xrightarrow{\Omega X} X$ from Example 2 in Sect. 9.4: EX is the space of paths beginning at the base point of X (we will constantly use this fibration in the future). Since the space EX is contractible, in E^∞ everything is trivial with the exception of $E^\infty_{00} = \mathbb{Z}$. By Hurewicz's theorem, $H_1(X) = \dots = H_{n-1}(X) = 0$, and hence there is nothing but zeroes in the vertical strip in E^2 formed by the 1st to $(n-1)$st columns. Furthermore, since $\pi_i(\Omega X) \cong \pi_{i+1}(X)$ (see Exercise 11 in Sect. 9.9), we also have $H_1(\Omega X) = \dots = H_{n-2}(\Omega X) = 0$, and the horizontal strip in E^2 formed by the 1st to $(n-2)$nd rows are also filled with zeroes. A diagram for the E^2-term (with some future differentials) is shown in Fig. 84.

The corner cell contains the group $E^2_{n,n-1}$. The only potentially nonzero differential from this cell is directed to the $(2n-2)$nd cell of the zeroth column; the only differential directed to this cell comes from the $2n$th cell in the zeroth row. Thus, the groups below the $(2n-2)$nd cell in the zeroth column and the groups to the left of the $2n$th cell of the zeroth row can be annihilated only by differentials acting from the zeroth row to the zeroth column (and they must be annihilated since the E^∞-term is zero). This shows that (actually, for all r) the differential $d^r_{r,0}\colon E^r_{r,0} \to E^r_{0,r-1}$ must be an isomorphism, and if $r \le 2n-2$, then this isomorphism connects the groups $E^r_{r,0} = E^2_{r,0} = H_r(X)$ and $E^r_{0,r-1} = E^2_{0,r-1} = H_{r-1}(\Omega X)$, which are, thus, isomorphic. This completes the proof.

Remark. This proof actually provides for $r \le 2n-2$, a canonical isomorphism $(d^r_{r,0})^{-1}: H_{r-1}(\Omega X) \to H_r(X)$. For $r = 2n-1$, our arguments fail only because of the differential $d^n_{n,n-1}: E^n_{n,n-1} \to E^n_{0,2n-2}$, so we have an isomorphism $d^{2n-1}_{2n-1,0}$ of $E^{2n-1}_{2n-1,0} = E^2_{2n-1,0} = H_{2n-1}(X)$ onto the quotient of $E^2_{0,2n-2} = H_{2n-2}(\Omega X)$ over the image of $d^n_{n,n-1}$. This gives us the right to consider $(d^{2n-1}_{2n-1,0})^{-1}$ as an *epimorphism* $H_{2n-2}(\Omega X) \to H_{2n-1}(X)$.

We will prove in the next lecture that this isomorphism and this epimorphism are actually induced by a certain continuous map, namely, by the map $\pi(X): \Sigma\Omega X \to X$ acting by the formula $\tau_X(s,t) = s(t)$ (where $s \in \Omega X$ and $t \in I$): The isomorphism and the epimorphism can be described as

$$H_{r-1}(\Omega X) \xrightarrow{\Sigma^{-1}} H_r(\Sigma\Omega X) \xrightarrow{(\pi_X)_*} H_r(X).$$

We take the liberty of using this statement right now, before proving it.

C: A Generalization of Freudenthal's Theorem

In Sect. 10.1, we proved Freudenthal's theorem, which states that the suspension homomorphism $\pi_r(S^n) \to \pi_{r+1}(S^{n+1}$ is an isomorphism for $r \ge 2n-2$ and an epimorphism for $r = 2n-1$. In that section we promised to prove later a similar statement where S^n is replaced by an arbitrary $(n-1)$-connected CW complex. We are going to do that now.

Theorem. *Let X be an $(n-1)$-connected CW complex. Then the suspension homomorphism $\Sigma: \pi_r(X) \to \pi_{r+1}(\Sigma X)$ is an isomorphism if $r \le 2n-2$ and is an epimorphism if $r = 2n-1$.*

Proof. In this proof, it will be convenient to use the base point version of the definition of the suspension (with the segment $x_0 \times I$ contracted to a point). Besides the map π_X defined above (that definition works for the base point suspension), we will consider the map $\iota_X: X \to \Omega\Sigma X$, $[\iota_X(x)](t) = (x,t)$ [recall that the vertices of the suspension, $(x,0)$ and $(x,1)$, are identified]. Obviously,

$$\Sigma X \xrightarrow{\Sigma\iota_X} \Sigma\Omega\Sigma X \xrightarrow{\pi_{\Sigma X}} \Sigma X$$

is the identity, $\mathrm{id}_{\Sigma X}$. According to the result of the preceding subsection, Sect. B (the statement whose proof was postponed to Lecture 23), $\pi_{\Sigma X}$ induces an isomorphism in homology of dimensions $\le 2n$ [if X is $(n-1)$-connected, then ΣX is n-connected]; hence, $\Sigma\iota_X$ induces an isomorphism in homology of dimensions $\le 2n$, and ι_X induces an isomorphism in homology of dimensions $\le 2n-1$ [the homomorphisms $(\Sigma\iota_X)_*$ and $(\iota_X)_*$ are the same up to a dimension shift by 1]. By Whitehead's theorem (Sect. 14.5), ι_X induces isomorphisms in homotopy groups of dimensions

$\leq 2n-2$ and an epimorphism in homotopy groups of dimension $2n-1$. Finally, consider the homomorphism

$$\pi_r(X) \xrightarrow{(\iota_X)_*} \pi_r(\Omega\Sigma X) \cong \pi_{r+1}(\Sigma X)$$

(where the last isomorphism arises from the homotopy sequence of the fibration $EY \xrightarrow{\Omega Y} Y$, $Y = \Sigma X$; see Example 2 in Sect. 9.4).

Lemma. *This homomorphism is just* $\Sigma: \pi_r(X) \to \pi_{r+1}(\Sigma X)$.

This is proved by a direct comparison of the definitions. We leave the details to the reader.

Remark 1. The proof given above does not use Freudenthal's theorem as given in Sect. 10.1. Thus, in particular, we get a new proof of that theorem.

Remark 2. The last results may serve as one more illustration of the Eckmann–Hilton duality described in Lecture 4. The operations Ω and Σ are dual to each other. The spaces X and ΣX have equal (co)homology groups (with a dimension shift by 1); their homotopy groups are the same in "stable dimensions," that is, in dimensions less than twice the connectivity of X. On the other hand, X and ΩX have the same homotopy groups (again, with a dimension shift), while their (co)homology groups are the same in stable dimensions.

Lecture 23 Additional Properties of Spectral Sequences of Fibrations

23.1 Continuous Maps and Homomorphisms of Spectral Sequences

For simplicity, we begin with the case of homology with coefficients in $\mathbb{Z}$. The cases of homology and cohomology with arbitrary coefficients are absolutely similar and do not even deserve separate consideration. Without saying this explicitly, we assume below throughout this lecture that the fibrations considered are homologically simple and their bases and fibers are path connected. (The reader will decide at every occasion whether this is really necessary.)

Let

$$\emptyset = X_{-1} \subset X_0 \subset X_1 \subset \ldots \subset X,\ \emptyset = X'_{-1} \subset X'_0 \subset X'_1 \subset \ldots \subset X'$$

be two spaces with filtrations. For the corresponding spectral sequences, we will use the notations E^r_{pq}, d^r_{pq} and $'E^r_{pq}, 'd^r_{pq}$. Let $f: X \to X'$ be a continuous map such that $f(X_p) \subset X'_p$ for every p. Such map induces, for all r, homomorphisms $f_\#: C_r(X) \to$

$C_r(X')$ compatible with boundary operations and filtrations; it also induces, for all r, p, q (including $r = \infty$), homomorphisms

$$f_* = (f_*)^r_{pq}: E^r_{pq} \to {}'E^r_{pq}$$

commuting with the differentials:

$${}'d^r_{pq} \circ (f_*)^r_{pq} = (f_*)^r_{p-r,q+r-1} \circ d^r_{pq}.$$

These homomorphisms are compatible with the statements of the Leray proposition in the best possible way:

(1) $(f_*)^1_{pq}: E^1_{pq} \to {}'E^1_{pq}$ is the same as

$$f_*: H_{p+q}(X_p, X_{p-1}) \to H_{p+q}(X'_p, X'_{p-1}).$$

(2) The homomorphism $(f_*)^{r+1}_{pq}$ is the homology homomorphism induced by the homomorphism $(f_*)^r_{pq}$ (compatible with the differentials d and $'d$).
(3) The map $(f_*)^\infty: \bigoplus_{p+q=m} E^\infty_{pq} \to \bigoplus_{p+q=m} {}'E^\infty_{pq}$ is induced by the map $f_*: H_m(X) \to H_m(X')$.

The proof of this is obvious.

All these properties of homomorphisms $(f_ast)^r_{pq}$ are briefly expressed in these words: $\{(f_*)^r_{pq}\}$ *is a homomorphism of the spectral sequence* $\{E^r_{pq}, d^r_{pq}\}$ *into the spectral sequence* $\{'E^r_{pq}, {}'d^r_{pq}\}$ (see Fig. 85).

Mark one obvious but important property of homomorphisms of spectral sequences: *If for some* r, *the homomorphism* $E^r \to {}'E^r$ *belonging to a homomorphism between spectral sequences is an isomorphism, then so are all homomorphisms* $E^s \to {}'E^s$ *with* $s > r$ *(including* $s = \infty$*).* Moreover, *if two homomorphisms between two spectral sequences coincide on* E^r *for some* r, *then they coincide on* E^s *for all* $s > r$.

MAIN EXAMPLE. Let (E, B, F, p) and (E', B', F', p') be two homologically simple fibrations with connected CW bases B, B', and let $f: E \to E'$ be a fiberwise

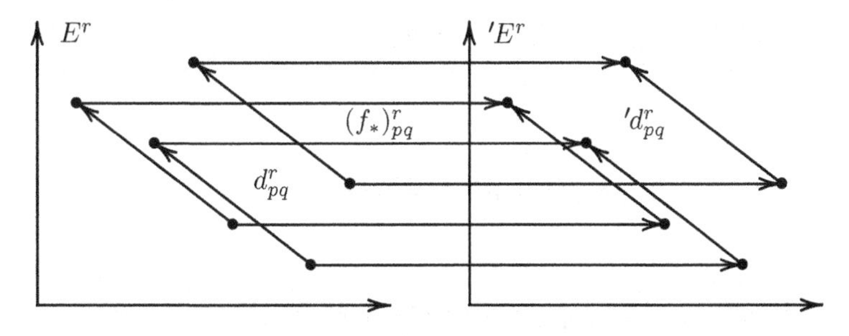

Fig. 85 Homomorphism between spectral sequences

continuous map. The latter means that there exists a continuous map $g: B \to B'$ such that the diagram

$$\begin{array}{ccc} E & \xrightarrow{f} & E' \\ \downarrow{\scriptstyle p} & & \downarrow{\scriptstyle p'} \\ B & \xrightarrow{g} & B' \end{array}$$

is commutative. For every point $x \in B$, the map f induces a map h of the fiber $p^{-1}(x)$ into the fiber $(p')^{-1}(g(x))$, which, because of the homological simplicity, induces homomorphisms $h_*: H_q(F) \to H_q(F')$ not depending on the choice of x.

By the cellular approximation theorem, the map g is homotopic to a cellular map, and by the covering homotopy property, this homotopy (rather the homotopy of $p \circ g$) may be lifted to a homotopy of the map f. The new map f is compatible with the filtrations in E and E', and we can assume that the maps f and g had these properties from the very beginning. Then a homomorphism f_* arises between the spectral sequences of the two fibrations.

Proposition. (1) *The homomorphism* $(f_*)^2_{pq}: H_p(B; H_q(F)) \to H_p(B'; H_q(F'))$ *coincides with the homomorphism induced by the maps g and h.*

(2) *For $r \geq 2$ (including $r = \infty$), the homomorphisms $(f_*)^r_{pq}$ do not depend on the choices of the cellular approximation of g and the mapping f compatible with filtrations.*

Proof of Part (1) The proof of part (1) is left to the reader (who will need to use the details of the construction of the isomorphism $E^2_{pq} \cong H_p(B; H_q(F))$); part (2) follows from part (1) in view of the preceding remark.

Corollary. *Starting from the E^2-term, the spectral sequence of a fibration does not depend on the CW structure of the base.*

Proof. If two fibrations differ only by a CW structure of the base B, we can apply the proposition to a (possibly noncellular, but continuous) map id_B.

Everything said in this section has obvious analogs for homology and cohomology with coefficients in an arbitrary Abelian group. The consideration of homologically nonsimple fibrations here and further in this lecture is left to the reader.

23.2 Zeroth Row and Zeroth Column

One can obtain important corollaries from the result of the previous section by applying them to simplest maps between fibrations, namely, to the maps

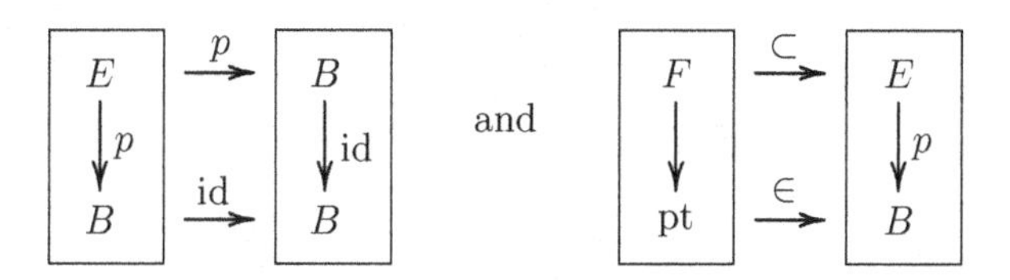

At the E^2-level, the induced homomorphisms of spectral sequences are schematically shown in Fig. 86.

The E^2-term of the fibration $B \xrightarrow{\text{pt}} B$, as well as all the subsequent terms, consists of one (zeroth) row which contains homology of B. The homomorphism f_*^2 is the identity on this row and is trivial on the remaining (shadowed) part of E^2. Similarly, for the fibration $F \xrightarrow{F} \text{pt}$, every term starting from E^2 consists of one (zeroth) column which contains homology of F, and homomorphism f_*^2 is an isomorphism on this column and zero elsewhere. Since all the differentials (of the homological spectral sequence) directed at the groups in the zeroth row, as well as all the differentials from the groups in the zeroth column, are trivial (see Fig. 87), the groups E^r_{m0} and E^r_{0m} are, correspondingly, chains of subgroups and quotient groups:

$$E^2_{p0} \supset E^3_{p0} \supset E^4_{p0} \supset \ldots \supset E^\infty_{p0};$$
$$E^2_{0q} \xrightarrow{\operatorname{Im} d} E^3_{0q} \xrightarrow{\operatorname{Im} d} E^4_{0q} \xrightarrow{\operatorname{Im} d} \ldots \longrightarrow E^\infty_{0q}.$$

There arise a monomorphism and an epimorphism

$$E^\infty_{p0} \to E^2_{p0} \text{ and } E^2_{0q} \to E^\infty_{0q}.$$

On the other hand, E^∞_{0q} is a subgroup of the group $H_q(E)$, and E^∞_{p0} is a quotient group of the group $H_p(E)$. Consider composed homomorphisms

$$H_q(F) = H_0(B; H_q(F)) = E^2_{0q} \to E^\infty_{0q} \subset H_q(E),$$
$$H_p(E) \to E^\infty_{p0} \subset E^2_{p0} = H_p(B; H_0(F)) = H_p(B).$$

Proposition. *These homomorphisms coincide with the homology homomorphisms induced by the inclusion $F \to E$ and the projection $E \to B$.*

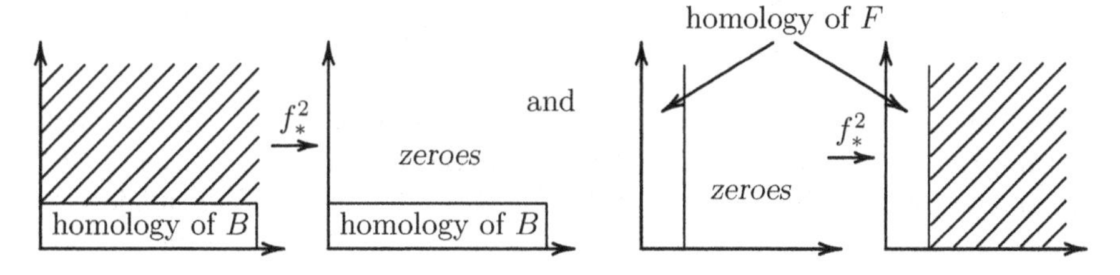

Fig. 86 The homomorphisms between the E^2-terms

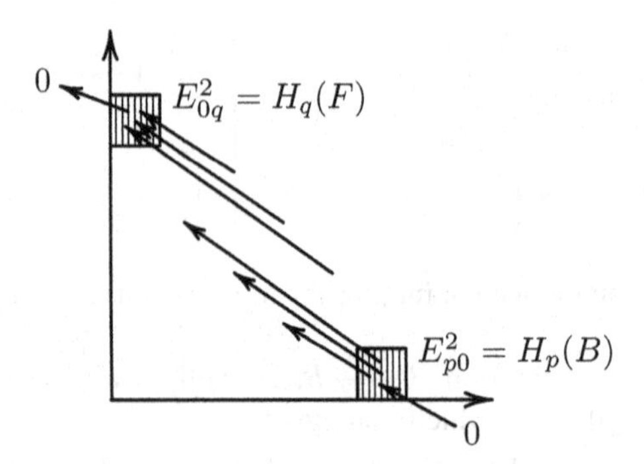

Fig. 87 Differentials at the zeroth row and the zeroth column

Indeed, these homomorphisms coincide with the homology maps induced by the maps between the fibered spaces described in the beginning of this section.

Precisely the same statements hold for homology groups with coefficients in an arbitrary Abelian group. In the cohomology case they take the following form. The spectral sequence provides a monomorphism and an epimorphism

$$E_\infty^{0q} \to E_2^{0q} \text{ and } E_2^{p0} \to E_\infty^{p0}.$$

On the other hand, E_∞^{p0} is a subgroup of $H^p(E;G)$ and E_∞^{q0} is a quotient group of $H^q(E;G)$. There arise compositions

$$H^p(B;G) = H^p(B;H^0(F;G)) = E_2^{p0} \to E_\infty^{p0} \subset H^p(E;G),$$
$$H^q(E;G) \to E_\infty^{0q} \subset E_2^{0q} = H^0(B;H^q(F;G)) = H^q(F;G).$$

Proposition. *These homomorphisms coincide with the cohomology homomorphisms induced by the projection $E \to B$ and the inclusion $F \to E$.*

23.3 Transgression

Consider differentials

$$d^m_{m0} \colon E^m_{m0} \to E^m_{0,m-1} \text{ and } d_m^{0,m-1} \colon E_m^{0,m-1} \to E_m^{m0}$$

of the homological and cohomological spectral sequences of a homologically simple fibration (E,B,F,p) with connected B and F. As we noticed before, E^m_{m0} and $E_m^{0,m-1}$ are subgroups of $H_m(E;G)$ and $H^{m-1}(E;G)$, while $E^m_{0,m-1}$ and E_m^{m0} are quotients of groups $H_{m-1}(F;G)$ and $H^m(E;G)$. Hence, our differentials have the form

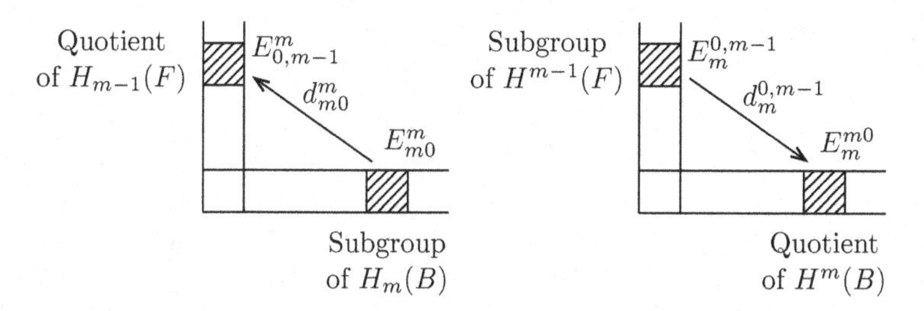

Fig. 88 Transgression

If A, B are Abelian groups, then a homomorphism of a subgroup of A into the quotient of B is called a *partial multivalued homomorphism of* A into B. For partial multivalued homomorphisms $A \to B$ we will sometimes use the notation $A \overset{h}{\dashrightarrow} B$. Notice that partial multivalued homomorphisms always have "inverses": For a partial multivalued homomorphism $A \supset C \xrightarrow{f} B/D$, its inverse is $B/p^{-1}\operatorname{Im} f \xrightarrow{f} C/\operatorname{Ker} f \subset A/\operatorname{Ker} f$ (where p is the projection $B \to B/D$). In particular, the inverse to a usual homomorphism $f: A \to B$ is the partial multivalued homomorphism $f^{-1}: \operatorname{Im} f \to A/\operatorname{Ker} f$.

Thus, our differentials are partial multivalued homomorphisms

$$H_m(B;G) \dashrightarrow H_{m-1}(F;G) \text{ and } H^{m-1}(F;G) \dashrightarrow H^m(B;G).$$

These homomorphisms are called, respectively, *homological* and *cohomological transgression* (see Fig. 88).

Elements of the domain of transgression are called *transgressive*. As far as we know, this term is used only in the cohomology case.

Theorem. *Homological and cohomological transgressions coincide, respectively, with the following compositions:*

$$H_m(B;G) = H_m(B, \mathrm{pt}; G) \overset{(p_*)^{-1}}{\dashrightarrow} H_m(E, F; G) \xrightarrow{\partial_*} H_{m-1}(F;G),$$
$$H^{m-1}(F;G) \xrightarrow{\delta^*} H^m(E, F; G) \overset{(p^*)^{-1}}{\dashrightarrow} H^m(B, \mathrm{pt}; G) = H^m(B;G).$$

Proof. We will consider only the case of homology and of $G = \mathbb{Z}$; all other cases are similar. We can assume that B has only one zero-dimensional cell. Elements of the group E^m_{m0} are represented by chains $c \in C_m(p^{-1}B^m) \subset C_m(E)$ whose boundaries belong to $C_{m-1}(p^{-1}(B^0)) = C_{m-1}(F)$, that is, by relative cycles of the pair $(p^{-1}(B^m), F)$. The identification of E^m_{m0} with a subgroup of the group $H_m(B)$ is done by the map which assigns to the class of c the homology class of the cycle $p_\#(c)$ of B. The differential d^m_{m0} takes this element of E^m_{m0} into the element of $E^m_{0,m-1}$ represented by the cycle $\partial c \in C_{m-1}(F)$. This is our statement.

23.4 Application: Three Exact Sequences

In Sect. 21.3, we encountered a situation when the information contained in a spectral sequence may be presented by an exact sequence. We will demonstrate in this section three more exact sequences which are equivalent to (or, at least, can be derived from) a certain spectral sequence of a fibration. Notice that at least two of these exact sequences had been discovered before the method of spectral sequences appeared in algebraic topology.

A: Gysin's Sequence

Let (E, B, S^n, p) be a homologically simple fibration with a spherical fiber (the condition of homological simplicity is equivalent to the condition of orientability: The fibers $p^{-1}(x)$ have orientations continuously depending on $x \in B$). The E^2-term of the homological spectral sequence of this filtration consists of two identical rows containing the homology of B (Fig. 89). Potentially nontrivial differentials are $d^{n+1}_{m0}\colon E^{n+1}_{m0} \to E^{n+1}_{m-n-1,n}$, that is, $H_m(B) \to H_{m-n-1}(B)$.

For every m, there are (at most) two nonzero groups E^∞_{pq} with $p+q=m$: $E^\infty_{m0} = \operatorname{Ker} d^{n+1}_{m0}$ and $E^\infty_{m-n,n} = \operatorname{Coker} d^{n+1}_{m+1,0}$; the second one is a subgroup of $H_m(E)$, while the first one is the corresponding quotient group. This can be written as a short exact sequence,

$$0 \to \operatorname{Coker} d^{n+1}_{m+1,0} \to H_m(E) \to \operatorname{Ker} d^{n+1}_{m0} \to 0,$$

which is the same as a five-term exact sequence

$$H_{m+1}(B) \xrightarrow{d^{n+1}_{m+1,0}} H_{m-n}(B) \longrightarrow H_m(E) \longrightarrow H_m(B) \xrightarrow{d^{n+1}_{m0}} H_{m-n-1}(B).$$

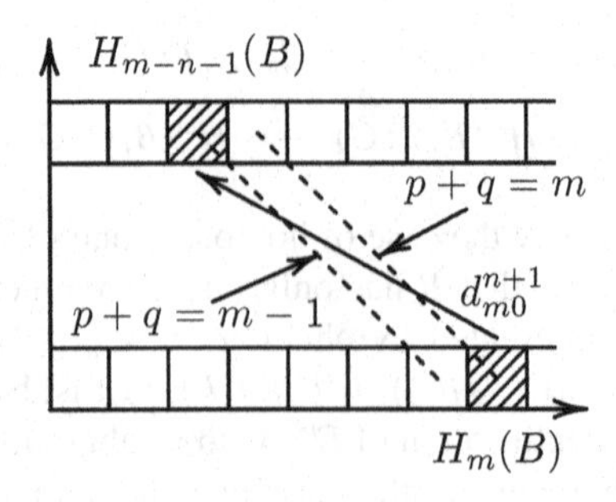

Fig. 89 Spectral sequence of a fibration with a spherical fiber

These five-term fragments may be merged into one infinite exact sequence,

$$\dots \xrightarrow{p_*} H_{m+1}(B) \xrightarrow{d} H_{m-n}(B) \xrightarrow{\ell} H_m(E) \xrightarrow{p_*} H_m(B) \xrightarrow{d} \dots$$

which is called the *(homological) Gysin sequence*. A similar sequence exists for homology with arbitrary coefficients.

EXERCISE 1. Prove that in the Gysin sequence, the homomorphism p_* is induced by the projection p (essentially, this is contained in the proposition of Sect. 23.2), the map ℓ ("lifting") assigns to a homology class of a cycle in B the homology cycle of its inverse image in E (in the smooth situation, when E and B are manifolds and p is smooth, this is $p^!$), and the map d is the $\frown$-product with the primary characteristic class $C \in H^{n+1}(B; \pi_n(S^n)) = H^{n+1}(B; \mathbb{Z})$ (see Sect. 18.5; it may be reasonable to postpone proving the last statement until Sect. 23.5 or the next lecture).

The *cohomological Gysin sequence* is defined in a similar way. It looks like this:

$$\dots \xleftarrow{p_*} H^{m+1}(B; G) \xleftarrow{d} H^{m-n}(B; G) \xrightarrow{\ell} H^m(E; G) \xleftarrow{p^*} H^m(B; G) \xleftarrow{d} \dots .$$

EXERCISE 2. Prove that in the cohomological Gysin sequence, the homomorphism p^* is induced by the projection p, the map ℓ is $p_!$) in the smooth situation, and the map d is the $\smile$-product with $C \in H^{n+1}(B; \mathbb{Z})$.

B: Wang's Sequence

Let $\xi = (E, S^n, F, p)$ be a fibration with a spherical base. (If $n \geq 2$, this fibration is automatically homologically simple; if $n = 1$, then we need to assume that the fibration is homologically simple, but in this case, the construction presented here requires some clarification.) The E^2-term of the homological spectral sequence consists of two identical columns, zeroth and nth; each contains homology of F. The groups in these columns are connected with differentials d^n (see Fig. 90).

Precisely as in Gysin's case, we get a short exact sequence

$$0 \to \operatorname{Coker} d^n_{n,m-n+1} \to H_m(E) \to \operatorname{Ker} d^n_{n,m-n} \to 0$$

and then develop it into a long exact sequence

$$\dots \xrightarrow{r} H_{m-n+1}(F) \xrightarrow{d} H_m(F) \xrightarrow{i_*} H_m(E) \xrightarrow{r} H_{m-n}(F) \xrightarrow{d} \dots .$$

This is the *homological Wang sequence*. There is also the *cohomological Wang sequence*, which is constructed similarly and has the following form:

$$\dots \xleftarrow{r} H^{m-n+1}(F; G) \xleftarrow{d} H^m(F; G) \xleftarrow{i^*} H^m(E; G) \xleftarrow{r} H^{m-n}(F; G) \xleftarrow{d} \dots$$

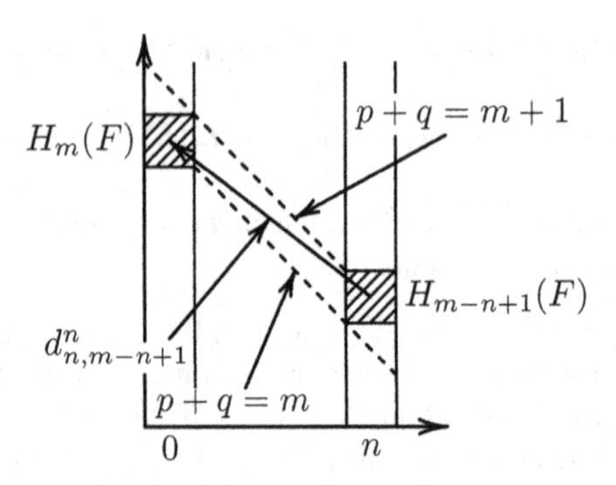

Fig. 90 Spectral sequence of a fibration with a spherical base

(it goes without saying that in the homological case we can also insert into all homology groups of Wang's sequence an arbitrary coefficient group G).

EXERCISE 3. Prove the following interpretation of the homomorphisms of the homological Wang sequence. The map i_* is just i_*, where i is the inclusion map $F \to E$. The operation r can be described as the "intersection" of a cycle of E with F; in the smooth situation, r can be described as $i_!$. To obtain a geometric description for d, consider the map $h: D^n \to D^n/S^{n-1} = S^n$ (this map can be described as a characteristic map for the n-dimensional cell of S^n). The fibration $h^*\xi$ with base D^n must be trivial. Thus, the canonical map $h^*\xi \to \xi$ provides a map $D^n \times F \to E$ which covers h. In particular, it restricts to the map $\widetilde{h}: S^{n-1} \times F \to F$, and d is $\alpha \longmapsto \widetilde{h}_*([S^{n-1}] \times \alpha)$ (you need to prove this).

EXERCISE 4. State and prove all the similar facts for the cohomological Wang sequence.

C: Serre's Sequence

Let us suppose now that the fibration (E, B, F, p) with a cellular base B has an additional property: For some n,

$$\pi_0(B) = \pi_1(B) = \ldots = \pi_{n-1}(B) = 0,$$
$$\pi_0(F) = \pi_1(F) = \ldots = \pi_{n-2}(F) = 0.$$

Then the E^2-term of the homological spectral sequence has the form familiar to us from Sect. 22.3.B (see Fig. 91).

Since $E^2_{pq} = 0$ for $p < n$ and for $q < n-1$ (with the exception of $E^2_{00} = \mathbb{Z}$), the same is true of E^∞_{pq}, and, for $m < 2n-1$, there is an exact sequence

$$0 \to E^\infty_{0m} \to H_m(E) \to E^\infty_{m0} \to 0.$$

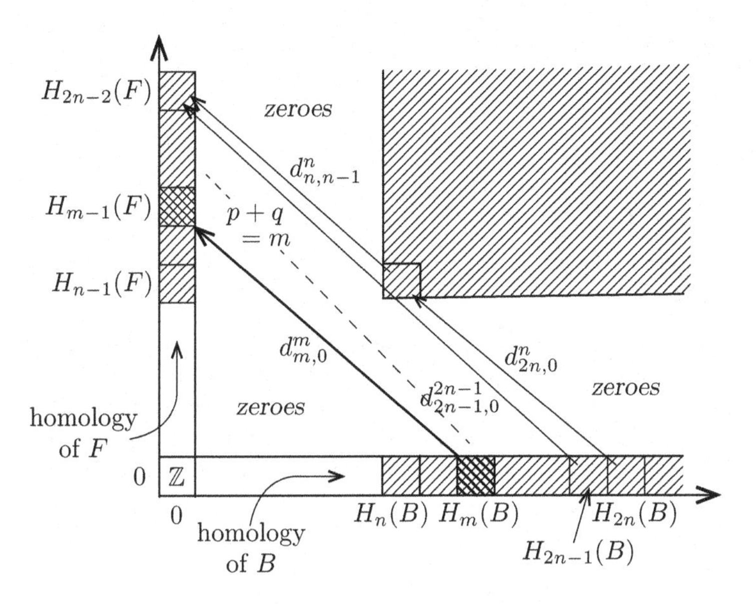

Fig. 91 The spectral sequence which implies the Serre exact sequence

But for $m \geq 2n-2$, $E^\infty_{m0} = \mathrm{Ker}[d^m_{m,0}\colon H_m(B) \to H_{m-1}(F)]$ and for $m \geq 2n-3$, $E^\infty_{0m} = \mathrm{Coker}[d^{m+1}_{m+1,0}\colon H_{m+1}(B) \to H_m(F)]$, which shows that for $m \geq 2n-3$ the short exact sequence above is equivalent to a five-term exact sequence

$$H_{m+1}(B) \to H_m(F) \to H_m(E) \to H_m(B) \to H_{m-1}(F).$$

In addition, $E^\infty_{0,2n-2}$ is a quotient of $H_{2n-2}(F)$ (which is factorized successively over the images of two differentials). All this leads to the following long exact sequence (the homomorphisms involved are known to us: They are p_*, i_*, where i is the inclusion map $F \to E$, and the transgression τ, which is, within this sequence, a genuine, not partial and multivalued, homomorphism):

$$H_{2n-2}(F) \xrightarrow{i_*} H_{2n-2}(E) \xrightarrow{p_*} H_{2n-2}(B) \xrightarrow{\tau} H_{2n-3}(F) \xrightarrow{i_*}$$
$$\dots \xrightarrow{p_*} H_n(B) \xrightarrow{\tau} H_{n-1}(F) \xrightarrow{i_*} H_{n-1}(E) \to 0.$$

This exact sequence is called the *Serre exact sequence*. It has a strong resemblance to the homotopy sequence of the same fibration, but unlike the homotopy sequence, it is *finite* (exists only in the "stable" dimensions).

EXERCISE 5. Prove that for $n-1 \le m \le 2n-2$, the diagram

$$\begin{array}{ccccccc} \pi_m(F) & \xrightarrow{i_*} & \pi_m(E) & \xrightarrow{p_*} & \pi_m(B) & \xrightarrow{\partial} & \pi_{m-1}(F) \\ \downarrow h & & \downarrow h & & \downarrow h & & \downarrow h \\ H_m(F) & \xrightarrow{i_*} & H_m(E) & \xrightarrow{p_*} & H_m(B) & \xrightarrow{\tau} & H_{m-1}(F), \end{array}$$

formed by homotopy and Serre sequences and Hurewicz homomorphisms, is commutative.

If the homology of E is trivial (at least up to the dimension $2n-2$), the Serre exact sequence implies isomorphisms between homologies of B and F with a shift of dimensions by 1; we already observed this phenomenon in Sect. 22.3.B. It should also be noted that the Serre sequence admits the transition to the homology with coefficients in an arbitrary Abelian group G and also has the following cohomology version:

$$H^{2n-2}(F;G) \xleftarrow{i^*} H^{2n-2}(E;G) \xleftarrow{p^*} H^{2n-2}(B;G) \xleftarrow{\tau}$$
$$H^{2n-3}(F;G) \xleftarrow{i^*} \dots \xleftarrow{p^*} H^n(B;G) \xleftarrow{\tau}$$
$$H^{n-1}(F;G) \xleftarrow{i^*} H^{n-1}(E) \leftarrow 0.$$

We conclude this section by proving a statement whose proof was promised in Sec. 22.3.B (and which was used in Sect. 22.3.C in the proof of a generalization of Freudenthal's theorem). Here we will prove an even stronger statement.

Proposition. *Transgression $H_m(X;G) \dashrightarrow H_{m-1}(\Omega X;G)$ is, for every m, a partial multivalued homomorphism inverse to the homomorphism*

$$H_{m-1}(\Omega X;G) \xrightarrow{\Sigma} H_m(\Sigma\Omega X;G) \xrightarrow{(\pi_X)_*} H_m(X;G).$$

Proof. Consider an auxiliary map $\varphi: C\Omega X \to EX$ (where C denotes the cone) which assigns to a loop $s: I \to X$ and a number $t \in I$ the "shortened loop" $u \mapsto s(tu)$. (This map illustrates the fact that the fiber ΩX is contractible in EX.) In addition to that, we consider the map $\pi'_X: C\Omega X \to X$ defined (like π_X) by the formula $(s,t) \mapsto s(t)$ and form a diagram

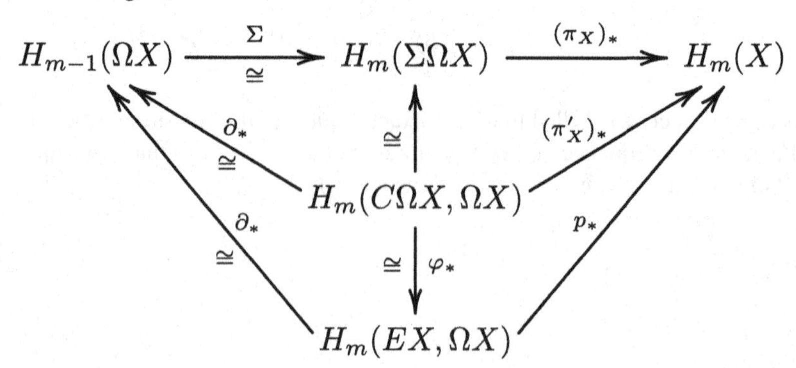

(we skip the notations for the coefficient group). Since the diagram is commutative, we see that the homomorphism $(\pi_X)_* \circ \Sigma: H_{m-1}(\Omega X) \to H_m(X)$ is inverse to the transgression $\partial_* \circ (p_*)^{-1}: H_m(X) \to H_{m-1}(\Omega X)$; this is precisely what we need.

D: An Application of the Serre Sequence: A Factorization Theorem for Relative Homotopy Groups

We know from Chap. 2 that relative homology groups, at least, for "good pairs," can be interpreted as absolute homology groups of the "quotient space": $H_q(X,A) \cong \widetilde{H}_q(X/A)$; the isomorphism is induced by the projection map $(X,A) \to (X/A, \text{pt})$. However, there is no similar result for homotopy groups. Now, we can state that the factorization theorem for homotopy groups holds in "stable dimensions"; that is, the following holds.

Proposition. *Suppose that for a CW pair (X,A), the homotopy groups $\pi_r(X), \pi_r(A)$ are trivial for $r < n$. Then the homomorphism*

$$\pi_q(X,A) \to \pi_q(X/A)$$

is an isomorphism for $q < 2n-2$ and is an epimorphism for $q = 2n-2$.

Proof. Recall that the inclusion map $i: A \to X$ is homotopy equivalent to a (Serre) fibration $p: \widetilde{A} \to X$ (this was done in a more general form in Sect. 9.7). The construction runs as follows. $\widetilde{A}$ is the space of paths $s: I \to X$ with $s(0) \in A$. The homotopy equivalence $\widetilde{A} \to A$ is established by the map $s \mapsto s(0)$. The map $p: \widetilde{A} \to X$ is defined by the formula $p(s) = s(1)$; this map $p: \widetilde{A} \to X$ is a Serre fibration. The fiber F of this fibration is the space of paths s with $s(0)$ being a fixed point $a \in X$; we assume below that $a \in A \subset X$. It follows from the homotopy sequence of this fibration that $\pi_r(F) = 0$ for $r < n-1$.

Now notice that the projection $X \to X/A$ takes every path from F into a loop of X/A, which yields a map $F \to \Omega(X/A)$, or, equivalently, $\Sigma F \to X/A$. This map induces homomorphisms $H_{q-1}(F) \to H_q(X/A)$. Consider a diagram composed of the homological sequence of the pair (X,A), the Serre sequence of the fibration $p: \widetilde{A} \to X$, and the homomorphism constructed above (plus a bunch of the identity homomorphisms):

$$\begin{array}{ccccccccc} H_q(A) & \to & H_q(X) & \to & H_q(X/A) & \to & H_{q-1}(A) & \to & H_{q-1}(X) \\ \| & & \| & & \uparrow & & \| & & \| \\ H_q(A) & \to & H_q(X) & \to & H_{q-1}(F) & \to & H_{q-1}(A) & \to & H_{q-1}(X). \end{array}$$

It is important to notice that the bottom sequence, and hence the whole diagram, exists only in stable dimensions.

A direct checking (which we leave to the reader) shows that this diagram is commutative. Then the five-lemma shows that the middle vertical homomorphism

is an isomorphism. Hence, our map $\Sigma F \to X/A$ induces, in stable dimensions, a homology isomorphism, and, according to Whitehead's theorem, it also induces, in the same dimensions, a homotopy group isomorphism. This is equivalent to our statement (we leave to the reader details related to the boundary case $q = 2n - 2$).

23.5 Transgression and the Characteristic Class

Let $\xi = (E, B, F, p)$ be a homologically simple fibration, and let $\pi = \pi_n(F)$ be the first nontrivial homotopy group of F. We suppose that $n \geq 2$ or, at least, that the group π is Abelian. In $H^n(F; \pi)$, there is the fundamental class $c(F)$ (it would be awkward to use a more traditional notation F_F for this class), which can be described as the cohomology class whose value on the homology class corresponding to an $\alpha \in \pi_n(X)$ is α [we can write $\langle c_F, h(\alpha)\rangle = \alpha$, where h is the Hurewicz homomorphism], or c_F is the characteristic class of the path fibration $EX \xrightarrow{\Omega X} X$. (This class was considered in Sect. 18.3 and Exercise 9 in Sect. 18.5.)

EXERCISE 6. Prove that the image of c_F with respect to the (cohomological) transgression equals the characteristic class C_ξ of the fibration ξ. [*Hint:* We need to prove that the fundamental class and the characteristic class have the same image under the homomorphisms

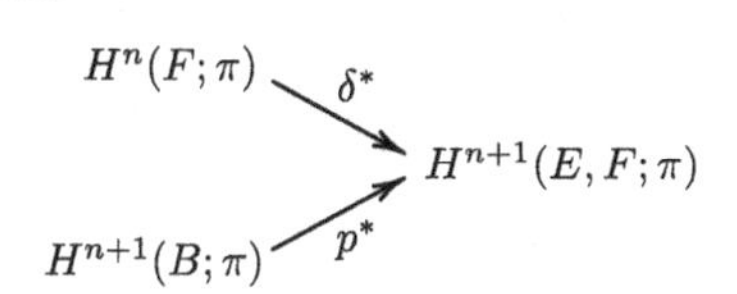

These images are equal, actually, to the first [$(n+1)$-dimensional] obstruction to extending the map id: $F \to F$ to a map $E \to F$.]

EXERCISE 7. Extend the result of Exercise 6 to homologically nonsimple fibrations.

Lecture 24 A Multiplicative Structure in a Cohomological Spectral Sequence

Up to now, we have alternated between considering homological and cohomological cases and did not see any significant difference between them. We have usually discussed the homological case in detail and for the cohomological case only pointed out the changes needed. However, the experience of the previous chapter shows that cohomology has a serious advantage over homology, because of the multiplicative structure (and some other structures, which we will consider in the next chapter). In this lecture, we will furnish cohomological spectral sequences with a multiplication.

24.1 Main Results: The Multiplicative Structure and Its Properties

Assume that the coefficient domain G is a ring, such as $\mathbb{Z}$, or a field. We will construct for every term E_r (starting with $r = 2$) of the cohomological spectral sequence of a fibration (E, B, F, p) (with a CW base) a multiplication with the following properties.

(1) The multiplication is bihomogeneous; that is, if $\alpha \in E_r^{pq}$ and $\beta \in E_r^{p'q'}$, then $\alpha\beta \in E_r^{p+p',q+q'}$.
(2) The differentials satisfy the product rule: If $\alpha \in E_r^{pq}$, $\beta \in E_r^{p'q'}$, then

$$d_r^{p+p,q+q'}(\alpha\beta) = (d_r^{pq}\alpha)\beta + (-1)^{p+q}\alpha(d_r^{p'q'}\beta).$$

(3) The multiplication in E_{r+1} is induced by the multiplication in E_r: If $\alpha, \beta \in E_{r+1}$ are represented by d_r-cycles $a, b \in E_r$, then $\alpha\beta$ is represented by the d_r-cycle ab.
(4) The multiplication in E_2 coincides with the multiplication in the cohomology of B with the coefficients in the cohomology ring of F. (This statement concerns the homologically simple case; in the case of homologically nonsimple fibration, it should involve a multiplication in the cohomology with coefficients in local systems; we leave the details to the reader.)
(5) The multiplication in E^∞ is adjoint to the multiplication in $H^*(E; G)$ in the following sense. If $a \in F^pH^m(E; G)$ and $b \in F^qH^n(E; G)$, then $ab \in F^{p+q}H^{m+n}(E; G)$, and if the elements $\alpha \in E_\infty^{p,m-p}$, $\beta \in E_\infty^{q,n-q}$, and $\gamma \in E_\infty^{p+q,m+n-p-q}$ are represented by a, b, and ab, then $\gamma = \alpha\beta$.

In connection with statement (5), we will introduce an algebraic notion which will also be useful in the future. Let A be a ring (not necessarily unitary), and let

$$A = F^{-1} \supset F^0A \supset F^1A \supset \ldots \supset F^nA \supset F^{n+1}A = 0$$

be a filtration of the Abelian group A. This filtration is called *multiplicative* if $(F^pA)(F^qA) \subset F^{p+q}A$ for all p and q. The group $\operatorname{Gr} A = \bigoplus(F^pA/F^{p+1}A)$ adjoint to A with respect to the multiplicative filtration as above has a natural structure of a graded ring: If $\alpha \in F^pA/F^{p+1}A$ and $\beta \in F^qA/F^{q+1}A$ and $a \in F^pA$, $b \in F^qA$ are representative of α and β, then $ab \in F^{p+q}A$ represents some element of $F^{p+q}A/F^{p+q+1}A$, and we take this element for $\alpha\beta$; it is obvious that this $\alpha\beta$ does not depend on the choice of a and b in α and β.

Notice that while the difference between Abelian groups G and $\operatorname{Gr} G$ in many respects may be regarded as insignificant, this is not the case for the rings A and $\operatorname{Gr} A$: The multiplication in $\operatorname{Gr} A$ is far less rich than the multiplication in A. For example, if the multiplication in $\operatorname{Gr} A$ is trivial (the product of any two elements is zero), then for A this means only that $(F^pA)(F^qA) \subset F^{p+q+1}A$; shift by 1 the filtration in any ring, and you will obtain a ring with this property. On the contrary, any statement showing a nontriviality of the multiplication in $\operatorname{Gr} A$ implies, as a rule, a similar statement for A. Here is an example.

EXERCISE 1. Prove that if $\mathrm{Gr}\,A$ has no zero divisors, then A has no zero divisors.

There is another statement of this kind which will be useful later.

Proposition. *Let A be a ring or an algebra over some field possessing a finite ($F^{n+1}A = 0$) multiplicative filtration, and let $\{x_i\}$ be a system of (multiplicative) generators of $\mathrm{Gr}\,A$ with $x_i \in F^{p_i}A/F^{p_i+1}A$. Further, let $\widetilde{x_i}$ be a representative of x_i in $F^{p_i}A \subset A$. Then $\{\widetilde{x_i}\}$ is a system of generators of A.*

Proof. Let $y \in F^pA \subset A$, and let $x \in F^pA/F^{p+1}A$ be the class of y. Then $x = P(x_i)$ is a polynomial (with coefficients in $\mathbb{Z}$ or in the ground field) in the generators x_i. The same polynomial in $\widetilde{x_i}$ differs from y by an element of a higher filtration: $z = y - P(\widetilde{x_i}) \in F^{p+1}A$. Apply the same construction to z, and we will get a polynomial $Q(\widetilde{x_i})$ such that $z - Q(\widetilde{x_i}) \in F^{p+2}A$, so $y - P(\widetilde{x_i}) - Q(\widetilde{x_i}) \in F^{p+2}A$. Proceeding in the same way, we obtain a polynomial in $\widetilde{x_i}$ whose difference with y belongs to $F^{n+1}A$, that is, equals 0.

24.2 The Construction of the Multiplication

We begin with a general algebraic construction in which a spectral sequence turns out to be multiplicative. Let A be a filtered differential graded ring; that is, the additive group of A is furnished by mutually compatible filtration $\{F_pA\}$, grading $A = \bigoplus_r A^r$, and a differential $d: A \to A$ of degree $+1$, and, in addition to the compatibility, the filtration is multiplicative, the grading is also multiplicative ($A^rA^s \subset A^{r+s}$), and the differential satisfies the product rule [$d(ab) = (da)b + (-1)^r a(db)$ for $a \in A^r$, $b \in A$]. In this situation, there is multiplication in the spectral sequence which satisfies conditions (1)–(3) above and also condition (5) with $H^n(E; G)$ replaced by the cohomology of A.

Let

$$\alpha \in E_r^{pq} = \frac{F^pA^{p+q} \cap d^{-1}(F^{p+r}A^{p+q+1})}{[F^{p+1}A^{p+q} \cap d^{-1}(F^{p+r}A^{p+q+1})] + [F^pA^{p+q} \cap d(F^{p-r+1}A^{p+q-1})]}$$

and $\beta \in E_r^{p'q'}$. Choose representative $a \in \alpha, b \in \beta$; thus,

$$a \in F^pA^{p+q}, da \in F^{p+r}A^{p+q+1}, \text{ and } b \in F^{p'}A^{p'+q'}, db \in F^{p'+r}A^{p'+q'+1}.$$

Then

$$ab \in F^{p+p'}A^{p+q+p'+q'},$$

$$(da)b \in F^{p+p'+r}A^{p+p'+q+q'+1},\ a(db) \in F^{p+p'+r}A^{p+p'+q+q'+1};$$

hence, $d(ab) = (da)b \pm a(db) \in F^{p+p'+r}A^{p+p'+q+q'+1}$, that is, $ab \in d^{-1}(F^{p+p'+r}A^{p+p'+q+q'+1})$, so ab represents an element of $E_r^{p+p',q+q'}$ which we take for $\alpha\beta$. Certainly, we need to check that this $\alpha\beta$ does not depend on the choice of $a \in \alpha$ and $b \in \beta$. For example, suppose that we make a different choice for a: Take some other $a' \in \alpha$. Then $a' = a+u+dv$, where $u \in F^{p+1}A^{p+q}$, $v \in F^{p-r+1}A^{p+q-1}$, and $a'b = ab + ub + (dv)b$. We have $ub \in F^{p+p'+1}A^{p+q+p'+q'}$ and $(dv)b = \pm d(vb) \pm v(db)$, $vb \in F^{p-r+1+p'}A^{p+q-1+p'+q'}$, $v(db) \in F^{p+p'+1}A^{p+q+p'+q'}$; that is, $a'b - ab$ belongs to the denominator in the definition of $E_r^{p+p',q+q'}$. Thus, the choice of a does not affect $\alpha\beta$; the same is true for the choice of b.

Thus, we have an (obviously, bilinear) multiplication $E_r^{pq} \times E_r^{p'q'} \to E_r^{p+p',q+q'}$. Properties (2) and (3) are obvious. Indeed, $d_r^{p+p',q+q'}(\alpha\beta)$ is represented by $d(ab)$ (where $a \in \alpha$, $b \in \beta$), and $d(ab) = (da)b+(-1)^{p+q}a(db)$; hence, $d_r^{p+p',q+q'}(\alpha\beta) = d_r^{pq}(\alpha)\beta + (-1)^{p+q}\alpha d_r^{p'q'}(\beta)$; this proves (2). Furthermore, if $\widehat{\alpha} \in E_{r+1}^{pq}$, $\widehat{\beta} \in E_{r+1}^{p'q'}$ are represented by d_r-cocycles $\alpha \in E_r^{pq}$, $\beta \in E_r^{p'q'}$, then a representative of $\widehat{\alpha}\widehat{\beta}$ in $F_{p+p'}A^{p+p'+q+q'}$ is also a representative of $\alpha\beta$ which implies (3). Property (5) (after an appropriate modification) is obvious.

It remains to do the last thing: to define an appropriate multiplication for cochains of the total space E of a fibration $\xi = (E,B,F,p)$; this multiplication must be compatible with the already existing structures: filtration, grading, and the differential (coboundary operator). To define this, we consider the diagram

$$\begin{array}{ccc} E & \xrightarrow{\Delta_E} & E \times E \\ \downarrow{\scriptstyle p} & & \downarrow{\scriptstyle p\times p} \\ B & \xrightarrow{\Delta_B} & B \times B, \end{array}$$

where Δ_B and Δ_E are diagonal maps. Then we consider a homotopy h_t connecting Δ_B with some cellular approximation Δ_B° and lift the homotopy $h_t \circ p$ to a homotopy of Δ_E. There arises a map $\Delta_E^\circ\colon E \to E \times E$ which is compatible with the filtrations in E and $E \times E$. After this, we define the "product" c_1c_2 of cochains $c_1 \in C^{n_1}(E;G)$ and $c_2 \in C^{n_2}(E;G)$ as $(\Delta_E^\circ)^\#(c_1 \times c_2) \in C^{n_1 \times n_2}(E;G)$. This product satisfies the product rule with respect to the usual coboundary operator in the cochain complex of E and induces the usual multiplication in the cohomology of E (since $\Delta_E^\circ \sim \Delta_E$). We leave to the reader the verification of the last necessary property of the resulting multiplication in the cohomological spectral sequence of the fibration ξ.

EXERCISE 2. Prove that if the fibration ξ is homologically simple, then the multiplication in the E_2-term satisfies property (5).

Note that property (5) shows that the multiplication in the E_2-term is associative and skew-commutative [the latter means that if $\alpha \in E_2^{pq}$ and $\beta \in E_2^{p'q'}$, then $\beta\alpha = (-1)^{(p+q)(p'+q')}\alpha\beta$]. After this, property (3) implies the same properties of the multiplication in E_r for $3 \leq r \leq \infty$. (Certainly, for $r = 0$ and 1, we have no reasons to expect that the multiplication in E_r is either associative or skew-commutative;

actually, the right choice of cellular approximations can make these multiplications associative; but they cannot be made skew-commutative, as we will see in Chap. 4.)

24.3 The First Application: The Cohomology of $SU(n)$

Now we can finish the computation started in Sect. 22.3.A.

Theorem. *There is a multiplicative isomorphism*

$$H^*(SU(n);\mathbb{Z}) \cong H^*(S^3 \times S^5 \times \ldots \times S^{2n-1};\mathbb{Z}).$$

Remark 1. A more common way to describe the preceding result is to say that $H^*(SU(n);\mathbb{Z})$ is an *exterior algebra* (over $\mathbb{Z}$) with $n-1$ generators of dimensions $3, 5, \ldots, 2n-1$. The latter means that there are generators $x_{2i-1} \in H^{2n-1}(SU(n);\mathbb{Z})$, $i = 2, 3, \ldots, n$, of the ring $H^{2n-1}(SU(n);\mathbb{Z})$ with the generating system of relations $x_i x_j = -x_j x_i$ for $i < j$ and $x_i^2 = 0$. A similar result holds for the cohomology of $SU(n)$ with coefficients in any field.

Remark 2. The theorem implies that all the differentials of the homological spectral sequence of the fibration $SU(n) \xrightarrow{SU(n-1)} S^{2n-1}$ are trivial (this was proved in Sect. 22.3.A for $n \leq 4$ and stated for all n). Indeed, for some n let some differentials be nontrivial, but let all the differentials be trivial for all smaller n. Then $H_*(SU(n-1)) = H_*(S^3 \times S^5 \times \ldots \times S^{2n-3})$ and the E^2-term of our spectral sequence is isomorphic to $H_*(S^3 \times S^5 \times \ldots \times S^{2n-1})$. But every nontrivial differential acts between free Abelian groups, so it must affect the total rank of the corresponding E^r-term. Hence, we would have had

$$\operatorname{rank} H_*(SU(n)) < \operatorname{rank} H_*(S^3 \times S^5 \times \ldots \times S^{2n-1}),$$

which contradicts the theorem, because the ranks of homology and cohomology groups are the same in all dimensions.

Proof of Theorem. We proceed by induction. For $n = 2$, the statement is correct. Suppose that $H^*(SU(n-1);\mathbb{Z})$ is an exterior algebra with generators $x_3, \ldots, x_{2n-3}$. Then the ring E_2 becomes an exterior algebra with generators $x_{2i-1} \in E_2^{0,2i-1}$, $i = 2, 3, \ldots, n-1$, and $y \in E_2^{2n-1,0}$ (see the diagram in Fig. 92). By the dimension arguments, $E_2 = E_3 = \ldots = E_{2n-1}$. Consider the differential d_{2n-1}. It is zero on the generators x_{2i-1} since it sends x_{2i-1} into $E_{2n-1}^{2n-1,(2i-1)-(2n-2)} = 0$; and more so, it is zero on y. But since d_{2n-1} satisfies the product rule, it is zero on any product of generators; that is, it is totally zero. Hence, $E^\infty = E^2$ is an exterior algebra with generators of (full) dimensions $3, 5, \ldots, 2n-1$.

That is not all, however. Our computation of E^∞ shows only that there is, for our n, an additive isomorphism $H^*(SU(n);\mathbb{Z}) \cong H^*(S^3 \times S^5 \times \ldots \times S^{2n-1})$; but

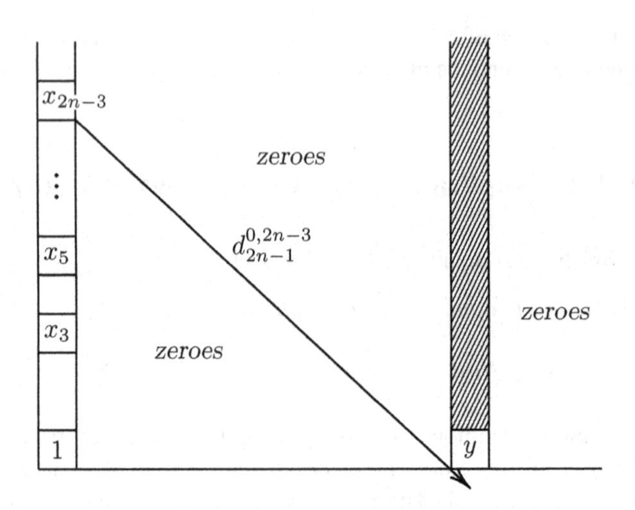

Fig. 92 Cohomological spectral sequence of the fibration $SU(n) \xrightarrow{SU(n-1)} S^{2n-1}$

for our induction we need to know that there is a multiplicative isomorphism. Take representatives of the generators $x_3, x_5, \ldots, x_{2n-3}, y$ in $H^*(SU(n); \mathbb{Z})$ (they have dimensions $3, 5, \ldots, 2n-3, 2n-1$); for the representatives of $x_3, \ldots, x_{2n-3}$ we preserve their notations, and for the representative of y we use the notation x_{2n-1}. By the proposition in Sect. 24.1, $x_3, x_5, \ldots, x_{2n-1}$ is a system of generators in $H^*(SU(n); \mathbb{Z})$. The relations $x_i x_j = -x_j x_i$, $x_i^2 = 0$ follow from the general properties of cohomology and the fact that $H^*(SU(n); \mathbb{Z})$ has no torsion. If this system of relations were not full, then $H^*(SU(n); \mathbb{Z})$ would have been obtained from $H^*(S^3 \times S^5 \times \ldots \times S^{2n-1})$ by an additional factorization which contradicts the equality $\operatorname{rank} H^*(SU(n); \mathbb{Z}) = \operatorname{rank} H^*(S^3 \times S^5 \times \ldots \times S^{2n-1})$.

EXERCISE 3. Prove that there is a multiplicative isomorphism

$$H^*(\mathbb{C}V(n,k); \mathbb{Z}) \cong H^*(S^{2(n-k)+1} \times S^{2(n-k)+3} \times \ldots \times S^{2n-1}; \mathbb{Z})$$

[see Sect. 1.8 for the definition of $\mathbb{C}V(n,k)$].

24.4 Cohomology of Other Classical Groups

A: Symplectic Groups

EXERCISE 4. Prove that there is a multiplicative isomorphism

$$H^*(Sp(n); \mathbb{Z}) \cong H^*(S^3 \times S^7 \times \ldots \times S^{4n-1}; \mathbb{Z}).$$

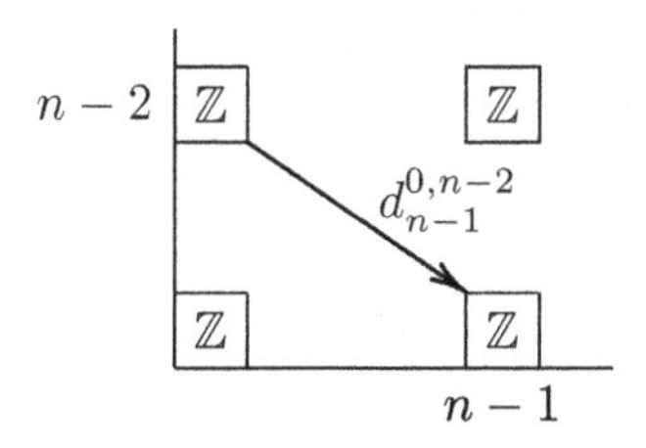

Fig. 93 Cohomological spectral sequence of the fibration $V(n,2) \xrightarrow{S^{n-2}} S^{n-1}$

B: Orthogonal Groups

Theorem. *If* $\mathbb{K} = \mathbb{Q}, \mathbb{R}$, *or* $\mathbb{C}$, *then*

$$H^*(SO(n);\mathbb{K}) = \begin{cases} H^*(S^3 \times S^7 \times \ldots \times S^{4m-1};\mathbb{K}), \text{ if } n = 2m+1, \\ H^*((S^3 \times S^7 \times \ldots \times S^{4m-5}) \times S^{2m-1};\mathbb{K}), \\ \qquad\qquad\qquad\qquad\qquad\qquad \text{if } n = 2m. \end{cases}$$

Proof. Since $SO(2) = S^1$ and $SO(3) = \mathbb{R}P^3$ (see Sect. 1.7), for $n = 2$ and 3 the statement is known to us. Assume that the statement is true for all $SO(n')$ with $n' < n$ and begin with the case when n is odd.

The map $SO(n) \to V(n,2)$, which assigns to an orthogonal matrix the 2-frame formed by its first two rows, is a fibration with the fiber $SO(n-2)$ [one can notice that $V(n,2) = SO(n)/SO(n-2)$; see Sect. 1.9].

Lemma. *If n is odd, then* $H^*(V(n,2);\mathbb{K}) = H^*(S^{2n-3};\mathbb{K})$.

Proof. Consider the cohomological spectral sequence of the fibration

$$V(n,2) \xrightarrow{S^{n-2}} S^{n-1}$$

(which is, actually, the fibration of the manifold of unit tangent vectors to S^{n-1} over S^{n-1}) with the coefficients in $\mathbb{Z}$ (Fig. 93).

The only differential which can be nontrivial is the transgression $d_{n-1}^{0,n-2}: \mathbb{Z} \to \mathbb{Z}$. But, according to Sect. 23.5 (see Exercise 6), the image of the generator of $E_{n-1}^{0,n-2}$ is the characteristic class of the fibration, that is, the Euler characteristic of the sphere S^{n-1} times the generator of $H^{n-1}(S^{n-1};\mathbb{Z})$ (see Proposition 2 of Sect. 18.5). Since $\chi(S^k) = 2$ if k is even and is 0 if k is odd, we conclude that $d_{n-1}^{0,n-2}$ is 0 if n is even and is a multiplication by 2 if n is odd. Thus, for n odd, the spectral sequence shows that

$$H^q(V(n,2);\mathbb{Z}) = \begin{cases} \mathbb{Z}, & \text{if } q = 0, 2n-3, \\ \mathbb{Z}_2, & \text{if } q = n-1, \\ 0 & \text{for all other } q. \end{cases}$$

This implies the statement of the lemma.

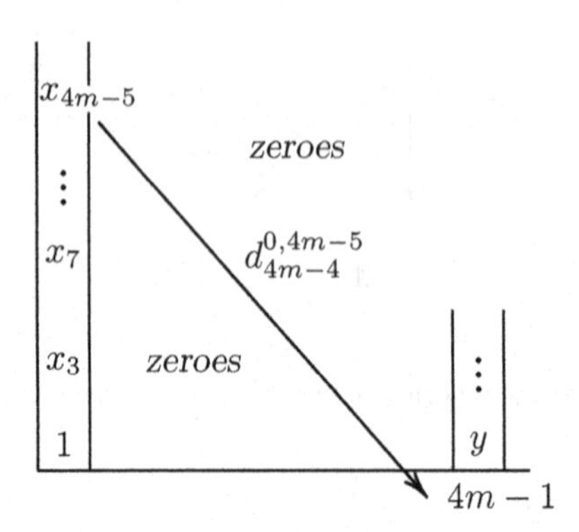

Fig. 94 Cohomological spectral sequence of the fibration $SO(2m+1) \xrightarrow{SO(2m-1)} V(2m+1,2)$

[By the way, we see also that if n is even, then $H^*(V(n,2);\mathbb{Z}) \cong H^*(S^{n-1} \times S^{n-2};\mathbb{Z})$, and similarly for the coefficients in $\mathbb{K}$; we do not need this statement—at least not now.]

Now let us return to the theorem. The cohomological spectral sequence of the fibration $SO(n) \xrightarrow{SO(n-2)} V(n,2)$ for $n = 2m+1$ (with coefficients in $\mathbb{K}$) is shown in Fig. 94. If we repeat the argumentation in Sect. 24.3 word for word, we obtain a multiplicative isomorphism

$$H^*(SO(2m+1);\mathbb{K}) \cong H^*(S^3 \times S^7 \times \ldots \times S^{4m-1};\mathbb{K}).$$

Note one more consequence of the triviality of the differentials in the last spectral sequence: The homomorphism $H^*(SO(2m+1);\mathbb{K}) \to H^*(SO(2m-1);\mathbb{K})$ induced by the inclusion map is onto.

Consider now the spectral sequence of the fibration

$$SO(2m) \xrightarrow{SO(2m-1)} S^{2m-1}$$

(Fig. 95).

The dimension argumentations do not prove the triviality of the differentials of this spectral sequence, but still these differentials are all trivial: The composition

$$H^*(SO(2m+1);\mathbb{K}) \longrightarrow H^*(SO(2m);\mathbb{K}) \xrightarrow{*} H^*(SO(2m-1);\mathbb{K})$$

induced by the composed inclusion $SO(2m-1) \to SO(2m) \to SO(2m+1)$ is (as was shown above) an epimorphism, and therefore so is the homomorphism marked by an asterisk. But the kernel of the differential considered is the image of the homomorphism $(*)$; thus, these differentials are all zero. This completes the proof.

Theorem. *There is an additive isomorphism*

$$H^*(SO(n);\mathbb{Z}_2) \cong H^*(S^1 \times S^2 \times S^3 \times \ldots \times S^{n-1};\mathbb{Z}_2).$$

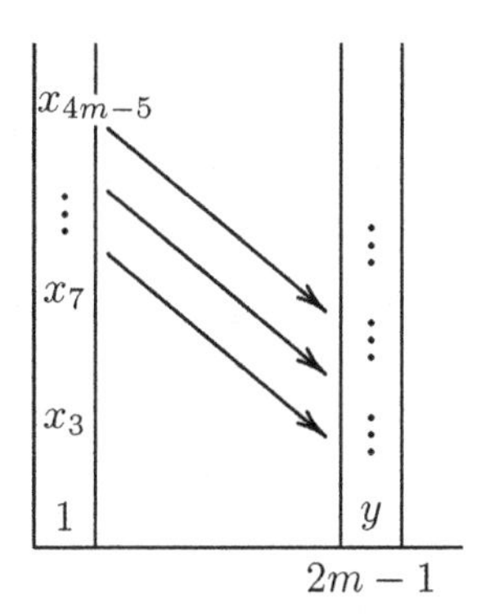

Fig. 95 Cohomological spectral sequence of the fibration $SO(2m) \xrightarrow{SO(2m-1)} S^{2m-1}$

Moreover, there exist elements $x_{n,i} = x_i \in H^i(SO(n); \mathbb{Z}_2), i = 1, 2, \dots, n-1$, *with the following properties:*

(1) *The monomials* $x_{i_1} x_{i_2} \dots x_{i_s}$, $1 \le i_1 < i_2 < \dots < i_s \le n-1$, *form an additive basis in* $H^*(SO(n); \mathbb{Z}_2)$.
(2) *The cohomology homomorphism induced by the inclusion map* $SO(n-1) \to SO(n)$ *takes* $x_{n,i}$ *with* $i < n-1$ *into* $x_{n-1,i}$.
(3) $x_{n,n-1}$ *is the image of the generator of the group* $H^{n-1}(S^{n-1}; \mathbb{Z}_2)$ *with respect to the homomorphism induced by the projection* $SO(n) \to S^{n-1}$.

Proof. Apply induction with respect to n. For $n = 2$ the statement is true. Consider the spectral sequence of the fibration

$$SO(n) \xrightarrow{SO(n-1)} S^{n-1}.$$

We want to prove that the differential (transgression)

$$d_{n-1}^{0,n-2} : H^{n-2}(SO(n-2); \mathbb{Z}_2) \to H^{n-2}(S^{n-1}; \mathbb{Z}_2)$$

takes x_{n-1} into 0 (all the rest is deduced in the usual way from the properties of the spectral sequences). To prove this, consider the map between fibrations,

$$\begin{array}{ccc} SO(n) & \longrightarrow & V(n,2) \\ \big\downarrow{\scriptstyle SO(n-1)} & & \big\downarrow{\scriptstyle S^{n-2}} \\ S^{n-1} & \xrightarrow{\text{id}} & S^{n-1} \end{array}$$

and the corresponding homomorphism between spectral sequences (Fig. 96).

It follows from the induction hypothesis that the homomorphism between the $E_2^{0,n-2}$-terms takes the generator of the group $H^{n-2}(S^{n-2}; \mathbb{Z}_2)$ into $x_{n-2} \in$

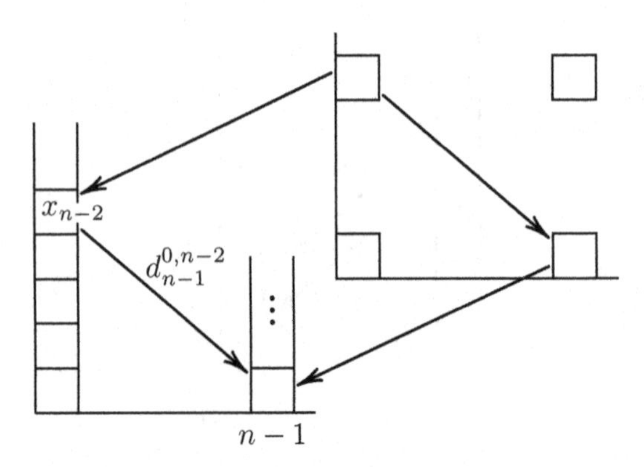

Fig. 96 The homomorphism between spectral sequences

$H^{n-2}(SO(n-1);\mathbb{Z}_2)$. But in the spectral sequence of the fibration $V(n,2) \to S^{n-1}$, our differential, as we know, is 0 (modulo 2). Hence, it is also 0 in our spectral sequence.

Remark 1. The theorem shows that $H^*(SO(2m+1);\mathbb{Z}_2) \not\cong H^*(Sp(2m+1);\mathbb{Z}_2)$ although, according to Sect. 24.4.A, $H^*(SO(2m+1);\mathbb{K}) \cong H^*(Sp(2m+1);\mathbb{K})$ for $\mathbb{K} = \mathbb{Q}, \mathbb{R}, \mathbb{C}$.

Remark 2. The theorem does not describe the ring $H^*(SO(n);\mathbb{Z}_2)$, since what x_i^2 is remains unknown. For example, $SO(3) = \mathbb{R}P^3$ (see Sect. 1.7), and the $\mathbb{Z}_2$-cohomology of $\mathbb{R}P^3$ is additively generated by $1, x, x^2, x^3$, $x \in H^1(\mathbb{R}P^3;\mathbb{Z}_2)$. Thus, $x_1^2 = x_2$ for $n = 3$ (and consequently for all $n \geq 3$). In fact, the following is true: *The ring $H^*(SO(n);\mathbb{Z}_2)$ is generated by $y_i \in H^{2i-1}(SO(n);\mathbb{Z}_2), i = 1, 2, \ldots, [n/2]$, with the defining system of relations (besides the commutativity relations) $y_i^{2^{k_i}} = 0$, where 2^{k_i} is such a power of 2 that $n \leq i \cdot 2^{k_i} < 2n$* (see Kac [50]). The reader can find some further information on the cohomology of Lie groups in the article by Fuchs [39].

24.5 Cohomology of the Loop Space of a Sphere

Theorem.

$$H^m(\Omega S^n;\mathbb{Z}) = \begin{cases} \mathbb{Z}, & \text{if } m \geq 0 \text{ is divisible by } n-1, \\ 0, & \text{if } m \text{ is not divisible by } n-1. \end{cases}$$

Moreover, there exist generators $x_k \in H^{k(n-1)}(\Omega S^n;\mathbb{Z})$ such that $x_k x_\ell = \alpha_{k,\ell} x_{k+\ell}$, where

$$\alpha_{k.\ell} = \begin{cases} \binom{k+\ell}{\ell}, & \text{if } n \text{ is odd,} \\ \binom{[(k+\ell)/2]}{[\ell/2]}, & \text{if } n \text{ is even, and at least one of } k,\ell \text{ is even,} \\ 0, & \text{if } n \text{ is even, and both } k,\ell \text{ are odd.} \end{cases}$$

Proof. The additive part follows directly from the exactness of the Wang sequence applied to the path fibration $ES^n \xrightarrow{\Omega S^n} S^n$: Since $H^m(ES^n;\mathbb{Z}) = 0$ for $m > 0$, the differentials $d_n^{m0}: E_n^{0m} \to E_n^{n,m-n}$; that is, $H^m(\Omega S^n;\mathbb{Z}) \to H^{m-n}(\Omega S^n;\mathbb{Z})$ are isomorphisms for $m > 0$, whence the result. If we choose some generators x_k in $E_n^{0,k(n-1)}$, then $E_n^{n,k(n-1)}$ is generated by sx_k, where s is the canonical generator in $E_n^{n0} = H^n(S^n;\mathbb{Z})$. If we require that $x_0 = 1$ and $d_n^{0,k(n-1)}x_k = sx_{k-1}$, the system $\{x_k \in H^{k(n-1)}(\Omega S^n;\mathbb{Z})\}$ is uniquely determined. Now, by the dimension arguments, we must have $x_k x_\ell = \alpha_{k,\ell} x_{k+\ell}$ for some $\alpha_{k,\ell} \in \mathbb{Z}$, and, from the multiplicative properties of the differentials,

$$\begin{aligned} d(x_k x_\ell) &= d(\alpha_{k,\ell} x_{k+\ell}) = \alpha_{k,\ell}\, s x_{k+\ell-1}, \text{ and} \\ d(x_k x_\ell) &= (dx_k)x_\ell + (-1)^{k(n-1)} x^k dx_\ell = s x_{k-1} x_\ell + (-1)^{k(n-1)} x_k s x_{\ell-1} \\ &= \left[\alpha_{k-1.\ell} + (-1)^{k(n-1)} \alpha_{k,\ell-1}\right] s x_{k+\ell-1} \end{aligned}$$

[here we use the fact that $x_k s = (-1)^{k(n-1)n} s x_k = s x_k$]. Thus,

$$\alpha_{k,\ell} = \begin{cases} \alpha_{k-1,\ell} + \alpha_{k,\ell-1}, & \text{if } n \text{ is odd,} \\ \alpha_{k-1,\ell} + (-1)^k \alpha_{k,\ell-1}, & \text{if } n \text{ is even.} \end{cases}$$

We leave for the reader's entertainment deducing the formula for $\alpha_{k,\ell}$ as stated in the theorem from these Pascal triangle–like relations.

24.6 One More Example (Demonstrating the Perfidy of the Multiplicative Adjointness)

Readers have certainly noticed that in the computation of cohomology of $SU(n), SO(n)$, and $Sp(n)$ we made the transition from the ring E_∞ to the cohomology ring with some cautiousness although these cohomology rings finally turned out to be isomorphic. This was not the case for $H^*(SO(n);\mathbb{Z}_2)$, but it does not seem surprising for the case of finite coefficient domain. However, this isomorphism does not hold even in the case of torsion-free integral cohomology, and even in the case of coefficients in $\mathbb{Q}$ (or $\mathbb{R}$, or $\mathbb{C}$), as the following example shows.

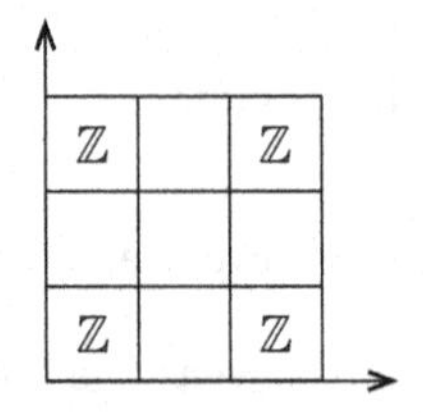

Fig. 97 The spectral sequence of the fibration $E \xrightarrow{S^2} S^2$

Suppose that we have a fibration with the base S^2 and the fiber S^2 (we do not specify which). The E_2-term of the cohomological spectral sequence of this fibration is shown in Fig. 97.

By the dimension arguments, all the differentials are trivial, and $E_\infty = E_2$. Thus, the E_∞-term is additively generated by $1 \in E_\infty^{00}$, $x \in E_\infty^{02}$, $y \in E_\infty^{20}$, and $xy \in E_\infty^{22}$. Accordingly, $H^*(E;\mathbb{Z})$ is additively generated by $1 \in H^0(E;\mathbb{Z}), \widetilde{x}, \widetilde{y} \in H^2(E;\mathbb{Z})$, and $\widetilde{z} \in H^4(E;\mathbb{Z})$. Since $E_\infty^{20} \subset H^2(E;\mathbb{Z}), \widetilde{y}$ is essentially the same as y, and hence it is well defined. However, $\widetilde{x}$ is a representative of the class $x \in E_\infty^{02} = H^2(E;\mathbb{Z})/E_\infty^{20}$ and it is defined up to a summand of the form $k\widetilde{y}$, where k is an integer. Since the ring E_∞ is adjoint to the ring $H^*(E;\mathbb{Z})$, we see that $\widetilde{y}^2 = 0. \widetilde{xy} = \widetilde{z}$, and $\widetilde{x}^2$ has a 0 image in the quotient group E_∞^{40} of the group $H^4(E;\mathbb{Z})$, which does not mean a thing, since $E_\infty^{40} = 0$. Thus, $\widetilde{x}^2 = \lambda\widetilde{z}$, where λ is an integer that is unknown to us. We can influence λ by adding to $\widetilde{x}$ a multiple of $\widetilde{y}$: $(\widetilde{x}+k\widetilde{y})^2 = \widetilde{x}^2 + 2k\widetilde{x}\widetilde{y} + k^2\widetilde{y}^2 = (\lambda+2k)\widetilde{z}$. Thus, we can shift λ by any even number, which shows that there may be only two essentially different cases: λ is even and λ is odd. [The rings with odd and even λ are not isomorphic: For λ odd, every element of $H^4(E;\mathbb{Z})$ is a square; for λ even, the generator of $H^4(E;\mathbb{Z})$ is not a square.] Let us show that both cases are represented by actual fibrations.

The case of even λ occurs for the trivial fibration; this case is not interesting, because for trivial fibrations we again have a multiplicative isomorphism $E_\infty \cong H^*(E;\mathbb{Z})$.

Let us now construct a nontrivial fibration with the base and the fiber homeomorphic to S^2. Choose a point $a \in \mathbb{C}P^2 - \mathbb{C}P^1$, take a ball $D^4 \subset \mathbb{C}P^2$ centered at a and disjoint from $\mathbb{C}P^1$, and put $X = \mathbb{C}P^2 - \operatorname{Int} D^4$; this is a four-dimensional manifold with the boundary S^3. Draw all (complex, projective) lines through a. They are disjoint in X, they cover X, and each of them intersects X by a two-dimensional disk and intersects $\mathbb{C}P^1$ in one point (the "center" of this disk). As a result, X becomes a fibered space with the base $\mathbb{C}P^1 = S^2$ and the fiber D^2; the boundary $\partial X = S^3$ is fibered with the base S^2 and the fiber $\partial D^2 = S^1$, and this is the Hopf fibration. Take two copies of X and attach them to each other according to the identity map of the boundary. We get a four-dimensional manifold which is known to us as the connected sum $E = \mathbb{C}P^2 \# \mathbb{C}P^2$ of two copies of $\mathbb{C}P^2$ (we dealt with connected sums in Sect. 17.10, Exercises 44 and 45 and an example between them; the construction of a connected sum is shown in Fig. 98).

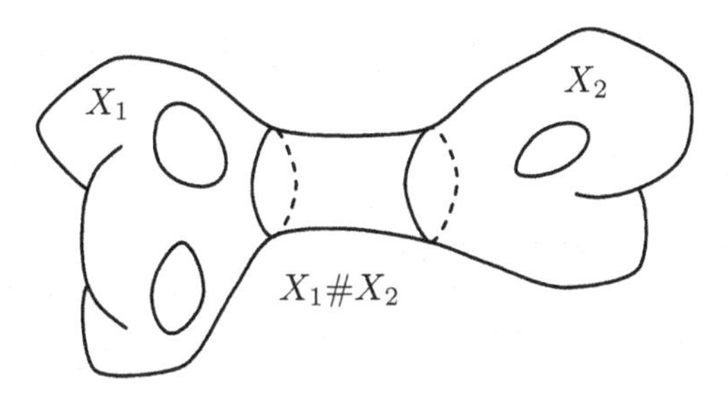

Fig. 98 The connected sum $X_1\#X_2$ of two manifolds, X_1 and X_2

The fibers of the two fibrations are also attached to each other, forming a fibration of E over S^2 with the fiber S^2.

The cohomology of the space E can be found, for example, using the Mayer–Vietoris sequence using the decomposition $E = X \cup X$ with $X \cap X = S^3$. The result is that the cohomology $H^*(E;\mathbb{Z})$ is additively generated by four classes, $1, x, y, z$ of dimensions $0, 2, 2, 4$. The two natural maps $E \to \mathbb{C}P^2$ (obtained by collapsing one of the Xs to a point) map $H^*(\mathbb{C}P^2;\mathbb{Z})$ onto subrings of $H^*(E;\mathbb{Z})$ generated, respectively, by $1, x, z$ and $1, y, z$, which implies that [in $H^*(E;\mathbb{Z})$] $x^2 = z$ and $y^2 = z$. Also, $xy = 0$ [the two-dimensional homology classes forming the basis in $H_2(E)$ are represented by the two complex projective lines $\mathbb{C}P^1 \subset X$, and these two are disjoint; their intersection number is 0]. We see that the cohomology ring of E is not isomorphic to that of $S^2 \times S^2$ (z is a square), which gives us an example of a nontrivial multiplicative adjointness of E_∞ to $H^*(E;\mathbb{Z})$.

Remark. If we replace in our construction $\mathbb{C}$ by $\mathbb{R}$, then X will become a Möbius band, and E will turn into a Klein bottle. So our E can be regarded as a "complex Klein bottle." Also, we can replace $\mathbb{C}$ by $\mathbb{H}$. This will give us an interesting nontrivial fibration $E \xrightarrow{S^4} S^4$.

In conclusion, we present several exercises concerning multiplicative cohomological spectral sequences.

EXERCISE 4. Let the homology of a connected closed orientable manifold X be known. Find the homology of the manifold of nonzero tangent vectors to X.

EXERCISE 5. Find the (already known to you) cohomology ring of $\mathbb{C}P^n$ using the spectral sequence of the Hopf fibration $S^{2n+1} \xrightarrow{S^1} \mathbb{C}P^n$.

EXERCISE 6. Find the ring of rational cohomology of $SU(n)/SO(n)$.

EXERCISE 7. Find the ring of rational cohomology of $SO(2n)/SU(n)$, at least for some small n (the first interesting case is $n = 3$).

Lecture 25 Killing Spaces Method for Computing Homotopy Groups

This lecture, as well as the two next lectures and a considerable part of subsequent chapters, will be devoted to computing homotopy groups. We begin with a quotation from an interview which Jean-Pierre Serre gave in Singapore in February 1985 (published in *Mathematical Intelligencer,* 1986, **8**, 8–13).

Q: Have you ever had the experience where you found a problem to be impossible to solve, and then after putting it aside for some time, an idea suddenly occurred leading to the solution?

A: Yes, of course this happens quite often. For instance, when I was working on homotopy groups ($\sim$1950), I convinced myself that, for a given space X, there should exist a fiber space E, with base X, which is contractible; such a space would indeed allow me (using Leray's methods) to do lots of computations on homotopy groups and Eilenberg–MacLane cohomology. But how to find it? It took me several weeks (a very long time, at the age I was then[1]...) to realize that the space of "paths" on X had all the necessary properties—if only I dared call it a "fiber space," which I did. This was the starting point of the loop space method in algebraic topology; many results followed quickly.

25.1 Killing Spaces Revisited

The initial Serre method of computing homotopy groups was based on considering multiple loop spaces. Suppose that we want to calculate the homotopy groups of some space X. Consider the sequence of spaces, $X, \Omega X, \Omega\Omega X, \Omega\Omega\Omega X, \dots$. Imagine that there is a way of calculating the homology of the loop space ΩY, provided that the homology of Y is known. For this, we can use the spectral sequence of the path fibration $EY \xrightarrow{\Omega Y} Y$. But the first nontrivial homology group of $\Omega^k X$ is, by Hurewicz's theorem, the same as its first nontrivial homotopy group, say, $\pi_r(\Omega^k X)$, which is, in turn, $\pi_{r+k}(X)$.

We will not directly use this method here. We prefer another method, also belonging to Serre: the method of killing spaces. (One can say that, roughly speaking, these two Serre's methods are closely related to each other and lead to similar results.)

Killing spaces were introduced in Sect. 11.9. Remember that the killing space $X|_{n+1}$ is constructed for X in a canonical way and has the property

[1]24 years—AF&DF.

$$\pi_q(X|_{n+1}) = \begin{cases} \pi_q(X), & \text{if } q > n, \\ 0, & \text{if } q \le n. \end{cases}$$

If $\pi_n(X) = \pi$ is the first nontrivial homotopy group of X, then $X|_{n+1}$ is related to X by two fibrations:

$$X \xrightarrow{X|_{n+1}} K(\pi, n), \; X|_{n+1} \xrightarrow{K(\pi,n-1)} X.$$

The first of them was constructed in Sect. 11.9: Its projection belongs to the homotopy class of maps corresponding to the fundamental class $F_X \in H^n(X; \pi)$. The projection of the second fibration is homotopic to the inclusion $X|_{n+1} \to X$ of the fiber of the first fibration; in other words, it is the fibration over X induced by the path fibration $EK(\pi, n) \xrightarrow{K(\pi,n-1)} K(\pi, n)$ with respect to the projection of the first fibration. In still other words, both fibrations may be described as parts of the following commutative diagram of four fibrations:

$$\begin{array}{ccc} X|_{n+1} & \xrightarrow{X|_{n+1}} & * \\ \Big\downarrow{\scriptstyle K(\pi,n-1)} & & \Big\downarrow{\scriptstyle K(\pi,n-1)} \\ X & \xrightarrow{X|_{n+1}} & K(\pi, n) \end{array}$$

(where all terms represent homotopy types; in particular, $*$ denotes a contractible space). We usually will use the second of these fibrations. In principle, if we know the cohomology of $K(\pi, n-1)$, then we can at least try to find the cohomology of $X|_{n+1}$, and hence the first nontrivial homotopy group of $X|_{n+1}$, which is the second nontrivial homotopy group of X. And so on.

25.2 First Application: A Computation of $\pi_{n+1}(S^n)$

At the moment, we know almost nothing of the cohomology of the Eilenberg–MacLane spaces. Still we know that $K(\mathbb{Z}, 2) = \mathbb{C}P^\infty$, and this makes it possible to find $H^*(S^3|_4; \mathbb{Z})$.

Theorem.

$$H^q(S^3|_4; \mathbb{Z}) \cong \begin{cases} \mathbb{Z}_m, & \text{if } q = 2m + 1,\ m = 2, 3, 4, \dots, \\ 0 & \text{for all other } q > 0. \end{cases}$$

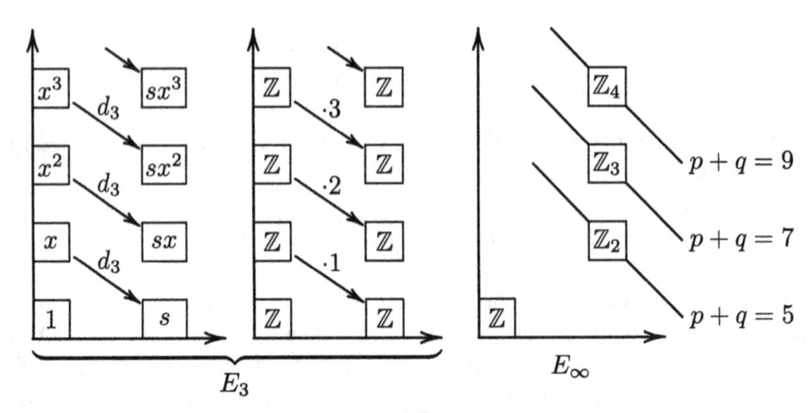

Fig. 99 Spectral sequence of the fibration $S^3|_4 \xrightarrow{K(\mathbb{Z},2)} S^3$

Proof. Consider the $\mathbb{Z}$-cohomology spectral sequence of the fibration $S^3|_4 \xrightarrow{K(\mathbb{Z},2)} S^3$ (see Fig. 99). Denote by x the (canonical) generator of the group $H^3(\mathbb{C}P^2;\mathbb{Z}) = H^2(K(\mathbb{Z},2);\mathbb{Z}) = E_2^{02}$ and denote by s the generator of the group $H^3(S^3;\mathbb{Z}) = E_2^{30}$. Then

the group $E_2^{0,2m} = \mathbb{Z}$ is generated by x^m,
the group $E_2^{3,2m} = \mathbb{Z}$ is generated by sx^m,
and all the other groups E_2^{pq} are trivial.

Obviously, $E_3 = E_2$ and $E_\infty = E_4$; but there are potentially nontrivial differentials $d_3 = d_3^{0,2m}: E_3^{0,2m} \to E_3^{3,2m-2}$, which are, for $m \geq 1$, $\mathbb{Z} \to \mathbb{Z}$. Since $H^2(S^3|_4;\mathbb{Z}) = H^3(S^3|_4;\mathbb{Z}) = 0$, the differential d_3^{02} must be an isomorphism; thus, $d_3(x) = \pm s$; we can assume that $d_3(x) = s$ (although this is not important). Then, by the product rule, $d_3(x^m) = mx^{m-1}d_3x = msx^{m-1}$; hence, the terms E_3 and E_∞ look like what is shown in Fig. 99. The theorem follows.

The formulas connecting integral homology and cohomology (Corollary 2 in Sect. 15.5) show that for $q > 0$

$$H_q(S^3|_4) = \begin{cases} \mathbb{Z}_m, & \text{if } q = 2m, \\ 0, & \text{if } q \neq 4, 6, 8, \ldots \end{cases}$$

and hence $\pi_4(S^3|_4) = \mathbb{Z}_2 \Rightarrow \pi_4(S^3) = \mathbb{Z}_2 \Rightarrow \pi_{n+1}(S^n) = \mathbb{Z}_2$ for $n \geq 3$. The reader who was able to get through the exercises in Sect. 10.5 already knows this fact. But even for that reader the easiness of the proof may be strong evidence of a big advantage of the method of spectral sequences. At the same time, we can see a great importance of the "Eilenberg–MacLane cohomology" $H^*(K(\pi,n);-)$. The rest of this chapter, as well as most of the next chapter, will be devoted mostly to this cohomology.

Lecture 26 Rational Cohomology of $K(\pi, n)$ and Ranks of Homotopy Groups

The computation of the Eilenberg–MacLane cohomology turned out to be hard work. Still, the problem was fully solved (for finitely generated Abelian π) in the 1950s, mostly by the French topologists A. Borel, H. Cartan, and J.-P. Serre. The easiest part of this work was the computation of the rational cohomology ring $H^*(K(\pi, n); \mathbb{Q})$. This lecture will be devoted to this computation and its corollaries. We begin with a useful general result.

26.1 The Case of Finite and Finitely Generated Group

A: The Main Result

Theorem. *If π is a finitely generated (finite) Abelian group, then for every finitely generated Abelian group G and every $n > 0$ and $q > 0$, the group $H^q(K(\pi, n); G)$ is finitely generated (finite).*

Proof. Because of the universal coefficients formula, we can restrict ourselves to the case when $G = \mathbb{Z}$.

For $n = 1$, the statement is true: We know the spaces $K(\mathbb{Z}, 1) = S^1, K(\mathbb{Z}_2, 1) = \mathbb{R}P^\infty, K(\mathbb{Z}_m, 1) = L_m^\infty$ (the infinite-dimensional lens space) for $m > 2$ and their cohomology:

q	0	1	2	3	4	5	6	$\cdots$
$H^q(S^1; \mathbb{Z})$	$\mathbb{Z}$	$\mathbb{Z}$	0	0	0	0	0	$\cdots$
$H^q(\mathbb{R}P^\infty; \mathbb{Z})$	$\mathbb{Z}$	0	$\mathbb{Z}_2$	0	$\mathbb{Z}_2$	0	$\mathbb{Z}_2$	$\cdots$
$H^q(L_m^\infty; \mathbb{Z})$	$\mathbb{Z}$	0	$\mathbb{Z}_m$	0	$\mathbb{Z}_m$	0	$\mathbb{Z}_m$	$\cdots$

Now assume the statement to be true for $K(\pi, n-1)$ and assume that it is wrong for $K(\pi, n)$. Let $H^m(K(\pi, n); \mathbb{Z})$ be an infinitely generated (infinite) cohomology group of $K(\pi, n)$ of the smallest positive dimension. Consider the $\mathbb{Z}$-cohomology spectral sequence of the fibration $* \xrightarrow{K(\pi,n-1)} K(\pi, n)$, where $*$ denotes the contractible space $EK(\pi, n)$ (see Fig. 100). It follows from the induction hypothesis, the universal coefficients formula, and the definition of m that the groups E_2^{pq} with $p < m$ (except E_2^{00}) are finitely generated (finite), while the group E_2^{m0} is infinitely generated (infinite). [See Fig. 100, where the light squares correspond to

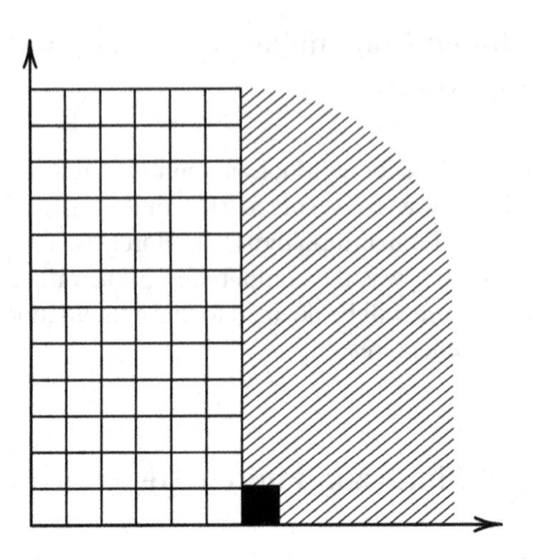

Fig. 100 The spectral sequence of the fibration $* \xrightarrow{K(\pi,n-1)} K(\pi,n)$

finitely generated (finite) groups, and the dark square corresponds to an infinitely generated (infinite) group.]

The groups $E_r^{p,q}$ with $p < m$ are obtained from E_2^{pq} by finitely many transitions to subgroups and quotients, so they (except E_r^{00}) are also finitely generated (finite). Since the quotient of an infinitely generated (infinite) group over a finitely generated (finite) subgroup is also an infinitely generated (infinite) group, all the groups

$$\begin{aligned} E_3^{m0} &= E_2^{m0}/d_2^{m-2,1}(E_2^{m-2,1}), \\ E_4^{m0} &= E_3^{m0}/d_3^{m-3,2}(E_3^{m-3,2}), \\ &\dots\dots\dots\dots\dots\dots \end{aligned}$$

are infinitely generated (infinite), which contradicts the equality $E_{m+1}^{m0} = E_\infty^{m0} = 0$. This completes the proof.

B: A Generalization: Classes of Abelian Groups

We say that a class $\mathcal{C}$ of Abelian groups is given if every Abelian group belongs or does not belong to $\mathcal{C}$, and (1) isomorphic groups belong or do not belong to $\mathcal{C}$ simultaneously; (2) if a group belongs to $\mathcal{C}$, then all its subgroups and all its quotients belong to $\mathcal{C}$; (3) if a subgroup H of a group G and the quotient G/H belong to $\mathcal{C}$, then G belongs to $\mathcal{C}$. This definition has numerous variations (see Serre [76] and subsequent works of different authors). Examples: finitely generated groups; finite groups; periodic groups; finite p-groups.

EXERCISE 1. Prove that if $\mathcal{C}$ is a class of finitely generated Abelian groups, $\pi \in \mathcal{C}$, and G is a finitely generated Abelian group, then $H^q(K(\pi,n);G) \in \mathcal{C}$ for all $n > 0, q > 0$.

C: Application

Theorem. *Let X be a simply connected topological space such that all homology groups $H_q(X)$, $q > 0$ are finitely generated (finite). Then all the homotopy groups $\pi_q(X)$ are finitely generated (finite).*

Remark. The requirement of X being simply connected is not unnecessary: For example, the homology groups of $S^1 \vee S^2$ are all finitely generated, while the group $\pi_2(S^1 \vee S^2)$ is not finitely generated. Still this requirement can be weakened to, say, the requirement of X being homotopically simple.

Lemma. *If all the homology groups of positive dimensions of the base and the fiber of a homologically simple fibration are finitely generated (finite), then so are all homology groups of positive dimension of the total space of this fibration. The same is true for homology and cohomology with any finitely generated coefficients.*

Proof of Lemma. Because of the universal coefficients formula, we can restrict ourselves to the integral homology. In the homology spectral sequence of our fibration, all the groups E^2_{pq} (except E^2_{00}) are finitely generated (finite). Since the transitions $E^2 \to E^3 \to E^4 \to \dots$ consist in taking subgroups and quotient groups, the same is true for E^∞_{pq}, and hence for the homology groups of the total space.

EXERCISE 2. Let $E \xrightarrow{F} B$ be a homologically simple fibration. Prove that if the homology groups of positive dimensions of two of the three spaces E, B, and F are finitely generated (finite), then the same is true for the homology groups of the third space.

Proof of Theorem. We successively apply the lemma to fibrations

$$X|_3 \xrightarrow{K(H_3(X),1)} X,\ X|_4 \xrightarrow{K(H_3(X|_3),2)} X|_3,\ X|_5 \xrightarrow{K(H_4(X|_4),3)} X|_4, \dots,$$

to prove that all homology groups of positive dimensions of the spaces $X|_3, X|_4, X|_5, \dots$ are finitely generated (finite). Hence, the groups $\pi_2(X) = H_2(X)$, $\pi_3(X) = H_3(X|_3)$, $\pi_4(X) = H_4(X|_4)$, $\pi_5(X) = H_5(X|_5), \dots$ are finitely generated (finite).

Corollary. *The homotopy groups $\pi_m(S^3)$ are finite for all $m \geq 4$.*

Proof. It was shown in Sect. 25.2 that all the groups $H_m(S^3|_4)$, $m > 0$, are finite. Hence, by the theorem, the groups $\pi_m(S^3|_4)$ are all finite, and $\pi_m(S^3|_4) = \pi_m(S^3)$ for all $m \geq 4$.

EXERCISE 3. Prove that if all homology groups of some positive dimension of a simply connected (or homotopically simple) space belong to a class $\mathcal{C}$ of finitely generated Abelian groups, then all the homotopy groups of this space also belong to the class $\mathcal{C}$.

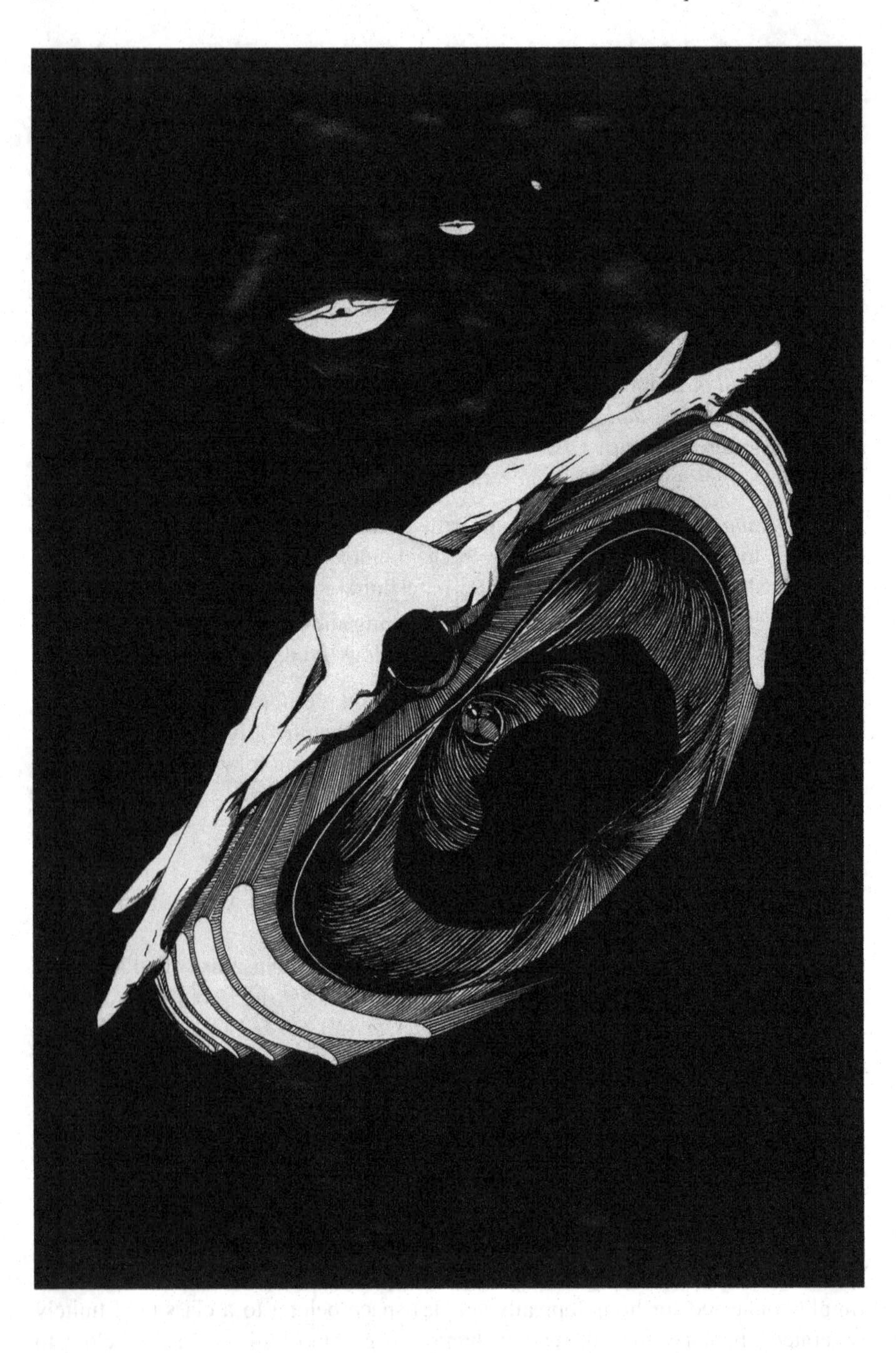

EXERCISE 4. Prove the following "Hurewicz $\mathcal{C}$-theorem": If for some simply connected space X, all homotopy groups $\pi_m(X)$ with $m < n$ belong to a class $\mathcal{C}$ (as in Exercise 3), then the groups $H_m(X)$ with $m < n$ also belong to $\mathcal{C}$ and the Hurewicz homomorphism $\pi_n(X) \to H_n(X)$ is a $\mathcal{C}$-isomorphism (that is, its kernel and cokernel belong to $\mathcal{C}$). Prove also the inverse statement where homology and homotopy groups are swapped.

Remark. It follows from the Hurewicz $\mathcal{C}$-theorem that for every prime p the order of the group $\pi_m(S^3)$ with $4 \leq m < 2p$ is not divisible by p. To prove that, we apply the Hurewicz $\mathcal{C}$-theorem to the space $S^3|_4$ and the class $\mathcal{C}$ of finite Abelian groups of the order not divisible by p. This statement will be radically generalized in Lecture 27 and then in Chap. 4.

A detailed exposition of the "$\mathcal{C}$-theory" is contained in Serre's article cited above.

26.2 The Computation of the Rings $\mathbf{H^*(K(\pi, n); \mathbb{Q})}$ for Finitely Generated Abelian Groups π

Let π be a finitely generated Abelian group. Then $\pi \cong \mathbb{Z} \oplus \ldots \oplus \mathbb{Z} \oplus \rho$, where ρ is a finite Abelian group. Accordingly, $K(\pi, n) = K(\mathbb{Z}, n) \times \ldots \times K(\mathbb{Z}, n) \times K(\rho, n)$, and, by Künneth's formula,

$$H^*(K(\pi, n); \mathbb{Q}) = \\ H^*(K(\mathbb{Z}, n); \mathbb{Q}) \otimes \ldots \otimes H^*(K(\mathbb{Z}, n); \mathbb{Q}) \otimes H^*(K(\rho, n); \mathbb{Q}).$$

By the theorem in Sect. 26.1.A, $H^*(K(\rho, n), \mathbb{Q}) = H^*(\text{pt}; \mathbb{Q})$ (if the integral homology groups of a topological spaces are finite, then its rational cohomology is trivial). Hence, we can remove the last factor in the formula for $H^*(K(\pi, n); \mathbb{Q})$ and all we need to compute is $H^*(K(\mathbb{Z}, n); \mathbb{Q})$.

Theorem.

$$H^*(K(\mathbb{Z}, n); \mathbb{Q}) = \begin{cases} \Lambda_{\mathbb{Q}}(x), \ \dim x = n, & \text{if } n \text{ is odd}, \\ \mathbb{Q}[x], \ \dim x = n, & \text{if } n \text{ is even}. \end{cases}$$

Let us explain the notation. If $\mathbb{K}$ is a field, then $\Lambda_{\mathbb{K}}(x_1, \ldots, x_m)$ denotes the exterior algebra with generators $x_1, \ldots, x_m$, that is, the algebra with these generators and relations $x_i x_j = -x_j x_i$ and $x_i^2 = 0$. The dimension of this algebra is 2^m, and an (additive) basis formed by monomials $x_{i_1} \ldots x_{i_s}$ with $1 \leq i_1 < \ldots < i_s \leq m$. Sometimes, the letter $\mathbb{K}$ denotes not a field, but a sufficiently good ring, like $\mathbb{Z}$; it may be omitted if the context allows that. The formula $H^*(K(\mathbb{Z}, n); \mathbb{Q}) = \Lambda_{\mathbb{Q}}(x)$, $\dim x = n$ has a very simple meaning:

$$H^*(K(\mathbb{Z}, n); \mathbb{Q}) = H^*(S^n; \mathbb{Q}).$$

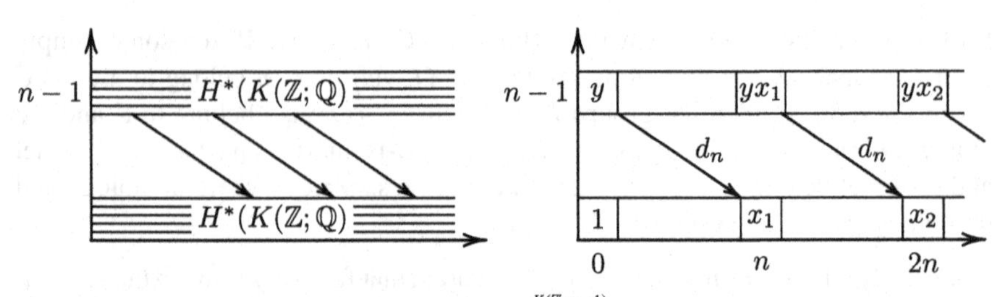

Fig. 101 The spectral sequence of the fibration $* \xrightarrow{K(\mathbb{Z},n-1)} K(\mathbb{Z},n)$, n is even

By $\mathbb{K}[x_1,\dots,x_m]$ we understand (as usual) the ring of polynomials in $x_1,\dots x_m$ with the coefficients in $\mathbb{K}$. Differently, this is a $\mathbb{K}$-algebra with generators $x_1,\dots,x_m$ and relations $x_ix_j = x_jx_i$. The formula $H^*(K(\mathbb{Z},n);\mathbb{Q})$ means that $H^q(K(\mathbb{Z},n);\mathbb{Q}) = \mathbb{K}$ for $q = 0,n,2n,3n,\dots$ and $H^q(K(\mathbb{Z},n);\mathbb{Q}) = 0$ for all other q and that $H^{kn}(K(\mathbb{Z},n);\mathbb{Q})$ is generated by x^k, where x is an arbitrarily chosen nonzero element of $H^n(K(\mathbb{Z},n);\mathbb{Q})$.

Proof of Theorem. For $n = 1$ (and $n = 2$) the statement is known to us, since $K(\mathbb{Z},1) = S^1$, $K(\mathbb{Z},2) = \mathbb{C}P^\infty$. We assume that the statement is true for $K(\mathbb{Z},n-1)$ and prove it for $K(\mathbb{Z},n)$. Consider the $\mathbb{Q}$-cohomological spectral sequence of the fibration

$$* \xrightarrow{K(\mathbb{Z},n-1)} K(\mathbb{Z},n)$$

and begin with the case when n is even. By the induction hypothesis, $H^*(K(\mathbb{Z},n-1);\mathbb{Q}) = H^*(S^{n-1};\mathbb{Q})$, and the $E_2 = E_n$-term of the spectral sequence looks like the diagram in Fig. 101, left. Since $E_\infty^{pq} = 0$ (except E_∞^{00}), all the differentials $d_n^{q-n,n}\colon E_n^{q-n,n} \to E_n^{q,0}$ with $q \neq 0$ must be isomorphisms; that is, for $q \neq 0$, $H^{q-n}(K(\mathbb{Z},n);\mathbb{Q}) \cong H^q(K(\mathbb{Z},n);\mathbb{Q})$ and

$$H^q(K(\mathbb{Z},n);\mathbb{Q}) = \begin{cases} \mathbb{Q} \text{ for } q = 0,n,2n,\dots, \\ 0 \ \text{ for all other } q. \end{cases}$$

Moreover, there exist such nonzero $y \in H^{n-1}(K(\mathbb{Z},n-1);\mathbb{Q}) = E_2^{0,n-1} = E_n^{0,n-1}$ and $x_k \in H^{kn}(K(\mathbb{Z},n);\mathbb{Q}) = E_2^{kn,0} = E_n^{kn,0}$ $(k = 1,2,3,\dots)$ that $d_n(y) = x_1$, $d_n(yx_1) = x_2$, and so on (see Fig. 101, right). But then

$$x_k = d_n(yx_{k-1}) = d_n(y)x_{k-1} = x_1x_{k-1},$$

from which $x_k = x_1^k$. This establishes an isomorphism between the rings $H^*(K(\mathbb{Z},n);\mathbb{Q})$ and $\mathbb{Q}[x_1]$.

Now let n be odd. Then $E_2^{0,k(n-1)} = \mathbb{Q}$ with a generator x^k, and $E_2^{0,q} = 0$ if q is not divisible by $n-1$. Furthermore, since $H^p(K(\mathbb{Z},n);\mathbb{Q}) = 0$ for $0 < p < n$ and $H^n(K(\mathbb{Z},n);\mathbb{Q}) = \mathbb{Q}$ (this is known to us, but also can be deduced from our spectral

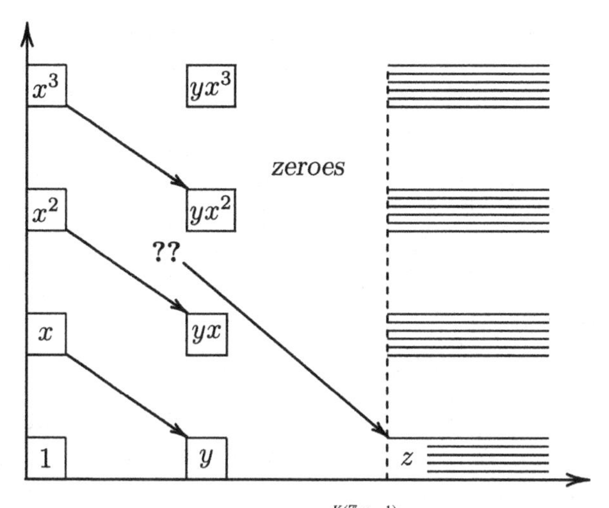

Fig. 102 The spectral sequence of the fibration $* \xrightarrow{K(\mathbb{Z},n-1)} K(\mathbb{Z},n)$, n is odd

sequence), then $E_2^{p,q} = 0$ for $0 < p < n$ and $E_2^{n,k(n-1} = \mathbb{Q}$ with the generator yx^k, where $y \in H^n(K(\mathbb{Z},n);\mathbb{Q}) = E_2^{n,0}$ is the image of x with respect to the differential d_n. Also, the rows with the numbers not divisible by $n-1$ consist of zeroes; hence, $E_2 = E_n$. We have

$$d_n(x^k) = kx^{k-1}d_n(x) = kyx^{k-1} \neq 0;$$

the differential d_n eliminates all the groups $E_n^{p,q}$ with $p = 0$ and n (except E_n^{00}).

Suppose now that there are nonzero elements of dimension greater than n in the ring $H^*(K(\mathbb{Z},n);\mathbb{Q})$, and that $z \in H^m(K(\mathbb{Z},n);\mathbb{Q})$ has, among them, the smallest dimension. But there are no elements in E_2 whose differential (of some number ≥ 2) could hit z (see Fig. 102). This contradicts the equality $E_\infty^{m,0} = 0$.

Corollary. *If* rank $\pi = r$, *then*

$$H^*(K(\pi,n);\mathbb{Q}) = \begin{cases} \Lambda_Q(x_1,\dots,x_r), \ \dim x_i = n, & \text{if } n \text{ is odd,} \\ \mathbb{Q}[x_1,\dots,x_r], \ \dim x_i = n, & \text{if } n \text{ is even.} \end{cases}$$

26.3 Ranks of the Homotopy Groups of Spheres

Theorem.

$$\operatorname{rank} \pi_q(S^n) = \begin{cases} 1, \text{ if } q = n, \text{ or } n \text{ is even and } q = 2n-1, \\ 0 \ \text{ in all other cases.} \end{cases}$$

Remark. We already know that $\pi_n(S^n) = \mathbb{Z}$ (Sect. 10.2) and that $\pi_{4n-1}(S^{2n})$ contains an element of an infinite order: the Whitehead square of the generator of $\pi_{2n}(S^{2n})$ (see Sect. 16.5). Now we are going to show that

$$\pi_{4n-1}(S^{2n}) = \mathbb{Z} \oplus \text{a finite group},$$

and that all the other groups $\pi_q(S^n)$ with $q \neq n$ are finite.

Lemma. *Let X be a simply connected space with finitely generated homology such that $H^*(X;\mathbb{Q}) \cong H^*(S^m;\mathbb{Q})$ for some odd m. Then all the groups $\pi_q(X)$ with $q \neq m$ are finite, and* rank $\pi_m(X) = 1$.

Proof of Lemma. Since $H^m(X;\mathbb{Q}) = \mathbb{Q}$, there exists an element $\gamma \in H^m(X;\mathbb{Z})$ of infinite order. Thus, there is a continuous map $f: X \to K(\mathbb{Z}, m)$ such that $f^*(\varphi) = \gamma$ [where $\varphi \in H^m(K(\mathbb{Z}, m);\mathbb{Z})$ is the fundamental class]. This shows that $f^*: H^m(K(\mathbb{Z}, m);\mathbb{Q})$ is an isomorphism, and hence $f^*: H^*(K(\mathbb{Z}, m);\mathbb{Q}) \to H^*(X;\mathbb{Q})$ is an isomorphism (in virtue of the theorem of Sect. 26.2 and assumptions concerning the cohomology of X).

Turn f into a homotopy equivalent fibration, and let F be the fiber of this fibration. Since the projection of this fibration induces an isomorphism in the rational cohomology, the $\mathbb{Q}$-cohomological spectral sequence shows that $H^q(F;\mathbb{Q}) = 0$ for any $q > 0$ (indeed, no differential of this spectral sequence can hit the zeroth row, and if m is the smallest positive dimension of a nonzero rational cohomology group of F, then $E_2^{0m} \neq 0$ must stay in E_∞, which contradicts the isomorphism above). By the theorem in Sect. 26.1.C, all the homotopy groups of F are finite. The homotopy sequence of our fibration shows that every homomorphism $f_*: \pi_q(X) \to \pi_q(K(\pi, m))$ has finite kernel and finite cokernel, which implies the statement of the lemma.

Proof of Theorem. For n odd, the theorem follows directly from the lemma. Let n be even. We already know that $\pi_n(S^n) = \mathbb{Z}$. Consider the fibration $S^n|_{n+1} \xrightarrow{K(\mathbb{Z},n-1)} S^n$. The initial term of its $\mathbb{Q}$-cohomological spectral sequence looks like what appears in Fig. 103 with the differential $d_n^{0,n-1}: E_n^{0,n-1} \to E_n^{n,0}$ killing both $E_n^{0,n-1}$ and $E_n^{n,0}$.

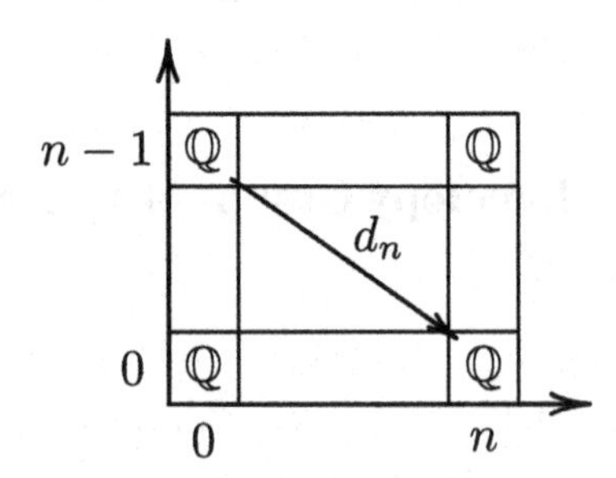

Fig. 103 The spectral sequence of the fibration $S^n|_{n+1} \xrightarrow{K(\mathbb{Z},n-1)} S^n$

We see that $H^*(S^n|_{n+1};\mathbb{Q}) \cong H^*(S^{2n-1};\mathbb{Q})$, and the lemma shows that

$$\operatorname{rank}\pi_q(S^n|_{n+1}) = \begin{cases} 1, \text{ if } q = 2n-1, \\ 0, \text{ if } q \neq 2n-1. \end{cases}$$

The theorem follows.

26.4 Theorem of H. Cartan and J.-P. Serre

We say that the rational cohomology of a space X forms a free skew-commutative algebra with generators $x_s \in H^{m_s}(X;\mathbb{Q})$ if these generators are not tied by any relations besides the relations of the skew commutativity: $x_s x_t = (-1)^{m_s m_t} x_t x_s$. In other words,

$$H^*(X;\mathbb{Q}) = \Lambda_{\mathbb{Q}}(\text{odd dimensional } x_s) \otimes \mathbb{Q}[\text{even dimensional } x_s].$$

Example 1. S^n with n odd.

Example 2. $S^n|_{n+1}$ with n even.

Example 3. $K(\pi, n)$ with a finitely generated Abelian group π.

Example 4. $SU(n), Sp(n), SO(n)$.

Example 5. ΩS^n.

Actually, it is true that the rational cohomology of an arbitrary H-space, in particular, of an arbitrary loop space, forms a free skew-commutative algebra. We do not prove it here; a relevant reference is Milnor and Moore's article [59].

Theorem (Cartan–Serre). *Let X be a simply connected space with finitely generated homology groups. Suppose that the rational cohomology of X is a free skew-commutative algebra,*

$$H^*(X;\mathbb{Q}) = \Lambda_{\mathbb{Q}}(x_1,\dots,x_m) \otimes \mathbb{Q}[y_1,\dots,y_\ell],$$

where the x_s are odd-dimensional rational cohomology classes of X and the y_t are even-dimensional rational cohomology classes of X. Then the rank of $\pi_q(X)$ equals the number of q-dimensional elements among $x_1,\dots,x_m,y_1,\dots,y_\ell$.

In other words, for a simply connected space X with a free skew-commutative rational cohomology algebra, there is a dimension-preserving bijection between free additive generators of homotopy groups and free multiplicative generators of a rational cohomology algebra.

Proof of Theorem. Since $H^*(X;\mathbb{Q}) = H^*(X;\mathbb{Z}) \otimes \mathbb{Q}$ and the homology groups of X are finitely generated, there exist nonzero integers $a_1,\dots,a_m,b_1,\dots,b_\ell$

such that the classes $a_1x_1, \dots, a_mx_m, b_1y_1, \dots, b_\ell y_\ell$ belong to the image of the homomorphism $H^*(X;\mathbb{Z}) \to H^*(X;\mathbb{Q})$ induced by the inclusion $\mathbb{Z} \to \mathbb{Q}$. With respect to this homomorphism, let $\widetilde{x}_1 \mapsto a_1x_1, \dots, \widetilde{y}_\ell \mapsto b_\ell y_\ell$. The classes $\widetilde{x}_s, \widetilde{y}_t$ give rise to (homotopically well-defined) continuous maps $X \to K(\mathbb{Z}, i_s)$, $X \to K(\mathbb{Z}, j_t)$, where $i_s = \dim x_s$, $j_t = \dim y_t$. Together, these maps determine a map

$$X \to Y = (\times_s K(\mathbb{Z}, i_s)) \times (\times_t K(\mathbb{Z}, j_t)), \qquad (*)$$

and this map induces an isomorphism in the rational cohomology (it follows from the theorem in Sect. 26.3 and the assumptions of the theorem that the rational cohomology algebras of X and Y are the same, and the mapping $(*)$ induces a bijection between the generators of these algebras). As in the proof of the lemma in Sect. 26.3, we can turn the map $(*)$ into a fibration, the projection of this fibration induces an isomorphism in the rational cohomology, and hence the fiber F of this fibration has trivial rational cohomology. Hence, the homotopy groups of F are all finite (see theorem in Sect. 26.1.C), the homotopy sequence of our fibration shows that every homomorphism $f_*\colon \pi_q(X) \to \pi_q(Y)$ has finite kernel and finite cokernel, and hence the groups $\pi_q(X)$ and $\pi_q(Y)$ have the same ranks, which is our theorem.

The Cartan–Serre theorem implies some statements already known to us (for example, the theorem of ranks of homotopy groups of odd-dimensional spheres) and also some new statements, such as the following one.

Corollary.

$$\pi_q(SU(n)) = \begin{cases} \mathbb{Z} \oplus \text{a finite group}, & \text{if } q = 3, 5, 7, \dots, 2n+1, \\ \text{a finite group} & \text{for all other } q; \end{cases}$$
$$\pi_q(SO(2m+1)) = \begin{cases} \mathbb{Z} \oplus \text{a finite group}, & \text{if } q = 3, 7, 11, \dots, 4m-1, \\ \text{a finite group} & \text{for all other } q; \end{cases}$$
$$\pi_q(SO(2m)) = \begin{cases} \mathbb{Z} \oplus \text{a finite group}, & \text{if } q = 3, 7, 11, \dots, 4m-5, 2m-1, \\ \text{a finite group} & \text{for all other } q; \end{cases}$$
$$\pi_q(Sp(n)) = \begin{cases} \mathbb{Z} \oplus \text{a finite group}, & \text{if } q = 3, 7, 11, \dots, 4n-1, \\ \text{a finite group} & \text{for all other } q. \end{cases}$$

Remark. The exactness of the homotopy sequences of the fibrations

$$SO(n) \xrightarrow{SO(n-1)} S^{n-1}, \ SU(n) \xrightarrow{SU(n-1)} S^{2n-1}, \ Sp(n) \xrightarrow{Sp(n-1)} S^{4n-1},$$

and the triviality of the homotopy groups $\pi_q(S^m)$ with $q < m$ imply the isomorphisms

$$\pi_q(SO(n-1)) \cong \pi_q(SO(n)) \text{ for } q < n-2,$$
$$\pi_q(SU(n-1)) \cong \pi_q(SU(n)) \text{ for } q < 2n-2,$$
$$\pi_q(Sp(n-1)) \cong \pi_q(Sp(n)) \text{ for } q < 4n-2$$

(induced by the corresponding inclusion maps). (In the "critical dimensions" $q = n-2, 2n-2$, and $4n-2$, these inclusion homomorphisms are epimorphisms.) Thus, for n large, the groups $\pi_q(SO(n)), \pi_q(SU(n))$, and $\pi_q(Sp(n))$ do not depend on n; these stable groups are denoted as $\pi_q(SO)$, $\pi_q(SU)$, and $\pi_q(Sp)$ [actually, these groups are homotopy groups of groups of infinite matrices, $SO = \bigcup_n SO(n)$, $SU = \bigcup_n SU(n)$, $Sp = \bigcup_n Sp(n)$]. Unlike the homotopy groups $\pi_q(SO(n))$, $\pi_q(SU(n))$, and $\pi_q(Sp(n))$ (which are known only partially), the stable homotopy groups are fully known:

$$\pi_q(U) = \begin{cases} \mathbb{Z} \text{ for odd } q, \\ 0 \text{ for even } q; \end{cases} \qquad \pi_q(SO) = \begin{cases} \mathbb{Z}, & \text{if } q \equiv -1 \bmod 4, \\ \mathbb{Z}_2, & \text{if } q \equiv 0, 1 \bmod 8, \\ 0 & \text{for all other } q; \end{cases}$$

$\pi_q(Sp) = \pi_{q+4}(SO)$ (compare with the last corollary). This fact was proved in the late 1950s by Raoul Bott, who applied methods of variational calculus in the large (this result is broadly known under the name of the *Bott periodicity*). A presentation of Bott's proof is contained in the last chapter of Milnor's book [57]. Another proof for the case of U is given in Chap. 6 of this book.

26.5 Comments About the Cartan–Serre Theorem

We have already noticed that the theorem on the ranks of homotopy groups of an odd-dimensional sphere is a corollary of the Cartan–Serre theorem. However, the case of an even-dimensional sphere is not covered by Cartan–Serre, since the cohomology ring of an even-dimensional sphere has an undesirable relation: The square of an even-dimensional generator is equal to 0. Still, we were able to compute the ranks of the homotopy groups of an even-dimensional sphere. A reader may create an impression that as soon as we know the rational cohomology ring of, say, a simply connected CW complex, we can find the ranks of its homotopy groups. This impression is wrong, however, as the following example shows. Take the bouquet of two two-dimensional spheres, $S^2 \vee S^2$.

EXERCISE 5. Using methods of this lecture, prove that rank $\pi_4(S^2 \vee S^2) = 2$ (the linearly independent elements of this group are $[[s_1, s_2], s_1]$ and $[[s_1, s_2], s_2]$, where s_1 and s_2 are the generators of the groups π_2 of the two spheres and the square brackets denote the Whitehead product; Sect. 10.5).

Take a spheroid $S^4 \to S^2 \vee S^2$ representing an infinite-order element of $\pi_4(S^2 \vee S^2)$ and attach a five-dimensional disk to $S^2 \vee S^2$ using this spheroid. We get a space X with a multiplicative isomorphism $H^*(X; \mathbb{Q}) \cong H^*(S^2 \vee S^2 \vee S^5; \mathbb{Q})$, but the ranks of homotopy groups (in particular, of the groups π_4) of X and $S^2 \vee S^2 \vee S^5$ are not the same. A question arises: Is it possible to express in terms of rational cohomology information about a (say, simply connected, CW) space sufficient for determining the ranks of homotopy groups? The answer is yes; however, for this purpose, we need to consider, in addition to the classical cohomological multiplication, a sequence of "higher multiplications," the *Massey products* (Massey products were briefly mentioned in Lecture 17, see Exercise 52.).

Here is a construction of the first of them. Let $\alpha \in H^p(X;G), \beta \in H^q(X;G), \gamma \in H^r(X;G)$ (where G is a ring) be cohomology classes such that $\alpha\beta = 0$ and $\beta\gamma = 0$. Let $a \in \alpha, b \in \beta, c \in \gamma$ be cocycles (it is not important what kind, maybe singular, maybe classical, or maybe differential forms; the only thing that matters is the existence for this kind of cochains of an associative bilinear product satisfying the product rule and inducing the standard cohomological multiplication). Then there are cochains e, f such that $\delta e = ab$ and $\delta f = bc$. Consider the cochain $ec-(-1)^p af$. It is a cocycle:

$$\delta(ec - (-1)^p af = (ab)c - (-1)^p(-1)^p a(bc) = 0,$$

and its cohomology class is called the Massey product of α, β, γ and is denoted as $\langle\alpha, \beta, \gamma\rangle$. Obviously, this cohomology class is not "well defined": Varying the choice of cochains e and f can result in adding to $\langle\alpha, \beta, \gamma\rangle$ an arbitrary class of the form $\alpha\lambda + \mu\gamma$ with $\lambda \in H^{q+r-1}(X;G)$, $\mu \in H^{p+q-1}(X;G)$. Thus, the Massey product is a partial multivalued operation.

EXERCISE 6. Compute the Massey products in the cohomology of the space $(S^2 \vee S^2) \cup D^5$ considered above.

EXERCISE 7. Compute the Massey products in the cohomology of the complement to the Borromean rings (compare Exercise 52 in Lecture 17).

It is possible to extend this construction to still "higher" Massey products. For example, for a quadruple of cohomology classes $\alpha, \beta, \gamma, \delta$ such that $\alpha\beta{=}0, \beta\gamma{=}0, \gamma\delta = 0, \langle\alpha, \beta, \gamma\rangle \ni 0, \langle\beta, \gamma, \delta\rangle \ni 0$, one can construct a cohomology class $\langle\alpha, \beta, \gamma, \delta\rangle$ with a still bigger indeterminacy than triple Massey products. And so on. A general algebraic description of Massey products is contained in the article by Fuchs and Weldon [41].

It turns out that the rational cohomology of a simply connected space, given with the multiplicative structure and the whole infinite sequence of Massey products, determines the ranks of homotopy groups (with Whitehead products and a sequence of Whitehead–Massey products in homotopy groups tensored with $\mathbb{Q}$).

A different approach (related to, but not based on Massey products) to rational homotopy types was developed by Dennis Sullivan as the minimal model theory. The main ingredient of this theory consists in assigning to a space a rational cochain complex furnished with an associative skew-commutative multiplication whose cohomology is the rational cohomology of the space (in Sullivan's original construction it is the complex of piecewise rational differential forms). A space is called *formal* if there exists a multiplicative homomorphism from the cohomology ring into a minimal model which assigns to every cohomology class a cocycle representing this class. It is true that for a formal space all the Massey products in rational cohomology are zero (but this does not imply formality). For formal spaces, a rational cohomology algebra determines ranks of homotopy groups (by a procedure similar to that of this lecture). Examples of formal spaces include loop spaces, spheres (both odd- and even-dimensional), Kähler manifolds, and symmetric spaces. See the article by Deligne, Griffiths, Morgan, and Sullivan [33]; there is also a famous book by Félix, Halperin, and Thomas [36].

Lecture 27 Odd Components of Homotopy Groups

In this lecture, p denotes an odd prime (the case of $p = 2$ will be considered in Chap. 4).

27.1 Cohomology $\mathbf{H^*(K(\pi, n); \mathbb{Z}_p)}$ for a Finite Group π of Order Not Divisible by p

Theorem. *If p' is a prime different from p, then*

$$H^*(K(\mathbb{Z}_{p'^s}; \mathbb{Z}_p) \cong H^*(\text{pt}; \mathbb{Z}_p).$$

This fact is, essentially, known to us: For $n = 1$, it follows from standard homology computations (see Chap. 2); in the general case it can be proved by induction based on spectral sequences of fibrations $* \xrightarrow{K(\mathbb{Z}_{p'^s}, n-1)} K(\mathbb{Z}_{p'^s}, n)$.

Corollary. *If π is a finite Abelian group whose order is not divisible by p, then every $H^q(K(\pi, n); \mathbb{Z})$, $q > 0$ is a finite Abelian group whose order is not divisible by p.*

This follows from the theorem and the universal coefficients formula. (See also Exercise 1 in Sect. 26.1.B.)

27.2 A Partial Computation of $H^*(K(Z_p, n); \mathbb{Z})$

We will do this computation up to dimension $n + 4p - 4$. (It will be clear from the computation why we restrict ourselves to this range of dimensions. We will continue this computation in Chap. 4.)

It will be convenient for us (and for the reader) to begin the computation before we state the final result.

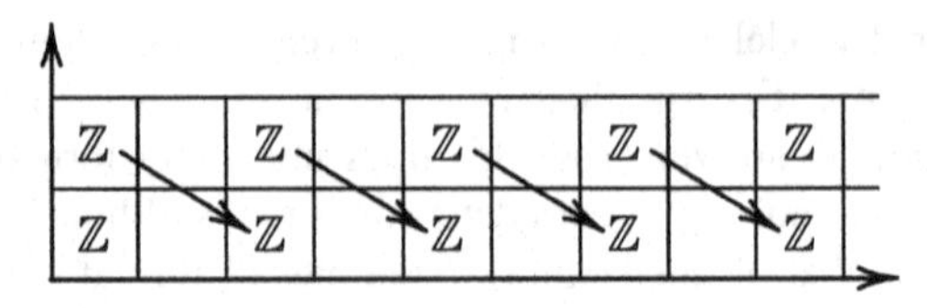

Fig. 104 The spectral sequence of the fibration $L_p^\infty \xrightarrow{S^1} \mathbb{C}P^\infty$

The cohomology of the space $K(\mathbb{Z}_p, 1) = L_p^\infty$ is known to us:

$$H^q(K(\mathbb{Z}_p, 1); \mathbb{Z}) = \begin{cases} \mathbb{Z}_p & \text{for an even } q > 0, \\ 0 & \text{for an odd } q. \end{cases}$$

It is easy to see that, in the positive dimensions, there is a multiplicative isomorphism

$$H^*(K(\mathbb{Z}_p, 1); \mathbb{Z}) \cong \mathbb{Z}_p[x],\ x \in H^2(K(\mathbb{Z}_p, 1); \mathbb{Z}).$$

(This can be proved, for example, in the following way. Consider the spectral sequence of the fibration $L_p^\infty \xrightarrow{S^1} \mathbb{C}P^\infty$ (see Fig. 104). From the product rule, or, still simpler, from our knowledge of the additive structure of cohomology of $K(\mathbb{Z}_p, 1)$, it follows that all the differentials are multiplications by p. Hence, in E_∞, there remains only the zeroth row; hence, the homomorphism $H^*(\mathbb{C}P^\infty; \mathbb{Z}) \to H^*(K(\mathbb{Z}_p, 1); \mathbb{Z})$ is onto. The multiplicative structure of $H^*(\mathbb{C}P^\infty; \mathbb{Z})$ implies our statement.)

Consider now the $\mathbb{Z}$-cohomological spectral sequence of the fibration $* \xrightarrow{K(\mathbb{Z}_p,1)} K(\mathbb{Z}_p, 2)$. Up to dimensions $\approx 4p$, this spectral sequence is sketched in Fig. 105. From this spectral sequence we see that, up to dimension $4p + 1$, the ring $H^*(K(\mathbb{Z}_p; \mathbb{Z})$ has the following structure. It has additive generators in dimensions $0, 3, 2p+1, 2p+2, 2p+4, 2p+5$; we denote them as $1, y_3, y_{2p+1}, y_{2p+2}, y_{2p+3}, y_{2p+4}$. These generators satisfy the relations $py_i = 0, y_3 y_{2p+1} = y_{2p+4}, y_3 y_{2p+2} = y_{2p+5}$.

Let us clarify this. The class y_3 is, by definition, the transgression image of x. From $d_3 x = y_3$, we have

$$d_3 x^m = m x^{m-1} y \begin{cases} \neq 0, & \text{if } m \text{ is not divisible by } p, \\ = 0, & \text{if } m \text{ is divisible by } p. \end{cases}$$

Since $d_3 x^p = 0$, the classes x^p and $x^{p-1} y_3$ remain in E_4, but they cannot survive in $E_\infty = 0$. They can be eliminated only by a differential which maps them into the zeroth row. Hence, there are y_{2p+1} and y_{2p+2} in the zeroth row such that $d_{2p+1} x^p = y_{2p+1}$ and $d_{2p-1}(x^{p-1} y_3) = y_{2p+2}$ [the last equality does not contradict the product rule, because there is neither y_3 nor x^{p-1} in E_{2p-2}, and $x^{p-1} y_3$ is not a product in

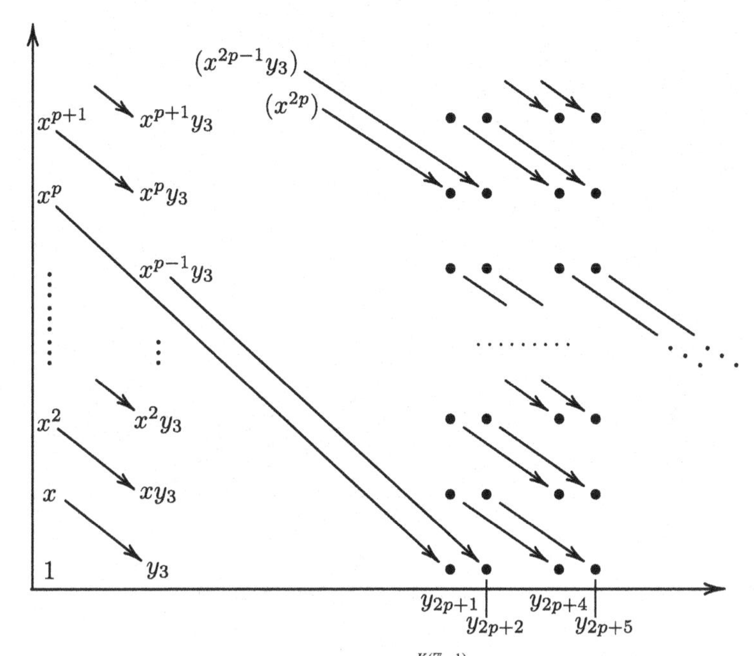

Fig. 105 The spectral sequence of the fibration $* \xrightarrow{K(\mathbb{Z}_p,1)} K(\mathbb{Z}_p, 2)$

E_{2p-1}]. The classes y_{2p+1} and y_{2p+2} must be present in E_2 (otherwise, how can they appear in E_{2p+1} and E_{2p-1}?). Furthermore, by the product rule,

$$d_3(xy_{2p+1}) = y_3 y_{2p+1} \text{ and } d_3(xy_{2p+2}) = y_3 y_{2p+2},$$

and these products cannot be zero, since otherwise xy_{2p+1} and xy_{2p+2} would have stayed in E_∞, which is impossible. We put

$$y_3 y_{2p+1} = y_{2p+4},\ y_3 y_{2p+2} = y_{2p+5}.$$

Finally, $d_3(x^m y_{2p+1}) = mx^{m-1} y_{2p+4}$, $d_3(x^m y_{2p+2}) = mx^{m-1} y_{2p+5}$, these differentials are not zero for $k < p$, and $x^p y_{2p+1}$ already has the full dimension $4p + 1$; moreover,

$$d_{2p+1}(x^{2p}) = 2x^p d_{2p+1} x^p = 2x^p y_{2p+1},$$
$$d_{2p-1}(x^{2p-1} y_3) = d_{2p-1}(x^p \cdot x^{p-1} y_3) = x^p y_{2p+2}.$$

Thus, there are no more elements of E_2 of full dimension $\leq 4p + 2$ not eliminated by the differentials d_3, d_{2p-1} and d_{2p+1}.

Now, we will not give the details of such an argumentation, instead restricting ourselves to a picture.

The spectral sequence of the fibration $* \xrightarrow{K(\mathbb{Z}_p,2)} K(\mathbb{Z}_p, 3)$ is shown in Fig. 106. We see that up to dimension $4p + 2$, the ring $H^*(K(\mathbb{Z}_p, 3);\ \mathbb{Z})$ is generated by the following elements:

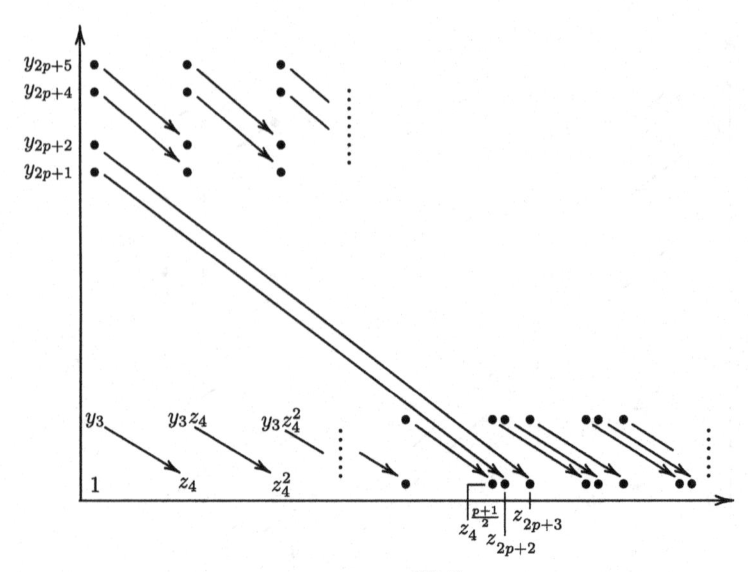

Fig. 106 The spectral sequence of the fibration $* \xrightarrow{K(\mathbb{Z}_p,2)} K(\mathbb{Z}_p,3)$

$$z_4, z_4^2, z_4^3, z_4^4, z_4^5, \ldots;$$
$$z_{2p+2}, z_{2p+2}z_4, z_{2p+2}z_4^2, \ldots;$$
$$z_{2p+3}, z_{2p+3}z_4, z_{2p+3}z_4^2, \ldots$$

($\dim z_m = m$). Notice that some of the elements listed have the same dimension:

$$\dim z_4^{k+\frac{p+1}{2}} = \dim z_{2p+2}z_4^k,$$

and it could happen that

$$pz_{2p+2} = rz_4^{\frac{p+1}{2}}, \ r \not\equiv 0 \bmod p,$$

but it does not happen, since in this case we would have had

$$0 = (pz_4)z_{2p+2} = z_4^{1+\frac{p+1}{2}} \neq 0.$$

Hence, all the elements listed have (additive) order p.

The spectral sequence of the fibration $* \xrightarrow{K(\mathbb{Z}_p,3)} K(\mathbb{Z}_p,4)$ is similar to that of the fibration $* \xrightarrow{K(\mathbb{Z}_p,1)} K(\mathbb{Z}_p,2)$; it is shown in Fig. 107. The generators of dimensions $4, 2p+2$, and $2p+3$ of the ring $H^*(K(\mathbb{Z}_p,3);\mathbb{Z})$ create in $H^*(K(\mathbb{Z}_p,4);\mathbb{Z})$ elements of dimensions $5, 2p+3$, and $2p+4$, and there arise also [because of the equality $d_5(z_4^p) = 0$] entirely new elements of dimensions $4p+1$ and $4p+2$.

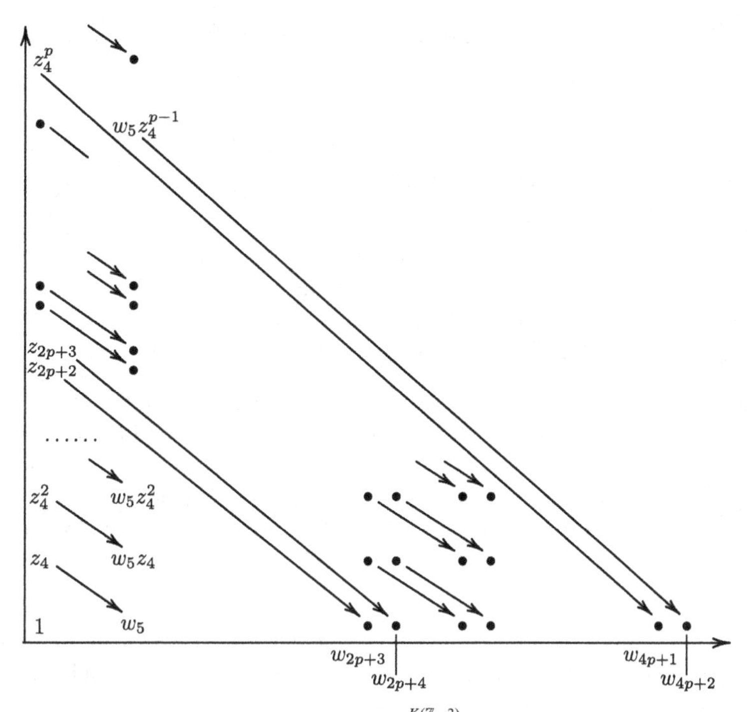

Fig. 107 The spectral sequence of the fibration $* \xrightarrow{K(\mathbb{Z}_p,3)} K(\mathbb{Z}_p, 4)$

Thus, $H^*(K(\mathbb{Z}_p, 4); \mathbb{Z})$ is additively generated in dimensions $\leq 4p + 2$ by the elements

$$1, w_5, w_{2p+3}, w_{2p+4}, w_{2p+3}w_5, w_{2p+4}w_5, w_{4p+1}, w_{4p+2}.$$

The further spectral sequences (alternatively, even and odd) are similar to those above. Moreover, the dimensions of the pth degrees of nonzero elements of E_2 will be too large, and so we will no longer need to care about them. Here is the final result.

Theorem. *For an even $n \geq 2$, the ring $H^*(K(\mathbb{Z}_p, n); \mathbb{Z})$ in dimensions $\leq n + 4p - 2$ is generated additively by the following elements of order p:*

$$\xi_{n+1}, \xi_{n+2p-1}, \xi_{n+2p}, \xi_{n+4p-3}, \xi_{n+4p-2},$$

and also $\xi_{n+1}\xi_{n+2p-1}$ and $\xi_{n+1}\xi_{n+2p}$ if their dimensions belong to our range. For an odd $n \geq 3$, this ring, in the same dimensions, is additively generated by the following elements of order p:

$$\eta_{n+1}^k\ (k \geq 1);\ \eta_{n+1}^k \eta_{n+2p-1}\ (k \geq 0);\ \eta_{n+1}^k \eta_{n+2p}\ (k \geq 0);\ \eta_{n+4p-3}, \eta_{n+4p-2}.$$

27.3 A Partial Computation of the Cohomology $K(\mathbb{Z}, n)$ mod p

Theorem. *In positive dimensions* $\leq n + 4p - 3$, *the ring* $H^*(K(\mathbb{Z}, n); \mathbb{Z}_p)$ *is isomorphic to the ring* $H^*(K(\mathbb{Z}_p, n-1); \mathbb{Z})$.

Indeed, there is a multiplicative isomorphism $\widetilde{H}^*(K(\mathbb{Z}, 2); \mathbb{Z}_p) \cong \widetilde{H}^*(K(\mathbb{Z}_p, 1); \mathbb{Z})$. On the other side, in the computation of $H^*(K(\mathbb{Z}_p, n);\ \mathbb{Z})$ (Sect. 27.2) we did not use anything beside the multiplicative structure of $H^*(K(\mathbb{Z}_p, 1); \mathbb{Z})$ and general properties of spectral sequences. Hence, this computation can be applied to the case of $H^*(K(\mathbb{Z}, n);\ \mathbb{Z}_p)$ without any changes.

Remark. We have to warn the reader that the multiplicative arguments alone do not provide a full computation of the rings $H^*(K(\mathbb{Z}_p, n);\ \mathbb{Z})$ and $H^*(K(\mathbb{Z}, n); \mathbb{Z}_p)$. Rather soon, there arises a situation when there are several different possibilities for a differential not contradicting the product rule. (A full computation requires using "cohomology operations," which we will study in the next chapter.) In particular, for $n \geq 3$, the rings $\widetilde{H}^*(K(\mathbb{Z}, n); \mathbb{Z}_p)$ and $\widetilde{H}^*(K(\mathbb{Z}_p, n-1); \mathbb{Z})$ are not isomorphic in all dimensions.

Let us now examine our knowledge of the cohomology $H^*(K(\mathbb{Z}, n);\ \mathbb{Z})$. According to results of Lecture 26, there are $\mathbb{Z}$-components in dimension n and, for n even, in dimensions $2n, 3n, \ldots$. In the transition to $H^*(K(\mathbb{Z}, n); \mathbb{Z}_p)$, the universal coefficients formula converts these $\mathbb{Z}$s into $\mathbb{Z}_p$s of the same dimensions, which agrees with our results. But we know some other $\mathbb{Z}_p$-components in $H^*(K(\mathbb{Z}, n); \mathbb{Z}_p)$, in particular, in dimensions $n + 2p - 2$ and $n + 2p - 1$. Since these components do not originate from any $\mathbb{Z}$-components in $H^*(K(\mathbb{Z}, n); \mathbb{Z})$, they have to arise from (the same) $\mathbb{Z}_{p^s}$-component in $H^{n+2p-1}(K(\mathbb{Z}, n)$ (as $\mathbb{Z}_{p^s} \otimes \mathbb{Z}$ and $\operatorname{Tor}(\mathbb{Z}_{p^s}, \mathbb{Z})$). We do not know what s is, but, actually, $s = 1$, and, moreover, elements of the order p^2 in $H^*(K(\mathbb{Z}, n); \mathbb{Z})$ first appear in dimension $n + 2p^2 - 1$. We leave the proof of this fact to the reader (who is recommended to begin by considering the integral cohomological spectral sequence of the fibration $* \xrightarrow{K(\mathbb{Z},2)} K(\mathbb{Z}, 3)$).

27.4 A Partial Computation of the p-Component of the Homotopy Groups of Spheres

Theorem. *If* $n \geq 3$, *then* $\pi_{n+2p-3}(S^n) \supset \mathbb{Z}_p$, $\pi_{n+4p-5}(S^n) \supset \mathbb{Z}_p$, *and the quotient*

$$\left[\bigoplus_{q=0}^{4p-5} \pi_{n+q}(S^n)\right] / (\mathbb{Z}_p \oplus \mathbb{Z}_p)$$

does not contain elements of order p.

We will prove a weaker statement here. Namely, we will prove that the p-component of the group $\bigoplus_{q=0}^{4p-7} \pi_{n+q}(S^n)$ is concentrated in $\pi_{n+2p-3}(S^n)$ and equals $\mathbb{Z}_p$. To prove the theorem in full, we will need to compute just one more differential. The reader will be able to fill this gap (and to proceed in computing the p-components of the groups $\pi_{n+q}(S^n)$ to $q \sim p^2$) after reading the next chapter.

Let us turn to a proof. Consider the case of odd $n \geq 7$. In this case, the E_2-term of the cohomological spectral sequence of the fibration

$$S^n|_{n+1} \xrightarrow{K(\mathbb{Z},n-1)} S^n$$

with the coefficients in $\mathbb{Z}_p$ looks as shown in Fig. 108, left. From this spectral sequence, the cohomology $H^*(S^n|_{n+1}; \mathbb{Z}_p)$ in dimensions $\leq n+4p-4$ is as follows:

0	$\cdots$	$n+2p-3$	$n+2p-2$	$\cdots$	$n+4p-5$	$n+4p-4$
$\mathbb{Z}_p$	zeroes	$\mathbb{Z}_p$	$\mathbb{Z}_p$	zeroes	$\mathbb{Z}_p$	$\mathbb{Z}_p$

We know that the integral cohomology of the space $S^n|_{n+1}$ is finite in all positive dimensions. It follows from this fact and from the absence, in the dimensions considered, of elements of order p^2 that the p-components of the integral cohomology of $S^n|_{n+1}$ look the following way:

0	$\cdots$	$n+2p-2$	$\cdots$	$n+4p-4$
$\mathbb{Z}_p$	zeroes	$\mathbb{Z}_p$	zeroes	$\mathbb{Z}_p$

The integral homology is the same, only the dimensions $n+2p-2$ and $n+4p-4$ should be turned into $n+2p-3$ and $n+4p-5$.

The transition to $S^n|_{n+2}$, $S^n|_{n+3}$, and so on, up to $S^n|_{n+2p-3}$, does not affect p-components of homology and cohomology, since the fibers of the fibrations $S^n|_{n+q} \to S^n|_{n+q-1}$ with $1 < q \leq 2p-3$ are $K(G,-)$-spaces, with G being a finite group of order prime to p; the positive-dimension homology and cohomology of these spaces are also finite groups of orders prime to p (see Sect. 27.1). In particular, the group $H_{2n+p-3}(S^n|_{2n+p-3})$ contains a component $\mathbb{Z}_p$ and no other p-components. Thus, $\pi_{n+q}(S^n)$ has no p-components for $q < 2p-3$ and the p-component of $\pi_{n+2p-3}(S^n)$ is $\mathbb{Z}_p$.

Next, we consider the fibration

$$S^n|_{n+2p-2} \xrightarrow{K(\pi_{n+2p-3}(S^n),n+2p-4)} S^n|_{n+2p-3}.$$

It follows from what was done before that $K(\pi_{n+2p-3}(S^n), n+2p-4) = K(\mathbb{Z}_p, n+2p-4) \times K(G, n+2p-4)$, where G is a finite group of order prime to p. The second factor does not affect p-components, so the p-components in the integral cohomological sequence of this fibration look as shown in Fig. 108, right.

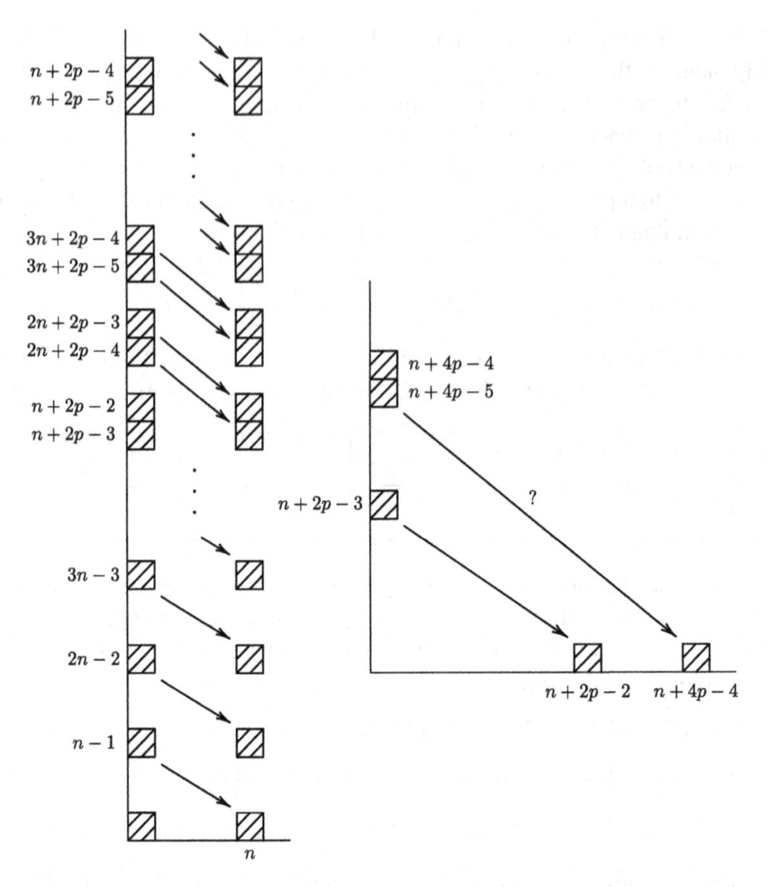

Fig. 108 Spectral sequences for computing p-components

This shows that the first p-component in $H^*(S^n|_{n+2p-2}; \mathbb{Z})$ can appear in dimension $n+4p-5$ (in homology, in dimension $n+4p-6$). We cannot say more, since we do not know the action of the differential labeled by the question mark. Actually, this differential is not trivial, and this is sufficient for the proof of the theorem. But the proof of this requires using Steenrod powers, which will appear in Chap. 4. What we can do now is notice that the transitions $S^n|_{n+2p-2} \to \dots \to S^n|_{n+4p-7}$ do not affect p-components, and hence the groups $\pi_{n+q}(S^n)$ with $2p - 3 < q < 4p - 6$ do not have p-components.

This settles the case of odd $n \geq 7$. The case $n = 3$ is easier, since we already know the homology of $S^3|_4$ (see Sect. 25.2). The case $n = 5$ is similar to the case $n \geq 7$, but the computations look slightly different, because $p(n - 1) = 4p$. Finally, the case of even n, which is not much different from the case of odd n but is sort of more cumbersome, is left to the reader.

Corollary. *The order of the stable homotopy group π_k^S (see Sect. 10.4) is not divisible by p for $1 \leq k < 2p - 3$ and $2p - 3 < k < 4p - 5$ and is divisible by p but not by p^2 for $k = 2p - 3$ and $4p - 5$.*

Notice that for the proof of this corollary, it is sufficient to know the theorem for the case of odd $n \geq 7$ (which was considered in detail above).

FYI. There is extensive information concerning the p-components of homotopy groups of spheres. Let us just mention the following result concerning *stable* homotopy groups.

Proposition. *All nontrivial p-components of π_k^S with $k \leq 2p(p-1)-1$ are contained in the following formula:*

$$\boxed{p\text{-component of } \pi_k^S} = \begin{cases} \mathbb{Z}_p, & \text{if } k = 2i(p-1) - 1, i = 1, \ldots, p-1, \\ \mathbb{Z}_p, & \text{if } k = 2p(p-1) - 2, \\ \mathbb{Z}_{p^2}, & \text{if } k = 2p(p-1) - 1. \end{cases}$$

The proof of this result, as well as of some further results, is contained in the classical book by Toda [86]. We can also refer the reader to the later book by Mosher and Tangora [63].

Chapter 4
Cohomology Operations

Lecture 28 General Theory

28.1 Definitions

Let n, q be two integers and let π, G be two Abelian groups. We say that a *cohomology operation* φ *of type* (n, q, π, G) is given if for every CW complex X a map $\varphi_X: H^n(X; \pi) \to H^q(X; G)$ is given and is natural with respect to X, in the sense that the diagram

$$\begin{array}{ccc} H^n(X;\pi) & \xrightarrow{\varphi_X} & H^q(X;G) \\ \uparrow{\scriptstyle f^*} & & \uparrow{\scriptstyle f^*} \\ H^n(Y;\pi) & \xrightarrow{\varphi_Y} & H^q(Y;G) \end{array}$$

is commutative for every continuous map $f: X \to Y$.

When no confusion is possible, we will abbreviate the notation φ_X to φ.

Notice that the map φ_X is not assumed to be a homomorphism of the group $H^n(X; \pi)$ into the group $H^q(X; G)$.

Since G is an Abelian group, the set of cohomology operations of the type (n, q, π, G) is an Abelian group. We denote this group by $\mathcal{O}(n, q, \pi, G)$.

28.2 Classification

Theorem. *There is a (canonical) isomorphism*

$$\mathcal{O}(n, q, \pi, G) \cong H^q(K(\pi, n); G).$$

A. Fomenko, D. Fuchs, *Homotopical Topology*, Graduate Texts in Mathematics 273,
DOI 10.1007/978-3-319-23488-5_4

This beautiful and unexpected statement is actually almost obvious, as we will immediately see.

Proof of Theorem. Recall that there is a natural bijection $H^n(X;\pi) \leftrightarrow \pi(X, K(\pi,n))$ between n-dimensional cohomology of X with coefficients in π and homotopy classes of continuous maps $X \to K(\pi,n)$ (see Sect. 18.3). The construction of this bijection uses a remarkable cohomology class $F_\pi \in H^n(K(\pi,n);\pi)$, which has several equivalent definitions; for example, the universal coefficients formula isomorphism $H^n(K(\pi,n);\pi) = \operatorname{Hom}(H_n(K(\pi,n)),\pi) = \operatorname{Hom}(\pi,\pi)$ connects F_π with id_π. Namely, for an $\alpha \in H^n(X;\pi)$, there exists a homotopically unique continuous map $h_\alpha: X \to K(\pi,n)$ such that $h_\alpha^*(F_\pi) = \alpha$. Our bijection is $\alpha \leftrightarrow [h_\alpha]$.

Let us assign to a cohomology operation $\varphi \in \mathcal{O}(n,q,\pi,G)$ the cohomology class $\varphi_{K(\pi,n)}(F_\pi) \in H^q(K(\pi,n);G)$. The formula $\varphi \mapsto \varphi_{K(\pi,n)}(F_\pi)$ determines a map of $\mathcal{O}(n,q,\pi,G)$ into $H^q(K(\pi,n);G)$, which is, obviously, a group homomorphism. We will show that it is actually an isomorphism.

Let us first prove that it is one-to-one. Let $\varphi,\psi \in \mathcal{O}(n,q,\pi,G)$ and let $\varphi_{K(\pi,n)}(F_\pi) = \psi_{K(\pi,n)}(F_\pi)$. Consider an X and an $\alpha \in H^n(X;\pi)$. Then $\alpha = h_\alpha^*(F_\pi)$ and

$$\begin{aligned}\varphi_X(\alpha) = \varphi_X(h_\alpha^*(F_\pi)) &= h_\alpha^*(\varphi_{K(\pi,n)}(F_\pi))\\ &= h_\alpha^*(\psi_{K(\pi,n)}(F_\pi)) = \psi_X(h_\alpha^*(F_\pi)) = \psi_X(\alpha);\end{aligned}$$

thus, $\varphi = \psi$.

Now let us prove that it is onto. Let $\gamma \in H^q(K(\pi,n);G)$ be arbitrary. We need to construct a cohomology operation $\varphi \in \mathcal{O}(n,q,\pi,G)$ such that $\varphi_{K(\pi,n)}(F_\pi) = \gamma$. For an arbitrary $\alpha \in H^n(X;\pi)$ put $\varphi_X(\alpha) = h_\alpha^*(\gamma) \in H^q(X;G)$. Since $h_{F_\pi} = \mathrm{id}_{K(\pi,n)}$, we have $\varphi_{K(\pi,n)}(F_\pi) = h_{K(\pi,n)}^*(\gamma) = \mathrm{id}^*\gamma = \gamma$. It remains to show that $\{\varphi_X\}$ is a cohomology operation. Let $f: X \to Y$ be a continuous map, and let $\beta \in H^n(Y;\pi)$ be arbitrary. Then $h_{f^*\beta} \sim h_\beta \circ f$ [since $(h_\beta \circ f)^*(F_\pi) = f^*(h_\beta^*(F_\pi)) = f^*\beta$] and $\varphi_X(f^*\beta) = h_{f^*\beta}^*(\gamma) = (h_\beta \circ f)^*(\gamma) = f^*(h_\beta^*(\gamma)) = f^*(\varphi_Y(\beta))$. This completes the proof.

Corollary. *A nonzero cohomology operation does not lower the dimension [that is, if $0 \neq \varphi \in \mathcal{O}(n,q,\pi,G)$, then $q \geq n$.]*

Indeed, $H^q(K(\pi,n);G) = 0$ for $q < n$, since $K(\pi,n)$ has no cells of dimension $< n$.

Here is an example of a cohomology operation which is not a homomorphism. Let π be a ring, and let n be an even number. The raising to a square $H^n(X;\pi) \to H^{2n}(X;\pi)$ is a cohomology operation which, in general, is not a homomorphism (it certainly is a homomorphism if $\pi = \mathbb{Z}_2$).

28.3 Examples

We know the cohomology $H^*(K(\pi, n); \mathbb{Q})$ for all finitely generated groups π. In particular, if π is finite, then $H^*(K(\pi, n); \mathbb{Q}) = 0$, so *there are no nontrivial cohomology operations from cohomology with finite coefficients into rational cohomology.* If n is odd, then $H^*(K(\mathbb{Z}, n); \mathbb{Q}) = H^*(S^n; \mathbb{Q})$, so *every nontrivial cohomology operation from odd-dimensional integral cohomology into rational cohomology preserves dimension and is a fixed rational number times the homomorphism induced by the inclusion* $\mathbb{Z} \to \mathbb{Q}$. If n is even, there also appear operations assigning to an integral cohomology class its powers (rationalized).

Consider now cohomology operations raising the dimension by one. Since $K(\mathbb{Z}, n)$ has no $(n+1)$-dimensional cells, *there are no nontrivial cohomology operations from integral cohomology into any other cohomology raising dimension by one.* The space $K(\mathbb{Z}_p, n)$ has precisely one $(n+1)$-dimensional cell (obtained by attaching D^{n+1} to S^n by a spheroid $S^n \to S^n$ of degree p). Thus, $H^{n+1}(K(\mathbb{Z}_p, n); \mathbb{Z}_p)$ is, at most, $\mathbb{Z}_p$, so there is, up to a multiplication by a constant, at most one cohomology operation of the type $(n, n+1, \mathbb{Z}_p, \mathbb{Z}_p)$. But we know such an operation: the Bockstein homomorphism (see Sect. 15.3). Thus, *the Bockstein homomorphism* (up to a multiplication by a constant) *is the only cohomology operation in mod p cohomology raising dimension by one.*

EXERCISE 1. Prove that any cohomology operation of type $(n, n+1, C, A)$ (where C and A are Abelian groups) is the connecting homomorphism in the coefficient exact sequence corresponding to some short exact sequence $0 \to A \to B \to C \to 0$.

28.4 Stable Cohomology Operations

A: Definition

A stable cohomology operation of type (r, π, G) is a sequence of cohomology operations $\varphi_n \in \mathcal{O}(n, n+r, \pi, G)$, $n = 1, 2, 3, \ldots$, such that for every X and every n the diagram

$$\begin{array}{ccc} H^n(X;\pi) & \xrightarrow{(\varphi_n)_X} & H^{n+r}(X;G) \\ \downarrow \Sigma & & \downarrow \Sigma \\ H^{n+1}(\Sigma X;\pi) & \xrightarrow{(\varphi_{n+1})_{\Sigma X}} & H^{n+r+1}(\Sigma X;G) \end{array}$$

(where Σ is the suspension isomorphism) is commutative.

H_*

The set (Abelian group) of all stable cohomology operations of type (r, π, G) is denoted as $\mathcal{S}tab\,\mathcal{O}(r, \pi, G)$.

For example, the different versions of the Bockstein homomorphism (Sect. 15.3) are stable cohomology operations of types $(1, \mathbb{Z}_p, \mathbb{Z})$ and $(1, \mathbb{Z}_p, \mathbb{Z}_p)$.

B: Relations with the Cohomology Sequence of a Pair

Theorem. *A stable cohomology operation is compatible with a cohomology sequence of a CW pair; that is, for every CW pair (X,A) and every stable cohomology operation $\varphi = \{\varphi_n\} \in \mathcal{S}tab\,\mathcal{O}(r, \pi, G)$ the diagram*

$$\begin{array}{ccccccc}
H^n(X;\pi) & \xrightarrow{i^*} & H^n(A;\pi) & \xrightarrow{\delta^*} & H^{n+1}(X,A;\pi) & \xrightarrow{j^*} & H^{n+1}(X;\pi) \\
\downarrow{\scriptstyle(\varphi_n)_X} & & \downarrow{\scriptstyle(\varphi_n)_A} & & \downarrow{\scriptstyle(\varphi_{n+1})_{(X,A)}} & & \downarrow{\scriptstyle(\varphi_{n+1})_X} \\
H^{n+r}(X;G) & \xrightarrow{i^*} & H^{n+r}(A;G) & \xrightarrow{\delta^*} & H^{n+r+1}(X,A;G) & \xrightarrow{j^*} & H^{n+r+1}(X;G)
\end{array}$$

is commutative.

(Here $(\varphi_{n+1})_{(X,A)}$ means $(\varphi_{n+1})_{(X/A)}$.)

Proof. We need to check the commutativity only for the middle square. But it is obvious that $\delta^*: H^n(A;\pi) \to H^{n+1}(X,A;\pi) = H^{n+1}(X/A;\pi)$ is the same as the composition

$$H^n(A;\pi) \xrightarrow{\Sigma} H^{n+1}(\Sigma A;\pi) \xrightarrow{p^*} H^{n+1}(X \cup CA;\pi) = H^{n+1}(X/A;\pi),$$

where p is the projection $X \cup CA \to (X \cup CA)/X = \Sigma A$ (and the last equality arises from the canonical homotopy equivalence $X \cup CA \sim X/A$) and similarly for $\delta^*: H^{n+r}(A;G) \to H^{n+r+1}(X,A;G)$. It remains to recall that a stable cohomology operation commutes with both p^* and Σ.

C: Relations with Transgression

Let us point out an important corollary of the last theorem. Let (E,B,F,p) be a fibration with a simply connected base, and let ϕ be a stable cohomology operation of the type (r,π,G). Suppose that the cohomology class $\alpha \in H^q(F;\pi) = E_2^{0q}$ is *transgressive*, that is, $d_2\alpha = d_3\alpha = \cdots = d_q\alpha = 0$. Then the class $\phi_q(\alpha) \in H^{q+r}(F;G) = E_2^{0,q+r}$ is also transgressive, that is, $d_2\varphi_q(\alpha) = d_3\varphi_q(\alpha) = \cdots = d_{q+r}\varphi_q(\alpha) = 0$. Moreover, if $\tau(\alpha) = d_{q+1}\alpha \in E_2^{q+1,0} = H^{q+1}(B;\pi)/\bigoplus_{s\leq q}\operatorname{Im} d_s$ contains $\beta \in H^{q+1}(B;\pi)$, then $\tau(\varphi_q(\alpha))$ contains $\varphi_{q+1}(\beta) \in H^{q+r+1}(B;G)$.

(Less precisely, but more sonorously, this can be expressed by the words *stable cohomology operations commute with transgression.*)

All this follows from the presentation of transgression as a composition

$$H^q(F;\pi) \xrightarrow{\delta^*} H^{q+1}(E,F;\pi) \overset{(p^*)^{-1}}{\dashrightarrow} H^{q+1}(B,\text{pt};\pi) = H^{q+1}(B;\pi)$$

(see Sect. 23.3) and the fact that φ commutes with both δ^* and p^*: The inclusion $\beta \in \tau(\alpha)$ means that $p^*(\beta) = \delta^*(\alpha)$; but then $p^*(\varphi(\beta)) = \varphi(p^*(\beta)) = \varphi(\delta^*(\alpha)) = \delta^*(\varphi(\alpha)$, which means that $\varphi(\beta) \in \tau(\varphi(\alpha))$.

D: Classification

A stable cohomology operation ϕ of the type (r,π,G) is a sequence

$$\varphi_n \in \mathcal{O}(n, r+n, \pi, G) = H^{r+n}(K(\pi,n);G).$$

The condition of commuting with Σ means that $f_n(\varphi_n) = \varphi_{n-1}$, where f_n is the composition

$$H^{r+n}(K(\pi,n);G) \xrightarrow{i_n^*} H^{r+n}(\Sigma K(\pi,n-1);G) \xrightarrow{\Sigma^{-1}} H^{r+n-1}(K(\pi,n-1);G),$$

where, in turn, the map $i_n\colon \Sigma K(\pi,n-1) \to K(\pi,n)$ is determined by the condition that $i_n^* F_\pi = \Sigma F_\pi$ [the two F_πs in the last formula lie, respectively, in $H^n(K(\pi,n);\pi)$ and $H^{n-1}(K(\pi,n-1);\pi)$]. Thus, in the language of algebra,

$$Stab\, \mathcal{O}(r,\pi,G) = \overleftarrow{\lim}\, (H^{r+n}(K(\pi,n);G), f_n),$$

the inverse (projective) limit of the sequence

$$\dots \xrightarrow{f_4} H^{r+3}(K(\pi,3);G) \xrightarrow{f_3} H^{r+2}(K(\pi,2);G) \xrightarrow{f_2} H^{r+1}(K(\pi,1);G)$$

[recall that the inverse limit $\overleftarrow{\lim}\, (G_n, f_n)$ of a sequence

$$\dots \xrightarrow{f_4} G_3 \xrightarrow{f_3} G_2 \xrightarrow{f_2} G_1$$

of groups and homomorphisms is the group of sequences $\{g_n \in G_n\}$ such that $f_n(g_n) = g_{n-1}$ for all n]. Our case is relatively simple: We know that if $n > r+1$, then the homomorphism $f_n\colon H^{r+n}(K(\pi,n);G) \to H^{r+n-1}(K(\pi,n-1);G)$

is an isomorphism (it is inverse to the transgression in the spectral sequence of the fibration $* \xrightarrow{K(\pi,n-1)} K(\pi,n)$). For this reason, $\overleftarrow{\lim}\,(H^{r+n}(K(\pi,n);G),f_n) = H^{r+N}(K(\pi,N);G)$ for N big enough (actually, for $N \geq r+1$).

We will say that a cohomology group (or a cohomology class) of $K(\pi,n)$ has a *stable dimension* if its dimension $< 2n$; otherwise, we say that it has a *nonstable dimension.*

E: The Algebra of Stable Operations (the Steenrod Algebra)

The multiplicative structure in $H^*(K(\pi,n);G)$ (where G is a ring) does not determine any multiplication for stable operations (a product of two cohomology classes of stable dimensions never has a stable dimension). However, the composition product turns $\bigoplus_{r\geq 0} \mathit{Stab}\,\mathcal{O}(r,G,G)$ into a graded ring, whether G is a ring or not. This ring of stable operations is a unitary associative ring, in general, not commutative. Notice that the cohomology $H^*(X;G)$ becomes a (graded) module over this ring, and all induced homomorphisms $f^*(Y;G) \to H^*(X;G)$ become module homomorphisms.

If G is a field, then the ring of stable cohomology operations becomes an algebra. If $G = \mathbb{Z}_p$, then this algebra is called the *Steenrod algebra* and is denoted as $\mathbb{A}_p$. In the next lectures, we will study the structure of the Steenrod algebra in all details (especially for $p = 2$).

Lecture 29 Steenrod Squares

29.1 An Introduction

We begin with a construction of some important elements of the Steenrod algebra $\mathbb{A}_2$ which are called *Steenrod squares.* Steenrod squares Sq^i are stable cohomology operations which are additive homomorphisms

$$\mathrm{Sq}^i\colon H^n(X;\mathbb{Z}_2) \to H^{n+i}(X;\mathbb{Z}_2).$$

They are defined for all $i \geq 0$ and possess the following properties (in addition to the properties required by the definition of stable cohomology operations):

(1)

$$\mathrm{Sq}^i\,\alpha = \begin{cases} 0, & \text{if } i > \dim\alpha, \\ \alpha^2, & \text{if } i = \dim\alpha, \\ \alpha, & \text{if i} = 0. \end{cases}$$

(2) The following *Cartan's multiplication formula* holds:

$$\mathrm{Sq}^i(\alpha \cdot \beta) = \sum_{p+q=i} \mathrm{Sq}^p \alpha \cdot \mathrm{Sq}^q \beta.$$

Remark. Consider the map $\mathrm{Sq} = \mathrm{Sq}^0 + \mathrm{Sq}^1 + \mathrm{Sq}^2 + \ldots : H^*(X; \mathbb{Z}_2) \to H^*(X; \mathbb{Z}_2)$ [thus, for $\alpha \in H^n(X; \mathbb{Z}_2)$, $\mathrm{Sq}\,\alpha = \alpha + \mathrm{Sq}^1 \alpha + \cdots + \mathrm{Sq}^{n-1} \alpha + \alpha^2$]. Cartan's formula means that $\mathrm{Sq}(\alpha \cdot \beta) = \mathrm{Sq}\,\alpha \cdot \mathrm{Sq}\,\beta$; that is, Sq is a multiplicative homomorphism.

Add to this that since Sq^1 is not zero ($\mathrm{Sq}^1 \alpha = \alpha^2$ for $\dim \alpha = 1$), it must be the Bockstein homomorphism (see Sect. 28.3).

29.2 Theorem of Existence and Uniqueness for Sq^i

We will prove the existence and uniqueness of stable cohomology operations which satisfy the properties listed above. Moreover, we will see that the uniqueness follows from stability and property (1), and then we will prove Cartan's formula as a theorem.

Denote the fundamental class in $H^n(K(\mathbb{Z}_2, n); \mathbb{Z}_2)$ by e_n. We need to define $\mathrm{Sq}^i e_n$ for all i and n. For $n = 1$ the definition is contained in property (1) [$\mathrm{Sq}^0 e_1 = e_1$, $\mathrm{Sq}^1 e_1 = e_1^2$, $\mathrm{Sq}^i e_1 = 0$ for $i > 1$]. Suppose that we have already defined $\mathrm{Sq}^i e_{n-1}$ for all i. Consider the spectral sequence of the fibration $* \xrightarrow{K(\mathbb{Z}_2, n-1)} K(\mathbb{Z}_2, n)$.

For $i < n - 1$, the transgression provides an isomorphism

$$d_{n+i}^{0,n+i-1} \colon H^{n+i-1}(K(\mathbb{Z}_2, n-1); \mathbb{Z}_2) \to H^{n+i}(K(\mathbb{Z}_2, n); \mathbb{Z}_2);$$

we take the image of $\mathrm{Sq}^i e_{n-1}$ for $\mathrm{Sq}^i e_n$. Hereby, $\mathrm{Sq}^i e_n$ is defined for $i < n - 1$. As to $i = n-1$, we have to observe the differential $d_n^{0,2n-2} \colon E_n^{0,2n-2} \to E_n^{n,n-1}$; however, $d_n^{0,2n-2}\,\mathrm{Sq}^{n-1} e_{n-1} = d_n^{0,2n-2}(e_{n-1}^2) = 2e_{n-1} d_n^{0,2n-2}(e_{n-1}) = 0$, so $\mathrm{Sq}^{n-1} e_{n-1}$ belongs to the domain of the differential

$$d_{2n-1}^{0,2n-2} \colon E_{2n-1}^{0,2n-2} \to E_{2n-1}^{2n-1,0} = H^{2n-1}(K(\mathbb{Z}_2, n); \mathbb{Z}_2),$$

and we take $d_{2n-1}^{0,2n-2}(\mathrm{Sq}^{n-1} e_{n-1})$ for $\mathrm{Sq}^{n-1} e_n$. Finally, we (have to) put $\mathrm{Sq}^n e_n = e_n^2$ and $\mathrm{Sq}^i e_n = 0$ for $i > n$.

Thus,

$$\mathrm{Sq}^i e_n \in H^{n+i}(K(\mathbb{Z}_2, n); \mathbb{Z}_2)$$

is defined for all i and n, and $\mathrm{Sq}^0 e_n = e_n$, $\mathrm{Sq}^n e_n = e_n^2$, $\mathrm{Sq}^i e_n = 0$ for $i > n$. It remains to check that the homomorphism

$$f_n \colon H^{n+i}(K(\mathbb{Z}_2, n); \mathbb{Z}_2) \to H^{n+i-1}(K(\mathbb{Z}_2, n-1); \mathbb{Z}_2)$$

(see Sect. 28.4.D) takes $\mathrm{Sq}^i e_n$ into $\mathrm{Sq}^i e_{n-1}$.

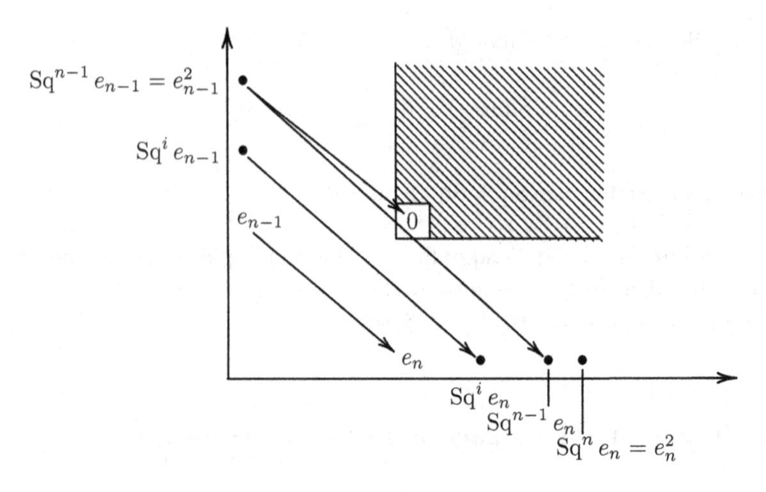

Fig. 109 Construction of Steenrod squares

For $i > n$, we have nothing to prove: $\mathrm{Sq}^i e_n$ and $\mathrm{Sq}^i e_{n-1}$ are both zeroes. For $i < n$, the statement is true, since f_n is inverse to the transgression in the spectral sequence of Fig. 109, and this transgression takes $\mathrm{Sq}^i e_{n-1}$ into $\mathrm{Sq}^i e_n$ by our construction. The remaining case is $i = n$. But $\mathrm{Sq}^n e_n = e_n^2$ and $\mathrm{Sq}^n e_{n-1} = 0$. The map $f_n: H^{2n}(K(\pi, n); \mathbb{Z}_2) \to H^{2n-1}(K(\pi, n-1); \mathbb{Z}_2)$ factors through $i_n^*: H^{2n}(K(\pi, n); \mathbb{Z}_2) \to H^{2n}(\Sigma K(\pi, n-1); \mathbb{Z}_2)$, and the latter takes e_n^2 into zero, since the multiplication in the cohomology of any suspension is trivial (see the proof below). Hence, $f_n(\mathrm{Sq}^n e_n) = f_n(e_n^2) = 0 = \mathrm{Sq}^n e_{n-1}$.

To finish the proof, notice that for any space X, the diagonal map $\Delta: \Sigma X \to \Sigma X \times \Sigma X$ is homotopic to a map taking ΣX into $\Sigma X \vee \Sigma X$: The homotopy $h_t: \Sigma X \to \Sigma X \times \Sigma X$ is defined by the formula $h_t(x) = (\varphi_t(x), \psi_t(x))$, where $\varphi_t, \psi_t: \Sigma X \to \Sigma X$ are two homotopies of the identity map ($\varphi_0 = \psi_0 = \mathrm{id}_{\Sigma X}$) such that $\varphi_1(C_1 X) = x_0, \psi_1(C_2 X) = x_0$, where $C_1 X, C_2 X$ are two cones composing ΣX and $x_0 \in \Sigma X$ is an (arbitrarily chosen) base point. On the other hand, the cross-product of any two cohomology classes of X of positive dimensions has zero restriction to $\Sigma X \vee \Sigma X$ (at least in the case when X is a CW complex). Hence, in the cohomology of ΣX, the cup-product of any two classes of positive dimensions is zero.

29.3 Proof of Cartan's Formula

We need to prove that for every CW complex X, every nonnegative integer i, and every $\alpha, \beta \in H^*(X; \mathbb{Z}_2)$,

$$\mathrm{Sq}^i(\alpha \cdot \beta) = \sum_{p+q=i} \mathrm{Sq}^p \alpha \cdot \mathrm{Sq}^q \beta. \qquad (*)$$

We can assume that $\alpha \in H^m(X;\mathbb{Z}_2), \beta \in H^n(X;\mathbb{Z}_2)$. There are some cases in which the formula $(*)$ is obvious. First, if $m = 0$ or $n = 0$, then the left-hand side and the right-hand side of the equality $(*)$ are simply the same (so we can assume that $m > 0$ and $n > 0$). Second, if $i > m+n$, then both parts of the equality are zeroes: If $p+q > m+n$ and both p and q are nonnegative, then either $p > m$ or $q > n$. Finally, if $i = m+n$, then $\mathrm{Sq}^i(\alpha \cdot \beta) = (\alpha \cdot \beta)^2$ and the only potentially nonzero summand on the right-hand side is $\mathrm{Sq}^m \alpha \cdot \mathrm{Sq}^n \beta = \alpha^2\beta^2$. Assume now that the formula $(*)$ has been proven for $i > m+n-s$ $(s > 0)$ and prove it for $i = m+n-s$.

The formula which we want to prove is equivalent to the equality

$$\mathrm{Sq}^{m+n-s}(\alpha \times \beta) = \sum_{p+q=m+n-s} \mathrm{Sq}^p \alpha \times \mathrm{Sq}^q \beta,$$

where $\alpha \in H^m(X;\mathbb{Z}_2), \beta \in H^n(Y;\mathbf{Z}_2)$ and X, Y are CW complexes. The cohomology classes in the equality are in $H^{2(m+n)-s}(X \times Y;\mathbb{Z}_2)$, but we have the right to replace the product $X \times Y$ by the smash product $X\#Y = X \times Y/X \vee Y$: Since $m > 0$ and $n > 0$, the class $\alpha \times \beta \in H^{m+n}(X \times Y;\mathbb{Z}_2)$ is the image of the class $\alpha \times \beta \in H^{m+n}(X\#Y;\mathbb{Z}_2)$ with respect to the cohomology homomorphism induced by the projection $X \times Y \to X\#Y$, so the formula for $X\#Y$ will imply the formula for $X \times Y$. And finally the most important remark: We do not need to prove our formula for arbitrary X, Y, α, β: It is sufficient to consider the case when $X = K(\mathbb{Z}_2, m)$ (which, to make our formulas shorter and better-looking, we will abbreviate to K_m), $Y = K_n$, and $\alpha = e_m, \beta = e_n$.

Consider the maps

$$K_m\#K_n \xleftarrow{i_m\#\,\mathrm{id}} (\Sigma K_{m-1})\#K_n = \Sigma(K_{m-1}\#K_n)$$
$$K_m\#K_n \xleftarrow{\mathrm{id}\,\#\,i_n} K_m\#(\Sigma K_{n-1}) = \Sigma(K_m\#K_{n-1})$$

[where i_m, i_n are determined by the relations $i_m^*(e_m) = \Sigma e_{m-1}, i_n^*(e_n) = \Sigma e_{n-1}$; compare with Sect. 28.4.D]. There arise cohomology homomorphisms

$$H^r(K_m\#K_n;\mathbb{Z}_2) \xrightarrow{(i_m\#\,\mathrm{id})^*} H^r(\Sigma(K_{m-1}\#K_n);\mathbb{Z}_2)$$
$$H^r(K_m\#K_n;\mathbb{Z}_2) \xrightarrow{(\mathrm{id}\,\#\,i_n)^*} H^r(\Sigma(K_m\#K_{n-1});\mathbb{Z}_2),$$

or

$$H^r(K_m\#K_n;\mathbb{Z}_2) \xrightarrow{\Sigma^{-1}(i_m\#\,\mathrm{id})^*} H^{r-1}(K_{m-1}\#K_n;\mathbb{Z}_2)$$
$$H^r(K_m\#K_n;\mathbb{Z}_2) \xrightarrow{\Sigma^{-1}(\mathrm{id}\,\#\,i_n)^*} H^{r-1}(K_m\#K_{n-1};\mathbb{Z}_2).$$

Note some properties of the last two maps. First, if α and β are cohomology classes of K_m and K_n, then these two homomorphisms map $\alpha \times \beta$ into, respectively, $f_m(\alpha) \times \beta$ and $\alpha \times f_n(\beta)$; in particular, they take $e_m \times e_n$ in, respectively, $e_{m-1} \times e_n$ and $e_m \times e_{n-1}$. Second, they both commute with all Steenrod squares (as well as f_m and f_n). And third, if some $\gamma \in H^r(K_m \# K_n; \mathbb{Z}_2)$ is annihilated by both homomorphisms and $r < 2(m+n)$, then $\gamma = 0$. The first two statements are already known to us, so let us prove the third. Let $\gamma = \sum(\alpha_i \times \beta_i)$. We can assume that the α_i are linearly independent, as well as the β_i. Indeed, if, say, some α_{i_0} is a linear combination of other α_i, then we can reduce the number of summands in the sum $\sum_i(\alpha_i \times \beta_i) = \gamma$ by distributing the summand $\alpha_{i_0} \times \beta_{i_0}$ among the other summands. Thus, if the number of summands in our presentation of γ is the minimum possible, then there cannot be any linear dependence between the α_i as well as between the β_i. Let $\gamma = \sum_i(\alpha_i \times \beta_i)$ be a presentation with this property. Our two maps take γ into, respectively, $\sum_i f_m(\alpha_i) \times \beta_i$ and $\sum_i \alpha_i \times f_n(\beta_i)$, and if both are zeroes, then $f_m(\alpha_i) = 0$ and $f_n(\beta_i) = 0$ for all i. But if a nonzero $\alpha \in H^r(K_m; \mathbb{Z}_2)$ is annihilated by f_m, then $r \geq 2m$ (the first cohomology class with this property is e_m^2), and similarly for $\beta \in H^r(K_n; \mathbb{Z}_2)$. Hence, if a nonzero γ is annihilated by both homomorphisms, then $\dim \gamma \geq 2(m+n)$.

Now, let us return to Cartan's formula. The difference

$$\mathrm{Sq}^{m+n-s}(e_m \times e_n) - \sum_{p+q=m+n-s} \mathrm{Sq}^p\, e_m \times \mathrm{Sq}^q\, e_n$$

is taken by our homomorphisms into, respectively,

$$\mathrm{Sq}^{m+n-s}(e_{m-1} \times e_n) - \sum_{p+q=m+n-s} \mathrm{Sq}^p\, e_{m-1} \times \mathrm{Sq}^q\, e_n,$$
$$\mathrm{Sq}^{m+n-s}(e_m \times e_{n-1}) - \sum_{p+q=m+n-s} \mathrm{Sq}^p\, e_m \times \mathrm{Sq}^q\, e_{n-1}.$$

Both are zeroes by the induction hypothesis [because $m+n-s = (m-1)+n-(s-1) = m+(n-1)-(s-1)$], and, according to the observation made above, this shows that

$$\mathrm{Sq}^{m+n-s}(e_m \times e_n) - \sum_{p+q=m+n-s} \mathrm{Sq}^p\, e_m \times \mathrm{Sq}^q\, e_n = 0$$

[since the dimension of the expression on the left-hand side is $2(m+n) - s < 2(m+n)$]. This completes the proof.

Lecture 30 The Steenrod Algebra

30.1 The Structure of the Steenrod Algebra A_2: The Statements

The Steenrod algebra $\mathbb{A} = \mathbb{A}_2$ is the algebra of all stable cohomology operations over the field $\mathbb{Z}_2$ where the multiplication is defined as the composition. In this section we will give a full algebraic description of $\mathbb{A}$; proofs of these statements will be given in subsequent sections.

It turns out that the algebra $\mathbb{A}$ is multiplicatively generated by Steenrod squares Sq^i; that is, every stable cohomology operation in $\mathbb{Z}_2$-cohomology is a linear combination of iterations of Steenrod squares.

However, the Steenrod squares do not form a free system of multiplicative generators: There are relations between them. In particular, an additive basis of the algebra $\mathbb{A}$ is not formed by all iterations of Steenrod squares, but only by iterations

$$\mathrm{Sq}^I = \mathrm{Sq}^{i_1}\,\mathrm{Sq}^{i_2}\ldots\mathrm{Sq}^{i_k}$$

for which the sequence $I = \{i_1, i_2, \ldots, i_k\}$ satisfies the conditions

$$i_1 \geq 2i_2, i_2 \geq 2i_3, \ldots, i_{k-1} \geq 2i_k$$

(such sequences are usually called *admissible sequences*, and the corresponding iterations of Steenrod squares are called *admissible iterations*).

The multiplicative structure of $\mathbb{A}$ is determined by the so-called Adem relations (see Adem [11]): If $a < 2b$, then

$$\mathrm{Sq}^a\,\mathrm{Sq}^b = \sum_c \binom{b-c-1}{a-2c} \mathrm{Sq}^{a+b-c}\,\mathrm{Sq}^c$$

[obviously, c in this sum varies from $\max(a-b+1, 0)$ to $\left\lfloor \frac{a}{2} \right\rfloor$]. Notice that all the iterations on Steenrod squares on the right-hand side of the Adem relation are admissible: If $a-2c \geq 0$ and $b-c-1 \geq 0$, then $a+b-c \geq 2c+c+1-c = 2c+1$. It is clear that using the Adem relations, one can reduce any iteration of Steenrod squares to a linear combination of admissible iterations. Indeed, let us order all the iterations of Steenrod squares lexicographically ($\{i_1, i_2, \ldots\} \succ \{j_1, j_2, \ldots\}$ if for some m, $i_1 = j_1, \ldots, i_{m-1} = j_{m-1}, i_m > j_m$). Consider any iteration $\mathrm{Sq}^{i_1}\,\mathrm{Sq}^{i_2}\ldots\mathrm{Sq}^{i_k}$. If (for some m) $i_m < 2i_{m+1}$, then replace $\mathrm{Sq}^{i_m}\,\mathrm{Sq}^{i_{m+1}}$ by the expression in the Adem relation. Then our iteration becomes a linear combination of lexicographically preceding iteration. Repeat this until we get a linear combination of admissible iterations.

We will discuss corollaries of the Adem relations after we prove them (Sect. 30.5).

In conclusion, let us formulate the main technical result of this lecture. To do so, we need one more definition. Let $I = \{i_1, i_2, \dots, i_k\}$ be an admissible sequence. The excess of I, $\operatorname{exc} I$, is defined by the formula

$$\operatorname{exc} I = (i_1 - 2i_2) + \dots + (i_{k-1} - 2i_k) + i_k = i_1 - (i_2 + \dots + i_k).$$

(In particular, the empty iteration is an admissible iteration of excess 0; Sq^i is an admissible iteration of excess i.)

Theorem. *$H^*(K(\mathbb{Z}_2, n); \mathbb{Z}_2)$ is a polynomial algebra with the generators $\mathrm{Sq}^I e_n$ for all admissible sequences $I = \{i_1, \dots, i_k\}$ with $\operatorname{exc} I < n$.*

This theorem belongs to J.-P. Serre [77]. It shows that all cohomology operations acting in the cohomology modulo 2 are linear combinations of products of admissible iterations of Steenrod squares, and these products are linearly independent. For example, all cohomology operations $H^2(X; \mathbb{Z}_2) \to H^5(X; \mathbb{Z}_2)$ are linear combinations of (linearly independent) operations $x \mapsto \mathrm{Sq}^2 \mathrm{Sq}^1 x$, $x \mapsto x \cdot \mathrm{Sq}^1 x$.

Serre's theorem will be proven in Sect. 30.3. The main technical tool of this proof (and many other proofs) is a theorem on spectral sequences, which we will prove in the next section.

30.2 Borel's Theorem

Suppose that there is a fibration (E, B, F, p) with a simply connected base and the following properties.

(1) $\widetilde{H}^*(E; \mathbb{Z}_2) = 0$.
(2) The algebra $H^*(F; \mathbb{Z}_2)$ has a system of *transgressive* generators $a_i \in H^{m_i}(F; \mathbb{Z}_2)$, $m_1 \leq m_2 \leq \dots$.
(3) Moreover, this system is *simple* in the sense that the monomials $a_{i_1} a_{i_2} \dots a_{i_k}$, $i_1 < i_2 < \dots < i_k$ form an additive basis in $H^*(F; \mathbb{Z}_2)$.

Theorem (Borel A. [22]). *If conditions* (1) – (3) *are met, then $H^*(B; \mathbb{Z}_2)$ is a polynomial algebra with generators $b_i \in H^{m_i+1}(B; \mathbb{Z}_2)$, where the b_i are arbitrary representatives of $\tau(a_i)$.*

Proof of Theorem. We will construct some abstract spectral sequence, $\{\widetilde{E}_r^{pq}, \widetilde{d}_r^{pq}\}$, with properties (1)–(3), and then we will show that this abstract spectral sequence coincides with the spectral sequence of our fibration.

Let $\widetilde{A} = H^*(F; \mathbb{Z}_2)$ and $\widetilde{B} = \mathbb{Z}_2[b_1, b_2, \dots]$, where the generators b_i correspond to the generators a_i of $\widetilde{A}$ and $\dim b_i = m_i + 1$ (where $m_i = \dim a_i$). Let $\widetilde{E}_2 = \widetilde{A} \otimes_{\mathbb{Z}_2} \widetilde{B}$ (with the natural bigrading).

Now, we define the differentials $\widetilde{d}_r$ as acting in $\widetilde{E}_2$. We put $\widetilde{d}_r b_i = 0$ for all r, $\widetilde{d}_r a_i = 0$ for $r \leq m_i$ (this means that every a_i is transgressive), and $d_{m_i+1} a_i = b_i$. Thus (if $i_1 \dots < i_k$, $j_1 < \dots < j_\ell$, $s_i > 0$),

$$\widetilde{d}_r(a_{i_1}\dots a_{i_k}\otimes b_{j_1}^{s_1}\dots b_{j_\ell}^{s_\ell}) = \begin{cases} a_{i_2}\dots a_{i_k}\otimes b_{i_1}b_{j_1}^{s_1}\dots b_{j_\ell}^{s_\ell}, & \text{if } i_1\le j_1 \text{ and } r=m_{i_1}+1,\\ 0 & \text{in all other cases.}\end{cases}$$

Obviously, $\widetilde{d}_r\circ\widetilde{d}_s = 0$ for any r and s, and for this reason the differentials $\widetilde{d}_r$ give rise to an (additive) spectral sequence. Namely, we put $\widetilde{E}_3 = \operatorname{Ker}\widetilde{d}_2/\operatorname{Im}\widetilde{d}_2$. Then, since $\widetilde{d}_3$ commutes with $\widetilde{d}_2$, we can consider $\widetilde{d}_3$ as acting in $\widetilde{E}_3$, then we put $\widetilde{E}_4 = \operatorname{Ker}\widetilde{d}_3/\operatorname{Im}\widetilde{d}_3$, and so on. We obtain an additive spectral sequence $\{\widetilde{E}_r,\widetilde{d}_r\}$, and, obviously $\widetilde{E}_\infty = \widetilde{E}_\infty^{00}$ (for every element α of the additive basis of $\widetilde{E}_2$, either $\widetilde{d}_r\alpha\neq 0$ for some r, or $\alpha\in\operatorname{Im}\widetilde{d}_s$ for some s).

Next, we want to show that the spectral sequence $\{\widetilde{E}_r,\widetilde{d}_r\}$ is actually multiplicative; that is,

$$\widetilde{d}_r(AA') = (\widetilde{d}_rA)A' + A(\widetilde{d}_rA') \qquad (*)$$

for any monomials A, A' in a_i and b_j (no minus signs, because everything is over $\mathbb{Z}_2$). Only one difficulty arises here: If

$$A = a_{i_1}\dots a_{i_k}\otimes b_{j_1}^{s_1}\dots b_{j_\ell}^{s_\ell},$$
$$A' = a_{i'_1}\dots a_{i'_{k'}}\otimes b_{j'_1}^{s'_1}\dots b_{j'_{\ell'}}^{s'_{\ell'}},$$

then the product AA' may contain an a_i with i less than both i_1 and i'_1 (this may arise from the square of an a_h, which appears in both A and A'). This does not happen, however, because for any a_i from our system of generators of $H^*(F;\mathbb{Z}_2)$, the square a_i^2 is a sum like $a_{h_1}+\dots+a_{h_m}$, not involving products $a_{g_1}\dots a_{g_s}$ with $s\ge 2$. This does not follow from the definition of a simple system of generators, but follows from the existence of a fibration with the fiber F and $\widetilde{H}^*(E;\mathbb{Z}_2) = 0$; indeed, if a_i^2 involves a monomial $a_{g_1}\dots a_{g_s}$ with $s\ge 2$, and g_1 is the smallest index in all such monomials, then $g_1 < i$ and in the spectral sequence of our fibration (actual, not artificial) $0 = d_{g_1}(a_i^2) = d_{g_1}(\dots+(a_{g_1}\dots a_{g_s})+\dots) = \dots+(b_{g_1}a_{g_2}\dots a_{g_s})+\dots\neq 0$, a contradiction. This shows that the squares of a_is from AA' involve only a_hs with $h > i$, and hence AA' does not contain a_is with $i < \min(i_1, i'_1)$. After this remark, checking the equality $(*)$ is immediate. [Indeed, if $i_1 < i'_1$, then $\widetilde{d}_{m_{i_1}+1}(A') = 0$ and $\widetilde{d}_{m_{i_1}+1}(AA') = (\widetilde{d}_{m_{i_1}+1}A)A'$, and if $i_1 = i'_1$, then $\widetilde{d}_{m_{i_1}+1}(AA') = 0$ and $(\widetilde{d}_{m_{i_1}+1}A)A' = A(\widetilde{d}_{m_{i_1}+1}A')$, so $(\widetilde{d}_{m_{i_1}+1}A)A' + A(\widetilde{d}_{m_{i_1}+1}A') = 0$; in all other cases, both sides of $(*)$ are zeroes.]

Almost nothing remains. We have two multiplicative spectral sequences, $\{E_r, d_r\}$ and $\{\widetilde{E}_r,\widetilde{d}_r\}$. Both of them have a trivial limit term, the E_2-terms have the same zeroth column, and there is a canonical multiplicative map of the zeroth row of $\widetilde{E}_2$ into the zeroth row of E_2. Suppose that this map is not an isomorphism; this means that either $H^*(B;\mathbb{Z}_2)$ has an element which is not a polynomial in b_is [let $c\in H^m(B;\mathbb{Z}_2)$ be such an element of the smallest dimension], or in $H^*(B;\mathbb{Z}_2)$ there is a relation between b_is [let $P(b_1, b_2,\dots) = 0$ be such a relation with the smallest possible dimension of the left-hand side]. In both cases we have an easy contradiction: In the first case, the spectral sequence $\{E_r, d_r\}$ has nothing to kill c, and it has to stay in E_∞; in the second case, the spectral sequence $\{\widetilde{E}_r, \widetilde{d}_r\}$ has nothing to kill $P(b_1, b_2, \dots)$, and it has to stay in $\widetilde{E}_\infty$.

This completes the proof of Borel's theorem.

30.3 Proof of Serre's Theorem

We use the induction with respect to n. The case $n = 1$ is simple: Only the empty sequence I has the excess 0, and $H^*(K(\mathbb{Z}_2, 1); \mathbb{Z}_2) = H^*(\mathbb{R}P^\infty; \mathbb{Z}_2) = \mathbb{Z}_2[e_1]$. Assume that $n \geq 2$ and $H^*(K(\mathbb{Z}_2, n-1); \mathbb{Z}_2) = \mathbb{Z}_2[\mathrm{Sq}^I e_{n-1} \mid \mathrm{exc}\, I < n-1]$ (the notation $\mathrm{exc}\, I$ means that I is admissible; we never state the admissibility of a sequence separately). As we have remarked before, the system of generators $\{\mathrm{Sq}^I e_{n-1}\}$ is not simple; to make it simple, we need to extend it to

$$\{\mathrm{Sq}^I e_{n-1}, (\mathrm{Sq}^I e_{n-1})^2, (\mathrm{Sq}^I e_{n-1})^4, \cdots \mid \mathrm{exc}\, I < n-1\},$$

which is the same as

$$\begin{aligned}\{&\mathrm{Sq}^I e_{n-1}, \mathrm{Sq}^{|I|+n-1} \mathrm{Sq}^I e_{n-1}, \mathrm{Sq}^{2(|I|+n-1)} \mathrm{Sq}^{|I|+n-1} \mathrm{Sq}^I e_{n-1}, \\ &\mathrm{Sq}^{4(|I|+n-1)} \mathrm{Sq}^{2(|I|+n-1)} \mathrm{Sq}^{|I|+n-1} \mathrm{Sq}^I e_{n-1}, \cdots \mid \mathrm{exc}\, I < n-1\}.\end{aligned} \tag{$**$}$$

Lemma. *The system* $(**)$ *is the same as* $\{\mathrm{Sq}^J e_{n-1} \mid \mathrm{exc}\, J < n\}$.

Proof of Lemma. If $I = \{i_1, \dots, i_k\}$, then, for

$$\begin{aligned} J = \{2^\ell(|I| + n - 1), 2^{\ell-1}(|I| + n - 1), \dots, 2(|I| + n - 1), \\ |I| + n - 1, i_1, \dots, i_k\}, \ \ell > 0, \end{aligned}$$

we have

$$\begin{aligned} \mathrm{exc}\, J = &\underbrace{\begin{matrix}[2^\ell(|I| + n - 1) - 2 \cdot 2^{\ell-1}(|I| + n - 1)] + \dots \\ \qquad + [2(|I| + n - 1) - 2 \cdot (|I| + n - 1)]\end{matrix}}_{\text{zero}} \\ &+ (|I| + n - 1 - 2i_1) + (i_1 - 2i_2) + \dots + (i_{k-1} - 2i_k) + i_k \\ &\qquad = |I| + n - 1 - |I| = n - 1. \end{aligned}$$

Since [in $(**)$] $\mathrm{exc}\, I < n - 1 < n$, we see that all members of the system $(**)$ have $\mathrm{exc} < n$.

Prove now that every $\mathrm{Sq}^J e_{n-1}$ with $\mathrm{exc}\, J < n$ is contained in $(**)$. Indeed, let $J = \{j_1, \dots, j_m\}$, $\mathrm{exc}\, J < n$. If $\mathrm{exc}\, J < n - 1$, then $\mathrm{Sq}^J e_{n-1}$ is contained in $(**)$. Let $\mathrm{exc}\, J = n - 1$. Consider the sequence

$$j_1 - 2j_2, j_2 - 2j_3, \dots, j_{m-1} - 2j_m, j_m.$$

This is a sequence of nonnegative numbers with a positive sum, $n - 1$. Let the first positive term of this sequence be $j_s - 2j_{s+1}$ (we put, if necessary, $j_{m+1} = 0$). Put $I = \{j_{s+1}, \dots, j_m\}$. Then

$$\begin{aligned} n - 1 = \operatorname{exc} J &= (j_s - 2j_{s+1}) + \dots + (j_{m-1} - 2j_m) + j_m \\ &= j_s - j_{s+1} - \dots - j_m = j_s - |I|, \end{aligned}$$

so $j_s = |I| + n - 1$, and

$$J = \{2^{s-1}(|I| + n - 1), \dots, 2(|I| + n - 1), |I| + n - 1, j_{s+1}, \dots, j_m\},$$

so $\mathrm{Sq}^J e_{n-1}$ is in $(**)$.

The lemma is proved, so let us return to the theorem. Consider the $\mathbb{Z}_2$-cohomological spectral sequence of the fibration $* \xrightarrow{K(\mathbb{Z}_2, n-1)} K(\mathbb{Z}_2, n)$. Since $e_{n-1} \in E_2^{0,n-1}$ is transgressive, and $\tau(e_{n-1}) = e_n \in E_2^{n0}$, all elements of the system $(**)$ are transgressive, and its image with respect to the transgression is represented by the system $\{\mathrm{Sq}^J e_n, \operatorname{exc} J < n\}$ (we use the lemma and the fact that all Steenrod squares commute with the transgression; see Sect. 28.4.C). Thus, Borel's theorem can be applied, and it shows that $H^*(K(\mathbb{Z}_2, n); \mathbb{Z}_2)$ is a polynomial algebra generated by $\mathrm{Sq}^J e_n$ for all admissible sequences J with $\operatorname{exc} J < n$, as was stated.

30.4 The Structure of the Steenrod Algebra (Modulo 2)

As was shown in Sect. 28.4, additively, $\mathbb{A}_2^q = \overleftarrow{\lim}_n H^{n+q}(K(\mathbb{Z}_2, n); \mathbb{Z}_2)$. Combining this with Serre's theorem, we arrive at the following result.

Theorem. *The Steenrod algebra $\mathbb{A}_2$ is additively generated by admissible iterations Sq^I (without restrictions on excesses).*

Thus, all stable cohomology operations in the cohomology with coefficients in $\mathbb{Z}_2$ are sums of iterations of Steenrod squares. Moreover, we see that there must exist many relations between these iterations, since an arbitrary iteration is equal to a sum of admissible iterations. We begin by studying these relations.

30.5 Relations

A: A Method of Finding Relations: the Splitting Principle

Suppose that we are given a certain (noncommutative) polynomial in Steenrod squares, $P(\mathrm{Sq}^1, \mathrm{Sq}^2, \dots)$. How can we prove that it is equal to zero? Theoretically, we need to prove the equality $P(\mathrm{Sq}^1, \mathrm{Sq}^2, \dots)x = 0$ for every cohomology class $x \in H^n(X; \mathbb{Z}_2)$ of every, say, CW complex X. Visibly, there arise two difficulties, and each of them seems to be a dead end. First, how do we observe *all* cohomology classes of *all* CW complexes? Second, even for an individual x, how do we compute $\mathrm{Sq}^I x$ for an iteration Sq^I of Steenrod squares?

The second difficulty does not exist in the case when x is a product of one-dimensional cohomology classes: Then $\mathrm{Sq}^I x$ can easily be found with Cartan's formula. And it turns out that we actually do not need anything else. Here is an explanation.

Let $\mathcal{P} = \underbrace{\mathbb{R}P^\infty \times \cdots \times \mathbb{R}P^\infty}_{N}$. Then $H^*(\mathcal{P}; \mathbb{Z}_2) = \mathbb{Z}_2[x_1, \dots, x_N]$ with the generators $x_i \in H^1(\mathcal{P}; \mathbb{Z}_2)$ coming from the ith factor $\mathbb{R}P^\infty$. Let $u = u_N = x_1 \dots x_N \in H^N(\mathcal{P}; \mathbb{Z}_2)$.

Theorem. *Let $\varphi = P(\mathrm{Sq}^1, \mathrm{Sq}^2, \dots) = \sum_j \mathrm{Sq}^{I_j}$ be a polynomial in Steenrod squares, and let $N \geq |I_j|$ for all j. If $\varphi(u_N) = 0$, then $\varphi = 0$ [that is, $\varphi(x) = 0$ for any $x \in H^*(X; \mathbb{Z}_2)$ and any X].*

Proof. We want to prove that if $q \leq N$, then the map $\eta: \mathbb{A}^q = \mathbb{A}_2^q \to H^{N+q}(\mathcal{P}; \mathbb{Z}_2)$, $\eta(\varphi) = \varphi(u)$, is a monomorphism. Let $\mathbb{B}^q = \eta(\mathbb{A}^q) \subset H^{N+q}(\mathcal{P}; \mathbb{Z}_2) = \mathbb{Z}_2[x_1, \dots, x_N]^{N+q}$. All we need is to prove that $\dim_{\mathbb{Z}_2} \mathbb{A}^q = \dim_{\mathbb{Z}_2} \mathbb{B}^q$; we will do this by a computation of both dimensions.

Let us begin with $\mathbb{B}^q$. First, for any (not necessarily even stable) cohomology operation $\psi \in \mathcal{O}(N, N+q, \mathbb{Z}_2, \mathbb{Z}_2)$, the polynomial $\psi(u) \in \mathbb{Z}_2[x_1, \dots, x_N]$ is symmetric; indeed, the map $\sigma: \mathcal{P} \to \mathcal{P}$ defined as a permutation of the factors $\mathbb{R}P^\infty$ takes u into U, and hence takes $\psi(u)$ into $\psi(u)$. Second, $\psi(u)$ is divisible by u. Indeed, let $\mathcal{P}_i$ be the product of $\mathbb{R}P^\infty$s obtained from $\mathcal{P}$ by deleting the ith factor. The inclusion map $\varepsilon_i: \mathcal{P}_i \to \mathcal{P}$ induces a map $\varepsilon_i^*: \mathbb{Z}_2[x_1, \dots, x_N] \to \mathbb{Z}_2[x_1, \dots \widehat{x_i} \dots, x_N]$, which acts to put $x_i = 0$. This map takes u into 0, hence takes $\psi(u)$ into 0; thus, $\psi(u)$ is divisible by every x_i; that is, it is divisible by u. In one word, $\psi(u) = u \cdot s(x_1, \dots, x_n)$, where s is a symmetric polynomial.

EXERCISE 1. Prove that if $q \leq N$, then $\mathrm{Sq}^q(u) = u \cdot e_q(x_1, \dots, x_N)$, where e_q is the qth elementary symmetric polynomial [if $q > N$, then, certainly, $\mathrm{Sq}^q(u) = 0$].

Let us call a polynomial from $\mathbb{Z}_2[x_1, \dots, x_N]$ *special* if every monomial in this polynomial has the form

$$x_1^{2^{k_1}} x_2^{2^{k_2}} \dots x_N^{2^{k_N}}$$

(in particular, any special polynomial is divisible by u).

Lemma. *$\mathbb{B}^q$ is the same as the set of symmetric special polynomials of degree $N+q$.*

Proof of Lemma. First, let us notice that if $\dim x = 1$, then $\mathrm{Sq}^i x^m = \binom{m}{i} x^{m+i}$; indeed, $\mathrm{Sq}\, x = x + x^2 = x(1+x)$, and hence $\mathrm{Sq}(x^m) = (\mathrm{Sq}\, x)^m = x^m(1+x)^m = \sum_{i=0}^{m} \binom{m}{i} x^{m+i}$. In particular,

$$\mathrm{Sq}^i\left(x^{2^k}\right) = \begin{cases} x^{2^{k+1}}, & \text{if } i = 2^k, \\ x^{2^k}, & \text{if } i = 0, \\ 0 & \text{in all other cases.} \end{cases}$$

Thus,

$$\mathrm{Sq}^i\left(x_1^{2^{k_1}} x_2^{2^{k_2}} \dots x_N^{2^{k_N}}\right) = \sum_{i_1+\dots+i_k=i} \mathrm{Sq}^{i_1}\left(x_1^{2^{k_1}}\right) \mathrm{Sq}^{i_1}\left(x_2^{2^{k_2}}\right) \dots \mathrm{Sq}^{i_1}\left(x_N^{2^{k_N}}\right)$$

is a special polynomial which shows that every Sq^I takes special polynomials into special polynomials. Since u is a special polynomial, we see that $\mathbb{B}^q$ consists of symmetric (see a notice above) special polynomials.

Let us prove now that every symmetric special polynomial of degree $N+q$ belongs to $\mathbb{B}^q$. The space of symmetric special polynomial of degree $N+q$ has a basis formed by the symmetrized monomials

$$\mathrm{Symm}\left(x_1^{2^k} \dots x_{n_1}^{2^k} x_{n_1+1}^{2^{k-1}} \dots x_{n_2}^{2^{k-1}} \dots x_{n_{k-1}+1}^2 \dots x_{n_k}^2 x_{n_k+1} \dots x_N\right), \tag{$*$}$$

where $1 \le n_1 \le n_2 \le \dots \le n_k \le N$ and

$$n_1(2^k - 1) + (n_2 - n_1)(2^{k-1} - 1) + \dots + (n_k - n_{k-1}) = q.$$

Let us order the monomials like the one in parentheses in $(*)$ lexicographically. Consider

$$\mathrm{Sq}^{2^{k-1}n_1} \mathrm{Sq}^{2^{k-2}n_2} \dots \mathrm{Sq}^{2n_{k-1}} \mathrm{Sq}^{n_k}(u).$$

This polynomial contains the monomial in $(*)$ plus an amount of symmetrized monomials which are lexicographically behind the monomial in $(*)$. [Indeed, when we apply Sq^{n_k} to u, we get precisely

$$\mathrm{Symm}(x_1^2 \dots x_{n_k}^2 x_{n_k+1} \dots x_N).$$

Then we apply $\mathrm{Sq}^2\, n_{k-1}$ to this and get

$$\mathrm{Symm}(x_1^4 \dots x_{n_{k-1}}^4 x_{n_{k-1}+1}^2 \dots x_{n_k}^2 x_{n_k+1} \dots x_N)$$

plus a variety of lexicographically smaller terms. And so on.] This shows that the systems (*) and $\mathrm{Sq}^I u$ are related by a lower triangular matrix with 1s on the diagonal, which provides the result we are looking for.

Thus, the lemma has been proven. Let us return to the theorem. The lemma shows that the dimension of $\mathbb{B}^q$ is the number of partitions of q such that every part in a power of 2 minus 1. On the other hand, the dimension of $\mathbb{A}^q$ is the number of admissible sequences I with $|I| = q$. Let $\{i_1, i_2, \dots, i_k\}$ be an admissible sequence and let $i_1 + i_2 + \dots + i_k = q$. Put $j_i = i_1 - 2i_2, j_2 = i_2 - 2i_3, \dots, j_{k-1} = i_{k-1} - 2i_k, j_k = i_k$. The statement that the given sequence is admissible means that all j_s are nonnegative. These nonnegative numbers, arbitrarily chosen, determine the admissible sequence. Express i in terms of j:

$$\begin{array}{lcl}
i_1 & = & j_1 + 2j_2 + \dots + 2^{k-3} j_{k-2} + 2^{k-2} j_{k-1} + 2^{k-1} j_k \\
i_2 & = & j_2 + \dots + 2^{k-4} j_{k-2} + 2^{k-3} j_{k-1} + 2^{k-2} j_k \\
\dots & \dots & \dots \\
i_{k-2} & = & j_{k-2} + 2j_{k-1} + 4j_k \\
i_{k-1} & = & j_{k-1} + 2j_k \\
i_k & = & j_k
\end{array}$$

Summing up these equalities, we get

$$\begin{aligned}
q &= j_1 + 3j_2 + \dots + (2^{k-2} - 1)j_{k-2} + (2^{k-1} - 1)j_{k-1} + (2^k - 1)j_k \\
&= \sum_{s=1}^{k} (2^s - 1)j_s.
\end{aligned}$$

Hence, the dimension of $\mathbb{A}^q$ is also the number of partitions of q such that every part in a power of 2 minus 1. Thus, $\dim \mathbb{A}^q = \dim \mathbb{B}^q$, which completes the proof of the theorem.

B: Examples of Relations

The previous theorem paves a road to finding relations in the Steenrod algebra. Here is the first example: Obviously,

$$\mathrm{Sq}^n(u) = \mathrm{Symm}(x_1^2 \dots x_2^n x_{n+1} \dots x_N)$$

and

$$\mathrm{Sq}^1\, \mathrm{Sq}^n(u) = (n+1)\, \mathrm{Symm}(x_1^2 \dots x_{n+1}^2 x_{n_2} \dots x_N).$$

Thus,

$$\mathrm{Sq}^1\,\mathrm{Sq}^n = \begin{cases} \mathrm{Sq}^{n+1}, & \text{if } n \text{ is even}, \\ 0, & \text{if } n \text{ is odd}. \end{cases}$$

Or,

$$\begin{aligned} \mathrm{Sq}^2\,\mathrm{Sq}^2(u) &= \mathrm{Sq}^2\,\mathrm{Symm}(x_1^2x_2^2x_3\dots x_N) \\ &= \mathrm{Symm}(x_1^4x_2^2x_3\dots x_n) + 6\,\mathrm{Symm}(x_1^2x_2^2x_3^2x_4^2x_5\dots x_N), \\ \mathrm{Sq}^3\mathrm{Sq}^1(u) &= \mathrm{Sq}^3\,\mathrm{Symm}(x_1^2x_2\dots x_N) \\ &= \mathrm{Symm}(x_1^4x_2^2x_3\dots x_n) + 4\,\mathrm{Symm}(x_1^2x_2^2x_3^2x_4^2x_5\dots x_N), \end{aligned}$$

which shows that $\mathrm{Sq}^2\mathrm{Sq}^2 = \mathrm{Sq}^3\mathrm{Sq}^1$.

C: Adem's Relations

We already mentioned in Sect. 30.1 that a complete system of relations between the Steenrod squares is formed by the so-called Adem's relation: If $a < 2b$, then

$$\mathrm{Sq}^a\,\mathrm{Sq}^b = \sum_{c=\max(a-b+1,0)}^{[a/2]} \binom{b-c-1}{a-2c} \mathrm{Sq}^{a+b-c}\,\mathrm{Sq}^c\,.$$

To prove this relation, we need only to check that the left- and right-hand sides take equal values on $u \in H^N(\mathcal{P};\mathbb{Z}_2)$ (where $N \geq a+b$). Both values can be calculated by means of Cartan's formula. The left-hand side takes u into

$$\sum_s \binom{a+b-3s}{b-a} \mathrm{Symm}(x_1^4\dots x_s^4x_{s+1}^2\dots x_{a+b-s}^2x_{a+b-s+1}\dots x_N),$$

while the right-hand side takes u into

$$\sum_c\sum_s \binom{b+a-3s}{b-s}\binom{b-c-1}{a-2c} \cdot \mathrm{Symm}(x_1^4\dots x_s^4x_{s+1}^2\dots x_{a+b-s}^2x_{a+b-s+1}\dots x_N).$$

To prove the Adem relations, we need to check the congruence

$$\binom{b+a-3c}{b-s} \equiv \sum_c \binom{b+a-3c}{c-s}\binom{b-c-1}{a-2c} \bmod 2,$$

which becomes, after the substitution $d = a-2s, e = b-s, f = c-s$, the congruence

$$\binom{d+e}{e} = \sum_{f=\max(0,d-e+1)}^{[d/2]} \binom{d+e}{f}\binom{e-f+1}{d-2f} \bmod 2,$$

which can be done by elementary means.

As we already mentioned in Sect. 30.1, the Adem system of relations is complete since it can reduce every iteration of Steenrod squares to the sum of admissible ones, and admissible iterations are linearly independent.

EXERCISE 2. Prove that the Steenrod squares $\mathrm{Sq}^1, \mathrm{Sq}^2, \mathrm{Sq}^4, \mathrm{Sq}^8, \mathrm{Sq}^{16}, \dots$ form a system of generators of the algebra $\mathbb{A}_2$, and this system is minimal. In other words, prove that Sq^n can be expressed as a polynomial in Sq^i with $0 < i < n$ if and only if n is not a power of 2.

EXERCISE 3. Find the defining system of relations between the generators of Exercise 2 (no satisfactory solution of this is known).

EXERCISE 4 (Bullet–MacDonald). Prove that the Adem relations may be presented as one identity,

$$P(s^2 + st)P(t^2) = P(t^2 + st)P(s^2),$$

where $P(u) = \sum \mathrm{Sq}^i u^i$.

30.6 Computing $\mathcal{O}(n, q, \mathbb{Z}, \mathbb{Z}_2)$

The computation of the cohomology of $K(\mathbb{Z}, n)$ modulo 2 was done by J.-P. Serre simultaneously with the computation of the cohomology of $K(\mathbb{Z}_2, n)$ modulo 2. No wonder: The two computations are essentially the same (induction with respect to n based on the Borel theorem). Here is the final result.

Theorem 1. *If $n \geq 2$, then the ring $H^*(K(\mathbb{Z}, n); \mathbb{Z}_2)$ is the ring of polynomials (with coefficients in $\mathbb{Z}_2$) of generators $\mathrm{Sq}^I \bar{e}_n$ where $\bar{e}_n \in H^n(K(\mathbb{Z}, n); \mathbb{Z}_2)$ is the generator and $I = (i_1, i_2, \dots, i_k)$ is an admissible sequence with $\operatorname{exc} I < n$ and $i_k > 1$ [the last inequality is the only difference between the results for $K(\mathbb{Z}, n)$ and $K(\mathbb{Z}_2, n)$].*

Taking the limit for $n \to \infty$ yields the following statement.

Theorem 2. *The vector space $\bigoplus_q \mathrm{Stab}\, \mathcal{O}(q, \mathbb{Z}, \mathbb{Z}_2)$ has a basis consisting of all operations Sq^I where $I = (i_1, i_2, \dots, i_k)$ is an admissible sequence with $i_k > 1$.*

The details of the proofs are left to the reader.

Remark 1. We consider Sq^I a cohomology operation from the integral cohomology to the cohomology modulo 2. Precisely, this means that in both statements Sq^I means $\mathrm{Sq}^I \circ \rho_2$, where ρ_2 is the reduction modulo 2.

Remark 2. One should not think that Sq^1 acts trivially in $H^*(K(\mathbb{Z}, n); \mathbb{Z}_2)$. It is true that $\mathrm{Sq}^1 \overline{e}_n = 0$, but, for example, $\mathrm{Sq}^1 \mathrm{Sq}^2 \overline{e}_n = \mathrm{Sq}^3 \overline{e}_n \neq 0$ for $n \geq 3$.

30.7 The Steenrod Algebra mod *p*

Let p be a prime number. For $p > 2$, as well as for $p = 2$, the only (up to a factor) cohomology operation in the cohomology modulo p raising the dimension by 1 is the Bockstein homomorphism β. However, while for $p = 2$ the operation $\beta = \mathrm{Sq}^1$ is just one of the Steenrod squares, for $p > 2$ this operation plays a very special role. The operations similar to other Steenrod squares also exist. Namely, there exists a unique stable cohomology operation $P^i_p \in \mathcal{S}tab\, \mathcal{O}(2i(p-1), \mathbb{Z}_p, \mathbb{Z}_p)$ such that $P^i_p x = x^p$ for $x \in H^{2i}(X; \mathbb{Z}_p)$. This operation is called the *(pth) Steenrod power*. In particular, $P^i_2 = \mathrm{Sq}^{2i}$. Precisely as for $p = 2$, $P^0_p = \mathrm{id}$ and $P^i_p x = 0$ if $\dim x < 2i$.

It turns out that the Steenrod algebra $\mathbb{A}_p$ is multiplicatively generated by the operations $\beta_p = \beta$ and P^i_p. We put

$$\mathrm{St}^k = \begin{cases} P^i_p & \text{for } k = 2i(p-1), \\ \beta_p \circ P^i_p & \text{for } k = 2i(p-1) + 1. \end{cases}$$

Thus, the operations St^k are defined for $k \equiv 0, 1 \bmod 2(p-1)$. Notice that for $p = 2$, $\mathrm{St}^k = \mathrm{Sq}^k$.

Let us describe the additive basis in $\mathbb{A}_p$ (over $\mathbb{Z}_p$). For a sequence $I = (i_1, i_2, \dots, i_k)$ of integers $\equiv 0, 1 \bmod 2(p-1)$, put $\mathrm{St}^I = \mathrm{St}^{i_1} \mathrm{St}^{i_2} \dots \mathrm{St}^{i_k}$. A sequence I is called admissible if

$$i_1 \geq p i_2,\ i_2 \geq p i_3, \dots, i_{k-1} \geq p i_k.$$

For an admissible I, we refer to St^I as to an *admissible iteration*. There is a theorem: *Admissible iterations* St^I *form an additive basis in the* $\mathbb{Z}_p$*-algebra* $\mathbb{A}_p$.

This theorem was first proved by H. Cartan in the mid-1950s. To the reader who wants to see a proof, we recommend reading the article by Postnikov [69].

The relations between the iterations St^I are generated by the *Adem relations* (see [12])

$$P_p^a P_p^b = \sum_{c=0}^{[a/p]} (-1)^{c+a} \binom{(p-1)(b-c)-1}{a-pc} P_p^{a+b-c} P_p^c;$$

$$P_p^a \beta_p P_p^b = \sum_{c=0}^{[a/p]} (-1)^{c+a} \binom{(p-1)(b-c)}{a-pc} \beta_p P_p^{a+b-c} P_p^c + \sum_{c=0}^{[(a-1)/p]} (-1)^{c+a+1} \binom{(p-1)(b-c)-1}{a-pc-1} P_p^{a+b-c} \beta_p P_p^c$$

(where $a < pb$).

30.8 Other Classification Results

We will need some new notations. Let $\mathbb{K}$ be a field. Denote as $\Lambda(m, \mathbb{K})$ the graded $\mathbb{K}$-algebra with the basis $\{1, x\}$ with $\deg x = m$, $x^2 = 0$ (this algebra is called the exterior algebra with the generator x). Furthermore, denote as $P(m, \mathbb{K})$ the $\mathbb{K}$-algebra with the basis $\{1, x = x^{(1)}, x^{(2)}, \dots\}$ with $\deg x^{(k)} = km$ and the multiplication defined by the formula $x^{(k)} \cdot x^{(\ell)} = \binom{k+\ell}{k} x^{(k+\ell)}$ (this algebra is called the algebra of *modified polynomials*). Of these algebras, we will form *graded tensor products*, which are defined as the usual tensor products with the multiplication acting by the formula $(a \otimes b) \cdot (c \otimes d) = (-1)^{\deg b \deg c} ac \otimes bd$ (in this section, we will abbreviate "graded tensor products" to "tensor products" since we will not consider any other tensor products).

Fix a prime p and a group $\Pi = \mathbb{Z}$ or $\mathbb{Z}_{p^s}$ and set up a definition of a *sequence of numbers satisfying the condition* (C_p). Let us be given a sequence of integers $I = (i_1, i_2, \dots, i_k)$. We say that I satisfies the condition (C_p) if

(1) $i_1 \geq pi_2$, $i_2 \geq pi_3, \dots, i_{k-2} \geq pi_{k-1}$, $i_{k-1} \geq 2(p-1)$;
(2) $i_k = 0$ if $\Pi = \mathbb{Z}$;
(3) $i_k = 0$ or 1 if $\Pi = \mathbb{Z}_{p^s}$;
(4) $i_\ell \equiv 0$ or $1 \bmod 2(p-1)$ for $1 \leq \ell \leq k$.

Theorem (Cartan [28]). *For $n \geq 1$ and a prime $p > 2$, the algebra $H^*(K(\Pi, n); \mathbb{Z}_p)$ (where $\Pi = \mathbb{Z}$ or $\mathbb{Z}_{p^s}$) is isomorphic to the tensor product of exterior algebras $\Lambda(m, \mathbb{Z}_p)$ with generators of odd degrees and algebras of usual polynomials with generators of even degrees. For $n \geq 2$ and $p = 2$, the algebra $H^*(K(\Pi, n); \mathbb{Z}_2)$ (where $\Pi = \mathbb{Z}$ or $\mathbb{Z}_{2^s}$) is isomorphic to the tensor product of algebras of usual polynomials. In both cases, the number of generators of degree $n + q$ is equal to the number of sequences I, with $|I| = q$ satisfying the condition (C_p) and the condition $p^i < (p-1)(n+q)$.*

Remark. This theorem implies all the preceding classification results. If $\Pi = \mathbb{Z}$, then the condition (C_p) implies that $i_k = 0$ and $i_{k-1} \geq 2(p-1)$, and if $p = 2$, then $i_{k-1} \geq 2$; that is, Sq^I does not contain Sq^1. If $p = 2$, then the condition $p^{i_1} < (p-1)(n+q)$ becomes the familiar condition $\operatorname{exc} I < n$.

It turns out that the homology algebra of $K(\Pi, n)$ can also be fully described [the multiplication in the homology of $K(\Pi, n)$ is induced by the structure of an H-space: $K(\Pi, n) = \Omega K(\Pi, n+1)$].

Theorem (Cartan [28]). *For $n \geq 1$ and a prime $p > 2$, the algebra $H_*(K(\Pi, n); \mathbb{Z}_p)$ (where $\Pi = \mathbb{Z}$ or $\mathbb{Z}_{p^s}$) is isomorphic to the tensor product of exterior algebras $\Lambda(m, \mathbb{Z}_p)$ with generators of odd degrees and algebras of modified polynomials with generators of even degrees. For $n \geq 2$ and $p = 2$, the algebra $H^*(K(\Pi, n); \mathbb{Z}_2)$ (where $\Pi = \mathbb{Z}$ or $\mathbb{Z}_{2^s}$) is isomorphic to the tensor product of algebras of modified polynomials. In both cases, the number of generators of degree $n+q$ is equal to the number of sequences I with $|I| = q$ satisfying the condition (C_p).*

Theorem of a Choice of a Basis (Cartan). *Let $\Pi = \mathbb{Z}$ and $\overline{e}_n \in H^n(K(\mathbb{Z}, n); \mathbb{Z}_p)$ be the generator. Then, for the generators of the exterior algebras and algebras of usual polynomials in the theorem concerning $H^*(K(\mathbb{Z}, n); \mathbb{Z}_p)$, one can take the classes $\mathrm{St}_p^I(\overline{e}_n)$ for all I satisfying the conditions (C_p) and $p^{i_1} < (p-1)(n+|I|)$.*

For the proofs of these results (as well as many other results) see the multivolume edition *Seminaire Henri Cartan* (which can be found in major mathematical libraries). See also Postnikov [69].

Lecture 31 Applications of Steenrod Squares

The general homotopy direction of this book forces us to consider Steenrod squares mainly as a tool for calculating homotopy groups. Indeed, Steenrod squares are very useful for this: In the next section, we will perform the calculation of the groups $\pi_{n+2}(S^n)$, and in the next chapter we will calculate the stable groups $\pi_{n+q}(S^n)$ for $q \leq 13$. One should not forget, however, that the homotopy calculations are, in some sense, a side effect of Steenrod's theory. To demonstrate this, we will display in this lecture an array of applications of Steenrod squares; our aim will be diversity rather than completeness.

31.1 Calculating Homotopy Groups

Here we assume that n is large.

From the spectral sequence of the fibration $S^n|_{n+1} \xrightarrow{K(\mathbb{Z}, n-1)} S^n$ (see Fig. 110) we obtain the following information about the cohomology of $S^n|_{n+1}$ modulo 2:

q	$n+1$	$n+2$	$n+3$	$n+4$
$H^q(S^n\|_{n+1};\mathbb{Z}_2)$	$\mathbb{Z}_2$	$\mathbb{Z}_2$	$\mathbb{Z}_2$	$\mathbb{Z}_2$
generators, relations	u	$\operatorname{Sq}^1 u$	v $(\operatorname{Sq}^2 u = 0)$	$\operatorname{Sq}^1 v$ $= \operatorname{Sq}^2 \operatorname{Sq}^1 u$

We know that $\pi_{n+1}(S^n) = \mathbb{Z}_2$. The integral cohomological spectral sequence of the fibration $S^n|_{n+2} \xrightarrow{K(\mathbb{Z}_2,n)} S^n|_{n+1}$ looks like Fig. 111.

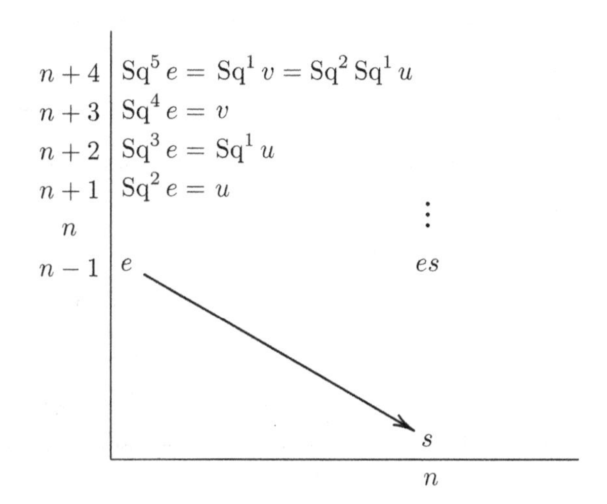

Fig. 110 The spectral sequence of the fibration $S^n|_{n+1} \xrightarrow{K(\mathbb{Z},n-1)} S^n$

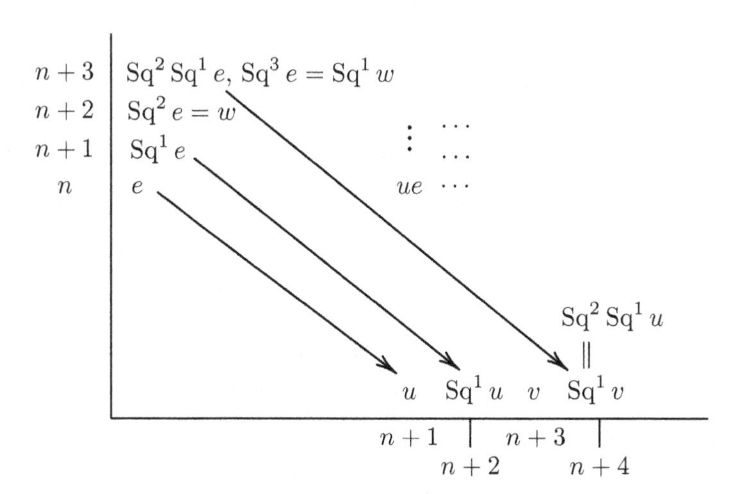

Fig. 111 The spectral sequence of the fibration $S^n|_{n+2} \xrightarrow{K(\mathbb{Z}_2,n)} S^n|_{n+1}$

For the cohomology of $S^n|_{n+2}$, we have

q	$n+2$	$n+3$
$H^q(S^n\|_{n+2};\mathbb{Z}_2)$	$\mathbb{Z}_2$	$\mathbb{Z}_2 \oplus \mathbb{Z}_2$
generators	w	v, $\mathrm{Sq}^1 w$

We already see that

$$H^{n+2}(S^n|_{n+2};\mathbb{Z}_2) = \mathbb{Z}_2,$$

and hence,

$$\pi_{n+2}(S^n) = H_{n+2}(S^n|_{n+2}) = (\mathbb{Z} \text{ or } \mathbb{Z}_{2^s}) \oplus (\text{a finite group of odd order}).$$

But the group $\pi_{n+2}(S^n)$ is finite (see Sect. 26.3) and does not contain elements of an odd order (Sect. 27.4). Thus,

$$\pi_{n+2}(S^n) = \mathbb{Z}_{2^s}.$$

Actually, the fact that $\mathrm{Sq}^1 w \neq 0$, that is, $b^{n+2}(w) \neq 0$, already implies that $s = 1$. But we will prove it using spectral sequences. The element $w \in H^{n+2}(S^n|_{n+2};\mathbb{Z}_2)$ gives rise to a map $S^n|_{n+2} \to K(\mathbb{Z}_2, n+2)$. Consider the fibration $E \xrightarrow{K(\mathbb{Z}_2,n+1)} S^n_{n+2}$ induced by the loop fibration $* \xrightarrow{K(\mathbb{Z}_2,n+1)} K(\mathbb{Z}_2, n+2)$. We can call the total space E of the induced fibration an "underkilling space": It follows from the homotopy sequence of the fibration that

$$\pi_q(E) = \begin{cases} 0, & \text{if } q < n+2, \\ \pi_{n+2}(S^n)/\mathbb{Z}_2, & \text{if } q = n+2, \\ \pi_q(S^n), & \text{if } q > n+2. \end{cases}$$

(Actually, as we shall see in a moment, E is the *killing* space.) Consider the spectral sequence of the fibration $E \xrightarrow{K(\mathbb{Z}_2,n+1)} S^n|_{n+2}$ (Fig. 112).

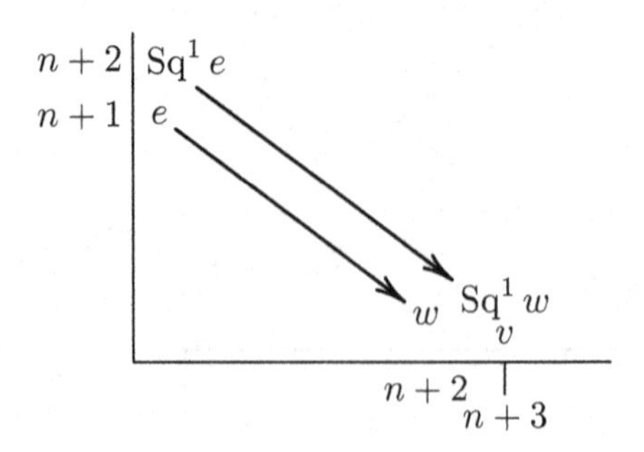

Fig. 112 The spectral sequence of the fibration $E \xrightarrow{K(\mathbb{Z}_2,n+1)} S^n|_{n+2}$.

We see from it (since $\mathrm{Sq}^1 w \neq 0$, we cannot avoid using that!) that $H^{n+2}(E;\mathbb{Z}_2) = 0$, hence $H_{n+2}(E;\mathbb{Z}_2) = 0$, hence $\pi_{n+2}(E)$ is a group of odd order, and hence it is 0, since $\pi_{n+2}(S^n)$ is a 2-group. Thus, E is a killing space $S^n|_{n+3}$ and $\pi_{n+2}(S^n|_{n+2}) = \pi_{n+2}(S^n) = \mathbb{Z}_2$.

By the way, we see also that $H_{n+3}(E;\mathbb{Z}_2) = H_{n+3}(S^n|_{n+3};\mathbb{Z}_2) = \mathbb{Z}_2$. Hence, the 2-component of the group $\pi_{n+3}(S^n)$ is a group of the form $\mathbb{Z}_{2^s}$ (with $s > 0$).

EXERCISE 1. Compute $\pi_{n+3}(S^n)$. [*Hint:* Consider the "underkilling space" E_1 with a fibration $E_1 \xrightarrow{K(\mathbb{Z}_2,n+2)} S^n|_{n+3}$. Show that $H_{n+3}(E_1;\mathbb{Z}_2)$ is again $\mathbb{Z}_2$. The same will be true for the next "underkilling space," E_2, $E_2 \xrightarrow{K(\mathbb{Z}_2,n+2)} E_1$, and only the third underkilling space, E_3, manages to kill the 2-component of $\pi_{n+3}(S^n)$. Hence, the 2-component of $\pi_{n+3}(S^n)$ is $\mathbb{Z}_8$. Furthermore, we know from Sect. 27.4 that the group $\pi_{n+3}(S^n)$ has a 3-component $\mathbb{Z}_3$ and has no p-components with $p > 3$. Thus, $\pi_{n+3}(S^n) = \mathbb{Z}_{24}$. (If you do this exercise, you will give an algebraic proof of the main result of one of the most difficult topological works of the pre-French epoch, that is, the work by Rokhlin [71] in which the group $\pi_{n+3}(S^n)$ was calculated by a geometric method.)]

31.2 Steenrod Squares and the Stiefel–Whitney Classes

A: A Formula for $\mathrm{Sq}^k\, w_m$

If we assign to a real vector bundle ξ the cohomology class $\mathrm{Sq}^k\, w_m(\xi)$ of its base, then we will get a new characteristic class of real vector bundles with values in $\mathbb{Z}_2$-cohomology. But there is no such thing as "new characteristic classes": We proved in Sect. 19.4.D that every characteristic class of real vector bundles with values in $\mathbb{Z}_2$-cohomology is a polynomial in the Stiefel–Whitney classes. What is this polynomial in our case? It turns out that the following formula holds:

$$\mathrm{Sq}^k\, w_m = \sum_{j=0}^{k} \binom{m+j-k-1}{j} w_{k-j} w_{m+j}.$$

We restrict ourselves to a draft of a proof given below; the reader may be able to reconstruct the details.

First, we check that the left- and right-hand sides of the equality take equal values on the fibration $\zeta \times \cdots \times \zeta$ with the base $\mathbb{R}P^N \times \cdots \times \mathbb{R}P^N$, where ζ is the (one-dimensional) Hopf bundle over $\mathbb{R}P^N$ and N and the number of factors are both sufficiently large. After this (or before this) we prove that no nonzero polynomial of Stiefel–Whitney classes takes a zero value on the fibration $\zeta \times \cdots \times \zeta$. (Notice that calculations for products of real projective spaces provide a key tool in the theories of both Stiefel–Whitney classes and Steenrod squares.) This completes the proof.

For example,

$$\mathrm{Sq}^1 w_m = w_1 w_m + (m-1) w_{m+1}.$$

This shows that if $w_1 = 0$, that is, if the vector bundle is orientable, then $\mathrm{Sq}^1 w_m = w_{m+1}$ for every even m. But Sq^1 is the Bockstein homomorphism. According to the direct description of the Bockstein homomorphism (see Sect. 15.3), this means that all odd-numbered Stiefel–Whitney classes of an orientable vector bundle are integral; that is, they are images of integral cohomology classes of order 2 with respect to the reduction homomorphism $\rho_2 \colon H^*(-;\mathbb{Z}) \to H^*(-;\mathbb{Z}_2)$.

B: A Digression: Thom Spaces and Thom Isomorphisms

Let ξ be a real vector bundle of dimension n with the base B and the total space E. Fix Euclidean structures in all fibers of ξ (such that the square-of-length function is continuous on the total space of ξ) and denote by $D(\xi)$ and $S(\xi)$ the spaces of unit ball and unit sphere bundles associated with ξ. The quotient space $T(\xi) = D(\xi)/S(\xi)$ is called the *Thom space* of ξ. Obviously, $T(\xi)$ contains B (the zero section of ξ) and is covered by n-dimensional spheres which have a common point and otherwise are disjoint; each of these spheres contains one point of B (see Fig. 113). Certainly these spheres are not canonically homeomorphic to S^n (since the fibers of ξ are not canonically homeomorphic to $\mathbb{R}^n$).

It is obvious that $T(\xi)$ does not depend, up to a homeomorphism, on the Euclidean structures in the fibers. Moreover, this homeomorphism may be chosen to be the identity on B and to preserve the decomposition of the Thom space into n-dimensional spheres.

EXERCISE 2. Prove that if the bundle ξ is trivial, then $T(\xi) = \Sigma^n(B \sqcup \mathrm{pt})$ (where Σ denotes the base point version of suspension; see Sect. 2.6).

EXERCISE 3. Prove the equality $T(\xi \times \eta) = T(\xi) \# T(\eta)$.

EXERCISE 4. Construct a homeomorphism between the Thom space of the Hopf bundle ζ over $\mathbb{R}P^n$ and $\mathbb{R}P^{n+1}$. Give a similar construction for the complex case.

EXERCISE 5. Give a geometric description of the Thom space of the tautological vector bundle over the Grassmannian.

If the base B is a CW complex, then $T(\xi)$ has a natural CW structure: The cells are the inverse images of cells of B in $D(\xi) - S(\xi)$ (their dimensions are n more than the dimensions of the cells in B) plus one more (zero-dimensional) cell obtained from $S(\xi)$. Moreover, if the vector bundle is oriented, then our correspondence assigns to

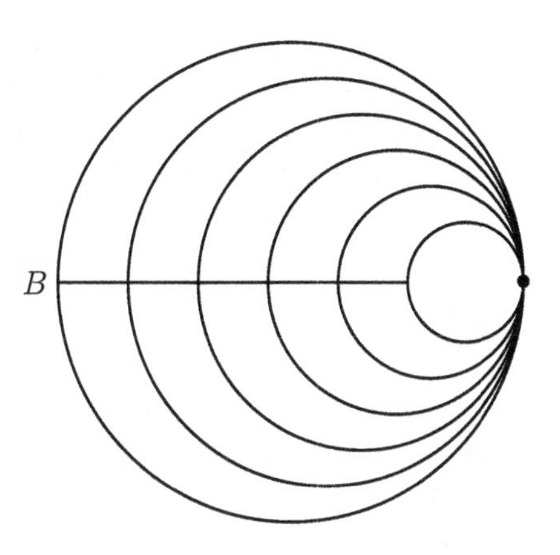

Fig. 113 The Thom space

oriented cells of B *oriented* cells of $T(\xi)$; moreover, the correspondence preserves the incidence numbers. Thus, for an oriented vector bundle with a CW base, there arise *Thom isomorphisms*

$$\mathbf{t}\colon \widetilde{H}_{q+n}(T(\xi);G) \xrightarrow{\cong} H_q(B;G)), \ \mathbf{t}\colon H^q(B;G) \xrightarrow{\cong} \widetilde{H}^{q+n}(T(\xi);G)$$

where G is an arbitrary coefficient group. Moreover, if $G = \mathbb{Z}_2$, then the assumption of orientability for ξ is not needed. Thom isomorphisms possess many naturality properties, of which we mention the commutative diagrams

$$\begin{array}{ccc} \widetilde{H}_{q+n}(T(f^*\xi);G) & \xrightarrow{T(f)_*} & \widetilde{H}_{q+n}(T(\xi);G) \\ \Big\downarrow{\scriptstyle \mathbf{t}} & & \Big\downarrow{\scriptstyle \mathbf{t}} \\ H_q(B';G) & \xrightarrow{f_*} & H_q(B;G) \end{array}$$

and

$$\begin{array}{ccc} \widetilde{H}^{q+n}(T(f^*\xi);G) & \xleftarrow{T(f)^*} & \widetilde{H}^{q+n}(T(\xi);G) \\ \Big\uparrow{\scriptstyle \mathbf{t}} & & \Big\uparrow{\scriptstyle \mathbf{t}} \\ H^q(B';G) & \xleftarrow{f_*} & H^q(B;G) \end{array},$$

which arise for an arbitrary continuous map $f: B' \to B$ (between CW complexes); $T(f): T(f^*\xi) \to T(\xi)$ is the naturally arising map.

The Thom isomorphisms have a different description in the *smooth case* when B is a closed oriented smooth manifold (say, of dimension m) and the oriented vector bundle ξ is smooth (that is, E is a smooth manifold and $p: E \to B$ is a smooth submersion; we assume also that the Euclidean metric used in the definition of $\mathbf{t}$ is smooth). Then the cohomological Thom isomorphism can be described, by means of Poincaré isomorphisms, as a composition

$$\begin{aligned} H^q(B;G) \xrightarrow[D^{-1}]{D} H_{m-q}(B;G) \xrightarrow{\cong} H_{m-q}(D(\xi);G) \\ \longrightarrow H^{m+n-m+q}(D(\xi), \partial D(\xi) = S(\xi); G) = \widetilde{H}^{n+q}(T(\xi);G) \end{aligned}$$

(where the middle arrow is induced by a homotopy equivalence), and a similar description exists for the homological Thom isomorphisms. (All the orientability assumptions may be dropped in the case of $G = \mathbb{Z}_2$.)

EXERCISE 6. Show that these description of Thom isomorphisms are equivalent to those given before.

Proposition. *Let ξ be an oriented vector bundle with a CW base B, let G be a ring, and let $\alpha_1 \in H^{q_1}(B;G)$, $\alpha_2 \in H^{q_2}(B;G)$. Then*

$$\mathbf{t}(\alpha_1 \smile \alpha_2) = \mathbf{t}(\alpha_1) \smile \alpha_2.$$

[The last cup-product has the following meaning:

$$\begin{aligned} \mathbf{t}(\alpha_1) \in H^{q_1+n}(D(\xi), S(\xi); G), \alpha_2 \in H^{q_2}(B;G) = H^{q_2}(D(\xi);G) \\ \Rightarrow \mathbf{t}(\alpha_1) \smile \alpha_2 \in H^{n+q_1+q_2}(D(\xi), S(\xi); G)\ .] \end{aligned}$$

Proof. For a smooth ξ with a compact oriented base of dimension m, it follows from the previous description of $\mathbf{t}$. Indeed, according to that description, for an $\alpha \in H^q(B;G)$, in $H_{m-q}(D(\xi);G)$,

$$[B] \frown \alpha = [D(\xi), S(\xi)] \frown \mathbf{t}(\alpha);$$

thus,

$$\begin{aligned} [D(\xi), S(\xi)] \frown \mathbf{t}(\alpha_1 \smile \alpha_2) = [B] \frown (\alpha_1 \smile \alpha_2) = ([B] \frown \alpha_1) \frown \alpha_2 \\ = ([D(\xi), S(\xi)] \frown \mathbf{t}(\alpha_1)) \frown \alpha_2 = [D(\xi), S(\xi)] \frown (\mathbf{t}(\alpha_1) \smile \alpha_2); \end{aligned}$$

that is, $\mathbf{t}(\alpha_1 \smile \alpha_2) = \mathbf{t}(\alpha_1) \smile \alpha_2$.

EXERCISE 7. Deduce the general case of the proposition from the smooth case. (*Hint:* For a compact B, the vector bundle ξ can be embedded into a vector bundle $\widetilde{\xi}$ with a closed oriented smooth base $\widetilde{B}$ such that the embedding $B \to \widetilde{B}$ induces an isomorphism in homology up to an arbitrarily high dimension. The transition to a noncompact case is straightforward.)

Corollary. *For an arbitrary* $\alpha \in H^q(B; G)$,

$$\mathbf{t}(\alpha) = \mathbf{t}(1) \smile \alpha.$$

The cohomology class $\mathbf{t}(1) \in H^n(T(\xi); G)$ is called the *Thom class* of the bundle ξ.

Remark. We can generalize the proposition to the case when $\alpha_1 \in H^{q_1}(B; G_1), \alpha_2 \in H^{q_2}$ and the cup-product $\alpha_1 \smile \alpha_2 \in H^{q_1+q_2}(B; G)$ is taken with respect to some pairing $G_1 \times G_2 \to G$ (see the end of Sect. 16.2). Using that, we can generalize the corollary to the case of an arbitrary Abelian group (not a ring) G considering the Thom class as belonging to $H^n(T(\xi); \mathbb{Z})$.

Let us do some final remark for the case when ξ is the normal bundle of a smooth submanifold Y of a smooth manifold X.

EXERCISE 8. Prove that in this case $T(\xi) = X/(X - U)$, where U is a tubular neighborhood of Y.

EXERCISE 9. Let X and Y be closed oriented, and let $\dim X - \dim Y = n$. Prove that the composition

$$H^q(Y; G) \xrightarrow{\mathbf{t}} H^{q+n}(T(\xi); G) = H^{q+n}(X/(X - U); G) \to H^{q+n}(X; G)$$

(where the last map is induced by the projection $X \to X/(X - U)$) coincides with $i_!$ (see Sect. 17.8), where i is the inclusion map $Y \to X$.

EXERCISE 10. Let X be an m-dimensional closed orientable manifold embedded in S^n, and let U be a tubular neighborhood of X in S^n. Prove that the diagram

$$\begin{array}{ccc} H^q(X; G) & \xrightarrow{\mathbf{t}} & H^{n-m+q}(S^n, S^n - U; G) \\ \downarrow D & & \downarrow D \\ H_{m-q}(X; G) = H_{m-q}(\overline{U}; G) & = & H_{m-q}(S^n - (S^n - U); G) \end{array}$$

(where $\mathbf{t}$ denotes the Thom isomorphism associated with the normal bundle of X in S^n) is commutative.

C: An Sq-definition of the Stiefel–Whitney Classes

Theorem. *For an arbitrary vector bundle* ξ *with a CW base,*

$$w_m(\xi) = \mathbf{t}^{-1}\,\mathrm{Sq}^m\,\mathbf{t}(1)$$

(the equality holds in $\mathbb{Z}_2$*-cohomology).*

EXERCISE 11. Prove this theorem. [*Hints:* A standard proof of this result (which can be found, for example, in the book by Milnor and Stasheff [60]) consists in systematically checking for the classes $\mathbf{t}^{-1}\,\mathrm{Sq}^m\,\mathbf{t}(1)$ the axioms of Stiefel–Whitney classes (mentioned in the beginning of Sect. 19.5). Another approach is based on the splitting principle. Since both sides of the equality in the theorem are characteristic classes of real vector bundles with values in the $\mathbb{Z}_2$-cohomology, it is sufficient to prove the equality for products $\zeta \times \dots \times \zeta$ of Hopf bundles. It is not hard to do this (using the results of Exercises 3 and 4).]

Corollary. *If* ξ *is the normal bundle of a submanifold* Y *of* X*, then*

$$i_!(w_m(\xi)) = \mathrm{Sq}^m\, i_!(1) = \mathrm{Sq}^m(D_X^{-1}[Y])$$

(where i *is the inclusion map* $Y \to X$*).*

D: The Wu Formula

Put $w = 1 + w_1 + w_2 + w_3 + \dots$ and $\mathrm{Sq} = 1 + \mathrm{Sq}^1 + \mathrm{Sq}^2 + \mathrm{Sq}^3 + \dots$ Notice that the operation $\mathrm{Sq}\colon H^*(-;\mathbb{Z}_2) \to H^*(-;\mathbb{Z}_2)$ is invertible: $(\mathrm{Sq})^{-1}\,\mathrm{Sq} = 1$, where

$$(\mathrm{Sq})^{-1} = 1 + \mathrm{Sq}^1 + \mathrm{Sq}^2 + \mathrm{Sq}^2\,\mathrm{Sq}^1 + (\mathrm{Sq}^4 + \mathrm{Sq}^3\,\mathrm{Sq}^1) + \dots.$$

Theorem (Wu). *Let* X *be a closed manifold. Then, for every* $\alpha \in H^*(X;\mathbb{Z}_2)$*,*

$$\langle (\mathrm{Sq})^{-1} w(X), D(\alpha) \rangle = \langle \mathrm{Sq}\,\alpha, [X] \rangle.$$

This "Wu formula" completely determines the Stiefel–Whitney classes of (the tangent bundle of) a closed smooth manifold. Its standard proof can be found in the books by Milnor and Stasheff [60] (Sect. 11) or Spanier [79] (Sect. 10 of Chap. 6).

EXERCISE 12. Prove the Wu formula following the plan outlined here. The tangent bundle of X is the same as the normal bundle of $\Delta(X)$ in $X \times X$, where $\Delta\colon X \to X \times X$ is the diagonal map. Hence,

$$\Delta_! w(X) = \mathrm{Sq}(D_{X\times X}^{-1}[\Delta(X)]).$$

Since $p \circ \Delta = \mathrm{id}$ (where p is the projection of $X \times X$ onto, say, the second factor), the last formula can be rewritten as

$$w(x) = p_! \,\mathrm{Sq}(D_{X\times X}^{-1}[\Delta(X)]).$$

But it may be deduced from the results of Lecture 17 (see Theorem 4 of Sect. 17.5 and Lemma 2 of Sect. 17.6) that $D_{X\times X}^{-1}[\Delta(X)] = \sum(\alpha_i^* \times \alpha_i)$, where $\{\alpha_i\}$ is a basis in the $\mathbb{Z}_2$-cohomology of X and $\{\alpha_i^*\}$ is the "dual" basis in the sense that $\langle \alpha_i^* \smile \alpha_j, [X]\rangle = \delta_{ij}$. It is true also that $p_!(\alpha \times \beta) = \langle \beta, [X]\rangle\alpha$. The reader will do the rest.

Corollary 1. *The Stiefel–Whitney classes of a smooth manifold are its homotopy invariants.*

(This result was mentioned in Sect. 19.6.E.)

Corollary 2 (The Stiefel Theorem). *Every closed orientable three-dimensional manifold is parallelizable.*

Proof. To prove that a closed orientable three-dimensional manifold X is parallelizable, it is sufficient to construct two linearly independent (at every point) vector fields on X, that is, to construct a section of the fibration $E \xrightarrow{V(3,2)} X$ associated with the tangent bundle of X. Since $V(3.2) = \mathbb{R}P^3$, we have $\pi_1(V(3,2)) = \mathbb{Z}_2$, $\pi_2(V(3,2)) = 0$. The first obstruction to the construction of such a section belongs to $H^2(X;\pi_1(V(3,2))) = H^2(X;\mathbb{Z}_2)$ and equals w_2. All we need is to prove that $w_2 = 0$, since the next obstruction belongs to the zero group $H^3(X;\pi_2(V(3,2)))$. To apply the Wu formula, let us determine the action of Steenrod squares with values in $H^3(X;\mathbb{Z}_2)$. The operation $\mathrm{Sq}^1 = b^2\colon H^2(X;\mathbb{Z}_2) \to H^3(X;\mathbb{Z}_2)$ is zero, because it is the Bockstein homomorphism, that is, a composition $H^2(X;\mathbb{Z}_2) \to H^3(X;\mathbb{Z}) \to H^3(X;\mathbb{Z}_2)$, and the first arrow is zero, since it is a homomorphism of a finite group into $H^3(X;\mathbb{Z}) = \mathbb{Z}$ (the last equality holds since X is orientable, and we may assume that X is connected). Furthermore, $\mathrm{Sq}^2\colon H^1(X;\mathbb{Z}_2) \to H^3(X;\mathbb{Z}_2)$ is zero, since Sq^2 is zero on the one-dimensional cohomology, and, the more so, $\mathrm{Sq}^3\colon H^0(X;\mathbb{Z}_2) \to H^3(X;\mathbb{Z}_2)$ is zero. Hence, $\langle \mathrm{Sq}\,\alpha, [X]\rangle = 0$ when $\dim\alpha \neq 3$; hence, $\mathrm{Sq}^{-1} w(X)$ can take nonzero values only in zero-dimensional cohomology, and hence $\mathrm{Sq}^{-1} w(X) \in H^0(X;\mathbb{Z}^2) \Rightarrow w(X) \in H^0(X;\mathbb{Z}^2) \Rightarrow w_2(X) = 0$.

EXERCISE 13. Prove that if X is an n-dimensional manifold, then $\overline{w}_m(X) = 0$ for $m > n-\alpha(n)$ where $\alpha(n)$ is the number of ones in the binary representation of n and $\overline{w}_m(X)$ is the mth Stiefel–Whitney class of the normal bundle of X in an Euclidean space. (This result belongs to W. S. Massey. The proof is quite involved but does not use anything not known to the reader.)

EXERCISE 14. Prove that the bound $n - \alpha(n)$ in Exercise 13 cannot be reduced. Namely, if $n = 2^{k_1} + \cdots + 2^{k_r}$ with $k_1, \ldots, k_r \in \mathbb{Z}$, $k_1 > \cdots > k_r \geq 0$, then $\overline{w}_{n-r}(\mathbb{R}P^{2^{k_1}} \times \cdots \times \mathbb{R}P^{2^{k_r}} \neq 0$.

FYI. The results in Exercises 13 and 14, together with Exercise 26 in Sect. 19.6, gave rise to a known *Massey conjecture* that every closed (actually, not necessarily closed) n-dimensional smooth manifold can be immersed into $\mathbb{R}^{2n-\alpha(n)}$ (and, as follows from Exercise 14, the dimension $2n - \alpha(n)$ cannot be reduced in this statement). The Massey conjecture was proved by Ralph Cohen [31].

31.3 Steenrod Squares and Second Obstructions

It is not surprising that Steenrod squares were first discovered by Norman Steenrod. But one should not think that the goal of the groundbreaking work of Steenrod was studying any cohomology operations. The work (Steenrod [81]) was devoted to an old homotopy problem. In the early 1930s, H. Hopf gave a homotopy classification of continuous maps of an n-dimensional CW complex X into the n-dimensional sphere. It turned out that these homotopy classes bijectively correspond to cohomology classes in $H^n(X;\mathbb{Z})$[1] (see Sect. 18.4); namely, the homotopy class of a map $f: X \to S^n$ is bijectively characterized by the cohomology class $f^*(s) \in H^n(X;\mathbb{Z})$, where $s \in H^n(S^n;\mathbb{Z})$ is the canonical generator. Steenrod extended this classification to maps $X \to S^n$, where $\dim X = n + 1$. His results are described in the following exercises.

EXERCISE 15. Let X be an $(n+2)$-dimensional CW complex, and let $\gamma \in H^n(X;\mathbb{Z})$. Prove that a map $f: X \to S^n$ such that $f^*(s) = \gamma$ exists if and only if $\mathrm{Sq}^2(\rho_2\gamma) = 0$ (where ρ_2 is the reduction modulo 2).

EXERCISE 16. Let X be an $(n+1)$-dimensional CW complex, and let $\gamma \in H^n(X;\mathbb{Z})$. Construct a bijection between homotopy classes of maps such that $f^*(s) = \gamma$ and the cokernel of the composition

$$H^{n-1}(X;\mathbb{Z}) \xrightarrow{\rho_2} H^{n-1}(X;\mathbb{Z}_2) \xrightarrow{\mathrm{Sq}^2} H^{n+1}(X;\mathbb{Z}_2).$$

EXERCISE 17 (This Exercise is independent of Exercises 15 and 16). Let $f: S^{n+1} \to S^n$, $n \geq 3$ be a spheroid representing a nonzero element in $\pi_{n+1}(S^n) \cong \mathbb{Z}_2$. Let $X = S^n \cup_f D^{n+2}$. Find the action of Steenrod squares in the $\mathbb{Z}_2$-cohomology of X.

It should be noted that the original construction of Steenrod squares is substantially different from the one presented above. Steenrod's construction is outlined in Exercises 18–20.

Let X be a CW complex. We pointed out before that the diagonal map $\Delta: X \to X \times X$ is not cellular. We choose for it a cellular approximation Δ_0 and consider the chain homomorphism

$$D_0 = \Delta_0^{\#}: C^r(X \times X;\mathbb{Z}_2) \to C^r(X;\mathbb{Z}_2).$$

[1] Certainly, Hopf did not formulate his result in terms of cohomology, which appeared in topology several years later.

Consider also the chain homomorphism $T: \mathcal{C}^r(X \times X; \mathbb{Z}_2) \to \mathcal{C}^r(X \times X; \mathbb{Z}_2)$ induced by the map $t: (x, y) \mapsto (y, x)$. Although $t \circ \Delta = \Delta$, the map $t \circ \Delta_0$ is only homotopic to Δ_0. Because of this, there arises a cochain homotopy

$$D_1: \mathcal{C}^r(X \times X; \mathbb{Z}_2) \to \mathcal{C}^{r-1}(X \times X; \mathbb{Z}_2),\ D_1 \circ \delta + \delta \circ D_1 = D_0 + D_0 \circ T.$$

EXERCISE 18. Prove that there exists a whole sequence

$$D_q: \mathcal{C}^r(X \times X; \mathbb{Z}_2) \to \mathcal{C}^{r-q}(X \times X; \mathbb{Z}_2),\ D_q \circ \delta + \delta \circ D_q = D_{q-1} + D_{q-1} \circ T.$$

EXERCISE 19. Prove that for a cocycle $c \in \mathcal{C}^r(X; \mathbb{Z}_2)$ the cochain $D_q(c \times c)$ is also a cocycle, and the cohomology class of this cocycle is determined by the cohomology class of c (does not depend either on the choice of c within a cohomology class or on the choice of the sequence D_q satisfying the conditions above). [By the way, it is wrong, in general, that if b and c are cocycles, then $D_q(b \times c)$ is a cocycle; to this effect, find a formula for the coboundary of $D_q(b \times c)$ in terms of coboundaries of b and c and homomorphisms D_q, D_{q-1}.]

EXERCISE 20. Prove that if c is a cocycle of the class $\gamma \in H^r(X; \mathbb{Z}_2)$, then $D_q(c \times c)$ is a cocycle of the class $\mathrm{Sq}^{r-q}\,\gamma$.

Remark 1. According to Steenrod's construction, the reason for the existence of Steenrod squares lies in the noncommutativity of multiplication of cellular (or classical) cochains. Actually, if the coefficients of a cohomology theory lie in $\mathbb{Z}$ or $\mathbb{Z}_2$, then this multiplication cannot be commutative (otherwise, Steenrod squares could not have existed). As we mentioned briefly at the end of Lecture 26, a (skew-) commutative ring of cochains can be constructed when the coefficients are taken in $\mathbb{Q}$.

Remark 2. If X is an ordered triangulated space, then the choice of operations D_q can be made quite explicit (in the spirit of the simplicial cup-product defined in Sect. 16.2 [the usual notation: $D_q(c_1 \times c_2) = c_1 \smile_i c_2$]. However, the existing formulas for $\smile_i$ do not look encouraging. The reader can think of something more appealing.

31.4 Nonexistence of Spheroids with an Odd Hopf Invariant

Theorem. *If n is not a power of 2, the group $\pi_{2n-1}(S^n)$ does not contain elements whose Hopf invariant is odd.* (See Remark 5 in Sect. 16.5.)

Remark. This theorem was proved by Hopf in the 1930s. However, the proof was complicated and did not seem convincing to topologists of that time; even some (false!) counterexamples to it were published by some prominent mathematicians.

Proof of Theorem. Let $\alpha \in \pi_{2n-1}(S^n)$ be an element with an odd Hopf invariant, and let $Y = S^n \cup_f D^{2n}$, where $f: S^{2n-1} \to S^n$ is a spheroid of the class α. Then

$$H^q(Y; \mathbb{Z}_2) = \begin{cases} \mathbb{Z}_2, \text{ if } q = 0, n, 2n, \\ 0 \quad \text{for all other } q, \end{cases}$$

and, by definition of the Hopf invariant, the squaring operation $H^n(Y; \mathbb{Z}_2) \to H^{2n}(Y; \mathbb{Z}_2)$ is nontrivial. But this operation is the same as Sq^n, and if n is not a power of 2, then Sq^n may be presented as a polynomial of Sq^i with $0 < i < n$ (see Exercise 2 in Sect. 30.5). But since $H^q(Y; \mathbb{Z}_2) = 0$ for $n < q < 2n$, any such polynomial is zero on $H^n(Y; \mathbb{Z}_2)$. This contradiction proves the theorem.

Remark. As we mentioned in Lecture 16, there are no elements with odd Hopf invariant in $\pi_{2n-1}(S^n)$ with $n > 8$. One of the possible proofs of this fact consists in a construction, which shows that the operations $\mathrm{Sq}^{16}, \mathrm{Sq}^{32}, \dots$, indecomposable within the class of usual ("primary") cohomology operations, are decomposable within the class of ("secondary") cohomology operation. We will not discuss this (historically, first) proof in detail although some explanations will be made in Chap. 5; a full presentation of this proof can be found in the book by R. E. Mosher and M. C. Tangora [63]. Another (remarkably simple) proof using K-theory will be presented in Chap. 6.

31.5 Lens Spaces

In this section we will use not Steenrod squares but only Bockstein homomorphisms, and only for $p > 2$.

Let p and q be relatively prime integers, $1 < q < p, p > 2$. Consider the transformation T of the sphere $S^3 \subset \mathbb{C}^2$ acting by the formula $T(z_1, z_2) = \left(z_1 e^{\frac{2\pi i}{p}}, z_2 e^{\frac{2piq}{p}}\right)$. Obviously, T generates a free action of the group $\mathbb{Z}_p$ in S^3. The quotient $S^3/\mathbb{Z}_p$ is denoted by $L(p, q)$ and is called the (three-dimensional) *lens space*. (We have encountered infinite-dimensional lens spaces before.) The question is, for which p, q, p', q' is the lens space $L(p, q)$ homeomorphic, or homotopy equivalent, to the lens space $L(p', q')$? Since $\pi_1(L(p, q)) = \mathbb{Z}_p$, we can assume from the very beginning that $p' = p$ (otherwise, the lens spaces cannot be either homeomorphic, or homotopy equivalent).

Theorem. *The lens spaces $L(p, q)$ and $L(p, q')$ are homotopy equivalent if and only if $q' \equiv k^2 q \bmod p$ for some integer k.*

We will sketch a proof of the "only if" part. The "if" part is less important; the reader can prove it using the obstruction theory (and the fact that the dimensions of lens spaces are small).

Obviously, $H^r(L(p, q); \mathbb{Z}_p) = \mathbb{Z}_p$ for $r = 0, 1, 2, 3$. Consider on S^3 a big circle arc joining the points $(1, 0)$ and $(e^{\frac{2\pi i}{p}}, 0)$; this arc determines an element of the group $\pi_1(L(p,q)) = H_1(L(p,q)) = \mathbb{Z}_p$. Denote by a the element of $H^1(L(p,q);\mathbb{Z}_p)$ which takes the value $1 \in \mathbb{Z}_p$ on this element.

Lemma.

$$\langle ab(a), [L(p,q)]\rangle \equiv q \bmod p$$

where b is the Bockstein homomorphism

$$H^1(L(p,q);\mathbb{Z}_p) \to H^2(L(p,q);\mathbb{Z}_p).$$

Proof of Lemma. The lens space has a natural cell decomposition which is obtained from the following $\mathbb{Z}_p$-invariant cell decomposition of S^3 with p cells in every dimension $0,1,2,3$. The circle $S^1 = \{(z,0) \mid |z| = 1\} \subset S^3$ is divided by the p points $e_k^0 = (e^{\frac{2\pi ki}{p}}, 0)$, $k = 0,1,\ldots,p-1$, into the union of p arcs; we denote the (open) arc joining e_k^0 with e_{k+1}^0 (throughout this section, we regard subscripts as residues modulo p) as e_k^1. Next, we put $e_k^2 = \{(z_1,z_2) \in S^3 \mid z_2 \in \mathbb{R}_{>0} \cdot e^{\frac{2\pi ki}{p}}\}$ and take for e_k^3 the domain between e_k^2 and e_{k+1}^2. We get the desired cell decomposition of S^3. (Some people say that this decomposition reminds them of an orange; if the reader finds this comparison helpful, we appreciate it.) The transformation T maps the cells $e_k^0, e_k^1, e_k^2, e_k^3$ homeomorphically onto $e_{k+1}^0, e_{k+1}^1, e_{k+q}^2, e_{k+q}^3$. Thus, our cell decomposition of S^3 gives rise to a cell decomposition of $L(p,q)$, with one cell in every dimension, e^0, e^1, e^2, e^3. All the cells in S^3 have natural orientations preserved by T; thus, the cells in $L(p,q)$ are also naturally oriented.

Obviously, in the cellular complex of S^3, $\partial e_k^2 = e_0^1 + \cdots + e_{p-1}^1$; thus, in the cellular complex of $L(p,q)$, $\partial e^2 = pe^1$ and $b[e^2] = [e^1]$ (here b is the homological Bockstein homomorphism). To finish the proof of the lemma, we need to check that the intersection index of the $\mathbb{Z}_p$-cycles e_1 and e_2 in $L(p,q)$ is q.

Take a small generic perturbation $\widetilde{e^1}$ of the cycle e^1. The inverse image of this perturbed curve in S^3 will be a T-invariant curve $\widetilde{S^1}$ close to S^1. We can think of this curve as located on the torus $|z_1| = \rho$, $|z_2| = \varepsilon$, where ε is a small positive number and $\rho = \sqrt{1-\varepsilon^2}$. Let this curve start at $(\rho,\varepsilon) \in e_0^2$. Since it is T-invariant, it goes to the point $(\rho e^{\frac{2\pi i}{p}}, \varepsilon e^{\frac{2\pi qi}{p}})$, then to the point $(\rho e^{\frac{4\pi i}{p}}, \varepsilon e^{\frac{4\pi qi}{p}})$, and so on. For example, we can choose the perturbation $\widetilde{e^1}$ in such a way that the parametric equation of $\widetilde{S^1}$ will be $z(t) = (\rho e^{ti}, \varepsilon e^{qti})$, $0 \le t < 2\pi$. Obviously, this curve intersects e_0^2 at q points (for $t = \dfrac{2\pi k}{q}$, $k = 0,1,\ldots,q-1$), and all the intersections are counted with the same sign (we can assume that it is $+$ just taking the right orientations). This shows that $\phi(e_1,e_2) = q$, as stated in the lemma.

Proof of Theorem. Consider also $L(p,q')$ and let a' be the same for $L(p,q')$ as a is for $L(p,q)$. Let

$$\varphi\colon L(p,q) \to L(p,q')$$

be a homotopy equivalence. Then

$$\varphi_*: H_3(L(p,q)) \to H_3(L(p,q'))$$

is an isomorphism, and hence $\varphi_*[L(p,q)] = \pm[L(p,q')]$. Further, let $\varphi^*(a') = ka$ (where k is a residue modulo p). Then

$$\begin{aligned} q' &= \langle a'b(a'), [L(p,q']\rangle = \pm\langle a'b(a'), \varphi_*[L(p,q)]\rangle \\ &= \pm\langle \varphi^*(a'b(a')), [L(p,q)] = \pm\langle kab(ka), [L(p,q)]\rangle \\ &= \pm k^2\langle ab(a), [L(p,q)]\rangle = \pm k^2 q, \end{aligned}$$

which is our statement.

For example, the lens spaces $L(5,1), L(5,2)$ are not homotopy equivalent (since $2 \not\equiv \pm k^2 \bmod 5$).

Add to this that if we are interested only in homotopy equivalences preserving or reversing orientation, then $\pm$ in our statement can be replaced by, respectively, $+$ or $-$. For example, there is no orientation reversing map (that is, simply no map of degree -1) of $L(3,1)$ (this fact was proved by H. Kneser long before the theory discussed in this section was created). Moreover, elementary number theory states that if p is prime and $p \equiv 3 \bmod 4$, then every q not divisible by p is congruent either to a square or to a negative square (for example, $1,2,3,4,5,6 \equiv 1^2, 3^2, -2^2, 2^2, -3^3, -1^2 \bmod 7$), but never to both; hence, in this case every two lens spaces $L(p,q), L(p,q')$ are homotopy equivalent, but for a homotopy equivalence its homotopy preserving or reversing is determined by p, q, q'. On the contrary, if $p \equiv 1 \bmod 4$, then every q is congruent modulo p to both a square and a negative square, or to neither. Thus, in this case, if lens spaces $L(p,q), L(p,q')$ are homotopy equivalent, then they can be related by both orientation preserving and orientation reversing homotopy equivalences.

And a final remark. Homotopy equivalent lens spaces are not necessarily homeomorphic. For example, the lens spaces $L(7,1), L(7,2)$ are homotopy equivalent, but not homeomorphic. This fact can be established with the help of the so-called Reidemeister–Franz torsion. The reader can find details in the book by Dubrovin, Fomenko, and Novikov [35] (Sect. 11 of Chap. 1, Problems 5–8), or in the book by de Rham, Maumary, and Kervaire [34].

Chapter 5
The Adams Spectral Sequence

Lecture 32 General Idea

32.1 Introduction

As we demonstrated in the previous lecture, the information about the action of stable cohomology operations may be used for computing stable homotopy groups. If we know the cohomology of the space X, we can find, relatively easily, the "stable part" of the cohomology of the first killing space of X, and then the same for the second, the third, and so on killing spaces. Hurewicz's theorem gives us, every time, the corresponding homotopy group. This method (the Serre method) does not permit us, however, to find the homotopy groups without dealing with other difficulties.

Let us imagine, for example, that we want to find stable homotopy groups of some space X and that we know (in the stable range of dimensions) the cohomology of X with any coefficients and with the action of all stable cohomology operations. For example, let us imagine that the first nontrivial homotopy group of X has dimension N and equals $\mathbb{Z}_2$. Consider the spectral sequence of the fibration $X|_N \xrightarrow{K(\mathbb{Z}_2,N-1)} X$.

To make the pictures more compact, we will draw the spectral sequences in two rows. The lower row is the same as the lower row of our old pictures, only we start not with the dimension 0, but from the positive dimension, where nontrivial cohomology first appears. The upper row is what used to be the left column, located in such a way that the cohomologies of the same dimension were above/below each other. The differential (the transgression) acts downward/rightward. The group adjoint to the cohomology group of the total space is now all in the column: The lower group of the column is a subgroup of the cohomology group, and the upper group is the quotient of the cohomology group over this subgroup. Certainly, our picture shows only the stable dimensions.

A. Fomenko, D. Fuchs, *Homotopical Topology*, Graduate Texts in Mathematics 273,
DOI 10.1007/978-3-319-23488-5_5

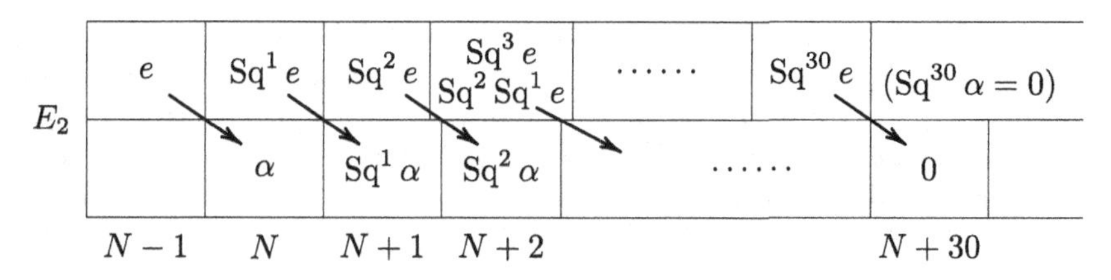

Fig. 114 The E_2-term of the stable spectral sequence

The spectral sequence of our fibration is shown in Fig. 114. In the upper row, there is the cohomology $K(\mathbb{Z}_2, N-1)$ modulo 2, which coincides, in the stable dimensions, with the Steenrod algebra $\mathbb{A}_2$. The differential takes e into the differential α, and for any stable cohomology operation φ, it takes $\varphi(e)$ into $\varphi(\alpha)$. After this differential, in the upper row there remain all $\varphi(e)$ such that $\varphi(\alpha) = 0$, and in the lower row there remain all cohomology classes of X which are not obtained from α by cohomology operations. Thus, the cohomology of $X|_N$ is fully known to us, but the action of the cohomology operations in this cohomology is known to us only partially. Let us imagine, for example, that $\mathrm{Sq}^{30}\,\alpha = 0$ and that in the Steenrod algebra there is a relation $\mathrm{Sq}^{20}\,\mathrm{Sq}^{30} = 0$ (probably there is no such relation, but it is not important to us). Then, in the upper row, there remains an element $f = \mathrm{Sq}^{30}\, e$, but in the cohomology of $X|_N$ this f is not an Sq^{30} of anything. What is $\mathrm{Sq}^{20} f$? This cannot be an element of the upper row, since $\mathrm{Sq}^{20}\,\mathrm{Sq}^{30} = 0$. But it is possible that $\mathrm{Sq}^{20} f = y \in H^*(X|_N; \mathbb{Z}_2)$ is not 0: f comes from $\mathrm{Sq}^{30}\, e \in H^{N+29}(K(\mathbb{Z}_2, N-1); \mathbb{Z}_2)$, and y comes from an element of $H^{N+49}(X; \mathbb{Z}_2)$ (Fig. 115).

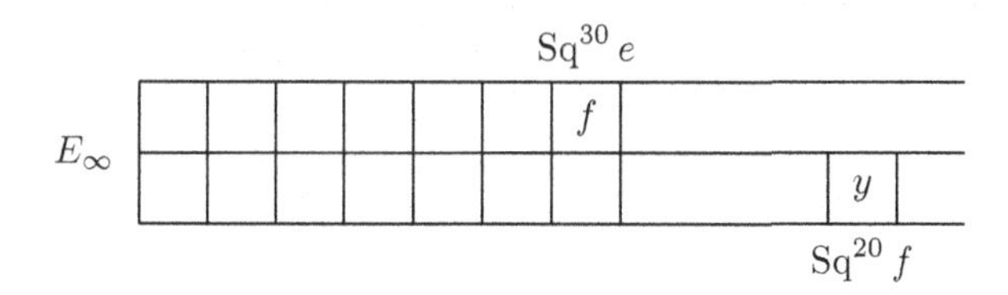

Fig. 115 The E_∞-term of the stable spectral sequence

Thus, we have no full information on the action of cohomology operations in the cohomology of $X|_N$, and hence we cannot find the cohomology of the next killing space.

The modern topology has no way to overcome this difficulty; the homotopy groups of spheres have not been calculated yet. But there is a possibility to "expose the difficulty in a pure form," namely, to collect all the computations related to stable homotopy groups into one spectral sequence. The initial term of this spectral sequence will be known, and all the difficulties will be concentrated in the computations of the differential. This is what the Adams spectral sequence is.

Certainly, the role of the Adams spectral sequence is not restricted to exposing difficulties: It also allows us to resolve part of them. Namely, the Adams spectral sequence may be furnished with an amount of additional structures (for example, the multiplicative structure) which do not show themselves at the level of usual killing spaces and which provide rich information on the action of the differentials.

32.2 The Serre Method and the Adams Method: A Comparison

The Serre method of computing homotopy groups consists in killing the cohomology groups ordered by their dimension: First, we kill the nth group, then pass to the $(n+1)$st, and so on. The Adams method also consists in killing cohomology groups, but in a different order. Take a space X. For example, let it be $(N-1)$-connected, and we want to find the p-components of its homotopy groups from the dimension N to the dimension $N+n$. At the first step we kill *all* the cohomology of X modulo p in these dimensions. One can do this in the following way. Every additive generator of $H^{N+q}(X;\mathbb{Z}_p)$ determines a map $X \to K(\mathbb{Z}_p, N+q)$. All these generators determine a map of X into the product $\Pi_i K(\mathbb{Z}_p, N+q_i)$. There arises a fibration

$$X(1) \xrightarrow{\Pi_i K(\mathbb{Z}_p, N+q_i-1)} X$$

The spectral sequence of this fibration looks as follows. In the upper row, there is a "free $\mathbb{A}_p$-module"; that is, the action of cohomology operations in this row, at least in the dimensions not very much exceeding N, is free: There are no relations which do not follow from the relations in the Steenrod algebra $\mathbb{A}_p$ itself. (See Fig. 116 for the case $p=2$.)

The differential provides an *epimorphism* of the upper row onto the lower row. In the E_∞-term, in the upper row there remains the kernel of the differential, and nothing remains in the lower row.

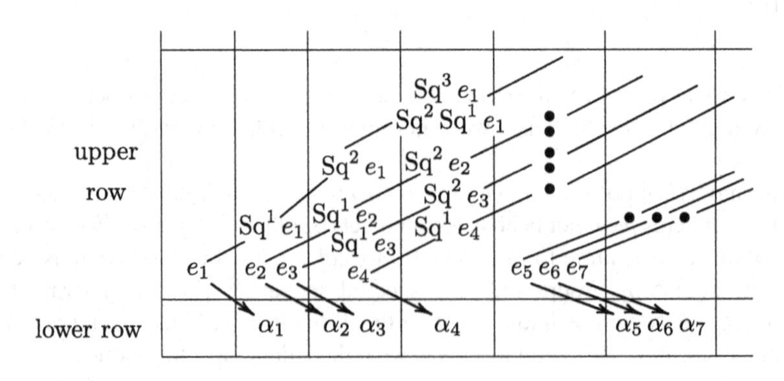

Fig. 116 The spectral sequence of the Adams killing

There arises a sequence of Adams killing spaces, $X(1), X(2), X(3), \ldots$, with a chain of fibrations

$$\ldots X(3) \to X(2) \to X(1) \to X$$

whose fibers are products of $K(\mathbb{Z}_p, n)$s.

Notice that the killing of all elements of $\mathbb{Z}_p$-cohomology of X, as well as all subsequent killing, may be made in a more economical way. For example, if we kill the element ξ and, say, $P^i_p\xi \neq 0$, we do not need to kill $P^i_p\xi$ separately: It will be killed automatically simultaneously with ξ. Speaking more algebraically, we should not kill all elements of some basis of the vector space over $\mathbb{Z}_p$, but rather all elements of some system of generators of an $\mathbb{A}_p$-module. In other words, consider an $\mathbb{A}_p$-module $H^*(X; \mathbb{Z}_p)$. Then there exists a *free* $\mathbb{A}_p$-module F_0 with an $\mathbb{A}_p$-epimorphism $F_0 \to H^*(X; \mathbb{Z}_p)$. [It is easy to understand: F_0 is a free module whose free generators bijectively correspond to the chosen generators of the module $H^*(X; \mathbb{Z}_p)$.] The kernel of this epimorphism is again an $\mathbb{A}_p$-module, but not necessarily free. We construct a free $\mathbb{A}_p$-module F_1 and an $\mathbb{A}_p$-epimorphism of it onto our kernel, in other words, an $\mathbb{A}_p$-homomorphism $F_1 \to F_0$ whose image is the kernel of the epimorphism $F_0 \to H^*(X; \mathbb{Z}_p)$. Repeating this construction, we get an exact sequence

$$\ldots \to F_3 \to F_2 \to F_1 \to F_0 \to H^*(X; \mathbb{Z}_p) \to 0$$

of $\mathbb{A}_p$-modules and $\mathbb{A}_p$-homomorphisms in which all the modules F_i are free. Such a sequence is called a *free resolution* of the module $H^*(X; \mathbb{Z}_p)$. Free resolutions have many remarkable properties, which we will discuss after we finish discussing "general idea."

32.3 The Spectral Sequence

Let us return to geometry. Our process is, in some sense, converging; that is, the cohomology of the spaces $X(k)$ becomes less and less and, again in some sense, completely vanishes at the limit. However, the Serre killing process was directly related (via Hurewicz's theorem) to the homotopy groups of the space. Namely, we always killed the cohomology of the smallest dimension, and every killing corresponded to some element of the homotopy group. It is not so, however, for the Adams killing.

Let us return to the example which we considered in the very beginning (Sect. 32.1): the $(N-1)$-connected space X, $\pi_N(X) = \mathbb{Z}_2$, the group $H_N(X; \mathbb{Z}_2)$ has a generator α, $\mathrm{Sq}^{30}\alpha = 0$, in the cohomology of X there remains an element $f = \mathrm{Sq}^{30} e$, and (thanks to an imaginary relation $\mathrm{Sq}^{20}\mathrm{Sq}^{30} = 0$) $\mathrm{Sq}^{20} f = y \in H^*(X; \mathbb{Z}_2)$ is not zero. If we had performed the Serre killing, we would not have had to separately kill y: It would have been automatically killed together with $f \in H^*(X|_N; \mathbb{Z}_2)$. The Adams killing, however, removes both α and y at the first step. Thus, the Adams method contains more killings than the Serre method. If we

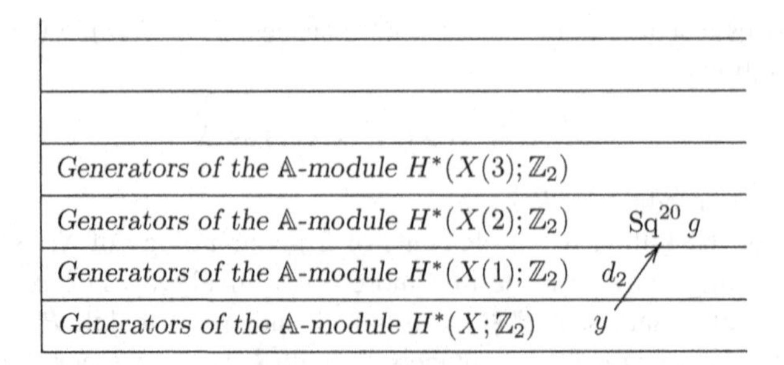

Fig. 117 An overview of the Adams spectral sequence

count the number of generators killed by the Adams method in some dimension $N + q$, we will get an upper estimate for the p-component (2-component in our example) of $\pi_{N+q}(X)$. This upper estimate is the initial term of the Adams spectral sequence. The differentials of this spectral sequence remove all the extra elements, according to the following scheme.

Observe that it is not only y that is an unnecessary killed element. In the cohomology of $X(1)$, as well as in the cohomology of $X|_N$, there is an element f. But in the cohomology of $X|_N$ we have $\mathrm{Sq}^{20} f = y$, while in the cohomology of $X(1)$, $\mathrm{Sq}^{20} f = 0$, which leads to the appearance of an extra element in the cohomology of $X(2)$. Namely, when we kill f by some q, then $\mathrm{Sq}^{20}\, q$ remains, and this $\mathrm{Sq}^{20}\, g$, which even does not appear in the Serre method, has to be killed in the third step of the Adams method. The initial term of the Adams spectral sequence is shown in Fig. 117. The second differential of the Adams spectral sequence will take y into the element coming from $\mathrm{Sq}^{20}\, g$ and will kill both of these extra elements.

The limit term of the Adams spectral sequence will be adjoint to the p-components of the stable homotopy groups of X.

Lecture 33 The Necessary Algebraic Material

33.1 Modules, Free Modules

Let A be an associative algebra (with the unit) over a field $\mathbb{K}$, and let it be $\mathbb{Z}$-graded:

$$A = \bigoplus_q A_q = \ldots \oplus A_{-1} \oplus A_0 \oplus A_1 \oplus \ldots, \quad A_r A_s \subset A_{r+s}.$$

A *left module* over A (or an A-module) is a vector space T over $\mathbb{K}$ which is $\mathbb{Z}$-graded, $T = \bigoplus_q T_q$, and is furnished with a bilinear map $A \times T \to T,\ (a, t) \mapsto at$ such that

(1) If $a \in A_q, t \in T_r$, then $at \in T_{q+r}$.
(2) $b(at) = (ba)t$ for $a, b \in A,\ t \in T$.

A right A-module is defined in a similar way (we will call left A-modules simply A-modules).

A subset U of a A-module T is called a *system of generators* if every $t \in T$ can be presented as a finite sum $a_1u_1 + \ldots + a_nu_n$ with $a_1, \ldots, a_n \in A$, $u_1, \ldots, u_n \in U$. A system of generators is called *homogeneous* if every $u \in U$ belongs to some T_q; in other words, $U \subset T^\circ$ where $T^\circ = \bigcup_q T_q$. If a presentation as above (with all u_i different) is unique for every (= for some) $t \in T$, then the system of generators U is called free, or a basis. A module which possesses a free system of generators is called *free*. For example, A itself is a free A-module with one generator.

An obvious construction extends an arbitrary graded set $U = \bigcup_q U_q$ to a free A-module $T = F_U \supset U$ with a basis U.

A *homomorphism* (or *module homomorphism* or *A-homomorphism*) of an A-module T into an A-module T' is a linear map $f: T \to T'$ such that $f(at) = af(t)$ for all $a \in A, t \in T$. A homomorphism $f: T \to T'$ is called *homogeneous of degree d* if $f(T_q) \subset T'_{q+d}$; a homogeneous homomorphism of degree 0 is called simply a homogeneous homomorphism. Kernels and images of homogeneous A-homomorphisms (of any degree) are (in the obvious sense) A-modules.

Here we often implicitly assume homomorphisms considered homogeneous (of degree 0).

Obviously, for every A-module T there is an exact sequence

$$0 \to K_T \to F_{T^\circ} \xrightarrow{\pi} T \to 0,$$

where π is the extension of the inclusion map $T^\circ \to T$. In particular, every A-module is an image of a (homogeneous) epimorphism of a free module. Notice that F_{T° in this construction can be replaced by F_U, where U is an arbitrary homogeneous system of generators of T.

33.2 Projective Modules

An A-module P is called *projective* if every diagram of the form

$$\begin{array}{ccccc} M & \to & N & \to & 0 \\ & & \uparrow & & \\ & & P & & \end{array}$$

with exact row can be extended to a commutative diagram

$$\begin{array}{ccccc} M & \to & N & \to & 0 \\ & \nwarrow & \uparrow & & \\ & & P. & & \end{array}$$

E3

In other words, any homomorphism of P into a quotient module M/R can be "lifted" to M, that is, factored as $P \to M \to M/R$ where the second arrow is the projection of a module onto the quotient module.

Proposition. *An A-module P is projective if and only if it is a direct summand of some free module.*

Proof. First, let us prove that every free module is projective (this is, actually, the only part of the proposition which will be used later). Let an A-module P be free; let us prove that it is projective. Consider the diagram

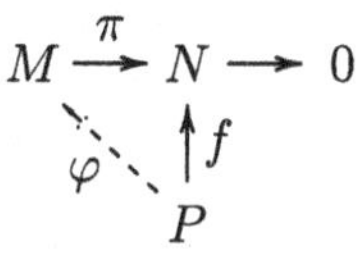

$$\begin{array}{ccccc} M & \xrightarrow{\pi} & N & \longrightarrow & 0 \\ & \nwarrow^{\varphi} & \uparrow f & & \\ & & P & & \end{array}$$

in which the maps π and f are given and the map φ needs to be defined. Let $Q \subset P$ be a basis of P. For every $q \in Q$ choose a $q' \in M$ such that $\pi(q') = f(q)$ (it exists, since π is onto). After it, extend the map $Q \to M$, $q \mapsto q'$ to an A-homomorphism $\varphi: P \to M$ (that is, put $\varphi\left(\sum_i a_i q_i\right) = \sum_i a_i q_i'$). Obviously, $\pi \circ \varphi = f$.

Next, let us prove that a direct summand of a projective module is projective. Let P be a projective module, and let P' be a direct summand of P. This means that there are module homomorphisms $\alpha: P \to P'$, $\beta: P' \to P$ such that $\alpha \circ \beta: P' \to P'$ is id. Consider the diagram shown on the left and add to it a couple of (solid) arrows as shown on the right.

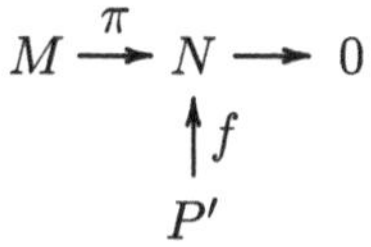

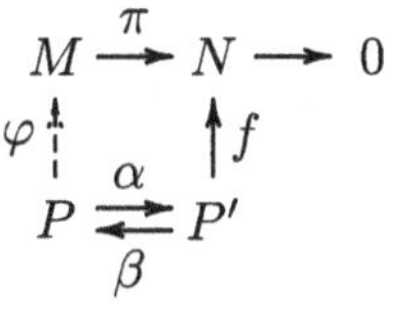

$$\begin{array}{ccccc} M & \xrightarrow{\pi} & N & \longrightarrow & 0 \\ & & \uparrow f & & \\ & & P' & & \end{array} \qquad\qquad \begin{array}{ccccc} M & \xrightarrow{\pi} & N & \longrightarrow & 0 \\ \varphi \uparrow & & \uparrow f & & \\ P & \underset{\beta}{\overset{\alpha}{\rightleftarrows}} & P' & & \end{array}$$

Since P is projective, we can apply the definition of the projectivity to the homomorphism $f \circ \alpha: P \to N$; we get a homomorphism $\varphi: P \to M$ such that $\pi \circ \varphi = f \circ \alpha$. Compose this equality with β: $\pi \circ (\varphi \circ \beta) = f \circ \alpha \circ \beta = f \circ \mathrm{id} = f$. This establishes the projectivity of P'.

Finally, we need to prove that every projective module is a direct summand of a free module. Let P be a projective module. As we have seen before, there is an exact sequence $F_P \xrightarrow{\pi} P \to 0$ where F_P is a free module. Since P is projective, this exact sequence can be extended to a diagram

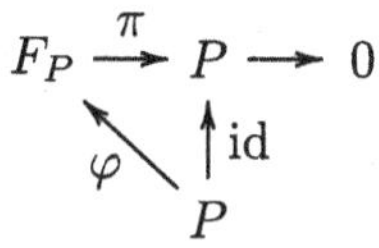

$$\begin{array}{ccccc} F_P & \xrightarrow{\pi} & P & \longrightarrow & 0 \\ & \nwarrow^{\varphi} & \uparrow \mathrm{id} & & \\ & & P & & \end{array}$$

which show that P is a direct summand of the free module F_P, $F_P = P \oplus \operatorname{Ker} \pi$.

33.3 Projective Resolutions

Let T be an arbitrary A-module. Then there exists an exact sequence

$$\ldots \to P_q \to P_{q-1} \to \ldots \to P_1 \to P_0 \to T,$$

where P_q are projective modules. Such a sequence is called a *projective resolution* of T. If all the modules P_q are free, then the sequence is called a *free resolution*. Here is a construction of a free resolution. For every module T we have an exact sequence

$$0 \to K_T \to F_T \to T \to 0$$

with the module F_T being free. Put $T_1 = K_T T_2 = K_{T_1}, T_3 = K_{T_2}, \ldots$. We have exact sequences

$$\begin{array}{l} 0 \to T_1 \to F_T \to T \to 0, \\ 0 \to T_2 \to F_{T_1} \to T_1 \to 0, \\ 0 \to T_3 \to F_{T_2} \to T_2 \to 0, \\ \ldots\ldots\ldots\ldots\ldots\ldots \end{array}$$

of which we can compose one long exact sequence:

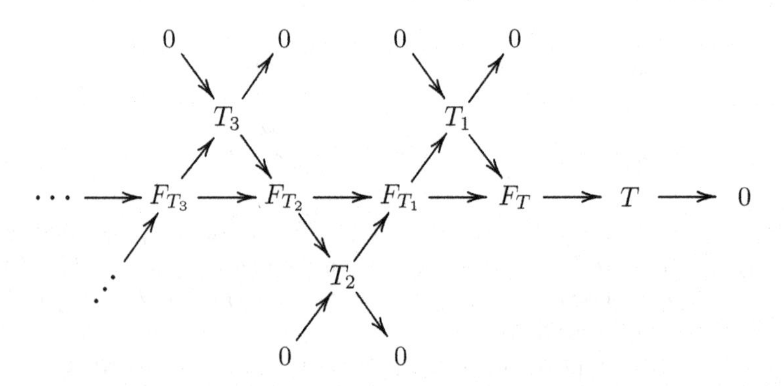

The sequence of horizontal arrows is our resolution.

Certainly, a free (the more so, a projective) resolution of a given A-module is not unique. In many cases, this creates a necessity of checking the independence of various constructions of the choice of a resolution. Usually, we will leave this checking to the reader, restricting ourselves to the remark that is always based on the following proposition which is sometimes called the fundamental lemma of homological algebra.

Theorem. *Let*

$$\begin{array}{c} \dots \longrightarrow P_2 \xrightarrow{d_2} P_1 \xrightarrow{d_1} P_0 \xrightarrow{\varepsilon} T \longrightarrow 0, \\ \dots \longrightarrow P_2' \xrightarrow{d_2'} P_1' \xrightarrow{d_1'} P_0' \xrightarrow{\varepsilon'} T' \longrightarrow 0 \end{array}$$

be projective resolutions of modules T, T', and let $f: T \to T'$ be a module homomorphism. Then the following holds.

(a) *There exist module homomorphisms $\varphi_i: P_i \to P_i'$, $i = 0, 1, 2, \dots$, which make the diagram*

$$\begin{array}{ccccccccc} \dots \longrightarrow & P_2 & \xrightarrow{d_2} & P_1 & \xrightarrow{d_1} & P_0 & \xrightarrow{\varepsilon} & T & \longrightarrow 0 \\ & \downarrow{\scriptstyle \varphi_2} & & \downarrow{\scriptstyle \varphi_1} & & \downarrow{\scriptstyle \varphi_0} & & \downarrow{\scriptstyle f} & \\ \dots \longrightarrow & P_2' & \xrightarrow{d_2'} & P_1' & \xrightarrow{d_1'} & P_0' & \xrightarrow{\varepsilon'} & T' & \longrightarrow 0 \end{array}$$

commutative.

(b) *If $\{\varphi_i'\}$ is another family of homomorphisms with the same property, then there exist module homomorphisms $D_i: P_i \to P_{i+1}'$, $i = 0, 1, 2, \dots$ such that, for $i = 0, 1, 2, \dots$,*

$$\varphi_i' - \varphi_i = D_{i-1} \circ d_i + d_{i+1}' \circ D_i$$

(in this formula, we mean that $D_{-1} = 0$.)

Proof of (a). First, let us construct φ_0. Since ε' is onto and P_0 is projective, the homomorphism $f \circ \varepsilon: P_0 \to T'$ can be factored through P_0', that is, exists a $\varphi_0: P_0 \to P_0'$ such that $\varepsilon' \circ \varphi_0 = f \circ \varepsilon$, as required. Suppose now that there are already $\varphi_j, j < i$ satisfying the commutativity condition. Then $\varphi_{i-2} \circ d_{1-1} \circ d_i = 0 \Rightarrow d_{i-1}' \circ \varphi_{i-1} \circ d_i = 0 \Rightarrow \operatorname{Im}(\varphi_{i-1} \circ d_i) \subset \operatorname{Ker} d_{i-1}' \Rightarrow \operatorname{Im}(\varphi_{i-1} \circ d_i) \subset \operatorname{Im} d_i'$. For this reason, we can consider the composition $\varphi_{i-1} \circ d_i$ as a homomorphism $P_i \to \operatorname{Im} d_i'$, and since P_i is projective and the map $d_i': P_i' \to \operatorname{Im} d_i'$ is onto, there exists a $\varphi_i: P_i \to P_i'$ such that $d_i' \circ \varphi_i = \varphi_{i-1} \circ d_i$. This completes the proof of (a).

Proof of (b). First, let us construct D_0. Since $\varepsilon' \circ \varphi_0' = \varepsilon' \circ \varphi_0 = f \circ \varepsilon, \varepsilon' \circ (\varphi_0' - \varphi_0) = 0 \Rightarrow \operatorname{Im}(\varphi_0' - \varphi_0) \subset \operatorname{Ker} \varepsilon' = \operatorname{Im} d_1'$—hence, $\varphi_0' - \varphi_0$ can be regarded as a map $P_0 \to \operatorname{Im} d_1'$—and since P_0 is projective and the map $d_1': P_1' \to \operatorname{Im} d_1'$ is onto, the homomorphism $\varphi_0' - \varphi_0$ can be factored through P_1'; that is, $\varphi_0' - \varphi_0 = d_1' \circ D_0$ for some $D_0: P_0 \to P_1'$. Suppose now that D_j with $j < i$ have been already constructed and satisfy the condition requested. Then $d_i' \circ (\varphi_i' - \varphi_i - D_{i-1} \circ d_i) = (\varphi_{i-1}' - \varphi_{i-1}) \circ d_i - d_i' \circ D_{i-1} \circ d_i = [\varphi_{i-1}' - \varphi_{i-1} - (D_{i-2} \circ d_{i-1} + d_i' \circ D_{i-1})] \circ d_i = 0$. Hence, $\operatorname{Im}(\varphi_i' - \varphi_i - D_{i-1} \circ d_i) \subset \operatorname{Ker} d_i' = \operatorname{Im} d_{i+1}'$ and since P_i is projective and $d_{i+1}': P_{i+1}' \to \operatorname{Im} d_{i+1}'$ is onto, the homomorphism $\varphi_i' - \varphi_i - D_{i-1} \circ d_i$ can be factored through P_{i+1}'; that is, $\varphi_i' - \varphi_i - D_{i-1} \circ d_i = d_{i+1}' \circ D_i$ for some $D_i: P_i \to P_{i+1}'$; that is, $\varphi_i' - \varphi_i = D_{i-1} \circ d_i + d_{i+1}' \circ D_i$. This is (b).

EXERCISE 1. Prove that this theorem (and, actually, its proof) hold in the following, more general, setting. (1) The modules P_i are projective, but the sequence $\ldots \to P_2 \to P_1 \to P_0 \to T \to 0$ is not necessarily exact: All we need is that it is a complex; that is, the compositions of the consecutive homomorphisms are zeroes. (2) The sequence $\ldots \to P_2' \to P_1' \to P_0' \to T' \to 0$ is exact, but the modules P_i' are not necessarily projective.

Notice finally that everything said in this section can be repeated word for word for right A-modules.

33.4 Tor and Ext

(Compare with Sect. 15.4.) Let M be a right A-module and let N be a left A-module. Fix a projective resolution

$$\ldots \to P_2 \to P_1 \to P_0 \to M \to 0$$

of M and apply to the portion $\ldots \to P_2 \to P_1 \to P_0$ of this resolution the functor $\otimes_A N$[1] with values in graded vector spaces over $\mathbb{K}$. We get a sequence

$$\ldots \to P_2 \otimes_A N \to P_1 \otimes_A N \to P_0 \otimes_A N \ (\to 0 \to 0 \to \ldots),$$

which, in general, is not exact but still is a complex. The degree of nonexactness of this sequence is measured by the homology of this complex, which is denoted as $\mathrm{Tor}_n^A(M,N)$:

$$\mathrm{Tor}_n^A(M,N) = \frac{\mathrm{Ker}(P_n \otimes_A N \to P_{n-1} \otimes_A N)}{\mathrm{Im}(P_{n+1} \otimes_A N \to P_n \otimes_A N)}.$$

If M and N are positively graded, then the space $\mathrm{Tor}_n^A(M,N)$ is graded: $\mathrm{Tor}_n^A(M,N) = \bigoplus_q \mathrm{Tor}_{n,q}^A(M,N)$, where

[1]Recall the definition of $M \otimes_A N$. Consider the vector space F generated by the set $M \times N$ and the subspace G of F spanned by all elements of F of the form $(m+m',n)-(m,n)-(m.,n)$, $(m,n+n')-(m,n)-(m,n')$, $(ma,n)-(m,an)$ for all $m,m' \in M$, $n,n' \in N$, $a \in A$. The quotient space F/G is $M \otimes_A N$ or, briefly, $M \otimes N$. The image of (m,n) in $M \otimes N$ is denoted as $m \otimes n$; thus, $m \otimes n$ is bilinear with respect to m and n and satisfies the relation $ma \otimes n = m \otimes an$. Notice that $M \otimes_A N$ is just a vector space, but if the algebra A is commutative, then it has a natural structure of an A-module: $a(m \otimes n) = ma \otimes n$. There is a generalization: If M is a B-A-bimodule and N is an A-C-bimodule, then $M \otimes_A N$ is a B-C-bimodule. Finally, notice that if M and N are *positively graded*, then the space $M \otimes_A N$ gets a grading: $(M \otimes_A N)_q$ is generated by the products $m \otimes n$ with $m \in M_r, n \in N_s, r+s=q$.

$$\operatorname{Tor}^A_{n,q}(M,N) = \frac{\operatorname{Ker}((P_n \otimes_A N)_q \to (P_{n-1} \otimes_A N)_q)}{\operatorname{Im}((P_{n+1} \otimes_A N)_q \to (P_n \otimes_A N)_q)}.$$

Notice that Tor^A_n is a (covariant) functor with respect to both variables M and N (we leave to the reader the work of relating the last statement with different kinds of homogeneity of homomorphisms).

EXERCISE 2. Prove that Tor is well defined: $\operatorname{Tor}^A_n(M,N)$ does not depend on the choice of a resolution.

EXERCISE 3. Prove that $\operatorname{Tor}^A_0(M,N) = M \otimes_A N$.

EXERCISE 4. If at least one of the modules M, N is projective, then $\operatorname{Tor}^A_n(M,N) = 0$ for all $n > 0$.

EXERCISE 5. If the algebra A is commutative, then $\operatorname{Tor}^A_n(M,N)$ has a natural structure of an A-module.

EXERCISE 6. If the algebra A is commutative, then $\operatorname{Tor}^A_n(M,N) \cong \operatorname{Tor}^A_n(N,M)$ (*Anti-hint:* This is not really easy).

A transition from the definition and properties of the Tor operation to those of the Ext operation is straightforward but still requires some changes in the initial settings. First of all, we need to assume that both modules M and N are left (or both are right, which will not make any difference). Again, we consider a projective resolution

$$\ldots \to P_2 \to P_1 \to P_0 \to M \to 0$$

of M and apply to the portion $\ldots \to P_2 \to P_1 \to P_0$ of this resolution not the functor $\otimes_A N$, but rather the functor $\operatorname{Hom}_A(-,N)$. At this moment we need some remarks concerning this functor.

The notation $\operatorname{Hom}_A(M,N)$ means the vector space of not necessarily homogeneous A-homomorphisms $f\colon M \to N$ satisfying the following, apparently artificial, condition: There exists a positive integer C such that for every q, $f(M_q) \subset \bigoplus_{r=q-C}^{q+C} N_r$ (we do not fix this C: It may be different for different fs). For such a homomorphism $f\colon M \to N$ and an arbitrary $m \in M$, we have $f(m) \in N = \bigoplus_r N_r$; that is,

$$f(m) = (\ldots, f(m)_{-2}, f(m)_{-1}, f(m)_0, f(m)_1, f(m)_2, \ldots),$$

where $f(m)_r \in N_r$ [with only finitely many $f(m)_q$ being different from 0]. Then we can present f as a finite sum $\ldots + f_{-2} + f_{-1} + f_0 + f_1 + f_2 + \ldots$ of homomorphisms where $f_s|_{M_q}$ is defined by the formula $f_s(m) = f(m)_{q+s}$. (Thus, $f_s = 0$ if $|s| > C$.) Obviously, f_s is a homogeneous homomorphism of degree s, and we have

$$\operatorname{Hom}_A(M,N) = \bigoplus_s \operatorname{Hom}^s_A(M,N),$$

where $\operatorname{Hom}^s_A(M,N)$ is the space of homogeneous homomorphisms $M \to N$ of degree s.

It is well known that the two-variable functor $\mathrm{Hom}_A(M,N)$ is covariant with respect to N and contravariant with respect to M. Thus, when we apply to our sequence the functor $\mathrm{Hom}_A(-,N)$, all the arrows change their directions, and this is what we get:

$$\dots \leftarrow \mathrm{Hom}_A(P_2,N) \leftarrow \mathrm{Hom}_A(P_1,N) \leftarrow \mathrm{Hom}_A(P_0,N)\ (\leftarrow 0 \leftarrow \dots).$$

The homology of this complex is Ext:

$$\mathrm{Ext}^n_A(M,N) = \frac{\mathrm{Ker}(\mathrm{Hom}_A(P_n,N) \to \mathrm{Hom}_A(P_{n+1},N))}{\mathrm{Im}(\mathrm{Hom}_A(P_{n-1},N) \to \mathrm{Hom}_A(P_n,N))}.$$

Like Tor, Ext is graded: $\mathrm{Ext}^n_A(M,N) = \bigoplus_s \mathrm{Ext}^{n,s}_A(M,N)$, where

$$\mathrm{Ext}^{n,s}_A(M,N) = \frac{\mathrm{Ker}(\mathrm{Hom}^s_A(P_n,N) \to \mathrm{Hom}^s_A(P_{n+1},N))}{\mathrm{Im}(\mathrm{Hom}^s_A(P_{n-1},N) \to \mathrm{Hom}^s_A(P_n,N))}.$$

Notice also that, like Hom, Ext is covariant in the second variable and is contravariant in the first variable.

EXERCISE 7. Prove that Ext is well defined: $\mathrm{Ext}^n_A(M,N)$ does not depend on the choice of a resolution.

EXERCISE 8. Prove that $\mathrm{Ext}^0_A(M,N) = \mathrm{Hom}_A(M,N)$.

EXERCISE 9. If the module M is projective, then $\mathrm{Ext}^n_A(M,N) = 0$ for any N and any $n > 0$.

EXERCISE 10. An alternative way of defining the Ext operation is to use *injective* resolutions of the module N. The definition of an injective module is a replica of the definition of the projective module with all arrows reversed. (In other words, the module I is injective if any homomorphism of a submodule of some module M into I can be extended to a homomorphism of M into I.) Modules possess *injective resolutions*; an injective resolution of a module N is an exact sequence $0 \to N \to I_0 \to I_1 \to I_2 \to \dots$ with all I_j injective. $\mathrm{Ext}^n_A(M,N)$ can be defined as the homology of the complex

$$(\dots 0 \to 0 \to)\ \mathrm{Hom}_A(M,I_0) \to \mathrm{Hom}_A(M,I_1) \to \mathrm{Hom}_A(M,I_2) \to \dots.$$

Reconstruct the details of this definition, prove that modules possess injective resolutions, and prove the equivalence of the two definitions of Ext.

EXERCISE 11. Abelian groups can be regarded as $\mathbb{Z}$-modules. Prove that for arbitrary Abelian groups A, B,

$$\operatorname{Tor}_n^{\mathbb{Z}}(A,B) = \begin{cases} \operatorname{Tor}(A,B) & \text{for } n = 1, \\ 0 & \text{for } n \geq 2, \end{cases}$$
$$\operatorname{Ext}_{\mathbb{Z}}^n(A,B) = \begin{cases} \operatorname{Ext}(A,B) & \text{for } n = 1, \\ 0 & \text{for } n \geq 2, \end{cases}$$

where on the right-hand sides Tor and Ext are taken in the sense of Sect. 15.4.

Lecture 34 The Construction of the Adams Spectral Sequence

34.1 Topological Adams Filtration

Let X be a topological space. Our goal is to find its stable homotopy groups $\pi_q^S(X) = \pi_{N+q}(\Sigma^N X)$, where $N \gg q$ (according to the generalized Freudenthal theorem, Sect. 23.3.C, this group does not depend on N). The main case: $X = S^0$ (the two-point space); then $\pi_q^S(X) = \pi_{N+q}(S^N) = \pi_q^S$.

Ahead, we fix a prime number p. We will denote the Steenrod algebra $\mathbb{A}_p$ simply as $\mathbb{A}$, and for the cohomology with coefficients in $\mathbb{Z}_p$ we will use the notation with the coefficient group omitted.

Since $\mathbb{A}$ acts in $\widetilde{H}^*$, we can consider $\widetilde{H}^*(X)$ as an $\mathbb{A}$-module; its grading is nonnegative (that is, all components of negative degrees are zero). Let us construct for this $\mathbb{A}$-module a free resolution. We already did this in Lecture 33. First, we take a free module B_0 with an epimorphism $B_0 \to \widetilde{H}^*(X)$. If we assume, for simplicity's sake, that X has finitely many cells in every dimension, then we can construct the module B_0, as well as all the modules which will appear in the future, with finitely many generators of each degree.

The kernel of the epimorphism above is, in general, not free, and we cover it with a new free module, B_2, and so on. We arrive at the exact sequence

$$0 \leftarrow \widetilde{H}^*(X) \leftarrow B_0 \leftarrow B_1 \leftarrow B_2 \leftarrow \dots$$

with all the modules B_i free.

The modules B_i are not cohomology modules for any topological space, because nonzero cohomology modules are never free [for example, in the cohomology modules there are always relations of the form $P_p^i(x) = 0$ for $i > (p-1)\dim x$]. The cohomology module with the smallest possible amount of relations is $\widetilde{H}^*(K(\mathbb{Z}_p, n))$. We want to "approximate" the modules B_i with modules of this kind.

Let N be a large number. The $\mathbb{A}$-module $\widetilde{H}^*(\Sigma^N X)$ is the same as $\widetilde{H}^*(X)$, only the grading is shifted by N. Let $\alpha_i \in H^{q_i}(X)$ be the images of the free generators of B_0, that is, the chosen generators of the $\mathbb{A}$-module $\widetilde{H}^*(X)$. Consider the maps $\Sigma^N X \to K(\mathbb{Z}_p, N + q_i)$ corresponding to the classes $\Sigma^N \alpha_i \in H^{N+q_i}(\Sigma^N X)$ for all i

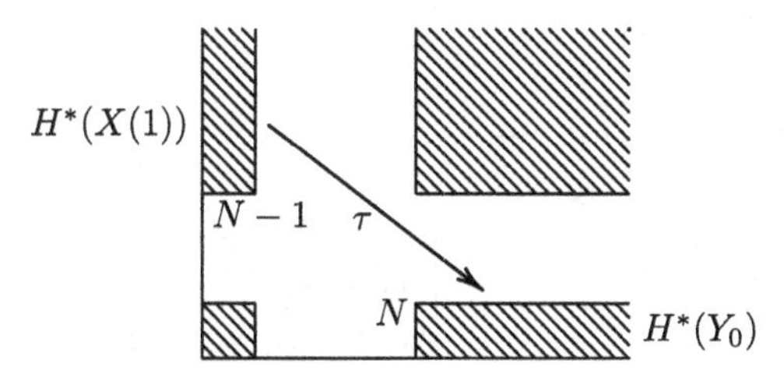

Fig. 118 The spectral sequence of the fibration $\Sigma^N X \xrightarrow{X(1)} Y_0$

such that $q_i < N$. All together, they define a map $\Sigma^N X \to Y_0 = \Pi_i K(\mathbb{Z}_p, N + q_i)$. The $\mathbb{A}$-module $H^*(Y_0)$ in dimensions from N to $2N$ coincides with the B_0-module in dimensions from 0 to N, and the map $\Sigma^N X \to Y_0$ induces a homomorphism $\widetilde{H}^*(Y_0) \to \widetilde{H}^*(\Sigma^N X)$ which coincides in the dimensions from N to $2N$ with the homomorphism $B_0 \to \widetilde{H}^*(X)$ in dimensions from 0 to N.

We can regard the map $\Sigma^N X \to Y_0$ as a fibration; denote the fiber of this fibration by $X(1)$ and put $X(0) = \Sigma^N X$. It would be not easy to fully describe $\widetilde{H}^*(X(1))$, but up to dimension $2N - 3$ the $\mathbb{A}$-module $\widetilde{H}^*(X(1))$, up to a dimension shift by 1, is the same as the $\mathbb{A}$-module $\mathrm{Ker}[\widetilde{H}^*(Y_0) \to H^*(\Sigma^N X)]$; it is seen from the spectral sequence of the fibration (Fig. 118).

The dimension shift occurs because the transgression raises the dimension by 1. Thus, the $\mathbb{A}$-module $\widetilde{H}^*(X(1))$ in dimensions from $N-1$ to $2N-3$ is isomorphic to the $\mathbb{A}$-module $\mathrm{Ker}[B_0 \to \widetilde{H}^*(X)]$ in dimensions from 0 to $N-2$ with a dimension shift by $N-1$.

Remark. Certainly, we can define $X(1)$ not as the fiber of the fibration $\Sigma^N X \to Y_0$, but also as the total space of the fibration $X(1) \xrightarrow{\Pi K(\mathbb{Z}_p, N+q_i-1)} \Sigma^N X$ induced by the fibration $* \xrightarrow{\Pi K(\mathbb{Z}_p, N+q_i-1)} \Pi K(\mathbb{Z}_p, N+q_i) = Y_0$ with respect to the map $\Sigma^N X \to Y_0$:

$$\begin{array}{ccc} X(1) & \longrightarrow & * \approx \mathrm{pt} \\ {\scriptstyle \Pi K(\mathbb{Z}_p, N+q_i-1)} \downarrow & & \downarrow {\scriptstyle \Pi K(\mathbb{Z}_p, N+q_i-1)} \\ \Sigma^N X & \xrightarrow{X(1)} & \Pi K(\mathbb{Z}_p, N+q_i) = Y_0 \end{array}$$

It is important that we have defined not only the space $X(1)$, but also the map $X(1) \to \Sigma^N X$, and both this space and this map are defined up to a homotopy equivalence.

Next, we do for $X(1)$ the same as we did for $\Sigma^N X$. In the $\mathbb{A}$-module $\widetilde{H}^*(X(1))$, we take a system of generators bijectively corresponding (up to dimension $2N - 3$) to the free generators of the $\mathbb{A}$-module B_1. Recall that there is an epimorphism $B_1 \to \mathrm{Ker}[B_0 \to \widetilde{H}^*(X)]$. The chosen generators differ from our previous generators by the dimension shift by $N - 1$. Let $\beta_i \in H^{N+r_i-1}(X(1))$ be these generators. They give rise to a map

$$X(1) \to Y_1 = \Pi K(\mathbb{Z}_p, N + r_i - 1).$$

The $\mathbb{A}$-modules $\widetilde{H}^*(Y_1)$ and B_1, up to the group $H^{2N-3}(Y_1)$, are the same with the dimension shift by $N-1$. We denote by $X(2)$ the fiber of the fibration homotopy equivalent to the map $X(1) \to Y_1$. Thus, we obtain a space $X(2)$ and a map $X(2) \to X(1)$; both are well defined up to a homotopy equivalence.

Proceeding in the same way, we arrive at a sequence of spaces and maps,

$$\ldots \to X(3) \to X(2) \to X(1) \to X(0) = \Sigma^N X.$$

In this sequence, $X(i)$ is the fiber of the fibration whose total space is (homotopy equivalent to) $X(i-1)$ and whose base is a product Y_{i-1} of the spaces $K(\mathbb{Z}_p, m)$.

Fix a dimension $n \ll N$. The $\mathbb{A}$-module $\widetilde{H}^*(Y_i)$ up to dimension $(N-i+1)+n$ and the $\mathbb{A}$-module B_i up to dimension n are the same, only the gradings differ by $N-i+1$. (Actually, they coincide much further, namely, $\widetilde{H}^*(Y_i)$ up to dimension $\sim 2N$ and B_i up to dimension $\sim N$, but it is not important to us, since in the near future we will take limits with both N and n going to ∞.)

On the other hand, we can consider the map $X(i+1) \to X(i)$ as a fibration. Its fiber will be the product Z_i of the same amount of $K(\mathbb{Z}_p, m)$s as Y_i, only the numbers m will be lower by 1. In our range of dimensions we can assume that $Z_i = \Omega Y_i$ and $Y_i = \Sigma Z_i$.

The $\mathbb{A}$-module $\widetilde{H}^*(Z_i)$ is isomorphic to B_i up do dimension $N-i+n$ with the dimension shift by $N-i$. Furthermore, $\widetilde{H}^*(X(i))$ coincides, up to dimension $N-i+n$ and with the dimension shift by $N-i$, with the kernel of the homomorphism $B_{i-1} \to B_{i-2}$ for $i \geq 2$ and with the kernel of the homomorphism $B_0 \to \widetilde{H}^*(X)$ for $i = 1$. Finally, the composition $Z_i \subset X(i+1) \to Y_{i+1}$ induces a homomorphism $\widetilde{H}^*(Y_{i+1}) \to \widetilde{H}^*(Z_i)$ which coincides, up to a dimension shift, with the homomorphism $B_{i+1} \to B_i$ from our resolution.

Notice also that all the spaces $X(i)$ are $(N-1)$-connected. Let us clarify this for the space $X(1)$. The difference in gradings of $\widetilde{H}^*(X(1))$ and $\operatorname{Ker}[B_0 \to \widetilde{H}^*(X)]$ is $N-1$. But the kernel is 0 in dimension 0: There are no relations between the generators of $\widetilde{H}^0(X)$. Thus, $H^{N-1}(X(1)) = 0$. Similarly, using the fact that $[\operatorname{Ker}(B_i \to B_{i-1})]_j = 0$ for $j \geq i$, we deduce that the space $X(i)$ is $(N-1)$-connected for all i.

Next, we want to replace the chain of maps between $X(i)$ by a filtration. For this, we have to turn all these maps into embeddings. Construct the cylinders of all these maps and attach them to each other as shown in Fig. 119. Denote by $X'(k)$ the part of the space obtained to the left of $X(k)$ [including $X(k)$]. There arises a chain of inclusion maps

$$\ldots \subset X'(k) \subset \ldots \subset X'(2) \subset X'(1) \subset X'(0),$$

which is homotopy equivalent to our chain of maps

$$\ldots \to X(3) \to X(2) \to X(1) \to X(0) = \Sigma^N X.$$

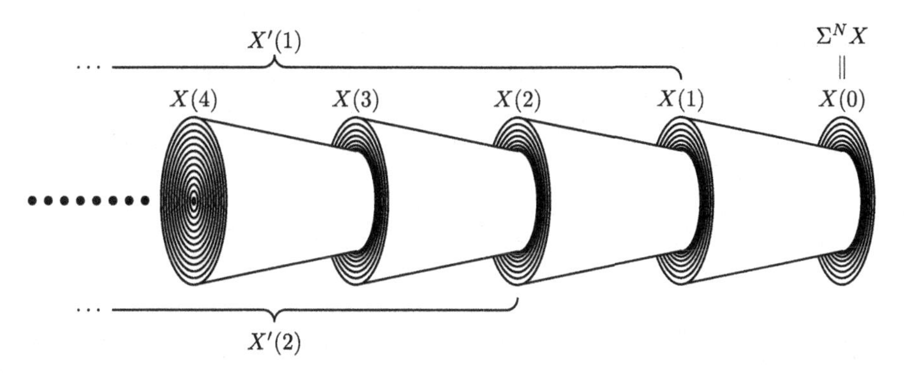

Fig. 119 The Adams filtration

In the future, we will not use the notation $X'(k)$—we will just write $X(k)$. We can do it since all spaces and maps we construct are defined up to a homotopy equivalence.

It will be convenient for us to put $\Sigma^N X = X(0) = X(-1) = X(-2) = \ldots$, so our filtration will be infinite both ways. Notice that the direction of this filtration is not the same as in Chap. 3: The numeration goes backward.

34.2 The Groups and Differentials of the Spectral Sequence

The Adams spectral sequence is constructed from the filtration defined in Sect. 34.1 in the same way as the Leray spectral sequence is constructed from the filtration defined in Sect. 22.1. The main difference is that now we will use homotopy groups rather than homology groups. In general, homotopy groups cannot be used to construct spectral sequences, roughly speaking, because of the absence of the equality $\pi_q(A, B) = \pi_q(A/B)$. However, as proven in Sect. 23.4, this equality holds in stable dimensions (we will not use this fact explicitly, but this is the reason why our constructions will go through).

Consider the inclusion map $(X(s), X(s+r)) \to (X(s+1-r), X(s+1))$, where $r \geq 1$, and introduce groups $E_r^{s,t}$ by the formula

$$E_r^{s,t} = \mathrm{Im}[\pi_{N+t-s}(X(s), X(s+r)) \to \pi_{N+t-s}(X(s+1-r), X(s+1))],$$

where $t \leq n - s$. One may wonder from where this restriction on t comes. Certainly, the group $E_r^{s,t}$ is defined by this formula for all r, s, and t. But its definition involves the arbitrarily chosen number N. How much does the group $E_r^{s,t}$ depend on N? The cohomology of the space $X(m)$ in dimensions from N to $N - m + n$ does not depend on N. If we replace N by a bigger number, M, then the space $X(m)$ will be replaced by a space $\widetilde{X}(m)$ that will be, up to dimension $M - m + n$, homotopy equivalent to $\Sigma^{M-N}X(m)$. This shows that under our restrictions on t, the group $E_r^{s,t}$ does not depend on N. If we take N and n sufficiently large, we obtain the groups $E_r^{s,t}$ well defined for all r, s, and t.

There is another, equivalent, way of defining the groups $E_r^{s,t}$. We put $E_r^{s,t} = G_r^{s,t}/D_r^{s,t}$,, where

$$\begin{aligned} G_r^{s,t} &= \operatorname{Im}[\pi_{N+t-s}(X(s), X(s+r)) \to \pi_{N+t-s}(X(s), X(s+1))] \\ D_r^{s,t} &= \operatorname{Im}[\pi_{N+t-s+1}(X(s+1-r), X(s)) \to \pi_{N+t-s}(X(s), X(s+1))], \end{aligned}$$

and the homomorphisms come from homotopy sequences of triples $(X(s), X(s+1), X(s+r))$, $(X(s+1-r), X(s), X(s+1))$. (This is the initial definition of Adams.) To check that the second definition makes sense and that the two definitions are equivalent, we consider the diagram

$$\begin{array}{ccccc} \pi_q(X(s), X(s+r)) & & & & \\ \uparrow f & \searrow g & & & \\ \pi_{q+1}(X(s+1-r), X(s)) & \xrightarrow{h} & \pi_q(X(s), X(s+1)) & \xrightarrow{k} & \pi_q(X(s+1-r), X(s+1)) \end{array}$$

where all the arrows come from homotopy sequences of triples. In this diagram, the triangle is commutative and the row is exact. We have

$$E_r^{s,t} = \operatorname{Im}(k \circ g),\ G_r^{s,t} = \operatorname{Im}(g),\ D_r^{s,t} = \operatorname{Im}(h).$$

Since $h = g \circ f$, $\operatorname{Im}(h) \subset \operatorname{Im}(g)$; thus, $D_r^{s,t} \subset G_r^{s,t}$. Furthermore,

$$\begin{aligned} E_r^{s,t} = \operatorname{Im}(k \circ g) &= \operatorname{Im}(g)/(\operatorname{Ker}(k) \cap \operatorname{Im}(g)) \\ &= \operatorname{Im}(g)/(\operatorname{Im}(h) \cap \operatorname{Im}(g)) = \operatorname{Im}(g)/\operatorname{Im}(h) = G_r^{s,t}/D_r^{s,t}, \end{aligned}$$

as stated.

Let us look more closely at the groups $E_r^{s,t}$. They are defined for $r \geq 1, s \geq 0$, $t \geq 0$; moreover, $E_r^{s,t} = 0$ for $t < s$ [this follows from the fact that all the spaces $X(s)$ are $(n-1)$-connected]. The term $E_r = \bigoplus_{s,t} E_r^{s,t}$ is shown in Fig. 120 (note the order of the axes!).

Let us look at how the group $E_r^{s,t}$ varies when r grows. Let us use the first definition of $E_r^{s,t}$. Then, starting from $r = s+1$ (we assume s and t fixed), the group $\pi_{N+t-s}(X(s+1-r), X(s+1))$ stabilizes: It becomes $\pi_{N+t-s}(\Sigma^N X, X(s+1))$. However, the group $\pi_{N+t-s}(X(s), X(s+r))$ has no reasons to stabilize. Still, it is true that

$$\operatorname{Im}[\pi_{N+t-s}(X(s), X(s+r)) \to \pi_{N+t-s}(\Sigma^N X, X(s+1))]$$

is a subgroup of

$$\operatorname{Im}[\pi_{N+t-s}(X(s), X(s+r+1)) \to \pi_{N+t-s}(\Sigma^N X, X(s+1))],$$

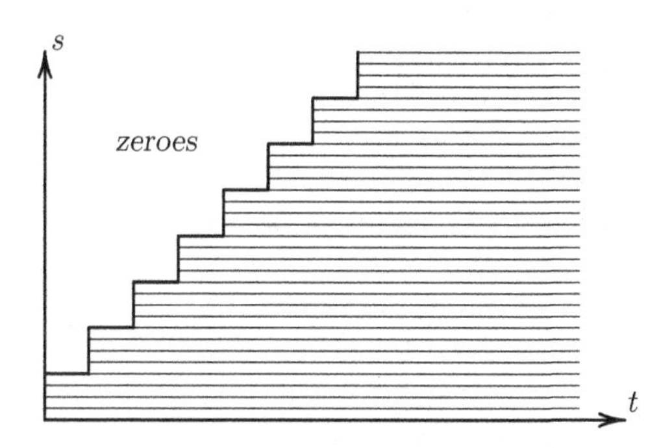

Fig. 120 A graphic presentation of a term of the Adams spectral sequence

since the second map is the composition of the first map and the map

$$\pi_{N+t-s}(X(s),X(s+r+1)) \to \pi_{N+t-s}(X(s),X(s+r)).$$

Thus, we can define the limit groups $E_\infty^{s,t} = \bigcap_{r>s} E_r^{s,t}$ and $E_\infty = \bigoplus_{s,t} E_\infty^{s,t}$.

It is equally easy to make this limit transition using the second description of the groups $E_r^{s,t}$.

Let us turn now to the differentials of the Adams spectral sequence. They will act as $d_r^{s,t}: E_r^{s,t} \to E_r^{s+r,t+r-1}$. Thus, the direction of the differentials is different from the direction of the differential of the Leray spectral sequence (see Fig. 121).

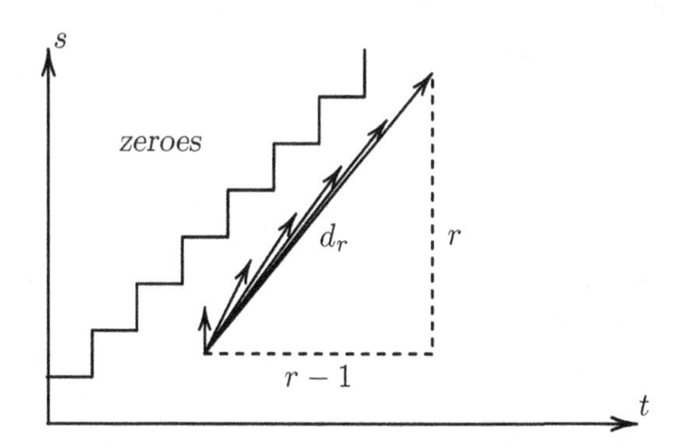

Fig. 121 Differentials of the Adams spectral sequence

Consider the triples $(X(s),X(s+r),X(s+2r))$ and $(X(s+1-r),X(s+1),X(s+r+1))$ and the diagram

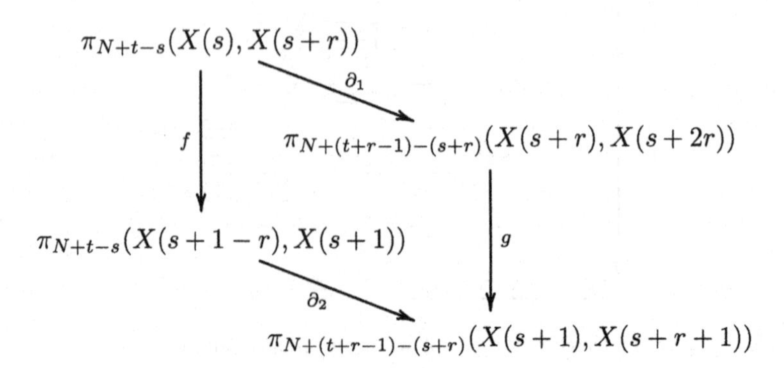

where ∂_1 and ∂_2 are connecting homomorphisms in homotopy sequences of our triples and f and g are homomorphisms used in the (first) definition of the terms of the Adams spectral sequence: $\mathrm{Im}(f) = E_r^{s,t}$ and $\mathrm{Im}(g) = E_r^{s+r,t+r-1}$. The commutativity of our diagram shows that ∂_2 maps $\mathrm{Im}(f)$ into $\mathrm{Im}(g)$. This is the differential $d_r^{s,t}: E_r^{s,t} \to E_r^{s+r,t+r-1}$.

It is obvious that $d_r^{s+r,t+r-1} \circ d_r^{s,t} = 0$ (basically, this means that $\partial \circ \partial = 0$). Now we are in a position to state and prove the main result of this chapter.

34.3 The Adams Theorem

Theorem. *Let X be a CW complex with finite skeletons, and let p be a prime number. Then there exists a spectral sequence $\{E_r^{s,t}, d_r^{s,t}: E_r^{s,t} \to E_r^{s+r,t+r-1} \mid r \geq 1, t \geq s \geq 0\}$, with the following properties.*

(1) *There is a canonical isomorphism*

$$E_2^{s,t} \cong \mathrm{Ext}_{\mathbb{A}}^{s,t}(\widetilde{H}^*(X); \mathbb{Z}_p).$$

Here $\mathbb{Z}_p$ is regarded as an $\mathbb{A}$-module with the trivial action of $\bigoplus_{q \geq 0} \mathbb{A}_q$ with the degree of the generator being 0.

(2) *There is a canonical isomorphism*

$$E_{r+1}^{s,t} = \mathrm{Ker}\, d_r^{s,t} / \mathrm{Im}\, d_r^{s-r,t-r+1}.$$

(3) *For $r > s$, $\mathrm{Im}\, d_r^{s-r,t-r+1} = 0$; thus, $E_{r+1}^{s,t} = \mathrm{Ker}\, d_r^{s,t} \subset E_r^{s,t}$; let $E_\infty^{s,t} = \bigcap_{r>s} E_r^{s,t}$. Claim: For $t > s$, there exists a chain of subgroups*

$$\ldots \subset B^{s+1,t+1} \subset B^{s,t} \subset \ldots \subset B^{1,t-s+1} \subset B^{0,t-s} \subset \pi_{t-s}^S(X)$$

such that $B^{s,t}/B^{s+1,t+1} = E_\infty^{s,t}$.

(4) *$\bigcap_{t-s=m} B^{s,t}$ is the subgroup of $\pi_m^S(X)$ consisting of all elements whose order is finite and is not divisible by p.*

Notice that the adjointness of the limit term of the spectral sequence to the stable homotopy group of X is substantially different from the adjointness in the Leray theory. For every $m \geq 0$, there are, in general, infinitely many groups $E_\infty^{s,t}$ with $0 \geq s, t-s = m$, as shown in the picture.

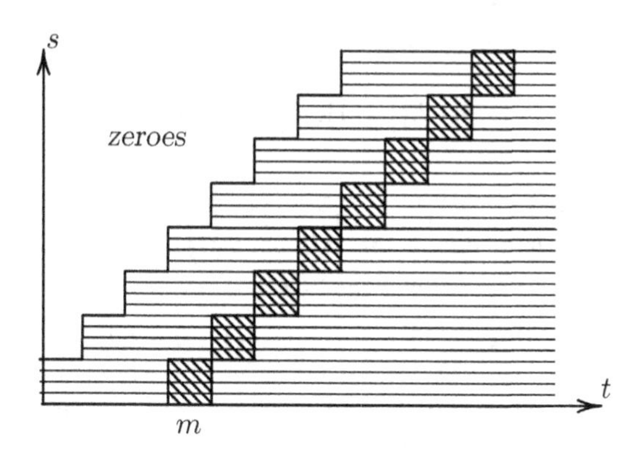

Fig. 122 The structure of the E_∞-term

The sum $E_\infty^{0,m} \oplus E_\infty^{1,m+1} \oplus E_\infty^{2,m+2} \oplus \ldots$ of E_∞ groups marked in Fig. 122 is adjoint to a quotient of the group $\pi_m^S(X)$ with respect to a, possibly infinite, filtration. The groups in the marked squares are all finite—actually, they are all finite sums of groups $\mathbb{Z}_p$; however, the group to which the sum is adjoint may have elements of infinite order. A typical example: The group $\mathbb{Z}_p \oplus \mathbb{Z}_p \oplus \mathbb{Z}_p \oplus \mathbb{Z}_p \oplus \ldots$ is adjoint to $\mathbb{Z}$ with respect to the filtration $\ldots \subset p^4\mathbb{Z} \subset p^3\mathbb{Z} \subset p^2\mathbb{Z} \subset p\mathbb{Z} \subset \mathbb{Z}$.

The proof of the Adams theorem (and some additional statements) is contained in Sects. 34.4–34.8.

34.4 Proof of Statements (1) and (2)

The groups $E_r^{s,t}$ and the differentials $d_r^{s,t}$ have already been defined. Let us clarify the structure of E_1 and E_2. By definition, $E_1^{s,t} = \pi_{N+t-s}(X(s), X(s+1))$. By the construction of $X(k)$, there is a fibration $X(s) \xrightarrow{X(s+1)} \Pi K_m$ [here K_m is an abbreviated notation for $K(\mathbb{Z}_p, m)$]. As we know from Sect. 9.8, the projection of a Serre fibration establishes an isomorphism between the relative homotopy groups of the total space modulo a fiber and the homotopy groups of the base. Hence,

$$E_1^{s,t} = \pi_{N+t-s}(X(s), X(s+1)) = \pi_{N+t-s}(\Pi K_m).$$

On the other hand, in the stable dimensions (that is, when $N \gg t-s$), $H^*(X(s), X(s+1))$ is the same as $\widetilde{H}^*(\Pi K_m)$, which is, again in the stable dimensions, a free module over the Steenrod algebra $\mathbb{A}$. We see that, again in the stable dimensions, the free generators of $H^*(X(s), X(s+1))$ bijectively correspond to the additive generators of $\bigoplus_t E_1^{s,t}$, which can be expressed as an isomorphism, with a grading shift by

$N-s$, $\bigoplus_t E_1^{s,t} = \mathrm{Hom}_{\mathbb{A}}(H^*(X(s), X(s+1)), \mathbb{Z}_p)$ (an $\mathbb{A}$-homomorphism of a free $\mathbb{A}$-module into $\mathbb{Z}_p$ is fully determined by the images of the generators, which are chosen arbitrarily). Finally, we know from Sect. 34.1 that in the stable dimensions the cohomology $H^*(X(s), X(s+1))$, with the same dimension shift, coincides, as an $\mathbb{A}$-module, with B_s. This leads to the final result: *Without* any grading shift and in *all* dimensions, there is an isomorphism

$$\bigoplus_t E_1^{s,t} = \mathrm{Hom}_{\mathbb{A}}(B_s, \mathbb{Z}_p).$$

Lemma. *The differential* $d_1^{s,t}: E_1^{s,t} \to E_1^{s+1,t}$ *coincides with the map* $\mathrm{Hom}_{\mathbb{A}}(B_s, \mathbb{Z}_p) \to \mathrm{Hom}_{\mathbb{A}}(B_{s+1}, \mathbb{Z}_p)$ *induced by the* $\mathbb{A}$*-homomorphism* $B_{s+1} \to B_s$ *from the resolution of* $\widetilde{H}^*(X)$ *constructed in Sect. 34.1.*

Proof of Lemma. The differential $d_1^{s,t}$ is the connecting homomorphism $\pi_{N+t-s}(X(s), X(s+1)) \to \pi_{n+t-s-1}(X(s+1), X(s+2))$ of the homotopy sequence of the triple $(X(s), X(s+1), X(s+2))$. To prove our statement, we need to prove that the connecting homomorphism $H^*(X(s+1), X(s+2)) \to H^{*+1}(X(s), X(s+1))$ of the same triple is the same as the homomorphism $B_{s+1} \to B_s$. In slightly different notations, this was done in Sect. 34.1 (where the homomorphism $B_{s+1} \to B_s$ was identified as induced by the mapping $Z_i \to Y_{i+1}$). Details (if there are any) are left to the reader.

By the lemma and the definition of Ext, the latter is the homology of the complex $\{E_1^{s,t}, d_1^{s,t}\}$. Thus, to finish the proof of statement (1), we need to prove statement (2). Let us do this.

First, we will construct a homomorphism $\mathrm{Ker}\, d_r^{s,t} \to E_{r+1}^{s,t}$. Consider the chain of inclusions:

$$X(s-r) \supset X(s+1-r) \supset X(s) \supset X(s+1) \supset X(s+r) \supset X(s+r+1).$$

Let $\alpha \in \mathrm{Ker}\, d_r^{s,t} \subset E_r^{s,t}$. As an element of $E_r^{s,t}$, α belongs to the image of the homomorphism $\pi_{N+t-s}(X(s), X(s+r)) \to \pi_{N+t-s}(X(s+1-r), X(s+1))$; in other words, α is an element of $\pi_{N+t-s}(X(s+1-r), X(s+1))$, which can be represented by a spheroid

$$F: (D^{N+t-s}, S^{N+t-s-1}) \to (X(s), X(s+r)).$$

Since $X(s+1-r) \subset X(s-r)$, we have $\alpha \mapsto \beta \in \pi_{N+t-s}(X(s-r), X(s+1))$ and to be an element of $E_{r+1}^{s,t} = \mathrm{Im}[\pi_{n+t-s}(X(s), X(s+r+1)) \to \pi_{N+t-s}(X(s-r), X(s+1))]$, it has to be representable by a spheroid $G: (D^{N+t-s}, S^{N+t-s-1}) \to (X(s), X(s+r+1))$. Is it? In general, it is not, but $\alpha \in \mathrm{Ker}\, d_r^{s,t}$; that is, α is annihilated by the boundary map $\pi_{N+t-s}(X(s+1-r), X(s+1)) \to \pi_{N+t-s-1}(X(s+1), X(s+r+1))$. This means that our spheroid F has the following property: The restriction $F|_{S^{N+t-s-1}}: S^{N+t-s-1} \to X(s+r)$, regarded as a map into $X(s-1)$, is homotopic to a map into $X(s+r+1)$. If we append this homotopy to the spheroid F, we will get

a homotopy of the spheroid F to a spheroid $G\colon (D^{N+t-s}, S^{N+t-s-1}) \to (X(s), X(s+r+1))$ within the class of spheroids of the pair $(X(s+1-r), X(s+1))$. Thus, α is represented by this spheroid G, hence, so is β, and hence $\beta \in E_{r+1}^{s,t}$. Our map $\alpha \mapsto \beta$ is a homomorphism $\operatorname{Ker} d_r^{s,t} \to E_{r+1}^{s,t}$, as required.

Next, we must prove that the homomorphism $\operatorname{Ker} d_r^{s,t} \to E_{r+1}^{s,t}$ constructed above is onto. The proof of this is more or less the previous proof read from the end to the beginning. An element $\beta \in E_{r+1}^{s,t}$ is an element of $\pi_{N+t-s}(X(s-r), X(s+1))$, which can be represented by a spheroid $G\colon (D^{N+t-s}, S^{N+t-s-1}) \to (X(s), X(s+r+1))$. The same spheroid represents a certain element $\alpha \in \pi_{N+t-s}(X(s+1-r), X(s+1))$ which belongs to the image of $\pi_{N+t-s}(X(s), X(s+r))$, that is, $\alpha \in E_r^{s,t}$. Moreover, since the boundary of the spheroid G is contained in $X(s+r+1)$, its class α is annihilated by the boundary homomorphism $\partial\colon \pi_{N+t-s}(X(s+1-r), X(s+1)) \to \pi_{N+t-s-1}(X(s+1), X(s+r+1))$, that is, $\alpha \in \operatorname{Ker} d_r^{s,t}$. This is precisely what we need.

It remains to check that the kernel of the epimorphism $\operatorname{Ker} d_r^{s,t} \to E_{r+1}^{s,t}$ is precisely $\operatorname{Im} d_r^{s-r,t-t+1}$.

First, let $\alpha \in \operatorname{Ker} d_r^{s,t}$. As we have seen before, this means that α is an element of $\pi_{N+t-s}(X(s+1-r), X(s+1))$ representable by a spheroid $F\colon (D^{N+t-s}, S^{N+t-s-1}) \to (X(s), X(s+r))$. Furthermore, α belongs to the kernel of the map $\operatorname{Ker} d_r^{s,t} \to E_{r+1}^{s,t}$ if and only if F, regarded as a spheroid of the pair $(X(s-r), X(s+1))$, is homotopic to the trivial spheroid. This means that there exists a homotopy $H\colon D^{N+t-s} \times I \to X(s-r)$ which is F on the lower base and maps the side surface and the upper base into $X(s+1)$ (see Fig. 123). This map H may be regarded as a relative $(N+t-s+1)$-dimensional spheroid of the pair $(X(s-r), X(s))$. This spheroid determines an element of $\pi_{N+t-s-1}(X(s-r), X(s))$ whose image in $\pi_{N+t-s+1}(X(s+1-2r), X(s+1))$ is some element β of $E_r^{s-r,t-s+1}$. Modulo $X(s+1)$, the boundary of the spheroid H is F. Thus, $d_r^{s-r,t-s+1}\beta = \alpha$, and $\alpha \in \operatorname{Im} d_r^{s-r,t-s+1}$, as required.

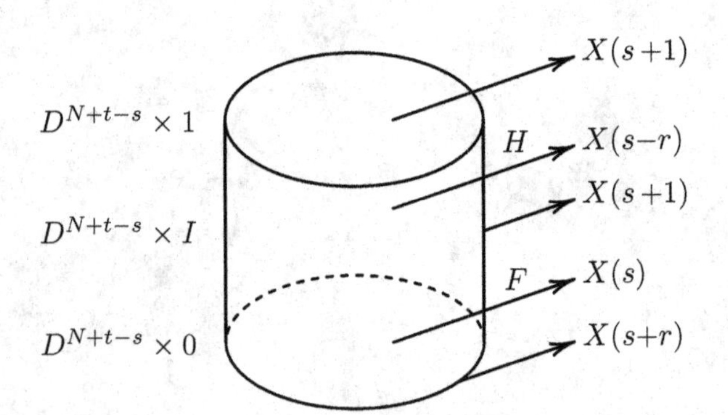

Fig. 123 A homotopy $H\colon (D^{N+t-s} \times I, S^{N+t-s-1} \times I) \to (X(s-r), X(s+1))$ of $F\colon (D^{N+t-s}, S^{N+t-s-1}) \to (X(s), X(s+r))$

Finally, we need to show that for every $\beta \in E_r^{s-r,t-r+1}$, the element $\alpha = d_r^{s-r,t-r+1}\beta$ of $\operatorname{Ker} d_r^{s,t} \subset E_r^{s,t}$ has a zero image in $E_{r+1}^{s,t}$. By definition, β is an element of $\pi_{N+t-s+1}(X(s+1-2r), X(s-r+1))$ represented by a relative spheroid of a pair $(X(s-r), X(s))$. Then α is represented by an absolute spheroid of $X(s)$ which is homotopic to zero in $X(s-r)$. Hence, it represents a zero element in $\pi_{N+t-s}(X(s-r), X(s+1))$, that is, a zero element of $E_{r+1}^{s,t}$.

This completes the proof of statement (2), and hence of statement (1).

34.5 A Digression: A Remark on the Resolutions

Let us describe the most convenient free resolution

$$\ldots \to B_3 \to B_2 \to B_1 \to B_0 \to \widetilde{H}^*(X)$$

of the $\mathbb{A}$-module $\widetilde{H}^*(X)$. First, we take a "minimal" system of generators in the module $\widetilde{H}^*(X)$. We begin with a basis (over $\mathbb{Z}_p$) in the first nontrivial space $\widetilde{H}^{q_1}(X)$. Then, in the second nontrivial space, $H^{q_2}(X)$, we take the image of $H^{q_1}(X)$ with respect to the action of $\mathbb{A}$, and take the basis in a complement to it. Then we do the same with the third nontrivial cohomology space, and so on. We take for B_0 the free $\mathbb{A}$-module on these generators, consider the kernel of the epimorphism $B_0 \to \widetilde{H}^*(X)$, choose generators in this kernel in the same way, get a B_1, and so on.

With a resolution constructed in this way, we will have all the homomorphisms $\mathrm{Hom}_{\mathbb{A}}(B_k, \mathbb{Z}_p) \to \mathrm{Hom}_{\mathbb{A}}(B_{k+1}, \mathbb{Z}_p)$ trivial [indeed, every $\mathbb{A}$-homomorphism $B_k \to \mathbb{Z}_p$ vanishes at all elements of the form $\sum_i \varphi_i a_i^{(k)}$, where $a_i^{(k)}$ are our generators of B_k and $\deg \varphi_i > 0$, and our homomorphism $B_{k+1} \to B_k$ takes all the generators of B_{k+1} into elements of such a form]. Thus, for our resolution, the complex $\{\mathrm{Hom}_{\mathbb{A}}(B_k, \mathbb{Z}_p)\}$ has a trivial differential, and, consequently,

$$\mathrm{Ext}^k_{\mathbb{A}}(\widetilde{H}^*(X), \mathbb{Z}_p) = \mathrm{Hom}_{\mathbb{A}}(B_k, \mathbb{Z}_p).$$

34.6 Proof of Statements (3) and (4): The Case of Finite Stable Homotopy Groups

In this section, we assume that all the stable homotopy groups of X are finite. Then the same is true for all the spaces $X(s)$.

Let us introduce a convenient notation. Every finite Abelian group is decomposed into the sum of two groups: "p-component" and "non-p-component." We will denote these groups as $\mathrm{comp}_p\, G$ and $\mathrm{comp}_{\overline{p}}\, G$, respectively. Thus, $G = \mathrm{comp}_p\, G \oplus \mathrm{comp}_{\overline{p}}\, G$, the order of $\mathrm{comp}_p\, G$ is a power of p, and the order of $\mathrm{comp}_{\overline{p}}\, G$ is not divisible by p.

Lemma 1. *The group* $\mathrm{comp}_{\overline{p}}\, \pi_{N+q}(X(s))$ *does not depend on* s.

Proof. According to Sect. 34.1, there is a fibration $X(s) \xrightarrow{X(s+1)} Y_s$. Since the base of this fibration is a product of spaces of the type $K(\mathbb{Z}_p, m)$, the homotopy groups of the base have no nontrivial non-p-components, and the homotopy sequence of our fibration yields the result.

Lemma 2. *For $q < n$, there exists an s_0 such that*

$$\operatorname{comp}_p \pi_{N+q}(X(s)) = 0$$

for $s > s_0$.

Proof. Let m be a smallest integer $< n$ for which

$$\operatorname{comp}_p H_{N+m}(X(s)) \neq 0.$$

Then m is also the smallest integer $< n$ for which

$$\operatorname{comp}_p \pi_{N+m}(X(s)) \neq 0$$

(strictly speaking, here we need the reference to the $\mathcal{C}$-theorem of Hurewicz, as presented in Sect. 26.1.B; but the fact is obvious: It is sufficient to consider the sequence of killing spaces for $X(s)$). The map $X(s) \to Y_s$ induces an epimorphism of the homotopy groups π_{N+m} [if $\operatorname{comp}_p \pi_{N+m}(X(s))$ is the sum of k cyclic groups of the form $\mathbb{Z}_{p^r}$, then $\pi_{N+m}(Y_s)$ is the sum of k groups $\mathbb{Z}_p$, and the map $\operatorname{comp}_p \pi_{N+m}(X(s)) \to \pi_{N+m}(Y_s)$ is the sum of projections $\mathbb{Z}_{p^r} \to \mathbb{Z}_p$]. Obviously, the group $\operatorname{comp}_p \pi_{N+m}(X(s+1))$ is the kernel of this map. Hence, the order of the last group is less than the order of $\operatorname{comp}_p \pi_{N+m}(X(s))$ (all homotopy groups considered have finite orders!). From this, if m' is the smallest integer $< n$ such that $\operatorname{comp}_p \pi_{N+m'}(X(s+1)) \neq 0$, then either $m' > m$ or $m' = m$, but the order of $\operatorname{comp}_p \pi_{N+m}(X(s+1))$ is less than the order of $\operatorname{comp}_p \pi_{N+m}(X(s))$. The result follows.

Lemma 3. *Let $q < n$. (1). If $0 \leq s < s'$, then $\pi_{N+q}(X(s), X(s'))$ is a p-group. (2). If s' is sufficiently large, then $\pi_{N+q}(X(s), X(s')) = \operatorname{comp}_p \pi_{N+q}(X(s))$.*

Proof. The statement follows from Lemmas 1 and 2 and the exactness of the homotopy sequence

$$\pi_{N+q}(X(s')) \to \pi_{N+q}(X(s)) \to \pi_{N+q}(X(s), X(s')) \\ \to \pi_{N+q-1}(X(s')) \to \pi_{N+q-1}(X(s)).$$

By Lemma 1, in this sequence both the first and last arrows are isomorphisms on $\operatorname{comp}_{\overline{p}}$; hence, both $\operatorname{Ker}[\pi_{N+q-1}(X(s')) \to \pi_{N+q-1}(X(s))]$ and $\operatorname{Coker}[\pi_{N+q}(X(s')) \to \pi_{N+q}(X(s))]$ are p-groups, and hence so is $\pi_{N+q}(X(s), X(s'))$. This is (1). Furthermore, if s' is large, then the p-components of $\pi_{N+q}(X(s'))$ and $\pi_{N+q-1}(X(s'))$ are trivial, and hence $\operatorname{Ker}[\pi_{N+q-1}(X(s')) \to \pi_{N+q-1}(X(s))] = 0$ and $\operatorname{Coker}[\pi_{N+q}(X(s')) \to \pi_{N+q}(X(s))] = \operatorname{comp}_p \pi_{N+q}(X(s))$, which implies (2).

We pass to statements (3) and (4) of the Adams theorem. Set

$$B^{s,t} = \operatorname{Im}[\pi_{N+t-s}(X(s)) \to \pi_{N+t-s}(X(0)) = \pi_{N+t-s}(\Sigma^N X) = \pi^S_{t-s}(X)]$$

(where N is large). We get a sequence of inclusions (a filtration)

$$\ldots \subset B^{s,t} \subset B^{s-1,t-1} \subset \ldots \subset B^{0,t-s}$$

in which $B^{0,t-s} = \operatorname{Im}[\pi_{N+t-s}(X(0)) \xrightarrow{\text{id}} \pi_{N+t-s}(X(0))] = \pi_{N+t-s}(\Sigma^N X) = \pi^S_{t-s}(X)$. We want to prove that $E^{s,t}_\infty = B^{s,t}/B^{s+1,t+1}$ and that

$$\bigcap_{t-s=m} B^{s,t} = \operatorname{comp}_{\overline{p}} \pi^S_m(X).$$

By our definition,

$$E^{s,t}_M = \operatorname{Im}[\pi_{N+t-s}(X(s), X(s+M)) \to \pi_{N+t-s}(\Sigma^N X, X(s+1))],$$

where $N \ll t-s$. By Lemma 3, $\pi_{N+t-s}(\Sigma^N X, X(s+1))$ is a p-group; if M is large, then $\pi_{N+t-s}(X(s), X(s+M)) = \operatorname{comp}_p \pi_{N+t-s}(X(s))$. Hence, for large M,

$$E^{s,t}_M = \operatorname{Im}[\pi_{N+t-s}(X(s)) \to \pi_{N+t-s}(\Sigma^N X, X(s+1))].$$

This does not depend on M, so this is, actually, $E^{s,t}_\infty$. The homomorphism $\pi_{N+t-s}(X(s)) \to \pi_{N+t-s}(\Sigma^N X, X(s+1))$ is a composition

$$\pi_{N+t-s}(X(s)) \to \pi_{N+t-s}(\Sigma^N X) \to \pi_{N+t-s}(\Sigma^N X, X(s+1))],$$

where the kernel of the second homomorphism is $\operatorname{Im}[\pi_{N+t-s}(X(s+1)) \to \pi_{N+t-s}(\Sigma^N X)]$, which is contained in the image of the first homomorphism. Hence,

$$\begin{aligned} E^{s,t}_\infty &= \frac{\operatorname{Im}[\pi_{N+t-s}(X(s)) \to \pi_{N+t-s}(\Sigma^N X)]}{\operatorname{Ker}[\pi_{N+t-s}(\Sigma^N X) \to \pi_{N+t-s}(\Sigma^N X, X(s+1))]} \\ &= \frac{\operatorname{Im}[\pi_{N+t-s}(X(s)) \to \pi_{N+t-s}(\Sigma^N X)]}{\operatorname{Im}[\pi_{N+t-s}(X(s+1)) \to \pi_{N+t-s}(\Sigma^N X)]} = \frac{B^{s,t}}{B^{s+1,t+1}}, \end{aligned}$$

which is statement (3) of the Adams theorem. Furthermore, for M large, by Lemmas 1 and 2,

$$\begin{aligned} B^{s+M,t+M} &= \operatorname{Im}[\pi_{N+t-s}(X(s+M) \to \pi_{N+t-s}(\Sigma^N X)] \\ &= \operatorname{comp}_{\overline{p}} \pi_{N+t-s}(\Sigma^N X) = \operatorname{comp}_{\overline{p}} \pi^S_{t-s}(X), \end{aligned}$$

which is statement (4) of the Adams theorem.

This completes the proof of the Adams theorem in the case of finite stable groups. It remains to prove statements (3) and (4) in the general case. This requires some preparation.

34.7 The Adams Spectral Sequence and Continuous Maps

Let us consider the behavior of the Adams spectral sequence with respect to continuous maps. Suppose that the previous construction of the Adams spectral sequence (including the special choice of resolutions) has been performed for spaces X, X'. Let $f: X \to X'$ be a continuous map. It induces the $\mathbb{A}$-homomorphism $f^*: \widetilde{H}^*(X') \to \widetilde{H}^*(X)$. The fundamental lemma of homological algebra (the theorem of Sect. 33.3) provides a (homotopically unique) homomorphism between the projective resolutions, that is, a commutative diagram

$$\begin{array}{ccccccccc}
\dots \xrightarrow{\beta_3} & B_2 & \xrightarrow{\beta_2} & B_1 & \xrightarrow{\beta_1} & B_0 & \xrightarrow{\beta_0} & \widetilde{H}^*(X) \\
& \uparrow{\scriptstyle \varphi_2} & & \uparrow{\scriptstyle \varphi_1} & & \uparrow{\scriptstyle \varphi_0} & & \uparrow{\scriptstyle f^*} \\
\dots \xrightarrow{\beta_3'} & B_2' & \xrightarrow{\beta_2'} & B_1' & \xrightarrow{\beta_1'} & B_0' & \xrightarrow{\beta_0'} & \widetilde{H}^*(X').
\end{array}$$

We want to convert this into a map between Adams filtrations, that is, a homotopy commutative diagram

$$\begin{array}{ccccccc}
\dots \longrightarrow & X(2) & \longrightarrow & X(1) & \longrightarrow & X(0) & = \Sigma^N X \\
& \downarrow & & \downarrow & & \downarrow{\scriptstyle \Sigma^N f} & \\
\dots \longrightarrow & X'(2) & \longrightarrow & X'(1) & \longrightarrow & X'(0) & = \Sigma^N X'.
\end{array}$$

The construction is as follows. Up to a dimension shift (by N), the $\mathbb{A}$-homomorphisms $\beta_0: B_0 \to \widetilde{H}^*(X)$ and $\beta_0': B_0' \to \widetilde{H}^*(X')$ are induced by maps $\Sigma^N X \to Y_0$ and $\Sigma^N X' \to Y_0'$, which can be regarded as fibrations with the fibers $X(1)$ and $X'(1)$. Since Y_0 and Y_0' are products of Eilenberg–MacLane spaces, the $\mathbb{A}$-homomorphism φ_0 is induced by the (homotopically well-defined) map $Y_0 \to Y_0'$, which forms (together with the map $\Sigma^N f$) a map between the fibrations $\Sigma^N X \to Y_0$ and $\Sigma^N X' \to Y_0'$. In particular, there arises a map between the fibers, $X(1) \to X'(1)$, which forms, together with $\Sigma^N f$ and the inclusion maps $X(1) \to \Sigma^N X = X(0)$ and $X'(1) \to \Sigma^N X' = X'(0)$, a homotopy commutative diagram

$$\begin{array}{ccc}
X(1) & \longrightarrow & X(0) \\
\downarrow & & \downarrow{\scriptstyle \Sigma^N f} \\
X'(1) & \longrightarrow & X'(0).
\end{array}$$

Next, we notice that, up to a dimension shift, $\widetilde{H}^*(X(1))$ and $\widetilde{H}^*(X'(1))$ are $\operatorname{Ker} \beta_0$ and $\operatorname{Ker} \beta_0'$. Since B_1 and B_1' are (again, up to a dimension shift) cohomology modules of spaces Y_1 and Y_1' (which are products of Eilenberg–MacLane spaces), the homomorphisms $\beta_1: B_1 \to \operatorname{Ker} \beta_0$ and $\beta_1': B_1' \to \operatorname{Ker} \beta_0'$ are induced by the maps

$X(1) \to Y_1$ and $X'(1) \to Y_1'$, which can be regarded as fibrations with the fibers $X(2)$ and $X'(2)$. The homomorphism φ_1 gives rise to a map $Y_1 \to Y_1'$, which, together with the already constructed map $X(1) \to X'(1)$, forms a map between these fibrations and, in particular, between their fibers. We obtain a map $X(2) \to X'(2)$, which, together with the map $X(1) \to X'(1)$ constructed before and the inclusion maps $X(2) \to X(1)$ and $X'(2) \to X'(1)$, form a homotopy commutative diagram

$$\begin{array}{ccc} X(2) & \longrightarrow & X(1) \\ \downarrow & & \downarrow \\ X'(2) & \longrightarrow & X'(1). \end{array}$$

Proceeding in the same way, we obtain the map between the Adams filtrations for X and X', as promised.

The map between the Adams filtrations induces homomorphisms between all relative homotopy groups involved in the construction of the Adams spectral sequences and, consequently, between the whole Adams spectral sequences. The meaning of this is presented in the following theorem.

Theorem. *A map $f: X \to X'$ induces a homomorphism of the Adams spectral sequence $\{E_r^{s,t}, d_r^{s,t}\}$ of X into the Adams spectral sequence $\{'E_r^{s,t}, 'd_r^{s,t}\}$ of X', that is, a set of homomorphisms $f_r^{s,t}: E_r^{s,t} \to {}'E_r^{s,t}$ such that*

(1) *These homomorphisms commute with the differentials; that is, the diagram*

$$\begin{array}{ccc} E_r^{s,t} & \xrightarrow{d_r^{s,t}} & E_r^{s+r,t+r-1} \\ \Big\downarrow f_r^{s,t} & & \Big\downarrow f_r^{s+r,t+r-1} \\ 'E_r^{s,t} & \xrightarrow{'d_r^{s,t}} & 'E_r^{s+r,t+r-1} \end{array}$$

is commutative.

(2) *The map $f_{r+1}: E_{r+1} \to {}'E_{r+1}$ is the map of the homology of the complex $\{E_r, d_r\}$ into the homology of the complex $\{'E_r, 'd_r\}$ induced by the map $f_r: E_r \to {}'E_r$.*

(3) *The homomorphism*

$$f_2^{s,t}: E_2^{s,t} = \mathrm{Ext}_{\mathbb{A}}^{s,t}(\widetilde{H}^*(X), \mathbb{Z}_p) \to {}'E_r^{s,t} = \mathrm{Ext}_{\mathbb{A}}^{s,t}(\widetilde{H}^*(X'), \mathbb{Z}_p)$$

is induced by the $\mathbb{A}$-homomorphism $f^*: \widetilde{H}^*(X') \to \widetilde{H}^*(X)$.

Let us clarify the last statement. The Ext functor is contravariant with respect to the first argument (and covariant with respect to the second argument). Namely, an $\mathbb{A}$-homomorphism $M_2 \to M_1$ gives rise to a (homotopically unique) homomorphism of a projective resolution of M_2 into a projective resolution of M_1. The operation $\mathrm{Hom}(-, N)$ reverses all the arrows, and the resulting homology homomorphism maps $\mathrm{Ext}_{\mathbb{A}}^{*,*}(M_1, N)$ into $\mathrm{Ext}_{\mathbb{A}}^{*,*}(M_2, N)$.

The theorem is obvious, and we can add to it one more statement, which is equally obvious but not fully proved, since we have not given a proof, in the general case, of the corresponding statement of the Adams theorem. (Certainly, we will not use this statement before we finish the proof of the Adams theorem.)

(4) *The limit map* $f_\infty^{s,t}: E_\infty^{s,t} \to {}'E_\infty^{s,t}$ *is induced by the map* $f_*: \pi_*^S(X) \to \pi_*^S(X')$.

Corollary. *The Adams spectral sequence of X, beginning from the* E_2*-term, depends only on the stable homotopy type of X.*

34.8 End of Proof of the Adams Theorem

In the general case, the homotopy groups $\pi_{N+q}(X(s))$ are not finite, but they are finitely generated (since X was assumed to be a CW complex with finite skeletons). The decomposition $G = \mathrm{comp}_p G \oplus \mathrm{comp}_{\overline{p}} G$ (see Sect. 34.6) in the case when G is finitely generated, but not necessarily finite, should be replaced by the decomposition $G = \mathrm{Comp}_p G \oplus \mathrm{comp}_{\overline{p}} G$, where $\mathrm{comp}_{\overline{p}} G$ is the same as before, the subgroup of G consisting of 0 and all elements of finite order prime to p, and $\mathrm{Comp}_p G \cong (G/\mathrm{comp}_{\overline{p}} G)$. Thus, $\mathrm{Comp}_p G$ is the sum of a group of the form $\mathbb{Z} \oplus \dots \oplus \mathbb{Z}$ and a finite group whose order is a power of p. (The decomposition $G = \mathrm{Comp}_p G \oplus \mathrm{comp}_{\overline{p}} G$ is not canonical, but this is not important for us.)

Lemma 1 of Sects. 34.6 (with its proof) remains valid for our new case; so does part (1) of Lemma 3. Lemma 2, however, should be modified. If the group $\pi_{N+q}(\Sigma^N X)$ is infinite, then we cannot expect that $\mathrm{Comp}_p \pi_{N+q}(X(s))$ will be zero for sufficiently large s; actually, all the groups $\pi_{N+q}(X(s)$ have the same rank. The role of Lemma 2 in our new setting will be played by the following proposition.

Proposition. *If the order of* $\alpha \in \pi_{N+q}(X(s))$ *is* ∞ *or a power of p, then, for M sufficiently large,* α *does not belong to the image of the homomorphism* $\pi_{N+q}(X(s+M)) \to \pi_{N+q}(X(s))$. *In other words,*

$$\bigcap\nolimits_M \mathrm{Im}[\pi_{N+q}(X(s+M)) \to \pi_{N+q}(X(s))] = \mathrm{comp}_{\overline{p}}\, \pi_{N+q}(X(s)).$$

Proof. First, notice that it is sufficient to prove the proposition in the case when $s = 0$, that is, $\alpha \in \pi_{N+q}(\Sigma^N X)$; indeed, the sequence $\dots \to X(s+2) \to X(s+1) \to X(s)$ is the Adams filtration for $X(s)$.

The main ingredient of the proof is the geometric construction presented in Fig. 124. Take the suspension ΣX and some positive integer k and then collapse to points $p^k - 1$ parallel copies of X. We obtain a map of ΣX into a garland of p^k copies of ΣX attached to each other by vertices. This garland is homotopy equivalent to the p^k-fold bouquet $\Sigma X \vee \dots \vee \Sigma X$. Then we apply the map of this bouquet onto ΣX, which is the identity on every ΣX in the bouquet. The composition map $h: \Sigma X \to \Sigma X$ may also be described as the smash product $X\#S^1 \to X\#S^1$ of the identity map $X \to X$ and a map $S^1 \to S^1$ of degree p^k.

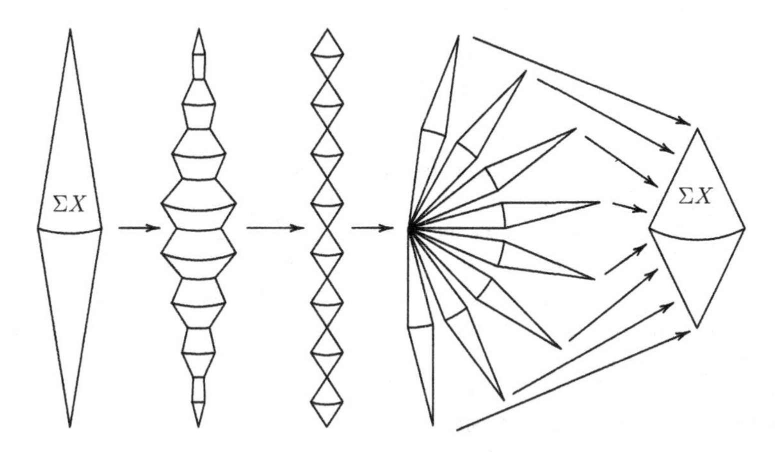

Fig. 124 A p^k-fold map $\Sigma X \to \Sigma X$

Then we take the cone X' of the map h constructed; that is, we attach the cone $C(\Sigma X)$ to ΣX by the map h of the base of the cone onto ΣX. There arises the inclusion map $\Sigma X \to X'$ of the base of the mapping cone into this cone; we will also need the $(N-1)$-fold suspension $\Sigma^N X \to \Sigma^{N-1} X'$ of this map.

Lemma. *The stable homotopy groups of the space X' are finite. (They are actually p-groups.)*

Proof of Lemma. Consider the homotopy sequence of the pair $(\Sigma^{N-1}X', \Sigma^N X)$:

$$\pi_{N+q+1}(\Sigma^{N-1}X', \Sigma^N X) \xrightarrow{\partial} \pi_{N+q}(\Sigma^N X) \to \pi_{N+q}(\Sigma^{N-1}X') \\ \longrightarrow \pi_{N+q}(\Sigma^{N-1}X', \Sigma^N X) \xrightarrow{\partial} \pi_{N+q-1}(\Sigma^N X).$$

In stable dimensions,

$$\begin{aligned}\pi_{N+q+1}(\Sigma^{N-1}X', \Sigma^N X) &= \pi_{N+q+1}(\Sigma^{N-1}X' / \Sigma^N X) \\ &= \pi_{N+q+1}(\Sigma^{N+1}X) = \pi_{N+q}(\Sigma^N X),\end{aligned}$$

and the map $\partial\colon \pi_{N+q+1}(\Sigma^{N-1}X', \Sigma^N X) = \pi_{N+q}(\Sigma^N X) \to \pi_{N+q}(\Sigma^N X)$ is induced by the map shown in Fig. 124. It is clear that this map is the multiplication by p^k (it is seen in Fig. 124 for the classes of spheroids of ΣX which are suspensions over spheroids of X; but in stable dimensions all spheroids in ΣX are homotopic to suspensions). It remains to notice that both the kernel and cokernel of the multiplication by p^k are finite p-groups; hence, $\pi_{N+q}(\Sigma^{N-1}X')$ is also a finite p-group. This proves the lemma.

Now return to the proposition. The map $\Sigma^N X \to \Sigma^{N-1}X'$ constructed above gives rise to a map between the Adams filtrations:

$$\begin{array}{ccccccc} \dots \to X(2) & \to & X(1) & \to & X(0) & = & \Sigma^N X \\ \downarrow & & \downarrow & & \downarrow & & \downarrow \\ \dots \to X'(2) & \to & X'(1) & \to & X'(0) & = & \Sigma^{N-1} X'. \end{array}$$

Assume now that k is so large that α is not divisible by p^k. The homotopy sequence above (in the proof of the lemma) shows that the kernel of the map $\pi_{N+q}(\Sigma^N X) \to \pi_{N+q}(\Sigma^{N-1} X')$ consists of elements divisible by p^k. Thus, α does not belong to this kernel, and hence its image $\beta \in \pi_{N+q}(\Sigma^{N-1} X')$ is not zero, but its order is a power of p. Since the stable homotopy groups of X' are finite, we can apply to X' Lemma 2 of Sect. 34.6, and it implies that for sufficiently large M, β does not belong to the image of the homomorphism $\pi_{N+q}(X'(M)) \to \pi_{N+q}(\Sigma^{N-1} X')$. Hence, α does not belong to the image of the homomorphism $\pi_{N+q}(X(M)) \to \pi_{N+q}(\Sigma^N X)$. This completes the proof of the proposition.

Now we can prove, in the general case, statements (3) and (4) of the Adams theorem. The proof repeats the proof in Sect. 34.6 with only one difference. We can no longer state, for M large, that

$$E_M^{s,t} = \mathrm{Im}[\pi_{N+t-s}(X(s), X(s+M)) \to \pi_{N+t-s}(\Sigma^N X, X(s+1))]$$

is the same as

$$\mathrm{Im}[\pi_{N+t-s}(X(s)) \to \pi_{N+t-s}(\Sigma^N X, X(s+1))].$$

We will prove instead that the intersection

$$\bigcap_{M \text{ large}} E_M^{s,t} = \bigcap_{M \text{ large}} \mathrm{Im}[\pi_{N+t-s}(X(s), X(s+M)) \to \pi_{N+t-s}(\Sigma^N X, X(s+1))]$$

is the same as

$$\mathrm{Im}[\pi_{N+t-s}(X(s)) \to \pi_{N+t-s}(\Sigma^N X, X(s+1))],$$

so it is still true that the last image is $E_\infty^{s,t}$, and the rest of the proof goes smoothly. Thus, we want to prove that if some $\beta \in \pi_{N+t-s}(\Sigma^N X, X(s+1))$ does not belong to the image of $\pi_{N+t-s}(X(s))$, then β also does not belong to the image of $\pi_{N+t-s}(X(s), X(s+M))$ for M sufficiently large. Consider the commutative diagram with an exact row,

$$\begin{array}{ccccc} & & \pi_{N+t-s}(X(s), X(s+M)) & \xrightarrow{\partial_1} & \pi_{N+t-s-1}(X(s+M)) \\ & & \downarrow{\scriptstyle \varphi_2} & & \downarrow{\scriptstyle \varphi_4} \\ \pi_{N+t-s}(X(s)) & \xrightarrow{\varphi_1} & \pi_{N+t-s}(X(s), X(s+1)) & \xrightarrow{\partial_2} & \pi_{N+t-s-1}(X(s+1)) \\ & & \downarrow{\scriptstyle \varphi_3} & & \| \\ & & \pi_{N+t-s}(\Sigma^N X(s), X(s+1)) & \xrightarrow{\partial_3} & \pi_{N+t-s-1}(X(s+1)) \end{array}$$

We want to prove that if $\beta \notin \mathrm{Im}(\varphi_3 \circ \varphi_1)$, then $\beta \notin \mathrm{Im}(\varphi_3 \circ \varphi_1)$ for large M. Let $\alpha = \partial_3(\beta) \in \pi_{N+t-s-1}(X(s+1)$. Then either $\alpha = 0$ or $\alpha \neq 0$, and the order of α is a power of p [because $\pi_{N+t-s}(\Sigma^N X, X(s+1))$ is a finite p-group].

Let $\alpha = 0$. If $\beta = \varphi_3\gamma$ for a $\gamma \in \pi_{N+t-s}(X(s), X(s+1))$, then $\partial_2\gamma = \alpha = 0$, hence $\gamma = \varphi_1\delta$ for some $\delta \in \pi_{N+t-s}(X(s))$, and hence $\beta = \varphi_3 \circ \varphi_1\delta \in\in (\varphi_3 \circ \varphi_1$, a contradiction. Hence, $\beta \notin \mathrm{Im}\,\varphi_3$; even more, $\beta \notin \mathrm{Im}(\varphi_3 \circ \varphi_2)$ for any M.

Let α be a nonzero element whose order is a power of p. Then, by the proposition, $\alpha \notin \mathrm{Im}\,\varphi_4$ for a large M. Then $\alpha \notin \mathrm{Im}(\varphi_4 \circ \partial_2) = \mathrm{Im}(\partial_3 \circ \varphi_3 \circ \varphi_2)$, and hence $\beta \notin \mathrm{Im}(\varphi_3 \circ \varphi_2$.

This completes the proof of the Adams theorem.

Lecture 35 Multiplicative Structures

The multiplicative structure in the cohomological Leray spectral sequence appeared because the cohomology groups of which this spectral sequence is made possess a multiplicative structure. The groups of which the Adams spectral sequence is made are homotopy groups which do not possess any multiplicative structure. Well, there certainly is the Whitehead product, but it is useless for the Adams spectral sequence since it is zero in the stable dimensions. For this reason, the Adams spectral sequence does not have, at least, an immediate multiplicative structure. Still, there is a way to introduce a multiplication in the Adams spectral sequence of a space X, at least in the case when X is a sphere; but this case is very important and interesting.

Let us begin with this case. If we imagine that there is a multiplicative structure in the Adams spectral sequence of the sphere, then the sum $\bigoplus_q \pi_q^S(S^0)$ should be a ring. What is the multiplication in this sum? It turns out that this multiplication exists, is very natural, and can be described in a purely geometric way.

35.1 The Composition Product in the Stable Homotopy Groups of Spheres

Let $\alpha \in \pi_k^S(S^0)$, $\beta \in \pi_\ell^S(S^0)$. We can regard α as an element of $\pi_{N+k}(S^N)$ and β as an element of $\pi_{N+k+\ell}(S^{N+k})$ (where $N \gg k, \ell$). Spheroids $a\colon S^{N+k} \to S^N$, $b\colon S^{N+k+\ell} \to S^{N+k}$ representing α and β can be composed to create a spheroid $a \circ b\colon S^{N+k+\ell} \to S^N$, which, in turn, belongs to some class in $\pi_{N+k+\ell}(S^N) = \pi_{k+\ell}^S(S^0)$. We denote this element as $\alpha \circ \beta$ and call it the *composition product* of α and β. It is obvious that the composition product is well defined; that is, $\alpha \circ \beta$ depends only on α and β.

There is another definition of the composition product; we will give this definition and then establish its equivalence to the previous definition. Let $\alpha \in \pi_{N+k} \to S^N$, $\beta \in \pi_{N+\ell}(S^N)$, and let $a\colon S^{N+k} \to S^N$ and $b\colon S^{N+\ell} \to S^N$ be spheroids of the classes α and β. We define $\alpha \circ \beta$ as $(-1)^{Nk}$ times the class of a spheroid $a\#b\colon S^{N+k}\#S^{N+\ell} \to S^N\#S^N$, that is, $S^{2N+k+\ell} \to S^{2N}$. Again, we have $\alpha \circ \beta \in \pi^S_{k+\ell}(S^0)$.

Proposition 1. *The two definitions of* $\circ$ *are equivalent.*

Proof. The map $(-1)^{Nk}(a\#b) = ((-1)^{Nk}a)\#b\colon S^{N+k}\#S^{N+\ell} \to S^N\#S^N$ can be presented as a composition of two maps,

$$\begin{array}{ccccc} S^N & \# & S^k & \# & S^{N+\ell} \\ \big\downarrow \mathrm{id} & & \big\downarrow \mathrm{id} & & \big\downarrow b \\ S^N & \# & S^k & \# & S^N \\ & \underbrace{\qquad\qquad}_{\big\downarrow (-1)^{Nk}a} & & & \big\downarrow \mathrm{id} \\ & S^N & & \# & S^N \end{array}$$

Now, supplement this composition with a map swapping the two copies of S^N in the second row. To make this map homotopic to the identity, we make one of the maps $S^N \to S^N$ not the identity, but rather $(-1)^N\mathrm{id}$. After that, we also swap the two copies of S^N, and again, to make the map homotopic to the identity, we turn one of the maps $S^N \to S^N$ into $(-1)^N\mathrm{id}$. Thus, our map is homotopic to the composition

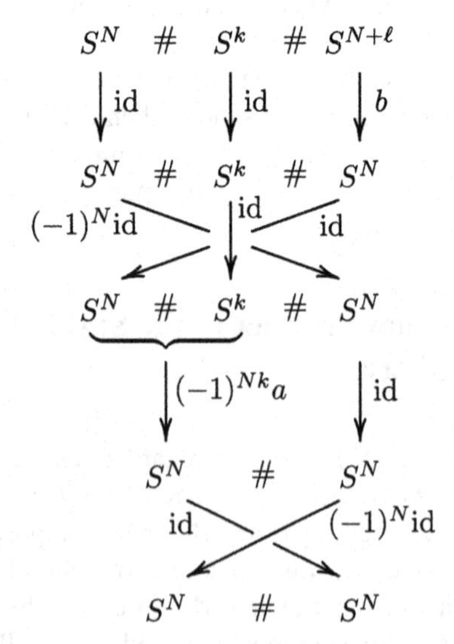

This composition is a map $S^N\#S^{N+k+l} \to S^N\#S^N$, which is the #-product of the identity map $S^N \to S^N$ and the composition

$$S^{N+k+\ell} \xrightarrow{\Sigma^k b} S^{N+k} = S^k\#S^N \xrightarrow{\text{swap}} S^N\#S^k \xrightarrow{(-1)^{Nk}a} S^N.$$

Since the degree of the swap map is Nk, this composition is homotopic to $a \circ b$, as described in the first definition. This completes the proof of Proposition 1.

Proposition 2. *The composition product is skew-commutative, that is,* $\beta \circ \alpha = (-1)^{k\ell}\alpha \circ \beta$.

Proof. Let $a\colon S^{N+k} \to S^N$ and $b\colon S^{N+\ell} \to S^N$ be spheroids of classes α and β. There is an obvious homotopy commutative diagram

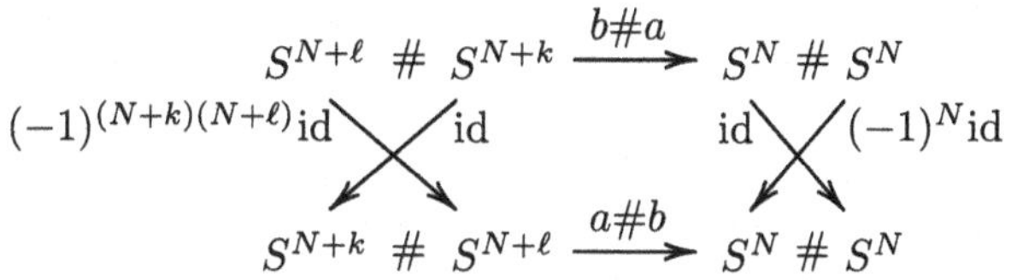

$$\begin{array}{ccc} S^{N+\ell} \# S^{N+k} & \xrightarrow{b\#a} & S^N \# S^N \\ (-1)^{(N+k)(N+\ell)}\mathrm{id} \swarrow\!\!\!\searrow \mathrm{id} & & \mathrm{id} \swarrow\!\!\!\searrow (-1)^N\mathrm{id} \\ S^{N+k} \# S^{N+\ell} & \xrightarrow{a\#b} & S^N \# S^N \end{array}$$

By Proposition 1, the classes $\beta \circ \alpha$ and $\alpha \circ \beta$ are represented by the spheroids $((-1)^{N\ell} b\#a$ and $((-1)^{Nk} a\#b$, and we have

$$\beta \circ \alpha = (-1)^{Nl}(-1)^{(N+k)(N+\ell)}(-1)^{Nk}(-1)^{N}\alpha \circ \beta = (-1)^{k\ell}\alpha \circ \beta.$$

Proposition 3. *The composition product is distributive:*

$$(\beta + \gamma) \circ \alpha = \beta \circ \alpha + \gamma \circ \alpha;\ \alpha \circ (\beta + \gamma) = \alpha \circ \beta + \alpha \circ \gamma.$$

Proof. Since we have already proven that the composition product is skew-commutative, it is sufficient to prove any one of these formulas, and we will prove the second one. Actually, we will prove a stronger statement, namely, that for any $\beta, \gamma \in \pi_m(S^n)$ and $\alpha \in \pi_n(X)$ (where X is arbitrary), the equality $\alpha \circ (\beta + \gamma) = \alpha \circ \beta + \alpha \circ \gamma$ holds. In other words, for any spheroids $b, c\colon S^m \to S^n$, and $a\colon S^n \to X$, the spheroids $a \circ (b + c), (a \circ b) + (a \circ c)\colon S^m \to X$ are homotopic, where $+$ stands for the sum of spheroids as defined in Sect.8.1. But it is obvious that these spheroids are not just homotopic—they are the same.

Since the associativity of the composition product is obvious, we can say that the direct sum $\pi_*^S(S^0) = \bigoplus_q \pi_q^S(S^0)$ acquires the structure of graded associative skew-commutative ring.

Remark. As we noticed in the proof of Proposition 3, the composition products exist not only in stable homotopy groups, and not only of spheres. However, even for a homotopy group of spheres, the composition product does not have any good algebraic properties. Certainly, it is associative: If $\alpha \in \pi_m(S^n), \beta \in \pi_\ell(S^m)$, and $\gamma \in \pi_k(S^\ell)$, then $(\alpha \circ \beta) \circ \gamma = \alpha \circ (\beta \circ \gamma)$. However, no kind of commutativity can exist: If $\alpha \in \pi_m(S^n)$ and $\beta \in \pi_\ell(S^m)$, then $\alpha \circ \beta$ cannot be equal to $\pm\beta \circ \alpha$ simply because the latter is not defined. The "right distributivity" $\alpha \circ (\beta + \gamma) = \alpha \circ \beta + \alpha \circ \gamma$ holds (as shown in the proof of Proposition 3), but the "left distributivity" may fail. Here is the simplest example: If $\iota \in \pi_2(S^2)$ is the class of the identity spheroid and $\alpha \in \pi^3(S^2)$ is the Hopf class, then $(2\iota) \circ \alpha$ is not 2α (which would have followed from the left distributivity), but rather 4α.

EXERCISE 1. Prove this using the Hopf invariant.

35.2 An Algebraic Digression

A: Definition and Examples

An associative unitary graded algebra $A = \bigoplus_m A^m$ over a field $\mathbb{K}$ is called a *Hopf algebra* if the following axioms hold.

(1) $A^m = 0$ for negative m, and $A^0 = \mathbb{K}$.
(2) There is defined a *diagonal map* or *comultiplication* $\Delta: A \to A \otimes_{\mathbb{K}} A$ which is a homomorphism between graded algebras (the multiplication in $A \otimes_{\mathbb{K}} A$ is defined by the rule $(\alpha \otimes \beta) \cdot (\alpha' \otimes \beta') = (-1)^{\deg \beta \deg \alpha'} \alpha\alpha' \otimes \beta\beta'$).
(3) If $\deg a = d > 0$, then $\Delta(a) = a \otimes 1 + \ldots + (-1)^d 1 \otimes a$, where $\ldots$ denotes a sum of tensor products of elements of positive degrees.
(4) The comultiplication Δ is *coassociative*; that is, the diagram

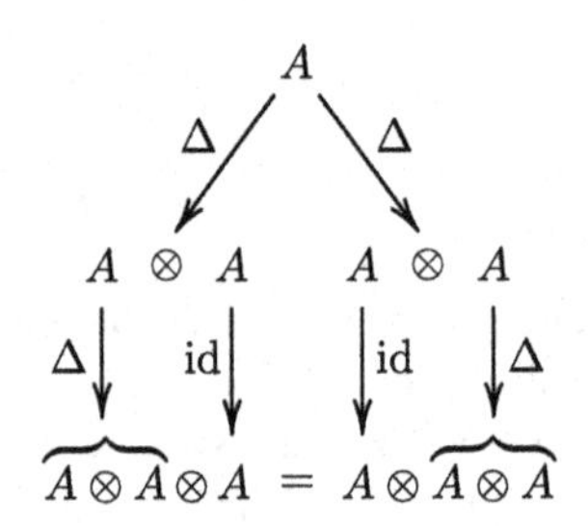

is commutative.

A remarkable property of the definition of a Hopf algebra is its symmetry with respect to the multiplication $\mu: A \otimes_{\mathbb{K}} A \to A$ ($\mu(a \otimes b) = ab$) and comultiplication $\Delta: A \to A \otimes_{\mathbb{K}} A$. This symmetry displays itself in the following fact.

EXERCISE 2. Prove that if A is a Hopf algebra, then the graded dual space $A^* = \bigoplus_m (A^m)^*$ is a Hopf algebra with respect to the multiplication $\Delta^*: A^* \otimes_{\mathbb{K}} A^* \to A^*$ and the comultiplication $\mu^*: A^* \to A^* \otimes_{\mathbb{K}} A^*$.

Details concerning Hopf algebras can be found in the article by Milnor and Moore [59].

Example 1. Cohomology (as well as homology) of an H-space X with coefficients in a field is a Hopf algebra: The multiplication and comultiplication are induced by the diagonal map $X \to X \times X$ and the product map $X \times X \to X$ (plus Künneth's formula).

Example 2. The Steenrod algebra is a Hopf algebra: The comultiplication is defined by the formula

$$\Delta(\beta) = \beta \otimes 1 - 1 \otimes \beta, \ \Delta(P_p^s) = \sum_{s+t=i} P_p^s \otimes P_p^t$$

(in particular, if $p = 2$, then $\Delta(\mathrm{Sq}^i) = \sum_{s+t=i} \mathrm{Sq}^s \otimes \mathrm{Sq}^t$).

B: Modules over Hopf Algebras and Their Tensor Products

If A is a Hopf algebra and B, C are A-modules, then the tensor product $B \otimes_{\mathbb{K}} C$ also has a natural structure of an A-module. Indeed, $B \otimes_{\mathbb{K}} C$ is naturally an $(A \otimes_{\mathbb{K}} A)$-module, and the algebra homomorphism $\Delta: A \to A \otimes_{\mathbb{K}} A$ turns it into an A-module.

Remark. The Hopf algebra structure on the Steenrod algebra is compatible with Künneth's formula. Namely, if X and Y are arbitrary spaces, then $\widetilde{H}^*(X\#Y; \mathbb{Z}_p) = H^*(X; \mathbb{Z}_p) \otimes \widetilde{H}^*(Y; \mathbb{Z}_p)$. Thus, there are two ways to furnish $\widetilde{H}^*(X\#Y; \mathbb{Z}_p)$ with an $\mathbb{A}$-module structure: First, the cohomology of any space has a structure of an $\mathbb{A}$-module; second, it is an $\mathbb{A}$-module, because it is the tensor product of two $\mathbb{A}$-modules. Cartan's formula shows that these two structures are the same.

Proposition. *For any Hopf algebra A, the A-module $A \otimes_{\mathbb{K}} A$ is free.*

Proof. Let $\{a_i\}$ be a $\mathbb{K}$-basis of A composed of homogeneous elements $a_i \in A^{m_i}$ such that $m_i \leq m_j$ for $i \leq j$. We will prove that $\{1 \otimes a_i\}$ is a free system of generators of $A \otimes_{\mathbb{K}} A$. First, let us prove that the system $\{1 \otimes a_i\}$ does generate $A \otimes_{\mathbb{K}} A$. For this, we need to prove that for every j, k, the product $a_j \otimes a_k$ can be expressed as a linear combination of elements $1 \otimes a_i$ with coefficients in A. If $a_j = 1$, it is true. Assume that our statement has been proven for all $a_{j'} \otimes a_{k'}$ with $\deg a_{j'} < \deg a_j$. Then it is also true for $a_j \otimes a_k$, since $a_j \otimes a_k - a_j(1 \otimes a_k) = a_j \otimes a_k - (a_j \otimes 1 + \ldots)(1 \otimes a_k)$ is a linear combination (over $\mathbb{K}$) of products $a_{j'} \otimes a_{k'}$ with $\deg a_{j'} < \deg a_j$. Next, let us prove that the system $\{1 \otimes a_i\}$ is linearly independent over A. Let $\sum_m b_m(1 \otimes a_{i_m}) = 0$ (where $b_m \in A$). Assume that $\deg a_{i_1} \leq \deg a_{i_2} \leq \deg a_{i_3} \leq \ldots$. Then the sum $\sum_m b_m(1 \otimes a_{i_m})$ contains the term $b_1 \otimes a_{i_1}$, which is not cancelled with anything in this sum; this is a contradiction.

Corollary. *If B and C are free A-modules, then $B \otimes_{\mathbb{K}} C$ is also a free A-module.*

Proof. A free A-module is a direct sum of modules isomorphic to A (with a possible grading shift).

C: Exts, Cohomology, and Related Multiplicative Structures

Let A be a Hopf algebra, and let M', M'', N', N'' be A-modules. We will define a multiplication (pairing)

$$\operatorname{Ext}_A^{s',t'}(M',N') \otimes_{\mathbb{K}} \operatorname{Ext}_A^{s'',t''}(M'',N'') \longrightarrow \operatorname{Ext}_A^{s'+s'',t'+t''}(M' \otimes_{\mathbb{K}} M'', N' \otimes_{\mathbb{K}} N'').$$

Let

$$\dots \xrightarrow{\partial_3'} P_2' \xrightarrow{\partial_2'} P_1' \xrightarrow{\partial_1'} P_0' \xrightarrow{\varepsilon'} M' \longrightarrow 0,$$
$$\dots \xrightarrow{\partial_3''} P_2'' \xrightarrow{\partial_2''} P_1'' \xrightarrow{\partial_1''} P_0'' \xrightarrow{\varepsilon''} M'' \longrightarrow 0$$

be free resolutions of the modules M', M''. We consider the following free resolution of the module $M' \otimes_{\mathbb{K}} M''$:

$$\dots \xrightarrow{\partial_3} (P_2' \otimes_{\mathbb{K}} P_0'') \oplus (P_1' \otimes_{\mathbb{K}} P_1'') \oplus (P_0' \otimes_{\mathbb{K}} P_2'') \xrightarrow{\partial_2}$$
$$(P_1' \otimes_{\mathbb{K}} P_0'') \oplus (P_0' \otimes_{\mathbb{K}} P_1'') \xrightarrow{\partial_1} (P_0' \otimes_{\mathbb{K}} P_0'') \xrightarrow{\varepsilon} M' \otimes_{\mathbb{K}} M'' \longrightarrow 0$$

where for $\alpha' \in P'i, \alpha'' \in P_j''$

$$\partial_{p+q}(\alpha \otimes \alpha') = \partial_p'(\alpha') \otimes \alpha'' + (-1)^i \alpha' \otimes \partial_j''(\alpha'')$$

and $\varepsilon = \varepsilon' \otimes \varepsilon''$. There is a natural homomorphism

$$\operatorname{Hom}_A(P_i', N') \otimes_{\mathbb{K}} \operatorname{Hom}_A(P_j'', N'') \to \operatorname{Hom}_A(P_i' \otimes_{\mathbb{K}} P_j'', N' \otimes_{\mathbb{K}} N''),$$

which yields, after the transition to the homologies of the Hom complexes, the promised pairing of Exts.

EXERCISE 3. If $M' = M'' = N' = N'' = \mathbb{K}$ (with the natural structure of A-modules), then the construction described above determines a bigraded ring structure in $\bigoplus_{s,t} \operatorname{Ext}_A^{s,t}(\mathbb{K}, \mathbb{K})$.

Definition. The spaces $\operatorname{Ext}_A^{s,t}(\mathbb{K}, \mathbb{K})$ are called *cohomology spaces of A* and are denoted by $H^{s,t}(A)$.

Thus, we have defined the (bigraded) cohomology algebra of a Hopf algebra.

EXERCISE 4. This is an associative skew-commutative algebra.

If in the construction above $M'' = M \neq \mathbb{K}$, then we get a homomorphism

$$\operatorname{Ext}_A^{s',t'}(\mathbb{K}, \mathbb{K}) \otimes_{\mathbb{K}} \operatorname{Ext}_A^{s'',t''}(M, \mathbb{K}) \to \operatorname{Ext}_A^{s'+s'',t'+t''}(M, \mathbb{K}).$$

Briefly, we can say that for every A-module M, $\mathrm{Ext}_A^{*,*}(M, \mathbb{K})$ is an $H^{*,*}(A)$-module. Furthermore, if $M \to N$ is an A-homomorphism, then the induced map

$$\mathrm{Ext}_A^{*,*}(N, \mathbb{K}) \to \mathrm{Ext}_A^{*,*}(M, \mathbb{K})$$

is an $H_A^{*,*}(A)$-homomorphism.

35.3 The Multiplicative Structure in the Adams Spectral Sequence

Theorem (Adams). *If $X = S^0$, then in the Adams spectral sequence, there arises a multiplication $E_r^{s,t} \otimes E_r^{s',t'} \to E_r^{s+s',t+t'}$ with the following properties.*

(1) *The multiplication is associative and skew-commutative.*
(2) *The multiplication $E_2^{s,t} \otimes E_2^{s',t'} \to E_2^{s+s',t+t'}$ coincides with the multiplication in the cohomology of the Steenrod algebra.*
(3) $d_r(uv) = (d_r u)v + (-1)^{s+t} u(d_r v)$ *(where $u \in E_r^{s,t}$).*
(4) *The multiplication in E_{r+1} is induced by the multiplication in E_r.*
(5) *The multiplication in E_∞ is adjoint to the composition multiplication $\pi_k^S(S^0) \otimes \pi_\ell^S(S^0) \to \pi_{k+\ell}^S(S^0)$.*

Proof. We will prove a more general statement, and for this purpose we begin with considering two arbitrary spaces, X' and X''. Let

$$\dots \longrightarrow B_2' \longrightarrow B_1' \longrightarrow B_0' \longrightarrow \widetilde{H}^*(X'),$$
$$\dots \longrightarrow B_2'' \longrightarrow B_1'' \longrightarrow B_0'' \longrightarrow \widetilde{H}^*(X'')$$

be free $\mathbb{A}$-resolutions of the $\mathbb{A}$-modules $\widetilde{H}^*(X')$ and $\widetilde{H}^*(X'')$. Let

$$\dots \longrightarrow X'(2) \longrightarrow X'(1) \longrightarrow X'(0) = \Sigma^{N'} X',$$
$$\dots \longrightarrow X''(2) \longrightarrow X''(1) \longrightarrow X''(0) = \Sigma^{N''} X''$$

be the corresponding Adams filtrations. Let us construct a filtration in $(\Sigma^{N'} X')\#(\Sigma^{N''} X'') = \Sigma^{N'+N''}(X'\#X'')$. Put

$$X(n) = \bigcup_{i+j=n} (X'(i)\#X''(j)) \text{ and } Y_n = X(n)/X(n+1);$$

obviously,

$$Y_n = \bigvee\nolimits_{i+j=n} [(X'(i)/X'(i+1))\#(X''(j)/X''(j+1))].$$

But the spaces $X'(i)/X'(i+1)$ and $X''(j)/X''(j+1)$ are homotopy equivalent, in stable dimensions, to the products Y'_i and Y''_j of the Eilenberg–MacLane spaces used in the construction of the Adams filtrations $\{X'(i)\}$ and $\{X''(j)\}$. Hence, $\widetilde{H}^*(Y_n) = \bigoplus_{i+j=n}[\widetilde{H}^*(Y'_i) \otimes \widetilde{H}^*(Y''_j)]$. We see that the filtration

$$\dots \longrightarrow X(2) \longrightarrow X(1) \longrightarrow X(0) = \Sigma^{N'+N''}(X'\#X'')$$

is the Adams filtration corresponding to the resolution

$$\begin{aligned}\dots \to (B'_2 \otimes B''_0) \oplus (B'_1 \otimes B''_1) \oplus (B'_0 \otimes B''_2) \to (B'_1 \otimes B''_0) \oplus (B'_0 \otimes B''_1)\\ \to (B'_0 \otimes B''_0) \to \widetilde{H}^*(X'\#X'') \to 0\end{aligned}$$

of $\widetilde{H}^*(X'\#X'')$. Consider the Adams spectral sequences for the spaces $X', X'', X'\#X''$ corresponding to these resolutions and filtrations. The multiplication

$$E_r^{s',t'}(X') \otimes E_r^{s'',t''}(X'') \to E_r^{s'+s'',t'+t''}(X'\#X'')$$

is defined in the following way. Let

$$\begin{aligned}\alpha' \in E_r^{s',t'}(X') = \operatorname{Im}[\pi_{N'+t'-s'}(X'(s'), X'(s'+r))\\ \to \pi_{N'+t'-s'}(X'(s'+1-r), X'(s'+1))],\\ \alpha'' \in E_r^{s'',t''}(X'') = \operatorname{Im}[\pi_{N''+t''-s''}(X''(s''), X''(s''+r))\\ \to \pi_{N''+t''-s''}(X''(s''+1-r), X''(s''+1))].\end{aligned}$$

Take elements of the homotopy groups $\pi_{N'+t'-s'}(X'(s'), X'(s'+r))$ and $\pi_{N''+t''-s''}(X''(s''), X''(s''+r))$ whose images are α' and α'' and choose spheroids, that is, maps of the cubes $I^{N'+t'-s'}$ and $I^{N''+t''-s''}$ into X' and X'' such that the cubes are mapped into $X'(s')$ and $X''(s'')$ and their boundaries are mapped into $X'(s''+r)$ and $X''(s''+r)$. Then we form the "product" of these maps, that is, a map

$$I^{N'+t'-s'} \times I^{N''+t''-s''} = I^{N'+N''+t'+t''-s'-s''} \to X' \times X'' \to X'\#X''.$$

This map takes the cube $I^{N'+N''+t'+t''-s'-s''}$ into $X'(s')\#X''(s'') \subset X(s+s')$ and its boundary into $[X'(s'+r)\#X''(s'')] \bigcup [X'(s')\#X(s''+r)]$ (see Fig. 125).

Hence, this map is a spheroid of a pair $(X(s'+s''), X(s+s''+r))$ and determines an element of

$$\pi_{N'+N''+t+t''-s'-s''}(X(s'+s''), X(s+s''+r)).$$

The image of this element in

$$\pi_{N'+N''+t+t''-s'-s''}(X(s'+s''+1-r), X(s+s''+1))$$

is an element of $E_r^{s'+s'',t'+t''}(X'\#X'')$, which is taken for the product $\alpha'\alpha''$.

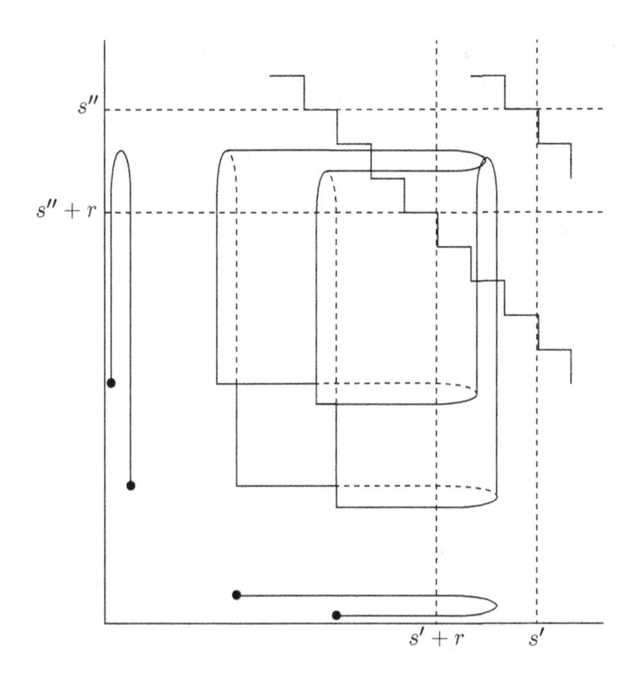

Fig. 125 Product of relative spheroids

The product rule

$$d_r^{s'+s'',t'+t''}(\alpha'\alpha'') = \left(d_r^{s',t'}\alpha'\right)\cdot\alpha'' + (-1)^{s'+t'}\alpha'\cdot\left(d_r^{s'',t''}\alpha''\right)$$

holds for this multiplication; for $r = 2$ the multiplication coincides with the multiplication

$$\begin{aligned}\operatorname{Ext}_{\mathbb{A}}^{s',t'}(\widetilde{H}^*(X'),\mathbb{Z}_p)\otimes\operatorname{Ext}_{\mathbb{A}}^{s'',t''}(\widetilde{H}^*(X''),\mathbb{Z}_p))\\ \to\operatorname{Ext}_{\mathbb{A}}^{s'+s'',t'+t''}(\widetilde{H}^*(X'\#X''),\mathbb{Z}_p)\end{aligned}$$

defined above (Sect. 35.2.C) and for $r = \infty$ turns into a multiplication

$$E_\infty^{**}(X')\otimes E_\infty^{**}(X'')\to E_\infty^{**}(X'\#X'')$$

adjoint to the multiplication $\pi_k^S(X')\otimes\pi_\ell^S(X'')\to\pi_{k+l}^S(X'\#X'')$.

We leave to the reader the verification of all these statements.

If $X' = X'' = S^0$, then the statement proved becomes that of the Adams theorem.

If $X' = S^0$ and $X'' = X$ is arbitrary, then our construction defines for $E_r^{**}(X)$ a structure of an $E_r^{**}(S^0)$-module compatible with differentials. For $r = 2$, this structure coincides with the structure of an $H^{**}(\mathbb{A})$-module in $\operatorname{Ext}_{\mathbb{A}}^{**}(\widetilde{H}^*(X),\mathbb{Z}_p)$ (see Sect. 35.2.C), and for $r = \infty$ it is adjoint to the structure of a $\pi_*^S(S^0)$-module in $\pi_*^S(X)$ defined by the composition product.

Lecture 36 An Application of the Adams Spectral Sequence to Stable Homotopy Groups of Spheres

We begin with a computation of the E_2-term of the Adams spectral sequence for the case of 2-components of stable homotopy groups of spheres, that is, the cohomology of the Steenrod algebra mod 2.

36.1 The Additive Structure of $\mathbf{E_2}$

Let us describe a free $\mathbb{A} = \mathbb{A}_2$-resolution

$$\dots \to B_2 \to B_1 \to B_0 \to \mathbb{Z}_2$$

of the $\mathbb{A}$-module $\widetilde{H}^*(S^0; \mathbb{Z}_2) = \mathbb{Z}_2$. For B_0 we can take $\mathbb{A}$. The epimorphism $B_0 \to \mathbb{Z}_2$ isomorphically maps $\mathbb{A}^0 = \mathbb{Z}_2$ onto $\mathbb{Z}_2$ and is zero on $\bigoplus_{q>0} \mathbb{A}^q = \widetilde{\mathbb{A}}$. To define B_1, we need a system of generators in the $\mathbb{A}$-module $\widetilde{\mathbb{A}}$. But we know such a system of generators: It is $\{\alpha_k = \mathrm{Sq}^{2^{k-1}}\}$. Thus, B_1 is a free $\mathbb{A}$-module whose generators have the same degrees as α_k; we denote them again as α_k; so the map $B_1 \to B_0$ is defined by the formula $\alpha_k \mapsto \mathrm{Sq}^{2^{k-1}}$. The (free) generators of B_2 correspond to (and have the same notation as) the generators β_k of the $\mathrm{Ker}(B_1 \to B_0)$, that is, to the generators of the module of relations between $\mathrm{Sq}^1, \mathrm{Sq}^2, \mathrm{Sq}^4, \mathrm{Sq}^8, \dots$ over $\mathbb{A}$. The space $\mathbb{A}^1$ is one-dimensional and is spanned by $\alpha_1 = \mathrm{Sq}^1$; no relations, so far. The space $\mathbb{A}^2$ is one-dimensional and is spanned by $\alpha_2 = \mathrm{Sq}^2$; the first relation arises: $\mathrm{Sq}^1\alpha_1 = 0$. The space $\mathbb{A}^3$ is two-dimensional with the basis $\mathrm{Sq}^3, \mathrm{Sq}^2\mathrm{Sq}^1$, that is, $\mathrm{Sq}^2\alpha_1$ and $\mathrm{Sq}^1\alpha_2$. There are no relations. The space $\mathbb{A}^4$ is also two-dimensional, with the basis $\mathrm{Sq}^4 = \alpha_3$, $\mathrm{Sq}^3\mathrm{Sq}^1 = \mathrm{Sq}^3\alpha^1$. But in the free $\mathbb{A}$-module B_1 there are two more additive generators: $\mathrm{Sq}^2\alpha_2$ and $\mathrm{Sq}^2\mathrm{Sq}^1\alpha_1$. In $\mathbb{A}^4$, they must be linear combinations of elements of our basis, and they are $\mathrm{Sq}^2\alpha_2 = \mathrm{Sq}^2\mathrm{Sq}^2 = \mathrm{Sq}^3\mathrm{Sq}^1 = \mathrm{Sq}^3\alpha_1$ and $\mathrm{Sq}^2\mathrm{Sq}^1\alpha_1 = \mathrm{Sq}^2\mathrm{Sq}^1\mathrm{Sq}^1 = 0$. Thus, two more relations arise: $\mathrm{Sq}^2\alpha_2 = \mathrm{Sq}^3\alpha_1$ and $\mathrm{Sq}^2\mathrm{Sq}^1\alpha_1 = 0$. This work can be continued indefinitely; in the following table, we show additive generators of $\widetilde{\mathbb{A}}$ and additive generators of the module of relations in $\widetilde{\mathbb{A}}$ [that is, in $\mathrm{Ker}(B_1 \to B_0)$] to degree 12; these relations are determined with the help of Adem's formulas.

Deg	Generators	Relations
1	α_1	
2	α_2	$\underline{\mathrm{Sq}^1\alpha_1 = 0}$
3	$\mathrm{Sq}^1\alpha_2, \mathrm{Sq}^2\alpha^1$	
4	$\alpha_3, \mathrm{Sq}^3\alpha_1$	$\underline{\mathrm{Sq}^2\alpha_2 = \mathrm{Sq}^3\alpha_1}$, $\mathrm{Sq}^2\mathrm{Sq}^1\alpha_1 = 0$
5	$\mathrm{Sq}^1\alpha_3, \mathrm{Sq}^4\alpha^1$	$\underline{\mathrm{Sq}^2\mathrm{Sq}^1\alpha_2 = \mathrm{Sq}^1\alpha_3 + \mathrm{Sq}^4\alpha_1}$, $\mathrm{Sq}^3\alpha_2 = 0$, $\mathrm{Sq}^3\mathrm{Sq}^1\alpha_1 = 0$

(continued)

Deg	Generators	Relations
6	$\mathrm{Sq}^2\,\alpha_3,\ \mathrm{Sq}^4\,\alpha_2,$ $\mathrm{Sq}^5\,\alpha_1$	$\mathrm{Sq}^5\,\alpha_1 = \mathrm{Sq}^3\,\mathrm{Sq}^1\,\alpha_2,$ $\mathrm{Sq}^4\,\mathrm{Sq}^1\,\alpha_1 = 0$
7	$\mathrm{Sq}^3\,\alpha_3,\ \mathrm{Sq}^5\,\alpha_2,$ $\mathrm{Sq}^4\,\mathrm{Sq}^2\,\alpha_1,\ \mathrm{Sq}^6\,\alpha_1$	$\mathrm{Sq}^6\,\alpha_1 = \mathrm{Sq}^2\,\mathrm{Sq}^1\,\alpha_3,$ $\mathrm{Sq}^4\,\mathrm{Sq}^1\,\alpha_2 = \mathrm{Sq}^5\,\alpha_2,\ \mathrm{Sq}^5\,\mathrm{Sq}^1\,\alpha_1 = 0$
8	$\alpha_4,\ \mathrm{Sq}^6\,\alpha_2$ $\mathrm{Sq}^5\,\mathrm{Sq}^2\,\alpha_1,\ \mathrm{Sq}^7\,\alpha_1$	$\underline{\mathrm{Sq}^4\,\alpha_3 = \mathrm{Sq}^7\,\alpha_1 + \mathrm{Sq}^6\,\alpha_2,}$ $\mathrm{Sq}^3\,\mathrm{Sq}^1\,\alpha_3 = \mathrm{Sq}^7\,\alpha_1, \mathrm{Sq}^5\,\mathrm{Sq}^1\,\alpha_2 = 0,$ $\mathrm{Sq}^5\,\mathrm{Sq}^2\,\alpha_1 = \mathrm{Sq}^4\,\mathrm{Sq}^2\,\alpha_2,$ $\mathrm{Sq}^4\,\mathrm{Sq}^2\,\mathrm{Sq}^1\,\alpha_1 = 0,\ \mathrm{Sq}^6\,\mathrm{Sq}^1\,\alpha_1 = 0$
9	$\mathrm{Sq}^1\,\alpha_4,$ $\mathrm{Sq}^6\,\mathrm{Sq}^1\,\alpha_2,$ $\mathrm{Sq}^7\,\alpha_2,\ \mathrm{Sq}^8\,\alpha_1$	$\underline{\mathrm{Sq}^4\,\mathrm{Sq}^1\,\alpha_3 = \mathrm{Sq}^1\,\alpha_4 + \mathrm{Sq}^7\,\alpha_2 + \mathrm{Sq}^8\,\alpha_1,}$ $\mathrm{Sq}^4\,\mathrm{Sq}^1\,\alpha_3 + \mathrm{Sq}^4\,\mathrm{Sq}^2\,\mathrm{Sq}^1\,\alpha_2,$ $+\,\mathrm{Sq}^6\,\mathrm{Sq}^1\,\alpha_1 = 0,$ $\mathrm{Sq}^5\,\alpha_3 = \mathrm{Sq}^7\,\alpha_2,\ \mathrm{Sq}^5\,\mathrm{Sq}^2\,\alpha_2 = 0,$ $\mathrm{Sq}^5\,\mathrm{Sq}^2\,\mathrm{Sq}^1\,\alpha_1 = 0,\ \mathrm{Sq}^7\,\alpha_1 = 0$
10	$\mathrm{Sq}^4\,\mathrm{Sq}^2\,\alpha_3,$ $\mathrm{Sq}^7\,\mathrm{Sq}^1\,\alpha_2$ $\mathrm{Sq}^7\,\mathrm{Sq}^2\,\alpha_1,$ $\mathrm{Sq}^8\,\alpha_2,\ \mathrm{Sq}^9\,\alpha_1$	$\underline{\mathrm{Sq}^2\,\alpha_4 = \mathrm{Sq}^4\,\mathrm{Sq}^2\,\alpha_3 + \mathrm{Sq}^8\,\alpha_2 + \mathrm{Sq}^7\,\mathrm{Sq}^2\,\alpha_1,}$ $\mathrm{Sq}^6\,\alpha_3 = \mathrm{Sq}^7\,\mathrm{Sq}^1\,\alpha_2,\ \mathrm{Sq}^5\,\mathrm{Sq}^1\,\alpha_3 = \mathrm{Sq}^9\,\alpha_1,$ $\mathrm{Sq}^6\,\mathrm{Sq}^2\,\alpha_2 = \mathrm{Sq}^6\,\mathrm{Sq}^3\,\alpha_1,\ \mathrm{Sq}^6\,\mathrm{Sq}^2\,\mathrm{Sq}^1\,\alpha_1 = 0,$ $\mathrm{Sq}^8\,\mathrm{Sq}^1\,\alpha_1 = 0$
11	$\mathrm{Sq}^3\,\alpha_4,\ \mathrm{Sq}^9\,\alpha_2,$ $\mathrm{Sq}^7\,\mathrm{Sq}^3\,\alpha_1$ $\mathrm{Sq}^8\,\mathrm{Sq}^2\,\alpha_1,$ $\mathrm{Sq}^{10}\,\alpha_1$	$\mathrm{Sq}^3\,\alpha_4 = \mathrm{Sq}^5\,\mathrm{Sq}^2\,\alpha_3 + \mathrm{Sq}^9\,\alpha_2,$ $\mathrm{Sq}^4\,\mathrm{Sq}^2\,\mathrm{Sq}^1\,\alpha_3 = \mathrm{Sq}^8\,\mathrm{Sq}^2\,\alpha_1 + \mathrm{Sq}^{10}\,\alpha_1$ $\mathrm{Sq}^6\,\mathrm{Sq}^1\,\alpha_3 = \mathrm{Sq}^8\,\mathrm{Sq}^1\,\alpha_2 + \mathrm{Sq}^9\,\alpha_2,$ $\mathrm{Sq}^6\,\mathrm{Sq}^2\,\mathrm{Sq}^1\,\alpha_2 + \mathrm{Sq}^9\,\alpha_2 + \mathrm{Sq}^8\,\mathrm{Sq}^1\,\alpha_2$ $= \mathrm{Sq}^7\,\mathrm{Sq}^3\,\alpha_1,\ \mathrm{Sq}^7\,\alpha_3 = 0,$ $\mathrm{Sq}^6\,\mathrm{Sq}^3\,\alpha_2 = 0,\ \mathrm{Sq}^7\,\mathrm{Sq}^2\,\alpha_2 = \mathrm{Sq}^7\,\mathrm{Sq}^3\,\alpha_1,$ $\mathrm{Sq}^6\,\mathrm{Sq}^3\,\mathrm{Sq}^1\,\alpha_1 = 0,\ \mathrm{Sq}^7\,\mathrm{Sq}^2\,\mathrm{Sq}^1\,\alpha_1 = 0,$ $\mathrm{Sq}^9\,\mathrm{Sq}^1\,\alpha_1 = 0,\ \mathrm{Sq}^2\,\mathrm{Sq}^1\,\alpha_4 = \mathrm{Sq}^{10}\,\alpha_1$
12	$\mathrm{Sq}^4\,\alpha_4,\ \mathrm{Sq}^8\,\alpha_3,$ $\mathrm{Sq}^9\,\mathrm{Sq}^1\,\alpha_2,$ $\mathrm{Sq}^{10}\,\alpha_2,$ $\mathrm{Sq}^8\,\mathrm{Sq}^3\,\alpha_1,$ $\mathrm{Sq}^9\,\mathrm{Sq}^2\,\alpha_1,$ $\mathrm{Sq}^{11}\,\alpha_1$	$\mathrm{Sq}^3\,\mathrm{Sq}^1\,\alpha_4 = \mathrm{Sq}^{11}\,\alpha_1,$ $\mathrm{Sq}^7\,\mathrm{Sq}^1\,\alpha_3 = \mathrm{Sq}^9\,\mathrm{Sq}^1\,\alpha_2,$ $\mathrm{Sq}^5\,\mathrm{Sq}^2\,\mathrm{Sq}^1\,\alpha_3 = \mathrm{Sq}^{11}\,\alpha_1 + \mathrm{Sq}^9\,\mathrm{Sq}^1\,\alpha_2,$ $\mathrm{Sq}^6\,\mathrm{Sq}^2\,\alpha_3 = \mathrm{Sq}^{11}\,\alpha_1 + \mathrm{Sq}^9\,\mathrm{Sq}^2\,\alpha_1 + \mathrm{Sq}^{10}\,\alpha_2$ $+\,\mathrm{Sq}^8\,\mathrm{Sq}^3\,\alpha_1 + \mathrm{Sq}^9\,\mathrm{Sq}_1\,\alpha_2,$ $\mathrm{Sq}^6\,\mathrm{Sq}^3\,\mathrm{Sq}^1\,\alpha_2 = \mathrm{Sq}^9\,\mathrm{Sq}^2\,\alpha_1 + \mathrm{Sq}^8\,\mathrm{Sq}^3\,\alpha_1,$ $\mathrm{Sq}^7\,\mathrm{Sq}^2\,\mathrm{Sq}^1\,\alpha_2 = \mathrm{Sq}^9\,\mathrm{Sq}^1\,\alpha_2,\ \mathrm{Sq}^7\,\mathrm{Sq}^3\,\alpha_2 = 0,$ $\mathrm{Sq}^8\,\mathrm{Sq}^2\,\alpha_2 = \mathrm{Sq}^8\,\mathrm{Sq}^3\,\alpha_1,\ \mathrm{Sq}^7\,\mathrm{Sq}^3\,\mathrm{Sq}^1\,\alpha_1 = 0,$ $\mathrm{Sq}^8\,\mathrm{Sq}^2\,\mathrm{Sq}^1\,\alpha_1 = 0,\ \mathrm{Sq}^{10}\,\mathrm{Sq}^1\,\alpha_1 = 0$

The generators of the module $\mathrm{Ker}(B_1 \to B_0)$ are underlined in the right column of the last diagram. Within our range of degrees there are six of them (of degrees 2, 4, 5, 8, 9, and 10). We denote them as $\beta_1, \dots, \beta_6$; thus, the homomorphism $B_2 \to B_1$ acts by the formulas

$$\begin{aligned}
&\beta_1 \mapsto \mathrm{Sq}^1\,\alpha_1,\\
&\beta_2 \mapsto \mathrm{Sq}^2\,\alpha_2 + \mathrm{Sq}^3\,\alpha_1,\\
&\beta_3 \mapsto \mathrm{Sq}^2\,\mathrm{Sq}^1\,\alpha_2 + \mathrm{Sq}^1\,\alpha_3 + \mathrm{Sq}^4\,\alpha_1,\\
&\beta_4 \mapsto \mathrm{Sq}^4\,\alpha_3 + \mathrm{Sq}^7\,\alpha_1 + \mathrm{Sq}^6\,\alpha_2,\\
&\beta_5 \mapsto \mathrm{Sq}^4\,\mathrm{Sq}^1\,\alpha_3 + \mathrm{Sq}^1\,\alpha_4 + \mathrm{Sq}^8\,\alpha_1 + \mathrm{Sq}^7\,\alpha_2,\\
&\beta_6 \mapsto \mathrm{Sq}^2\,\alpha_4 + \mathrm{Sq}^4\,\mathrm{Sq}^2\,\alpha_3 + \mathrm{Sq}^8\,\alpha_2 + \mathrm{Sq}^7\,\mathrm{Sq}^2\,\alpha_1.
\end{aligned}$$

The next table shows the bases in the module $\mathrm{Ker}(B_1 \to B_0)$ and in the module of relations in this module [which is the same as $\mathrm{Ker}(B_2 \to B_1)$]. Thus, the left column of this table is the same as the right column in the previous column, only everything is expressed in terms of β_k.

Deg	Generators	Relations
2	β_1	
3		$\mathrm{Sq}^1\,\beta_1 = 0$
4	β_2, $\mathrm{Sq}^2\,\beta_1$	
5	β_3, $\mathrm{Sq}^3\,\beta_1$, $\mathrm{Sq}^1\,\beta_2$	$\mathrm{Sq}^2\,\mathrm{Sq}^1\,\beta_1 = 0$
6	$\mathrm{Sq}^4\,\beta_1$, $\mathrm{Sq}^1\,\beta_3$	$\mathrm{Sq}^2\,\beta_2 = \mathrm{Sq}^4\,\beta_1 + \mathrm{Sq}^1\,\beta_3$, $\mathrm{Sq}^3\,\mathrm{Sq}^1\,\beta_1 = 0$
7	$\mathrm{Sq}^5\,\beta_1$, $\mathrm{Sq}^2\,\beta_3$, $\mathrm{Sq}^2\,\mathrm{Sq}^1\,\beta_2$	$\mathrm{Sq}^3\,\beta_2 = \mathrm{Sq}^5\,\beta_1$, $\mathrm{Sq}^4\,\mathrm{Sq}^1\,\beta_1 = 0$
8	β_4, $\mathrm{Sq}^6\,\beta_1$, $\mathrm{Sq}^4\,\beta_2$, $\mathrm{Sq}^4\,\mathrm{Sq}^2\,\beta_1$, $\mathrm{Sq}^3\,\mathrm{Sq}^1\,\beta_2$, $\mathrm{Sq}^3\,\beta_3$	$\mathrm{Sq}^2\,\mathrm{Sq}^1\,\beta_3 = \mathrm{Sq}^3\,\mathrm{Sq}^1\,\beta_2 + \mathrm{Sq}^6\,\beta_1$, $\mathrm{Sq}^5\,\mathrm{Sq}^1\,\beta_1 = 0$
9	β_5, $\mathrm{Sq}^7\,\beta_1$, $\mathrm{Sq}^5\,\mathrm{Sq}^2\,\beta_1$, $\mathrm{Sq}^5\,\beta_2$, $\mathrm{Sq}^2\,\beta_4$,	$\mathrm{Sq}^7\,\beta_1 = \mathrm{Sq}^3\,\mathrm{Sq}^1\,\beta_3$, $\mathrm{Sq}^5\,\beta_2 = \mathrm{Sq}^4\,\mathrm{Sq}^1\,\beta_2$, $\mathrm{Sq}^4\,\mathrm{Sq}^2\,\mathrm{Sq}^1\,\beta_1 = 0$, $\mathrm{Sq}^6\,\mathrm{Sq}^1\,\beta_1 = 0$

(continued)

Deg	Generators	Relations
10	β_6, $Sq^8\,\beta_1$, $Sq^5\,\beta_2$, $Sq^2\,\beta_4$, $Sq^6\,Sq^2\,\beta_1$, $Sq^4\,Sq^2\,\beta_2$.	$Sq^1\,\beta_5 = Sq^5\,\beta_3 + Sq^4\,Sq^1\,\beta_3 + Sq^8\,\beta_1$, $Sq^4\,Sq^1\,\beta_3 = Sq^4\,Sq^2\,\beta_2 + Sq^6\,Sq^2\,\beta_1$, $Sq^5\,Sq^1\,\beta_2 = 0$, $Sq^5\,Sq^2\,\beta_1 = 0$, $Sq^7\,Sq^1\,\beta_1 = 0$
11	$Sq^9\,\beta_1$, $Sq^7\,Sq^2\,\beta_1$, $Sq^6\,Sq^3\,\beta_1$, $Sq^7\,\beta_2$, $Sq^6\,Sq^1\,\beta_2$, $Sq^2\,Sq^1\,\beta_4$, $Sq^6\,\beta_3$, $Sq^4\,Sq^2\,\beta_3$, $Sq^1\,\beta_6$, $Sq^3\,\beta_4$, $Sq^2\,\beta_5$,	$Sq^2\,Sq^1\,\beta_4 = Sq^4\,Sq^2\,Sq^1\,\beta_2$, $+ Sq^7\,\beta_2 + Sq^6\,\beta_3$, $Sq^5\,Sq^2\,\beta_2 = Sq^9\,\beta_1 + Sq^7\,Sq^2\,\beta_1$, $Sq^5\,Sq^1\,\beta_3 = Sq^9\,\beta_1$, $Sq^6\,Sq^2\,Sq^1 = 0$, $Sq^8\,Sq^1\,\beta_1 = 0$
12	$Sq^{10}\,\beta_1$, $Sq^8\,Sq^2\,\beta_1$, $Sq^7\,Sq^3\,\beta_1$, $Sq^8\,\beta_2$, $Sq^5\,Sq^2\,Sq^1\,\beta_2$, $Sq^7\,Sq^1\,\beta_2$, $Sq^6\,Sq^1\,\beta_3$, $Sq^3\,Sq^1\,\beta_4$, $Sq^3\,\beta_5$, $Sq^2\,\beta_6$, $Sq^5\,Sq^2\,\beta_3$.	$Sq^4\,\beta^4 = Sq^2\,\beta_6 + Sq^3\,\beta_5$, $+ Sq^3\,Sq^1\,\beta_4 + Sq^8\,\beta_2$, $Sq^4\,Sq^2\,Sq^1\,\beta_3 = Sq^5\,Sq^2\,Sq^1\,\beta_2$ $+ Sq^{10}\,\beta_1 + Sq^8\,Sq^2\,\beta_1$, $Sq^7\,\beta_3 = Sq^5\,Sq^2\,Sq^1\,\beta_2 + Sq^3\,Sq^1\,\beta_4$, $Sq^6\,Sq^2\,\beta_2 = Sq^6\,Sq^1\,\beta_3 + Sq^7\,Sq^3\,\beta_1$, $Sq^2\,Sq^1\,\beta_5 = Sq^{10}\,\beta_1$, $Sq^6\,Sq^3\,Sq^1\,\beta_1 = 0$, $Sq^7\,Sq^2\,Sq^1\,\beta_1 = 0$, $Sq^9\,Sq^1\,\beta_1 = 0$

Within our range of degrees, the module $\mathrm{Ker}(B_2 \to B_1)$ has five generators $\gamma_1, \ldots, \gamma_5$ (underlined in the right column of the last diagram), of degrees 4, 6, 10, 11, and 12. We attribute the same notations to the free generators of B_3. Thus, the homomorphism $B_3 \to B_2$, in our degrees, acts by the formulas

$$\begin{aligned}
&\gamma_1 \mapsto Sq^1\,\beta_1,\\
&\gamma_2 \mapsto Sq^2\,\beta_2 + Sq^1\,\beta_3 + Sq^4\,\beta_1,\\
&\gamma_3 \mapsto Sq^1\,\beta_5 + Sq^5\,\beta_3 + Sq^8\,\beta_1 + Sq^4\,Sq^1\,\beta_3,\\
&\gamma_4 \mapsto Sq^6\,\beta_3 + Sq^4\,Sq^2\,Sq^1\,\beta_2 + Sq^7\,\beta_2 + Sq^2\,Sq^1\,\beta_4,\\
&\gamma_5 \mapsto Sq^2\,\beta_6 + Sq^3\,\beta_5 + Sq^3\,Sq^1\,\beta_4 + Sq^4\,\beta_4 + Sq^8\,\beta_2.
\end{aligned}$$

The next table shows the bases in the module $\mathrm{Ker}(B_2 \to B_1)$ (expressed in γ_k) and in the module of relations in this module, which is the same as $\mathrm{Ker}(B_3 \to B_2)$.

Deg	Generators	Relations
3	γ_1	
4		$\underline{\mathrm{Sq}^1\,\gamma_1 = 0}$
5	$\mathrm{Sq}^2\,\gamma_1$	
6	$\gamma_2,\ \mathrm{Sq}^3\,\gamma_1$	$\mathrm{Sq}^2\,\mathrm{Sq}^1\,\gamma_1 = 0$
7	$\mathrm{Sq}^1\,\gamma_2,\ \mathrm{Sq}^4\,\gamma_1$	$\mathrm{Sq}^3\,\mathrm{Sq}^1\,\gamma_1 = 0$
8	$\mathrm{Sq}^2\,\gamma_2,\ \mathrm{Sq}^5\,\gamma_1$	$\mathrm{Sq}^4\,\mathrm{Sq}^1\,\gamma_1 = 0$
9	$\mathrm{Sq}^2\,\mathrm{Sq}^1\,\gamma_2,\ \mathrm{Sq}^3\,\gamma_2,$ $\mathrm{Sq}^6\,\gamma_1\ \mathrm{Sq}^4\,\mathrm{Sq}^2\,\gamma_1.$	$\mathrm{Sq}^5\,\mathrm{Sq}^1\,\gamma_1 = 0$
10	$\gamma_3,\ \mathrm{Sq}^3\,\mathrm{Sq}^1\,\gamma_2,\ \mathrm{Sq}^4\,\gamma_2,$ $\mathrm{Sq}^7\,\gamma_1,\ \mathrm{Sq}^5\,\mathrm{Sq}^2\,\gamma_1.$	$\mathrm{Sq}^6\,\mathrm{Sq}^1\,\gamma_1 = 0,\ \mathrm{Sq}^4\,\mathrm{Sq}^2\,\mathrm{Sq}^1\,\gamma_1 = 0$
11	$\gamma_4,\ \mathrm{Sq}^1\,\gamma_3,\ \mathrm{Sq}^5\,\gamma_2,$ $\mathrm{Sq}^6\,\mathrm{Sq}^2\,\gamma_1,\ \mathrm{Sq}^8\,\gamma_1.$	$\underline{\mathrm{Sq}^4\,\mathrm{Sq}^1\,\gamma_2 = \mathrm{Sq}^1\,\gamma_3,}$ $\underline{+\,\mathrm{Sq}^5\,\gamma_2 + \mathrm{Sq}^8\,\gamma_1,}$ $\mathrm{Sq}^7\,\mathrm{Sq}^1\,\gamma_1 = 0,\ \mathrm{Sq}^5\,\mathrm{Sq}^2\,\mathrm{Sq}^1\,\gamma_1 = 0$
12	$\gamma_5,\ \mathrm{Sq}^1\,\gamma_4,\ \mathrm{Sq}^2\,\gamma_3,$ $\mathrm{Sq}^4\,\mathrm{Sq}^2\,\gamma_2,\ \mathrm{Sq}^6\,\gamma_2,$ $\mathrm{Sq}^6\,\mathrm{Sq}^3\,\gamma_1,\ \mathrm{Sq}^7\,\mathrm{Sq}^2\,\gamma_1,$ $\mathrm{Sq}^9\,\gamma_1$	$\mathrm{Sq}^5\,\mathrm{Sq}^1\,\gamma_2 = \mathrm{Sq}^0\,\gamma_1,$ $\mathrm{Sq}^6\,\mathrm{Sq}^2\,\mathrm{Sq}^1\,\gamma_1 = 0,\ \mathrm{Sq}^8\,\mathrm{Sq}^1\,\gamma_1 = 0$

We see that the module $\mathrm{Ker}(B_3 \to B_2)$ has two generators (underlined in the preceding table), of degrees 4 and 11. We denote them as δ_1 and δ_2 and use the same notation for the free generators of the module B_4. The map $B_4 \to B_3$ (within our range of degrees) acts according to the formulas

$$\begin{aligned}
&\delta_1 \mapsto \mathrm{Sq}^1\,\gamma_1,\\
&\delta_2 \mapsto \mathrm{Sq}^1\,\gamma_3 + \mathrm{Sq}^4\,\mathrm{Sq}^1\,\gamma_2 + \mathrm{Sq}^5\,\gamma_2 + \mathrm{Sq}^8\,\gamma_1.
\end{aligned}$$

Furthermore, it is clear that, up to degree 12, all relations between δ_1 and δ_2 have the form $[\ldots]\,\mathrm{Sq}^1\,\delta_1 = 0$, and so, up to degree 12, the module $\mathrm{Ker}(B_4 \to B_3)$ has only one generator, $\mathrm{Sq}^1\,\delta_1$, and its degree is 5. Thus, the free module B_5 has (within our range of degrees) only a generator of degree 5. In a similar way, free modules $B_6, \ldots, B_{12}$ have, in degrees not exceeding 12, one generator each; the degrees of these generators are $6, \ldots, 12$. The maps $B_k \to B_{k-1}$, $k = 6, \ldots, 12$ take each of these generators into Sq^1 of the previous generator. The modules B_k with $k > 12$ do not have any nonzero components of degree ≤ 12.

As we have shown, $\mathrm{Ext}^s_{\mathbb{A}}(\mathbb{Z}_2, \mathbb{Z}_2) = \mathrm{Hom}_{\mathbb{A}}(B_s, \mathbb{Z}_2)$, so we know all the additive generators of the cohomology $H^{s,t}(\mathbb{A})$ of the Steenrod algebra for $t \leq 12$. In other words, we know the additive structure of the E_2-term of the Adams spectral sequence for $t \leq 12$ (Fig. 126).

					*							
				δ_1							δ_2	
			γ_1			γ_2				γ_3	γ_4	γ_5
		β_1		β_2	β_3			β_4	β_5	β_6		
	α_1	α_2		α_3				α_4				
1												

Fig. 126 The E_2-term of the Adams spectral sequence for S^0

Now, we make the following general observation. Obviously, the minimal degree of a relation in the $\mathbb{A}$-module $\operatorname{Ker}(B_k \to B_{k-1})$ is at least one more than the minimal degree of a generator of this module. This shows that the minimal degree of a generator of B_k exceeds, at least by one, the minimal degree of a generator of B_{k-1}. This observation, however, gives nothing for our spectral sequence: We already know that the module B_k has a generator of degree k and has no generators of smaller degrees. In particular, we cannot guarantee that even the diagonal $t-s=1$ does not contain infinitely many elements. But we can say more if we consider the Adams spectral sequence for the Serre killing space $S^n|_{n+1}$.

The $\mathbb{A}$-module of the cohomology of $S^n|_{n+1}$ can be found from the spectral sequence of the fibration $S^n|_{n+1} \xrightarrow{K(\mathbb{Z},n-1)} S^n$. In the stable dimensions, this is the same as the cohomology of $K(\mathbb{Z}, n-1)$ with $H^{n-1}(K(\mathbb{Z}, n-1)) = \mathbb{Z}_2$ deleted. We know that, up to a dimension shift by $n-1$, this $\mathbb{A}$-module is the same as the module $\widetilde{\mathbb{A}}$ factorized over the right ideal generated by Sq^1. Thus, the zeroth row of the Adams spectral sequence for $S^n|_{n+1}$ is the same, up to a dimension shift, as the first row of the Adams spectral sequence for S^0, only the generator α_1 is deleted. It is easy to show that, also in the rows above, the spectral sequences for $S^n|_{n+1}$ and S^0 differ only by a dimension shift (by 1 downward and by $n-1$ to the right) and by deleting the diagonal $1, \alpha_1, \beta_1, \gamma_1, \delta_1, \ldots$.

EXERCISE 1. Prove this.

The transition from S^0 to $S^n|_{n+1}$ turns the diagram in Fig. 126 into the following diagram in Fig. 127.

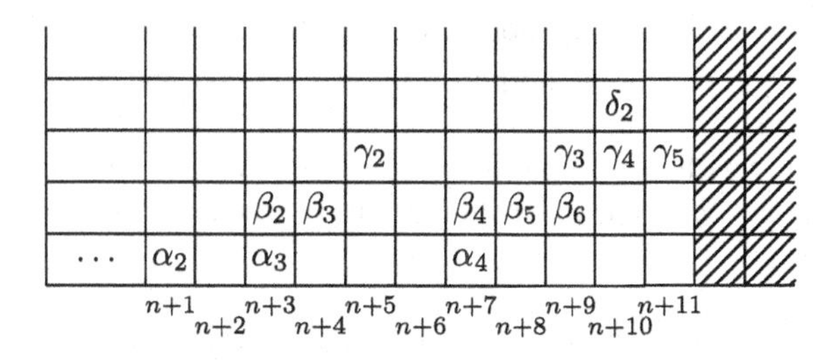

Fig. 127 The E_2-term of the Adams spectral sequence for $S^n|_{n+1}$

Let us now apply to this spectral sequence our observation regarding the minimal degree of a generator of the module B_k. We see that in the spectral sequence for S^0 the second-lowest degree of a generator in the kth row grows with k growing. Since in the fifth row the second-lowest degree of a generator is at least 13, we deduce that all the nontrivial terms $E_2^{s,t}$ with $0 < t - s \leq 7$ are shown in Fig. 126. This yields some preliminary results regarding the orders of the stable homotopy groups of spheres.

Consider possible differentials. By the dimension arguments, within the diagram in Fig. 126, there are two possibilities for a nontrivial differential: First, α_2 can be mapped by a differential into one of $\gamma_1, \delta_1, \ldots$; second, there is a possible differential $\beta_6 \mapsto \delta_2$. The first, however, does not happen, since this would have implied that $\pi_{n+1}(S^n) = 0$, while we know that it is $\mathbb{Z}_2$. Actually, $d_2\beta_6$ is also 0, but we cannot prove this before we study the multiplicative structure in our spectral sequence. But even without this, we see now that the orders of the 2-components of the stable groups $\pi_{n+k}(S^n)$ with $k = 1, 2, 3, 4, 5, 6, 7$ are 2, 2, 8, 1, 1, 2, 16 or 8.

36.2 The Multiplicative Structure

It is clear that we can obtain from a resolution

$$\ldots \xrightarrow{\partial_3} B_2 \xrightarrow{\partial_2} B_1 \xrightarrow{\partial_1} B_0 \xrightarrow{\partial_0} \mathbb{Z}_2 \longrightarrow 0$$

of the $\mathbb{A}$-module $\mathbb{Z}_2$ a resolution of the module $\operatorname{Ker} \partial_k$ simply by cutting the resolution of $\mathbb{Z}_2$ in the appropriate place:

$$\ldots \xrightarrow{\partial_{k+4}} B_{k+3} \xrightarrow{\partial_{k+3}} B_{k+2} \xrightarrow{\partial_{k+2}} B_{k+1} \xrightarrow{\partial_{k+1}} \operatorname{Ker} \partial_k \longrightarrow 0.$$

Hence, $\operatorname{Ext}_{\mathbb{A}}^{s,t}(\operatorname{Ker} \partial_k, \mathbb{Z}_2) = \operatorname{Ext}_{\mathbb{A}}^{s+k+1,t}(\mathbb{Z}_2, \mathbb{Z}_2)$ and the action of $H^{**}(\mathbb{A})$ in both Exts is the same.

Suppose that we want to compose the "multiplication table" by, for example, the element $\operatorname{Sq}^1 = \alpha_1$ of $\operatorname{Ker} \partial_0 = \widetilde{\mathbb{A}}$. Consider the $\mathbb{A}$-homomorphism $\widetilde{\mathbb{A}} \to \mathbb{Z}_2$, which takes α_1 into $1 \in \mathbb{Z}_2$. This homomorphism lowers the degrees by 1. Hence, the induced homomorphism $\operatorname{Ext}_{\mathbb{A}}^{**}(\mathbb{Z}_2, \mathbb{Z}_2) \to \operatorname{Ext}_{\mathbb{A}}^{**}(\widetilde{\mathbb{A}}, \mathbb{Z}_2)$ has the bidegree $(0, 1)$. Thus, we have a homomorphism $\operatorname{Ext}_{\mathbb{A}}^{s,t}(\mathbb{Z}_2, \mathbb{Z}_2) \to \operatorname{Ext}_{\mathbb{A}}^{s,t+1}(\widetilde{\mathbb{A}}, \mathbb{Z}_2) = \operatorname{Ext}_{\mathbb{A}}^{s+1,t+1}(\mathbb{Z}_2, \mathbb{Z}_2)$. This homomorphism takes 1 into α_1 and, hence, an arbitrary element ξ into $\xi\alpha_1$.

To determine this homomorphism between the Exts, we need to construct the corresponding map between the resolutions, that is, the commutative diagram

$$\begin{array}{ccccccccc}
\cdots \xrightarrow{\partial_3} & B_2 & \xrightarrow{\partial_2} & B_1 & \xrightarrow{\partial_1} & B_0 & \xrightarrow{\partial_0} & \mathbb{Z}_2 & \longrightarrow 0 \\
& \uparrow f_3 & & \uparrow f_2 & & \uparrow f_1 & & \updownarrow f_0 & \\
\cdots \xrightarrow{\partial_4} & B_3 & \xrightarrow{\partial_3} & B_2 & \xrightarrow{\partial_2} & B_1 & \xrightarrow{\partial_1} & \widetilde{\mathbb{A}} & \longrightarrow 0
\end{array}$$

in which $f_0(\alpha_1) = 1$ and every f_i lowers degrees by 1. Since the modules $B_1, B_2, \dots$ are free, it is sufficient to define $f_1, f_2, \dots$ on the generators.

Begin with f_1. The map ∂_1 takes the generators α_k of B_1 into the elements $\mathrm{Sq}^{2^{k-1}}$ of $\widetilde{\mathbb{A}}$ (which are also denoted by α_k). Hence, $f_0 \circ \partial_1(\alpha_1) = 1$ and $f_0 \circ \partial_1(\alpha_k) = 0$ for $k > 1$. Since $B_0 = \widetilde{A}$ and $\partial_0(1) = 1$, $\partial_0(\widetilde{\mathbb{A}}) = 0$, we can put $f_1(\alpha_1) = 1, f_1(\alpha_k) = 0$ for $k > 1$. [Notice that the homomorphisms f_i are not determined by our conditions uniquely; for example, we could take for $f_1(\alpha_k)$ with $k > 1$ an arbitrary element of $B_0 = \mathbb{A}$ of the appropriate degree; we choose these $f(\alpha_k)$ in the most convenient way, taking care only of the commutativity of the diagram.]

Turn now to f_2. Using the description of the homomorphism $B_2 \to B_1$ (denoted now as ∂_2) given in Sect. 36.1 and the fact that f_1 takes α_1 to 1 and kills α_k with $k > 1$, we conclude that $f_1 \circ \partial_2 = \partial_1 \circ f_2$ takes $\beta_1, \beta_2, \beta_3, \beta_4, \beta_5, \beta_6$ into $\mathrm{Sq}^1, \mathrm{Sq}^3, \mathrm{Sq}^4, \mathrm{Sq}^7, \mathrm{Sq}^8, \mathrm{Sq}^7 \mathrm{Sq}^2$. Thus, we should take for $f(\beta_1), \dots, f(\beta_6)$ arbitrary elements of B_1, which are taken by ∂_1 into the combination of the Steenrod squares listed. Since ∂_1 takes $\alpha_1, \alpha_2, \alpha_3, \alpha_4$ into $\mathrm{Sq}^1, \mathrm{Sq}^2, \mathrm{Sq}^4, \mathrm{Sq}^8$, we can put $f_2(\beta_1) = \alpha_1, f_2(\beta_3) = \alpha_3, f_2(\beta_5) = \alpha_4$, and $f_2(\beta_6) = \mathrm{Sq}^7 \alpha_2$. Since $\mathrm{Sq}^3 = \mathrm{Sq}^1 \mathrm{Sq}^2$ and $\mathrm{Sq}^7 = \mathrm{Sq}^3 \mathrm{Sq}^4$ (Adem's formulas), we can also put $f_2(\beta_2) = \mathrm{Sq}^1 \alpha_2$ and $f_2(\beta_4) = \mathrm{Sq}^3 \alpha_3$.

To determine f_3, we find, using the formulas known to us, the expressions for $f_2 \circ \partial_3(\gamma_i)$:

$$\begin{aligned}
f_2 \circ \partial_3(\gamma_1) &= \mathrm{Sq}^1 \alpha_1, \\
f_2 \circ \partial_3(\gamma_2) &= \mathrm{Sq}^1 \alpha_3 + \mathrm{Sq}^2 \mathrm{Sq}^1 \alpha_2 + \mathrm{Sq}^4 \alpha_1, \\
f_2 \circ \partial_3(\gamma_3) &= \mathrm{Sq}^1 \alpha_4 + \mathrm{Sq}^5 \alpha_3 + \mathrm{Sq}^4 \mathrm{Sq}^1 \alpha_3 + \mathrm{Sq}^8 \alpha_1, \\
f_2 \circ \partial_3(\gamma_4) &= \mathrm{Sq}^6 \alpha_3 + \mathrm{Sq}^7 \mathrm{Sq}^1 \alpha_2, \\
f_2 \circ \partial_3(\gamma_5) &= \mathrm{Sq}^3 \alpha_4 + \mathrm{Sq}_5 \mathrm{Sq}^2 \alpha_3 + \mathrm{Sq}^9 \alpha_2,
\end{aligned}$$

and then take for $f_3(\gamma_1), \dots, f_3(\gamma_5)$ elements of B_2 whose ∂_2-images are the right-hand sides of the last formulas. We can find such elements of B_2 using the expressions for $\partial_2(\beta_i)$ from Sect. 36.1 and Adem's formulas; a possible choice of these five elements (of degrees 2, 5, 9, 10, 11) is β_1; β_3; $\beta_5 + \mathrm{Sq}^1 \beta_4$; $\mathrm{Sq}^1 \beta_5 + \mathrm{Sq}^2 \beta_4 + \mathrm{Sq}^8 \beta_1$; $\mathrm{Sq}^1 \beta_6$. We take those for $f_3(\gamma_i)$.

Finally, $f_3 \circ \partial_4(\delta_1) = \mathrm{Sq}^1 \beta_1 = \partial_3(\gamma_1)$ and $f_3 \circ \partial_4(\delta_2) = \mathrm{Sq}^1 \beta_5 + \mathrm{Sq}^4 \mathrm{Sq}^3 \beta_3 + \mathrm{Sq}^5 \beta_3 + \mathrm{Sq}^8 \beta_1 = \partial_3(\gamma_3)$, and we can put $f_4(\delta_1) = \gamma_1$ and $f_4(\delta_2) = \gamma_3$.

Here is a full diagram of the action of homomorphisms f_i in degrees ≤ 12:

$f_1: B_1 \to B_2$	$f_2: B_2 \to B_3$	$f_3: B_3 \to B_4$	$f_4: B_4 \to B_5$
$\alpha_1 \mapsto 1$	$\beta_1 \mapsto \alpha_1$	$\gamma_1 \mapsto \beta_1$	$\delta_1 \mapsto \gamma_1$
$\alpha_2 \mapsto 0$	$\beta_2 \mapsto \mathrm{Sq}^1 \alpha_2$	$\gamma_2 \mapsto \beta_2$	$\delta_2 \mapsto \gamma_3$
$\alpha_3 \mapsto 0$	$\beta_3 \mapsto \alpha_3$	$\gamma_3 \mapsto \beta_5 + \mathrm{Sq}^1 \beta^4$	
$\alpha_4 \mapsto 0$	$\beta_4 \mapsto \mathrm{Sq}^3 \alpha_3$	$\gamma_4 \mapsto \mathrm{Sq}^1 \beta_5 + \mathrm{Sq}^2 \beta_4 + \mathrm{Sq}^8 \beta_1$	
$\alpha_5 \mapsto 0$	$\beta_5 \mapsto \alpha_4$	$\gamma_5 \mapsto \mathrm{Sq}^1 \beta_6$	
	$\beta_6 \mapsto \mathrm{Sq}^7 \alpha_2$		

Now consider the homomorphisms

$$f_k^*: \mathrm{Hom}_{\mathbb{A}}(B_{k-1}, \mathbb{Z}_2) \to \mathrm{Hom}_{\mathbb{A}}(B_k, \mathbb{Z}_2).$$

The additive generators of $\mathrm{Hom}_{\mathbb{A}}(B_k, \mathbb{Z}_2)$ correspond to the free generators of the module B_k and are denoted by the same letters: Each takes the chosen generator to 1 and takes all other generators, as well as the images of all the generators (including the chosen one) with respect to Steenrod squares and their compositions, to zero. This gives the following description of f_k^*:

f_1^*	f_2^*	f_3^*	f_4^*
$1 \mapsto \alpha_1$	$\alpha_1 \mapsto \beta_1$	$\beta_1 \mapsto \gamma_1$	$\gamma_1 \mapsto \delta_1$
	$\alpha_2 \mapsto 0$	$\beta_2 \mapsto 0$	$\gamma_2 \mapsto 0$
	$\alpha_3 \mapsto \beta_3$	$\beta_3 \mapsto \gamma_2$	$\gamma_3 \mapsto \delta_2$
	$\alpha_4 \mapsto \beta_5$	$\beta_4 \mapsto 0$	$\gamma_4 \mapsto 0$
		$\beta_5 \mapsto \gamma_3$	$\gamma_5 \mapsto 0$
		$\beta_6 \mapsto 0$	

Since, with our choice of generators, $\mathrm{Ext}_{\mathbb{A}}^k(\mathbb{Z}_2, \mathbb{Z}_2)$ is not different from $\mathrm{Hom}_{\mathbb{A}}(B_k, \mathbb{Z}_2)$ (see Sect. 34.5), our homomorphism $\xi \mapsto \xi\alpha_1$ acts on the generators in E_2 by the same formulas. Thus,

$$\alpha_1^2 = \beta_1, \alpha_1\alpha_3 = \beta_3, \alpha_1\alpha_4 = \beta_5, \alpha_1\beta_1 = \gamma_1,$$
$$\alpha_1\beta_3 = \gamma_2, \alpha_1\beta_5 = \gamma_3, \alpha_1\gamma_1 = \delta_1, \alpha_1\gamma_3 = \delta_2,$$

and all the other products of α_1 with the generators in Fig. 126 are zeroes (which also follows directly from the grading argumentations). Notice that the relations $\alpha_1^2 = \beta_1, \alpha_1\beta_1 = \gamma_1, \alpha_1\gamma_1 = \delta_1$ form a beginning of an infinite sequence, which shows that for every s, the generator of $E_2^{s,s} \cong \mathbb{Z}_2$ is the α_1^s.

In a similar way, one can find products of generators of E_2 with α_2, α_3, and any other generator. The computation (which is way easier than the computation above) shows that

$$\alpha_2^2 = \beta_2, \alpha_2\alpha_4 = \beta_6, \alpha_2\beta_2 = \gamma_4, \alpha_2\beta_6 = \gamma_5, \alpha_3^2 = \beta_4, \alpha_3\beta_4 = \gamma_5,$$

and all other pairwise products of generators listed in Fig. 126 are zeroes.

Below, we will use classical notations for cohomology classes of the Steenrod algebra. The classes $\alpha_1, \alpha_2, \alpha_3, \dots$ are usually denoted as $h_0, h_1, h_2, \dots$ (and they are called Hopf classes). With these notations, the part of the E_2-term of the Adams spectral sequence for S^0 known to us (and presented in Fig. 126) looks like what appears in Fig. 128.

Here a is a new multiplicative generator. By the way, we can now state that $d_2(\beta_6) = d_2(h_1h_3) = 0$, and so the stable homotopy group $\pi_{n+7}(S^n)$ has order 16, not 8.

We see that the problem of the computation of the stable homotopy groups of spheres falls into two parts: the computation of the cohomology of the Steenrod algebra and the computation of the differentials of the Adams spectral sequence. The first part is reduced to a purely mechanical work (or to a computer program) and can be done up to any degree. In the book *Stable Homotopy Theory* of Adams (see Adams [2]) the result of this computation is presented for $t - s \leq 17$. Thus, the diagram in Fig. 128 can be considerably extended; the additional part of this diagram is shown in Fig. 129. Since obtaining this diagram requires only boring but automatic work, we may consider it as established. To get this diagram we would have to investigate the Steenrod algebra not to degree 12, as we did, but to degree 27. In particular, we see that $E_2^{10,t} = 0$ for $10 < t \leq 27$, and hence, for $0 < t - s \leq 17$, the E_2 contains no nontrivial elements not shown in Figs. 128 and 129.

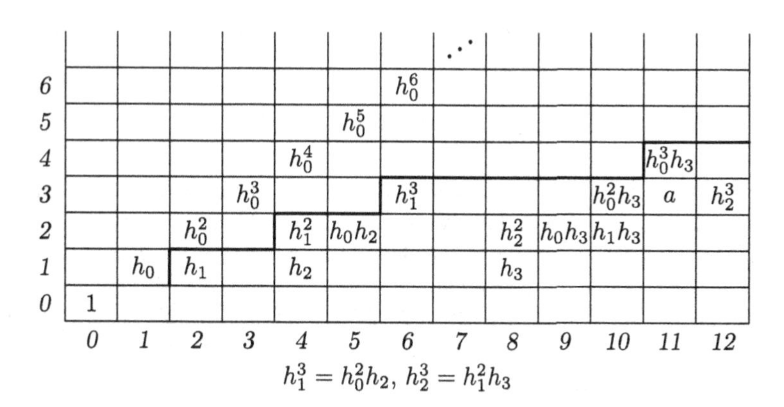

Fig. 128 Multiplicative structure of the E_2-term of the Adams spectral sequence for S^0

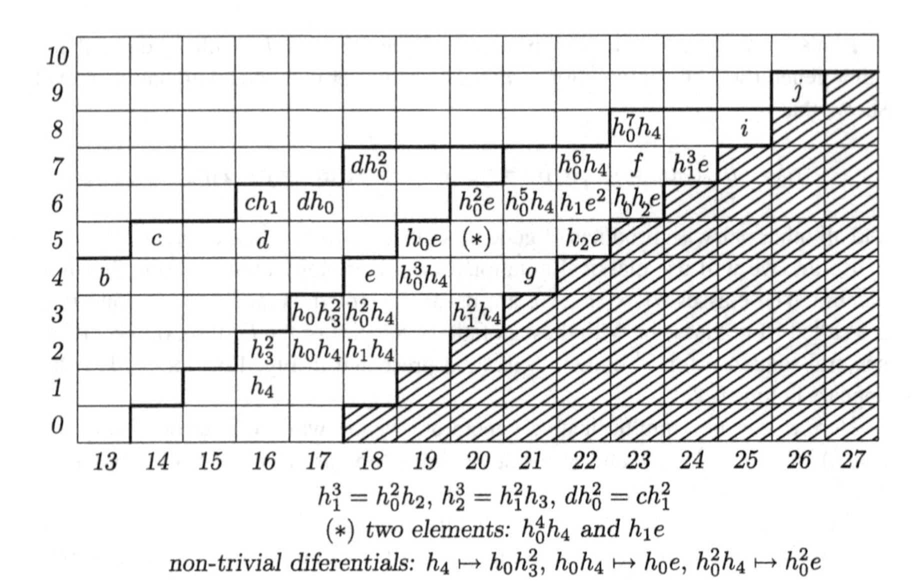

Fig. 129 Multiplicative structure of the E_2-term of the Adams spectral sequence for S^0 (continuation)

Figures 128 and 129 show three polygonal lines. The right one bounds the domain $t - s \leq 17$, and the middle one bounds the domain $t - s \leq 13$. The meaning of the remaining (left) polygonal line will be explained in Sect. 36.4.

The grading arguments show that all the differentials of the multiplicative generators $h_0, h_1, h_2, h_3, a, b, c, d, e, f, i, j$ are zeroes. Hence, $E_\infty^{s,t} = E_2^{s,t}$ for $t - s \leq 13$, and we can derive the following statement concerning the first 13 stable homotopy groups of spheres.

The orders of 2-components of the groups $\pi_r^S(S^0)$ for $r = 1, 2, \ldots, 13$ are as follows:

$$2, 2, 8, 1, 1, 2, 16, 4, 8, 2, 8, 1, 1.$$

The elements $1, h_0, h_0^2, h_0^3, \ldots$ generate a group $\mathbb{Z}_2 \oplus \mathbb{Z}_2 \oplus \mathbb{Z}_2 \oplus \ldots$ adjoint to $\pi_0^S(S^0) = \mathbb{Z}$. The filtration in this group must be $\mathbb{Z} \supset 2\mathbb{Z} \supset 4\mathbb{Z} \supset 8\mathbb{Z} \supset \ldots$. Thus, $h_0^r \in E_2^{r,r}$ represents in $\pi_0^S(S^0) = \mathbb{Z}$ the element 2^r. From this we see that if for some $u \in E_2^{s,t}$, $h_0^r u \neq 0$ and u represents some $\xi \in \pi_{t-s}^S(S^0)$, then $2^r\xi \neq 0$ in $\xi \in \pi_{t-s}^S(S^0)$. In particular, we see from Figs. 128 and 129 that $\pi_3^S(S^0)$, $\pi_7^S(S^0)$, and $\pi_{11}^S(S^0)$ are cyclic groups of orders 8, 16, and 8. This gives us full information about the groups $\pi_r^S(S^0)$ with $r \leq 13$, except the cases $r = 8, 9$.

The group $\pi_8^S(S^0)$ has order 4. It has a subgroup $\mathbb{Z}_2$ generated by some element u represented by $a \in E_\infty^{3,11}$, and the quotient of $\pi_8^S(S^0)$ over this subgroup is also a group $\mathbb{Z}_2$ generated by $h_1h_3 \in E_\infty^{2,10}$. Choose a representative $v \in \pi_8^S(S^0)$ for this generator. Then there are two possibilities: $2v = 0$ or $2v \neq 0$. But since $h_0(h_1h_3) \in$

$E_\infty^{3,11}$ is zero, this class $2v$ must have filtration > 11. But there are no elements of filtration > 11 in $\bigoplus_s E_\infty^{s,s+8}$; thus, $2v = 0$ and $\pi_8^S(S^0) \cong \mathbb{Z}_2 \oplus \mathbb{Z}_2$. [It is possible to prove in a different way that $2v = 0$. Since v represents h_1h_3, we can take for v the product of elements of $\pi_1^S(S^0)$ and $\pi_7^S(S^0)$, which represent h_1 and h_3. But $\pi_1^S(S^0) \cong \mathbb{Z}_2$, and thus the first of these two factors has order 2. Hence, the product has order 2.]

Finally, the group $\pi_9^S(S^0)$ has order 8. In $\bigoplus_s E_\infty^{s,s+9}$, there are three generators: h_2^3, b, and c. The representatives of the generators b and c generate in $\pi_9^S(S^0)$ a subgroup isomorphic to $\mathbb{Z}_2 \oplus \mathbb{Z}_2$; the proof is the same as for $\pi_8^S(S^0)$. The quotient of $\pi_9^S(S^0)$ over this subgroup is $\mathbb{Z}_2$, which leaves us with two possibilities for $\pi_9^S(S^0)$: $\mathbb{Z}_2 \oplus \mathbb{Z}_2 \oplus \mathbb{Z}_2$ or $\mathbb{Z}_4 \oplus \mathbb{Z}_2$. But the generator of the quotient $\pi_9^S(S^0)/\mathbb{Z}_2 \oplus \mathbb{Z}_2)$ represents $h_2^3 \in E_\infty^{3,12}$; so we can take for this generator $\alpha \circ \alpha \circ \alpha$, where α is the generator of $\pi_3^S(S^0)$, and it has order 2, since $\alpha \circ \alpha \in \pi_6^S(S^0)$ already has order 2. Thus, $\pi_9^S(S^0) \cong \mathbb{Z}_2 \oplus \mathbb{Z}_2 \oplus \mathbb{Z}_2$, and we now know the 2-components of all groups $\pi_r^S(S^0)$ with $r \leq 13$:

$$\mathbb{Z}_2,\ \mathbb{Z}_2,\ \mathbb{Z}_8,\ 0,\ 0,\ \mathbb{Z}_2,\ \mathbb{Z}_{16},\ \mathbb{Z}_2 \oplus \mathbb{Z}_2,\ \mathbb{Z}_2 \oplus \mathbb{Z}_2 \oplus \mathbb{Z}_2,\ \mathbb{Z}_2,\ \mathbb{Z}_8,\ 0,\ 0.$$

We can also derive some results from the multiplicative structure in our spectral sequence. If $\eta \in \pi_1^S(S^0)$ is the generator, then η^2 generates $\pi_2^S(S^0)$, and $\eta^3 = 4\alpha$, where α is the generator of $\pi_3^S(S^0)$. The group $\pi_6^S(S^0)$ is generated by α^2. The elements of the group $\pi_7^S(S^0)$ are not decomposable into compositions; denote the generator of this group by β. Then the group $\pi_8^S(S^0)$ is generated by $\eta\beta$ and an indecomposable generator γ. The group $\pi_9^S(S^0)$ is generated by $\alpha^3 = \eta^2\beta$ and two more generators, one of which is indecomposable, and the other may be equal to $\eta\gamma$. The group $\pi_{10}^S(S^0)$ is generated by the product of one of these generators by η, and the group $\pi_{11}^S(S^0)$ is generated by an indecomposable element.

In addition to all this, we remark that the computation of 2-components of 14th and further stable homotopy groups of spheres encounters a difficulty: The grading arguments no longer guarantee the triviality of differentials. And, indeed, nontrivial differentials appear at the first opportunity. It turns out that $d_2(h_4) = h_0h_3^2$, $d_3(h_0h_4) = h_0i$, and $h_3(h_0^2h_4) = h_0^2i$.

36.3 The Odd Components

Within our dimension range, the only necessity for considering the Adams spectral sequence arises for $p = 3$. Indeed, we already know (see Sect. 27.4) that the first p-component in the stable homotopy groups of spheres is $\mathbb{Z}_p \subset \pi_{2p-3}^S(S^0)$ and the second is $\mathbb{Z}_p \subset \pi_{4p-5}^S(S^0)$; thus, besides $\mathbb{Z}_5 \subset \pi_7^S(S^0)$ and $\mathbb{Z}_7 \subset \pi_{11}^S(S^0)$, the groups $\pi_n^S(S^0)$ with $n \leq 13$ do not have p-components with $p > 3$.

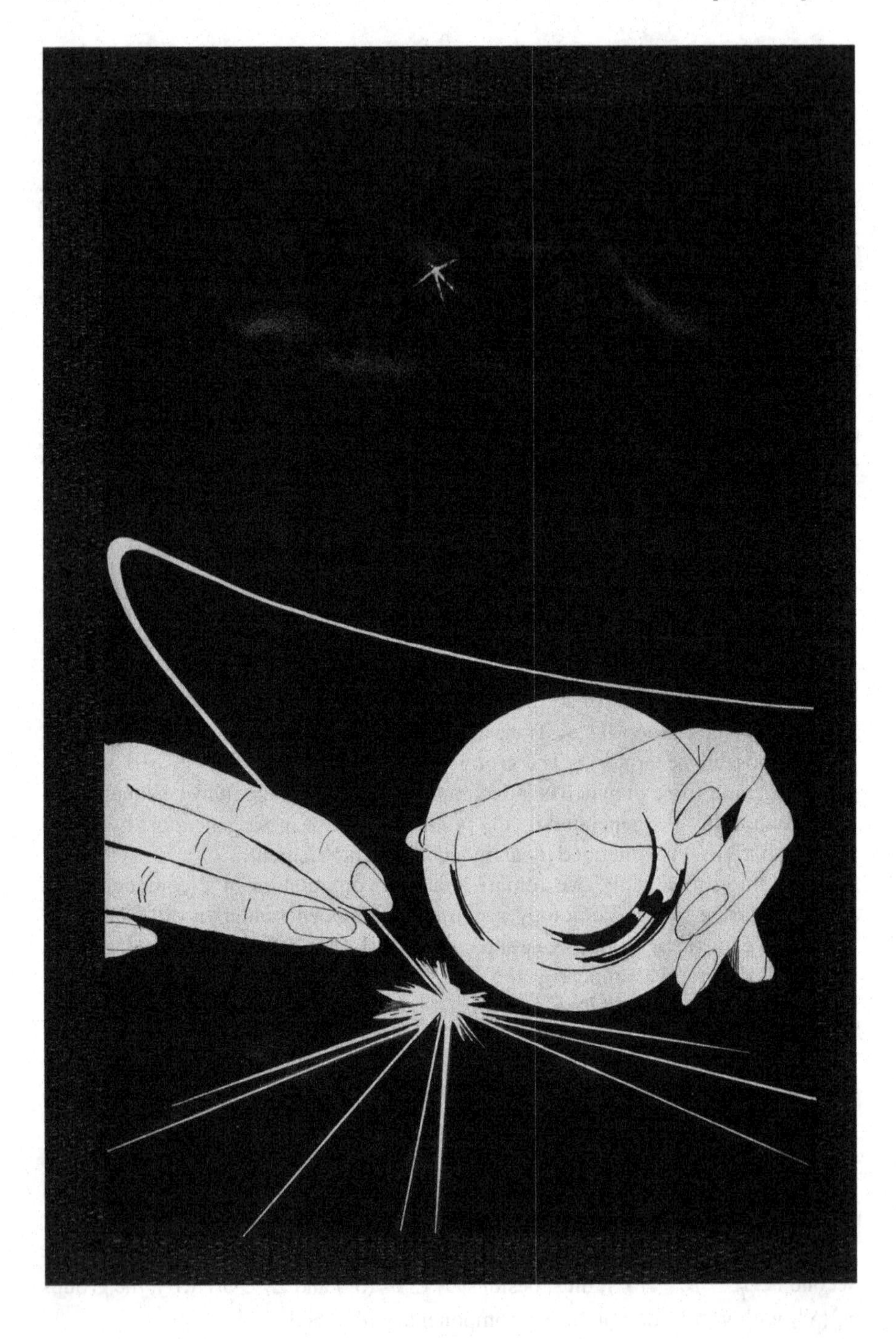

The 3-components are as follows:

$$0,\ 0,\ \mathbb{Z}_3,\ 0,\ 0,\ 0,\ \mathbb{Z}_3,\ 0,\ 0,\ \mathbb{Z}_3,\ \mathbb{Z}_9,\ 0,\ \mathbb{Z}_3$$

(it will be a good exercise for the reader to prove this result using the Adams spectral sequence for $p = 3$).

The composition multiplication in these groups is trivial. Thus, we have

$$\begin{aligned}
\pi_n(S^n) &= \mathbb{Z} && \text{for } n \geq 1,\\
\pi_{n+1}(S^n) &= \mathbb{Z}_2 && \text{for } n \geq 3,\\
\pi_{n+2}(S^n) &= \mathbb{Z}_2 && \text{for } n \geq 4,\\
\pi_{n+3}(S^n) &= \mathbb{Z}_{24} && \text{for } n \geq 5,\\
\pi_{n+4}(S^n) &= 0 && \text{for } n \geq 6,\\
\pi_{n+5}(S^n) &= 0 && \text{for } n \geq 7,\\
\pi_{n+6}(S^n) &= \mathbb{Z}_2 && \text{for } n \geq 8,\\
\pi_{n+7}(S^n) &= \mathbb{Z}_{240} && \text{for } n \geq 9,\\
\pi_{n+8}(S^n) &= \mathbb{Z}_2 \oplus \mathbb{Z}_2 && \text{for } n \geq 10,\\
\pi_{n+9}(S^n) &= \mathbb{Z}_2 \oplus \mathbb{Z}_2 \oplus \mathbb{Z}_2 && \text{for } n \geq 11,\\
\pi_{n+10}(S^n) &= \mathbb{Z}_6 && \text{for } n \geq 12,\\
\pi_{n+11}(S^n) &= \mathbb{Z}_{504} && \text{for } n \geq 13,\\
\pi_{n+12}(S^n) &= 0 && \text{for } n \geq 14,\\
\pi_{n+13}(S^n) &= \mathbb{Z}_3 && \text{for } n \geq 15.
\end{aligned}$$

36.4 The Adams Theorems on the Initial Term of His Spectral Sequence

It is shown that homological algebra can be applied to stable homotopy theory. In this application, we deal with $\mathbb{A}$-modules, where $\mathbb{A}$ is the mod p Steenrod algebra. Obtaining a concrete geometrical result by this method usually involves work of two different sorts. To illustrate this, we consider the spectral sequence

$$\operatorname{Ext}^{s,t}_{\mathbb{A}}(\widetilde{H}^*(Y;\mathbb{Z}_p), \widetilde{H}^*(X;\mathbb{Z}_p)) \Longrightarrow {}_p\pi^S_*(X, Y).$$

Here each group $\operatorname{Ext}^{s,t}$ which occurs in the E_2-term can be effectively computed; the process is purely algebraic. However, no such effective method is given for computing the differentials d_r in the spectral sequence or to determine the group extension by which ${}_p\pi^S_*(X, Y)$ is built up from the E_∞-term; these are topological problems.

A mathematical logician might be satisfied with this account: an algorithm is given for computing E_2; to find the maps d_r still requires intelligence. The practical mathematician, however, is forced to admit that the intelligence of mathematicians is an asset at least as reliable as their willingness to do large amounts of tedious mechanical work[2]. In fact, when a chance has arisen to show that such a differential d_r is nonzero, it has been regarded as an interesting problem, and duly solved. However, the difficulty of actually computing groups $\mathrm{Ext}_{\mathbb{A}}^{s,t}(L,M)$ has remained the greatest obstacle to the method.

As we can see from this quotation from Adams' article (Adams [8]), its author did not think that the existence of the algorithm for computing the cohomology is equivalent to its computing. Adams devoted several works to this subject. Here we will formulate his most important results.

Theorem of the First Three Rows (Adams [1]).

(1) *The group* $E_2^1 = \bigoplus_t E_1^{s,t}$ *is additively generated by the elements* $h_i \in E_2^{1,1+2^i}$, $i = 0, 1, 2, \ldots$.
(2) *The group* $E_2^2 = \bigoplus_t E_2^{s,t}$ *is additively generated by the linearly independent elements* $h_i h_j$, $0 \le i \le j$, $j \ne i+1$; *the products* $h_i h_{i+1}$ *are all equal to* 0.
(3) *In the group* $E_2^3 = \bigoplus_t E_3^{s,t}$ *the following relations hold:* $h_{i+2}h_i^2 = h_{i+1}^3$, $h_{i+2}^2 h_i = 0$. *The set of products* $h_i h_j h_k$, $0 \le i \le j \le k$ *with the products* $h_i h_j h_{j+1}$, $h_i h_{i+1} h_k$, $h_i^2 h_{i+2}$, *and* $h_i h_{i+2}^2$ *deleted is linearly independent in* E_2^3.

Notice that part (3) does not give a full description of E_2^3: It can contain elements which cannot be expressed in terms of h_is. We know one such element: This is $a \in E_2^{3,11}$.

Triviality Theorem (Adams [7]). $E_2^{s,t} = 0$ *for* $s < t < f(s)$, *where*

$$\begin{aligned} f(4n) &= 12n - 1,\ n > 0,\\ f(4n+1) &= 12n + 2,\ n \ge 0,\\ f(4n+2) &= 12n + 4,\ n \ge 0,\\ f(4n+3) &= 12n + 6,\ n \ge 0. \end{aligned}$$

The part of the E_2-term which is zero by this theorem is marked (by the left polygonal line) in Figs. 128 and 129.

Periodicity Theorem (Adams [7]). *For every k, there exists a neighborhood* N_k *of the line* $t = 3s$ *in which the groups* $E_2^{s,t}$ *are periodic with the period* $(2^{k+2}, 3 \cdot 2^{k+2})$. *The union of these neighborhoods* N_k *is a domain* $s < t < g(s)$, *where* $g(s)$ *satisfies the inequalities* $4s \le g(s) \le 6s$.

[2]This was written in the precomputer era.

36.5 A Final Remark

We conclude this lecture with an example of an Adams spectral sequence with a nontrivial differential.

Let $X = K(\mathbb{Z}_4, n)$, where n is large enough. The stable homotopy groups of X up to dimension $\approx 2n$ are $\pi_n^S(X) = \mathbb{Z}_4$, $\pi_i^S(X) = 0$ for $i \neq n$. The $\mathbb{A}$-module $\widetilde{H}^*(X; \mathbb{Z}_2)$ is isomorphic to $\widetilde{H}^*(K(\mathbb{Z}, n); \mathbb{Z}_2) \otimes \widetilde{H}^*(K(\mathbb{Z}, n+1); \mathbb{Z}_2)$; that is, it has two generators, $\alpha \in H^n(X; \mathbb{Z}_2)$ and $\beta \in H^{n+1}(X; \mathbb{Z}_2)$, with $\mathrm{Sq}^1 \alpha = 0$, $\mathrm{Sq}^1 \beta = 0$. This shows that in the E_2-term of the Adams spectral sequence for X, there are two nontrivial diagonals, as shown in Fig. 130.

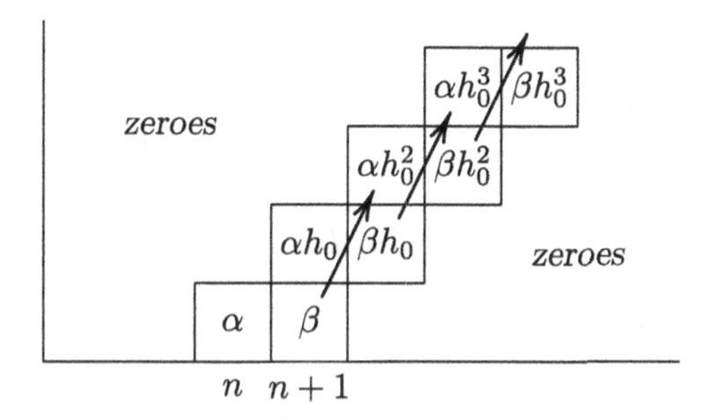

Fig. 130 The Adams spectral sequence for $X = K(\mathbb{Z}_4, n)$

Obviously, $d_2(\beta h_0^k) = \alpha h_0^{k+2}$ for $k = 0, 1, 2, \ldots$: Otherwise, the order of the group $\pi_n^S(X)$ could not equal 4.

Lecture 37 Partial Cohomology Operation

In our presentation of the Adams spectral sequence, we avoid using partial cohomology operations; we did not even mention the term although it could be appropriate under the title "General idea." Still, it would be fair to say that the notion of a partial cohomology operation is a true base for the Adams method.

37.1 The Construction of Partial Cohomology Operations

Let

$$\sum\nolimits_{i=1}^{m} \beta_i \alpha_i$$

be a relation in the Steenrod algebra $\mathbb{A}_p$, where α_i, β_i are stable cohomology operations of degrees, respectively, $q_i, n - q_i$. For every N, the operation α_i determines a map

$$\widetilde{\alpha}_i: K(\mathbb{Z}_p, N) \to K(\mathbb{Z}_p, N + q_i),$$

and all these operations together determine a map

$$\widetilde{\alpha}: K(\mathbb{Z}_p, N) \to \prod_i K(\mathbb{Z}_p, N + q_i).$$

Consider the induced fibration

$$\begin{array}{ccc} E & \longrightarrow & * \\ \Big\downarrow{\scriptstyle \prod K(\mathbb{Z}_p, N+q_i-1)} & & \Big\downarrow{\scriptstyle \prod K(\mathbb{Z}_p, N+q_i-1)} \\ K(\mathbb{Z}_p, N) & \xrightarrow{\widetilde{\alpha}} & \prod K(\mathbb{Z}_p, N + q_i). \end{array}$$

The $\mathbb{Z}_p$-cohomology spectral sequence of this fibration is presented in Fig. 131. We see that the element $\sum \beta_i u_i \in E_2^{0,N+n-1}$ remains in E_∞ and determines, because of this, in the group $H^{N+n-1}(E; \mathbb{Z}_p)$ a coset of the subgroup $\mathrm{Im}[H^{N+n-1}(K(\mathbb{Z}_p, N) \to H^{N+n-1}(E; \mathbb{Z}_p)]$. Choose some element v of this coset. It is clear from the definition that the image of v with respect to the homomorphism

$$H^{N+n-1}(E; \mathbb{Z}_p) \to H^{N+n-1}(\prod K(\mathbb{Z}_p, N + q_i - 1); \mathbb{Z}_p)$$

is $\sum_i \beta_i u_i$.

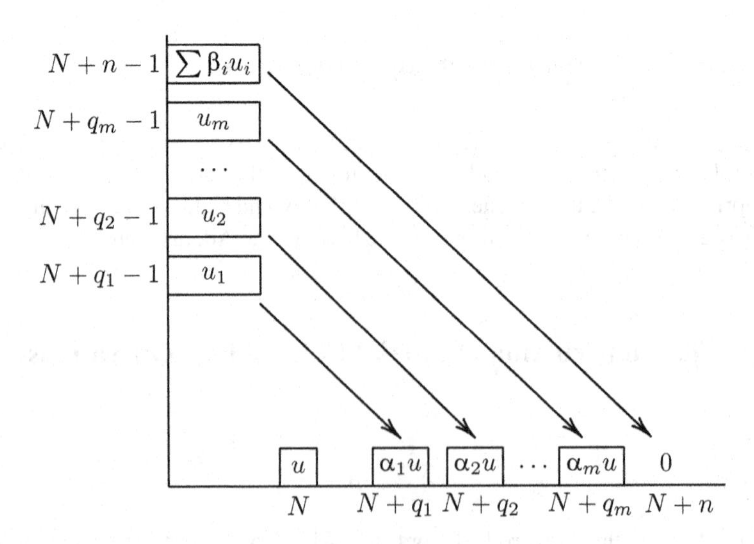

Fig. 131 The spectral sequence of the fibration $E \to K(\mathbb{Z}_p, N)$

For an arbitrary X, we define a (natural with respect to X) homomorphism

$$\bigcap_{i=1}^{m} \operatorname{Ker}\left[\alpha_i : H^N(X;\mathbb{Z}_p) \to H^{N+q-i}(X;\mathbb{Z}_p)\right]$$
$$\longrightarrow \frac{H^{N+n-1}(X;\mathbb{Z}_p)}{\sum\limits_{i=1}^{m} \operatorname{Im}\left[\beta_i : H^{N+q_i+1}(X;\mathbb{Z}_p) \to H^{N+n-1}(X;\mathbb{Z}_p)\right]}.$$

This is our *secondary operation.*

Let $\xi \in H^N(X;\mathbb{Z}_p)$, and let $\alpha_i(\xi) = 0$ for $i = 1, 2, \ldots, m$. Consider the map $\widetilde{\xi}: X \to K(\mathbb{Z}_p, N)$ corresponding to the cohomology class ξ. Its composition with the map $\widetilde{\alpha}$ is homotopy trivial, and hence the fibration induced over X by the fibration $E \to K(\mathbb{Z}_p, N)$ is trivial [it coincides with the fibration induced by the fibration $* \to \prod K(\mathbb{Z}_p, N + q_i)$ via the map $\widetilde{\alpha} \circ \widetilde{\xi}$]. Hence, it has a section

$$\begin{array}{ccc} X \times \prod K(\mathbb{Z}_p, N + q_i - 1) \approx E' & \longrightarrow & E \\ \downarrow\uparrow & & \downarrow \\ X & \xrightarrow{\widetilde{\xi}} & K(\mathbb{Z}_p, N). \end{array}$$

The composition of this section with the upper horizontal map is a map $\widetilde{\eta}: X \to E$ which forms, with the other maps of the diagram, the commutative triangle

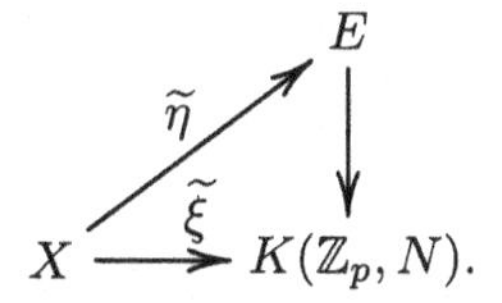

Obviously, $\eta^* u = \xi$. We take for $\varphi(\xi) \in H^{N+n-1}(X;\mathbb{Z}_p)$ the cohomology class $\widetilde{\eta}^*(v)$. It is not uniquely defined, since the section used for the construction of the map $\widetilde{\eta}$ was not uniquely defined. How not unique is this section? Its existence was derived from the equality $E' = X \times \prod K(\mathbb{Z}_p, N + q_i - 1)$, and the section itself was defined by the formula $x \mapsto (x, *)$. Any other section acts according to the formula $x \mapsto (x, \zeta(x))$, where $\zeta = (\zeta_1, \ldots, \zeta_m): X \to \prod K(\mathbb{Z}_p, N + q_i - 1)$ is some map.

EXERCISE 1. This change of section gives rise to the change of the element $\varphi(\xi)$ by the element $\varphi(\xi) + \sum_{i=1}^{m} \beta_i \zeta_i^*(u_i)$, where $u_i \in H^{N+q_i-1}(K(\mathbb{Z}_p, N + q_i - 1);\mathbb{Z}_p)$ are the fundamental classes.

This shows that $\beta(\xi)$ is defined uniquely as an element of the quotient group indicated above.

Thus, we have constructed a secondary operation φ. It is a family of partial multivalued homomorphisms $H^N(X; \mathbb{Z}_p) \dashrightarrow H^{N+n-1}(X; \mathbb{Z}_p)$ (*partial* because they are defined only on the intersection of the kernels of the operations α_i; *multivalued* because they are defined modulo the images of the operations β_i).

Notice that the secondary operation ϕ is uniquely determined by the relation $\sum \beta_i \alpha_i = 0$ up to addition of the usual (primary) cohomology operations [this nonuniqueness arises from the nonuniqueness of the class $v \in H^{N+n-1}(E; \mathbb{Z}_p)$; see above].

EXERCISE 2. Formulate and prove the properties of the secondary operations, including their naturality and stability.

The best-known example of a secondary cohomology operation is the so-called second Bockstein homomorphism. It corresponds to the relation $\beta^2 = 0$ and thus is defined on the kernel and takes values modulo the image of the usual, "first" Bockstein homomorphism. If $\xi \in H^N(X; \mathbb{Z}_p)$, then, to define $\beta(\xi)$, we take a mod p cocycle x in the class ξ, and then choose an integral cochain, $\widetilde{x}$, which is projected onto x by reducing modulo p. The $\beta(\xi)$ is represented by the cocycle $\frac{1}{p}\delta\widetilde{x}$, and the equality $\beta(\xi) = 0$ means that there exists an integral cochain $\widetilde{y}$ such that $\frac{1}{p}\delta\widetilde{x} \equiv \delta\widetilde{y} \bmod p$. In other words, $\delta(\widetilde{x} - p\widetilde{y})$ is divisible by p^2, and we define $\beta_2(\xi) \in H^{N+1}(X; \mathbb{Z}_p)$ as the cohomology class of the cocycle $\frac{1}{p^2}\delta(\widetilde{x} - p\widetilde{y})$.

EXERCISE 3. What are the space E and the cohomology class v which correspond, according to the constructions of this section, to the relation $\beta^2 = 0$?

Similar to the second Bockstein homomorphism, one can construct the "third Bockstein homomorphism" using the cocycle $\frac{1}{p^3}\delta(\widetilde{x} - p\widetilde{y} - p^2\widetilde{z})$, and then fourth, fifth, etc. Bockstein homomorphisms. These are examples of ternary, quaternary, quinary, etc. cohomology operations. For example, a ternary operation corresponds to a relation like $\sum \beta_i \varphi_i = 0$, where the β_i are primary cohomology operations and the φ_i are secondary cohomology operations. A reader who is interested in higher cohomology operations can construct a theory of n-ary cohomology operations with an arbitrary n.

If one does not restrict this theory to stable operations, and considers arbitrary coefficient groups, then it becomes true that the cohomology with the action of all this higher cohomology operations fully determines the homotopy type of a simply connected space. The reader can try to make this statement formal and precise, but in this form it becomes more or less tautological. A more interesting statement of this kind is discussed in Sect. 37.4.

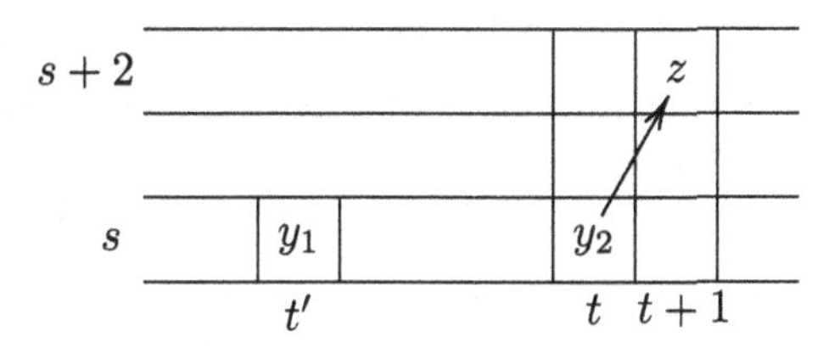

Fig. 132 A differential in the Adams spectral sequence arising from a secondary cohomology operation

37.2 Higher Cohomology Operations and Differentials of the Adams Spectral Sequence

We will not discuss in detail the relations between higher cohomology operations and differentials of the Adams spectral sequence. Instead, we restrict ourselves to a simple remark.

Suppose that in the Adams spectral sequence for some space X, there are elements $y_1 \in E_2^{s,t'}, y_2 \in E_2^{s,t}$, and $z \in E_2^{s+2,t+1}$ (see Fig. 132). Then y_1 and y_2 are generators of the $\mathbb{A}$-module $\widetilde{H}^*(X(s);\mathbb{Z}_p)$. Assume that $y_2 = \varphi(y_1)$ [better to say $y_2 \in \varphi(y_1)$], where φ is a secondary cohomology operation constructed from some relation $\sum \beta_i\alpha_i = 0$ in the Steenrod algebra.

By the construction of secondary cohomology operations, $\alpha_i y_1 = 0$. These are relations in the $\mathbb{A}$-module $\widetilde{H}^*(X(s);\mathbb{Z}_p)$. If we apply the operation β_i to the relation $\alpha_i y_1 = 0$ and then sum up all the results, then we get the relation $0 = 0$. Thus, we have a relation in the module of relations. Assume that this relation is one of generators in the module of relations in $\widetilde{H}^*(X(s);\mathbb{Z}_p)$ and that this generator is z (indeed, it has the same degree as z).

EXERCISE 4. Prove that in the Adams spectral sequence $d_2 y_2 = z$.

37.3 Higher Cohomology Operations and Homotopy Groups of Spheres

The E_2-term of the Adams spectral sequence for S^0 is related to the primary, secondary, and so on cohomology operations in the obvious way. The elements in the first row correspond to primary operations; the elements in the second row correspond to the relations in the Steenrod algebra, that is, to the secondary operations, and so on. On the other hand, elements of the stable homotopy groups of spheres arise from the Adams spectral sequence. What is the relation between (primary and higher) cohomology operations and stable homotopy groups of spheres?

Am element α of $\pi_{N+k}(S^N)$ $(k \geq 0)$ is a class of spheroids $S^{N+k} \to S^n$. Attach D^{N+k+1} to S^N by such a spheroid. We get a CW complex whose homotopy type depends only on α; we denote this CW complex by X_α. Obviously, X_α has nontrivial cohomology groups in two positive dimensions: $H^N(X_\alpha;\mathbb{Z}_p) \cong \mathbb{Z}_p$, $H^{N+k+1}(X_\alpha;\mathbb{Z}_p) \cong \mathbb{Z}_p$.

EXERCISE 5. Prove that the higher cohomology operation corresponding to the element of the E_2-term of the Adams spectral sequence which becomes in E_∞ an element represented by α acts nontrivially in the cohomology of X_α. (*Hint:* Consider the Adams spectral sequences for X_α and S^N and the homomorphism of the first into the second arising from the inclusion map $S^N \to X_\alpha$.

Consider, for example, the elements $h_1, h_2, h_3, \ldots$ of the first row of the mod 2 Adams spectral sequence for S^0. Those which survive to E_∞ correspond to the elements α of the groups $\pi_{N+1}(S^N)$, $\pi_{N+3}(S^N), \pi_{N+7}(S^N), \ldots$ such that in the corresponding spaces X_α the operations $\mathrm{Sq}^2, \mathrm{Sq}^4, \mathrm{Sq}^8, \ldots$ act nontrivially. Adams proved that the elements h_i with $i \geq 4$ do not survive to E_∞: They are killed by differentials. For example, $d_2 h_4 = h_0 h_3^2 \neq 0$. This shows that a space of the form $S^N \cup D^{N+q}$ with a nontrivial action of Sq^q exists only for $q = 1, 2, 4, 8$. We explained in Sect. 31.4 that this is equivalent to the fact that a group $\pi_{4n-1}(S^2 n)$ contains an element with odd Hopf invariant only for $n = 1, 2$, and 4.

Let us mention a more classical (and actually equivalent) approach to the problem of the odd Hopf invariant (known as the Frobenius conjecture). As explained in Sect. 31.4, the reason why the operation Sq^q cannot act nontrivially in the space $S^N \cup D^{N+q}$, if q is not a power of 2, is the fact that for such a q, the Steenrod squares Sq^q are expressed as sums of compositions of Steenrod squares Sq^r with $r < q$ (which act trivially in $S^N \cup D^{N+q}$). However, $\mathrm{Sq}^{16}, \mathrm{Sq}^{32}, \ldots$ are not decomposable into compositions. Why are there no spaces $S^N \cup D^{N+16}$, $S^N \cup D^{N+32}, \ldots$ with nontrivial actions of $\mathrm{Sq}^{16}, \mathrm{Sq}^{32}, \ldots$? The answer is simple: $\mathrm{Sq}^{16}, \mathrm{Sq}^{32}, \ldots$ are decomposable, but not through Steenrod squares, but rather through higher cohomology operations. The reader can try to prove that for Steenrod squares Sq^{2^r}, the existence of such a decomposition is equivalent to the nontriviality of some differentials of the Adams spectral sequence for the elements h_r of the E_2-term.

37.4 Postnikov Towers

We already know that the homotopy groups of a space (even of a CW complex) do not fully determine its homotopy type. There are two exceptions: when all the homotopy groups are trivial, and when there is only one nontrivial homotopy group.

Even in the case of two nontrivial homotopy groups it is not necessarily so. Indeed, let Π_1 and Π_2 be two Abelian groups, and let n_1 and n_2 be two integers with $1 < n_1 < n_2$ (to avoid unnecessary complications, we do not consider the case $n_1 = 1$). Suppose that a CW complex X has trivial homotopy groups of all dimensions except n_1 and n_2, and $\pi_{n_1}(X) = \Pi_1$, $\pi_{n_2}(X) = \Pi_2$. Then there arises a (homotopically unique) map $X \to K(\Pi_1, n_1)$ which may be considered a fibration, and the fiber is $K(\Pi_2, n_2)$. In the cohomology spectral sequence of this fibration with coefficients in Π_2, the fundamental class $u_2 \in H^{n_2}(K(\Pi_2, n_2); \Pi_2) = E_2^{0,n_2}$ is transgressive and is taken by the transgression into some element f_1 of $E_2^{n_2+1,0} = H^{n_2+1}(K(\Pi_1, n_1); \Pi_2)$. This f_1, on one hand, fully determines the fibration, and hence (the homotopy type of) X, and on the other hand it can be regarded as a (primary) cohomology operation $H^{n_1}(-; \Pi_1) \to H^{n_2+1}(-; \Pi_2)$. Thus, the homotopy type of a space X with two nontrivial homotopy groups is determined by these groups plus one primary cohomology operation, which is called the *(first) Postnikov factor* of X.

Suppose now that Y is a space (a CW complex) with three nontrivial homotopy groups, $\pi_{n_1}(Y) = \Pi_1$, $\pi_{n_2}(Y) = \Pi_2$, $\pi_{n_3}(Y) = \Pi_3$, where $1 < n_1 < n_2 < n_3$. The capping operation (described in Sect. 11.9) may turn Y into a space $X \supset Y$ with only two nontrivial homotopy groups, $\pi_{n_1}(X) = \Pi_1$, $\pi_{n_2}(X) = \Pi_2$, such that the homomorphisms $\pi_q(Y) \to \pi_q(X)$ induced by the inclusion map $Y \to X$ are the identity maps for $q = n_1, n_2$. If we turn this inclusion map into the homotopy equivalent fibration, then the fiber of this fibration will be $K(\Pi_3, n_3)$. The transgression in the cohomology spectral sequence of this fibration takes the fundamental class $u_3 \in H^{n_3}(K(\Pi_3, n_3); \Pi_3)$ into some class $f_2 \in H^{n_3+1}(X; \Pi_3)$ and this class fully determines the fibration and hence Y (provided that X is known).

We explained in Sect. 37.1 that the class f_2 can be regarded as a secondary cohomology operation defined on the kernel of the primary cohomology operation f_1. Let us repeat this explanation using the data of the current construction. Let W be some space (some CW complex), and let $\xi \in H^{n_1}(X; \Pi_1)$. The class ξ gives rise to a map $h: W \to K(\Pi_1, n_1)$. If we assume that $f_1(\xi) = 0$, then the fibration over W induced by the fibration $X \xrightarrow{K(\Pi_2, n_2)} K(\Pi_1, n_1)$ via the map h is trivial, and h can be lifted to a map $\widetilde{h}: W \to X$; we put $f_2(\xi) = \widetilde{h}^*(f_2) \in H^{n_3+1}(W; \Pi_3)$ (we use the same notations for cohomology operations and cohomology classes which determine them). Thus, f_2 is a partial multivalued operation; it is partial, because it is defined only on the kernel of f_1, and it is multivalued because the choice of $\widetilde{h}$ is not unique. The operation f_2 is called the *second Postnikov factor* of W.

EXERCISE 6. Prove that the values of f_2 are defined up to the image of the primary cohomology operation $H^{n_2}(-; \Pi_2) \to H^{n_3+1}(-; \Pi_3)$, which is the first Postnikov factor of the killing space $W|_{n_2}$.

This construction may be repeated infinitely many times. The result is schematically shown in Fig. 133.

The notation $K(\Pi_1, n_1; \Pi_2, n_2; \ldots)$ means a space X with nontrivial homotopy groups $\pi_{n_1}(X) = \Pi_1$, $\pi_{n_2}(X) = \Pi_2, \ldots$. As we know, the homotopy groups do not determine the homotopy type, so these notations do not fully specify the spaces. The space W appears as the limit of the vertical sequence. The cohomology classes u_i are the fundamental classes, and τ means the transgressions. The transgression images $f_1, f_2, f_3, \ldots$ are cohomology operations: f_1 is a primary operation, f_2 is a secondary operation defined on the kernel of f_1, f_3 is a ternary operation defined on the kernel of f_2, and so on. These cohomology operations are called *Postnikov factors*.

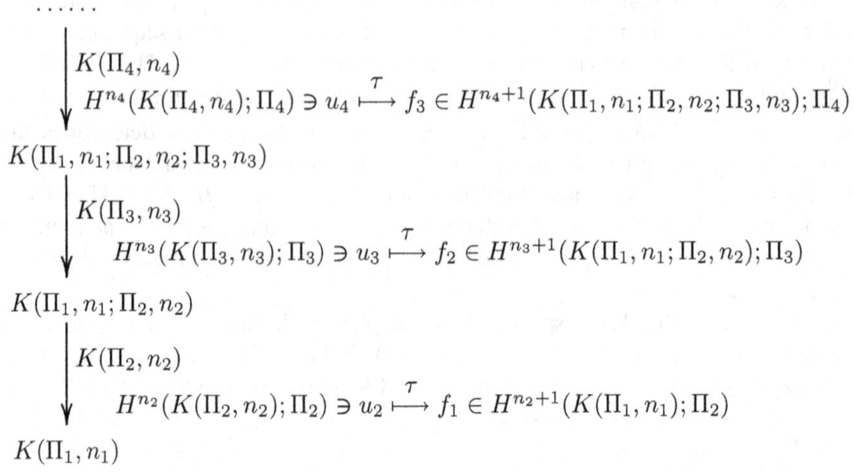

$$
\begin{array}{l}
\cdots\cdots \\
\Big\downarrow \begin{array}{l} K(\Pi_4, n_4) \\ H^{n_4}(K(\Pi_4, n_4); \Pi_4) \ni u_4 \stackrel{\tau}{\longmapsto} f_3 \in H^{n_4+1}(K(\Pi_1, n_1; \Pi_2, n_2; \Pi_3, n_3); \Pi_4) \end{array} \\
K(\Pi_1, n_1; \Pi_2, n_2; \Pi_3, n_3) \\
\Big\downarrow \begin{array}{l} K(\Pi_3, n_3) \\ H^{n_3}(K(\Pi_3, n_3); \Pi_3) \ni u_3 \stackrel{\tau}{\longmapsto} f_2 \in H^{n_3+1}(K(\Pi_1, n_1; \Pi_2, n_2); \Pi_3) \end{array} \\
K(\Pi_1, n_1; \Pi_2, n_2) \\
\Big\downarrow \begin{array}{l} K(\Pi_2, n_2) \\ H^{n_2}(K(\Pi_2, n_2); \Pi_2) \ni u_2 \stackrel{\tau}{\longmapsto} f_1 \in H^{n_2+1}(K(\Pi_1, n_1); \Pi_2) \end{array} \\
K(\Pi_1, n_1)
\end{array}
$$

Fig. 133 The Postnikov tower

Thus, according to the Postnikov theory, *the homotopy type of a simply connected CW complex is determined by the homotopy groups plus a sequence of cohomology operations, a primary one, a secondary one, and so on, such that every operation from this sequence in defined on the kernel of the previous one.*

EXERCISE 7. Reconstruct the details of this construction (including the limit transition).

EXERCISE 8. Prove that the first Postnikov factor of S^n (with $n \geq 3$) is Sq^2. Try to describe the second Postnikov factor.

Chapter 6
K-Theory and Other Extraordinary Cohomology Theories

Lecture 38 General Theory

38.1 Introduction

K-theory emerged as an independent part of topology in the late 1950s, when the limits of the possibilities of the methods based on spectral sequences and cohomology operations (and studied in Chaps. III–V) became visible. Progress in homotopy topology considerably slowed down, the leading topologists got involved in cumbersome calculations, the results became less and less impressive, and to obtain them one had to combine a great inventiveness with a readiness to do a huge amount of tedious work. To get these activities revived a radical method was needed. And it was found! It consisted of replacing cohomology as the basic homotopy invariant by an entirely new object, the so-called K-functor.

The history of K-theory is enlightening. The period of its active development was unbelievably short: The first works in K-theory were published in 1959, and as early as 1963 most of its results were completed (including all the results described below, with the exception of the proof of the Adams conjecture). And it is remarkable that all works in K-theory of this period belong to four authors: Frank Adams, Michael Atiyah, Raoul Bott, and Friedrich Hirzebruch. There is no name of Alexander Grothendieck on this list, for he has no works in topology; but according to common opinion, it was Grothendieck who had first conceived of the key ideas of K-theory that found their applications not only in topology.

The topological applications of K-theory obtained during its "heroic period" were really impressive. Let us mention only a simple proof of the nonexistence of division algebras in dimensions different from 1, 2, 4, 8; the precise computation of the maximal number of linearly independent vector fields on the sphere of an arbitrary dimension; the computation of the order of the image of the "J-homomorphism" $\pi_n(SO) \to \pi_n^S$; theorems of nonembeddability and nonimmersibility of various manifolds into Euclidean spaces; as well as various

A. Fomenko, D. Fuchs, *Homotopical Topology*, Graduate Texts in Mathematics 273,
DOI 10.1007/978-3-319-23488-5_6

integrality theorems for rational linear combinations of characteristic classes. The solution of the problem of the index of an elliptic operator found in 1963 by Atiyah and Singer would have been totally impossible without K-theory.

The majority of topologists became aware of the existence of K-theory only after its construction had mainly been completed and its most remarkable applications had mostly been obtained. Nevertheless, they found work to do. They included K-theory in the frames of a general theory of "extraordinary cohomology." It turned out that K-theory is a rather modest representative of the family of extraordinary cohomologies. The main reason for that is as follows. As the reader remembers, for the usual cohomology (for example, with the coefficients in $\mathbb{Z}$) the following Whitehead theorem holds: If X and Y are simply connected CW complexes, and if a continuous map $f: X \to Y$ induces an isomorphism $f^*: H^*(Y; \mathbb{Z}) \to H^*(X; \mathbb{Z})$, then f is a homotopy equivalence (see Sect. 14.5). K-theory also assigns to a (finite) CW complex a graded group that depends on it in a functorial way, but there is no analog of Whitehead's theorem: It is quite possible that $f^*: K^*(Y) \to K^*(X)$ is an isomorphism, but f is not a homotopy equivalence. Some extraordinary theories were conceived (or, better to say, recollected) that have all the advantages of K-theory and are free of this flaw.

The best known of these theories is the cobordism theory, which exists in a variety of versions (the same is true for K-theory, by the way). For cobordisms, the analog of Whitehead's theorem is valid; thus, from the point of view of homotopy theory, cobordism theory is not worse than cohomology. The topologists were able to construct for cobordisms an analog of the Adams spectral sequence, the so-called Adams–Novikov spectral sequence, which was repeatedly applied to homotopy calculations. A theory of cobordism-valued characteristic classes proved to be very useful. At last the assertion that cobordisms are better than K-theories had a precise statement: The (complex) K-theory may be embedded into the (complex) cobordism theory as a direct summand. We pay a tribute to cobordism theories by devoting a separate lecture to them (Lecture 44).

Nevertheless, one has to admit that from the viewpoint of applications, K-theory considerably surpasses all other extraordinary cohomology theories. Thus, K-theory is the main subject of this chapter. By necessity, we restrict ourselves almost entirely to the complex case.

38.2 Definitions

A: The Groups $K(X)$

Let X be a finite CW complex. Usually we assume that X has a base point, x_0. Denote by $F(X)$ the set of equivalence classes of complex vector bundles with the base X. If X is connected, then $F(X) = \coprod_{n \geq 0} F_n(X)$, where $F_n(X)$ is the set of classes of n-dimensional vector bundles with the base X. If X is not connected, then we do not require that the dimensions of the fibers over different components be the same; in this case we define the dimension of the bundle as the dimension of the fiber over the base point, if there is one.

There are two binary operations in $F(X)$: direct sum and tensor product. With respect to addition (direct summation), $F(X)$ is a commutative semigroup with zero; the multiplication is commutative, associative, distributive with respect to addition, and possesses the identity element.

There is a general operation in algebra which converts a commutative semigroup with zero into a commutative group (the so-called Grothendieck group). Namely, let F be a commutative semigroup with zero. Consider the set of all formal differences $a - b, a \in F, b \in F$ and introduce the following equivalence relation in this set: $(a - b) \sim (c - d)$ if there exists some $e \in F$ such that $a + d + e = b + c + e$. Obviously, this relation is reflexive and symmetric.

EXERCISE 1. Check that it is also transitive.

Denote by $G(F)$ the set of equivalence classes of formal differences in F, and define an addition operation in $G(F)$ by the formula $\{a-b\} + \{c-d\} = \{(a+c) - (b+d)\}$.

EXERCISE 2. Check that $G(F)$ with this operation is a group.

This is *the Grothendieck group of the semigroup F*. The formula $a \mapsto \{a - 0\}$ defines a map $F \to G(F)$.

EXERCISE 3. Check that this map is a homomorphism [with respect to the addition operations in F and $G(F)$].

EXERCISE 4. This map is not necessarily an injection: Give an example.

EXERCISE 5. Let G be an Abelian group. Prove that any homomorphism $F \to G$ is uniquely decomposed into a composition of the canonical map $F \to G(F)$ and a group homomorphism $G(F) \to G$.

EXERCISE 6. Prove that the last property of the group $G(F)$ and the homomorphism $F \to G(F)$ uniquely determines both.

Let us return to the vector bundles.

Definition. $K(X) = G(F(X))$. The elements of the group $K(X)$ are sometimes called *virtual vector bundles* over X.

A continuous map $f: X \to Y$ defines, in the obvious way, the "induced homomorphism" $f^*: K(Y) \to K(X)$. Homotopic continuous maps always induce equal homomorphism; thus, $K(X)$ is a homotopy invariant of X.

Example. $F(\text{pt}) = \mathbb{Z}_{\geq 0}, K(\text{pt}) = \mathbb{Z}$; the same holds for any contractible space.

The homomorphism $\dim: F(X) \to \mathbb{Z}_{\geq 0}$ determines a homomorphism $\dim: K(X) \to \mathbb{Z}$. This is seen from Exercise 5 but may be described directly: $\dim\{a - b\} = \dim a - \dim b$ (check that it is well defined). Notice that the dimension of a virtual vector bundle may be negative.

Proposition 1. (i) *For any virtual vector bundle* $\alpha \in K(X)$ *there exist a usual vector bundle* a *and an integer* N *such that* $\alpha = \{a - N\}$. *(Recall that in the theory of vector bundles* N *denotes the standard trivial vector bundle of dimension* N, $X \times \mathbb{C}^N \to X$.*)*

(ii) *Virtual vector bundles* $\alpha = \{a - N\}$ *and* $\beta = \{b - M\}$ *are equal if and only if* $\dim a = \dim b$ *and the vector bundles* a *and* b *are stably equivalent (see the definition in Sect. 19.1).*

The proof is based on the following lemma.

Lemma. *For any vector bundle* a *over (a finite* CW *complex)* X *there exists a vector bundle* $\overline{a}$ *over* X *such that the sum* $a \oplus \overline{a}$ *is trivial.*

Proof of Lemma. According to Sect. 19.4, $a = f^*\zeta$, where ζ is the tautological bundle over an appropriate complex Grassmannian and f is a continuous map of X into this Grassmannian. Hence, we need to prove the lemma only in the case when a itself is the tautological bundle ζ over the Grassmannian $\mathbb{C}G(N; n)$. Denote by $\zeta^\perp$ the "dual tautological bundle," whose fiber over $x \in \mathbb{C}G(N; n)$ is the orthogonal complement $x^\perp$ to x in $\mathbb{C}^N$. The fiber of $\zeta \oplus \zeta^\perp$ over x is $x \oplus x^\perp = \mathbb{C}^N$. Hence, the bundle $\zeta \oplus \zeta^\perp$ is trivial.

Proof of Proposition. (i) Let $\alpha = \{c - d\}$, and let $\overline{d}$ be a bundle such that $d \oplus \overline{d} = N$. Then the virtual bundle $\{c - d\}$ is equivalent to $\{(c \oplus \overline{d}) - (d \oplus \overline{d})\} = \{a - N\}$, where $a = c \oplus \overline{d}$.

(ii) The equality $\{a - N\} = \{b - M\}$ means that $a \oplus M \oplus c = b \oplus N \oplus c$ for some c. Choose a $\overline{c}$ such that $c \oplus \overline{c} = K$. Then $a \oplus M \oplus K = a \oplus M \oplus c \oplus \overline{c} = b \oplus N \oplus c \oplus \overline{c} = b \oplus N \oplus K$; hence, a is stably equivalent to b. Conversely, let the bundles a and b be stably equivalent, which means that $a \oplus K = b \oplus L$ for some K and L. Hence, $a \oplus K \oplus M \oplus N = b \oplus L \oplus M \oplus N$, and since $\dim(a - N) = \dim(b - M)$ and $\dim(a + K) = \dim(b + L)$, then $K + N = L + M$, and therefore $a \oplus M \oplus c = b \oplus N \oplus c$, where $c = K + N = L + M$, which shows that $a - N$ is equivalent to $b - M$.

B: The Groups $\widetilde{K}(X)$

Definition. $\widetilde{K}(X) = \operatorname{Ker}[\dim\colon K(X) \to \mathbb{Z}]$.

Since $K(\mathrm{pt}) = \mathbb{Z}$ and $\dim\colon K(\mathrm{pt}) \to \mathbb{Z}$ is an isomorphism, our definition is equivalent to the following definition:

$$\widetilde{K}(X) = \operatorname{Ker}[K(X) \to K(x_0)],$$

where x_0 is the base point of X.

Obviously, $K(X) = \widetilde{K}(X) \oplus \mathbb{Z}$ always.

Theorem. $\widetilde{K}(X)$ *is the set of classes of stably equivalent vector bundles over* X.

Proof. The preceding proposition implies that any element α of $\widetilde{K}(X)$ is represented by the difference $a - \dim a$ where a is a vector bundle, and a is determined by α up to stable equivalence. At the same time, the difference $a - \dim a$ determines for any vector bundle a a certain element of $\widetilde{K}(X)$.

Set

$$BU = \mathbb{C}G(\infty,\infty) = \lim_{n\to\infty}\lim_{N\to\infty}\mathbb{C}G(N,n),\ U = \lim_{n\to\infty} U(n).$$

The above theorem has the following important

Corollary. $\widetilde{K}(X) = \pi(X, BU)$.

Proof. We need to prove that classes of stably equivalent vector bundles over (a finite CW complex) X correspond one-to-one to homotopy classes of maps $X \to BU$. According to Sect. 19.4, a vector bundle (of dimension n) over X gives rise to a continuous map $X \to \mathbb{C}G(\infty,n)$ and since $\mathbb{C}G(\infty,n) \subset \mathbb{C}G(\infty,\infty)$ also gives rise to a continuous map $X \to \mathbb{C}G(\infty,\infty)$. But it is obvious that the maps $X \to \mathbb{C}G(\infty,n_1), X \to \mathbb{C}G(\infty,n_2)$, which correspond to two vector bundles over X, are homotopic in $\mathbb{C}G(\infty,\infty)$ if and only if these two vector bundles are stably equivalent.

Homotopy Lemma. *For every r, there is an isomorphism $\pi_r(U) = \pi_{r+1}(BU)$. The same is true if BU and U are replaced by $BU(n) = \mathbb{C}G(\infty,n)$ and $U(n)$.*

Let us prove the assertion concerning $BU(n)$ and $U(n)$ (the part concerning BU and U may be obtained by considering the obvious limits). We have a fibration

$$\mathbb{C}V(\infty,n) \xrightarrow{U(n)} \mathbb{C}G(\infty,n),\ \text{that is } EU(n) \xrightarrow{U(n)} BU(n)$$

with $EU(n)$ being contractible. [The reader may prove the contractibility of $\mathbb{C}V(\infty,n)$ as an exercise; for our purposes, however, it is sufficient to know that $\pi_r(\mathbb{C}V(\infty,n)) = 0$ for any G, and this follows from the equality $\pi_r(\mathbb{C}V(N,n)) = 0$ for $\leq 2(N-n)$; see Sect. 19.3.] The isomorphism needed arises from the exactness of the homotopy sequence of this fibration.

EXERCISE 7. Prove the following stronger form of the homotopy lemma: The spaces ΩBU and U are homotopy equivalent, and the spaces $\Omega BU(n)$ and $U(n)$ are homotopy equivalent.

EXERCISE 8. A further strengthening of the homotopy lemma: The spaces above are homotopy equivalent as H-spaces.

APPLICATION. $\widetilde{K}(S^r) \cong \pi_r(BU) = \pi_{r-1}(U)$. *In particular,* $\widetilde{K}(S^2) = \pi_1(U) = \mathbb{Z}$.

Notice that the group $\widetilde{K}(S^2)$ is generated by the stable class of the Hopf bundle ζ over $S^2 = \mathbb{C}P^1$. [In other words, $\widetilde{K}(S^2)$ is generated by the class of the virtual bundle $\zeta - 1$.] Indeed, the Hopf bundle ζ is not stably trivial since its first Chern class $c_1(\zeta)$

does not vanish (see Sect. 19.5); hence, ζ determines a nonzero element of $\widetilde{K}(S^2)$. But this element cannot be divisible, since it is always true that $c_1(\xi_1 \oplus \xi_2) = c_1(\xi_1) + c_1(\xi_2)$ and hence $c_1(n\xi) = nc_1(\xi)$, which shows that c_1 of a divisible vector bundle is divisible, while $c_1(\zeta)$ is the generator of the group $H^2(S^2; \mathbb{Z}) \cong \mathbb{Z}$, which is not divisible.

Furthermore, $K(S^2) = \widetilde{K}(S^2) \oplus \mathbb{Z} = \mathbb{Z} \oplus \mathbb{Z}$. This group is generated by ζ and 1. Thus, we have an isomorphism

$$K(S^2) \longrightarrow \mathbb{Z} \oplus \mathbb{Z},\ \xi \mapsto (c_1(\xi), \dim \xi)$$

[we regard $c_1(\xi)$ as an integer, because $H^2(S^2; \mathbb{Z}) \cong \mathbb{Z}$]. In particular, the bundles $(\zeta \otimes \zeta) \oplus 1$ and $\zeta \oplus \zeta$ determine the same element of $K(S^2)$ (actually, these two vector bundles are equivalent), and hence we have the relation

$$\zeta^2 = 2\zeta - 1$$

in $K(S^2)$ [and $(\zeta - 1)^2 = 0$ in $\widetilde{K}(S^2)$].

C: The Groups $K^q(X, A)$ with $q \leq 0$

Let (X, A) be a finite CW pair (a finite CW complex and a CW subcomplex of it). For $q \leq 0$, put

$$\begin{aligned} K^q(X, A) &= \widetilde{K}(\Sigma^{-q}(X/A)), \\ K^q(X) &= K^q(X, \emptyset), \\ \widetilde{K}^q(X) &= K^q(X, x_0), \end{aligned}$$

where x_0 is the base point. Obviously,

$$K^0(X) = \widetilde{K}(X/\emptyset) = \widetilde{K}(X \sqcup \mathrm{pt}) = K(X)$$

[$\widetilde{K}(X \sqcup \mathrm{pt})$ is the group of virtual bundles over $X \sqcup \mathrm{pt}$ of dimension 0 over pt, which is simply the group of virtual bundles over X],

$$\widetilde{K}^0(X) = \widetilde{K}(X/x_0) = \widetilde{K}(X).$$

(Notice, however trivial it is, that the quotient $X/\emptyset$ consists of X and one more isolated point: The elements of the set $X/\emptyset$ are elements of the difference $X - \emptyset = X$ and one more element $\{\emptyset\}$.)

Example. $K^q(\mathrm{pt}) = \widetilde{K}^q(\mathrm{pt}/\emptyset) = \widetilde{K}^q(\mathrm{pt} \sqcup \mathrm{pt}) = \widetilde{K}(\Sigma^{-q}(S^0)) = \widetilde{K}(S^{-q}) = \pi_{-q}(BU)$.

In K-theory, the groups $K^q(X,A)$ with $q > 0$ are, certainly, also studied, but just to define them we need the so-called Bott periodicity, and we postpone any discussion of this group until Sect. 38.3.

D: K-Sequences of Pairs

Lemma. *The functor $\widetilde{K}$ is half-exact. This means that for any (finite CW) pair (X,A), the sequence*

$$\widetilde{K}(X/A) \longrightarrow \widetilde{K}(X) \longrightarrow \widetilde{K}(A)$$

of homomorphisms induced by the inclusion $A \to X$ and the projection $X \to X/A$ is exact.

The easiest way to prove it is to use the equality $\widetilde{K}(Y) = \pi(Y, BU)$.

Now for a pair (X,A), consider the following diagram:

$$\begin{array}{ccccc} A \longrightarrow X \longrightarrow & X/A & & & \\ \| & \uparrow & & & \\ X \longrightarrow & X \cup CA & \longrightarrow & \Sigma A & \\ & \| & & \uparrow & \\ & X \cup CA & \longrightarrow & CX \cup CA & \longrightarrow \Sigma X. \end{array}$$

The vertical arrows of this diagram denote homotopy equivalences: the factorization over CA and the factorization over CX. All the rows have the form $B \to Y \to Y/B$. This lets us identify all the $\widetilde{K}$-groups for the spaces in any column and get the sequence

$$\widetilde{K}(A) \longleftarrow \widetilde{K}(X) \longleftarrow \widetilde{K}(X/A) \longleftarrow \widetilde{K}(\Sigma A) \longleftarrow \widetilde{K}(\Sigma X),$$

which is exact by the lemma. Notice that the first, second, and fourth arrows in this sequence are induced, correspondingly, by the inclusion map $A \to X$, by the projection $X \to X/A$, and by the inclusion map $\Sigma A \to \Sigma X$. Precisely in the same way, using the pair $(\Sigma X, \Sigma A)$ instead of the pair (X,A), we construct the exact sequence

$$\widetilde{K}(\Sigma A) \longleftarrow \widetilde{K}(\Sigma X) \longleftarrow \widetilde{K}(\Sigma X/\Sigma A) \longleftarrow \widetilde{K}(\Sigma^2 A) \longleftarrow \widetilde{K}(\Sigma^2 X),$$

and, furthermore, the exact sequence

$$\begin{aligned} \widetilde{K}(\Sigma^q A) \longleftarrow \widetilde{K}(\Sigma^q X) \longleftarrow & \widetilde{K}(\Sigma^q X/\Sigma^q A) \\ \longleftarrow & \widetilde{K}(\Sigma^{q+1} A) \longleftarrow \widetilde{K}(\Sigma^{q+1} X) \end{aligned}$$

for any $q \geq 0$. All these sequences may be combined into one long exact sequence

$$\begin{aligned} \widetilde{K}(A) \longleftarrow \widetilde{K}(X) \longleftarrow \widetilde{K}(X/A) \longleftarrow \dots \\ \longleftarrow \widetilde{K}(\Sigma^q X/\Sigma^q A) \longleftarrow \widetilde{K}(\Sigma^{q+1} A) \longleftarrow \widetilde{K}(\Sigma^{q+1} X) \longleftarrow \dots \end{aligned}$$

If we replace in this sequence the spaces X and A first by the spaces $X \sqcup \mathrm{pt}$ and $A \sqcup \mathrm{pt}$, and then by the spaces X/B and A/B, where B is a CW subcomplex of A, then we get (using the definitions from Sect. 38.2.C) exact sequences

$$\begin{aligned} K^0(A) \longleftarrow K^0(X) \longleftarrow K^0(X,A) \longleftarrow K^{-1}(A) \longleftarrow \dots \\ \longleftarrow K^{-q}(X) \longleftarrow K^{-q}(X,A) \longleftarrow K^{-q-1}(A) \longleftarrow K^{-q-1}(X) \longleftarrow \dots \end{aligned}$$

and

$$\begin{aligned} K^0(A,B) \longleftarrow K^0(X,B) \longleftarrow K^0(X,A) \longleftarrow K^{-1}(A,B) \longleftarrow \dots \\ \longleftarrow K^{-q}(X,B) \longleftarrow K^{-q}(X,A) \longleftarrow K^{-q-1}(A,B) \longleftarrow K^{-q-1}(X,B) \longleftarrow \dots. \end{aligned}$$

These are the K-sequences of pairs and triples.

Naturally, continuous maps between (finite CW) pairs induce homotopy invariant homomorphisms of all groups K^q (in the opposite direction), and these homomorphisms are compatible with all the K-sequences of pairs and triples.

Remark. The geometric construction, which we used for the definition of the K-sequences of pairs and triples, may be continued. As a result we will get a sequence of continuous maps

$$\begin{aligned} A \longrightarrow X \longrightarrow X/A \dashrightarrow \Sigma A \longrightarrow \Sigma X \\ \longrightarrow \Sigma(X/A) \dashrightarrow \Sigma^2 A \longrightarrow \dots, \end{aligned}$$

where dotted arrows are defined up to a homotopy. This sequence is called the *Puppe sequence.* The Puppe sequence is homotopy exact in the sense that any three-term fragment of this sequence is homotopy equivalent to a fragment of the form $B \to Y \to Y/B$.

E: Attempts to Generalize K to the Case of Infinite CW Complexes

There is no satisfactory generalization of K-theory to the case of infinite CW complexes. Let us consider several possible versions of the definition of K-groups in this case.

VERSION 1. $K(X)$ is the Grothendieck group of the semigroup of the equivalence classes of vector bundles.

VERSION 2. $\widetilde{K}(X) = \pi(X, BU)$, $K(X) = \widetilde{K}(X) \oplus \mathbb{Z}$.

VERSION 3. $K(X) = \varprojlim K(Y)$, where the limit is taken with respect to all finite CW subcomplexes of X.

Version 1 looks the most natural, but nothing is known about such a "K-functor": It is unlikely that it possesses any good properties. It is not even clear how to compute it, for example, in the case $X = \mathbb{C}P^\infty$. Versions 2 and 3 appear to make more sense; the corresponding K-functors are denoted by the symbols k and $\mathcal{K}$, respectively. The biggest advantage of $\mathcal{K}$-theory is its relatively easy computability: actually, the methods of computing groups $K(X)$, developed in Lecture 39 ahead, if applied to an infinite CW complex X, yield precisely $\mathcal{K}(X)$. However, there are no exact sequences of pairs and triples in this theory: It is not hard to construct an example of an infinite CW pair (X, Y) with $\mathcal{K}(X/Y) = 0$, $\mathcal{K}(Y) = 0$, but $\mathcal{K}(X) \neq 0$ (the reader may try to construct such an example after reading Lecture 39). On the contrary, the k-functor is (half) exact, but it is very difficult to compute its values. Indeed, let us try to understand the difference between k and $\mathcal{K}$. There is an obvious map $k \to \mathcal{K}$ [a continuous map $X \to BU$ restricts to a map $Y \to BU$, and hence to an element of $\widetilde{K}(Y)$ for any finite CW subcomplex Y of X; these elements compose an element of $\varprojlim K(Y) = \widetilde{\mathcal{K}}(X)$].

EXERCISE 9. Prove that this map $\widetilde{k}(X) \to \widetilde{\mathcal{K}}(X)$ is an epimorphism.

What is the kernel of this map? This kernel consists of classes of maps $X \to BU$ which are homotopic to a constant on any finite CW subspace $Y \subset X$, but still not homotopic to a constant on the whole X. Is such an odious thing possible? It turns out that yes!

Definition. A continuous map of a CW complex X into some topological space is called *a phantom map* if it is not homotopic to a constant, but its restriction to any finite CW subspace of X is homotopic to a constant.

The first example of a phantom, rather a complicated one, was constructed by J. F. Adams and G. Walker (see [10]). Simpler examples were found later, and now an example of a phantom map $\mathbb{C}P^\infty \to S^3$ is known (the reader may try to construct such a map). Relationships between k- and $\mathcal{K}$-theories were first studied by Atiyah and Hirzebruch, and later by V. Buchstaber and A. Mishchenko [26]. We will mention two of Buchstaber and Mishchenko's results. (1) *If all the odd-numbered Betti numbers of X are 0, then $k(X) = \mathcal{K}(X)$.* (2) *Let $X = K(\mathbb{Z}, 3)$. Then $\widetilde{\mathcal{K}}(X) = 0$, while $\widetilde{k}(X) = \widehat{\mathbb{Z}}/\mathbb{Z}$, where $\widehat{\mathbb{Z}}$ is the completion of the group $\mathbb{Z}$ with respect to all finite index subgroups.* (In other words, $\widehat{\mathbb{Z}}$ is the subgroup of the infinite product $\mathbb{Z}_2 \times \mathbb{Z}_3 \times \mathbb{Z}_4 \times \dots$ consisting of such sequences $(a_2, a_3, a_4, \dots)$ that $a_{qr} \equiv a_q \bmod q$ for any integers q, r; the group $\mathbb{Z}$ is embedded into $\widehat{\mathbb{Z}}$ by the formula $n \mapsto (\mathrm{res}_2\, n, \mathrm{res}_3\, n, \mathrm{res}_4\, n, \dots)$; in other words, $\mathbb{Z} \subset \widehat{\mathbb{Z}}$ is the group of stabilizing sequences; it is clear that it is considerably less than $\widehat{\mathbb{Z}}$.)

38.3 Bott Periodicity

A: The Statement of Bott's Theorem

As we already noticed, $K(S^2) = \mathbb{Z} \oplus \mathbb{Z}$, and the generators of the two groups $\mathbb{Z}$ are the classes 1 and ζ of the trivial bundle and the Hopf bundle. Let X be a finite CW complex, and let $\alpha_1, \alpha_2 \in K(X)$. Then we can consider the virtual bundle

$$(\alpha_1 \otimes 1) + (\alpha_2 \otimes \zeta) \in K(X \times S^2).$$

Theorem (Bott). *For any finite* CW *complex X, the mapping*

$$K(X) \oplus K(X) \longrightarrow K(X \times S^2),\ (\alpha_1, \alpha_2) \mapsto (\alpha_1 \otimes 1) + (\alpha_2 \otimes \zeta)$$

is an isomorphism.

We postpone the proof of this theorem until Sect. 38.3.C. Now we will reduce its statement to a more convenient to us (and, actually, more habitual) form.

Consider the K-sequence of the pair $(X \times S^2, X \vee S^2)$:

$$\cdots \to K^{-1}(X \times S^2) \to K^{-1}(X \vee S^2) \to K(X \times S^2, X \vee S^2)$$
$$\to K(X \times S^2) \xrightarrow{*} K(X \vee S^2)$$

(recall that $K^0 = K$). Obviously, $K(X \vee S^2) = K(X) \oplus K(S^2)$ [actually, $K(X \vee S^2) = \mathbb{Z} \oplus \widetilde{K}(X \vee S^2) = \mathbb{Z} \oplus \widetilde{K}(X) \oplus \widetilde{K}(S^2)$, $K(X) = \mathbb{Z} \oplus \widetilde{K}(X), K(S^2) = \mathbb{Z} \oplus \widetilde{K}(S^2)$]. Furthermore, any $a \in K(X)$ is the image with respect to the homomorphism $*$ of $p_1^* a \in K(X \times S^2)$ and any $b \in K(S^2)$ is the image of $p_2^* b \in K(X \times S^2)$. Hence, $*$ is an epimorphism, and in a similar way the map $K^{-1}(X \times S^2) \to K^{-1}(X \vee S^2)$ from the above sequence is an epimorphism. Hence, $K(X \times S^2, X \vee S^2)$ is simply the kernel of the homomorphism $*$. According to Bott's theorem, any element of $K(X \times S^2)$ has the form $(\alpha_1 \otimes 1) + (\alpha_2 \otimes \zeta)$. Rewrite the last sum as $(\alpha \otimes (\zeta - 1)) + (\beta \otimes \zeta)$. The restriction of this virtual bundle to $X \subset X \times S^2$ is equal to $\alpha \cdot \dim(\zeta - 1) + \beta \cdot \dim 1 = \beta$; the restriction of this virtual bundle to $S^2 \subset X \times S^2$ is equal to $\dim \alpha \cdot (\zeta - 1) + \dim \beta \cdot 1$. Thus, our virtual bundle belongs to the kernel of the homomorphism $*$ if and only if $\beta = 0$ and $\dim \alpha = 0$, that is, if it has the form $\alpha \otimes (\zeta - 1)$ with $\alpha \in \widetilde{K}(X)$. We see that the map

$$\alpha \mapsto \alpha \otimes (\zeta - 1)$$

establishes an isomorphism

$$\widetilde{K}(X) \longrightarrow K(X \times S^2, X \vee S^2) = \widetilde{K}((X \times S^2)/(X \vee S^2)) = \widetilde{K}(\Sigma^2 X).$$

Thus, we obtain

Corollary. *There is an isomorphism that is natural with respect to X:*

$$\widetilde{K}(X) \cong \widetilde{K}(\Sigma^2 X).$$

In the case $X = S^q$, this implies [since $\widetilde{K}(S^i) = \pi_i(BU)$]

Corollary of corollary. *For any q,*

$$\pi_q(BU) \cong \pi_{q-2}(BU),$$

and hence $\pi_q(U) \cong \pi_{q-2}(U)$.

It was this theorem that was originally proved by Bott and which is known as (unitary) Bott periodicity. Since $\pi_1(U) = \mathbb{Z}$ and $\pi_2(U) = \pi_2(SU) \oplus \pi_2(S^1) = \pi_2(SU) \oplus \pi_2(S^1) = 0$,

$$\pi_i(U) = \begin{cases} \mathbb{Z} \text{ for } i \text{ odd}, \\ 0 \text{ for } i \text{ even}, \end{cases}$$

and

$$\pi_i(BU) = \widetilde{K}(S^i) = \begin{cases} \mathbb{Z} \text{ for } i \text{ even}, \\ 0 \; for \; i \text{ odd}. \end{cases}$$

B: The Completion of the Construction of K-Theory

We postpone the proof of Bott's theorem a little more and show how it can be used to complete the construction of K-theory. The corollary of Bott's theorem given in this section shows that for any (finite CW) pair (X,A) and any $q \leq 0$, there is a natural [with respect to continuous maps between pairs and to connecting homomorphisms $K^q(X,A) \leftarrow K^{q-1}(A;B)$] isomorphism

$$K^q(X,A) \cong K^{q-2}(X,A).$$

We use this isomorphism to define $K^q(X,A)$ for all q: By definition,

$$K^q(X,A) = K^{q-2N}(X,A), \; N > \max(0,q)$$

[the groups $K^{q-2N}(X,A)$ with different large N are identified by means of the Bott periodicity isomorphism]. We get a 2-periodic extraordinary cohomology theory (the precise meaning of this term will be explained in Sect. 38.5), that is, a rule that assign to a finite CW pair (X,A) a sequence of groups $K^q(X,A), q \in \mathbb{Z}$

and a sequence of homomorphisms $K^q(A) = K^q(A, \emptyset) \to K^{q+1}(X, A)$, and to a continuous map $(X, A) \to (Y, B)$ between two finite CW pairs a sequence of homomorphisms $K^q(Y, B) \to K^q(X, A)$, and all these groups and homomorphisms are homotopy invariant and satisfy suitable conditions of the commutativity of diagrams and the exactness of sequences (see the details in Sect. 38.5). In addition to this, compatible with all these structures are isomorphisms $K^q(X, A) = K^{q-2}(X, A)$. (The last feature illustrates the important difference between K and H, since $H^q = 0$ for $q < 0$.)

Now we can apply to K-theory all the standard corollaries of the exactness of the sequences of pairs and triples. In particular, there are suspension isomorphisms

$$\Sigma\colon \widetilde{K}^q(X) \cong \widetilde{K}^{q+1}(\Sigma X)$$

[the exactness of the $\widetilde{K}$-sequence of the pair (CX, X)

$$\begin{array}{ccccccc} \widetilde{K}^q(CX) & \longrightarrow & \widetilde{K}^q(X) & \longrightarrow & K^{q+1}(CX, X) & \longrightarrow & \widetilde{K}^{q+1}(CX) \\ \| & & & & & & \| \\ 0 & & & & & & 0 \end{array}$$

implies that $\widetilde{K}^q(X) \cong K^{q+1}(CX, X) = \widetilde{K}^{q+1}(\Sigma X)$].

EXERCISE 10. Define a natural multiplication in K-theory,

$$K^q(X, A) \otimes K^r(X, A) \longrightarrow K^{q+r}(X, A).$$

C: Proof of Bott's Theorem

We will prove that Bott's map $K(X) \oplus K(X) \to K(X \times S^2)$, $(\alpha_1, \alpha_2) \mapsto (\alpha_1 \otimes 1) + (\alpha_2 \otimes \zeta)$ is an epimorphism. The proof that it is a monomorphism is based on similar ideas, and we will restrict ourselves to a short sketch of this proof (which will probably be sufficient for the reader). The full proof of Bott's theorem may be found in the books by Atiyah [16], Karoubi [51], Mishchenko [61], J. Schwartz [73], and Husemoller [49]). All these proofs, like the proof presented below, are based on K-theory; the original proof given by Bott was based on differential geometry and the Morse theory; it is presented in the book *The Morse Theory* by J. Milnor (Milnor [57]).

Consider a vector bundle ξ over $X \times S^2$. The base $X \times S^2$ splits into the union of two copies of $X \times D^2$. Over each of the two pieces the vector bundle is the product of some vector bundle over X (we denote this by α) and D^2. The two pieces of the base are attached along $X \times S^1$ (S^1 is the equator of the sphere S^2); for any $z \in S^1$ the restrictions $\xi|_{X \times D^2}$ are attached by an isomorphism $u(z)\colon \alpha \to \alpha$ which covers the identity of the base X. Hence, for each $x \in X$ we have an automorphism depending on z, $u(x, z)\colon \alpha_x \to \alpha_x$, of the fiber α_x of the vector bundle α over the point x. Hence, our vector bundle is determined by the pair α, u; we use the notation $\xi = [\alpha, u]$.

EXERCISE 11. Prove that

(i) $\zeta = [1, z^{-1}]$.
(ii) $[\alpha_1, u_1] \oplus [\alpha_2, u_2] = [\alpha_1 \oplus \alpha_2, u_1 \oplus u_2]$.
(iii) $[\alpha_1, u_1] \otimes [\alpha_2, u_2] = [\alpha_1 \otimes \alpha_2, u_1 \otimes u_2]$.

Take the Fourier series for the function $u(x, z)$ with respect to z (we regard z as a point in $S^1 = \{|z| = 1\} \subset \mathbb{C}^2$):

$$u(x, z) = \sum_{k \in \mathbb{Z}} u_k(x) z^k.$$

Here $u_k(x)$ are endomorphisms of the vector space α_x which may be degenerate.

It is clear that the vector bundle ξ will not be affected if we replace u by a sufficiently close map. Hence, we may assume that u is a trigonometric polynomial:

$$u(x, z) = \sum_{k=-N}^{N} u_k(x) z^k.$$

The tensor multiplication of the function u by z corresponds to the tensor multiplication of the vector bundle ξ by ζ^{-1} (see Exercise 11). Replace the vector bundle ξ by $\xi' = \xi \otimes \zeta^{-N}$. Then $\xi' = [\alpha, u']$, where

$$u' = \sum_{k=0}^{m} u_k(x) z^k, \ m = 2N.$$

Let $\xi'' = \xi' \oplus (m\alpha \otimes I) = [\alpha'', u'']$, where $\alpha'' = (m+1)\alpha$ and $u'' = u' \oplus I \oplus \cdots \oplus I$. Thus, the matrix u'' has the form

$$\begin{pmatrix} u' & & & \text{zeroes} \\ & I & & \\ & & \ddots & \\ \text{zeroes} & & & I \end{pmatrix}$$

Multiply this matrix from the left and from the right by, respectively,

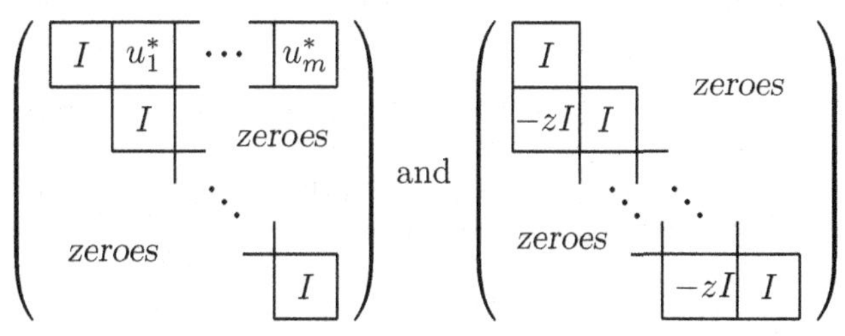

This multiplication does not change the bundle (both matrices are homotopic as maps $X \times S^1 \to \{\text{matrices}\}$ to the constant identity matrix) and it reduces the matrix $u''(x,z)$ to the form $a(x) + b(x)z$, where $a(x)$ and $b(x)$ are, respectively,

$$\begin{pmatrix} u_0 & u_1 & \cdots & u_m \\ & I & \text{zeroes} & \\ & & \ddots & \\ \text{zeroes} & & & I \end{pmatrix} \quad \text{and} \quad \begin{pmatrix} -I & & \text{zeroes} \\ & \ddots & \\ \text{zeroes} & & -I \end{pmatrix}$$

Lemma from Linear Algebra. *Let a and b be two endomorphisms of a complex vector space V with $\operatorname{Ker} a \cap \operatorname{Ker} b = 0$. Consider $c_{\lambda,\mu} = \lambda a + \mu b$ $(\lambda, \mu \in \mathbb{C}, (\lambda,\mu) \neq (0,0))$, and let $(\lambda_i : \mu_i) \in \mathbb{C}P^1$, $i = 1, \ldots, m$, all be pairwise different solutions of the equation $\det c_{\lambda,\mu} = 0$. Then there exist direct decompositions*

$$V = A_1 \oplus A_2 \oplus \ldots A_m, \; V = B_1 \oplus B_2 \oplus \ldots B_m$$

such that $c_{\lambda,\mu}(A_i) \subset B_i$ for all λ, μ and $A_i \xrightarrow{c_{\lambda,\mu}} B_i$ is an isomorphism for $(\lambda : \mu) \neq (\lambda_i : \mu_i)$.

The proof is left to the reader.

Let $c_{\lambda,\mu}(x) = \lambda a(x) + \mu b(x) \colon \alpha''_x \to \alpha''_x$. Since the endomorphism $a(x) + zb(x)$ with $|z| = 1$ is invertible, the equation $\det c_{\lambda,\mu}(x) = 0$ has no solutions (λ,μ) with $|\lambda|/|\mu| = 1$. The lemma implies that the decompositions $\alpha''_x = \beta_x \oplus \gamma_x$ and $\alpha''_x = \overline{\beta}_x \oplus \overline{\gamma}_x$ exist such that $c_{\lambda,\mu}(x)(\beta_x) \subset \overline{\beta}_x$, $c_{\lambda,\mu}(x)(\gamma_x) \subset \overline{\gamma}_x$, and $c_{\lambda,\mu}(x) \colon \beta_x \cong \overline{\beta}_x$ for $|\lambda|/|\mu| \geq 1$, $c_{\lambda,\mu}(x) \colon \gamma_x \cong \overline{\gamma}_x$ for $|\lambda|/|\mu| \leq 1$. In particular,

$$\beta_x \xrightarrow{a(x)} \overline{\beta}_x \text{ and } \gamma_x \xrightarrow{b(x)} \overline{\gamma}_x$$

are isomorphisms. Denote by $d_{\lambda,\mu}(x)$ and $e_{\lambda,\mu}(x)$ maps

$$\beta_x \xrightarrow{c_{\lambda,\mu}(x)} \overline{\beta}_x \xrightarrow{a(x)^{-1}} \beta_x \text{ and } \gamma_x \xrightarrow{c_{\lambda,\mu}(x)} \overline{\gamma}_x \xrightarrow{b(x)^{-1}} \gamma_x.$$

It follows from the preceding remark that $d_{\lambda,\mu}(x)$ is invertible for $|\lambda|/|\mu| \geq 1$ and $e_{\lambda,\mu}(x)$ is invertible for $|\lambda|/|\mu| \leq 1$. Assume that X is connected (this assumption is harmless). Then $\beta_x, \gamma_x, \overline{\beta}_x, \overline{\gamma}_x$ compose vector subbundles $\beta, \gamma, \overline{\beta}, \overline{\gamma}$ of the vector bundle α'', and

$$\xi'' = [\beta, d_{1,z}] \oplus [\gamma, e_{1,z}].$$

But $d_{1,z}$ is homotopic to $d_{1,0}$ and $e_{1,z}$ is homotopic to $e_{0,z}$ within the class of invertible maps (the homotopies are $d_{1,tz}$ and $e_{t,z}$). Thus,

$$\begin{aligned} \xi'' &= [\beta, d_{1,0}] \oplus [\gamma, e_{0,z}] \\ &= [\beta, 1] \oplus [\gamma, z] = (\beta \otimes 1) \oplus (\gamma \otimes \zeta^{-1}). \end{aligned}$$

Finally, we have

$$(\xi \otimes \zeta^{-N}) \oplus (m\alpha \otimes 1) = (\beta \otimes 1) \oplus (\gamma \otimes \zeta^{-1});$$

that is,

$$\xi = (((\beta - m\alpha) \otimes 1) \oplus (\gamma \otimes \zeta^{-1})) \otimes \zeta^N.$$

This completes the proof of $K(X) \oplus K(X) \to K(X \times S^2)$ being an epimorphism.

The fact that it is a monomorphism may be proved in a very similar way. We simply need a relative version of the preceding theorem: If (X, Y) is a finite CW pair, ξ is a vector bundle over $X \times S^2$, and f is an isomorphism

$$\xi|_Y \longrightarrow (\alpha \otimes 1) \oplus (\beta \otimes \zeta),$$

where α and β are vector bundles over Y, then there exists an isomorphism

$$\xi \oplus N \longrightarrow (\widetilde{\alpha} \otimes 1) \oplus (\widetilde{\beta} \otimes \zeta) \oplus N$$

(where $\widetilde{\alpha}$ and $\widetilde{\beta}$ are vector bundles over X) compatible with the preceding isomorphism. Once proved, the last assertion may be applied to the pair $(X \times I, X \times \{0, 1\})$, which yields the monomorphism theorem. Another way to prove it is to use easily available information on the homotopy groups of BU (see Exercise 12).

EXERCISE 12. Deduce the fact that $K(X) \oplus K(X) \to K(X \times S^2)$ is a monomorphism from the fact that it is an epimorphism in the following way. The epimorphness implies that the map $\pi_i(BU) \to \pi_{i+2}(BU)$ is an epimorphism for all i. But $\pi_1(BU) = 0$, $\pi_2(BU) = \mathbb{Z}$. Hence, $\pi_i(BU) = 0$ for any odd i, and either the map $\pi_i(BU) \to \pi_{i+2}(BU)$ is always an isomorphism, or the group $\pi_i(BU)$ is finite for sufficiently large i. The latter contradicts the rank computations: $\operatorname{rank} \pi_i(BU) = \operatorname{rank} \pi_{i-1}(U)$ is equal to 1 for any even i (see Sect. 26.4). Hence, it is an isomorphism, which easily implies that $K(X) \oplus K(X) \to K(X \times S^2)$ is an isomorphism for any X (the last implication is contained, implicitly, in the next section).

38.4 Chern Character

A: Chern Character in K-Theory

In Sect. 19.5.D we define the Chern character, which is a characteristic class of complex vector bundles taking values in $H^{\text{even}}(X; \mathbb{Q}) = H^0(X; \mathbb{Q}) \oplus H^2(X; \mathbb{Q}) \oplus \ldots$, where X is the base of the bundle; notation: $\xi \mapsto \operatorname{ch} \xi$. This characteristic class is additive and multiplicative:

$$\mathrm{ch}(\xi_1 \oplus \xi_2) = \mathrm{ch}\,\xi_1 + \mathrm{ch}\,\xi_2,\ \mathrm{ch}(\xi_1 \otimes \xi_2) = \mathrm{ch}\xi_1\mathrm{ch}\xi_2.$$

The first property lets us extend ch to the K-theory:

$$\mathrm{ch}\colon K(X) \longrightarrow H^{\mathrm{even}}(X;\mathbb{Q});$$

the second property makes this K-theory's character multiplicative. Furthermore, this definition obviously gives rise to a definition of the character

$$\mathrm{ch}\colon K^q(X,A) \longrightarrow \begin{cases} H^{\mathrm{even}}(X,A;\mathbb{Q}) \text{ for } q \text{ even}, \\ H^{\mathrm{odd}}(X,A;\mathbb{Q}) \text{ for } q \text{ odd}; \end{cases}$$

$H^{\mathrm{odd}} = H^1 \oplus H^3 \oplus \dots$). The following notation is convenient:

$$\mathcal{H}^q(X,A) = \begin{cases} H^{\mathrm{even}}(X,A;\mathbb{Q}) \text{ for } q \text{ even}, \\ H^{\mathrm{odd}}(X,A;\mathbb{Q}) \text{ for } q \text{ odd}. \end{cases}$$

This $\mathcal{H}$ looks like K in some respects: It has exact sequences of pairs and triples, but also it is 2-periodic: $\mathcal{H}^q = \mathcal{H}^{q+2}$. The character ch becomes then a sequence of homomorphisms

$$\mathrm{ch}\colon K^q(X,A) \longrightarrow \mathcal{H}^q(X,A),$$

which commutes with the induced homomorphisms and connecting homomorphisms of the sequences of pairs and triples.

EXERCISE 13. Prove that ch commutes with the 2-periodicity isomorphisms $K^q(X,A) \to K^{q+2}(X,A)$ and $\mathrm{id}\colon \mathcal{H}^q(X,A) \to \mathcal{H}^{q+2}(X,A)$.

EXERCISE 14. Prove that this map ch is multiplicative (cf. Exercise 10).

B: Chern Character $\mathbf{ch}_{\mathbb{Q}}$

Let

$$\mathrm{ch}_{\mathbb{Q}} = \mathrm{ch} \otimes \mathbb{Q}\colon K^q(X,A) \otimes \mathbb{Q} \longrightarrow \mathcal{H}^q(X,A) \otimes \mathbb{Q} = \mathcal{H}^q(X,A).$$

Theorem. *The map* $\mathrm{ch}_{\mathbb{Q}}$ *is an isomorphism for any* $\mathbb{Q}$ *and any (finite* CW*) pair* (X,A). *In particular,*

$$K(X) \otimes \mathbb{Q} \cong H^{\mathrm{even}}(X;\mathbb{Q}),$$

and the rank of the group $K(X)$ is equal to the sum of the even Betti numbers of X. [For example, $\mathrm{rank}\, K(S^2) = 2$*, as we already know.]*

Proof. It is sufficient to consider the case when $A = \emptyset$ [since $K^q(X,A) = \widetilde{K}^q(X/A)$ and $\mathcal{H}^q(X,A) = \widetilde{\mathcal{H}}^q(X/A)$].

The statement of the theorem is obvious if $\dim X = 0$, that is, if X is a finite set. Since there are suspension isomorphisms for both $K \otimes \mathbb{Q}$ and K, and $\mathrm{ch}_{\mathbb{Q}}$ commutes with these isomorphisms, then $\mathrm{ch}_{\mathbb{Q}}$ is an isomorphism also for the (multiple) suspensions over finite sets, that is, for bouquets of spheres of equal dimensions. We proceed by induction.

Assume that $\dim X = n > 1$ and that for CW complexes of dimension less than n the theorem is right. Let X^{n-1} be the $(n-1)$-skeleton of X; then X/X^{n-1} is a bouquet of n-dimensional spheres. Consider the commutative ladder composed of the $\mathcal{H}$- and K-sequences of the pair (X, X^{n-1}) and the appropriate homomorphisms $\mathrm{ch}_{\mathbb{Q}}$:

$$\begin{array}{ccccc} K^{q-1}(X^{n-1}) \otimes \mathbb{Q} & \longrightarrow & K^q(X, X^{n-1}) \otimes \mathbb{Q} & \longrightarrow & K^q(X) \otimes \mathbb{Q} \\ \Big\downarrow \mathrm{ch}_{\mathbb{Q}} & & \Big\downarrow \mathrm{ch}_{\mathbb{Q}} & & \Big\downarrow \mathrm{ch}_{\mathbb{Q}} \\ \mathcal{H}^{q-1}(X^{n-1}) & \longrightarrow & \mathcal{H}^q(X, X^{n-1}) & \longrightarrow & \mathcal{H}^q(X) \end{array}$$

$$\begin{array}{ccccc} \longrightarrow & K^q(X^{n-1}) \otimes \mathbb{Q} & \longrightarrow & K^{q+1}(X, X^{n-1}) \otimes \mathbb{Q} \\ & \Big\downarrow \mathrm{ch}_{\mathbb{Q}} & & \Big\downarrow \mathrm{ch}_{\mathbb{Q}} \\ \longrightarrow & \mathcal{H}^q(X^{n-1}) & \longrightarrow & \mathcal{H}^{q+1}(X, X^{n-1})\,. \end{array}$$

All the vertical homomorphisms, with the exception of the middle one, are isomorphisms: For two of them it follows from the inequality $\dim X^{n-1} < n$; for the other two it is true because X/X^{n-1} is a bouquet of n-dimensional spheres. It remains to apply the five-lemma.

38.5 Extraordinary Homology and Cohomology

A: Eilenberg–Steenrod Axioms

Let us return to Chap. 2. In that chapter we assign to every finite CW pair (actually, to every topological pair, but it is not important to us now) (X, A) a sequence of groups $H_q(X,A), q \in \mathbb{Z}$ [we put $H_q(X,A) = 0$ for $q < 0$] and homomorphisms $\partial_*: H_q(X,A) \to H_{q-1}(A)$ [we put $H_r(Y) = H_r(Y, \emptyset)$] and to every continuous map

$y\colon (X,A) \to (Y,B)$ a sequence of homomorphisms $f_*\colon H_q(X,A) \to H_q(Y,B)$ such that $\mathrm{id}_* = \mathrm{id}$, $(f \circ g)_* = f_* \circ g_*$, $f_* \circ \partial_* = \partial_* \circ f_*$. In addition, the following axioms hold:

HOMOTOPY AXIOM. If $f \sim g$, then $f_* = g_*$.

EXACTNESS AXIOM. The sequences of pairs are exact. (Corollary: The sequences of triples are exact.)

FACTORIZATION AXIOM. The projection $(X,A) \to (X/A, \mathrm{pt})$ induces for all q isomorphisms $H_q(X,A) \cong H_q(X/A, \mathrm{pt})$ ($\overset{\mathrm{def}}{=} \widetilde{H}_q(X/A)$).

DIMENSION AXIOM. $H_q(\mathrm{pt}) = 0$ for $q \neq 0$.

These axioms are called *Eilenberg–Steenrod axioms*. If we add to these axioms the statement that $H_0(\mathrm{pt}) = \mathbb{Z}$, then the following uniqueness theorem will hold: *The theory $\{H_q, \partial_*, f_q\}$ satisfying all the axioms above is unique and coincides with the theory of usual (singular) homology*. Actually, this theorem was proven in Lecture 13, where we calculated the homology of CW complexes (and CW pairs) using only the properties of homology listed above.

Drop the requirement $H_0(\mathrm{pt}) = \mathbb{Z}$ and call the group $G = H_0(\mathrm{pt})$ the coefficient group. Then a similar uniqueness theorem states that a theory satisfying all the Eilenberg–Steenrod axioms coincides with the theory of usual homology with the coefficients in G.

The Eilenberg–Steenrod axiom system of cohomology is set up in a similar way, and it satisfies the same uniqueness theorems.

B: Extraordinary Theories

Suppose that for every finite CW pair (X,A) there is defined in some way a sequence of Abelian groups $h_q(X,A)$ and homomorphisms $\partial_*\colon h_q(X,A) \to h_{q-1}(A)$ [we put $h_r(Y) = h_r(Y, \emptyset)$] and for every continuous map $f\colon (X,A) \to (Y,B)$ there is defined a sequence of homomorphisms $f_*\colon h_q(X,A) \to h_q(Y,B)$ such that $\mathrm{id}_* = \mathrm{id}$, $(f \circ g)_* = f_* \circ g_*$, $f_* \circ \partial_* = \partial_* \circ f_*$. Suppose now that the homotopy, exactness, and factorization axioms hold, but the dimension axiom may fail. In this case we say that $\{h_q, \partial_*, f_q\}$ is an *extraordinary (or generalized) homology theory*. An *extraordinary (or generalized) cohomology theory* is defined in a similar way.

K-theory is an example of an extraordinary cohomology theory. $\mathcal{H}$ is another example. Some other examples are considered in Sect. 38.5.D and Lecture 44.

EXERCISE 15. Prove that $h_q(X_1 \sqcup X_2) = h_q(X_1) \oplus h_q(X_2)$ for any extraordinary homology theory h (and that the same is true for extraordinary cohomology).

C: Generalized Uniqueness Theorem

Let h and k be two extraordinary cohomology theories (or extraordinary homology theories; in the homological case all that is said here is valid with appropriate modifications: One should lower some indices, reverse some arrows, etc.). We say that a homomorphism $\varphi: h \to k$ is given if for every finite CW pair (X, A) there are homomorphisms

$$\varphi = \varphi_{(X,A)}: h^q(X, A) \longrightarrow k^q(X, A)$$

such that for any (X, A), (Y, B), and $f: (X, A) \to (Y, B)$, the diagrams

$$\begin{array}{ccc} h^q(X,A) & \xleftarrow{\delta^*} & h^{q-1}(A) \\ \big\downarrow{\scriptstyle\varphi} & & \big\downarrow{\scriptstyle\varphi} \\ k^q(X,A) & \xleftarrow{\delta^*} & k^{q-1}(A) \end{array} \quad \text{and} \quad \begin{array}{ccc} h^q(X,A) & \xleftarrow{f^*} & h^q(Y,B) \\ \big\downarrow{\scriptstyle\varphi} & & \big\downarrow{\scriptstyle\varphi} \\ k^q(X,A) & \xleftarrow{f^*} & k^q(Y,B) \end{array}$$

are commutative.

Theorem. *Let $\varphi: h \to k$ be a homomorphism of the theory h into the theory k such that for any q*

$$\varphi_{(\mathrm{pt},\emptyset)}: h^q(\mathrm{pt}) \longrightarrow k^q(\mathrm{pt})$$

is an isomorphism. Then

$$\varphi_{(X,A)}: h^q(X, A) \longrightarrow k^q(X, A)$$

is an isomorphism for any q and (X, A); in particular, in this case $h \cong k$.

Proof. The proof repeats that of the theorem in Sect. 38.4.B (where $\mathrm{ch}_{\mathbb{Q}}$ plays the role of φ); we only need to prove that $\varphi_{(X,A)}$ is an isomorphism if $\dim X = 0$. This follows from Exercise 15 but can also be proved by induction over the number of points in X. Let X consist of $n \geq 2$ points. Then $X = Y/\emptyset$, where Y consists of $n - 1$ points. The h- and k-sequences of the pair (X, pt) can be arranged in the commutative ladder

$$\begin{array}{ccccccccc} h^{q-1}(\mathrm{pt}) & \to & h^q(X,\mathrm{pt}) & \to & h^q(X) & \to & h^q(\mathrm{pt}) & \to & h^{q+1}(X,\mathrm{pt}) \\ \big\downarrow{\scriptstyle\varphi} & & \big\downarrow{\scriptstyle\varphi} & & \big\downarrow{\scriptstyle\varphi} & & \big\downarrow{\scriptstyle\varphi} & & \big\downarrow{\scriptstyle\varphi} \\ k^{q-1}(\mathrm{pt}) & \to & k^q(X,\mathrm{pt}) & \to & k^q(X) & \to & k^q(\mathrm{pt}) & \to & k^{q+1}(X,\mathrm{pt}), \end{array}$$

in which all the maps φ except the middle one are isomorphisms [since $h^r(X, \mathrm{pt}) = h^r(Y/\emptyset, \mathrm{pt}) = h^r(Y, \emptyset) = h^r(Y)$]. Hence, the middle homomorphism φ is also an isomorphism.

Remark. If $h^q(\mathrm{pt}) \cong k^q(\mathrm{pt})$, but the isomorphism is not induced by any map $h \to k$, then the theories h and k may be nonisomorphic. For example,

$$K^q(\mathrm{pt}) \cong \mathcal{H}^q_{\mathbb{Z}}(\mathrm{pt}),$$

where $\mathcal{H}^q_{\mathbb{Z}}(X, A) = H^{\text{even (odd)}}(X, A; \mathbb{Z})$, but, as we will prove in Lecture 39,

$$K(\mathbb{R}P^n) \ncong H^{\text{even}}(\mathbb{R}P^n; \mathbb{Z}).$$

D: Examples of Extraordinary Theories

Along with the complex K-theory there exists the real K-theory. We do not deal with it here although in some cases this will cost us a little weakening of the result (see, for example, the remarks in Sects. 41.3 and 42.5). The role similar to that of BU is played in the real K-theory by the space $BSO = G_+(\infty, \infty)$ [in another version of it by the space $BO = G(\infty, \infty)$], and the role of the Bott 2-periodicity is played by the real Bott 8-periodicity $\pi_i(SO) \cong \pi_{i+8}(SO)$. See the details in Atiyah [16] or Karoubi [51].

There also exists a general construction that generalizes both complex and real K-theories, as well as the usual cohomology (and homology).

Definition. A *spectrum*, or an Ω-*spectrum*, is a sequence of CW complexes W_i and continuous maps $f_i \colon W_i \to \Omega W_{i+1}$ $(i \geq 1)$.

Notice that for any topological spaces A, B with base points there exists a natural one-to-one correspondence between continuous maps $A \to \Omega B$ and $\Sigma A \to B$ (we mean the suspension operation for spaces with base points; see Sect. 6.2): For a map $f\colon A \to \Omega A$, the corresponding map $\overline{f}\colon \Sigma f \to B$ is defined by the formula $\overline{f}(a, t) = [f(a)](t)$. In particular, $\pi(A, \Omega B) = \pi(\Sigma A, B)$.

Definition. Let (X, A) be a CW pair, and let $\mathcal{W} = \{W_i, f_i\}$ be an arbitrary Ω-spectrum. The homology and cohomology groups of the pair (X, A) with the coefficients in $\mathcal{W}$ are defined by the formulas (in which $q \in \mathbb{Z}$)

$$h^q(X, A; \mathcal{W}) = \varinjlim_N \pi(\Sigma^N(X/A), W_{N+q}),$$

$$h_q(X, A; \mathcal{W}) = \varinjlim_N \pi_{N+q}(W_N \# (X/A)).$$

CLARIFICATION. The maps

$$\pi(\Sigma^N(X/A), W_{N+q}) \longrightarrow \pi(\Sigma^{N+1}(X/A), W_{N+1+q})$$
$$= \pi(\Sigma^N(X/A), \Omega W_{N+1+q})$$

that give rise to the first limit are induced by the maps $f_{N+q}\colon W_{N+q} \to \Omega W_{N+q+1}$. The maps

$$\pi_{N+q}(W_N\#(X/A)) \longrightarrow \pi_{N+1+q}(W_{N+1}\#(X/A))$$

that give rise to the second limit are defined by the map $\Sigma W_N \to W_{N+1}$ corresponding to f_N, as the composition

$$\pi_{N+q}(W_N\#(X/A)) \xrightarrow{\Sigma} \pi_{N+1+q}(\Sigma(W_N\#(X/A)))$$
$$= \pi_{N+1+q}(\Sigma W_N\#(X/A)) \longrightarrow \pi_{N+1+q}(W_{N+1}\#(X/A)).$$

EXERCISE 16. Define induced and connecting homomorphisms for the groups $h^q(X,\ A;\mathcal{W})$ and $h_q(X,A;\mathcal{W})$ in such a way that they form an extraordinary cohomology theory and an extraordinary homology theory.

Examples. The usual homology and cohomology (with coefficients in G) correspond to the *Eilenberg–MacLane spectrum* in which W_i is $K(G,i)$ and f_i is the standard homotopy equivalence $K(G,i) \to \Omega K(G,i+1)$. (It is known to us for cohomology; for homology this is an exercise.) The complex K-theory corresponds to the periodic Ω-spectrum $U, BU, U, BU, \ldots$ with the homotopy equivalences $U \to \Omega BU$ and $BU \to \Omega U$ (the first equivalence was announced in Exercise 7; the second is one of the forms of Bott periodicity). The homology in this spectrum, the so-called K-homology, is also very important, but we will not consider it. The real K-theory is defined by a 8-periodic spectrum, in which

$$W_{8r} = BSO,\ W_{8r-1} = SO, \ldots, W_{8r-k} = \Omega^k BSO \text{ for } k < 8.$$

One more example: The spherical spectrum $\mathcal{S}$ in which $W_i = S^i$ and the map $S^i \to \Omega S^{i+1}$ correspond to the standard homeomorphism $\Sigma S^i \to S^{i+1}$. Cohomology groups are stable homotopy groups $\pi_q^S(X,A)$ and the so-called stable cohomotopy groups

$$\pi_S^q(X,A) = \varinjlim_N (\Sigma^{N+q}(X/A), S^N).$$

Finally, remark that an arbitrary theory of extraordinary homology or cohomology is obtained by the procedure described above from some Ω-spectrum. This can easily be deduced from the so-called Brown representability theorem, which can be found, for example, in Spanier [79].

Lecture 39 Calculating *K*-Functor: Atiyah–Hirzebruch Spectral Sequence

39.1 The Construction of the Atiyah–Hirzebruch Spectral Sequence

A: The Statement

We will present a construction that assigns to a finite CW space X a family of Abelian groups $E_r^p = E_r^p(X)$ and homomorphisms $d_r^p\colon E_r^p \to E_r^{p+r}$, $p \geq 0, r \geq 2$ with the following properties.

(1) $E_2^p = H^p(X;\mathbb{Z})$.
(2) $E_{r+1}^p = \operatorname{Ker} d_p^r / \operatorname{Im} d_r^{p-r}$ (we mean that $\operatorname{Im} d_r^{p-r} = 0$ for $r > p$).
(3) $d_r^p = 0$ and $E_{r+1}^p = E_r^p$ for any even r and any p.

Since the CW complex X is finite, (1) and (2) imply also that $d_r^p = 0$ and $E_{r+1}^p = E_r^p$ for sufficiently large r. These groups E_r^p, which do not depend on r, we denote by E_∞^p.

(4) $E_\infty^p = \dfrac{\operatorname{Ker}[K^p(X) \to K^p(X^{p-2})]}{\operatorname{Ker}[K^p(X) \to K^p(X^p)]}$.

Let us explain what this means. Put

$$^{(r)}K^p(x) = \operatorname{Ker}[K^p(X) \longrightarrow K^p(X^r)].$$

These groups compose filtrations (Bott's periodicity lets us assume that $p = 0$ or 1)

$$K^0(X) = {}^{(-1)}K^0(X) \supset {}^{(0)}K^0(X) \supset \cdots \supset {}^{(n)}K^0(X) = 0,$$
$$K^1(X) = {}^{(-1)}K^1(X) \supset {}^{(0)}K^1(X) \supset \cdots \supset {}^{(n)}K^1(X) = 0,$$

where $n = \dim X$. Notice that if X is connected, then ${}^{(0)}K^0(X) = \widetilde{K}(X)$.

Lemma. *For r even,*

$$^{(r)}K^0(X) = {}^{(r+1)}K^0(X), \quad {}^{(r-1)}K^1(X) = {}^{(r)}K^1(X).$$

Proof. Consider the fragment of the K-sequence of the triple (X, X^{r+1}, X^r):

$$\begin{array}{ccccc} K^0(X,X^{r+1}) & \longrightarrow & K^0(X,X^r) & \longrightarrow & K^0(X^{r+1},X^r) \\ & & & & \| \\ & 0 & = \quad \widetilde{K}^{-r-1}(\vee S^0) & = & \widetilde{K}^0(\vee S^{r+1}) \\ & & \downarrow & & \\ & & \text{by Bott's periodicity} & & \end{array}$$

The exactness implies that the natural map

$$K^0(X, X^{r+1}) \longrightarrow K^0(X, X^r)$$

is epimorphic. Consider now the following fragment of the commutative ladder composed of the K-sequences of the pairs (X, X^r) and (X, X^{r+1}):

$$\begin{array}{ccccc} K^0(X, X^r) & \xrightarrow{\alpha} & K^0(X) & \xrightarrow{\beta} & K^0(X^r) \\ \downarrow{\scriptstyle\varepsilon} & & \| & & \\ K^0(X, X^{r+1}) & \xrightarrow{\gamma} & K^0(X) & \xrightarrow{\delta} & K^0(X^{r+1}). \end{array}$$

The epimorphness of ε implies that $\operatorname{Im}\alpha = \operatorname{Im}\gamma$, but $\operatorname{Im}\alpha = \operatorname{Ker}\beta$, $\operatorname{Im}\gamma = \operatorname{Ker}\delta$. Hence, $\operatorname{Ker}\beta = \operatorname{Ker}\delta$, which is our statement in the case of K^0.

The statement for K^1 is proven in the same way.

The lemma implies that our filtrations actually have the form

$$\begin{aligned} K^0(X) &= {}^{(0)}K^0(X) \supset {}^{(2)}K^0(X) \supset {}^{(4)}K^0(X) \supset \dots \\ K^1(X) &= {}^{(-1)}K^1(X) \supset {}^{(1)}K^1(X) \supset {}^{(3)}K^1(X) \supset \dots, \end{aligned}$$

and statement (4) shows that the graded groups

$$E_\infty^0 \oplus E_\infty^2 \oplus E_\infty^4 \oplus E_\infty^6 \oplus \dots,$$

$$E_\infty^1 \oplus E_\infty^3 \oplus E_\infty^5 \oplus E_\infty^7 \oplus \dots$$

are adjoint to these filtrations.

(5) A continuous map $f: X \to Y$ induces (depending only on the homotopy class of f) homomorphisms $f^*: E_r^p(Y) \to E_r^p(X)$ which satisfy the following conditions: (a) the diagrams

$$\begin{array}{ccc} E_r^p(Y) & \xrightarrow{d_r^p} & E_r^{p+r}(Y) \\ \downarrow{\scriptstyle f^*} & & \downarrow{\scriptstyle f^*} \\ E_r^p(X) & \xrightarrow{d_r^p} & E_r^{p+r}(X) \end{array}$$

are commutative; (b) $f^*: E_{r+1}^p(Y) \to E_{r+1}^p(X)$ coincides with the homology homomorphism induced by $f^*: E_r^p(Y) \to E_r^p(X)$; (c) $f^*: E_2^p(Y) \to E_2^p(X)$ coincides with f^*: $H^p(Y; \mathbb{Z}) \to H^p(X; \mathbb{Z})$; (d) $f^*: E_\infty^p(Y) \to E_\infty^p(X)$ is induced by the homomorphism $f^*: K^*(Y) \to K^*(X)$.

Notice that (5) contains the topological, and even homotopical, invariance of the spectral sequence.

(6) The spectral sequence is multiplicative: All $E_r^* = \oplus_p E_r^p$ are graded rings, the differentials d_r^p satisfy the product formula, the multiplication in E_2^* coincides with the multiplication in $H^*(X;\mathbb{Z})$, the multiplication in E_∞^* is adjoint to the multiplication in $K^*(X)$ (see Exercise 10 in Lecture 38).

(7) All the differentials d_r^p have finite order.

We can prove this property even before constructing the spectral sequence. Obviously, $\operatorname{rank} E_{r+1}^p \leq \operatorname{rank} E_r^p$, and the equality holds if and only if d_r^p and d_{p+r}^p both have finite order. Hence, if some differential has infinite order, then

$$\operatorname{rank}(K^0(X) \oplus K^1(X)) = \operatorname{rank} E_\infty^* < \operatorname{rank} E_2^* = \operatorname{rank} H^*(X;\mathbb{Z}),$$

which contradicts the theorem in Sect. 38.4.

Let us tensor our spectral sequence with $\mathbb{Q}$. Item (7) will imply that all the differentials will become zero, and we will have $E_2^* \otimes \mathbb{Q} = E_\infty^* \otimes \mathbb{Q}$. Hence, we get some isomorphisms $K^0(X) \otimes \mathbb{Q} \cong H^{\text{even}}(X;\mathbb{Q}), K^1(X) \otimes \mathbb{Q} \cong H^{\text{odd}}(X;\mathbb{Q})$.

(8) These isomorphisms coincide with $\text{ch}_\mathbb{Q}$ (see Sect. 38.4).

B: Construction of the Spectral Sequence

Put

$$E_r^{p,q} = \frac{\operatorname{Ker}[K^{p+q}(X^{p+r-1}, X^{p-r}) \to K^{p+q}(X^{p-1}, X^{p-r})]}{\operatorname{Ker}[K^{p+q}(X^{p+r-1}, X^{p-r}) \to K^{p+q}(X^{p}, X^{p-r})]}$$

and define the differential $d_r^{p,q}\colon E_r^{p,q} \to E_r^{p+r,q-r+1}$ in the following way. Consider the commutative diagram

$$\begin{array}{ccc} K^{p+q}(X^{p+r-1}, X^{p-r}) & \xrightarrow{\alpha} & K^{p+q}(X^p, X^{p-r}) \\ \Big\downarrow{\scriptstyle \delta^*} & & \Big\downarrow{\scriptstyle \delta^*} \\ K^{p+q+1}(X^{p+2r-1}, X^{p+r-1}) & \xrightarrow{\beta} & K^{p+q+1}(X^{p+2r-1}, X^p). \end{array}$$

The composition $\delta^* \circ \alpha = \beta \circ \delta^*\colon K^{p+q}(X^{p+r-1}, X^{p-r}) \to K^{p+q+1}(X^{p+2r-1}, X^p)$ may be regarded as a map

$$K^{p+q}(X^{p+r-1}, X^{p-r}) / \operatorname{Ker}\alpha \to \operatorname{Im}\beta.$$

But, by definition, $E_r^{p,q}$ is a subgroup of the group $K^{p+q}(X^{p+r-1}, X^{p-r}) / \operatorname{Ker}\alpha$, while $E_r^{p+r,q-r+1}$ is a quotient of the group

$$\operatorname{Im}\beta = \operatorname{Ker}[K^{p+q+1}(X^{p+2r-1}, X^p) \to K^{p+q+1}(X^{p+r-1}, X^p)].$$

Hence, the last map induces some homomorphism

$$E_r^{p,q} \longrightarrow E_r^{p+r,q-r+1}.$$

This is $d^{p,q}$.

It is easy to check that $d_r^{p+r,q-r+1} \circ d_r^{p,q} = 0$ and that

$$E_{r+1}^{p,q} = \operatorname{Ker} d_r^{p,q} / \operatorname{Im} d_r^{p-r,q+r+1}.$$

With this we have already constructed a certain spectral sequence, but it is not the one we were to construct, for the sequence constructed consists of groups with two superscripts, while the proposed sequence consists of groups with one superscript. Let us compute $E_1^{p,q}$ and $E_2^{p,q}$. By definition,

$$\begin{aligned} E_1^{p,q} &= \frac{\operatorname{Ker}[K^{p+q}(X^p, X^{p-1}) \to K^{p+q}(X^{p-1}, X^{p-1})]}{\operatorname{Ker}[K^{p+q}(X^p, X^{p-1}) \to K^{p+q}(X^p, X^{p-1})]} \\ &= K^{p+q}(X^p, X^{p-1}) = \widetilde{K}^{p+q}\Big(\bigvee_{p\text{-cells}} S^p\Big) \\ &= \bigoplus\nolimits_{p\text{-cells}} \widetilde{K}^{p+q}(S^p) = \bigoplus\nolimits_{p\text{-cells}} K^q(\mathrm{pt}) = C^p(X; K^q(\mathrm{pt})). \end{aligned}$$

It is easy to check (we leave the details to the reader) that

$$d_1^{p,q} = \delta \colon C^p(X; K^q(\mathrm{pt})) \longrightarrow C^{p+1}(X; K^q(\mathrm{pt}))$$

and hence

$$E_2^{p,q} = H^p(X; K^q(\mathrm{pt})).$$

Since

$$K^q(\mathrm{pt}) = \begin{cases} \mathbb{Z} & \text{for } q \text{ even}, \\ 0 & \text{for } q \text{ odd}, \end{cases}$$

the second term of our spectral sequence has the following structure; the odd rows contain only zeroes, while the even rows contain the integral cohomology of X:

...
$H^*(X;\mathbb{Z})$
zeroes
$H^*(X;\mathbb{Z})$
zeroes
$H^*(X;\mathbb{Z})$
zeroes
$H^*(X;\mathbb{Z})$
...

Moreover, Bott's 2-periodicity shows that our spectral sequence is 2-periodic with respect to q: $E_r^{p,q} = E_r^{p,q+2}$ and $d_r^{p,q} = d_r^{p,q+2}$ for all $p, q, r < \infty$ and $E_\infty^{p,q} = E_\infty^{p,q+2}$. Therefore, we can identify all the groups $E_r^{p,2s}$ into one group E_r^p and all the differentials $d_r^{p,2s}: E_r^{p,2s} \to E_r^{p+r,2s-r+1}$ with r odd into one differential $d_r^p : E_r^P \to E_r^{p+r}$ (the rest of $E_r^{p,q}$ and $d_r^{p,q}$ are equal to 0). We arrive at the spectral sequence we need. By the way, the unexpected property (3) becomes obvious.

We leave the reader to check the remaining properties.

The spectral sequence constructed is called the *Atiyah–Hirzebruch spectral sequence* (Atiyah and Hirzebruch [19]).

The Atiyah–Hirzebruch spectral sequence has the obvious relative version $H^*(X, A; \mathbb{Z}) \Rightarrow K^*(X, A)$ and the equally obvious reduced version $\widetilde{H}^*(X; \mathbb{Z}) \Rightarrow \widetilde{K}^*(X)$ (when related to a spectral sequence, the notation $G \Rightarrow H$ means that the initial, usually the second, term of the spectral sequence is G, while its limit term is adjoint to H with respect to some filtration).

EXERCISE 1. Prove that the order of the torsion in the groups $K^0(X), K^1(X)$ divides the order of the torsion in the groups $H^{\text{even}}(X; \mathbb{Z}), H^{\text{odd}}(X; \mathbb{Z})$, respectively.

C: Generalizations

The Atiyah–Hirzebruch spectral sequence [rather its $(E_r^{p,q}, d_r^{p,q})$-version of Sect. 39.1.B] exists for any extraordinary homology or cohomology theory; it has the form

$$H_*(X; h_*(\text{pt})) \Rightarrow h_*(X) \text{ or } H^*(X; h^*(\text{pt})) \Rightarrow h^*(X).$$

A further generalization of this spectral sequence provides a "spectral sequence of a fibration (E, B, F, p)":

$$H_*(B; h_*(F)) \Rightarrow h_*(E) \text{ or } H^*(B; h^*(F)) \Rightarrow h^*(E).$$

(The construction of both spectral sequences is similar to the previous construction.)

The second spectral sequence may be nontrivial even for a trivial fibration. Still, it is possible to obtain a K-analog of Künneth's formula (see Atiyah [14]).

39.2 Examples of Calculations

A: $K^*(\mathbb{C}P^n)$

The second term of the Atiyah–Hirzebruch spectral sequence has the form

$\mathbb{Z}$	0	$\mathbb{Z}$	0	$\mathbb{Z}$	...	0	$\mathbb{Z}$
0	1	2	3	4	...	$2n-1$	$2n$

All the differentials are zeroes by the dimension argumentations. The general properties of the adjointness operation imply that

$$K^0(\mathbb{C}P^n) = \underbrace{\mathbb{Z} \oplus \cdots \oplus \mathbb{Z}}_{n+1 \text{ summands}}$$
$$K^1(\mathbb{C}P^n) = 0.$$

We will prove the following detailed explanation of this statement.

Theorem. *Multiplicatively,* $K^0(\mathbb{C}P^n) = \mathbb{Z}[\gamma]/(\gamma^{n+1})$, *where* $\gamma = \zeta - 1$ (ζ *is the Hopf bundle).*

Lemma. *The image of the homomorphism*

$$\begin{array}{ccc} \text{ch}\colon \widetilde{K}(S^{2n}) & \longrightarrow \widetilde{H}^*(S^{2n};\mathbb{Q}) = & H^{2n}(S^{2n};\mathbb{Q}) \\ \| & & \| \\ \mathbb{Z} & & \mathbb{Q} \end{array}$$

coincides with $H^{2n}(S^{2n};\mathbb{Z}) = \mathbb{Z} \subset \mathbb{Q} = H^{2n}(S^{2n};\mathbb{Q})$.

Proof of Lemma. For $n = 1$ this is known to us; the general case follows by induction because of the commutativity of the following diagram:

$$\begin{array}{ccc} \widetilde{K}^0(X) & \xrightarrow{\text{ch}} & \widetilde{H}^{\text{even}}(X;\mathbb{Q}) \\ \downarrow{\scriptstyle \Sigma^2} & & \downarrow{\scriptstyle \Sigma^2} \\ \widetilde{K}^2(X) & \xrightarrow{\text{ch}} & \widetilde{H}^{\text{even}}(X;\mathbb{Q}) \\ \downarrow{\scriptstyle \text{Bott}} & & \| \\ \widetilde{K}^0(X) & \xrightarrow{\text{ch}} & \widetilde{H}^{\text{even}}(X;\mathbb{Q}). \end{array}$$

Proof of Theorem. Let $x \in H^2(\mathbb{C}P^2;\mathbb{Z})$ be the generator. Let us compute the Chern characters:

$$\begin{aligned} \text{ch}\,\zeta &= e^{c_1(\zeta)} = e^x = 1 + x + \frac{x^2}{2} + \ldots; \\ \text{ch}\,\gamma &= \text{ch}\,\zeta - 1 = x + \frac{x^2}{2} + \ldots; \\ \text{ch}\,\gamma^k &= x^k + \frac{k}{2}x^{k+1} + \ldots; \\ \text{ch}\,\gamma^n &= x^n,\ \text{ch}\,\gamma^{n+1} = 0. \end{aligned}$$

This computation shows that $\gamma^{n+1} = 0$ and $1, \gamma, \gamma^n$ are linear independent in $K(\mathbb{C}P^n)$, and hence [since $K(\mathbb{C}P^n) = \mathbb{Z}^n$], they generate $K(\mathbb{C}P^n)$ over $\mathbb{Q}$. Show that they generate $K(\mathbb{C}P^n)$ over $\mathbb{Z}$. We know this for $\mathbb{C}P^1$; assume that it is true for $\mathbb{C}P^{n-1}$. Let $\alpha \in K(\mathbb{C}P^n)$, and let $\alpha = r_0 + r_1\gamma + \cdots + r_n\gamma^n$. Then $\alpha|_{\mathbb{C}P^{n-1}} = r_0 + r_1\gamma + \cdots + r_{n-1}\gamma^{n-1}$, and hence $r_0, \ldots, r_{n-1} \in \mathbb{Z}$. Therefore, $r_0 + r_1\gamma + \cdots + r_{n-1}\gamma^{n-1} \in K(\mathbb{C}P^n)$, and hence $r_n\gamma^n \in K(\mathbb{C}P^n)$. Furthermore, since $r_n\gamma^n|_{\mathbb{C}P^{n-1}} = 0$, the exactness of the K-sequence of the pair $(\mathbb{C}P^n, \mathbb{C}P^{n-1})$ implies that $r_n\gamma^n$ belongs to the image of the homomorphism $\widetilde{K}(S^2n) = K(\mathbb{C}P^n, \mathbb{C}P^{n-1}) \to K(\mathbb{C}P^n)$. Hence, $\mathrm{ch} r_n\gamma^n$, which is $r_n x^n$, is equal to the Chern character of the image of some element of $\widetilde{K}(S^2n)$, which belongs to $H^{2n}(\mathbb{C}P^n; \mathbb{Z})$ according to the lemma. Hence, $r_n \in \mathbb{Z}$.

B: $K^*(\mathbb{R}P^n)$

The initial term of the reduced Atiyah–Hirzebruch spectral sequence for $\mathbb{R}P^n$ with $n = 2m, 2m+1$ has the form

$$\begin{array}{ccccc} K^0(X, X^{r+1}) & \longrightarrow & K^0(X, X^r) & \longrightarrow & K^0(X^{r+1}, X^r) \\ & & & & \| \\ 0 & = & \widetilde{K}^{-r-1}(\vee S^0) & = & \widetilde{K}^0(\vee S^{r+1}) \\ \downarrow & & & & \\ \text{by Bott's periodicity} & & & & \end{array}$$

All the differentials are trivial by the dimension argumentations, and we have

$$K^1(\mathbb{R}P^n) = \begin{cases} 0 & \text{for } n \text{ even}, \\ \mathbb{Z} & \text{for } n \text{ odd}; \end{cases}$$

the order of the group $\widetilde{K}(\mathbb{R}P^n)$ is equal to $2^{[n/2]}$.

Theorem. *$\widetilde{K}(\mathbb{R}P^n)$ is a cyclic group of order* $2^{[n/2]}$.

Proof. Since the inclusion $\mathbb{R}P^{2m} \to \mathbb{R}P^{2m+1}$ induces, as is seen from the Atiyah–Hirzebruch spectral sequence, an isomorphism between the groups $\widetilde{K}^0$, it is sufficient to prove the theorem for $n = 2m$. The natural embedding $\mathbb{R}P^{2m} \to \mathbb{C}P^{2m}$ is homotopic, by the cellular approximation theorem, to some cellular map $f : \mathbb{R}P^{2m} \to \mathbb{C}P^m$. It is easy to see that the induced cohomology homomorphism

$$f^* : H^*(\mathbb{C}P^m; \mathbb{Z}) \to H^*(\mathbb{R}^{2m}; \mathbb{Z})$$

is an epimorphism. Property (5) of the Atiyah–Hirzebruch spectral sequence implies then that

$$f^* : \widetilde{K}(\mathbb{C}P^m) \to \widetilde{K}^0(\mathbb{R}P^{2m})$$

is also an epimorphism, which implies, in virtue of the last theorem, that $\widetilde{K}^0(\mathbb{R}P^{2m})$ is spanned by $\beta, \beta^2, \dots, \beta^m$, where $\beta = f^*\gamma$. But $\gamma = \zeta - 1$. It is easy to see that $f^*\zeta = \mathbb{C}\zeta_{\mathbb{R}}$, the complexification of the real Hopf bundle over $\mathbb{R}P^{2m}$. It is also obvious that $(\zeta_{\mathbb{R}})^2 = 1$ (actually, the square of any real one-dimensional vector bundle is equal to 1, for if $\dim \mu = 1$, then $\mu \otimes \mu = S^2\mu$, and a section of $S^2\mu$ is an Euclidean structure in μ, which always exists). Hence, $(\mathbb{C}\zeta_{\mathbb{R}})^2 = 1$, $\beta^2 = (\mathbb{C}\zeta_{\mathbb{R}} - 1)^2 = -2\mathbb{C}\zeta_{\mathbb{R}} + 2 = -2\beta$, $\beta^3 = (-2\beta)\beta = -2\beta^2 = 4\beta$, $\dots$, $\beta^m = (-2)^{m-1}\beta$. This shows that an arbitrary element of $\widetilde{K}^0(\mathbb{R}P^{2m})$ is a multiple of β; that is, the group $\widetilde{K}^0(\mathbb{R}P^{2m})$ is cyclic.

Corollary of Proof. *Any complex vector bundle over $\mathbb{R}P^{2m}$ or $\mathbb{R}P^{2m+1}$ is stably equivalent to the vector bundle $k\mathbb{C}\zeta_{\mathbb{R}}$, and the vector bundles $k\mathbb{C}\zeta_{\mathbb{R}}$ and $l\mathbb{C}\zeta_{\mathbb{R}}$ are stably equivalent if and only if $k \equiv l \bmod 2^m$.*

C: Other Computations of K-Functors

Other computations of K-functors in particular, for Grassmann manifolds and flag manifolds, are contained in Chapter IV of Karoubi's book [51] (see Sect. 38.3.C). By the way, Karoubi does not use in his computations the Atiyah–Hirzebruch spectral sequence, so the reader may try to improve his computations. For the computation of the K-functor of the classical Lie groups, see Atiyah [15].

39.3 Differentials of the Atiyah–Hirzebruch Spectral Sequence

The first (possibly) nontrivial differential of the Atiyah–Hirzebruch spectral sequence is

$$d_3 = d_3^n : H^n(X;\mathbb{Z}) \to H^{n+3}(X;\mathbb{Z}).$$

EXERCISE 2. Show that d_3 is a stable cohomology operation.

Although we have not calculated $H^*(K(\mathbb{Z},n);\mathbb{Z})$, the reader can easily deduce from the results of Lecture 27 that

$$H^{n+3}(K(\mathbb{Z},n);\mathbb{Z}) = \mathbb{Z}_2 \ (n \geq 3).$$

Thus, there exists a unique nontrivial stable cohomology operation which maps integral cohomology into integral cohomology and raises dimensions by 3. This operation acts as

$$H^n(X;\mathbb{Z}) \xrightarrow{\rho_2} H^n(X;\mathbb{Z}_2) \xrightarrow{\mathrm{Sq}^2} H^{n+2}(X;\mathbb{Z}_2) \xrightarrow{\beta} H^{n+3}(X;\mathbb{Z}).$$

After we reduce it mod 2, this operation becomes $\mathrm{Sq}^1\mathrm{Sq}^2 = \mathrm{Sq}^3$, and it is natural to denote it by $\widetilde{\mathrm{Sq}}^3$.

Theorem. $d_3 = \widetilde{\mathrm{Sq}}^3$.

Proof. Since the only alternative to this equality is $d_3 = 0$, it is sufficient to give an example when this differential is nontrivial.

Take a map $\alpha: S^5 \to S^4$ not homotopic to a constant and consider the space $S^4 \cup_\alpha D^6$. Since $\pi_5(S^4) = \mathbb{Z}_2$, the map $\beta: S^5 \xrightarrow{\lambda} S^5 \xrightarrow{\alpha} S^4$ with $\deg\lambda = 2$ is homotopic to 0, and there exists a map $\gamma: D^6 \to S^4$ with $\gamma|_{S^5} = \beta$. This map, combined with the map

$$D^6 = CS^5 \xrightarrow{C\lambda} CS^5 = D^6 \xrightarrow{\subset} S^4 \cup_\alpha D^6,$$

yields a map $\delta: (D^6 \cup D^6) = S^6 \to S^4 \cup_\alpha D^6$, and we set $X = (S^4 \cup_\alpha D^6) \cup_\delta D^7$. For X, we need to know only that $X \supset S^4$ and that the cohomology of X is as follows:

q	<4	4	5	6	7	>7
$\widetilde{H}^q(X;\mathbb{Z})$	0	$\mathbb{Z}$	0	0	$\mathbb{Z}_2$	0
$\widetilde{H}^q(X;\mathbb{Z}_2)$	0	$\mathbb{Z}_2$	0	$\mathbb{Z}_2$	$\mathbb{Z}_2$	0

Moreover, the operation $\mathrm{Sq}^3: H^4(X;\mathbb{Z}_2) \to H^7(X;\mathbb{Z}_2)$ is not trivial. [To check this notice that the operation $\mathrm{Sq}^2: H^4 \to H^6$ (with coefficients in $\mathbb{Z}_2$) is not trivial in $S^4 \cup_\alpha D^6$ (as we know from Lecture 31), and the Bockstein homomorphism $H^6 \to H^7$ (again with coefficients in $\mathbb{Z}_2$) is not trivial in X/S^4.]

If the differential $d_3^4: E_3^4(X) \to E_3^7(X)$ of the Atiyah–Hirzebruch spectral sequence were trivial, then the homomorphism $\widetilde{K}(X) \to \widetilde{K}(S^4)$ induced by the embedding $S^4 \to X$ would have been an isomorphism. In particular, the map $\varphi: S^4 \to BU$ corresponding to the generator of $\widetilde{K}(S^4) = \mathbb{Z}$ would have been extendable to a continuous map $\psi: X \to BU$. But the map $\varphi^*: H^4(BU;\mathbb{Z}_2) \to H^4(S^4;\mathbb{Z}_2)$ is not trivial. [Indeed, the Chern character of the generator ξ of $\widetilde{K}(S^4)$ is the generator of $H^4(S^4;\mathbb{Z})$; at the same time, $\mathrm{ch}^2(\xi) = \frac{1}{2}c_1^2(\xi) - c_2(\xi) = -c_2(\xi)$; hence, $\varphi^*(H^4(BU;\mathbb{Z})) \ni -c_2(\xi) = 1 \in H^4(S^4;\mathbb{Z})$, and hence $\varphi^*: H^4(BU;\mathbb{Z}_2) \to H^4(X;\mathbb{Z}_2)$ must be nontrivial, which contradicts the commutativity of the diagram

$$\begin{array}{ccc} H^4(BU;\mathbb{Z}_2) & \xrightarrow{\psi^*} & H^4(X;\mathbb{Z}_2) \\ \downarrow{\scriptstyle \mathrm{Sq}^3} & & \downarrow{\scriptstyle \mathrm{Sq}^3} \\ 0 = H^7(BU;\mathbb{Z}_2) & \xrightarrow{\psi^*} & H^7(X;\mathbb{Z}_2).] \end{array}$$

Corollary of Proof. *There exist simply connected finite* CW *complexes* X, Y *and a continuous map* $f: X \to Y$ *which is not a homotopy equivalence, but which induces an isomorphism* $f^*: K^*(Y) \to K^*(X)$. (This fact was promised and commented on in Sect. 37.1.)

Proof. One can take X from the previous proof and put $Y = S^4$. It is easy to show that the map $S^4 \to S^4$ of degree 2 may be extended to a map $X \to S^4$; the extended map is f. The computations above imply that $f^*: \widetilde{K}^*(Y) \to \widetilde{K}^*(X)$ is an isomorphism, and f is not a homotopy equivalence, since $f^*: H^4(Y; \mathbb{Z}_2) \to H^4(X; \mathbb{Z}_2)$ is not an isomorphism.

Notice that there exists only partial information about the subsequent differentials. It is known that for any prime p the differential d_r with $r < 2p - 1$ has order not divisible by p, while the differential d_{2p-1} has order not divisible by p^2. See the details in the article by Buchstaber [24].

Lecture 40 The Adams Operations

40.1 Definition and Main Properties

A: Introduction

A cohomology operation in K-theory (and in any extraordinary cohomology theory as well) is defined precisely in the same way as the usual cohomology operation. Namely, it is a class of homomorphisms

$$\alpha_X: K(X) \longrightarrow K(X)$$

[or $K^q(X) \to K^q(X)$, but we prefer to ignore this opportunity] defined for all finite CW complexes X and satisfying the condition of the commutativity of diagrams

$$\begin{array}{ccc} K(X) & \xrightarrow{\alpha_X} & K(X) \\ \Big\downarrow{\scriptstyle f^*} & & \Big\downarrow{\scriptstyle f^*} \\ K(Y) & \xrightarrow{\alpha_Y} & K(Y) \end{array}$$

for all continuous maps $f: Y \to X$. An approach to the classification of such operations is suggested by the general idea of Sect. 28.2, which reduces the classification problem to the problem of the calculation of the K-functor for BU (with simultaneous overcoming of the difficulties related to BU being an infinite CW complex; see Sect. 38.2.E). All these problems have been long solved; the results related to the complex case are presented in Sect. 4.7 of Karoubi's book (see Sect. 38.3.C), while the real case is studied in the article by Anderson [13].

We systematically ignore the real case. As to the complex case, we restrict ourselves to considering the most important specific cohomology operations in the K-theory. The readers of Karoubi's book know that these operations exhaust the whole variety of the cohomology operation in the complex K-theory—in the same sense in which the Steenrod squares and Steenrod powers exhaust the entire variety of ordinary cohomology operations.

B: The Definition of Adams Operations

A cohomology operation in K-theory should assign to a virtual vector bundle with the base X another virtual vector bundle with the same base X. If we drop the adjective "virtual" here, then we will have no difficulties in constructing such operations. Indeed, linear algebra provides a lot of ways to assign in a natural way a vector space to another vector space, and these linear algebra operations usually give rise to operations for vector bundles: One can associate with a vector bundle ξ a vector bundle $\xi \otimes \xi$, or $\Lambda^2\xi$, or $S^2\xi$, or one can take any complex linear representation (of some dimension N) of the group $GL(n, \mathbb{C})$ [or $U(n)$] and assign, with the help of this representation, some N-dimensional complex vector bundle to a given n-dimensional complex vector bundle. The common flaw of all these constructions is the lack of additivity: $(\xi_1 \oplus \xi_2) \otimes (\xi_1 \oplus \xi_2) \neq (\xi_1 \otimes \xi_1) \oplus (\xi_2 \otimes \xi_2)$, etc. Therefore, these constructions cannot be extended to the K-theory. Adams' construction provides a way (unique, in a sense: See the section of Karoubi's book mentioned above) to compose from these operations an additive combination.

Let $e_1, e_2, e_3, \ldots$ be the elementary symmetric polynomials [that is, $e_k(x_1, \ldots, x_m) = \sum_{1 \leq i_1 < \cdots < i_k \leq m} x_{i_1} \ldots x_{i_k}, m \gg k$], and let $N_1, N_2, N_3, \ldots$ be the Newton symmetric polynomials ($N_k(x_1, \ldots, x_m) = x_1^k + \cdots + x_m^k$). Let

$$
\begin{array}{ll}
N_1 = P_1(e_1), & P_1(t_1) = t_1, \\
N_2 = P_2(e_1, e_2) & P_2(t_1, t_2) = t_1^2 - 2t_2, \\
N_3 = P_3(e_1, e_2, e_3), & P_3(t_1, t_2, t_3) = t_1^3 - 3t_1t_2 + 3t_3 \\
\ldots\ldots\ldots\ldots & \ldots\ldots\ldots\ldots
\end{array}
$$

For a complex vector bundle (not virtual!), put

$$\psi^k(\xi) = P_k(\xi, \Lambda^2\xi, \Lambda^3\xi, \ldots).$$

For instance,

$$
\begin{aligned}
&\psi^1(\xi) = \xi, \\
&\psi^2(\xi) = \xi \otimes \xi - 2\Lambda^2\xi, \\
&\psi^3(\xi) = \xi \otimes \xi \otimes \xi - 3\xi \otimes \Lambda^2\xi + 3\Lambda^3\xi, \\
&\ldots\ldots\ldots\ldots
\end{aligned}
$$

According to this definition, $\psi^k(\xi)$ is a virtual vector bundle.

Lemma. $\psi^k(\xi \oplus \eta) = \psi^k(\xi) \oplus \psi^k(\eta)$.

Proof. Divide the variables $x_1, \dots, x_m$ into two groups, $y_1, \dots, y_r$ and $z_1, \dots, z_s$ (r, s, and $m = r + s$ are very large). Obviously,

$$N_k(x) = N_k(y) + N_k(z), \quad e_k(x) = \sum_{u+v=k} e_u(y)e_v(z),$$

where e_0 is 1. Hence,

$$\begin{aligned} P_k(e_1(y), e_2(y), \dots) &+ P_k(e_1(z), e_2(z), \dots) \\ &= P_k\left(\sum_{u+v=1} e_u(y)e_v(z), \sum_{u+v=2} e_u(y)e_v(z), \dots\right); \end{aligned}$$

that is,

$$P_k(\alpha_1, \alpha_2, \dots) + P_k(\beta_1, \beta_2, \dots) = P_k\left(\sum_{u+v=1} \alpha_u\beta_v, \sum_{u+v=2} \alpha_u\beta_v, \dots\right)$$

(where $\alpha_0 = \beta_0 = 1$). Substitute $\Lambda^u\xi$ and $\Lambda^v\eta$ for α_u and β_v and use the equality

$$\bigoplus_{u+v=w} \Lambda^u\xi \otimes \Lambda^v\eta = \Lambda^w(\xi \oplus \eta).$$

We arrive at the equality

$$P_k(\xi, \Lambda^2\xi, \dots) + P_k(\eta, \Lambda^2\eta, \dots) = P_k(\xi \oplus \eta, \Lambda^2(\xi \oplus \eta), \dots);$$

that is, $\psi^k(\xi) + \psi^k(\eta) = \psi^k(\xi \oplus \eta)$.

The lemma lets us extend the definition of the operation ψ^k to virtual bundles. We get a cohomology operation

$$\psi^k: K(X) \to K(Y)$$

(it is obvious that ψ^k commutes with the homomorphisms f^*). By construction, ψ^k is additive:

$$\psi^k(\xi + \eta) = \psi^k(\xi) + \psi^k(\eta).$$

C: The Properties of the Operations ψ^k

Theorem. (1) $\dim \psi^k(\xi) = \dim \xi$.
(2) $\psi^k(\xi\eta) = \psi^k(\xi)\psi^k(\eta)$.
(3) $\psi^k \circ \psi^l(\xi) = \psi^{kl}(\xi)$.
(4) $\mathrm{ch}^q\, \psi^k(\xi) = k^q\, \mathrm{ch}^q\, \xi$.
(5) *If ξ is a geometric (that is, not virtual) one-dimensional vector bundle, then* $\psi^k(\xi) = \xi^k$.
(6) *If p is a prime, then* $\psi^p(\xi) \equiv \xi^p \bmod p$.
(7) *Put* $\psi^0(\xi) = \dim \xi$, $\psi^{-k}(\xi) = \psi^k(\overline{\xi})$ $(k > 0)$. *Properties* (1)–(5) *hold for the operations ψ^k, $k \in \mathbb{Z}$ defined in this way.*

Proof. (1) Since the operations ψ^k and dim are both additive, we need only to consider the case when ξ is a geometric vector bundle. In this case we have $(n = \dim \xi)$:

$$\begin{aligned}\dim \psi^k(\xi) &= P_k(\dim \xi, \dim \Lambda^2\xi, \dots, \dim \Lambda^k\xi) = P_k\left(n, \binom{n}{2}, \dots, \binom{n}{k}\right)\\ &= P_k(e_1(\underbrace{1, \dots, 1}_{n}), e_2(\underbrace{1, \dots, 1}_{n}), \dots, e_k(\underbrace{1, \dots, 1}_{n}))\\ &= N_k(\underbrace{1, \dots, 1}_{n}) = \underbrace{1^k + \dots + 1^k}_{n} = n.\end{aligned}$$

(5) If ξ is a one-dimensional geometric vector bundle, then $\Lambda^r\xi = 0$ for $r > 1$, and $\psi^k(\xi) = P_k(\xi, 0, 0, \dots, 0) = \xi^k$.

(6) This follows from $N_p \equiv (e_1)^p \bmod p$.

The most natural proof of statements (2)–(4) is based on the informal "splitting principle" that was mentioned in Sect. 19.5. This principle states that in proofs of various formulas involving complex vector bundles and their characteristic classes one can restrict oneself to the case when these bundles are split into the sums of one-dimensional bundles. Technically this is based on the fact that the natural embedding $B(U(1) \times \dots \times U(1)) = (\mathbb{C}P^\infty)^n \to BU(n) = \mathbb{C}G(\infty, n)$ induces a monomorphism both in cohomology (this is essentially known to us) and in K-theory (this is deduced by means of the Atiyah–Hirzebruch spectral sequence). We, however, will give shorter proofs, which use the splitting principle implicitly.

(4) The formula $\xi \mapsto (\mathrm{ch}^q\, \psi^k\xi - k^q\, \mathrm{ch}^q\, \xi)$ defines a characteristic class of complex vector bundles with values in $2q$-dimensional rational cohomology. According to the results of Lecture 19 [see part (iii) of the theorem in Sect. 19.4.E)], this class may be represented as a (rational) polynomial in Chern classes. Also, it was proven in Sect. 19.5 that any nonzero polynomial in Chern classes takes a nonzero value on the vector bundle $\underbrace{\zeta \times \dots \times \zeta}_{n}$ with the base $(\mathbb{C}P^\infty)^n$, where n is sufficiently large. But

$$\begin{aligned}\operatorname{ch}\psi^k(\underbrace{\zeta\times\dots\times\zeta}_{n})\\ &=\operatorname{ch}\psi^k(p_1^*\zeta\oplus\dots\oplus p_n^*\zeta)=\operatorname{ch}[\psi^k(p_1^*\zeta)\oplus\dots\oplus\psi^k(p_n^*\zeta)]\\ &=\operatorname{ch}[(p_1^*\zeta^k)\oplus\dots\oplus(p_n^*\zeta^k)]=p_1^*(\operatorname{ch}\zeta)^k+\dots+p_n^*(\operatorname{ch}\zeta)^k\\ &=p_1^*e^{kx}+\dots+p_n^*e^{kx}=\sum_{i=1}^n e^{kx_i}=\sum_{q=0}^\infty\frac{k^q}{q!}\sum_{i=1}^n x_i^q.\end{aligned}$$

[Notations: p_i is the projection of $(\mathbb{C}P^\infty)^n$ onto the ith factor; $x\in H^2(\mathbb{C}P^\infty;\mathbb{Z})$ and $x_1,\dots,x_n\in H^2((\mathbb{C}P^\infty)^n;\mathbb{Z})$ are the cohomological generators.] Hence,

$$\operatorname{ch}^q\psi^k(\zeta\times\dots\times\zeta)=\frac{k^q}{q!}\sum_{i=1}^n x_i^q,$$

in particular,

$$\operatorname{ch}^q(\zeta\times\dots\times\zeta)=\frac{1}{q!}\sum_{i=1}^n x_i^q,$$

whence

$$\operatorname{ch}^q\psi^k(\zeta\times\dots\times\zeta)=k^q\operatorname{ch}(\zeta\times\dots\times\zeta),$$

which completes the proof.

(3) It is sufficient to prove the formula in the case when ξ is the tautological vector bundle over the Grassmannian. Since the cohomology of the Grassmannian is torsion-free, then its K-functor is also torsion-free (this is seen from the Atiyah–Hirzebruch spectral sequence; compare with Exercise 1 in Sect. 39.1). Hence, the Chern character is a monomorphism (its kernel is always finite), and we need only to prove that

$$\operatorname{ch}(\psi^k\circ\psi^l(\xi))=\operatorname{ch}(\psi^{kl}(\xi)),$$

which follows from (4):

$$\operatorname{ch}^q(\psi^k\circ\psi^l(\xi))=k^q\operatorname{ch}^q(\psi^l(\xi))=k^ql^q\operatorname{ch}^q\xi=(kl)^q\operatorname{ch}^q\xi=\operatorname{ch}^q(\psi^{kl}(\xi)).$$

(2) The proof is similar to the previous one: It is sufficient to consider the case when ξ and η are the vector bundles over the product of two Grassmannians that are pull-backs of the tautological vector bundles over the factors. The cohomology of the product of two (complex) Grassmannians is again torsion-free, and we have the right to apply the Chern character to both sides of the equality to be proved. Thus, all we need is the following immediate calculation:

$$\begin{aligned}\mathrm{ch}^q\,\psi^k(\xi\eta) &= k^q\,\mathrm{ch}^q(\xi\eta) = k^q \sum_{s+t=q} \mathrm{ch}^s\,\xi\,\mathrm{ch}^t\,\eta \\ &= \sum_{s+t=q} (k^s\,\mathrm{ch}^s\,\xi)(k^t\,\mathrm{ch}^t\,\eta) = \sum_{s+t=q} (\mathrm{ch}^s\,\psi^k\xi)(\mathrm{ch}^t\,\psi^k\eta) \\ &= \mathrm{ch}^q[(\psi^k\xi)(\psi^k\eta)].\end{aligned}$$

(7) All we need to prove this is $\mathrm{ch}^q\,\overline{\xi} = (-1)^q\,\mathrm{ch}^q\,\xi$, and $\psi^k\overline{\xi} = \overline{\psi^k\xi}$. The first is known [follows from $c_1(\zeta) = -c_1(\overline{\zeta})$], and the second is proved by comparing the Chern characters of the two sides of the equality.

Corollary from (1). $\psi(\widetilde{K}(X)) \subset \widetilde{K}(X)$. *Thus,* ψ^k *may be regarded as an operation* $\widetilde{K}(X) \to \widetilde{K}(X)$.

Corollary from (3). $\psi^k \circ \psi^l = \psi^l \circ \psi^k$.

40.2 A Short Proof of the Nonexistence of Spheroids with Odd Hopf Invariant

(Adams and Atiyah [9]) Let $\lambda \in \pi_{4n-1}(S^{2n})$. Attach the ball D^{4n} to the sphere S^{2n} by means of a spheroid $S^{4n-1} \to S^{2n}$ of the class λ. The resulting space X has the following cohomology:

$$H^q(X;\mathbb{Z}) = \begin{cases} \mathbb{Z} \text{ for } q = 0, 2n, 4n, \\ 0 \text{ otherwise.} \end{cases}$$

Let $a \in H^{2n}(X;\mathbb{Z})$ and $b \in H^{4n}(X;\mathbb{Z})$ be the natural generators. Then $a^2 = hb$, where $h \in \mathbb{Z}$. The number h is called the Hopf invariant of the class λ. The following theorem was stated and commented on in this book several times (see, in particular, Sects. 16.5 and 31.4).

Theorem. *If* $n \neq 1, 2, 4$, *then there are no elements with odd Hopf invariant in* $\pi_{4n-1}(S^{2n})$.

Proof. Since there are always elements with the Hopf invariant 2 in $\pi_{4n-1}(S^{2n})$ and since the Hopf invariant is additive (see Sect. 16.5), it is sufficient to prove that for $n \neq 1, 2, 4$ there are no elements with the Hopf invariant 1 in $\pi_{4n-1}(S^{2n})$. Let $h = 1$, that is, $a^2 = b$.

Compute $\widetilde{K}(X)$. Since $\widetilde{K}^1(S^{2n}) = \widetilde{K}^1(S^{4n}) = 0$, the $\widetilde{K}$-sequence of the pair (X, S^{2n}) has the form

$$0 \longrightarrow \widetilde{K}(S^{4n}) \longrightarrow \widetilde{K}(X) \longrightarrow \widetilde{K}(S^{2n}) \longrightarrow 0,$$

and its exactness implies $\widetilde{K}(X) \cong \widetilde{K}(S^{2n}) \oplus \widetilde{K}(S^{4n}) = \mathbb{Z} \oplus \mathbb{Z}$. There are canonical generators $\alpha \in \widetilde{K}(S^{2n}), \beta \in \widetilde{K}(S^{4n})$. Take generators $\widetilde{\alpha}, \widetilde{\beta}$ in $\widetilde{K}(X)$: $\widetilde{\beta}$ is defined canonically, as the image of β, while $\widetilde{\alpha}$ is an arbitrary inverse image of α and is defined up to adding a multiple of $\widetilde{\beta}$. Since $\operatorname{ch}\alpha = a$, $\operatorname{ch}\beta = b$, then $\operatorname{ch}\widetilde{\beta} = b$ and $\operatorname{ch}\widetilde{\alpha} = a + eb$, where e is some rational number unknown to us. (This number is not important for the current proof, but it will play a crucial role in Lecture 41.) Since

$$\operatorname{ch}\widetilde{\alpha}^2 = (\operatorname{ch}\widetilde{\alpha})^2 = (a+eb)^2 = a^2 = \operatorname{ch}\widetilde{\beta},$$

then $\widetilde{\alpha}^2 = \widetilde{\beta}$. Furthermore, property (4) of the operations ψ^k implies

$$\psi^k\widetilde{\alpha} = k^n\widetilde{\alpha} + \mu_k\beta \ (\mu_k \in \mathbb{Z}),\ \psi^k\widetilde{\beta} = k^{2n}\widetilde{\beta}.$$

The congruence $\psi^2\widetilde{\alpha} = \widetilde{\alpha}^2 \bmod 2$ implies that μ_2 is odd (this is the crucial point of the proof!). Hence,

$$\begin{aligned}\psi^2\psi^3\widetilde{\alpha} &= \psi^2(3^n\widetilde{\alpha} + \mu_3\widetilde{\beta}) = 6^n\widetilde{\alpha} + 3^n\mu_2\widetilde{\beta} + 2^{2n}\mu_3\widetilde{\beta},\\ \psi^3\psi^2\widetilde{\alpha} &= \psi^3(2^n\widetilde{\alpha} + \mu_2\widetilde{\beta}) = 6^n\widetilde{\alpha} + 2^n\mu_3\widetilde{\beta} + 3^{2n}\mu_2\widetilde{\beta},\end{aligned}$$

which implies, since $\psi^2\psi^3 = \psi^3\psi^2$, that

$$3^n\mu_2 + 2^{2n}\mu_3 = 2^n\mu_3 + 3^{2n}\mu_2;\ \mu_2 3^n(3^n - 1) = \mu_3 2^n(2^n - 1).$$

Since μ_2 is odd, $3^n - 1$ is divisible by 2^n, which is possible only for $n = 1, 2, 4$.

[Indeed, obviously $3^{2s} \equiv 1 \bmod 8$ and $3^{2s+1} \equiv 3 \bmod 8$. Therefore, if $n = 2^l m$ with m odd, then (for $l \geq 1$)

$$\begin{aligned}3^n - 1 &= (3^{2^{l-1}m} + 1)(3^{2^{l-2}m} + 1)\dots(3^m + 1)(3^m - 1)\\ &= \underbrace{2 \cdot \dots \cdot 2}_{l-1} \cdot 4 \cdot 2 \cdot \{\text{odd number}\} = 2^{l+2} \cdot \{\text{odd number}\};\end{aligned}$$

that is, for n even the number of factors 2 in the prime factorization of $3^n - 1$ is $l + 2$ if n is even and 1 in n is odd. Hence, the divisibility of $3^n - 1$ by 2^n implies $n = 2^l m \leq l + 2$; if $m = 1$, then this inequality holds for $l \leq 2$; if $m \geq 3$, then this inequality never holds.]

Recall that the existence in $\pi_{4n-1}(S^{2n})$ of an element with the Hopf invariant odd is equivalent to the parallelizability of the sphere S^{2n-1}. Hence, the problem considered above has a natural extension: What is the maximal possible number of pointwise linear independent tangent vector fields on the sphere S^m? If the sphere S^m is parallelizable, then this number is equal to m; if it is not parallelizable, one can say only that this number is less than m. What is this number? The answer is contained in the following result.

Theorem. *Let $n = (2a+1)2^{4b+c}$, where a, b, c are nonnegative integers and $c \leq 3$. Put $\rho(n) = 2^c + 8b$. Then on the sphere S^{n-1} there exists $\rho(n) - 1$ linear independent vector fields and there is no $\rho(n)$ linear independent vector fields.*

The first part of this theorem is classical; it has been proved by Radon and Hurwitz; the proof consists in a direct construction and is not related in any way to homotopy topology. The second part [that there are no $\rho(n)$ vector fields] was proved by Adams, who reduced it to the computation of the operations ψ^k in the K-functor of "truncated projective spaces" $\mathbb{R}P^s/\mathbb{R}P^t$. The reader can find the details in the original article by Adams (see Adams [3]); this paper is very famous because it is the paper where the operations ψ^k were first introduced. In what is for the reader a more convenient, self-contained way, this result (both Radon–Hurwitz's and Adams' parts) is presented in the book [73] of J. Schwartz (mentioned in Sect. 38.3.C).

Lecture 41 J-functor

41.1 Definition and Relations to Homotopy Groups of Spheres

A: Definition

Let ξ and η be two complex vector bundles of the same dimension with the same base, and let $S(\xi)$ and $S(\eta)$ be the corresponding sphere bundles. The vector bundles ξ and η are called *homotopy equivalent* or *J-equivalent* if there exists a fiberwise map $S(\xi) \to S(\eta)$ that covers the identity of the base and has degree 1 on each fiber.

Let X be a finite CW complex. An element $\alpha \in \widetilde{K}(X)$ is called *J-trivial* if $\alpha = \{\xi - \eta\}$ and the vector bundles ξ and η are J-equivalent. Obviously, J-trivial virtual vector bundles form a subgroup of $\widetilde{K}(X) \subset K(X)$; we denote this subgroup by $T(X)$ and put

$$J(X) = K(X)/T(X), \; \widetilde{J}(X) = \widetilde{K}(X)/T(X).$$

Thus, $\widetilde{J}(X) = \operatorname{Ker}[\dim: J(X) \to \mathbb{Z}], J(X) = \mathbb{Z} \oplus \widetilde{J}(X)$.

J-functor was thoroughly studied in the four articles "*On* $J(X)$, I–IV" of Adams ([4]–[7]).

EXERCISE 1. Prove that $T(X) \subset \widetilde{K}(X)$ consists of classes of stably equivalent vector bundles over X such that an arbitrary vector bundle of sufficiently large dimension from this class is J-equivalent to a trivial bundle.

A continuous map $f: X \to Y$ induces homomorphisms $f^*: J(Y) \to J(X)$ and $f^*: \widetilde{J}(Y) \to \widetilde{J}(X)$, which makes J and $\widetilde{J}$ functors. We will see, however, that the functor $\widetilde{J}$ *is not half-exact*; that is, the sequence

$$\widetilde{J}(X/A) \longrightarrow \widetilde{J}(X) \longrightarrow J(A),$$

induced by the inclusion map $A \to X$ and the projection $X \to X/A$, does not have to be exact.

B: J-Homomorphism and J-Functor of the Sphere

The action of the group $U(n)$ in S^{2n-1} determines a map $U(n) \to \text{Maps}(S^{2n-1} \to S^{2n-1})$. If we compose this map with the suspension map $\Sigma\colon \text{Maps}(S^{2n-1} \to S^{2n-1}) \to \text{Maps}(S^{2n} \to S^{2n})$ we will get a map of the group $U(n)$ into the space of maps $S^{2n} \to S^{2n}$ which fix some point (any of the two poles), that is, a map

$$U(n) \to \Omega^{2n} S^{2n}.$$

The corresponding homomorphism of homotopy groups

$$\pi_i(U(n)) \to \pi_i(\Omega^{2n} S^{2n}) \to \pi_{2n+i}(S^{2n})$$

is called the *J-homomorphism* (or Whitehead J-homomorphism). We will deal mainly with the *stable J-homomorphism* ($n \gg i$) in the case of odd i:

$$J\colon \pi_{2k-1}(U(n)) = \mathbb{Z} \to \pi_{2n+2k-1}(S^{2n}) = \pi^S_{2k-1}.$$

The image of this homomorphism is a cyclic subgroup of the group π^S_{2k-1}, which, as we will see, can have a big order (in the case of even k). [Actually, there is also the image of the "real J-homomorphism" $J\colon \pi_{4k-1}(SO(n)) = \mathbb{Z} \to \pi^S_{4k-1}$, which may be slightly (two times) bigger than the image of the complex J-homomorphism. We will comment on the difference between these two images in Lecture 42.]

Let us relate now the J-homomorphism to $\widetilde{J}(S^m)$.

Theorem. $\widetilde{J}(S^m) = \text{Im}[J\colon \pi_{m-1}(U) \to \pi^S_{m-1}]$.

Essentially, this fact is obvious. An n-dimensional complex vector bundle over S^m is determined by the "attaching map" $S^{m-1} \to U(n)$, which is used for gluing together two trivial bundles over the hemispheres $D^m_\pm \subset S^m$ along the equator $S^{m-1} = \partial D^m_\pm$; J-triviality of the resulting vector bundle means precisely that the attaching map is homotopic to the constant within the space of maps $(S^{2n-1} \to S^{2n-1}) \supset U(N)$.

C: J-Functor as an Image of the K-Functor in a Half-Exact Functor

Consider for a finite CW complex X the set of all oriented sphere bundles with the base X, that is, locally trivial fibrations over X with the fibers homeomorphic to a sphere and compatibly oriented. Again, if X is disconnected, we allow the fibers over different components to have different dimensions. Two sphere bundles, $p_1\colon E_1 \to X$ and $p_2\colon E_2 \to X$, are called equivalent if there exists a continuous map $E_1 \to E_2$ which for each $x \in X$ maps $p_1^{-1}(x)$ onto $p_2^{-1}(x)$ with the degree $+1$.

EXERCISE 2. Check that this is an equivalence relation.

Denote the set of classes of all sphere bundles over X by $S(X)$. The dimension of a sphere bundle is defined as the dimension of the fiber over the base point plus 1; equivalent sphere bundles have equal dimensions.

For sphere bundles there is an addition operation: The sum of two sphere bundles is defined with the help of the "fiberwise join":

$$E_1 *_X E_2 = \frac{\{(y_1, y_2, t) \in E_1 \times E_2 \times I | p_1(y_1) = p_2(y_2)\}}{(y_1, y_2, t) \sim (y_1', y_2', t) \begin{cases} \text{if } y_1 = y_1', t = 0 \\ \text{or } y_2 = y_2', t = 1 \end{cases}}.$$

This operation makes $S(X)$ a commutative semigroup. The corresponding Grothendieck group is denoted by $\Pi(X)$. There is an obvious map dim: $\Pi(X) \to \mathbb{Z}$. It is clear that

$$\Pi(X) = \mathbb{Z} \oplus \widetilde{\Pi}(X),$$

where $\widetilde{\Pi}(X) = \operatorname{Ker} \dim$.

EXERCISE 3. Prove that $\widetilde{\Pi}(X)$ is the group of classes of stably equivalent sphere bundles. (Two sphere bundles are called stably equivalent if their multiple suspensions are equivalent. The suspension over a sphere bundle is defined as a fiberwise suspension; in other words, the multiple suspension is the sum with the trivial sphere bundle.)

A continuous map $f: X \to Y$ induces in a natural way the homomorphisms $f^*: \Pi(Y) \to \Pi(X), f^*: \widetilde{\Pi}(Y) \to \widetilde{\Pi}(X)$, which depend only on the homotopy class of f.

EXERCISE 4. Prove that $\widetilde{\Pi}$ is a half-exact functor; that is, the sequence

$$\widetilde{\Pi}(X/A) \longrightarrow \widetilde{\Pi}(X) \longrightarrow \widetilde{\Pi}(A)$$

is exact for any finite CW pair (X, A).

There is a natural (functorial) homomorphism $K(X) \to \Pi(X)$ (it assigns to a vector bundle the corresponding sphere bundle). Obviously, *$J(X)$ is the image of $K(X)$ in $\Pi(X)$.*

Theorem.

$$\widetilde{\Pi}(S^n) = \begin{cases} 0 & \text{for } n = 1, \\ \pi_{n-1}^S & \text{for } n > 1. \end{cases}$$

Proof. A sphere bundle of dimension N over S^n is determined by the attaching map $S^{n-1} \to \text{Map}_1(S^{N-1} \to S^{N-1})$, where Map_1 denotes the topological space of maps of degree 1. But $\mathrm{Map}_1(S^{N-1} \to S^{N-1})$ is a fibration over S^{N-1} with the fiber $(\Omega^{N-1}S^{N-1})_0$ (the subscript 0 refers to a connected component). Hence, if $N \gg n > 1$, then homotopy classes of attaching maps compose the group $\pi_{n-1}(\Omega^{N-1}S^{N-1}) = \pi_{n-1}^S$.

Corollary. *The group $\widetilde{\Pi}(S^n)$ is finite for any n.*

Theorem. *The group $\widetilde{\Pi}(X)$ is finite for any finite* CW *complex X.*

Proof. Obviously, $\widetilde{\Pi}(\mathrm{pt}) = 0$. Assume that the statement is true for CW complexes that contain fewer cells than X. Let Y be a CW complex obtained from X by removing one cell of maximal dimension, say, n. Then we have an exact sequence

$$\widetilde{\Pi}(S^n) \longrightarrow \widetilde{\Pi}(X) \longrightarrow \widetilde{\Pi}(Y)$$

in which the first and last groups are finite; hence, $\widetilde{\Pi}(X)$ is also finite.

Corollary. *The group $\widetilde{J}(X)$ is finite for any finite* CW *complex X.*

Indeed, $\widetilde{J}(X) \subset \widetilde{\Pi}(X)$.

D: Some Properties of the Functor $\widetilde{\Pi}$

The half-exact functor $\widetilde{\Pi}$ gives rise to a "half of a cohomology theory":

$$\Pi(X,A) = \widetilde{\Pi}(\Sigma^{-q}(X/A)) \text{ for } q \leq 0.$$

However, the absence of the Bott periodicity for Π makes it difficult to define groups $\Pi^q(X,A)$ with $q > 0$. Nevertheless, for Π (and for any half-exact functor as well), there exists an analog of the Atiyah–Hirzebruch spectral sequence: The groups $E_r^{p,q}$ are defined for $p \geq 0$ and $q \leq 0$, and $E_2^{p,q} = H^p(X; \widetilde{\Pi}(S^{-q}))$; the groups $E_\infty^{p,q}$ with $p + q = r \leq 0$ compose a group adjoint to $\Pi^r(X)$ (with respect to some filtration). (See Fig. 134.)

Notice also that if $X = \Sigma Y$, then

$$\widetilde{\Pi}(X) = \pi(Y, \Omega^N S^N);$$

that is, $\widetilde{\Pi}(X) = \widetilde{P}^0(Y) = \widetilde{P}^1(X)$, where P is the extraordinary "cohomotopy cohomology theory" (see Sect. 38.5.D). Hence, for any finite CW pair (X,A)

$$\Pi(X,A) = P^{q+1}(X,A) \text{ for } q \leq -1.$$

This lets us say that the functor Π is close to the cohomotopy theory with the dimensions shifted by 1. Unfortunately, it is not known whether the equality $\Pi^0(X,A) = P^1(X,A)$ holds for any finite CW pair (X,A).

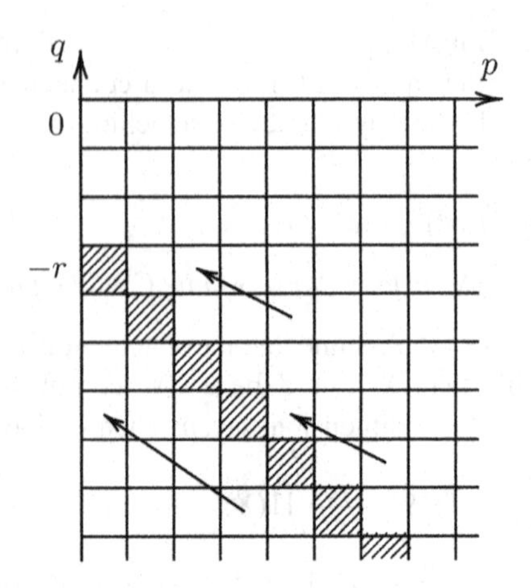

Fig. 134 The Atiyah–Hirzebruch spectral sequence for Π

41.2 The Adams Conjecture

A: Introduction

Although it has been long proven, the following statement is still called the Adams conjecture.

Theorem. *For any finite* CW *complex* X *and any* $\alpha \in K(X)$ *and* $k \in \mathbb{Z}$, *there exists* N *such that*

$$k^N(\psi^k(\alpha) - \alpha) \in T(X).$$

Adams first stated this conjecture in the article "On $J(X)$, I" (Adams [4]). He himself proved it for one-dimensional geometric bundles. After that the Adams conjecture constantly attracted the attention of the leading topologists, but it was proved only in 1970. In that year, first Quillen [70] announced a proof of the conjecture (his detailed paper was published later), and then Sullivan obtained another proof, which used, among other things, some of Quillen's ideas. Sullivan's proof with some preparatory material and further results was presented in his MIT lectures (1970), notes of which were broadly available. (These notes were published in a Russian translation as a book entitled *Geometric Topology*, in 1975. Thirty years later, in 2005, an English-language book named *Geometric Topology: Localization, Periodicity, and Galois Symmetry*, was published by Springer; see Sullivan [83].) Both proofs were long and complicated. However, in 1975 Becker and Gottlieb [21] were able to find a relatively simple proof of Adams' conjecture, based on the notion of the transfer; this proof is given later in this section.

B: The Case of One-Dimensional Vector Bundle

If α is a one-dimensional geometric vector bundle, then Adams' conjecture is a corollary of the following Adams–Dold theorem.

Theorem. *Let ξ_1, ξ_2 be two complex vector bundles of the same dimension with a finite* CW *base X. Suppose that there exists a fiberwise map $S(\xi_1) \to S(\xi_2)$ that covers the identity map of X and has degree $k > 0$ on each fiber [$S(\xi)$, as usual, denotes the sphere bundle associated with the complex vector bundle ξ]. Then for some M and N the vector bundles $k^N\xi_1 \oplus M$ and $k^N\xi_2 \oplus M$ are J-equivalent.*

First, let us show how this result implies the Adams conjecture for geometric one-dimensional bundles. Let α be such a bundle. According to property (5) of the Adams operations, $\psi^k(\alpha) = \alpha^k$. Furthermore, it is obvious that there exists a map $S(\alpha) \to S(\alpha^k)$ of degree k on each of the fibers: It is defined in the fibers by the formula $z \mapsto z^k$. Hence, by the Adams–Dold theorem, $k^N\xi_1 \oplus M$ is J-equivalent to $k^N\xi_2 \oplus M$, whence $k^N(\psi^k(\alpha) - \alpha) \in T(X)$.

Proof of the Adams–Dold Theorem. First, let us reduce the general case of the theorem to the case when the vector bundle ξ_2 is trivial. Assume that the theorem has already been proven in this case, and pick a vector bundle η such that the sum $\eta \oplus \xi_2$ is trivial. The degree k fiberwise map $S(\xi_1) \to S(\xi_2)$ becomes, after adding η, a degree k fiberwise map $S(\eta \oplus \xi_1) \to S(\eta \oplus \xi_2)$. By our assumption, for some N and M, the vector bundle $k^N(\eta \oplus \xi_1) \oplus M$ is J-equivalent to the vector bundle $k^N(\eta \oplus \xi_2) \oplus M$. But then $k^N(\eta \oplus \xi_1) \oplus M \oplus k^N\xi_2$ is J-equivalent to $k^N(\eta \oplus \xi_2) \oplus M \oplus k^N\xi_2$; that is, $k^N\xi_1 \oplus M'$ is J-equivalent to $k^N\xi_1 \oplus M'$, where $M' = M \oplus k^N(\eta \oplus \xi_2)$. Thus, the theorem is valid in the general case.

Consider now the case when ξ_2 is trivial. Let ξ be a vector bundle of dimension n with base X, and let E be the space of the associated sphere bundle. Let $f\colon E \to X \times S^{2n-1}$ be a degree k fiberwise map. Denote by lE the space of the sphere bundle, associated with $l\xi$; thus, lE is the space of the sphere bundle over X with the fiber S^{2ln-1}, whose fiber over $x \in X$ is the l-fold join of the fiber of the bundle $E \to X$ over x. Furthermore, denote by lf the map $lE \to X \times S^{2ln-1}$, which is the fiberwise l-fold join of the map f. We will prove the following.

There exist a positive integer N and a fiberwise map $g\colon k^NE \to X \times S^{2k^Nn-1}$, which has the degree 1 *on each fiber and such that the diagram*

$$\begin{array}{ccc} k^NE & \xrightarrow{g} & X \times S^{2k^Nn-1} \\ \| & & \big\downarrow{\scriptstyle \mathrm{id}_X \times h} \\ k^NE & \xrightarrow{k^Nf} & X \times S^{2k^Nn-1}, \end{array}$$

in which h is a map of degree k^N, is fiberwise homotopy commutative.

This is, certainly, sufficient for us; actually, the existence of g is all we need.

We will prove our statement by induction with respect to the number of cells in X. Let X be obtained by attaching an m-dimensional cell to a CW complex Y of dimension $\leq m$; assume that the statement has already been proven for Y (with N replaced with some N'). Choose a characteristic map $(D^m, S^{m-1}) \to (X, Y)$ of the cell $X - Y$ and take the induced bundle over D^m. Of course, this bundle is trivial. Thus, our problem is reduced to the following one. We have a fiberwise homotopy commutative diagram

$$\begin{array}{ccc} S^{m-1} \times S^{M-1} & \xrightarrow{g'} & S^{m-1} \times S^{M-1} \\ \Big\| & & \Big\downarrow {\scriptstyle \mathrm{id}_{S^{m-1}} \times h'} \\ S^{m-1} \times S^{M-1} & \xrightarrow{f'} & S^{m-1} \times S^{M-1}, \end{array}$$

of fiberwise [over S^{m-1}] maps, in which f' has degree $k^{N'}$ on the fibers, g' has degree 1 on fibers, and h' has degree $k^{N'}$. Let f' be extended to a fiberwise map $F: D^m \times S^{M-1} \to D^m \times S^{M-1}$. We want to extend, for some N'', the map $k^{N''} g': S^{m-1} \times S^{\bar{M}-1} \to S^{m-1} \times S^{\bar{M}-1}$, where $\bar{M} = k^{N''} M$, to a map $G: D^m \times S^{\bar{M}-1} \to D^m \times S^{\bar{M}-1}$ such that the diagram

$$\begin{array}{ccc} D^m \times S^{\bar{M}-1} & \xrightarrow{G} & D^m \times S^{\bar{M}-1} \\ \Big\| & & \Big\downarrow {\scriptstyle \mathrm{id}_{D^m} \times \bar{h}} \\ D^m \times S^{\bar{M}-1} & \xrightarrow{F} & D^m \times S^{\bar{M}-1}, \end{array}$$

where $\bar{h}$ has degree k^N, $N = N' + N''$, is fiberwise homotopy commutative. In other words, we have the following problem. Denote by $\mathcal{G}(M, r)$ the space of maps $S^{M-1} \to S^{M-1}$ of degree r. The map g' determines, via the formula $g'(x, y) = (x, [\gamma(x)](y))$, some map $\gamma: S^{m-1} \to \mathcal{G}(M, 1)$, and we need to prove that it has the following property. The composition

$$S^{m-1} \xrightarrow{\gamma} \mathcal{G}(M, 1) \xrightarrow{h' \circ} \mathcal{G}(M, k^{N'}), \tag{$*$}$$

where the second arrow is the composition with a fixed map $h': S^{m-1} \to S^{m-1}$ of degree $k^{N'}$, is homotopic to a constant map. But γ represents a certain element c of the group

$$\pi_{m-1}(\mathcal{G}(M, 1)) = \pi^S_{m-1},$$

and the composition $(*)$ being homotopic to a constant means precisely that $k^{N'} c = 0$.

Homotopy Lemma. *If a spheroid $\alpha: S^l \to \mathcal{G}(M,1)$, $M \gg l$, represents some $a \in \pi_l^S$, then the spheroid $\underbrace{\alpha * \cdots * \alpha}_{q}: S^l \to \mathcal{G}(qM,1)$,*

$$\begin{aligned}\alpha * \cdots * \alpha(x) \\ = \left[\alpha(x) * \cdots * \alpha(x): S^{M-1} * \cdots * S^{M-1} \to S^{M-1} * \cdots * S^{M-1}\right],\end{aligned}$$

represents $qa \in \pi_l^S$.

Proof of Lemma. We can assume that α takes the base point of S^l to the identity map. Replace the map $\alpha * \cdots * \alpha$ by a homotopic map $\alpha_1 * \cdots * \alpha_q$, where $\alpha_i = \alpha \circ \beta_i$, $\beta_i: S^l \to S^l$ is a map which maps into the base point the complement to a small ball $d_i \subset S^l$ and stretches d_i onto the whole sphere S^l; the balls d_i are supposed to be disjoint. Obviously, $\alpha_1 * \cdots * \alpha_q$ is the sum of q spheroids, of which each represents in π_l^S the same element, as α.

Now we can complete the proof of the theorem. Put $N'' = N'$. The lemma implies that the map $\underbrace{\gamma * \cdots * \gamma}_{k^{N'}}$ is homotopic to a constant. But this map is induced by $k^{N'} g'$; hence, $k^{N'} g'$ may be extended to a map G; for $\bar{h}$ we take $\underbrace{h * \cdots * h}_{k^{N'}}$.

C: The Theorem of Becker and Gottlieb on the Existence of Transfer

The general case of Adams' conjecture may be reduced to its one-dimensional case via a pure geometrical result of Becker and Gottlieb.

Let G be a compact Lie group [a compact subgroup of the group $GL(N, \mathbb{R})$], and let M be a closed smooth manifold with a smooth action of the group G. Further, let $p: E \to X$ be a smooth fiber bundle with the fiber M and the structure group G.

Denote for a topological space Y by Y^+ the disjoint union $Y \sqcup \mathrm{pt}$. It is important for us that

$$\Sigma^N Y^+ = (Y \times S^n)/(Y \times \mathrm{pt});$$

that is, an N-fold suspension over Y^+ is the union of N-dimensional spheres passing through the points of Y and attached to each other in one point (see Fig. 135).

Notice also that $\Sigma^N Y^+$ is also the Thom space of the N-dimensional trivial vector bundle over Y (see Sect. 31.2).

Theorem (Becker and Gottlieb [21]). *For a sufficiently large N, there exists a map*

$$t: \Sigma^N X^+ \to \Sigma^N E^+$$

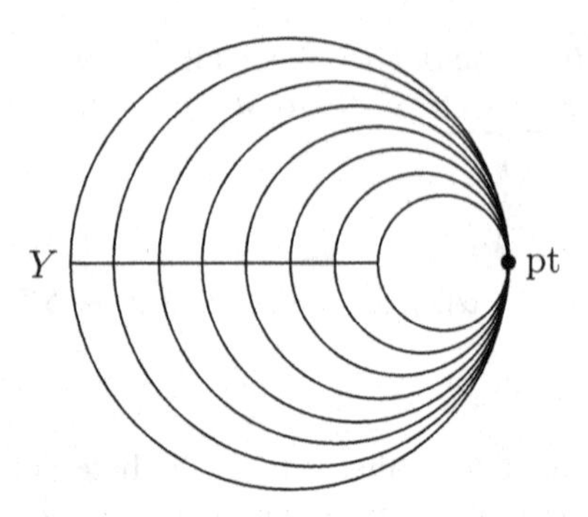

Fig. 135 $\Sigma^N Y^+$

such that the composition

$$\Sigma^N X^+ \xrightarrow{t} \Sigma^N E^+ \xrightarrow{\Sigma^N p} \Sigma^N X^+$$

maps each of the N-dimensional spheres of which $\Sigma^N X^+$ is made (see above) into the same sphere with degree $\chi(M)$ (the Euler characteristic of the manifold M).

The map t is called the *transfer*.

Example. Let $p: E \to X$ be a finite covering. One can assume that $E \subset X \times \mathbb{R}^N$ and p is the restriction of the projection $X \times \mathbb{R}^N \to X$. The normal bundle to E in $X \times \mathbb{R}^N$ is trivial (normal spaces to E are projected isomorphically onto the factor $\mathbb{R}^N$). Denote by D an open ball in $\mathbb{R}^N$ such that $E \subset X \times D$, and by U a narrow tubular neighborhood of E in $X \times \mathbb{R}^N$. A map

$$\begin{array}{ccc} (X \times \mathbb{R}^N)/(X \times (\mathbb{R}^N - D)) & \longrightarrow & (X \times \mathbb{R}^N)/((X \times \mathbb{R}^N) - U) \\ \| & & \| \\ \Sigma^N X^+ & & \Sigma^N E^+ \end{array}$$

arises. We take this map for t; check that the requirements of the theorem are met.

Proof of Theorem. The construction in the general case generalizes the construction of the last example.

Lemma. *Let M be a closed smooth manifold with a smooth action of the group G. Then there exist a linear representation of the group G in some $\mathbb{R}^N$ and an equivariant (that is, commuting with the operators from G) C^∞ embedding $M \to \mathbb{R}^N$.*

Proof of Lemma. Let L be the space of all C^∞-functions on M. The formula $gf(x) = f(gx), g \in G, f \in L, x \in M$ defines a representation of the group G in L. As is known from representation theory (see, e.g., Kirillov [52], Sect. 9.2), finite-dimensional subrepresentations of L span a dense subspace of L; in other words, for any function $f \in L$ there exists an arbitrary function close (in C^∞-topology) to it, $\bar{f} \in L$, which is contained in a finite-dimensional subrepresentation of L. Take an arbitrary embedding $M \to \mathbb{R}^k$; let $f_1, \dots, f_k: M \to \mathbb{R}$ be the coordinate functions. Choose functions $\bar{f}_1, \dots, \bar{f}_k$ which are contained in finite-dimensional subrepresentations $\bar{L}_1, \dots, \bar{L}_k$ of L and which are so close to $f_1, \dots, f_k$ that the map $M \to \mathbb{R}^k$ with the coordinate functions $\bar{f}_1, \dots, \bar{f}_k$ is still an embedding. Let $g_{i1}, \dots, g_{in_i}$ be a basis of the space $\bar{L}_i$ ($n_i = \dim \bar{L}_i$). These functions compose a map $M \to \mathbb{R}^{n_i}$, which is actually an equivariant map $M \to \bar{L}_i'$ (the latter is the dual of $\bar{L}_i$). All together, the maps $g_1, \dots, g_k$ compose an equivariant map $M \to \bar{L}_1' \times \dots \times \bar{L}_k'$, which is an embedding because $\bar{f}_1, \dots, \bar{f}_k$ are among the linear combinations of its coordinate functions.

Now fix an equivariant map of the manifold M into the space $\mathbb{R}^N$ of a representation of the group G. Since the group G is compact, the action of G in $\mathbb{R}^N$ is orthogonal with respect to an Euclidean structure (take any symmetric positive definite bilinear structure in $\mathbb{R}^n$ and average it with respect to G). Take an open ε-neighborhood T of M in $\mathbb{R}^N$ (with respect to this Euclidean structure, with a sufficiently small ε); since the action of G is orthogonal, this neighborhood is G-invariant. Embed $\mathbb{R}^N$ now into $S^N = \mathbb{R}^N \cup \infty$ and extend the action of G in $\mathbb{R}^N$ to S^N by $g(\infty) = \infty$ for all $g \in G$. The quotient $S^N/(S^N - T)$ is the Thom space of the normal bundle ν of M in $\mathbb{R}^N$; thus, we have constructed an equivariant map $S^N \to T(\nu)$. Furthermore, the bundle ν is embedded into the sum $\nu \oplus \tau$, where τ is the tangent bundle of M, and this defines a map $T(\nu) \to T(\nu \oplus \tau)$, which is certainly also equivariant. Since $\nu \oplus \tau = N$, then $T(\nu \oplus \tau) = \Sigma^N M^+$. Thus, we have constructed an equivariant map $S^N \to \Sigma^N M^+$ [the action of the group G in $\Sigma^N M^+ = (M \times S^N)/(M \times \mathrm{pt})$ is induced by the actions of G in M and S^N].

Let us show that the degree of the map

$$S^N \longrightarrow \Sigma^N M^+ \longrightarrow S^N$$

is equal to $\chi(M)$. The map acts like this. The points from $S^N - T$ go to $\infty \in S^N$; a point $x \in T$ is mapped into the vector $v \in \mathbb{R}^N \subset S^N$ parallel to the line through X perpendicular to M, directed from M to x and having length $\varphi(l)$, where l is the distance from x to M, and φ is a monotone function $[0, \varepsilon) \to [0, \infty)$ such that $\varphi(0) = 0, \varphi'(0) = 1$, and $\lim_{t \to \varepsilon} \varphi(t) = \infty$ (see Fig. 136).

Let us compute the degree of this map. Find the inverse image of a vector $v_0 \in \mathbb{R}^n$. This inverse image consists of points $x \in S^N$ at the distance $l = \varphi^{-1}(\|v_0\|)$ from M such that the direction of the perpendicular from x to M coincides with the direction of the vector $-v_0$. Consider on M the field of vectors of length l parallel to $-v_0$, and project this field orthogonally onto M. We obtain a tangential vector field on M whose singularities correspond exactly to the points of the inverse image of v_0 (see Fig. 137).

It is easy to understand that the signs, with which the points of the inverse image are counted when the degree of the map is computed, coincide with the signs with

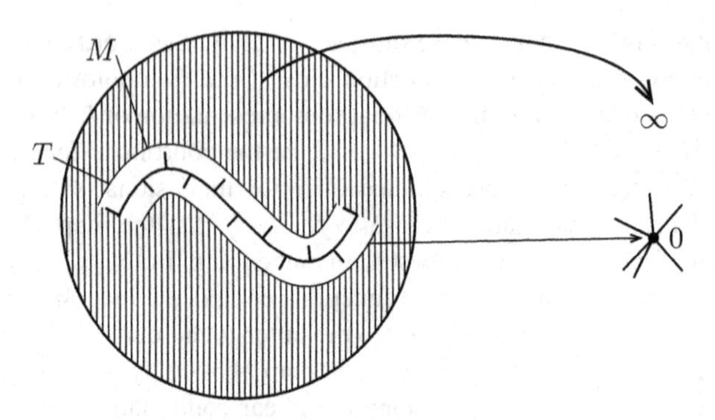

Fig. 136 The map $S^N \to \Sigma^N M^+ \to S^n$

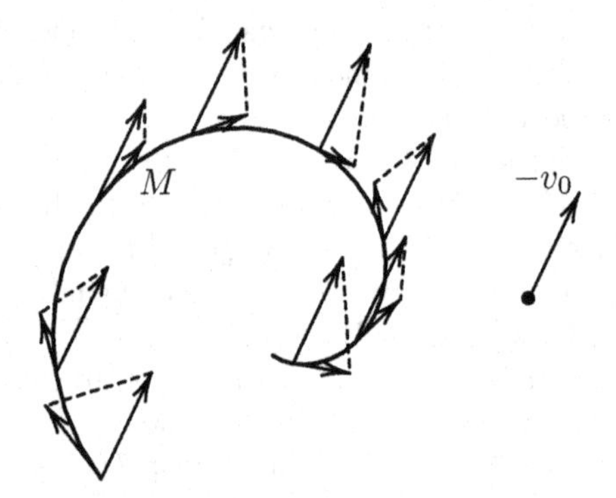

Fig. 137 Computing $\deg(S^n \to S^n)$

which the corresponding singularities of the vector field are counted when the Euler characteristic is computed. Hence, the degree of our map $S^N \to S^N$ equals the Euler characteristic of M.

For each fiber of the fibration $p: E \to X$ its homeomorphism with M is given up to a transformation from G. Since our construction is equivariant, it may be applied to the fibers of the fibration $p: E \to X$. As a result, we obtain for each point $x \in X$ a map $S^N = S_x^N \to \Sigma^N(p^{-1}(x))^+$ that takes the point ∞ into the base point of the suspension $\Sigma^N(p^{-1}(x))^+$. All together these maps define a map of the space composed of the spheres S_x^N glued together in the point ∞ into the space composed of the suspensions $\Sigma^N(p^{-1}(x))^+$ glued together in the base point. Obviously, both of these spaces are Thom spaces: The first one is the Thom space of the vector bundle ξ over X with the fiber $\mathbb{R}^N$ associated with the fibration $p: E \to X$ by means of our representation of G in $\mathbb{R}^N$; the second is the Thom space of the fibration $p^*\xi$ over E (also with the fiber $\mathbb{R}^N$). The composition of this map $T(\xi) \to T(p^*\xi)$ with the natural projection $T(p^*\xi) \to T(\xi)$ has degree $\chi(M)$ on each of the N-dimensional spheres of which the Thom space $T(\xi)$ is made. If the vector bundle ξ is trivial, this completes the proof. If it is nontrivial, then we find a vector bundle η over X (which does not have to be related in any way to G) such that the sum $\xi \oplus \eta$ is trivial; after it we extend our map in the obvious way to a map

$$T(\xi \oplus \eta) \longrightarrow T(p^*(\xi \oplus \eta)).$$

[When we pass from $T(\xi)$ to $T(\xi \oplus \eta)$, each of the spheres S_x^N, of which the space $T(\xi)$ is made, is multiplied tensorially by the augmented fiber $\eta_x^* = \eta_x \cup \infty$ of the bundle η over x; similarly, when we pass from $T(p^*\xi)$ to $T(p^*(\xi \oplus \eta))$, the part of the space $T(p^*\xi)$ which lies over the fiber $p^{-1}(x)$ of the fibration $p: E \to X$ is multiplied tensorially by η_x^*. The "extension" mentioned above is reduced to the tensor product of the map $S_x^N \to \Sigma^N(p^{-1}(x))^+$ constructed above with η_x^*.]

The last map is our t. This completes the proof of the Becker–Gottlieb theorem.

Corollary. *Let $p: E \to X$ denote the same as in the Becker–Gottlieb theorem, and let $\chi(M) = 1$. Then $p^*: P(X) \to P(E)$ is a monomorphism for any half-exact functor P.*

Proof. The composition $\Sigma^N p^+ \circ t: \Sigma^N X^+ \to \Sigma^N X^+$ maps each of the N-dimensional spheres, of which $\Sigma^N X^+$ is made, onto itself with degree $\chi(M) = 1$. Hence, it maps each of the cells of $\Sigma^N X^+$ onto itself with degree 1, which implies that the cohomology homomorphism

$$t^* \circ (\Sigma^N p^+)*: H^*(\Sigma^N X^+; A) \to H^*(\Sigma^N X^+; A)$$

is an isomorphism for any coefficient group A. Hence,

$$(\Sigma^N p^+)_*: H^*(X; A) \to H^*(E; A)$$

is a monomorphism onto a direct summand. The Atiyah–Hirzebruch spectral sequence shows then that the same is true for any half-exact functor P.

D: Direct Image (The Case of Coverings)

As we know, a continuous map $X \to Y$ induces a homomorphism $K(Y) \to K(X)$. However, even for usual (ordinary) homology and cohomology we encountered the situation when a continuous map induces homological and cohomological homomorphism which has a "wrong direction" (see Sect. 17.7). A similar phenomenon in the K-theory will be discussed in all generality in Lecture 42, but now we will consider it in the simplest situation.

Let $\pi: Y \to X$ be a finite covering, and let ξ be a (complex) vector bundle of dimension n over Y. Then a vector bundle $\pi_!\xi$ of dimension hn over X arises, where h is the number of sheets of the covering: The fiber of $\pi_!\xi$ over $x \in X$ is the sum of fibers of ξ over the points of $\pi^{-1}(x)$. The operation $\pi_!$ is clearly additive; hence, it gives rise to a homomorphism $\pi_!: K(Y) \to K(X)$. The equality $\dim(\pi_!\xi) = h \dim \xi$ shows that $\pi_!(\widetilde{K}(Y)) \subset \widetilde{K}(X)$. It is clear also that the homomorphism $\pi_!$ commutes with ψ^k and takes J-trivial virtual bundles into J-trivial virtual bundles.

E: Proof of the Adams Conjecture in the General Case

We need to prove that

$$k^N(\psi^k(\xi) - \xi) \in T(X)$$

for any $\xi \in K(X)$ and k and for N sufficiently large. We may assume that ξ is a geometric (not virtual) bundle; let $\dim \xi = n$. Consider the principal bundle $p: E \to X$ with the fiber $U(n)$ associated with ξ. Distinguish the following subgroups of the group $U(n)$:

T, the group of diagonal matrices;
N, the normalizer of T, that is, the group of unitary matrices which preserve the union of coordinate axes;
L, the subgroup of N consisting of matrices that preserve the nth coordinate axis.

Obviously, $T \subset L \subset N$, and $|N/T| = n!$ and $|L/T| = (n-1)!$. Consider the sequence of fibrations

$$E \xrightarrow[p_1]{T} E/T \xrightarrow[p_2]{(n-1)!} E/L \xrightarrow[p_3]{n} E/N \xrightarrow[p_4]{U(n)/N} X.$$

Lemma. $\chi(U(n)/N) = 1$.

Actually, this is quite a general fact: The Euler characteristic of the quotient of a simple compact Lie group G over its maximal torus T is equal to the order $|W|$ of the Weyl group W, the normalizer N of the torus T is an $|W|$-fold covering of T, and hence $\chi(G/N) = 1$. In the case of $U(n)$ it may be easily checked directly. Namely, $G/T = U(n)/T$ is the flag manifold in $\mathbb{C}^n$, and this manifold has a CW decomposition with precisely $n!$ cells, all of which are even-dimensional, hence $\chi(U(n)/T) = n!$. Another way of proving this consists in considering chain of fibrations

$$U(n)/T_n \xrightarrow{U(n-1)/T_{n-1}} \mathbb{C}P^{n-1}, \ldots, U(3)/T_3 \xrightarrow{U(2)/T_2} \mathbb{C}P^2,\ U(2)/T_2 = \mathbb{C}P^1$$

[where T_k denotes the group of diagonal matrices in $U(k)$]. Since the Euler characteristic of the fibered space is the product of those of the base and the fiber (see Sect. 21.1), it follows that

$$\chi(U(n)/T) = \prod_{k=1}^{n-1} \chi(\mathbb{C}P^k) = \prod_{k=1}^{n-1}(k+1) = n!.$$

Finally, $\chi(U(n)/N) = 1$ since there is an $n!$-fold covering $U(n)/T \to U(n)/N$.

According to the results of Sect. 41.2.C (see the corollary), the projection $p_4: E/N \to X$ induces a monomorphism $\Pi(X) \to \Pi(E/N)$. Hence, it is sufficient to prove the J-triviality of $k^N(\psi^k - 1)\eta$, where $\eta = p_4^*\xi$. A point of the space E/N is a pair (x, α), where α is an unordered system of n pairwise orthogonal lines in the fiber ξ_x of the bundle ξ. A point of the space E/L is a triple (x, α, l), where x and α are as before, and l is one of the lines from α. Consider the one-dimensional bundle ω over E/L whose fiber over the point (x, α, l) is the line l. Obviously, $(p_3)_!\omega = \eta$. Hence, $(p_3)_![k^N(\psi^k - 1)\omega] = k^N(\psi^k - 1)\eta$. But the virtual bundle $k^N(\psi^k - 1)\omega$ is J-trivial according to Sect. 41.2.B, and hence $k^N(\psi^k - 1)\eta$ is also J-trivial (Sect. 41.2.D). This completes the proof of Adams' conjecture.

41.3 An Application to the Homotopy Groups of Spheres

In this section we will study the order of the cyclic group

$$\widetilde{J}(S^{2n}) = \operatorname{Im}(J: \pi_{2n-1}(U) \longrightarrow \pi_{2n-1}^S) \subset \pi_{2n-1}^S.$$

A: *Upper Bound*

We know that $\widetilde{K}(S^{2n}) = \mathbb{Z}$ and $\psi^k: \widetilde{K}(S^{2n}) \to \widetilde{K}(S^{2n})$ is the multiplication by k^n. Thus, Adams' conjecture implies that for any k and sufficiently large N (which depends on k and n)

$$k^N(k^n - 1)\widetilde{K}(S^{2n}) \subset T(S^{2n}).$$

Denote by d_n the largest integer with the following property: For any k there is an N such that d_n divides $k^N(k^n - 1)$.

Theorem. $|\widetilde{J}(S^{2n})| \leq d_n$.

This follows directly from Adams' conjecture.

What is d_n?

Let us find d_2. The greatest common divisor of the numbers $2^N(2^2 - 1) = 2^N \cdot 3$ and $3^N(3^2 - 1) = 3^N \cdot 8$ equals 24. On the other hand, it is easy to check that the number $k^N(k^n - 1)$ is always divisible by 24. Hence, $d_2 = 24$.

Let us find d_3. The greatest common divisor of the numbers $2^N(2^3 - 1) = 2^N \cdot 7$ and $3^N(3^3 - 1) = 3^N \cdot 26$ equals 2. The number $k^N(k^3 - 1)$ is always even. Hence, $d_3 = 2$.

Let us find d_4. The greatest common divisor of the numbers $2^N(2^4 - 1) = 2^N \cdot 15$ and $3^N(3^4 - 1) = 3^N \cdot 80$ equals 240. It is not hard to check that the number $k^N(k^n - 1)$ is always divisible by 240. Hence, $d_4 = 240$.

One can easily check that $d_k = 2$ for any odd k. For even k the numbers d_k are expressed in terms of so-called Bernoulli numbers. The Bernoulli number $B_s, s \geq 1$ is defined by the power series

$$\frac{x}{1-e^{-x}} = 1 + \frac{x}{2} + \sum_{s=1}^{\infty}(-1)^{s-1}B_s\frac{x^{2s}}{(2s)!}.$$

The following table contains several first Bernoulli numbers.

s	1	2	3	4	5	6	7	8
B_s	$\frac{1}{6}$	$\frac{1}{30}$	$\frac{1}{42}$	$\frac{1}{30}$	$\frac{5}{66}$	$\frac{691}{2730}$	$\frac{7}{6}$	$\frac{3617}{510}$
denominator of $B_s/4s$	24	240	504	480	264	65520	24	16320

Bernoulli numbers are studied in textbooks in number theory. At the same time, as the reader will see in this and upcoming sections, these numbers are very important for topology, and topologists made considerable contributions to their study. The information about Bernoulli numbers relevant for topology is contained in Sect. 2 of Adams' "On J(X). II" (see Adams [5]). In particular, the following theorem is proved there.

Theorem. *The number d_s is equal to the denominator of the irreducible fraction equal to $B_s/4s$.*

(See the third line of the preceding table.)

Thus, the numbers in the third line of the table provide upper bounds for the orders of images of the J-homomorphism in π_{4s-1}^S.

B: Lower Bound

Let $\lambda \in \pi_{2N+2n-1}(S^{2N})$. Attach to S^{2N} the ball D^{2N+2n} by means of the spheroid $S^{2N+2n-1} = \partial D^{2N+2n} \to S^{2N}$ of the class λ. We will obtain the space X_λ for which

$$H^q(X_\lambda; \mathbb{Z}) = \begin{cases} \mathbb{Z} \text{ for } q = 0, 2N, 2N+2n, \\ 0 \text{ otherwise.} \end{cases}$$

Denote by a and b the natural generators of the groups $H^{2N}(X_\lambda; \mathbb{Q})$ and $H^{2N+2n}(X_\lambda; \mathbb{Q})$.

Since $X_\lambda / S^{2N} = S^{2N+2n}$, there is an exact sequence

$$0 \longrightarrow \underset{\underset{\mathbb{Z}}{\parallel}}{\widetilde{K}(S^{2N+2n})} \longrightarrow \widetilde{K}(X_\lambda) \longrightarrow \underset{\underset{\mathbb{Z}}{\parallel}}{\widetilde{K}(S^{2N})} \longrightarrow 0,$$

which shows that $\widetilde{K}(X_\lambda) \cong \mathbb{Z} \oplus \mathbb{Z}$. One of the generators of this group, $\widetilde{\beta}$, is defined canonically as the image of the natural generator β of the group $\widetilde{K}(S^{2N+2n})$. The other generator, $\widetilde{\alpha}$, is defined up to adding a multiple of $\widetilde{\beta}$; it is the inverse image of the natural generator α of $\widetilde{K}(S^{2N})$. Since $\operatorname{ch} \alpha = a$ and $\operatorname{ch} \beta = b$,

$$\begin{aligned} \operatorname{ch} \widetilde{\alpha} &= a + \widetilde{e} b, \widetilde{e} \in \mathbb{Q} \\ \operatorname{ch} \widetilde{\beta} &= b \end{aligned}$$

(compare with Sect. 39.2). Since $\widetilde{\alpha}$ is defined only up to adding a multiple of $\widetilde{\beta}$, the number $\widetilde{e}$ is defined only mod 1. Let $e = e(\lambda) \in \mathbb{Q}/\mathbb{Z}$ be the corresponding residue. This residue is called *Adams' e-invariant* of λ.

EXERCISE 5. Prove that $e \colon \pi_{2N+2n-1}(S^{2N}) \to \mathbb{Q}/\mathbb{Z}$ is a homomorphism.

EXERCISE 6. Prove that $e(\Sigma\lambda) = e(\lambda)$.

Example. Let $\gamma \in \pi_7(S^4)$ be the class of the Hopf map $S^7 \to S^4$. Let us compute $e(\gamma)$.

Obviously, $X_\gamma = \mathbb{H}P^2$, the quaternionic projective plane. Consider in $K(\mathbb{H}P^2)$ the class of the tautological bundle $\zeta_{\mathbb{H}}$ regarded as a two-dimensional complex vector bundle. The Chern classes of the bundle $\zeta_{\mathbb{H}}$ are known to us: $c_1(\zeta_{\mathbb{H}}) = 0$ [because $H^2(\mathbb{H}P^2; \mathbb{Z}) = 0$], $c_2(\zeta_{\mathbb{H}}) = a$ (this is the Euler class of $\zeta_{\mathbb{H}}$, that is, the first obstruction to a section of the corresponding sphere bundle), and $c_i(\zeta_{\mathbb{H}}) = 0$ for $i > 2$ (because $\dim_{\mathbb{C}} \zeta_{\mathbb{H}} = 2$). The components of the Chern character are expressed in terms of the Chern classes. In particular,

$$\begin{aligned} \operatorname{ch}_2 \zeta_{\mathbb{H}} &= \frac{1}{2}(c_1^2 - 2c_2), \\ \operatorname{ch}_4 \zeta_{\mathbb{H}} &= \frac{1}{24}(c_1^4 - 4c_1^2 c^2 + 2c_2^2 + 4c_1c_3 + 4c_4). \end{aligned}$$

Hence, $\operatorname{ch}_2 \zeta_{\mathbb{H}} = -a$, $\operatorname{ch}_4 \zeta_{\mathbb{H}} = \dfrac{a^2}{12} = \dfrac{b}{12}$, that is,

$$\underset{\underset{\dim_{\mathbb{C}} \zeta_{\mathbb{H}}}{\parallel}}{\operatorname{ch} \zeta_{\mathbb{H}} = 2} - a + \frac{1}{12} b.$$

Hence, for $\widetilde{\alpha}$ we can take $2 - \zeta_{\mathbb{H}}$, and $e(\gamma) = -\dfrac{1}{12}$.

This shows that the class $\Sigma^N\gamma$ generates in $\pi_{N+7}(S^{N+4})$ a subgroup of order at least 12. (Actually, this order is equal to 24—see Chap. 5.)

This example shows a way to obtain lower bounds for the order of the image of the J-homomorphism: to compute the e-invariant for the J-image of the group $\pi_{2s-1}(U)$ and to take the denominator.

EXERCISE 7. Let ξ be a geometric vector bundle representing an element α of the group $\widetilde{K}(S^{2n})$, and let $\dim_{\mathbb{C}}\xi = N$. Then the Thom space $T(\xi)$ has the homotopy type X_λ, where $\lambda = J(\alpha) \in \pi_{2N+2n-1}(S^{2N})$.

This statement shows that for finding lower bounds of the order of the image of the J-homomorphism, one needs to know how to compute the Chern character for the elements of the K-functor of the Thom space. This is the subject of Lecture 42.

Lecture 42 The Riemann–Roch Theorem

42.1 The General Riemann–Roch Theorem

A: The Orientability of a Vector Bundle with Respect to a Cohomology Theory

Let h be an extraordinary (possibly, ordinary) cohomology theory (see Sect. 38.5). Let $\xi = (p\colon E \to X)$ be a (complex or real) vector bundle. Consider the Thom space $T(\xi)$. As we know (see Sect. 31.2.B) there is an isomorphism

$$H^*(X;G) \cong H^*(T(\xi);G),$$

shifting the dimensions by $n = \dim_{\mathbb{R}}\xi$, where G is $\mathbb{Z}_2$ in the general case and an arbitrary Abelian group for an oriented (in particular, complex) vector bundle ξ. The following question arises: Is there a similar isomorphism for the theory h? The answer is suggested by the results mentioned above: The existence of the isomorphism depends on a certain "orientability" condition which varies with the theory h.

We restrict ourselves here to the case of *multiplicative* theories; that is, we assume that $h(X)$ is, for any X, a graded ring with the identity element $1 \in h^0(X)$; moreover, we assume that there exists a multiplication

$$h^p(X,A) \otimes h^q(X,B) \to h^{p+q}(X, A \cup B).$$

All these multiplications should be compatible with the homomorphisms induced by continuous maps (we left the precise statements of axioms to the reader). In particular, all $h^*(X,A)$ become modules over the ring $h^*(\mathrm{pt})$ (because $h^*(\mathrm{pt})$ is naturally mapped into $h^*(X)$: The homomorphism is induced by the map $X \to \mathrm{pt}$).

The suspension isomorphism $h^q(\mathrm{pt}) \to \widetilde{h}^{q+n}(S^n)$ gives rise to a canonical generator $s_h^n \in \widetilde{h}^n(S^n)$ of the $h^*(\mathrm{pt})$-module $h^*(S^n)$: It is the image of $1 \in h^0(\mathrm{pt})$.

EXERCISE 1. For any X, the suspension isomorphism $\Sigma^n\colon \widetilde{h}^q(X) \to \widetilde{h}^{q+n}(\Sigma^n X)$ is expressed via s_h^n:

$$\begin{array}{ccccccc}
\widetilde{h}^q(X) = h^q(X,\mathrm{pt}) & & h^q(X\times S^n, S^n) & & h^{q+n}(X\times S^n, X\vee S^n) & = & \widetilde{h}^{q+n}(\Sigma^n X) \\
\Cup & & \Cup & & \Cup & & \Cup \\
\alpha & \longmapsto & p_1^*\alpha & \left.\vphantom{\begin{array}{c}a\\a\\a\end{array}}\right\} \longmapsto & (p_1^*\alpha)(p_2^* s_h^n) & = & \Sigma^n\alpha \\
s_h^n & \longmapsto & p_2^* s_h^n & & & & \\
\Cap & & \Cap & & & & \\
\widetilde{h}^n(S^n) = h^q(S^n,\mathrm{pt}) & & h^n(X\times S^n, S^n) & & & &
\end{array}$$

Definition. A (real or complex) vector bundle ξ is called *orientable with respect to the theory h*, or simply *h-orientable*, if there exists an element $u \in \widetilde{h}^n(T(\xi))$, $n = \dim_{\mathbb{R}} \xi$ such that for any $x \in X$ the homomorphism $\widetilde{h}^n(T(\xi)) \to \widetilde{h}^n(S^n)$ induced by the embedding of the augmented fiber $\xi_x^* = \xi_x \cup \infty \approx S^n$ into $T(\xi)$ takes u into εs_h^n, where ε is an invertible element of the ring $h^0(\mathrm{pt})$. Any u with this property is called an *h-orientation of the vector bundle ξ*.

Obviously, any vector bundle is $H^*(\ ;\mathbb{Z}_2)$-orientable, and the $H^*(\ ;\mathbb{Z})$-orientability is the orientability in the usual sense. We will investigate here the notion of orientability with respect to K-theory. We will see that any complex vector bundle is K-orientable, while for real vector bundles the necessary and sufficient condition of K-orientability is the usual orientability plus vanishing an integral analog of the Stiefel–Whitney class w_3.

B: Thom Isomorphism

Suppose that a vector bundle $\xi = (p\colon E \to X)$ is h-orientable and fix an h-orientation $u \in \widetilde{h}^n(T(\xi))$, $n = \dim_{\mathbb{R}} \xi$. Denote by E' the complement to the zero section in E; obviously, $\widetilde{h}^r(T(\xi)) = h^r(E, E')$. The map

$$\mathbf{t}\colon h^q(X) = h^q(E) \xrightarrow{\cdot u \in h^n(E,E')} h^{q+n}(E,E') = \widetilde{h}^{q+n}(T(\xi))$$

is called the *Thom homomorphism*. Sometimes instead of $\mathbf{t}$ we will use more detailed notations $\mathbf{t}_h$ or $\mathbf{t}_h^{\xi}$.

Remark. Obviously, $\mathbf{t}(\alpha\beta) = \alpha\mathbf{t}(\beta) = \mathbf{t}(\alpha)\beta\ [= \alpha\beta u]$.

Theorem. *The Thom homomorphism is an isomorphism.*

Proof. For $A \subset X$ there is an obvious relative Thom homomorphism

$$\mathbf{t}: h^q(X,A) \longrightarrow \widetilde{h}^{q+n}(T(\xi), T(\xi|_A)),$$

and the diagram

$$\begin{array}{ccccc}
h^{q-1}(A) & \longrightarrow & h^q(X,A) & \longrightarrow & h^q(X) \\
\downarrow \mathbf{t} & & \downarrow \mathbf{t} & & \downarrow \mathbf{t} \\
\widetilde{h}^{q+n-1}(T(\xi|_A)) & \longrightarrow & \widetilde{h}^{q+n}(T(\xi), T(\xi|_A)) & \longrightarrow & \widetilde{h}^{q+n}(T(\xi))
\end{array}$$

$$\begin{array}{ccccc}
\longrightarrow & h^q(A) & \longrightarrow & h^{q+1}(X,A) \\
& \downarrow \mathbf{t} & & \downarrow \mathbf{t} \\
\longrightarrow & \widetilde{h}^{q+n}(T(\xi|_A)) & \longrightarrow & \widetilde{h}^{q+n+1}(T(\xi), T(\xi|_A))
\end{array} \qquad (*)$$

is commutative. If $X - A$ is one cell of, say, dimension m, then, independently of the bundle ξ, the second and fifth homomorphisms $\mathbf{t}$ of the diagram $(*)$ become

$$\widetilde{h}^q(S^m) \to \widetilde{h}^{q+n}(S^{m+n}),\ \widetilde{h}^{q+1}(S^m) \to \widetilde{h}^{q+n+1}(S^{m+n}),$$

and Exercise 1 shows that they are just Σ^n; in particular, they are isomorphisms. Assuming by induction that for $\xi|_A$ the homomorphism $\mathbf{t}$ has been already proven to be an isomorphism, we get the same for ξ by applying the five-lemma to the commutative ladder $(*)$.

In this section, we will use the term "Thom isomorphism" instead of "Thom homomorphism."

C: The Riemann–Roch Theorem

Let h and k be two multiplicative cohomology theories, and let $\tau: h \to k$ be a multiplicative map taking $1 \in h^0(\text{pt})$ into $1 \in k^0(\text{pt})$. (We do not assume that τ is homogeneous with respect to the gradings.) Let ξ be a vector bundle oriented with respect to both theories, and let X be the base of ξ. Consider the diagram

$$\begin{array}{ccc}
h^*(X) & \xrightarrow{\tau} & k^*(X) \\
\downarrow \mathbf{t}_h^\xi & & \downarrow \mathbf{t}_k^\xi \\
\widetilde{h}^*(T(\xi)) & \xrightarrow{\tau} & \widetilde{k}^*(T(\xi)).
\end{array}$$

Let us ask whether this diagram is commutative. The examples which are known to us show that it is not, in general. Indeed, let $h = k = H^*(\ ;\mathbb{Z}_2)$ and let $\tau = \mathrm{Sq} = 1 + \mathrm{Sq}^1 + \mathrm{Sq}^2 + \dots$ (the multiplicativity of this map is stated by Cartan's formula; see Lecture 29). As we know,

$$\mathbf{t}^{-1} \circ \mathrm{Sq} \circ \mathbf{t}(1) \in H^*(X;\mathbb{Z}_2)$$

is not 1, but rather $w(\xi) = 1 + w_1(\xi) + w_2(\xi) + \dots$, the total Stiefel–Whitney class of the vector bundle ξ (see Sect. 31.2).

Definition.

$$\mathcal{T}_\tau(\xi) = (\mathbf{t}_k^\xi)^{-1} \circ \tau \circ \mathbf{t}_h^\xi(1) \in k^*(X)$$

is called the *Todd class* of the vector bundle ξ (with respect to $\tau: h \to k$).

This is a characteristic class of vector bundles which are simultaneously h-oriented and k-oriented. The previous example shows that $\mathcal{T}_{\mathrm{Sq}} = w$.

Theorem. *For any* $\alpha \in h^*(X)$,

$$(\mathbf{t}_k^\xi)^{-1} \circ \tau \circ \mathbf{t}_h^\xi(\alpha) = \tau(\alpha)\mathcal{T}_\tau(\xi).$$

Proof. $\tau(\mathbf{t}_h^\xi(\alpha)) = \tau(\alpha \cdot \mathbf{t}_h^\xi(1)) = \tau(\alpha)\cdot\tau(\mathbf{t}_h^\xi(1)) = \tau(\alpha)\cdot\mathbf{t}_k^\xi\mathcal{T}_\tau(\xi) = \mathbf{t}_k^\xi(\tau(\alpha)\cdot\mathcal{T}_\tau(\xi))$.

It is this obvious statement that is called the general Riemann–Roch theorem. It stops being trivial as soon as we consider specific examples, since then it includes an explicit description of the class of orientable bundles and an explicit computation of $\mathcal{T}_\tau$. We will see this in the rest of this section, where the Chern character ch will be taken for τ. But first we will point out a corollary of the general Riemann–Roch theorem.

D: Direct Image in Extraordinary Cohomology

(We leave to the reader to compare the material of this subsection with that of Sect. 41.2.D.) A smooth closed manifold is called h-(co)orientable if its normal bundle in a sphere of a large dimension is h-orientable. Let X and Y be two smooth closed h-oriented manifolds, and let $f: X \to Y$ be a continuous map. For a large (much larger than $\dim X$ and $\dim Y$) integer N, fix a smooth embedding $\varphi: Y \to S^N$ and approximate the composition $\varphi \circ f: X \to S^N$ by a smooth embedding $\psi: X \to S^N$. The manifold $\psi(X)$ is contained in a narrow tubular neighborhood V of $\varphi(Y)$; choose a still more narrow tubular neighborhood U of $\psi(X)$, such that it is contained in V. Consider the natural map

$$\begin{array}{ccc} S^N/(S^N - U) & \xrightarrow{g} & S^N/(S^N - V) \\ \| & & \| \\ T(\nu(X)) & & T(\nu(Y)), \end{array}$$

where $\nu(X)$ and $\nu(Y)$ are the normal bundles of the manifolds Y and X in S^N. Since the vector bundles $\nu(X)$ and $\nu(Y)$ are h-orientable, there arises a map

$$h^*(X) \xrightarrow{\mathbf{t}_h} \widetilde{h}^*(T(\nu(X))) \xrightarrow{g^*} \widetilde{h}^*(T(\nu(Y))) \xrightarrow{\mathbf{t}_h^{-1}} h^*(Y).$$

EXERCISE 2. This map is determined by the map f, even by the homotopy class of the map f.

EXERCISE 3. If h is the theory of ordinary cohomology (with arbitrary coefficients, if the manifolds X and Y are orientable, and with the coefficients in $\mathbb{Z}_2$ in the general case), then this map coincides with the map $f_!$ defined in Sect. 17.7.

In the general case, this map is also denoted by $f_!$. Obviously,

$$f_!(h^q(X)) \subset h^{q+n-m}(Y),$$

where $m = \dim X, n = \dim Y$. The element $f_!\alpha$ of $h^*(Y)$ is called the *direct image* of $\alpha \in h^*(X)$.

EXERCISE 4. Prove that for any $\alpha \in h^*(X), \beta \in h^*(Y)$

$$f_!(f^*(\beta)\alpha) = \beta f_!(\alpha).$$

Let h and k be two multiplicative cohomology theories, let $\tau: h \to k$ be a multiplicative map, and let X be a smooth closed manifold which is both h-orientable and k-orientable. The Todd class of the normal bundle $\nu(X)$ of X with respect to τ, that is,

$$(\mathbf{t}_k^{\nu(X)})^{-1} \circ \tau \circ \mathbf{t}_h^{\nu(X)}(1) \in k^*(X),$$

is called the *Todd class* of the manifold X and is denoted by $\mathcal{T}(X)$ or $\mathcal{T}_\tau(X)$. The Riemann–Roch theorem has the following corollary (which is also sometimes called the Riemann–Roch theorem).

Theorem. *Let X and Y be smooth closed manifolds, which are simultaneously h-oriented and k-oriented, and let $f: X \to Y$ be continuous maps. Then the Todd classes $\mathcal{T}(X)$ and $\mathcal{T}(Y)$ measure the degree of noncommutativity of the diagram*

$$\begin{array}{ccc} h^*(X) & \xrightarrow{\tau} & k^*(X) \\ \Big\downarrow{\scriptstyle f_!} & & \Big\downarrow{\scriptstyle f_!} \\ h^*(Y) & \xrightarrow{\tau} & k^*(Y). \end{array}$$

More precisely, for any $\alpha \in h^*(X)$,

$$f_!(\tau(\alpha) \cdot \mathcal{T}(X)) = \tau(f_!(\alpha)) \cdot \mathcal{T}(Y).$$

Proof. Apply to both sides of the last equality the isomorphism $\mathbf{t}_k^{\nu(Y)}$:

$$\begin{aligned}
&\mathbf{t}_k^{\nu(Y)}[f_!(\tau(\alpha) \cdot \mathcal{T}(X))] \\
&\quad = \mathbf{t}_k^{\nu(Y)}[(\mathbf{t}_k^{\nu(Y)})^{-1} g^* \mathbf{t}_k^{\nu(X)}(\tau(\alpha) \cdot (\mathbf{t}_k^{\nu(X)})^{-1} \tau \mathbf{t}_h^{\nu(X)}(1))] \\
&\quad = g^* \mathbf{t}_k^{\nu(X)}(\tau(\alpha) \cdot (\mathbf{t}_k^{\nu(X)})^{-1} \tau \mathbf{t}_h^{\nu(X)}(1)) = g^*(\tau(\alpha) \cdot \tau \mathbf{t}_h^{\nu(X)}(1)) \\
&\quad = \tau g^*(\alpha \cdot \mathbf{t}_h^{\nu(X)}(1)) = \tau g^* \mathbf{t}_h^{\nu(X)}(\alpha);
\end{aligned}$$

$$\begin{aligned}
&\mathbf{t}_k^{\nu(Y)}[\tau(f_!(\alpha)) \cdot \mathcal{T}(Y)] \\
&\quad = \mathbf{t}_k^{\nu(Y)}[\tau((\mathbf{t}_h^{\nu(Y)})^{-1} g^* \mathbf{t}_h^{\nu(X)}(\alpha)) \cdot ((\mathbf{t}_k^{\nu(Y)})^{-1} \tau(\mathbf{t}_h^{\nu(Y)}(1))] \\
&\quad = \tau[((\mathbf{t}_h^{\nu(Y)})^{-1} g^* \mathbf{t}_h^{\nu(X)}(\alpha)) \cdot \mathbf{t}_h^{\nu(Y)}(1)] = \tau g^* \mathbf{t}_h^{\nu(X)}(\alpha).
\end{aligned}$$

[We always use the multiplicative property of the Thom isomorphism: $\alpha \cdot \mathbf{t}(\beta) = \mathbf{t}(\alpha) \cdot \beta = \mathbf{t}(\alpha) \cdot \beta$.]

Example. Let $h = k = H^*(\ ;\mathbb{Z}_2), \tau = \mathrm{Sq}, X$ be an arbitrary smooth closed manifold, and let Y be the point. The description of the direct image given in Sect. 17.7 shows that the homomorphism

$$f_!\colon H^*(X;\mathbb{Z}_2) \to H^*(Y;\mathbb{Z}_2) = \mathbb{Z}_2$$

acts as $x \mapsto \langle x, [X]\rangle$. Furthermore, $\mathcal{T}(X) = \mathbf{t}^{-1}\mathrm{Sq}\,\mathbf{t}(1) = 1 + \bar{w}_1(X) + \bar{w}_2(X) + \dots$ is the total normal Stiefel–Whitney class of X (see Sect. 30.2C), that is, $\mathcal{T}(X) = w(X)^{-1}$, where $w(X)$ is the total (tangential) Stiefel–Whitney class of the manifold X. (And, certainly, $\mathcal{T}(Y) = 1$ and Sq in $H^*(Y;\mathbb{Z}_2) = \mathbb{Z}_2$ is the identity map.) Apply the Riemann–Roch theorem:

$$\langle \mathrm{Sq}\,x \smile w(X)^{-1}, [X]\rangle = \langle x, [X]\rangle.$$

Put $\mathrm{Sq}\,x \smile w(X)^{-1} = \mathrm{Sq}\,\alpha$. Then $\mathrm{Sq}\,x = \mathrm{Sq}\,\alpha \smile w(X), x = \alpha \smile \mathrm{Sq}^{-1}w(X)$, and, finally,

$$\langle \mathrm{Sq}\,\alpha, [X]\rangle = \langle \alpha \smile \mathrm{Sq}^{-1}w(X), [X]\rangle = \langle \mathrm{Sq}^{-1}w(X), D(\alpha)\rangle,$$

where D is Poincaré isomorphism. This is the Wu formula (see Sect. 31.2.D).

42.2 The Riemann–Roch Theorem in K-Theory for Complex Vector Bundles

The goals of this section are to prove that complex vector bundles are orientable, and possess a canonical orientation, with respect to the complex K-theory and to compute the corresponding Todd class.

Let $\xi = (p\colon E \to X)$ be a complex vector bundle with a finite CW base X. Put

$$\begin{aligned}\lambda_0(\xi) &= 1 \oplus \Lambda^2\xi \oplus \Lambda^4\xi \dots,\\ \lambda_1(\xi) &= \xi \oplus \Lambda^3\xi \oplus \Lambda^5\xi \dots,\\ \lambda(\xi) &= \lambda_0(\xi) - \lambda_1(\xi) \in \widetilde{K}(X).\end{aligned}$$

Lemma. *If the vector bundle ξ possesses a nonvanishing section, then $\lambda_0(\xi) \sim \lambda_1(\xi)$. Moreover, there is a natural construction which assigns to a nonvanishing section of ξ an equivalence $\lambda_0(\xi) \longleftrightarrow \lambda_1(\xi)$.*

CONSTRUCTION. A nonvanishing section of the vector bundle ξ is the same as a splitting $\xi = 1 \oplus \eta$. But $\Lambda^r(1 \oplus \eta) = \oplus_{s+t=r}(\Lambda^s 1 \otimes \Lambda^t\eta) = \Lambda^r\eta \oplus \Lambda^{r-1}\eta$. Hence,

$$\left.\begin{array}{l}\lambda_0(1 \oplus \eta)\\ \lambda_1(1 \oplus \eta)\end{array}\right\} = 1 \oplus \eta \oplus \Lambda^2\eta \oplus \Lambda^3\eta \oplus \Lambda^4\eta \oplus \Lambda^5\eta \oplus \dots$$

Consider the vector bundle $p^*\xi$ over E. This bundle has a natural section: The fiber of the bundle $p^*\xi$ over a point $y \in E$ is naturally identified with the fiber $\xi_{p(y)} \ni y$ of the bundle ξ over $p(y)$, and we assign to the point $y \in E$ the point $y \in \xi_{p(y)} = (p^*\xi)_y$. This section vanishes over the zero section $X \subset E$ and nowhere vanishes over $E' = E - X$. Hence, the preceding lemma provides the equivalence between the restrictions to E' of $\lambda_0(p^*\xi)$ and $\lambda_1(p^*\xi)$, and therefore the equivalence between the restrictions to E' of their complex conjugates, $\lambda_0(p^*\bar{\xi})$ and $\lambda_1(p^*\bar{\xi})$. This lets us consider

$$\lambda(p^*\bar{\xi}) = \lambda_0(p^*\bar{\xi}) - \lambda_1(p^*\bar{\xi})$$

as an element of $K(E, E') = \widetilde{K}(T(\xi))$. Denote this element by $u(\xi)$.

[Let us explain the last step of the construction of $u(\xi)$. If ξ_1 and ξ_2 are two vector bundles over X with a given equivalence over $A \subset X$, then the difference $\xi_1 - \xi_2$ may be regarded as an element of $K(X, A)$. Namely, let η be a bundle over X such that the sum $\xi_2 \oplus \eta$ is trivial; fix a trivialization $\xi_2 \oplus \eta \longleftrightarrow N$. Then $\xi_1 - \xi_2 = (\xi_1 \oplus \eta) - (\xi_2 \oplus \eta) = (\xi_1 \oplus \eta) - N$, and for the bundle $\xi_1 \oplus \eta$ a trivialization is given over A. This lets us glue all the fibers of $\xi_1 \oplus \eta$ over A and get a vector bundle over X/A.]

Theorem. (i) *The element $u(\xi)$ of $\widetilde{K}(T(\xi))$ is a K-orientation of the bundle ξ; this K-orientation is natural in the sense that if ξ' is another complex vector bundle of the same dimension (possibly over a different base) and a map $\xi' \to \xi$ maps* *isomorphically each fiber of the bundle ξ' onto some fiber of the bundle ξ, then the corresponding map $T(\xi') \to T(\xi)$ induces a map $\widetilde{K}(T(\xi)) \to \widetilde{K}(T(\xi'))$ that takes $u(\xi)$ into $u(\xi')$.*

To state the second part of the theorem, we consider the symmetric power series

$$F(x_1, \ldots, x_n) = \prod_{i=1}^{n} \frac{1 - e^{-x_i}}{x_i}$$

and define a power series $G(y_1, \ldots, y_n)$ by the formula

$$F(x_1, \ldots, x_n) = G(e_1(x_1, \ldots, x_n), \ldots, e_n(x_1, \ldots, x_n)),$$

where $e_i(x_1, \ldots, x_n)$ is the ith elementary symmetric function. A computation shows that

$$\begin{aligned} G(y_1, \ldots, y_n) &= 1 - \frac{1}{2}y_1 + \frac{1}{12}(2y_1^2 - y_2) - \frac{1}{24}(y_1^3 - y_1y_2) \\ &\quad + \frac{1}{720}(6y_1^4 - 9y_1^2y_2 + 2y_2^2 - y_1y_3 + y_4) + \ldots. \end{aligned}$$

(ii) *The Todd class*

$$\mathcal{T}(\xi) = (\mathbf{t}^{\xi}_{H^*(\ ;\mathbb{Q})})^{-1} \operatorname{ch} u(\xi) \in H^*(X; \mathbb{Q})$$

corresponding to the orientation $u(\xi)$ is equal to $G(c_1(\xi), \ldots, c_n(\xi))$.

Proof. Let us show first that (ii) implies (i). We must show that the restriction of the class $u(\xi)$ to the augmented fiber of the bundle ξ, that is, to $S^{2n} \subset T(\xi)$, is the generator of the group $\widetilde{K}(S^{2n}) = \mathbb{Z}$. This restriction is $u(\xi_0)$, where ξ_0 is the (trivial) n-dimensional bundle over the point. But all the Chern classes of ξ_0 are trivial, hence

$$(\mathbf{t}^{\xi_0}_{H^*(\ ;\mathbb{Q})})^{-1} \operatorname{ch} u(\xi_0) = G(c_1(\xi_0), \ldots, c_n(\xi_0)) = 1,$$

and $\operatorname{ch} u(\xi_0) = \mathbf{t}^{\xi_0}_{H^*(\ ;\mathbb{Q})}(1)$ is the generator of the group $H^{2n}(S^{2n}; \mathbb{Z})$. Hence, $u(\xi_0) \in \widetilde{K}(S^{2n})$ is the generator.

The uniqueness part of (i) is obvious.

Now prove (ii). A priori, $\mathcal{T}(\xi)$ is a characteristic class of ξ; that is, homogeneous components of $\mathcal{T}(\xi)$ are rational polynomials of Chern classes. Hence, it is sufficient to prove our formula for $\mathcal{T}(\xi)$ in the case when ξ is the direct product $\zeta \times \cdots \times \zeta$ of a number of Hopf bundles over some $\mathbb{C}P^N$. Furthermore, it is obvious that $\mathcal{T}(\xi_1 \oplus \xi_2) = \mathcal{T}(\xi_1) \oplus \mathcal{T}(\xi_2)$ [this follows from the obvious equality $\lambda(\xi_1 \oplus \xi_2) = \lambda(\xi_1)\lambda(\xi_2)$]. Hence, we can restrict ourselves to the case when ξ is just the Hopf bundle ζ over $\mathbb{C}P^N$. In this case our formula takes the form $\mathcal{T}(\zeta) = \dfrac{1 - e^{-x}}{x}$, where $x \in H^2(\mathbb{C}P^N; \mathbb{Q})$ is the generator. Observe now that $T(\zeta) = \mathbb{C}P^{N+1}$. Indeed, fix a

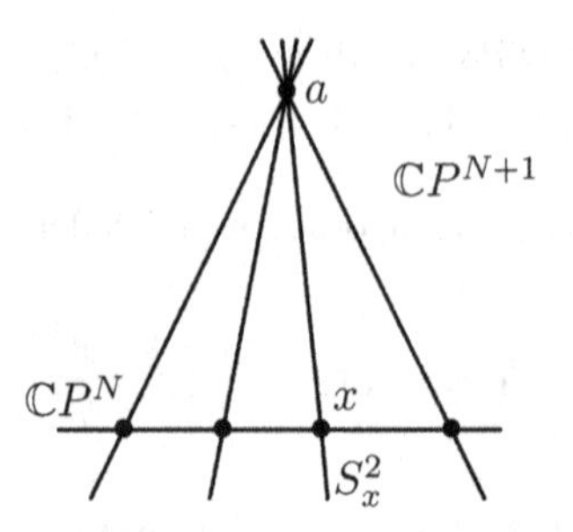

Fig. 138 The Thom space of the Hopf bundle

point $a \in \mathbb{C}P^{N+1} - \mathbb{C}P^N$ and draw all possible (complex projective) lines through a (see Fig. 139). Each of these lines intersects $\mathbb{C}P^N$ in its own point; that is, $\mathbb{C}P^{N+1}$ is made of two-dimensional spheres $S_x^2, x \in \mathbb{C}P^N$, all attached to each other in one common point. We see that $\mathbb{C}P^{N+1}$ is the Thom space of a certain one-dimensional complex vector bundle whose total space is $\mathbb{C}P^{N+1} - a$. This is the normal bundle of $\mathbb{C}P^N$ in $\mathbb{C}P^{N+1}$, that is, ζ (Fig. 138).

Let us show now that $u(\zeta) \in \widetilde{K}(\mathbb{C}P^{N+1})$ is the virtual bundle $1 - \overline{\zeta}$ over $\mathbb{C}P^{N+1}$ (we use the notation ζ for the Hopf bundle over $\mathbb{C}P^N$ and $\mathbb{C}P^{N+1}$ as well). Indeed, a point of $\mathbb{C}P^N$ is a line through the origin in $\mathbb{C}^{N+1}$, a point of $\mathbb{C}P^{N+1}$ is a line through the origin in $\mathbb{C}^{N+2}$, and is it natural to take for a the line perpendicular to $\mathbb{C}^{N+1}$. The projection $p: \mathbb{C}P^{N+1} - a \to \mathbb{C}P^N$ is induced by the orthogonal projection of $\mathbb{C}^{N+2}$ onto $\mathbb{C}^{N+1}$. The fiber of the bundle ζ over $l \in \mathbb{C}R^N$ is the line $l \subset \mathbb{C}P^{N+1}$ itself; the fiber of the bundle $p^*\zeta$ over $\widetilde{l} \in \mathbb{C}P^{N+1} - a$ is the projection of the line $\widetilde{l}$ in $\mathbb{C}^{N+1}$, which is naturally identified with $\widetilde{l}$. Thus, the lifted bundle $p^*\zeta$ is $\zeta|_{\mathbb{C}P^{N+1}-a}$. The isomorphism between $\lambda_1(p^*\zeta) = p^*\zeta$ and $\lambda_0(p^*\zeta) = 1$ over the complement to $\mathbb{C}P^N$ consists in the projection of the fibers of the bundle $\zeta|_{\mathbb{C}P^{N+1}-a}$ onto the line $a \subset \mathbb{C}^{N+2}$. Hence, when we pass from $\mathbb{C}P^{N+1} - a$ to $T(\zeta) = \mathbb{C}P^{N+1}$, then $\lambda(p^*\zeta)$ becomes $1-\zeta$, which shows that $u(\zeta) = 1-\overline{\zeta}$. Thus, $\operatorname{ch} u(\zeta) = \operatorname{ch}(1-\overline{\zeta}) = 1-e^{-x}$.

Finally, it is obvious that the cohomological Thom isomorphism $\mathbf{t}_H: H^*(\mathbb{C}P^N; \mathbb{Z}) \to \widetilde{H}^*(\mathbb{C}P^{N+1}; \mathbb{Z})$ acts as $x^k \mapsto x^{k+1}$. Therefore,

$$\mathcal{T}(\zeta) = \mathbf{t}_H^{-1}(1 - e^{-x}) = \frac{1 - e^{-x}}{x}.$$

This completes the proof of the theorem.

Thus, in our case the Riemann–Roch theorem takes the following form.

Theorem. *For an n-dimensional complex vector bundle ξ over a finite* CW *base X, there exists a natural (with respect to ξ) Thom isomorphism $\mathbf{t}_K^\xi: K(X) \to \widetilde{K}(T(\xi))$ such that*

$$\operatorname{ch}(\mathbf{t}_K^\xi \alpha) = \mathbf{t}_H^\xi(\operatorname{ch} \alpha \cdot \mathcal{T}(\xi)),$$

where $\alpha \in k(X)$, $\mathbf{t}_H^\xi$ is the cohomological Thom isomorphism, and

$$\mathcal{T}(\xi) = G(c_1(\xi), \dots, c_n(\xi)),$$

where G is the power series defined by the formula

$$G(e_1(x_1, \dots), e_2(x_1, \dots), \dots) = \prod \frac{1 - e^{-x_i}}{x_i}.$$

42.3 Application: The Computation of the e-Invariant

Recall the definition of the e-invariant of the class $\lambda \in \pi_{2N+2n-1}(S^{2N})$: We construct the space $X_\lambda = S^{2N} \cup_f D^{2N+2n}$, where $f\colon S^{2N+2n-1} \to S^{2N}$ is a spheroid of the class λ; next we take in $\widetilde{K}(X_\lambda) \cong \mathbb{Z} \oplus \mathbb{Z}$ an arbitrary element $\widetilde{\alpha}$ whose restriction to $S^{2N} \subset X$ is the standard generator of the group $\widetilde{K}(S^{2N}) = \mathbb{Z}$; then

$$\operatorname{ch} \widetilde{\alpha} = a + \widetilde{e} b,$$

where $a \in H^{2N}(X_\lambda; \mathbb{Q}), b \in H^{2N+2n}(X_\lambda; \mathbb{Q})$ are the standard generators, and $\widetilde{e} \in \mathbb{Q}$; the residue $e(\lambda) \in \mathbb{Q}/\mathbb{Z}$ modulo 1 is determined by the class λ; this residue is the e-invariant of the class λ.

Our goal is to calculate $e(\lambda)$ in the case when λ is the J-image of the generator of the group $\widetilde{K}(S^{2n}) = \mathbb{Z}$. We noticed in Lecture 41 (see Exercise 7) that in this case X_λ is the Thom space $T(\xi)$, where ξ is the vector bundle over S^{2n} representing the generator of $\widetilde{K}(S^{2n})$. Notice, in addition to that, that we can take $u(\xi) \in \widetilde{K}(S^{2n})$ for $\widetilde{\alpha}$, for the restriction of $u(\xi)$ to S^{2N} is the generator of $\widetilde{K}(S^{2N})$, by the definition of the orientation. According to the Riemann–Roch theorem,

$$\operatorname{ch} u(\xi) = \mathbf{t}_H^\xi \mathcal{T}(\xi) = \mathbf{t}_H^\xi G(c_1(\xi), c_2(\xi), \dots).$$

But the classes $c_i(\xi)$ with $0 < i < n$ are all zero (because the corresponding cohomology groups of the sphere are all zero), and $c_n(\xi)$ may be easily calculated:

$$\operatorname{ch} \xi = \dim \xi + s_{2n} \in H^*(S^{2n}; \mathbb{Q}),$$

where s_{2n} is the generator of the group $H^{2n}(S^{2n}; \mathbb{Z})$. Let us find the coefficient with which c_n appears in ch_n. By construction,

$$\operatorname{ch}_n = \frac{1}{n!} P_n(c_1, \dots, c_n),$$
$$\sum_{i=1}^n x_i^n = P_n(e_1(x_1, \dots, x_n), \dots, e_n(x_1, \dots, x_n)).$$

Substitute $x_i = \varepsilon_n^i$, where ε_n is the primitive root of degree n of 1, in the last equality. We will have $e_1(x) = \cdots = e_{n-1}(x) = 0, e_n(x) = \pm 1, \sum x_i^N = n$. This shows that c_n appears in P_n with the coefficient $\pm n$, and, hence,

$$s_n = \mathrm{ch}_n(\xi) = \frac{1}{n!}(\pm n)c_n(\xi),$$

$$c_n(\xi) = \pm(n-1)!s_n.$$

Thus,

$$\mathrm{ch}\,\widetilde{\alpha} = \mathrm{ch}\,u(\xi) = \mathbf{t}_H^{\xi}(1 \pm (n-1)!\mu s_n) = a \pm (n-1)!\mu b,$$

where μ is the coefficient with which $c_n(\xi)$ appears $G(c_1(\xi), c_2(\xi), \dots)$; thus,

$$e = \pm(n-1)!\mu.$$

It is easy to show that if n is odd and greater than 1, then $\mu = 0$. If $n = 2m$ is even, then it turns out that $\mu = \pm\dfrac{B_m}{2m!}$ [for $m = 1, 2$, it is seen from the partial formula for the series $G(y_1, y_2, \dots)$ given in Sect. 44.2 above]. To prove this, consider the class $\mathcal{T}^{-1}(\xi) = \mathcal{U}(c_1(\xi), c_2(\xi), \dots)$ where

$$\mathcal{U}(e_1(x), e_2(x), \dots) = \prod \frac{x_i}{1 - e^{-x_i}}.$$

Obviously, for our ξ this class is equal to

$$a \pm (n-1)\mu b.$$

Furthermore, it can be shown by elementary means that in the series $\mathcal{U}$ the coefficients in the terms with $e_1(x)^k$ and $e_k(x)$ coincide for any k (see the proof in Hirzebruch's book [46], Remark 7.2 in Sect. 1.7 of Chap. 1). And it is easy to find the coefficient in the term with $e_1(x)^{2m}$: Put $x_1 = z, x_2 = \cdots = 0$; then $e_1(x) = z, e_2(x) = \cdots = 0$,

$$\begin{array}{l} \displaystyle\prod \frac{x_i}{1 - e^{-x_i}} = \frac{z}{1 - e^{-z}} = 1 + \frac{1}{2}z + \sum_{s=1}^{\infty}(-1)^{s-1}\frac{B_s}{(2s)!}z^{2s}, \\ \qquad\qquad \| \\ \mathcal{U}(e_1(x), e_2(x), \dots) = \mathcal{U}(z, 0, \dots); \end{array}$$

that is, the coefficient at $e_1(x)^{2m}$ is equal to $\pm\dfrac{B_m}{(2m)!}$.

Thus, if $n = 2m$, then

$$e(\lambda) = \pm(2m-1)!\frac{B_m}{(2m)!} = \pm\frac{B_m}{2m}.$$

We have proven

Theorem. $|\widetilde{J}(S^{4m})|$ *is greater than or equal to the denominator of the irreducible fraction equal to* $B_m/2m$.

Let us compare this result with the result of Sect. 41.3. We proved there that $|\widetilde{J}(S^{4m})|$ is less than or equal to the denominator of $B_m/4m$. Since the numerator of B_m is always odd (it is an easy result from number theory), we see that our upper bound for $|\widetilde{J}(S^{4m})|$ is always twice the lower bound. Since in both theorems "greater than or equal to" should be read as "divisible by," the number $|\widetilde{J}(S^{4m})|$ must be equal either to the denominator of $B_m/2m$ or to the denominator of $B_m/4m$.

Actually, the situation is as follows (see Karoubi's book). If m is even, then the order of the group $\widetilde{J}(S^{4m})$ is equal to the denominator of $B_m/2m$; but in this case the homomorphism

$$\mathbb{Z} = \pi_{4m-1}(U) \to \pi_{4m-1}(SO) = \mathbb{Z}$$

induced by the embedding $U \to SO$ is the multiplication by 2, which implies that the J-image of the group $\pi_{4m-1}(SO)$ has order equal to the denominator of $B_m/4m$. If m is odd, then the above homomorphism is an isomorphism, but in this case $|\widetilde{J}(S^{4m})|$ coincides with its upper bound, namely, with the denominator of $B_m/4m$. Thus, for any m, the group π^S_{4m-1} contains a cyclic subgroup of order equal to the denominator of $B_m/4m$. It is also known that this subgroup of π^S_{4m-1} is always a direct summand.

42.4 The Riemann–Roch Theorem in K-Theory for Spinor Vector Bundles

Not only complex vector bundles are orientable with respect to the (complex) K-theory. Actually, the following proposition holds.

Theorem. *An orientable real vector bundle* ξ *is* K*-orientable if and only if* $w_3^{\mathbb{Z}}(\xi) = 0$, *where* $w_3^{\mathbb{Z}}$ *is the characteristic class corresponding to the nontrivial element of the group* $H^3(G_+(\infty, n); \mathbb{Z}) \cong \mathbb{Z}_2$. *An equivalent condition: The Stiefel–Whitney class* $w_2(\xi) \in H^2(X; \mathbb{Z}_2)$ *(where* X *is the base of* ξ*;* X *is supposed to be a finite* CW *complex) is integral, that is, belongs to the image of the reducing* mod 2 *homomorphism* $\rho_2 \colon H^2(X; \mathbb{Z}) \to H^2(X; \mathbb{Z}_2)$. *A choice of a* K*-orientation is equivalent to a choice of the inverse image* $\widetilde{w}_2(\xi) \in H^2(X; \mathbb{Z})$ *of the class* $w_2(\xi)$. *[*K*-orientation of a complex vector bundle* ξ *corresponds to* $\widetilde{w}_2(\xi) = c_1(\xi)$.*]*

Here we will restrict ourselves to the construction of a K-orientation for a real vector bundle with a chosen class $\widetilde{w}_2$. The reader may find a complete proof of the preceding theorem in Karoubi's book.

A: The Group Spin$^{\mathbb{C}}$

The construction of the K-orientation of a complex vector bundle given in Sect. 42.2 was based on the following observation: The natural representation of the group $U(n)$ in $\Lambda^*\mathbb{C}^n$ splits into the sum $\Lambda^{\text{even}}\mathbb{C}^n \oplus \Lambda^{\text{odd}}\mathbb{C}^n$ of subrepresentations, and these two subrepresentations are isomorphic as representations of the smaller group $U(n-1) \subset U(n)$. We will see now that there exist much bigger [than $U(n)$] groups which have representations in $\Lambda^*\mathbb{C}^n$ with similar properties.

The vector space $\Lambda^*\mathbb{C}^n$ has a natural Hermitian structure:$\Lambda^k\mathbb{C}^n$ with different k that are orthogonal to each other, and the length of $v_1 \wedge \cdots \wedge v_k$ is the (nonnegative) volume of the parallelepiped spanned by $v_1, \dots, v_k$. Due to this, $\operatorname{End}\Lambda^*\mathbb{C}^n$ possesses a natural conjugation $*$ with $*^2 = 1$. For $v \in \mathbb{C}^n$ define $F_v \in \operatorname{End}\Lambda^*\mathbb{C}^n$ as the operator of the exterior multiplication by v, and put $\varphi_v = F_v + F_v^*$. (F_v^* is "the differentiation with respect to v.") Obviously, $\varphi_v^* = \varphi_v$, $\varphi_v^2 = \|v\|^2$ (indeed, $F_vF_v^* + F_v^*F_v = \|v\|^2$, $F_v^2 = 0, F_v^{*\,2} = 0$), and $v \mapsto \varphi_v$ is a linear embedding $\mathbb{C}^n \to \operatorname{End}\Lambda^*\mathbb{C}^n$.

Denote by $\operatorname{Pin}^{\mathbb{C}}(2n)$ the subset of $\operatorname{End}\Lambda^*\mathbb{C}^n$ consisting of endomorphisms of $\Lambda^*\mathbb{C}^n$ that satisfy the following three conditions.

(i) $x^* \circ x = 1$ (in other words, x is a unitary transformation of $\Lambda^*\mathbb{C}^n$; in particular, the endomorphism x is invertible).
(ii) The endomorphism x is homogeneous (of degree 0 or 1) with respect to the grading $\Lambda^*\mathbb{C}^n = \Lambda^{\text{even}}\mathbb{C}^n \oplus \Lambda^{\text{odd}}\mathbb{C}^n$.
(iii) For any $v \in \mathbb{C}^n$ there exists a $w \in \mathbb{C}^n$ such that $x\varphi_v x^* = \varphi_w$.

Obviously, $\operatorname{Pin}^{\mathbb{C}}(2n)$ is a Lie group.

Denote by $\operatorname{Spin}^{\mathbb{C}}(2n)$ the subset of the group $\operatorname{Pin}^{\mathbb{C}}(2n)$ that consists of those x for which $x(\Lambda^{\text{even}}\mathbb{C}^n) \subset \Lambda^{\text{even}}\mathbb{C}^n$, $x(\Lambda^{\text{odd}}\mathbb{C}^n) \subset \Lambda^{\text{odd}}\mathbb{C}^n$. Obviously, $\operatorname{Spin}^{\mathbb{C}}(2n)$ is an index-2 subgroup of $\operatorname{Pin}^{\mathbb{C}}(2n)$.

For each $x \in \operatorname{Pin}^{\mathbb{C}}(2n)$, define a real linear transformation of $\tau: \mathbb{C}^n \to \mathbb{C}^n$ by the formula

$$\tau(v) = \begin{cases} w & \text{if } x \in \operatorname{Spin}^{\mathbb{C}}(2n) \\ -w & \text{if } x \notin \operatorname{Spin}^{\mathbb{C}}(2n), \end{cases}$$

where w is defined in condition (iii) above. This transformation is obviously orthogonal ($\|w\|^2 = \varphi_w^2 = x\varphi_v x^* x\varphi_v x^* = x\varphi_v^2 x^* = x\|v\|^2 x^* = \|v\|^2 xx^* = \|v\|^2$). Hence, we obtain a homomorphism

$$\tau: \operatorname{Pin}^{\mathbb{C}}(2n) \to O(2n).$$

Proposition 1. *This is an epimorphism whose kernel is the group* $S^1 \subset \mathrm{Spin}^{\mathbb{C}}(2n)$ *of multiplication by complex numbers of absolute value* 1*. The group* $\mathrm{Spin}^{\mathbb{C}}(2n)$ *is precisely* $\tau^{-1}(SO(2n))$.

Proof. Obviously, $\varphi_v \in \mathrm{Pin}^{\mathbb{C}}(2n) - \mathrm{Spin}^{\mathbb{C}}(2n)$ for any unit vector $v \in \mathbb{C}^n$. Furthermore, $\tau(\varphi_v)$ is in this case the reflection of $\mathbb{C}^n$ in the real hyperplane orthogonal to v. [Indeed, $-\varphi_v\varphi_v\varphi_v^* = -\varphi_v^3 = -\varphi_v$; $-\varphi_v\varphi_{iv}\varphi_v^* = -(F_v+F_v^*)(F_{iv}+F_{iv}^*)(F_v+F_v^*) = -i(F_v+F_v^*)(F_v-F_v^*)(F_v+F_v^*) = -i(F_v^*F_vF_v^*-F_vF_v^*F_v) = -i(F_v^*-F_v) = \varphi_{iv}$; if v and w are complex orthogonal, then $-\varphi_v\varphi_w\varphi_v^* = -\varphi_v\varphi_w\varphi_v = \varphi_v^2\varphi_w = \varphi_w$.] Hence, the image of the homomorphism τ contains all the reflections in hyperplanes, and hence it contains the whole group $O(2n)$.

Prove now that the kernel of the homomorphism τ consists of the multiplications by complex numbers of absolute value 1. Let $x \in \mathrm{Pin}^{\mathbb{C}}(2n)$ belong to $\mathrm{Ker}\,\tau$.

Suppose first that $x \in \mathrm{Spin}^{\mathbb{C}}\Lambda^*\mathbb{C}^n$. Then $x \in \mathrm{Ker}\,\tau$ means that $\varphi_v x = x\varphi_v$ for any $v \in \mathbb{C}^n$, or, equivalently, for any $v \in \mathbb{C}^n$ with $\|v\| = 1$. Denote by $\mathbb{C}_v^{n-1}$ the (complex) orthogonal complement to v. Then

$$\Lambda^*\mathbb{C}^n = (1+v)\wedge\Lambda^*\mathbb{C}_v^{n-1} \oplus (1-v)\wedge\Lambda^*\mathbb{C}_v^{n-1},$$

and φ_v is id on the first of these two summands and $-$id on the second. Since x commutes with φ_v, the summands $(1\pm v)\wedge\Lambda^*\mathbb{C}_v^{n-1}$ should be invariant with respect to x [for example, if $\alpha \in (1+v)\wedge\Lambda^*\mathbb{C}_v^{n-1}$, then $\varphi_v x(\alpha) = x\varphi_v(\alpha) = x\varphi_v(\alpha) = x(\alpha)$, whence $x(\alpha) \in (1+v)\wedge\Lambda^*\mathbb{C}_v^{n-1}$]. We can write

$$x((1+v)\wedge a + (1-v)\wedge b) = (1+v)\wedge y(a) + (1-v)\wedge z(b),$$

where y and z are endomorphisms of the space $\Lambda^*\mathbb{C}_v^{n-1}$. It is checked automatically that actually $y, z \in \mathrm{Spin}^{\mathbb{C}}(2n-2)$; moreover, they belong to the kernel of the homomorphism $\mathrm{Spin}^{\mathbb{C}}(2n-2) \to SO(2n-2)$ similar to τ. By induction, we can assume that y and z are multiplications by complex numbers β_v and γ_v with $|\beta_v| = |\gamma_v| = 1$. Furthermore,

$$\begin{aligned} x(1) &= x\left((1+v)\frac{1}{2} + (1-v)\frac{1}{2}\right) \\ &= (1+v)\frac{\beta_v}{2} + (1-v)\frac{\gamma_v}{2} \\ &= \frac{\beta_v+\gamma_v}{2} + \frac{\beta_v-\gamma_v}{2}v. \end{aligned}$$

Since $x(1)$ does not depend on v, we see first that $\beta_v - \gamma_v = 0$, that is, $\beta_v = \gamma_v$, and second that $\beta_v + \gamma_v$ does not depend on v. Hence, $\beta_v = \gamma_v = \beta$, $|\beta| = 1$, and x is the multiplication by β.

The remaining case $x \notin \mathrm{Spin}^{\mathbb{C}}(2n)$ is still easier, since in this case

$$x((1+v)\wedge a + (1-v)\wedge b) = (1+v)\wedge y(b) + (1-v)\wedge z(a),$$

and $y, z \in (\mathrm{Pin}^{\mathbb{C}}(2n-2) - \mathrm{Spin}^{\mathbb{C}}(2n-2)) \cap \mathrm{Ker}\,\tau$, which implies by induction that $y = z = 0$, and hence $x = 0$.

Finally, the part of the proposition that has been already proven shows that the group $\mathrm{Pin}^{\mathbb{C}}(2n)$ is generated by the transformations φ_v and multiplications by constants of absolute value 1; it is clear that a word of these generators belongs to either $\mathrm{Spin}^{\mathbb{C}}(2n)$ or $\tau^{-1}SO(2n)$ if and only if it involves an even number of φ_vs. This implies the last part of the proposition.

Remark 1. A similar real construction gives rise to the groups $\mathrm{Pin}^{\mathbb{R}}(k)$ and $\mathrm{Spin}^{\mathbb{R}}(k)$. The group $\mathrm{Spin}^{\mathbb{R}}(k)$ is a twofold covering of $SO(k)$; the group $\mathrm{Pin}^{\mathbb{R}}(k)$ consists of two components, one of which is $\mathrm{Spin}^{\mathbb{R}}(k)$. The group $\mathrm{Spin}^{\mathbb{R}}(2n)$ is contained in $\mathrm{Spin}^{\mathbb{C}}(2n)$, and the projections τ of these two groups onto $SO(2n)$ are compatible with each other.

Remark 2. The group $U(n)$ is included in $\mathrm{Spin}^{\mathbb{C}}(2n)$ as the group of transformations which preserve not only the grading $\Lambda^*\mathbb{C}^n = \Lambda^{\mathrm{even}}\mathbb{C}^n \oplus \Lambda^{\mathrm{odd}}\mathbb{C}^n$, but also a finer grading $\Lambda^*\mathbb{C}^n = \oplus_q \Lambda^q\mathbb{C}^n$. The restriction of the projection $\tau\colon \mathrm{Spin}^{\mathbb{C}}(2n) \to SO(2n)$ to $U(n)$ coincides with the natural embedding $U(n) \to SO(2n)$.

Thus, we have constructed a group $\mathrm{Spin}^{\mathbb{C}}(2n)$ and its (unitary) representation in $\Lambda^*\mathbb{C}^n$, and this representation is decomposed into the sum of two (actually, irreducible) subrepresentations: $\Lambda^*\mathbb{C}^n = \Lambda^{\mathrm{even}}\mathbb{C}^n \oplus \Lambda^{\mathrm{odd}}\mathbb{C}^n$.

Proposition 2. *$\Lambda^{\mathrm{even}}\mathbb{C}^n$ and $\Lambda^{\mathrm{odd}}\mathbb{C}^n$ are isomorphic as representations of the group* $\mathrm{Spin}^{\mathbb{C}}(2n-2) \subset \mathrm{Spin}^{\mathbb{C}}(2n)$.

Proof. The embedding $\mathrm{Spin}^{\mathbb{C}}(2n-2) \to \mathrm{Spin}^{\mathbb{C}}(2n)$ is induced by the embedding $\mathbb{C}^{n-1} \to \mathbb{C}^n$. Let $v \in \mathbb{C}^n$ be a unit vector orthogonal to $\mathbb{C}^{n-1}$. Then the operator $\varphi_v\colon \Lambda^{\mathrm{even}}\mathbb{C}^n \to \Lambda^{\mathrm{odd}}\mathbb{C}^n$ is a $\mathrm{Spin}^{\mathbb{C}}(2n-2)$-isomorphism.

In conclusion, consider the cohomology of the group $\mathrm{Spin}^{\mathbb{C}}(2n)$.

Proposition 3. *If $n > 1$, then*

$$H^1(\mathrm{Spin}^{\mathbb{C}}(2n); \mathbb{Z}) = \mathbb{Z};\ H^2(\mathrm{Spin}^{\mathbb{C}}(2n); \mathbb{Z}) = 0.$$

Lemma. *If $n > 1$, then the fibration* $\mathrm{Spin}^{\mathbb{C}}(2n) \to SO(2n)$ *(with the fiber S^1) is nontrivial.*

Proof of Lemma. The group $\mathrm{Spin}^{\mathbb{C}}(2n)$ contains the twofold covering $\mathrm{Spin}^{\mathbb{R}}(2n)$ of the group $SO(2n)$. If the fibration of the lemma is trivial, then the composition

$$\mathrm{Spin}^{\mathbb{R}}(2n) \longrightarrow \mathrm{Spin}^{\mathbb{C}}(2n) \xrightarrow{=} SO(2n) \times S^1 \xrightarrow{\text{projection}} S^1$$

maps the points of each fiber of the twofold covering $\mathrm{Spin}^{\mathbb{R}}(2n) \to SO(2n)$ into the opposite points of the circle. In this case we would have a commutative diagram

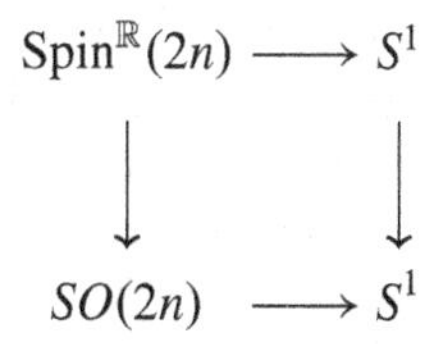

whose vertical arrows are twofold coverings. But the map $SO(2n) \to S^1$ should be homotopic to constant since $H^1(SO(2n); \mathbb{Z}) = 0$, which contradicts the nontriviality of the covering $\mathrm{Spin}^{\mathbb{R}}(2n) \to SO(2n)$.

Proof of Proposition 3. The fibration $\mathrm{Spin}^{\mathbb{C}}(2n) \to SO(2n)$ is simple since its fibers have canonical orientations as cosets of the group S^1. The E_2-term of the spectral sequence of this fibration looks like

$\mathbb{Z}$	0	$\mathbb{Z}_2$	$\cdots$
$\mathbb{Z}$	0	$\mathbb{Z}_2$	$\cdots$

and the differential $d_2^{0,1}: E_2^{0,1} = \mathbb{Z} \to E_2^{2,0} = \mathbb{Z}_2$ cannot be trivial, since its image contains the first (and the last) obstruction to a section (see Sect. 18.5). Thus, the spectral sequence implies our statement.

B: Spinor Vector Bundles and Their Characteristic Classes

A *spinor structure*, or, more specifically, a complex spinor structure or $\mathrm{Spin}^{\mathbb{C}}$-*structure*, in an oriented real $2n$-dimensional vector bundle is, by definition, a reduction of its structure group $SO(2n)$ to $\mathrm{Spin}^{\mathbb{C}}(2n)$. Let us explain this. As we said in Sect. 19.1.E, an oriented real $2n$-dimensional vector bundle over X may be given by an open covering $\{U_i\}$ of X and a family of continuous maps $\varphi_{ij}: U_i \cap U_j \to SO(2n)$ that satisfy some conditions [namely, $\varphi_{ij}(x)\varphi_{jk}(x)\varphi_{ki}(x) = 1$ for $x \in U_i \cap U_j \cap U_k$]. A reduction is defined by a family of liftings $\Pi_{ij}: U_i \cap U_j \to \mathrm{Spin}^{\mathbb{C}}(2n)$ of the maps φ_{ij} such that the condition above remains valid. The reader will reconstruct the notion of equivalence for reductions.

Vector bundles whose structure group is reduced to $\mathrm{Spin}^{\mathbb{C}}(2n)$ are called spinor bundles. (Notice that a complex spinor bundle is not, in general, a complex bundle.)

To proceed further we need some knowledge of general classifying spaces. For any topological group G there exists a CW complex BG such that the fibrations with a finite CW base X and a structure group G are in a one-to-one correspondence with the homotopy classes of maps $X \to BG$. This BG is unique up to a homotopy equivalence. It is called a *classifying space for* G. Here are some other properties of classifying spaces: There exists a contractible topological space EG with a (right) free action of G, such that $BG = EG/G$; in particular, there is a principal G-bundle $EG \to BG$; a group homomorphism $G \to H$ defines (up to a homotopy) a map $BG \to BH$; moreover, if $G \to H$ is an epimorphism with the kernel F, then there is a fibration $BG \to BH$ with the fiber BF. All this is easy to prove—the most difficult thing is to construct the space BG; this difficulty is resolved by using *Milnor's construction* in which EG is the infinite join of the group G:

$$EG = \ldots((G * G) * G) * G \ldots.$$

The classifying spaces for some groups are already known to us:

$$BU(n) = \mathbb{C}G(\infty, n), BO(n) = G(\infty, n), BSO(n) = G_+(\infty, n),$$

and if π is a discrete group, then $B\pi = K(\pi, 1)$. See Sect. 4.4 of the book by Fuchs and Rokhlin [40] for further details.

Thus, there exists a classifying space $B\mathrm{Spin}^{\mathbb{C}}(2n)$ such that all complex spinor bundles with a finite CW base X are in one-to-one correspondence with the homotopy classes of maps $X \to B\mathrm{Spin}^{\mathbb{C}}(2n)$, and the characteristic classes of complex spinor bundles correspond to the cohomology of the space $B\mathrm{Spin}^{\mathbb{C}}(2n)$. The group fibration

$$\mathrm{Spin}^{\mathbb{C}}(2n) \xrightarrow{S^1} SO(2n)$$

gives rise to a fibration

$$B\mathrm{Spin}^{\mathbb{C}}(2n) \xrightarrow{BS^1} BSO(2n)$$

in which $BSO(2n) = G_+(\infty, 2n)$, $BS^1 = BU(1) = \mathbb{C}G(\infty, 1) = \mathbb{C}P^\infty = K(\mathbb{Z}, 2)$. One can say more: The introduction of a complex spinor structure into an oriented $2n$-dimensional real vector bundle corresponds to a lifting of a map $X \to BSO(2n)$ to a map $X \to B\mathrm{Spin}^{\mathbb{C}}(2n)$.

It may be deduced in the usual way from Proposition 3 and the spectral sequences of the fibrations

$$E\mathrm{Spin}^{\mathbb{C}}(2n) \xrightarrow{\mathrm{Spin}^{\mathbb{C}}(2n)} B\mathrm{Spin}^{\mathbb{C}}(2n), \ ESO(2n) \xrightarrow{SO(2n)} BSO(2n)$$

that $H^i(BSO(2n); \mathbb{Z}) = 0$ for $i = 1, 2$, $H^3(BSO(2n); \mathbb{Z}) = \mathbb{Z}_2$, and $H^3(B\mathrm{Spin}^{\mathbb{C}}(2n); \mathbb{Z}) = 0$. The E_2-term of the spectral sequence of the fibration

$$B\mathrm{Spin}^{\mathbb{C}}(2n) \xrightarrow{K(\mathbb{Z},2)} BSO(2n)$$

looks like

0				
$\mathbb{Z}$	0	0	$\mathbb{Z}_2$	$\cdots$
0	0	0	0	$\cdots$
$\mathbb{Z}$	0	0	$\mathbb{Z}_2$	$\cdots$

and, since $H^3(B\mathrm{Spin}^{\mathbb{C}}(2n);\mathbb{Z}) = 0$, the nontrivial group $E_2^{3,0}$ cannot survive until E_∞, which means that the differential $d_3^{0,2} : E_3^{0,2} = \mathbb{Z} \to E_3^{3,0} = \mathbb{Z}_2$ must be nontrivial; the image of this differential should contain the only nonzero element of $H^3(BSO(2n);\mathbb{Z})$, which we denote by $w_3^{\mathbb{Z}}$. Now we can classify complex spinor structures in oriented even-dimensional real vector bundles. An oriented real $2n$-dimensional vector bundle ξ over X corresponds to a map $f_\xi\colon X \to BSO(2n)$; a spinor structure is a lifting of this map into $B\mathrm{Spin}^{\mathbb{C}}(2n)$, that is, a section of a fibration $Y \to X$ induced by the fibration $B\mathrm{Spin}^{\mathbb{C}}(2n) \to BSO(2n)$ by means of the map f_ξ:

$$\begin{array}{ccc} Y & \longrightarrow & B\mathrm{Spin}^{\mathbb{C}}(2n) \\ \Big\downarrow{\scriptstyle K(\mathbb{Z},2)} & & \Big\downarrow{\scriptstyle K(\mathbb{Z},2)} \\ X & \xrightarrow{f_\xi} & BSO(2n). \end{array}$$

The only obstruction to this section is the class $f_\xi^* w_3^{\mathbb{Z}}$, that is, the characteristic class of the bundle ξ that corresponds to $w_3^{\mathbb{Z}}$. Since $w_3^{\mathbb{Z}} = \beta w_2$, where $w_2 \in H^2(BSO(2n);\mathbb{Z}_2)$, we arrive at the following result.

Proposition 4. *An oriented real vector bundle ξ with a finite* CW *base X possesses a complex spinor structure if and only if $w_3^{\mathbb{Z}}(\xi) = 0$, or, equivalently, the class $w_2(\xi)$ is integral; that is, it belongs to the image of the reduction homomorphism $\rho_2\colon H^2(X;\mathbb{Z}) \to H^2(X;\mathbb{Z}_2)$. Moreover, the classes of equivalent complex spinor structures in ξ are in one-to-one correspondence with elements of the set $\rho_2^{-1}(w_2(\xi)) \subset H^2(X;\mathbb{Z})$.*

EXERCISE 5. Prove the last statement of Proposition 4 (*Hint:* difference cochains).

EXERCISE 6. A complex vector bundle ξ has a canonical complex spinor structure because of the inclusion $U(n) \subset \mathrm{Spin}^{\mathbb{C}}(2n)$. Prove that this structure corresponds to the class $c_1(\xi) \subset \rho_2^{-1}(w_2(\xi))$.

Proposition 5. *$H^*(B\mathrm{Spin}^{\mathbb{C}}(2n))$ is the ring of polynomials of the following variables: $c\,(\dim c = 2)$, $p_1, \dots, p_n\,(\dim p_i = 4i)$, $\chi\,(\dim \chi = 2n)$. Thus, rational characteristic classes of complex spinor bundles are rational polynomials of Pontryagin and Euler characteristic classes of this bundle regarded as an oriented* *real vector bundle without spinor structure, and the characteristic class c, which is the image in the rational cohomology of the integral cohomology class from* $\rho_2^{-1}(w_2(\xi))$*, which determines the spinor structure.*

The first statement follows from the spectral sequence of the fibration

$$B\mathrm{Spin}^{\mathbb{C}} \xrightarrow{K(\mathbb{Z},2)} BSO(2n)$$

and the computation of the rational cohomology of the space $BSO(2n)$ from Lecture 19. The rest of Proposition 5 is left to the reader as an exercise.

As we noticed several times, complex vector bundles are also spinor bundles. Therefore, characteristic classes of spinor bundles are also characteristic classes of complex vector bundles, and hence they should be expressed in terms of Chern classes. It is easy to find these expressions by computing the classes c, p_i, χ for the vector bundle $\zeta \times \cdots \times \zeta$ over $\mathbb{C}P^n \times \cdots \times \mathbb{C}P^n$ (n factors). The computation shows

$$\begin{aligned} c(\zeta \times \cdots \times \zeta) &= x_1 + \cdots + x_n, \\ p_i(\zeta \times \cdots \times \zeta) &= e_i(x_1^2, \ldots, x_n^2), \\ \chi(\zeta \times \cdots \times \zeta) &= x_1 \ldots x_n, \end{aligned}$$

where the e_i are elementary symmetric polynomials. This lets us express c, p_i, χ directly via c_i:

$$c = c_1,\ p_i = c_i^2 + c_{i-1}c_{i+1} + \cdots + c_{2i},\ \chi = c_n.$$

However, this direct expression is less useful than the above indirect one.

C: K-Orientation of Spinor Bundles, Todd Class, and Riemann–Roch Theorem

The K-orientation of a spinor bundle $\xi = (p: E \to X)$ is constructed in the same way as the K-orientation of a complex vector bundle. The unitary representation of the group $\mathrm{Spin}^{\mathbb{C}}(2n)$ in $\Lambda^{\mathrm{even}}\mathbb{C}^n$ and $\Lambda^{\mathrm{odd}}\mathbb{C}^n$ lets us associate with ξ vector bundles with the base X and the fibers $\Lambda^{\mathrm{even}}\mathbb{C}^n$ and $\Lambda^{\mathrm{odd}}\mathbb{C}^n$. Denote these bundles by $\lambda_0(\xi)$ and $\lambda_1(\xi)$. The lifted bundles $p^*\lambda_0(\xi)$ and $p^*\lambda_1(\xi)$ over E are canonically isomorphic over $E' = E - X$ by Proposition 2; hence, the difference $p^*\lambda_0(\xi) - p^*\lambda_1(\xi)$ may be regarded as an element of $K(E, E') = \widetilde{K}(T(\xi))$; denote this element by $u(\xi)$.

Theorem. (i) *$u(\xi)$ is a K-orientation of ξ.*
(ii) *The Todd class*

$$\mathcal{T}(\xi) = \mathbf{t}_H^{-1}\,\mathrm{ch}\,u(\xi) \in H^{\mathrm{even}}(X; \mathbb{Q})$$

is equal to

$$e^{c/2}A(p_1(\xi), p_2(\xi), \dots),$$

where the power series A is determined by the formula

$$A(e_1(x_1^2, x_2^2, \dots), e_2(x_1^2, x_2^2, \dots), \dots) = \prod \frac{\sinh(x_i/2)}{x_i/2}.$$

Remark 1. The characteristic class $A(p_1(\xi), p_2(\xi), \dots)$ is called the (*reduced*) *Atiyah–Hirzebruch class* of the real vector bundle ξ.

Remark 2. The K-orientation of a complex vector bundle ξ that was introduced in Sect. 41.2 coincides with the K-orientation in the spinor sense of the vector bundle $\bar{\xi}$.

The proof of this theorem is essentially known to us. As in the case of the similar theorem in Sect. 41.2, the first part follows from the second; as to $\mathcal{T}(\xi)$, it is computed in the following way. First, if the bundle ξ is complex, then our $\mathcal{T}(\xi)$ coincides with $\mathcal{T}(\bar{\xi})$ of Sect. 41.2. Second, the class $\mathcal{T}(\xi)$ is stable, that is, $\mathcal{T}(\xi \oplus 2) = \mathcal{T}(\xi)$. Hence, $\mathcal{T}(\xi)$ cannot depend on the Euler class χ. These properties of $\mathcal{T}(\xi)$ determine this class completely [since a symmetric polynomial cannot have two different expressions in terms of $e_1(x_1, \dots)$ and $e_i(x_1^2, \dots)$]. It remains to use the equality

$$\frac{1 - e^{-x}}{x} = e^{-x/2}\frac{\sinh(x/2)}{x/2}.$$

Here is a computation of several first terms of the series $A(p)$ and $A(p)^{-1}$:

$$A(p) = 1 + \frac{1}{24}p_1 + \frac{1}{5760}(3p_1^2 + 4p_2) + \dots,$$
$$A(p)^{-1} = 1 - \frac{1}{24}p_1 + \frac{1}{5760}(7p_1^2 - 4p_2) + \dots.$$

Now we combine the general theorems of Sect. 41.1 with the results of the calculations above to arrive at the main results of this section: Riemann–Roch theorems for Thom isomorphisms and direct images.

Theorem. *Let ξ be a complex spinor vector bundle with a finite* CW *base X. Then for any $\alpha \in K(X)$,*

$$\mathrm{ch}(\mathbf{t}_H^{\xi}(\mathrm{ch}\,\alpha \cdot e^{c(\xi)/2}A(p_1(\xi), p_2(\xi), \dots).$$

Theorem. *Let X, Y be manifolds whose normal bundles $\nu(X), \nu(Y)$ in a sphere of a large dimension are furnished with a complex spinor structure; let $f\colon X \to Y$ be a continuous map. Then for any $\alpha \in K(X)$,*

$$f_!(\operatorname{ch}\alpha \cdot e^{-c(\nu(X))}A(\bar{p}_i(X)) = \operatorname{ch}(f_!\alpha) \cdot e^{-c(\nu(Y))}A(\bar{p}_i(Y))$$

[$\bar{p}_i(X) = p_i(\nu(X))$, $\bar{p}_i(Y) = p_i(\nu(Y))$ are normal Pontryagin classes of manifolds X, Y].

The reader may find a number of other propositions of this kind in the book by Karoubi, Chap. 5, Sect. 4.

42.5 First Application: Integrality Theorems

The last theorem is highly nontrivial even in the case when Y is a point.

Theorem. *Let X be a closed smooth oriented manifold with integral $w_2(X)$, and let $c \in H^2(X;\mathbb{Z})$ be some class whose image in $H^2(X;\mathbb{Z}_2)$ is $w_2(X)$. Then for any $\alpha \in K(X)$, the value of the class*

$$e^{-c/2}\operatorname{ch}\alpha \cdot A^{-1}(p_1(X), p_2(X), \dots)$$

on the fundamental class $[X]$ of the manifold X is integral.

Proof. Apply the last theorem of the last section to the map $f: X \to \text{pt}$. We have

$$\begin{aligned}\mathbb{Z} \in \dim f_!\alpha &= \operatorname{ch} f_!\alpha = (\operatorname{ch} f_!\alpha)\cdot \mathcal{T}(\text{pt})\\ &= f_!(\operatorname{ch}\alpha \cdot e^{-c(\nu(X))/2}A(\bar{p}_1(X), p_2(X), \dots)\\ &= f_!(e^{-c/2}\cdot \operatorname{ch}\alpha \cdot A^{-1}(p_1(X), p_2(X), \dots)\\ &= \langle e^{-c/2}\cdot \operatorname{ch}\alpha \cdot A^{-1}(p_1(X), p_2(X), \dots), [X]\rangle\end{aligned}$$

[we have used the equality $A(p) = a^{-1}(\overline{p})$ which follows from the obvious multiplicativity of the class $\grave{A}$: $A(\xi \oplus \eta) = A(\xi)A(\eta)$].

Put in the theorem $\alpha = 1$ (the case when $\alpha \neq 1$ is also very interesting, as we will see in the next section). We get a theorem of divisibility for characteristic numbers of the manifold with the class w_2 being integral.

Examples. Let X be a closed oriented manifold, and let $c \in H^2(X;\mathbb{Z})$ be an arbitrary cohomology class whose image in the cohomology mod 2 is w_2. If $\dim X = 4$, then $p_1 - 3c^2$ is divisible by 24. If $\dim X = 6$, then $c^3 - cp^1$ is divisible by 48. If $\dim X = 8$, then $15c^4 + 30p_1c^2 + 7p_1^2 - 4p_2$ is divisible by 5760. [The divisibility holds in the group $H^{\dim X}(X;\mathbb{Z})$.]

If we use the real K-functor instead of the complex K-functor, then we can get a small but essential improvement of these results. Namely, the orientability of an orientable real vector bundle with respect to the real K-theory is equivalent to the reducibility of the structure group—not to the group $\text{Spin}^{\mathbb{C}}$, but to the smaller group $\text{Spin}^{\mathbb{R}}$. The latter means that the class w_2 needs not just to be integral but rather to be 0. On the other hand, the real Todd class is equal to one half of the complex Todd class, which lets us enhance the divisibility theorems in the case $w_2 = 0$ by the additional factor 2.

Examples. Let X be a closed oriented manifold with $w_2 = 0$. If $\dim X = 4$, then p_1 is divisible by 48. If $\dim X = 8$, then $7p_1^2 - 4p_2$ is divisible by 11520.

At first glance, this supplement does not look very essential. To illustrate the opposite, consider the case when X is a closed simply connected four-dimensional manifold with $w_2(X)$. The Wu formulas for Stiefel–Whitney classes (see Sect. 30.2) easily imply that the last statement is equivalent to the triviality of the homomorphism $\mathrm{Sq}^2: H^2(X;\mathbb{Z}_2) \to H^4(X;\mathbb{Z}_2)$, which is the plain squaring. In the language of integral homology this means that the square of an arbitrary two-dimensional cohomology class is even; that is, the diagonal entries of the matrix of the intersection form $H_2(X;\mathbb{Z}_2) \times H_2(X;\mathbb{Z}_2) \to \mathbb{Z}$ are all even. There is a term in algebra which describes this situation: an *even form*. There exists Serre's theorem, which states that the signature of an even unimodular form is divisible by 8. But it is known (see Sect. 19.5) that the signature of this form is equal to $\langle p_1(x), [X]\rangle/3$. Hence, our complex integrality theorem only repeats the algebraic result, while the more subtle real theorem gives rise to the following highly important result.

Theorem. *The signature of the intersection form of a closed simply connected smooth four-dimensional manifold X with $w_2(X) = 0$ is divisible by* 16.

This is the famous Rokhlin's theorem that was proved in 1952. It shows that not any integral unimodular quadratic form can be the intersection form of a smooth closed simply connected manifold; this fact has numerous applications in topology. Notice that until 1982 there were no known integral unimodular quadratic forms which could not be intersection forms of smooth closed four-dimensional manifolds but which were not prohibited by Rokhlin's theorem. Such forms became known in a great variety after the work of Donaldson, who proved that if the intersection form of a smooth closed 4-manifold is positive definite, then it can be reduced to the sum of squares over $\mathbb{Z}$.

42.6 Second Application: Theorems of Nonembeddability

The Riemann–Roch theorem and the integrality theorem have the following generalization. Let $\tau: h \to k$ be a multiplicative homomorphism of one cohomology theory into another, and let ξ be a vector bundle which is oriented with respect to the theory k but, in general, is not even orientable with respect to h. Let $u_k \in \widetilde{k}(T(\xi))$ be a k-orientation, and let $u_h \in \widetilde{h}(T(\xi))$ be arbitrary. These u_k and u_h give rise to, correspondingly, the *Thom isomorphism* $\mathbf{t}_k$ and the *Thom homomorphism* $\mathbf{t}_h$, and it should be remarked that the latter, as well as the former, possesses the multiplicative property.

Theorem. *For any $\alpha \in h(X)$ (where X is the base of the bundle ξ),*

$$\mathbf{t}_k^{-1} \circ \tau \circ \mathbf{t}_h(\alpha) = \tau(\alpha)\mathcal{T}(\xi),$$

where $\mathcal{T}(\xi) = \mathbf{t}_k^{-1} \circ \tau \circ \mathbf{t}_h(1) \in k(X)$.

The proof is as before (see Sect. 41.1).

The statements of the Riemann–Roch theorem for the direct image and of the integrality theorem are precisely as above, and we do not repeat them.

Consider the following example. The group $SO(2k)$ has a unitary representation in $\Lambda^*\mathbb{C}^{2k}$. There is an equivariant operator

$$*: \Lambda^r\mathbb{C}^{2k} \to \Lambda^{2k-r}\mathbb{C}^{2k},$$

in particular,

$$*: \Lambda^k\mathbb{C}^{2k} \to \Lambda^k\mathbb{C}^{2k}.$$

The space $\Lambda^k\mathbb{C}^{2k}$ is decomposed into the direct sum $\Lambda^k_+\mathbb{C}^{2k} \oplus \Lambda^k_-\mathbb{C}^{2k}$ of eigenspaces of $*$ corresponding to the eigenvalues 1 and -1. Set

$$T_0 - T_1 = \mathbb{C} - \Lambda^1\mathbb{C}^{2k} + \Lambda^2\mathbb{C}^{2k} - \cdots + (-1)^k\Lambda^k_+\mathbb{C}^{2k}.$$

This is a "difference" between two unitary representations of the group $SO(2k)$. It is easy to understand that these two representations are equivalent as representations of $SO(2k-1)$. Hence, for any $2k$-dimensional oriented vector bundle $\xi = (p: E \to X)$, we obtain, in complete analogy with previous constructions, an element

$$v(\xi) = T_0 p^*\xi - T_1 p^*\xi \in K(E, E - X) = \widetilde{K}(T\xi).$$

Denote by $B(\xi)$ the counterpart of the Todd class for $v(\xi)$ and ch. A computation (a replica of the computation above) shows that

$$B(\xi) = \frac{1}{2}\left[\prod_{i=1}^{k} \frac{e^{x_i} - e^{-x_i}}{x_i} + \prod_{i=1}^{k} \frac{(e^{x_i} - 1)(e^{-x_1} - 1)}{x_1}\right],$$

$$p_i(\xi) = e_i(x_1^2, \ldots, x_k^2), \ \chi(\xi) = x_1 \ldots x_k.$$

EXERCISE 7. Using this computation, show that $v(\xi)$, in general, is not an orientation.

All the results of this section taken together give rise to the following result.

Theorem. *The number*

$$\langle \operatorname{ch} \alpha \cdot B(\nu(X)), [X] \rangle$$

is integral for any $\alpha \in K(X)$.

Let us introduce now some notation. For $z \in H^{\text{even}}(x;\mathbb{Q})$, $z = z_0 + z_1 + \dots$, $z_i \in H^{2i}(X;\mathbb{Q})$, set $z(t) = \Sigma_q t^q z_q$; recall that if $z = \operatorname{ch}\alpha$ and $t \in \mathbb{Z}$, then $z^{(t)} = \operatorname{ch}\psi^t\alpha$ (see Sect. 40.1.C). For a closed $2n$-dimensional manifold X, a class $c \in H^2(X;\mathbb{Q})$, and a virtual bundle $\alpha \in K(X)$, put

$$H(t) = \langle e^{c/2} \cdot (\operatorname{ch}\alpha)^{(t)} \cdot A^{-1}(p_1(X), p_2(X), \dots), [X] \rangle.$$

This is a polynomial of degree n in t with rational coefficients; its coefficient at t^n is equal to $\langle \operatorname{ch}_n \alpha, [X] \rangle$. The theorem of Sect. 41.5 [with the remark that $(\operatorname{ch}\alpha)^{(t)} = \operatorname{ch}\psi^t\alpha$] shows that the polynomial $H(t)$ takes integral values at integral points. In addition, the last integrality theorem yields the following result.

Theorem. *If the manifold X admits a smooth embedding in S^{2n+2k}, then the number*

$$2^{n+k-1} H(1/2)$$

is integral.

Proof. As we proved in Sect. 19.6, the Euler class of the normal bundle of a manifold embedded into a sphere is always equal to zero. Hence, in the expression for $B(\nu(X))$, the second term, namely,

$$\prod \frac{(e^{x_i} - 1)(e^{-xi} - 1)}{x_i},$$

equals 0: It is divisible by $x_1 \dots x_k$. Hence, in the expression for $B(\nu(X))$ the second term may be removed, and we see that

$$\operatorname{ch}\alpha \cdot B(\nu(X)) = (-1)^k 2^{n+k-1} H(1/2).$$

[Indeed,

$$\frac{\operatorname{ch}\alpha}{2} \prod \frac{e^{x_i} - e^{-x_i}}{x_i} \quad \text{and} \quad 2^{n+k-1} (\operatorname{ch}\alpha)^{1/2} \prod \frac{\sinh(x_i/2)}{x_i/2}$$

have the same terms of degree n (where the formal variables x_i have degree 2).]

Now we can prove a mighty nonembeddability theorem.

Theorem. *Let X be a smooth closed manifold of dimension $2n > 2$ with $w_3^{\mathbb{Z}} = 0$, and let there exist an $\alpha \in K(X)$ such that the integer $\langle n! \operatorname{ch}_n \alpha, [X] \rangle$ is odd. Then X cannot be smoothly embedded in $S^{4n-\delta(n)}$, where $\delta(n)$ is the number of digits 1 in the binary presentation of the number n.*

Proof. Consider the polynomial $H(t)$ that corresponds to an arbitrary c from $\rho_2^{-1}(w_2(X))$ and the α from the theorem. According to the remark above, this polynomial takes integral values for integral t. Hence,

$$H(t) = a_n \frac{t(t-1)\dots(t-n+1)}{n!} + a_{n-1}\frac{t(t-1)\dots(t-n+2)}{(n-1)!} + \dots + a_2\frac{t(t-1)}{2} + a_1 t + a_0$$

with integral a_i. Since $a_n = \langle n!\, \mathrm{ch}_n\, \alpha, [X]\rangle$ (see above), then a_n is odd. Furthermore, according to the previous theorem, $2^{n+k-1}H(1/2)$ is an integer. On the other hand, the expression for H given above shows that

$$2^{n+k-1}H(1/2) = 2^{k-1}(a_n b + nc)/n!,$$

where b is odd and c is even. But it is well known that the prime decomposition of $n!$ contains precisely $n - \delta(n)$ factors 2. Hence, $k - 1 \geq n - \delta(n)$, which completes the proof.

In the statement of the theorem, the condition of the existence of α holds, in particular, if there exists such $d \in H^2(X;\mathbb{Z})$ that $\langle d^n, [X]\rangle$ is odd: In this case we take for α the stable class of the one-dimensional vector bundle ξ with $c_1(\xi) = d$, and we will have $\mathrm{ch}\,\xi = e^{c_1(\xi)} = e^d$ and $\mathrm{ch}_n\,\xi = d^n/n!$. In particular, the theorem holds for complex projective spaces:

Corollary. *$\mathbb{C}P^n$ cannot be embedded in $S^{4n-2\delta(n)}$.*

To estimate the strength of this result, let us compare it with the nonembeddability theorems of Sect. 19.6. For example, according to the results of Lecture 19, $\mathbb{C}P^{1000}$ cannot be embedded in S^{2046}, and according to the results here it cannot be embedded in S^{3988}. (Let us add that the general results of differential topology provide an embedding of $\mathbb{C}P^{1000}$ into S^{3999}.) All the results of this section belong to Atiyah and Hirzebruch; they are contained in their article [18].

42.7 Conclusion: The Origin of the Name

The term "Riemann–Roch theorem" is known in algebraic geometry. In the most classical way, it refers to the formula

$$r(-D) - i(D) = d(D) - g + 1,$$

where D is a divisor on a Riemannian surface of the genus g, $d(D)$ is the degree of D, $r(-D)$ is the dimension of the space of meromorphic functions with the divisor greater than or equal to $-D$, and $i(D)$ is the dimension of the space of meromorphic 1-forms with the divisor greater than or equal to D. This theorem may be regarded as the computation of the Euler characteristic for the sheaf of holomorphic line bundle over X determined by the divisor D. The initial version of the integrality theorem of Sect. 42.5 was a generalization of the classical Riemann–Roch theorem; see Atiyah and Hirzebruch [17, 19]:

Theorem. *Let X be a smooth complex projective algebraic variety, let c be its first Chern class, and let ξ be a holomorphic bundle over X. Then the value of the cohomology class*

$$e^{c/2} \cdot \operatorname{ch} \xi \cdot A^{-1}(p_1(X), p_2(X), \dots)$$

on [X] is equal to the Euler characteristic of the sheaf of holomorphic sections of ξ.

When this theorem was translated from the holomorphic setting into the differential one, the Euler characteristic disappeared, only the integrality statement remained. But the theorem was still called a Riemann–Roch theorem, and even the more general theorems of Sect. 41.1 were given this label.

It should be mentioned that of the creators of K-theory only Adams was a "pure topologist"; Bott was known by his works in differential geometry and global calculus of variations; Atiyah, Hirzebruch, and especially Grothendieck were famous in algebraic geometry. This showed itself in the terminology accepted in K-theory. Besides "Riemann–Roch theorem," the term "K-theory" has an origin in algebraic geometry. Moreover, Atiyah and Hirzebruch called the polynomial $H(t)$ of Sect. 42.6 the Hilbert polynomial of the variety X, for it is the Hilbert polynomial of X in the case when X is a complex projective algebraic variety, c is its first Chern class, and α is the restriction to X of the Hopf bundle over the ambient complex projective space.

Lecture 43 The Atiyah–Singer Formula: A Sketch

The Atiyah–Singer formula for the index of an elliptic operator is one of the most significant mathematical results of the second half of the 20th century. Its importance is not restricted to its numerous applications in mathematics and physics. It reminded experts in analysis of the necessity of studying topology and to topologists of the necessity of studying analysis, and in this way it promoted the rebirth of mathematics as a united discipline.

There are books devoted to the Atiyah–Singer theorem (the best known is Palais [66]). The goals of this lecture are more modest: to stimulate the reader's interest and to give a preliminary acquaintance with the subject.

43.1 Elliptic Operators and Their Indices

Let U be a domain in $\mathbb{R}^n$, let $\mathcal{C}^\infty(U)$ be the space of complex-valued $\mathcal{C}^\infty$-functions on U, and let

$$D: \mathcal{C}^\infty(U) \to \mathcal{C}^\infty$$

be a differential operator of order k. [The last assumption can be easily formulated in the coordinate language, but it also has an invariant description: If functions $f, g \in C^\infty(U)$ have, at some point $x \in U$, equal k-jets, then $Df(x) = Dg(x)$.]

[Let us recall the definition of a jet. If X, Y are manifolds and $x \in X$, $y \in Y$, then a k-jet with the source x and target (value) y is defined as a class of C^∞-maps $f\colon (X,x) \to (Y,y)$ with respect to the following equivalence relation: $f \sim g$ if $\|f(z)-g(z)\| = O(\|z-x\|)^k$, where the norms are taken with respect to arbitrary local coordinate systems at x and y.]

Thus, we assume that D is an operator of order k (and is not an operator of order $k-1$). The *symbol* (or *principal symbol*) σ_D of D is the function on the space T^*U of the cotangent bundle $\tau^*(U)$ of the domain U. In coordinates, the symbol has the following description: If

$$D = \sum_{k_1+\dots+k_n=k} \varphi_{k_1\dots k_n}(x)\frac{\partial^k}{\partial x_1^{k_1}\dots\partial x_n^{k_n}} + \text{ terms of orders } < k,$$

then

$$\sigma_D(x_1,\dots,x_n;\xi_1,\dots,\xi_n) = \sum_{k_1+\dots+k_n=k} \varphi_{k_1\dots k_n}(x)\xi_1^{k_1}\dots\xi_n^{k_n},$$

where $\xi_i = dx_i$ are coordinates in the cotangent space T^*X. It is important that the symbol also has a coordinate-free description. Namely, let us identify the space T^*X with the space $J^1(U,\mathbb{R})_0$ of 1-jets with the value 0. Then

$$\sigma_D(x,y) = \frac{1}{k!}Df^k(x),$$

where $y \in J^1(U,\mathbb{R})_0$ and f is an arbitrary representative of the jet y.

This invariant definition of the symbol makes possible its generalization to differential operators $C^\infty(X) \to C^\infty(X)$, where X is an arbitrary smooth manifold; in this case the symbol is a function on the space T^*X of the cotangent bundle $\tau^*(X)$ of the manifold X. A further generalization: Let ξ, η be complex vector bundles with the base X. The statement that

$$D\colon C^\infty(\xi) \to C^\infty(\eta)$$

[where $C^\infty(\xi)$ and $C^\infty(\eta)$ are spaces of smooth sections] is a differential operator of order k has an obvious meaning. Let $p\colon T^*X \to X$ be the projection of the fibration $\tau^*(X)$. Then the symbol of the operator D is the vector bundle map

$$\sigma_D\colon p^*\xi \to p^*\eta,$$

which is defined by the formula

$$\sigma_D(y, z) = \left(y, \frac{1}{k!}[D(f^k g)](x)\right),$$

where $y \in I(X, \mathbb{R})_0$ is a cotangent vector to X at the point $x \in X, f \in C^\infty(X)$ is a representative of the jet y, $z \in \xi_x$ is a point of the fiber of the bundle ξ over x, and g is a section of the bundle ξ such that $g(x) = z$. Thus, the symbol σ_D maps linearly the fiber of the bundle $p^*\xi$ over a cotangent vector of the manifold X into the fiber of the bundle $p^*\eta$ over the same cotangent vector. Obviously, over the zero section $X \subset T^*X$ the symbol σ_D vanishes.

The operator D is called *elliptic* if for every nonzero cotangent vector y the operator σ_D yields an isomorphism $(p^*\xi)_y \to (p^*\eta)_y$.

The following statement is proved in analysis (see Palais' book):

Theorem. *An elliptic operator on a compact manifold X has a finite-dimensional kernel and a finite-dimensional cokernel.*

An operator with finite-dimensional kernel and finite-dimensional cokernel is called a *Fredholm operator*. The index $\operatorname{ind} D$ of a Fredholm, in particular, elliptic, operator D is defined by the formula

$$\operatorname{ind} D = \dim \operatorname{Ker} D - \dim \operatorname{Coker} D.$$

This definition is justified by the fact that in the process of a deformation of an operator in the class of Fredholm operators the dimensions of kernel and cokernel may vary, but the index stays unchanged. This fact led to a general problem formulated in the 1950s by I. M. Gelfand: Express the index of an elliptic operator in topological terms.

The Atiyah–Singer formula provided a solution of this general Gelfand's problem.

Before stating the formula, we will consider several examples.

43.2 Examples

All examples considered here and in Sect. 43.4 assume some knowledge of the de Rham theory not presented in this book. A reader who is not familiar with this theory will have either to skip this section and Sect. 43.4 or, what is better, to study this theory. It is presented in many books, for example, in *Differential Forms in Algebraic Topology* by Bott and Tu [23], or in *Modern Geometry. Methods of Homology Theory* by Dubrovin, Fomenko, and Novikov [35].

A. Let X be a closed oriented smooth $2n$-dimensional manifold with a fixed Riemannian metric (that is, with fixed Euclidean structures in the fibers of the tangent bundles depending smoothly on a point of the base). Denote by $\Omega^k X$ the

space of C^∞ exterior differential forms of degree k. There are operators

$$\begin{aligned} d&: \Omega^k X \to \Omega^{k+1} X, \\ *&: \Omega^k X \to \Omega^{2n-k} X, \\ \delta&: \Omega^k X \to \Omega^{k-1} X, \end{aligned}$$

where d is the exterior differential, $*$ is a fiberwise operator defined by the formula

$$[*\varphi](v_1, \dots, v_{2n-k}) = \varphi(w_1, \dots, w_k) \cdot \frac{\mathrm{vol}(v_1, \dots, 2v_{n-k}, w_1, \dots, w_k)}{\mathrm{vol}(w_1, \dots, w_k)},$$

where $v_1, \dots, v_{2n-k}$ are arbitrary tangent vectors to X and $w_1, \dots, w_k$ are arbitrary linearly independent tangent vectors orthogonal to $v_1, \dots, v_{2n-k}$ (check that the right-hand side of this formula does not depend on the choice of $w_1, \dots, w_k$), and $\delta = * \circ d \circ *$. Obviously, d and δ are adjoint operators.

Consider the operator

$$d + \delta: \Omega^* X \to \Omega^* X.$$

EXERCISE 1. Prove that $d + \delta$ is an elliptic operator.

There is no sense in computing the index of this operator: It is self-adjoint, its kernel and cokernel are adjoint to each other, hence, they have the same dimension, and the index is zero. To straighten the things, let us consider the operator

$$(d + \delta)|_{\Omega^{\mathrm{even}} X}: \Omega^{\mathrm{even}} X \to \Omega^{\mathrm{odd}} X.$$

Obviously, the form $\varphi \in \Omega^{2k} X$ belongs to the kernel of the operator $d + \delta$ if and only if $d\varphi = 0$ and $\delta\varphi = 0$. Differential forms with these properties are called *harmonic*. There is a well-known *Hodge theorem*: Every de Rham cohomology class contains one and only one harmonic form (that is, for every closed form ψ, there exists a unique harmonic form φ such that $\psi - \varphi \in \mathrm{Im}\, d$). This lets us (in virtue of de Rham's theorem) make an identification

$$\mathrm{Ker}(d + \delta)|_{\Omega^{\mathrm{even}} X} = H^{\mathrm{even}}(X; \mathbb{R}).$$

The cokernel $\mathrm{Coker}(d + \delta)|_{\Omega^{\mathrm{even}} X}$ is the kernel of the adjoint operator, which is nothing but $(d + \delta)|_{\Omega^{\mathrm{odd}} X}: \Omega^{\mathrm{odd}} X \to \Omega^{\mathrm{even}} X$. Hence,

$$\mathrm{Coker}(d + \delta)|_{\Omega^{\mathrm{even}} X} = H^{\mathrm{odd}}(X; \mathbb{R})$$

and

$$\mathrm{ind}(d + \delta)|_{\Omega^{\mathrm{even}} X} = \dim H^{\mathrm{even}}(X; \mathbb{R}) - \dim H^{\mathrm{odd}}(X; \mathbb{R}) = \chi(X).$$

B. Obviously, $(*|_{\Omega^k X})^2 = (-1)^k$. Consider the operator

$$\alpha = * \cdot i^{k(k-1)-n} \colon \Omega^k X \to \Omega^{2n-k} X.$$

Obviously, $\alpha^2 = 1$. Put

$$\Omega^*_{\pm} X = \{\varphi \in \Omega^* X \mid \alpha(\varphi) = \pm\varphi\}.$$

EXERCISE 2. Prove that $(d + \delta)\left(\Omega^*_{\pm} X\right) \subset \Omega^*_{\mp} X$.

EXERCISE 3. Prove that if n is odd, then the index of the operator $(d + \delta)|_{\Omega^*_{+} X}$ is 0, and if n is even, then this index is equal to the signature of the intersection form of the manifold X.

43.3 The Formula

Let

$$D \colon \mathcal{C}^{\infty}(\xi) \to \mathcal{C}^{\infty}(\eta)$$

be an elliptic operator, and let

$$\sigma_D \colon p^*\xi \to p^*\eta$$

be its symbol (we use notations from Sect. 43.1). Since σ_D is an isomorphism over the complement to the zero section, we can consider the difference $p^*\eta - p^*\xi$ as an element from $\widetilde{K}(T(\tau^*(X)))$ (compare with Sects. 42.2, 42.4, and 42.6). We denote this element by Σ_D and put

$$\operatorname{ch} D = \mathbf{t}_H^{-1} \operatorname{ch} \Sigma_D \in \begin{cases} H^{\text{even}}(X; \mathbb{Q}), & \text{if } \dim X \text{ is even,} \\ H^{\text{odd}}(X; \mathbb{Q}), & \text{if } \dim X \text{ is odd.} \end{cases}$$

Theorem.

$$\operatorname{ind} D = \left\langle (\operatorname{ch} D \cdot \mathcal{T}(\mathbb{C}\tau(X))^{-1}), [X] \right\rangle,$$

where $\mathcal{T}(\mathbb{C}\tau(X))$ denotes the Todd class (see Sect. 41.2) of the complexification of the tangent bundle. In other words,

$$\mathcal{T}(\mathbb{C}\tau(X))^{-1} = U(p_1(X), p_2(X), \dots),$$

where the power series U is defined by the formula

$$U(e_1(x_i^2), e_2(x_i^2), \dots) = -\prod \frac{x_i^2}{(1-e^{-x_i})(1-e^{x_i})}.$$

There is some apparent similarity between this theorem and the Riemann–Roch theorem. In any case, the integrality of the right-hand side can be established by methods developed in Lecture 42 (see Sect. 42.6). But certainly the main thing in this theorem is the fact that this right-hand side is equal to the index of the operator D.

We will not discuss here the proof of this theorem, we will just present a general plan. It is not hard to prove a "cobordism invariance" of index: If the bundles ξ and η over X can be extended to bundles $\widetilde{\xi}$ and $\widetilde{\eta}$ over a compact manifold W with the boundary X, and the elliptic operator D can be extended to an elliptic operator $\widetilde{D}: C^\infty(\widetilde{\xi}) \to C^\infty(\widetilde{\eta})$, then the index of the operator D equals zero. In addition to that, the index of the elliptic operator is additive (if $X = X_1 \coprod X_2$ and $D = D_1 \coprod D_2$, then $\operatorname{ind} D = \operatorname{ind} D_1 + \operatorname{ind} D_2$) and multiplicative (if $X = X_1 \times X_2$ and $D = D_1 \otimes D_2$, then $\operatorname{ind} D = \operatorname{ind} D_1 \cdot \operatorname{ind} D_2$). All this lets us reduce the general problem of computing the index to the problem of computing the index of several explicitly defined operators (in the spirit of the examples in Sect. 43.2), and this can be done without serious difficulties.

43.4 Back to Examples

1. The symbol of the operator $d+\delta$ (considered as defined on the whole space Ω^*X) has the following description. Over a cotangent vector $v \in T_x^*X$ to the manifold X at the point x, it is the operator $\varphi_v: (\Lambda^*T_x^*X)\otimes\mathbb{C} \to (\Lambda^*T_x^*X)\otimes\mathbb{C}$ (see Sect. 42.4). Let $D = (d+\delta)|_{\Omega^{\text{even}}X}$. The comparison of the construction of Σ_D with the construction in Sect. 42.2 shows that Σ_D is precisely the canonical K-orientation $u(\mathbb{C}\tau^*(X)) \in \widetilde{K}(T(\mathbb{C}\tau^*(X)))$ of the complex vector bundle $\mathbb{C}\tau^*(X)$ restricted from $T(\mathbb{C}\tau^*(X))$ to $T(\tau^*(X))$.

EXERCISE 4. Let ξ be an arbitrary oriented (in the usual sense) real vector bundle, let $\mathbb{C}\xi$ be its complexification, and let $\eta: T(\xi) \to T(\mathbb{C}\xi)$ be the embedding induced by the canonical embedding $\xi \to \mathbb{C}\xi$. Then, for every $\alpha \in H^*(T(\mathbb{C}\xi); G)$, the following equality holds:

$$\left(\mathbf{t}_H^{\xi}\right)^{-1} \eta^*\alpha = \left(\mathbf{t}_H^{\mathbb{C}\xi}\right)^{-1}(\alpha)\cdot\chi(X).$$

Let us now apply the Atiyah–Singer formula.

$$\begin{aligned}
\mathrm{ind}_D &= \langle (\mathrm{ch}_D \cdot \mathcal{T}(\mathbb{C}\tau^*(X))^{-1}), [X]\rangle \\
&= \langle ((\mathbf{t}_H^{\tau^*(X)})^{-1} \,\mathrm{ch}\, \Sigma_D \cdot \mathcal{T}(\mathbb{C}\tau^*(X))^{-1}), [X]\rangle \\
&= \langle ((\mathbf{t}_H^{\tau^*(X)})^{-1} \,\mathrm{ch}\, \eta^* u(\mathbb{C}\tau^*(X)) \cdot \mathcal{T}(\mathbb{C}\tau^*(X))^{-1}), [X]\rangle \\
&= \langle ((\mathbf{t}_H^{\tau^*(X)})^{-1} \eta^* \,\mathrm{ch}\, u(\mathbb{C}\tau^*(X)) \cdot \mathcal{T}(\mathbb{C}\tau^*(X))^{-1}), [X]\rangle \\
&= \langle (\chi(\tau^*(x)) \cdot (\mathbf{t}_H^{\tau^*(X)})^{-1} \,\mathrm{ch}\, u(\mathbb{C}\tau^*(X)) \cdot \mathcal{T}(\mathbb{C}\tau^*(X))^{-1}), [X]\rangle \\
&= \langle (\chi(\tau^*(x)) \cdot \mathcal{T}(\mathbb{C}\tau^*(X)) \cdot \mathcal{T}(\mathbb{C}\tau^*(X))^{-1}, [X]\rangle \\
&= \langle \chi(\tau^*(X)), [X]\rangle = \chi(X).
\end{aligned}$$

Thus, the calculation based on the Atiyah–Singer formula yields the same result as the direct calculation in Sect. 43.2.A.

2. The calculation of the index of the elliptic operator considered in Sect. 43.2.B based on the Atiyah–Singer formula provides an expression of this index, that is, of the signature of the manifolds, via the Pontryagin numbers. The reader can check that the formulas arising are precisely those obtained in Sect. 19.6.D.

3. Let X be a complex manifold, and let η be a holomorphic vector bundle with the base X. Consider the space $\Omega^{p,q}(X;\eta)$ of differential forms of type p,q on X with the coefficients in sections of the bundle η and "anti-holomorphic differential" $d''\colon \Omega^{p,q}(X;\eta) \to \Omega^{p,q+1}(X;\eta)$. By Dolbeault's theorem (see, for example, Chern's book [30]) the cohomology of the sheaf $C^\omega(\eta)$ of germs of holomorphic sections of the bundle η can be calculated by the complex

$$\cdots \to \Omega^{p,q}(X;\eta) \xrightarrow{d''} \Omega^{p,q+1}(X;\eta) \to \cdots.$$

From this it can be easily deduced that the index of the operator $(d'' + \delta'')|_{\Omega^{0,\mathrm{even}}(X;\eta)}$ is just the Euler characteristic $\sum(-1)^r \dim H^r(X;\eta)$ of the sheaf $C^\omega(\eta)$. The calculation of this index using the Atiyah–Singer formula yields an expression of this Euler characteristic in terms of topological invariants, certainly the same as stated in the Atiyah–Hirzebruch theorem (see Sect. 42.7).

4. Notice in conclusion that all the integrality theorems from Lecture 42, including the theorems in Sect. 42.6, can be deduced from the Atiyah–Singer formula applied to an appropriate elliptic operator.

Lecture 44 Cobordisms

Not unlike the Atiyah–Singer theorem, the cobordism theory is the subject of whole books (see, for example, Stong [82]). In this lecture, we restrict ourselves to defining cobordisms, listing their major properties, and describing their relations to the K-theory.

44.1 Definitions

A: Bordisms

Let X be a topological space. A *(closed) n-dimensional singular manifold* of X is, by definition, a pair (M,f) where M is a closed smooth n-dimensional manifold and f is a continuous map of M into X. Singular manifolds (M_1,f_1) and (M_2,f_2) are called *bordant* if there exists a compact manifold W with the boundary $\partial W = M_1 \sqcup M_2$ and a continuous map $F\colon W \to X$ such that $F|_{M_1} = f_1$ and $F|_{M_2} = f_2$. The set of classes of bordant n-dimensional singular manifolds of X is denoted by $\Omega_n^O(X)$. The operation of the disjoint summation $\coprod$ of singular manifolds makes this set an Abelian group [in which, by the way, every element has the order 2 (or 1): the bordism $2(M,f) \sim (\emptyset, -)$ is provided by the manifold $W = M \times I$ and the composition of the projection $M \times I \to M$ and f]. A continuous map $X \to Y$ induces, in an obvious way, homomorphisms $\Omega_n^O(X) \to \Omega_n^O(Y)$, and, putting

$$\Omega_n^O(X) = 0 \text{ for } n < 0,$$
$$\widetilde{\Omega}_n^O(X) = \operatorname{Ker}(\Omega_n^O(X) \to \Omega_n^O(\mathrm{pt})),$$
$$\Omega_n^O(X,A) = \widetilde{\Omega}_n^O(X/A),$$

we compose of groups Ω_n^O a theory of extraordinary homology, which is called the *theory of (unoriented) bordisms*.

If we equip these definitions with orientations, we arrive at the definition of *oriented bordisms*, Ω_n^{SO}. Emphasize that in this definition both manifolds M and W must be oriented, and in the definition of a bordism we should assume, in addition to the equality $\partial W = M_1 \sqcup M_2$, that the orientation of M_1 must be compatible with that of W, and the orientation of M_2 must be opposite the orientation compatible with that of W. The operation of inversion in the group Ω_n^{SO} corresponds to the operation of reversion of the orientation. It does not follow from anything, and, as we will see later, it is not true that all elements of the group $\Omega_n^{SO}(X)$ must have order 2.

One more version of bordisms will arise if we assume that the manifolds M and W are endowed with compatible stable almost complex structures; that is, the bundles $\tau(M) \oplus N$ and $\tau(W) \oplus N$ (with some N) are equipped with structures of complex vector bundles. The corresponding notation is $\Omega_n^U(X)$.

We should mention that in modern topology many other versions of bordism (and cobordism) theories are considered. Some of them will be mentioned here. For further information on these theories, see the book by Stong mentioned above.

B: Thom Spectra

The bordism theories fit into the general scheme of constructing extraordinary homology theories using spectra (see Sect. 38.5.D).

Denote by $MO(n)$ the Thom space of the tautological vector bundle ξ_n over $BO(n) = G(\infty, n)$, and by $MSO(n)$ and $MU(n)$ the Thom spaces of the tautological bundles $\xi_n^+$and $\xi_n^{\mathbb{C}}$ over $BSO(n) = G_+(\infty, n)$ and $BU(n) = \mathbb{C}G(\infty, n)$. We will construct maps $\Sigma MO(n) \to MO(n+1)$, $\Sigma MSO(n) \to MSO(n+1)$, and $\Sigma^2 MU(n) \to MU(n+1)$. The restriction of the bundle ξ_{n+1} to $G(\infty.n)$ is, obviously, $\xi_n \oplus 1$; the inclusion map $\xi_n \oplus 1 \to \xi_{n+1}$ determines a map $T(\xi_n \oplus 1) \to T(\xi_{n+1})$. But it is clear that $T(\eta \oplus 1) = \Sigma T(\eta)$ for an arbitrary real vector bundle η (the same is true for complex vector bundles with Σ replaced by Σ^2). Thus, we obtain a map $\Sigma MO(n) \to M)(n+1)$; the other two maps are constructed in a similar way.

The spaces $MO(n), MSO(n), MU(n)$, with the maps constructed above, compose the spectra

$$\begin{gathered} \dots, MO(n), MO(n+1), \dots \\ \dots, MSO(n), MSO(n+1), \dots \\ \dots, MU(n), \Sigma MU(n), MU(n+1), \Sigma MU(n+1), \dots \end{gathered}$$

[the maps comprising the last spectrum are id $\Sigma MU(n)$ and our map $\Sigma(\Sigma MU(n)) = \Sigma^2 MU(n) \to MU(n+1)$]. These spectra are called *Thom spectra* and are denoted as MO, MSO, and MU.

Theorem. *The bordism theories coincide with the extraordinary homology theories with the coefficients in the Thom spectra. In other words,*

$$\begin{aligned} \widetilde{\Omega}_n^O(X) &= \lim \pi_{n+N}(X\#MO(N)), \\ \widetilde{\Omega}_n^{SO}(X) &= \lim \pi_{n+N}(X\#MSO(N)), \\ \widetilde{\Omega}_n^U(X) &= \lim \pi_{n+2N}(X\#MU(N)). \end{aligned} \qquad (*)$$

Before proving this theorem, it is appropriate to make the following remark. In the course of this book we have encountered Thom spaces in several different contexts: in connection with characteristic classes (Lecture 31), in connection with orientations of vector bundles (Lecture 42), etc. But all these things are byproducts of the Thom theory, whose main result consists in the equalities $(*)$ in the case of $X = S^0$. In this case, the equalities $(*)$ take the form

$$\begin{aligned} \widetilde{\Omega}_n^O(S^0) = \Omega^O(\text{pt}) &= \lim \pi_{n+N}(S^0\#MO(N)) = \lim \pi_{n+N}(MO(N)), \\ \Omega^{SO}(\text{pt}) &= \lim \pi_{n+N}(MSO(N)), \\ \Omega^U(\text{pt}) &= \lim \pi_{n+N}(MU(N)). \end{aligned}$$

Notice also that for sufficiently large N the groups $\pi_{n+N}(MO(N))$, $\pi_{n+N}(MSO(N))$, and $\pi_{n+2N}(MU(N))$, do not depend on N (prove that!), so in all these equalities we can drop the lim symbol.

But what is $\Omega_n^\Gamma(\text{pt})$ (for $\Gamma = O, SO, U$)? A singular manifold of a point is a pair consisting of a manifold M (possibly, with some structure) and a map $M \to \text{pt}$. But there is only one map of M into pt, so we do not need to specify this map.

The notation $\Omega_n^\Gamma(\mathrm{pt})$ is commonly abbreviated to Ω_n^Γ. Elements of the group Ω_n^O are classes of closed smooth abstract (that is, not embedded or mapped anywhere) n-dimensional manifolds with respect to the following equivalence relation: $M_1 \sim M_2$ (M_1 is *cobordant* to M_2) if there exists a compact manifold W with $\partial W = M_1 \sqcup M_2$. The sum, which is also the difference (all elements have order 2), is defined as a disjoint union. The direct product operation makes $\Omega_*^O = \bigoplus_n \Omega_n^O$ into a graded ring.

In Sect. 19.6.C we formulated Thom's results, which provided a classification of smooth closed manifolds up to a cobordism. Some partial results on a cobordism classification for oriented manifolds are formulated in Sect. 19.6.D (closed oriented manifolds M_1 and M_2 are oriented cobordant if there exists a compact oriented manifold W with $\partial W = M_1 \sqcup (-M_2)$ where $-M_2$ denotes the manifold M_2 with the opposite orientation). Now we can add to these cases the case of closed stably almost complex manifolds with an appropriate cobordism relation.

Let us prove that $\Omega_n^O = \pi_{n+N}(MO(n))$. First, we will construct a map $\Omega_n^O \to \pi_{n+N}(MO(N))$. Let X be a closed n-dimensional manifold. Embed it into S^{n+N}, where N is sufficiently large. By the results of Sect. 19.4, the normal bundle ν of the manifold X in S^{n+N} can be mapped into the tautological bundle ξ_N over $BO(N) = G(\infty, N)$, and this map is homotopically unique. There arises a map of the Thom space $T(\nu)$ into $T(\xi_N) = MO(N)$. But $T(\nu) = S^{n+N}/(S^{n+N} - U(X))$, where $U(X)$ is a tubular neighborhood of X in S^{n+N}. The composition

$$S^{n+N} \to S^{n+N}/(S^{n+N} - U(X)) = T(\nu) \to MO(N)$$

determines an element of the group $\pi_{n+N}(MO(N))$, which, as the reader can check, depends only on (N and) the class of the manifold X in Ω_n^O; thus, we obtain a map $\Omega_n^O \to \pi_{n+N}(MO(N))$. Next, let us construct a map $\pi_{n+N}(MO(N)) \to \Omega_n^O$. An element of the group $\pi_{n+N}(MO(N))$ is represented by a spheroid $\varphi\colon S^{n+N} \to MO(N)$. We can assume that the image of this map is contained in the Thom space of the tautological bundle over $G(M, N)$ with some large M; to avoid cumbersome notations, we will denote this finite-dimensional Thom space again by $MO(N)$. Furthermore, the map $\varphi\colon S^{n+N} \to MO(N)$ can be assumed smooth in the complement to the inverse image of the base point ∞ of the space $MO(N)$ ($MO(N) - \infty$ is a smooth manifold). Finally, a small perturbation of this map makes it generic with respect to the manifold $G(M, N) \subset MO(N)$ in the following sense: If $\varphi(x) \in G(M, N)$, then the image with respect to (the differential of) the map φ of the tangent space T_xS^{n+N} composes, in the sum with $T_{\varphi(x)}G(M, N)$, the whole space $T_{\varphi(x)}MO(N)$. This property of the map φ is called transversally regular, or t-regular (where t may be regarded as an abbreviation of the word "transversally" or of the name of Thom), with respect to $G(M, N)$ (see Fig. 139).

The transverse regularity implies that the inverse image $\varphi^{-1}(G(M, N) = X \subset S^{n+N}$ is a smooth submanifold of the sphere S^{n+N} whose codimension in the sphere is equal to the codimension of $G(M, N)$ is $MO(N)$), that is, is equal to N; thus, $\dim X = n$. (For details related to the t-regularity, see the books *Beginner's Course*

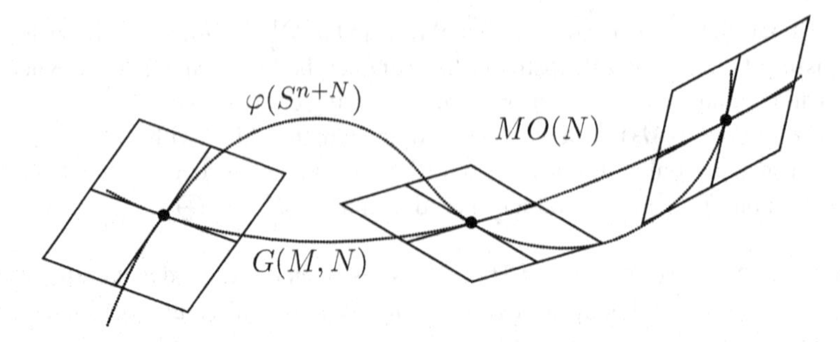

Fig. 139 t-regular map

in Topology by Fuchs and Rokhlin [40] or *Differential Topology* by Hirsch [45].) It is not hard to prove (again, we leave the details to the reader) that the class of the manifold X in Ω_n^O depends only on the original element of the group $\pi_{n+N}(MO(N))$, so we get a map $\pi_{n+N}(MO(N)) \to \Omega_n^O$. And we also leave to the reader the proof of the facts that the maps $\Omega_n^O \longleftrightarrow \pi_{n+N}(MO(N))$ constructed are mutually inverse homomorphisms.

Proofs of the equalities

$$\Omega_n^{SO} = \pi_{n+N}(MSO(N)) \text{ and } \Omega_n^U = \pi_{n+2N}(MU(N))$$

are absolutely similar (one needs only to notice that furnishing the tangent bundle of a manifold by an orientation or a stable complex structure is equivalent to endowing with an orientation or a complex structure the normal bundle of the manifold in a sphere of a sufficiently large dimension).

This completes the proof in the case $X = S^0$. The transition to the general case does not meet any serious difficulties, and we will restrict ourselves to a general outline. Let (M,f) be a closed n-dimensional singular manifold of the space X. Let us embed M into S^{n+N}. As before, we construct a map $\varphi\colon S^{n+N} \to MO(N)$ which takes the complement of a tubular neighborhood to M in S^{n+N} to the base point. The map $f\colon M \to X$ is extended, in an obvious way, to a map $\widetilde{f}\colon U \to X$, and the maps φ and $\widetilde{f}$ compose a map $S^{n+N} \to (X \times MO(N))/X$,

$$s \mapsto \begin{cases} \text{base point}, & \text{if } s \notin U, \\ (\widetilde{f}(s), \varphi(s)), & \text{if } s \in U. \end{cases}$$

The composition of this map with the projection $(X \times MO(N))/X \to X\#MO(N)$ determines an element of $\pi_{n+N}(X\#MO(N))$, and it is not hard to show that this gives rise to a map $\Omega_n^O(X) \to \pi_{n+N}(X\#MO(N))$, in particular, $\widetilde{\Omega}_n^O(X) \to \pi_{n+N}(X\#MO(N))$. The fact that it is an isomorphism can be proved directly, or deduced from the general theory of Sect. 38.5. (Consider Ω_*^O and homology with coefficients in the Thom spectrum MO as two different extraordinary homology theories; one can check that the construction above determines a homomorphism between these two theories, and then apply the uniqueness theorem from Sect. 38.5.C.)

The oriented and complex cases are similar.

C: Cobordisms

The construction of Sect. 38.5 gives not only extraordinary homology, but also extraordinary cohomology. The extraordinary cohomology theories corresponding to the Thom spectra are called theories of *cobordisms*. The notations are $\Omega_O^* = \bigoplus \Omega_O^n$, $\Omega_{SO}^* = \bigoplus \Omega_{SO}^n$, $\Omega_U^* = \bigoplus \Omega_U^n$. Thus,

$$\begin{aligned} \Omega_O^n(X,A) &= \pi(\Sigma^N(X/A), MO(n+N)), \\ \Omega_{SO}^n(X,A) &= \pi(\Sigma^N(X/A), MSO(n+N)), \\ \Omega_U^n(X,A) &= \pi(\Sigma^{2N}(X/A), MU(n+N)). \end{aligned}$$

In particular,

$$\Omega_O^n(\mathrm{pt}) = \pi(S^N, MO(n+N)) = \pi_N(MO(n+N) = \Omega_{-n}^O(\mathrm{pt}),$$

and, similarly,

$$\Omega_{SO}^n(\mathrm{pt}) = \Omega_{-n}^{SO}(\mathrm{pt}),\ \Omega_U^n(\mathrm{pt}) = \Omega_{-n}^U(\mathrm{pt}).$$

Thus, in particular, $\Omega_\Gamma^n(\mathrm{pt}) = 0$ for $n > 0, \Gamma = O, SO, U$.

EXERCISE 1. Cobordisms possess natural multiplicative structures: Reconstruct the details.

EXERCISE 2. Let X be a closed manifold of dimension m. Then $\Omega_O^n(X)$ is the group of equivalence classes of pairs (M, F) where M is a closed manifold of dimension $m - n$ and $f: M \to X$ is a continuous map. The groups $\Omega_{SO}^n(X)$ and $\Omega_U^n(X)$ are described in the same way, but with an additional condition that the virtual bundle $f^*\tau(X) - \tau(M)$ (over M) is furnished, respectively, with an orientation or a stable complex structure. In particular,

$$\begin{aligned} \Omega_n^O(X) &= \Omega_O^{m-n}(X) && \text{always}, \\ \Omega_n^{SO}(X) &= \Omega_{SO}^{m-n}(X), && \text{if } X \text{ is oriented}, \\ \Omega_n^U(X) &= \Omega_U^{m-n}(X), && \text{if the tangent bundle } \tau(X), \\ & && \text{is furnished by a stable complex structure} \end{aligned}$$

("Poincaré isomorphisms").

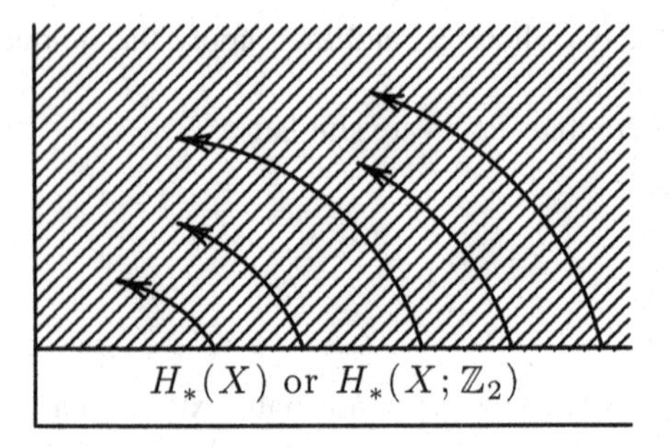

Fig. 140 Atiyah–Hirzebruch spectral sequence for Ω^O and Ω^{SO}

D: A Relation to the Usual Homology and Cohomology

The bordisms of all kinds are related to each other and to the homology by natural maps which compose the diagram

$$\begin{array}{ccccc} \Omega_n^U(X) & \longrightarrow & \Omega_n^{SO}(X) & \longrightarrow & \Omega_n^O(X) \\ \downarrow & & \downarrow & & \downarrow \\ H_n(X) & = & H_n(X) & \longrightarrow & H_n(X;\mathbb{Z}_2). \end{array}$$

The upper horizontal arrows are defined by partial or full forgetting additional structure, the lower horizontal arrow is the reduction mod 2, and the vertical arrows act according to the formula $(M,f) \mapsto f_*[M]$. These vertical homomorphisms have a transparent geometric sense. Namely, the class $\alpha \in H_n(X)$ belongs to the image of the homomorphism $\Omega_n^{SO}(X) \to H_n(X)$ if and only if it is represented by a cycle realizable as an oriented singular manifold. Moreover, if X is a manifold of dimension $> 2n$, then this singular deformation may be turned, by a small perturbation, into an embedded manifold. Thus, our homomorphism is directly related to the problem of realizing homology classes by (nonsingular) submanifolds, which we discussed in Sect. 17.2. A similar sense (but without orientations and with coefficients in $\mathbb{Z}_2$) can be attributed to the homomorphism $\Omega_n^O(X) \to H_n(X;\mathbb{Z}_2)$.

Notice also that the homomorphisms considered are also closely related with the Atiyah–Hirzebruch spectral sequence, more precisely, with its differentials. Namely, also, at the moment we know nothing of the group of bordisms of a point (we will study them in the next section); it is clear a priori that $\Omega_n^O(\text{pt}) = \Omega_n^{SO}(\text{pt}) = 0$ for $n < 0$, and $\Omega_0^O(\text{pt}) = \mathbb{Z}_2$, $\Omega_0^{SO}(\text{pt}) = \mathbb{Z}$. Therefore, the Atiyah–Hirzebruch spectral sequence for $\Omega_*^O(X)$ and $\Omega_*^{SO}(X)$ looks like what is shown in Fig. 140.

The kernels of differentials acting from $H_*(X)$ and $H_*(X;\mathbb{Z}_2)$ are quotient groups of $\Omega_*^{SO}(X)$ and $\Omega_*^O(X)$; this is how our homomorphisms $\Omega_*^{SO}(X) \to H_*(X)$ and $\Omega_*^O(X) \to H_*(X;\mathbb{Z}_2)$ are defined. Thus, the obstruction to realizability of a homology class by a singular manifold, in particular, by a submanifold, lies in the nontriviality of differentials of the Atiyah–Hirzebruch spectral sequence. See the details in two works of Buchstaber [24] and [25]. We restrict ourselves to a remark that all the differentials of the Atiyah–Hirzebruch spectral sequences have finite order, and, consequently, the homomorphisms $\Omega_*^{SO}(X) \to H_*(X)$ and $\Omega_*^U(X) \to H_*(X)$ have finite cokernels (for finite CW-complexes X).

E: Homotopy Sufficiency

As we noticed earlier (see Sects. 38.1 and 39.3), K-theory is not homotopy sufficient in the sense that a map of a simply connected finite CW complex X into a simply connected finite CW-complex Y inducing an isomorphism $K^*(X) \cong K^*(Y)$ does not have to be a homotopy equivalence. Contrary to this, the theories $\Omega_*^{SO}(X)$ and $\Omega_*^U(X)$ [as well as $\Omega_{SO}^*(X)$ and $\Omega_U^*(X)$] are homotopy sufficient.

Theorem. *If a continuous map $f: X \to Y$ of one simply connected CW complex to another one is such that one of the homomorphisms*

$$\begin{aligned} f_*&: \Omega_*^{SO}(X) \to \Omega_*^{SO}(Y), \\ f_*&: \Omega_*^U(X) \to \Omega_*^U(Y), \\ f^*&: \Omega_{SO}^*(Y) \to \Omega_{SO}^*(X), \\ f^*&: \Omega_U^*(Y) \to \Omega_U^*(X) \end{aligned}$$

is an isomorphism, then f is a homotopy equivalence.

Proof. Let us begin with the case of $f_*: \Omega_*^{SO}(X) \to \Omega_*^{SO}(Y)$. It is sufficient to prove that f induced an isomorphism in the (integral) homology. Furthermore, if we turn the map f into an embedding homotopy equivalent to this map and then use the exactness of the homology sequence of a pair, we arrive at the following, equivalent statement: If for a finite CW complex Z, $\widetilde{\Omega}_*^{SO}(Z) = 0$, which is the same as $\Omega_*^{SO}(Z) = \Omega_*^{SO}(\mathrm{pt})$, then $\widetilde{H}_*(Z) = 0$. We can assume that Z is connected. Suppose that $\widetilde{H}_*(Z) \neq 0$; let q be the smallest positive number such that $H_q(Z)$ is not zero. Then, the Hurewicz homomorphism $\pi_q(Z) \to H_q(Z)$ is an isomorphism, and hence there exists a nontrivial homology class $\alpha \in H_q(Z)$ which is represented by a spheroid $S^q \to Z$. But S^q is an orientable q-dimensional manifold, so this spheroid may be regarded as an oriented singular manifold. This singular manifold represents an element of $\widetilde{\Omega}_q^{SO}(Z)$, nontrivial since it has a nontrivial image in homology. Thus, $\widetilde{\Omega}_q^{SO}(Z) \neq 0$, a contradiction. The case of Ω^U is the same: A sphere has a stably trivial normal bundle, so the spheroid as above also represents a nontrivial element of $\widetilde{\Omega}_q^U(Z)$. The reduction of cases of f^* to the cases of f_* can be based on Exercise 2; we leave the details to the reader.

44.2 Computations

A: Ω_*^O

Theorem 1. *Ω_*^O is the ring of polynomials (over $\mathbb{Z}_2$) of variables x_i,* $\dim x_i = i$, *where i runs through positive integers not of the form $2^s - 1$, $s \geq 1$. For x_i, one can take the class of an arbitrary smooth closed i-dimensional manifold X such that*

$$\langle P_i(w_1, \ldots, w_i), [X] \rangle \neq 0,$$

where P_i is the (reduced modulo 2) polynomial which expresses the symmetric function $\sum x_k^i$ via the elementary symmetric functions (see Sect. 40.1.B) and $[X]$ is the fundamental homology class mod 2 *of X.*

This theorem belongs to Thom. Its proof runs as follows. We need to find the homotopy groups $\pi_{n+N}(MO(N)), n \ll N$. It is clear a priori that all elements of these groups have order 2. The mod 2 cohomology of $MO(N)$ is known to us: The Thom isomorphism connects it with the mod 2 cohomology of $BO(N) - G(\infty, N)$ calculated in Lecture 13 (Sect. 13.8.C). It can be easily shown with the help of the multiplication

$$H^*(MO(N); \mathbb{Z}_2) \times H^*(BO(N); \mathbb{Z}_2) \to H^*(MO(N); \mathbb{Z}_2)$$

that the cohomology $H^*(MO(N); \mathbb{Z}_2)$ is generated additively by all elements of the form $uw_{i_1}, \ldots, uw_{i_r}, 1 \leq i_1 \leq \cdots \leq i_r$, where $u \in H^N(MO(N); \mathbb{Z}_2)$ is the Thom class and $w_i \in H^i(BO(N); \mathbb{Z}_2)$ are the Stiefel–Whitney classes. The formulas

$$\operatorname{Sq}^k w = \sum \binom{m-k}{j} w_{k-j} w_{m+j} \text{ and } \operatorname{Sq}^k u = uw_k$$

(see Sect. 31.2) can be used to compute the action of the Steenrod squares in $H^*(MO(N); \mathbb{Z}_2)$. This computation yields the following result: Up to dimension $2N$, the cohomology $H^*(MO(N); \mathbb{Z}_2)$ is a free module over the Steenrod algebra $\mathbb{A}_2$, and the generators of this free module may be explicitly found (we recommend to the reader to do this computation). [Notice the fact that the $\mathbb{A}_2$-module $H^*(MO(N); \mathbb{Z}_2)$ is free has a simple proof involving almost no calculations; see Stong's book.]

Let us use the Adams spectral sequence (Chap. 5). We see that in the initial term of this spectral sequence only the first row is different from zero, and the nonzero elements in this row correspond to the free generators of the $\mathbb{A}_2$-module $H^*(MO(N); \mathbb{Z}_2)$. (The other rows contain relations between generators, relations between relations, and so on, but here are no such things, since the module is free.) Hence, in the Adams spectral sequence, up to a certain dimension, all the differentials are trivial, and this leads to an immediate computation of the homotopy groups. [Actually, this shows that up to dimension $\sim 2N$ the space $MO(N)$ is homotopy equivalent to a product of $K(\mathbb{Z}_2, N + k)$s.]

This computation implies the fact (already known to us from Lecture 19) that the class of a manifold in Ω_*^O is fully determined by its Stiefel–Whitney numbers (we leave to the reader the deduction of this fact from Theorem 1). In Sect. 19.6.D, we also noticed that the Stiefel–Whitney numbers are not independent and provided examples of relations between them. Now we can give a full list of these relations.

Theorem 2. *Let $\varphi: H^n(BO(N); \mathbb{Z}_2) \to \mathbb{Z}_2$ be a homomorphism. There exists a smooth closed n-dimensional manifold M such that for every $\alpha \in H^n(BO(N); \mathbb{Z}_2)$,*

$$\langle \alpha_M, [M] \rangle = \varphi(\alpha),$$

where α_M is a characteristic class of (the tangent bundle of) the manifold M, if and only if for every $\beta \in H^*(BO(N); \mathbb{Z}_2)$,

$$\varphi(\operatorname{Sq} \beta + \beta \operatorname{Sq}^{-1} w) = 0$$

(where $\operatorname{Sq} = 1 + \operatorname{Sq}^1 + \operatorname{Sq}^2 + \dots$ and $w = 1 + w_1 + w_2 + \dots$).

(The reader can prove this theorem for an exercise or find the proof in Stong's book.)

Let us look at what this theorem means for small dimensions. A simple computation shows that

$$\operatorname{Sq}^{-1} w = 1 + w_1 + (w_1^2 + w_2) + w_1 w_2 + (w_2^2 + w_1 w_3 + w_4) + \dots.$$

For a $\beta \in H^*(BO(N); \mathbb{Z}_2)$, we denote by $F_q(\beta)$ the component of dimension q of $\operatorname{Sq} \beta + \beta \operatorname{Sq}^{-1} w$. The last formula gives

β	$F_0(\beta)$	$F_1(\beta)$	$F_2(\beta)$	$F_3(\beta)$	$F_4(\beta)$
1	1	w_1	$w_1^2 + w_2$	$w_1 w_2$	$w_2^2 + w_1 w_3 + w_4$
w_1	0	0	0	$w_1^3 + w_1 w_2$	$w_1^2 w_2$
w_1^2	0	0	0	w_1^3	$w_1^2 w_2$
w_2	0	0	0	w_3	$w_1^2 w_2$
w_1^3	0	0	0	0	0
w_3	0	0	0	0	0
$w_1 w_2$	0	0	0	0	$w_1 w_3 + w_1^2 w_2$

By Theorem 2, every polynomial in the qth column of this table takes zero value on the fundamental class mod 2 of every smooth closed q-dimensional manifold, and this gives all relations between Stiefel–Whitney numbers. In particular, all Stiefel–Whitney numbers of one- and three-dimensional manifolds are equal to zero (for one-dimensional manifolds it is obvious; for three-dimensional manifolds it was proven by Rokhlin several years before Thom's work); if $\dim X = 2$, then $w_1^2[X] = w_2[X]$ (we already know this), and if $\dim X = 4$, then $w_1^2 w_2[X] = w_1 w_3[X] = 0$ and $w_2^2[X] = w_4[X]$.

In conclusion, we will give one of the possible descriptions of multiplicative generators x_i of the ring Ω_*^O (see Theorem 1). If i is even, then $x_i = \mathbb{R}P^i$; if $i = 2^p(2q+1) - 1, p \geq 1, q \geq 1$, then x_i is a submanifold of the product $\mathbb{R}P^m \times \mathbb{R}P^n$, $m = 2^{p+1}q, n = 2^p$, which is determined, in the homogeneous coordinates $(u_0 : \ldots : u_m), (v_0 : \ldots : v_n)$, by the equation $F(u, v) = 0$, where F is an arbitrary bilinear form of rank n. (This system was found in 1965 by Milnor; another system of generators was found in 1956 by Dold.)

B: Ω_*^{SO}

Theorem 3. *$\Omega_*^{SO} \otimes \mathbb{Q}$ is the ring of polynomials (over $\mathbb{Q}$) of variables of degree $4i$. More precisely: The homomorphism $\Omega^{SO} \to \mathbb{Z}[t_1, t_2, \ldots]$, $\deg(t_i) = 4i$ determined by the formula*

$$M \mapsto \sum_{i_1 \leq \cdots \leq i_r} p_{i_1} \ldots p_{i_r}[M] t_{i_1} \ldots t_{i_r}$$

has, in every dimension, a finite kernel and finite cokernel.

This fact, which has already been mentioned in Sect. 19.6.D, may be formulated in the following way. (1) If the dimension of an oriented manifold M is not divisible by 4, then, for some k, the manifold kM is oriented cobordant to zero. (2) If oriented manifolds M_1 and M_2 have equal dimensions divisible by 4 and have equal Pontryagin numbers, then, for some k, the manifolds kM_1 and kM_2 are oriented cobordant. (3) Every set $\{p_{i_1 \ldots i_r} \mid i_1 + \cdots + i_r = t\}$ of integers is proportional to the set of Pontryagin numbers of some oriented $4t$-dimensional manifold. Theorems 4 and 6, stated below, imply that in statements (1) and (2) one can take $k = 2$.

A proof of Theorem 3 is relatively simple: It is based on Cartan–Serre's theorem (Sect. 26.4). First of all, we already know that

$$\operatorname{rank} H_s(BSO(N)) = \begin{cases} 0, & \text{if } s \text{ is not divisible by } 4, \\ \text{the number of partitions of } t & \\ \quad \text{into sums of natural numbers,} & \text{if } s = 4t. \end{cases}$$

(Sect. 13.8.C). By the Thom isomorphism, the group $H_{N+s}(MSO(N))$ has the same rank. If we restrict ourselves to dimensions $< 2N$, we can assume that the ring of the rational cohomology of $MSO(N)$ is free (relations may appear in dimensions $\geq 2N$). Hence, for $s \ll N$, the rank of the group $\Omega_s^{SO} = \pi_{N+s}(MSO(N))$ coincides with the rank of the group $H_s(BSO(N))$. It is easy to see that the rank of the group

Ω_s^{SO} calculated this way is as stated in Theorem 3. It remains to notice that the Pontryagin numbers $p_{i_1\dots i_r}$ are linearly independent (see Sect. 19.6.D) and hence they determine a monomorphism of the space $\Omega_*^{SO}\otimes]\mathbb{Q}$ into $\mathbb{Q}[t_1, t_2, \dots]$. Since the dimensions of homogeneous components are the same, this monomorphism has to be an isomorphism.

Below, we give a survey of further results. Most are proved in Stong's book; see also bibliography in that book.

Theorem 4 (Milnor, Averbuch). *The ring Ω_*^{SO} does not have elements of an odd order.*

The most elegant proof of this result uses the Adams spectral sequence modulo an odd prime.

Theorem 5 (Milnor, Novikov). *The ring $\Omega_*^{SO}/\,\mathrm{Tors}$ is isomorphic to the ring of integral polynomials of variables of dimension $4i$, $i = 1, 2, 3, \dots$.*

(This theorem may be regarded as a more precise version of Theorem 3.)

Theorem 6 (Rokhlin, Wall). *The ring Ω_*^{SO} does not have elements of order 4.*

Theorem 7. *The quotient ring $\Omega_*^{SO}/2\Omega_*^{SO}$ may be described in the following way. Let W be a subring of the ring $\mathbb{Z}_2[x_i,\ i = 1, 2, \dots, \dim x_i = i]$ generated by x_{2k}, x_{2k-1} $(k \neq 2^j)$ and $(x_{2^j})^2$, and let $\partial\colon W \to W$ be the differential defined by the formulas $\partial x_{2k} = x_{2k-1}$, $\partial x_{2k-1} = 0$, $\partial\,(x_{2^j})^2 = 0$. Then $\Omega_*^{SO}/2\Omega_*^{SO}$ is isomorphic to $\mathrm{Ker}\,\partial$, and the image of $\mathrm{Tors}\,\Omega_*^{SO}$ in $\mathrm{Ker}\,\partial$ coincides with $\mathrm{Im}\,\partial$.*

Because of the absence in Ω_*^{SO} of the 4-torsion (Theorem 6), it is sufficient, for the proof of Theorem 7, to know $H^*(MSO(N);\mathbb{Z}_2)$ as an $\mathbb{A}_2$-module. The computation of this module was done by Novikov. He proved that the $\mathbb{A}_2$-module $H^*(MSO(N);\mathbb{Z}_2)$ coincides (up to dimension $\sim 2N$) with the $\mathbb{Z}_2$-cohomology of a product of some spaces of the type $K(\mathbb{Z}_2, N+k)$ and $K(\mathbb{Z}, N+k)$. Notice that even in stable dimensions the space $MSO(N)$ does not have a homotopy type of a product of Eilenberg–MacLane spaces: The Postnikov factors (see Sect. 37.4) are not zero, but have odd orders.

Theorem 8. *The homomorphism $\Omega_*^{SO} \to \Omega_*^O$ of forgetting orientation maps isomorphically the ring $\Omega_*^{SO}/2\Omega_*^{SO}$ onto the subring $\mathrm{Ker}\,\partial \subset \Omega_*^O = \mathbb{Z}_2[x_2, x_4, x_5, \dots]$, where x_i, $i \neq 2^j - 1$ are (appropriately chosen) generators of the ring Ω_*^O* (see Theorem 1).

As we already noted, Theorem 3 shows that an arbitrary set

$$\{p_{1_1, i_2, \dots} \mid i_1 \le i_2 \le \dots,\ i_1 + i_2 + \dots = t\}$$

becomes, after multiplication by some integer, the set of Pontryagin numbers of some $4t$-dimensional manifold. It remains to find this integer, which can be done by the following statement:

Theorem 9 (Stong). *A set $\{p_{i_1,i_2,\dots} \mid i_1 + i_2 + \dots = 4t\}$ is the set of Pontryagin numbers of a smooth oriented closed $4t$-dimensional manifold if and only if this does not contradict the integrality theorems from Sect.* 42.5.

In conclusion, we present a useful corollary from theorems given above.

Corollary. *For $n \le 3$, every smooth closed oriented n-dimensional manifold is the boundary of a smooth compact oriented $(n+1)$-dimensional manifold. Every smooth closed oriented four-dimensional manifold is oriented cobordant to a connected sum of some number of manifolds $\mathbb{C}P^2$ or $-\mathbb{C}P^2$. The sign is determined by the sign of the signature of a given manifold, and the number of summands is equal to the absolute value of the signature.*

(These results belong to Rokhlin; they were obtained way before the general theorems concerning the structure of Ω_*^{SO}.)

C: Ω_*^U

Theorem 10 (Milnor, Novikov). *The ring Ω_*^U is isomorphic to the ring of integral polynomials of generators of dimension $2i$, $i = 1, 2, \dots$. The generators are represented by some complex projective algebraic manifolds.*

Notice that after a tensor multiplication by $\mathbb{Q}$, this theorem becomes a corollary of the Cartan–Serre theorem (compare with Theorem 3). Moreover, Theorem 10 shows that the Chern numbers fully determine the class of a stably almost complex manifold in Ω_*^U.

Theorem 11 (Stong, Hattori). *A set $\{c_{i_1,i_2,\dots} \mid i_1 + i_2 + \dots = t\}$ is the set of Chern numbers of a smooth closed stably almost complex $2t$-dimensional manifold if and only if this does not contradict the integrality theorems from Sect. 42.5.*

Theorem 12 (Milnor, Novikov). *The forgetful homomorphism $\Omega_*^U \to \Omega_*^{SO}$ maps the ring Ω_*^U epimorphically onto $\Omega_*^{SO}/\,\mathrm{Tors}$. The forgetful homomorphism $\Omega_*^U \to \Omega_*^O$ maps the ring Ω_*^U epimorphically onto the subring of Ω_*^O consisting of the squares of elements of the ring Ω_*^O.*

D: Other Types of Cobordisms

Historically, the first computation of cobordism groups was done by Pontryagin, who proved, in the 1930s, that the ring Ω_*^{fr} of cobordism classes of cobordisms of "framed" manifolds, that is, manifolds with trivialized normal bundles, coincides with the (composition) ring of stable homotopy groups of spheres. (Pontryagin's work also contains a corresponding nonstable statement. See the details in his book [68].)

There exist computations of cobordism rings of many other types (see survey of results in Chap. 4 of Stong's book). Let us mention the computation of SU-cobordisms by Conner, Floyd, and Wall, and the computation of spinor cobordisms by Anderson, Brown, and Peterson (in both cases, important preliminary results belong to Novikov).

44.3 Relations to K-Theory

In accordance with the general arrangements of this chapter, we restrict ourselves to the complex version of K-theory. For a more complete and detailed presentation, see Conner and Floyd [32].

A: The Homomorphism μ

We are going to define a multiplicative map

$$\mu\colon \Omega_U^* \to K^*$$

of the theory of unitary cobordisms into the (complex) K-theory. Let $\alpha \in \widetilde{\Omega}_U^n(X)$, and let $M \gg N \gg |n| + \dim X$. This α can be represented by a map

$$f\colon \Sigma^{2N-n}X \to T(\xi),$$

where ξ is the (complex) tautological bundle over $\mathbb{C}G(M,N)$. According to Sect. 42.2, the bundle ξ possesses a canonical K-orientation $u(\xi) \in \widetilde{K}(T(\xi))$; put

$$\mu(\alpha) = f^*(u(\xi)) \in \widetilde{K}(\Sigma^{2N-n}X) = \widetilde{K}^{n-2N}(X) = \widetilde{K}^n(X).$$

Obviously, this $\mu(\alpha)$ is well defined.

EXERCISE 3. Check that μ is a multiplicative map of the theory Ω_U^* into the K-theory.

EXERCISE 4. Show that the homomorphism

$$\mu\colon \Omega_U^{-2n}(\mathrm{pt}) \to K^{-2n}(\mathrm{pt}) = \mathbb{Z}$$

assigns to the class of a $2n$-dimensional closed stably almost complex manifold M its "Todd genus"

$$\mathcal{T}(M) = \langle \mathcal{T}(\tau(M)), [M] \rangle ,$$

where $\mathcal{T}(\tau(M)) = G(c_1(M), c_2(M), \dots)$, where, in turn,

$$G(e_1(x_1, x_2, \dots), e_2(x_1, x_2, \dots), \dots) = \prod_i \frac{1 - e^{-x_i}}{x_i}.$$

EXERCISE 5. Describe the composition map

$$\Omega_U^* \xrightarrow{\mu} K \xrightarrow{\mathrm{ch}} H^*(-;\mathbb{Q}).$$

B: Chern Classes with Values in Cobordisms

Theorem. *There exists a unique function which assigns to a complex vector bundle ξ with a finite CW base X the "Chern classes" $\sigma_i(\xi) \in \Omega_U^{2i}(X)$, $i = 1, \dots, \dim \xi$ which possess the following properties (we put $\sigma(\xi) = 1 + \sigma_1(\xi) + \dots + \sigma_{\dim \xi}(\xi)$):*

(1) $\sigma(f^*\xi) = f^*\sigma(\xi)$.
(2) $\sigma(\xi \oplus \eta) = \sigma(\xi)\sigma(\eta)$.
(3) *If ξ is the (one-dimensional) Hopf bundle over $\mathbb{C}P^n$, then $\sigma(\xi) = 1 + \gamma$, where $\gamma \in \Omega_U^2(\mathbb{C}P^n)$ is the cobordism class represented by $\mathbb{C}P^{n-1}$* (see Exercise 2).

What is presented below is a sketch of a proof. The reader can find all the details in the book by Conner and Floyd cited above.

The construction which proves both existence and uniqueness of the classes σ_i is as follows. If the bundle ξ is one-dimensional, then it is induced by the Hopf bundle over $\mathbb{C}P^N$ with N large enough [indeed, $BU(1) = \mathbb{C}P^\infty$], and the class $\sigma(\xi)$ is determined by condition (3). Assume that cobordism Chern classes with required properties have already been constructed for bundles of dimension $< n$, and let $\dim \xi = n$. Consider the fibration $p: \widehat{X} \to X$ with the fiber $\mathbb{C}P^{n-1}$ associated with ξ [points of $\widehat{X}$ are passing through the origin lines in the fibers of ξ]. The bundle $p^*\xi$ splits canonically into the sum of a one-dimensional bundle and an $(n-1)$-dimensional complement: The fiber $(p^*\xi)_\ell$, where $\ell \in \widehat{X}$ is a line in the fiber ξ_x of ξ $(x = p(\ell) \in X)$, coincides with ξ_x and canonically splits as $\ell \oplus \ell^\perp$. Thus, in virtue of the induction hypothesis and condition (2), the cobordism Chern classes of $p^*\xi$ are defined. By condition (1), we must have

$$p^*(\sigma(\xi)) = \sigma(p^*(\xi)),$$

and our goal will be achieved if we show that

$$p^*: \Omega_U^*(X) \to \Omega_U^*(\widehat{X})$$

is a monomorphism. But this follows from a general fact established by Dold.

Lemma. *Let h be a multiplicative extraordinary cohomology theory and let $\pi: E \to X$ be a locally trivial fibration with the fiber F, where X and F are finite CW complexes, and X is connected. Let $i: F \to E$ be the inclusion map. Furthermore, let $h^*(F)$ be a free $h^*(\mathrm{pt})$-module with generators $i^*c_1, \dots, i^*c_n$ for some $c_1, \dots, c_n \in h^*(E)$. Then $h^*(E)$ is a free $h^*(X)$-module with generators $c_1, \dots, c_n$; in other words, for every $a \in h^*(E)$, there exist unique $a_1, \dots, a_n \in h^*(X)$ such that*

$$a = \pi^*(a_1)c_1 + \dots + \pi^*(a_n)c_n.$$

In particular, $\pi^*: h^*(X) \to h^*(E)$ is a monomorphism.

We leave the proof of this lemma to the reader. It can be easily deduced from the cobordism spectral sequence of the fibration π. Another plan, which is realized in the work of Conner and Floyd, consists in first proving the statement for trivial bundles, and then using multiple Mayer–Vietoris–like sequences (see Sect. 12.7).

Dold's lemma can be applied to our fibration $p: \widehat{X} \to X$. Indeed, it is easy to prove with the help of the Atiyah–Hirzebruch spectral sequence that $\Omega_U^*(\mathbb{C}P^{n-1})$ is a free $\Omega_U^*(\mathrm{pt})$-module with the generators $1, \gamma, \gamma^2, \dots, \gamma^{n-1}$, where $\gamma \in \Omega^2(\mathbb{C}P^{n-1})$ is the element described in condition (3) of our theorem. In particular, $\gamma = \sigma_1(\zeta)$, where ζ is the Hopf bundle over $\mathbb{C}P^{n-1}$. But the fibration ζ is the restriction of some one-dimensional bundle over $\widehat{X} \supset \mathbb{C}P^{n-1}$, namely, the one-dimensional summand of the splitting of the bundle $p^*\xi$ into a sum. Hence, the class γ is the image of some class $\widetilde{\gamma} \in \Omega_U^2(\widehat{X})$ (the first cobordism Chern class of the one-dimensional bundle mentioned above) with respect to the homomorphism induced by the inclusion map $\mathbb{C}P^{n-1} \to \widehat{X}$, and the classes $\gamma^2, \dots, \gamma^{n-1}$ are images of the classes $\widetilde{\gamma}^2, \dots, \widetilde{\gamma}^{n-1}$.

This completes the proof of the theorem.

C: Interaction of σ and μ

Since

$$\sigma_1(\xi \oplus \eta) = \sigma_1(\xi) + \sigma_1(\eta),$$

there arises a homomorphism

$$\sigma_1: K(X, A) \to \Omega_U^2(X, A).$$

Denote be σ_1^0 the composition

$$\begin{aligned} K(X,A) &= \widetilde{K}(X/A) = \widetilde{K}^{-2}(X/A) \\ &= \widetilde{K}(\Sigma^2(X/A)) \xrightarrow{\sigma_1} \widetilde{\Omega}^2(\Sigma^2(X/A)) = \widetilde{\Omega}_U^0(X/A) = \Omega_U^0(X,A). \end{aligned}$$

EXERCISE 6. Show that the composition

$$K(X,A) \xrightarrow{\sigma_1^0} \Omega_U^0(X,A) \xrightarrow{\mu} K(X,A)$$

is $-\operatorname{id}$. (*Hint:* It is sufficient to check this for $X = \mathrm{pt}, A = \emptyset$.)

Corollaries (From Exercise 6).

(i) *There exists a natural additive direct summand embedding*

$$K(X,A) \to \Omega_U^0(X,A).$$

(ii) *There is a natural multiplicative* $\mathbb{Z}_2$*-graded isomorphism*

$$K^*(X,A) = \Omega_U^*(X,A) \otimes_{\Omega_U^*(\mathrm{pt})} \mathbb{Z},$$

where the action of $\Omega_U^*(\mathrm{pt})$ *in* $\mathbb{Z}$ *is determined by the Todd genus* $\Omega_U^*(\mathrm{pt}) \to \mathbb{Z}$. (See Exercise 4.)

44.4 Cohomology Operations in Cobordisms and the Adams–Novikov Spectral Sequence

As we stated before (see the introductory phrase of Lecture 40), cohomology operations may be defined in any extraordinary cohomology theory. Here we will consider cohomology operations in the theory Ω_U.

Examples of cohomology operations. In our constructions of cohomology operations α we will restrict ourselves to the description of the map

$$\alpha_X : \Omega_U^*(X) \to \Omega_U^*(X)$$

for the case when X is a closed stably almost complex manifold. It is sufficient, since an arbitrary finite CW complex Y can be represented as a retract of such a manifold X (for example, of the Cartesian square of the double of the closure of its neighborhood in an Euclidean space), and then $\Omega_U^*(Y)$ will be a direct summand in $\Omega_U^*(X)$. In this case, we will use the interpretation of cobordisms given in Exercise 2.

We will consider two ways of constructing cohomology operations. The first one will work, actually, for any multiplicative cohomology theory. It consists in a fixation of some element a of the ring $\Omega_U^*(\mathrm{pt})$ and assigning to an arbitrary X the operation of the multiplication by a:

$$\Omega_U^*(X) \xrightarrow{\cdot a} \Omega_U^*(X).$$

Let us give a geometric description of this construction. If an element $a \in \Omega_U^{-2r}(\mathrm{pt})$ is represented by a $2r$-dimensional manifold A, and the element of $\Omega_U^*(X)$ is represented by a map $f: Z \to X$ (with an appropriate structure, as indicated in Exercise 2), then the image of this element of $\Omega_U^*(X)$ with respect to our cohomology operation is represented by the composition

$$Z \times A \xrightarrow{\text{projection}} Z \xrightarrow{f} X.$$

This operation reduces the dimension by $2r$.

Another way to construct cobordism cohomology operations assumes that a sequence $\omega = (\omega_1, \omega_2, \dots)$ of nonnegative integers with finitely many nonzero terms is fixed. Consider again an element of the ring $\Omega_U^*(X)$ represented by a map $f: Z \to X$ and a complex vector bundle ξ over Z stably equivalent to the virtual bundle $f^*\tau(X) - \tau(Z)$. According to Sect. 44.3, there is a characteristic class

$$\sigma^\omega(\xi) = \sigma_1(\xi)^{\omega_1}\sigma_2(\xi)^{\omega_2}\cdots \in \Omega_U^{2|\omega|}(Z),$$

where $|\omega| = \omega_1 + 2\omega_2 + \dots$. This class is represented by some map $g: W \to Z$ (again, with an appropriate structure). The image of our element of $\Omega_U^*(X)$ with respect to our cohomology operation is represented by the composition

$$W \xrightarrow{g} Z \xrightarrow{f} X.$$

This operation is denoted as s_ω; it raises the dimension by $2|\omega|$.

All the cobordism cohomology operations considered above are, obviously, stable (that is, they commute with the suspension isomorphism).

B: The Landweber–Novikov Algebra

Denote by $\mathbb{A}_U$ the algebra of all stable cohomology operations in Ω_U^*. The following theorem (which we formulate here without a proof) was proved in 1967 independently by Landweber and Novikov.

Theorem. (i) *The subgroup of the algebra $\mathbb{A}_U$ generated by the operations s_ω is closed with respect to the composition; that is, this subgroup is, actually, a subalgebra.*

(ii) *The map $\Omega_U^* \otimes S \to \mathbb{A}_U$ defined by the formula $\alpha \otimes s \mapsto \alpha \circ s$ is a monomorphism and has a dense image.*

The algebra S is called the Landweber–Novikov algebra.

The following interpretation of the Landweber–Novikov algebra was found by Buchstaber and Shokurov (see [27]). Denote by $\mathrm{Diff}_1(\mathbb{Z})$ the group of formal diffeomorphisms of a line of the form

$$t \mapsto t + x_1 t^2 + x_2 t^3 + \dots, \ x_1, x_2, \dots \in \mathbb{Z}.$$

Notice that the group $\mathrm{Diff}_1(\mathbb{Z})$ acts in the ring $P = \mathbb{Z}[x_1, x_2, \dots]$ of functions on itself by left translations. Denote by $\mathcal{S}$ the algebra of all $\mathrm{Diff}_1(\mathbb{Z})$-invariant formal differential operators in P, that is, formal power series in $\frac{\partial}{\partial x_1}, \frac{\partial}{\partial x_2}, \dots$ with coefficients in $\mathbb{Z}[x_1, x_2, \dots]$ which take elements from P into elements of P and commute with the transformations from $\mathrm{Diff}_1(\mathbb{Z})$. For an arbitrary sequence $\omega = (\omega_1, \omega_2, \dots)$ as above, define an operator $D_\omega \in \mathcal{S}$ by the formula

$$P(x \circ y) = \sum\nolimits_\omega D_\omega P(x) y^\omega,$$

where $x = (x_1, x_2, \dots)$, $y = (y_1, y_2, \dots)$, $\circ$ is the operation in $\mathrm{Diff}_1(\mathbb{Z})$, and y^ω means $y_1^{\omega_1} y_2^{\omega_2} \dots$. For example,

$$D_{(1,0,0,\dots)} = \frac{\partial}{\partial x_1} + \sum_{k=2}^{\infty} k x_{k-1} \frac{\partial}{\partial x_k},$$
$$D_{(0,1,0,\dots)} = \frac{\partial}{\partial x_2} + \sum_{k=3}^{\infty} (k-1) x_{k-2} \frac{\partial}{\partial x_k}$$

$$D_{(2,0,0,\dots)} = \frac{1}{2}\frac{\partial^2}{\partial x_1^2} + \sum_{k=2}^{\infty} k x_{k-1} \frac{\partial^2}{\partial x_1 \partial x_k}$$
$$+ \frac{1}{2} \sum_{k=2}^{\infty}\sum_{\ell=2}^{\infty} k\ell x_{k-1} x_{\ell-1} \frac{\partial^2}{\partial x_k \partial x_\ell} + \frac{1}{2} \sum_{k=3}^{\infty} (k-1)(k-2) x_{k-2} \frac{\partial}{\partial x_k}.$$

The following theorem is the main result of Buchstaber and Shokurov.

Theorem. *The correspondence $s_\omega \leftrightarrow D_\omega$ yields an isomorphism between the rings S and $\mathcal{S}$.*

The isomorphism $\mathcal{S} \cong S$ makes P an S-module, and this module also has a cobordism interpretation. Namely, denote by $\Omega_U^r(\mathbb{Z})$ the part of the tensor product $\Omega_U^r \otimes \mathbb{Q}$ which is distinguished by the condition that all the Chern numbers are integers. Thus, $\Omega_U^r \subset \Omega_U^r(\mathbb{Z}) \subset \Omega_U^r \otimes \mathbb{Q}$. It follows from the results of Sect. 44.2.C that the group $\Omega_U^r(\mathbb{Z}) / \Omega_U^r$ is finite. The action of the operations from S is naturally extended from Ω_U^* to $\Omega_U^*(\mathbb{Z})$.

Theorem. *The isomorphism between S and $\mathcal{S}$ can be extended to an isomorphism between the S-module $\Omega_U^*(\mathbb{Z})$ and the $\mathcal{S}$-module P.*

C: The Adams–Novikov Spectral Sequence (Novikov [65])

The homotopy sufficiency of the theory of unitary cobordisms (see Sect. 44.1.E) creates a principal possibility for computing stable homotopy groups for a space with known cobordism groups (with the action of operations from $\mathbb{A}_U$). This possibility finds its realization in the cobordism analogy of the Adams spectral sequence, the so-called Adams–Novikov spectral sequence. It is defined for an arbitrary finite CW complex X, has the initial term $\mathrm{Ext}_{\mathbb{A}_U}(\Omega_U^*(X), \Omega_U^*)$, and converges to the stable homotopy groups of X. In the case on the one-point X, we obtain a spectral sequence

$$\mathrm{Ext}_{\mathbb{A}_U}(\Omega_U^*, \Omega_U^*) \Longrightarrow \pi_*^S,$$

converging to the homotopy groups of spheres. The main difficulty in applications of this spectral sequence stems from the fact that not too much is known about its initial term. Let us mention some results concerning this term.

Part (ii) of the Landweber–Novikov theorem shows that

$$\mathrm{Ext}_{\mathbb{A}_U}(\Omega_U^*, \Omega_U^*) = \mathrm{Ext}_S(\mathbb{Z}, \Omega_U^*),$$

and the Buchstaber–Shokurov theorem allows us to replace S in this equality by $\mathcal{S}$. It is not hard to prove that

$$\mathrm{Ext}_{\mathcal{S}}(\mathbb{Z}, P) = 0.$$

Buchstaber suggested that to compute $\mathrm{Ext}_{\mathcal{S}}(\mathbb{Z}, \Omega_U^*)$ one should use the filtration

$$0 = N_0 \subset N_1 \subset \cdots \subset N_\infty = \Omega_U^*(\mathbb{Z}),$$

where $N_0 = \Omega_U^*$, $N_i^r = \{\sigma \in \Omega_U^r(\mathbb{Z}) \mid s_\omega \sigma \in N_{i-1}^{r+2|\omega|} \text{ for } |\omega| > 0\}$. The quotients N_i^r/N_{i-1}^r are finite and trivial as S-modules. (Currently, N_1/N_0 and N_2/N_1 are known; see Shokurov [78].) The Buchstaber filtration gives rise to a trigraded spectral sequence

$$\{E_r^{s,t,q}, s \geq 0, q \geq 0, s+t \geq 0,\ d_r^{s,t,q}\colon E_r^{s,t,q} \to E_r^{s-r,t+r-1,q}\}$$

in which $E_1^{0,*,*} = \mathrm{Ext}_S^*(\mathbb{Z}, \Omega_U^*)$, $E_\infty = E_\infty^{0,0,0} = \mathbb{Z}$, and, for $s > 0$,

$$E_1^{s,t,*} = \mathrm{Ext}_S^{-(s+t),*}(\mathbb{Z}, N_s/N_{s-1}).$$

Since N_s/N_{s-1} is a trivial S-module, we have (by Künneth's formula)

$$\begin{aligned}\operatorname{Ext}_S^{-(s+t),q}(\mathbb{Z}, N_s/N_{s-1}) = \bigoplus_{q_1+q_2=q} \Big\{ \Big[\operatorname{Ext}_S^{-(s+t),q_1}(\mathbb{Z},\mathbb{Z}) \otimes (N_s^{-q_2}/N_{s-1}^{-q_2}) \Big] \\ \oplus \operatorname{Tor}\Big(\operatorname{Ext}_S^{-(s+t)+1,q_1}(\mathbb{Z},\mathbb{Z}), (N_s^{-q_2}/N_{s-1}^{-q_2}) \Big) \Big\}.\end{aligned}$$

Thus, to compute the initial term of the Buchstaber spectral sequence, we need to know $\operatorname{Ext}_S(\mathbb{Z},\mathbb{Z})$, and this problem, in virtue of the equality $S = S$, may be regarded as a problem from the cohomology theory of infinite-dimensional Lie groups and Lie algebras. Within this theory, it is possible to prove that

$$\operatorname{rank} \operatorname{Ext}_S^{u,v}(\mathbb{Z},\mathbb{Z}) = \begin{cases} 1, & \text{if } v = 3u^2 \pm u, \\ 0 & \text{otherwise} \end{cases}$$

(see the book by Fuchs [38]).

Captions for the Illustrations

Page viii. *Alexander sphere, the so-called "wild sphere" or "horned sphere"*. This sphere separates three-dimensional Euclidean space into the union of two regions. One of them is homeomorphic to a 3-ball, the other is nonsimply connected. This embedding of 2-sphere in 3-space is "wild," i.e., nonlocally flat, in the infinite set of points, which is the Cantor set. Let us recall that the regular smooth embedding of a 2-sphere separates 3-space into the union of two simply connected regions

Page xi. *The first step in the infinite process of construction of Alexander sphere.* We see two "fingers," and at their end we see two new "fingers" that are "almost linked" in 3-space but do not touch one another. On the foreground we see the process of construction of "killing spaces" (see Chap. 3)

Page 9. *Two-dimensional surfaces in three-dimensional space.* On the left, like leafs of a fern, grow projective planes (projective 2-spaces). In the foreground is the Möbius strip in the form of a "crossed cap." We can also see the surfaces of high genus, i.e., 2-spheres with a large number of handles. The surfaces are not the manifolds. They are homeomorphic to a 2-sphere with three identified points

Page 14. *Bifurcations of two-dimensional surfaces inside three-dimensional manifolds. I.* Two tori are "glued" along the common nontrivial cycle and are transformed into one torus. Such bifurcations appear in the integrable Hamiltonian systems with two degrees of freedom

Page 15. *Bifurcations of two-dimensional surfaces inside three-dimensional manifolds. II*

Page 32. *Topological zoo.* At the right upper corner, we see the construction of a topological complex with a trivial one-dimensional homology group. But in this complex, there is a pair of limit points, where arbitrary small neighborhood for each of these two points has a nontrivial one-dimensional homology group. We then see the procedure of turning inside out the two-dimensional torus with a hole in ambient three-dimensional Euclidean space. As a result, we obtain again the 2-torus with a hole, but the parallel and meridian of the torus change places (switch positions). The inner side of the torus becomes the outer side. Then, at the left bottom corner, we see the so-called Antoine necklace. This is obtained as the intersection of the consequence of embedded solid tori. This "necklace" is totally disconnected compact perfect metric space and consequently is homeomorphic to the Cantor set. Then we see minimal surface (soap film) with boundary. It is interesting to note that this surface can be continuously mapped on its boundary curve (circle) in such a way that the mapping on the boundary is identical. But the boundary circle is not a deformational retract of the surface! This minimal surface is the connected sum of a standard Möbius strip with the so-called "triple Möbius strip," which is not a manifold. In the center of the drawing, we see 2-adic solenoid

Page 43. *Two-dimensional surfaces.* Two spheres with handles are shown as well as projective planes and "crossed caps," i.e., representations (models) of the Möbius strip in three-dimensional Euclidean space, such that its boundary becomes a flat circle. Almost everywhere such a "model" is a smooth immersion of a Möbius strip in 3-space, except for two points (branching points)

Page 55. *Simplicial (cell) approximation theorem.* Arbitrary continuous mapping of polyhedra can be continuously deformed into simplicial mapping. The idea of the proof is as follows. Using the small perturbation of the mapping in the image space, we can "clean out" a small ball. Then the images of the simplexes can be moved into some sub-polyhedrons formed by the simplices of the same dimension or of lower dimensions

Page 61. *Action of fundamental group on the higher homotopy groups. I.* The element of some homotopy group is represented as spheroid in ambient space, i.e., by a continuous mapping of the sphere. The element of a fundamental group is realized by some "loop." Then, from the spheroid the thin tube grows, moving along the "loop" until it reaches the initial point of the loop. Thus, each spheroid is replaced by a new spheroid. This procedure determines the automorphism of the fundamental group

Page 77. *Action of fundamental group on the higher homotopy groups. II.* Each topological space has some homotopy invariants. Among them, the homotopy groups have an important place. The first of these groups is called the fundamental group. Their elements are the classes of homotopic "loops." This group acts by automorphisms on the other homotopic groups

Page 78. *Knots in three-dimensional manifolds.* The theory of classification of the knots and analysis of their invariants is one of the most complicated fields in modern mathematics

Page 89. *Two-dimensional torus covers twice the Klein bottle.* This mapping is called two-sheeted covering

Page 105. *Fiber bundles (foliated spaces or fibrations).* These spaces can be considered as a union of fibers, i.e., of subspaces that are "similar." They can be homeomorphic to some fixed topological space. The fibers are parametrized by points of some other space called "base of fibration"

Page 120. *Dehn twisting along the circles (cycles) on surfaces.* These operations are the automorphisms of two-dimensional surface

Page 130. *Turning inside out of two-dimensional sphere in three-dimensional Euclidean space.* It is possible to interchange the outer and inner surfaces of two-dimensional sphere in three-dimensional Euclidean space by using a smooth regular deformation (without singularities). Important to note that we allow the self-intersection of the 2-sphere in the process of turning inside out. This deformation—family of smooth regular immersions—is nontrivial and here only one of its steps is shown (see the figure from the right)

Page 136. *It is possible to turn inside out the two-dimensional sphere in ambient three-dimensional Euclidean space in the class of smooth immersions.* It is clear that it is impossible to turn inside out the smooth circle embedded in Euclidean plane in the class of smooth immersions. Some singularities appear during the deformation when we try to contract the infinitely small loops. On the other hand, we can turn inside out 2-sphere in ambient three-dimensional Euclidean space in the class of smooth regular immersions. The rough scheme of deformation is shown in the drawing. We can see the nine steps of regular homotopy. The position number IX is "symmetrical," and at this moment 2-sphere appears to be in the small neighborhood of the immersed in 3-space two-dimensional projective plane. This immersion is usually called the Bolyai surface. We can then turn inside out the 2-sphere near projective plane and repeat the process of deformation in the opposite direction

Page 159. *The space, which is locally homologically disconnected (in Chech's sense) in dimension one.* This is a two-dimensional locally compact Hausdorff space with a trivial one-dimensional homology group (two-dimensional homology group is also equal to zero). On the other hand, the space contains two remarkable points. The first at the left bottom corner and the second located "in infinity." Their arbitrary open neighborhood (different from the whole complex, of course) has a nontrivial one-dimensional homology group. The complex is glued from an infinite number of "shells." The hole of each "shell" is glued by the scroll of the next "shell."

Page 164. *Cellular subdivision, cellular chains, and cellular homotopy*

Page 167. *Simplicial subdivisions and triangulations of polyhedra.* Simplicial complexes are obtained from the standard simplices as a result of "regular gluing" along their faces. Simplicial spaces are cell complexes

Page 177. *2-adic solenoid.* Let us take the solid torus, i.e., "filled torus." We then take a second solid torus and embed it inside the first one, twisting twice along the axis of the first torus. This process is then repeated with a third solid torus, and this goes on. Let us consider the "limit" of these solid tori. This limit is called 2-adic solenoid. It turns out that the "finite 2-adic twistings" appear in modern Hamiltonian and symplectic topology. They describe the important properties of some Liouville integrable differential equations. See details in the book by A.T. Fomenko. "Symplectic Geometry". Second revised edition. – Gordon and Breach, 1995

Page 198. *Foliations of three-dimensional manifolds. I.* We see the transformation of two tori into one as a result of gluing-bifurcation along the nontrivial cycle (circle). Such bifurcations appear in symplectic topology. The regular fibers are diffeomorphic to 2-tori, and the singular fibers are constructed in a more complicated way

Page 199. *Foliations of three-dimensional manifolds. II.* Transformation of one torus again into one torus in the neighborhood of a singular fiber of the type $(1, 2)$ inside the three-dimensional Seifert fibration

Page 216. *Between two maxima of a smooth function on the two-dimensional surface there always exists the saddle point.* Let us consider a smooth Morse function (i.e., with nondegenerate critical points) on 2-surface and assume that this function has two local maxima. Then "between them" there always exists a saddle point, i.e., "mountain pass." Let us consider the rubber thread on the surface and fix its ends in the maxima points. The thread starts to deform along the surface and stops soon. It then passes through the saddle point

Page 234. *Fantasy on the subject of fractals, analogues of Cantor sets, and noninteger Hausdorff dimension*

Page 254. *How to unclasp one's fingers in three-dimensional Euclidean space. I.* The 2-sphere with two handles can be embedded in three-dimensional Euclidean space in such a way that the handles will be "linked." But using the regular isotopy (in the class of smooth embeddings), we can deform the surface into a standard position and obtain the standard "pretzel." Several consecutive steps of such a deformation are shown in the next figures

Page 255. *How to unclasp one's fingers in three-dimensional Euclidean space. II*

Page 256. *How to unclasp one's fingers in three-dimensional Euclidean space. III*

Page 257. *How to unclasp one's fingers in three-dimensional Euclidean space. IV*

Page 258. *How to unclasp one's fingers in three-dimensional Euclidean space. V*

Page 265. *Hopf fibration and the subdivision of a 3-sphere into the union of two solid tori.* Let us take two solid tori and glue their boundaries (two tori) by a diffeomorphism, changing the parallel and meridian of the torus. We obtain 3-sphere. Let us foliate each of these two solid tori into the union of circles of a type (1,1), i.e., the circles go once along the parallel and once along the meridian. This is the well-known Hopf fibration of a 3-sphere, with the circle as the fiber

Page 296. *Fibrations with singularities.* We see here the "tangent bundle" of a circle with one singular point. The base here is a circle, embedded into the 2-plane and having one singular point, where the tangent line is not determined. If the circle is regularly embedded, then its tangent bundle is homeomorphic to a cylinder

Page 306. *Spectral sequences and the orbits of group actions. I.* The symmetry groups play an important role in many mathematical and physical problems. If the Lie group acts on some topological space, then this action determines the foliation into the union of the orbits of action. The topology of such foliations can be analyzed with the help of spectral sequences

Page 307. *Spectral sequences and the orbits of group actions. II.* The orbit of group action is the set of points obtained from the one point under action by all elements of the group of transformation. Different orbits can have different types, topologies, volumes, and dimensions. The volumes of the orbits determine the function on the space of orbits. This function plays an important role in the minimal surface theory and does not vary under the action of the symmetry groups

Page 320. *The terms of spectral sequence.* The calculation of the homology and cohomology groups of the topological space can be realized in some cases if we can represent the space as foliated space. Then we can calculate the infinite sequence of the "tables" – the terms of spectral sequence. They are connected by differential operators, which allow us to calculate "the limit table" that gives us the necessary information about homology groups (cohomology groups) of foliated space

Page 348. *Differentials in spectral sequences.* The homomorphisms-differentials in each term-table of a spectral sequence act approximately in the direction shown in the picture

Page 351. *The space of paths on the cell complex.* For each cell complex X, there always exists the contractible fibrated space with the base X. This is the space of all paths starting at the same point on the complex X

Page 353. *Method of "killing spaces."* This procedure is used to calculate homotopy groups for topological foliated spaces (Serre method)

Page 370. *Bott periodicity and Grassmannian manifolds.* Periodicity theorem for stable homotopy groups of classical Lie groups is proved with the help of the bundles of geodesics connecting the pairs of fixed points in symmetric spaces. The middle points of these geodesics, which connect the points E and –E on unitary group U(n), form the Grassmannian manifold

Page 392. *Spectral sequences, cohomological operations, and homotopy groups*

Page 396. *Method of killing spaces and spectral sequences*

Page 403. *Three-dimensional lens spaces.* These 3-manifolds are obtained by factorization of three-dimensional sphere with respect to the special actions of finite Abelian groups

Page 430. *Differentials in spectral sequences*

Page 436. *Consecutive terms (tables) of spectral sequence.* Each such table is endowed by differential (by special homomorphism), which allows us to calculate the next term (table)

Page 443. *The terms of spectral sequence*

Page 453. *Homotopy groups of spheres.* The calculation of these groups is a very complicated problem

Page 484. *Homotopy groups of topological spaces.* The elements of homotopy groups are the classes of homotopic mappings of a sphere into topological space. The problem of effective calculation of homotopy groups is very complicated

Page 526. The problem of effective algorithmic calculation of homotopy groups of spheres is not yet finally solved

Page 534. *Spectral sequences and differentials (homomorphisms)*

Page 550. *Homotopy of the Riemannian surface of algebraic function.* We see here a "three-dimensional model" for the smooth deformation of Riemannian 2-surface of the function $w=[(z-a)(z-b)(z-c)(z-d)]^{1/2}$ in four-dimensional ambient Euclidean space (two-dimensional complex space). The 2-surface is diffeomorphic to a torus if all polynomial roots a, b, c, d are different. If the same roots coincide (i.e., multiple roots appear), then the Riemannian surface is deformed in such a way that some disappearing cycles appear

Page 552. *Fantasy on the theme of K-theory, spinor foliations, and differentials in spectral sequences*

Page 580. *The kernels of differentials in spectral sequences and cobordisms*

Page 618. *Homotopy groups of spheres*

References

ADAMS J.F.

1. *On the non-existence of elements of Hopf invariant one.* Ann. Math. **72** (1960), 20–104.
2. *Stable homotopy theory.* Springer Lecture Notes, **3** (1961).
3. *Vector fields on spheres.* Ann. Math., **75** (1962), 603–632.
4. *On J(X), I.* Topology, **2** (1963), 181–195.
5. *On J(X), II.* Topology, **3** (1965), 137–171.
6. *On J(X), III.* Topology, **3** (1965), 193–222.
7. *On J(X), IV.* Topology, **5** (1966), 21–71.
8. *A periodicity theorem in homological algebra.* Proc. Cambr. Phil. Soc., **61** (1966).

ADAMS J.F. AND ATIYAH M.F.

9. *K-theory and the Hopf invariant.* Quarterly J. Math., **17** (1966), 31–38.

ADAMS J.F. AND WALKER G.

10. *An example in homotopy theory.* Proc. Cambr. Phil. Soc., **60** (1969), 699–700.

ADEM J.

11. *The iteration of Steenrod squares in algebraic topology.* Proc. Nat. Acad. Sci., **38** (1952), 720–726.
12. *Relations on iterations of reduced powers.* Proc. Nat. Acad. Sci., **39** (1953), 636–638.

ANDERSON D.W.

13. *The real K-theory of classifying spaces.* Proc. Nat. Acad. Sci. USA, **51** (1964), 634–636.

ATIYAH M.F

14. *Vector bundles and Künneth formula.* Topology, **1** (1961), 245–248.
15. *On the K-theory of compact Lie groups.* Topology, **4** (1965), 95–99.
16. *K-theory.* Addison-Wesley, 1989.

ATIYAH M.F. AND HIRZEBRUCH F.

17. *Riemann–Roch theorems for differentiable manifolds.* Bull. Amer. Math. Soc., **65** (1959) 276–281.
18. *On the non-embeddability of differential manifolds.* Colloque de topologie, Lille, 1959.
19. *Vector bundles and homogeneous spaces.* Proc. Symp. Pure Math., III, Amer. Math. Soc. (1961), 7–38.

A. Fomenko, D. Fuchs, *Homotopical Topology*, Graduate Texts in Mathematics 273,
DOI 10.1007/978-3-319-23488-5

BARRATT M.G. AND PRIDDY S.B.

20. *On the homology of non-connected monoids and their associated groups.* Comment. Math. Helv., **47** (1972), 1–14.

BECKER J. AND GOTTLIEB D.

21. *The transfer map and fiber bundles.* Topology, **14** (1975), 1–12.

BOREL A.

22. *Sur la cohomologie des espaces fibrés principaux et des espaces homogènes de groupes de Lie compacts.* Ann. Math., (2) **57** (1953), 115–207.

BOTT R. AND TU L.W.

23. *Differential Forms in Algebraic Topology.* Springer, 1982.

BUCHSTABER V.M.

24. *Modules of differentials of the Atiyah–Hirzebruch spectral sequence.* Math. of the USSR—Sbornik, **7** (1969), 299–313.
25. *Modules of differentials of the Atiyah–Hirzebruch spectral sequence II.* Math. of the USSR—Sbornik, **12** (1970), 59–75.

BUCHSTABER V.M. AND MISHCHENKO A.S.

26. *K-theory on the category of infinite cell complexes.* Math. of the USSR –Izvestiya, **2** (1968), 515–566.

BUCHSTABER V.M. AND SHOKUROV A.V.

27. *The Landweber–Novikov algebra and formal vector fields on the line.* Funct. Anal. Appl., **12** (1978), 159–168.

CARTAN H.

28. *Sur les groupes d'Eilenberg–MacLane $H(\Pi, n)$.* Proc. Nat. Acad. Sci., **40** (1954), 467–471, 704–707.

CARTAN H. AND EILENBERG S.

29. *Homological Algebra.* Princeton Univ. Press, 1956.

CHERN S.-S.

30. *Complex Manifolds.* Univ. of Chicago, 1956

COHEN R.

31. *The immersion conjecture for differentiable manifolds.* Ann. Math., **122** (1985), 237–328.

CONNER P.E. AND FLOYD E.E.

32. *The relation of cobordism to K-theories.* Springer Lecture Notes in Math., **28** (1966).

DELIGNE P., GRIFFITHS P.A., MORGAN J.W., AND SULLIVAN D.

33. *Real homotopy theory of Kähler manifolds.* Invent. Math, **29** (1965), 245–274.

DE RHAM G., MAUMARY S., AND KERVAIRE M.A..

34. *Torsion et type simple d'homotopie.* Springer Lecture Notes, **48** (1967).

DUBROVIN B.A., FOMENKO A.T., AND NOVIKOV S.P.

35. *Modern Geometry—Methods and Applications: Part III: Introduction to Homology Theory.* Springer, 1990.

FÉLIX Y., HALPERIN S., AND THOMAS J.-C.

36. *Rational Homotopy Theory.* Springer, 2001.

FUCHS D.B.

37. *Quillenization and bordisms.* Funct. Anal. and Appl., **8** (1984), 31–36.
38. *Cohomology of Infinite-Dimensional Lie Algebras.* Consultants Bureau, 1986.
39. *Classical manifolds.* Encyclopaedia Math. Sci. **24**, Topology II, 197–252, Springer, 2004.

FUCHS D.B. AND ROKHLIN V.A.

40. *Beginner's Course in Topology.* Springer, 1984.

FUCHS D.B. AND WELDON L.L.

41. *Massey brackets and deformations.* J. Pure and Appl. Alg., **156** (2001), 215–229.

GABRIELOV A.M., GELFAND I.M., AND LOSIK M.V.

42. *Combinatorial calculation of characteristic classes.* Funct. Anal. Appl. **9** (1975), 103–116, 186–202.

GINZBURG V.L.

43. *On the number of pre-images of a point with respect to a continuous map.* (Russian) Uspekhi Mat. Nauk, **41**, no. 2, 197–189.

HILTON P.

44. *Suspension theorem and generalized Hopf invariant.* Proc. London Math Soc., (3), **1** (1951), 462–493.

HIRSCH M.W.

45. *Differential Topology.* Springer, 1976.

HIRZEBRUCH F.

46. *Topological Methods in Algebraic Geometry.* Grundlehren Math. Wissenschaften, **131**, Springer, 1978.

HOCHSCHILD G.P. AND SERRE J.-P.

47. *The cohomology of group extensions.* Trans. Amer. Math. Soc., **74** (1949), 110–134.
48. *Cohomology of Lie algebras.* Ann. Math., **57** (1953), 591–603.

HUSEMOLLER D.H.

49. *Fibre Bundles.* Graduate Texts in Math., **20**, Springer, 1994.

KAC V.G.

50. *Torsion of cohomology of compact Lie groups and Chow rings of reductive algebraic groups.* Invent. Math., **80** (1985), 69–80.

KAROUBI M.

51. *K-Theory, an Introduction.* Springer, 1978.

KIRILLOV A.A.

52. *Elements of the Theory of Representations.* Springer, 1976.

LERAY J.

53. *Structure de l'anneau d'homologie de la représentation.* C. R. Acad. Sci. Paris, **222** (1946), 1419–1422.

MILNOR J.W.

54. *Link Groups.* Ann. Math., **59** (1954), 177–195.
55. *Construction of universal bundles II.* Ann. Math., **63** (1956), 430–436.

56. *On spaces having the homotopy type of a CW-complex.* Trans. Amer. Math Soc., **90** (1959), 272–280.
57. *Morse Theory.* Princeton Univ. Press, 1963.

MILNOR J.W. AND HUSEMOLLER D.H.

58. *Symmetric Bilinear Forms.* Ergebnisse Math. und Grenzgebiete, **73**, Springer, 1973.

MILNOR J.W. AND MOORE J.C.

59. *On the structure of Hopf algebras.* Ann. Math., **81** (1965), 211–264.

MILNOR J.W. AND STASHEFF J.D.

60. *Characteristic Classes.* Princeton Univ. Press, 1974.

MISHCHENKO A.S.

61. *Vector Bundles and Their Applications.* Kluwer, 1968.

MITCHELL W.J.

62. *Defining the boundary of a homology manifold.* Proc. Amer. Math. Soc., **110** (1990), 509–513.

MOSHER R.E. AND TANGORA M.C.

63. *Cohomology Operations and Applications in Homotopy Theory.* Harper&Row, 1968.

MUNKRES J.R.

64. *Elementary Differential Topology.* Annals of Math Studies, **54**, Princeton, 1966.

NOVIKOV S.P.

65. *The methods of algebraic topology from the point of view of cobordism theories.* Math. USSR—Izvestiya, **1** (1967), 827–913.

PALAIS R.

66. *Seminar on the Atiyah–Singer Index Theorem.* Princeton Univ. Press, 1965.

PONTRYAGIN L.S.

67. *Über the topologische Struktur der Lieschen Gruppen.* Comment. Math. Helv., **13/4** (1940/1941), 277–283.
68. *Smooth manifolds and their applications in homotopy theory.*, *Amer. Math. Soc. Translations, Ser. 2,* **11** (1959), 1–114.

POSTNIKOV M.M.

69. *On a theorem of Cartan.* Russian Math. Surveys, **21** (1966), 25–36.

QUILLEN D.

70. *The Adams conjecture.* Topology, **10** (1971), 67–80.

ROKHLIN V.A.

71. *Classification of mappings of an $(n+3)$-dimensional sphere into an n-dimensional sphere.* (Russian). Dokl. AN SSSR, **81** (1951), 19–22.

ROKHLIN V.A. AND SCHWARZ A.S.

72. *The combinatorial invariance of Pontryagin classes* (Russian). Dokl. AN SSSR, **114** (1957), 490–493.

SCHWARTZ J.T.

73. *Differential Geometry and Topology.* Gordon and Breach, 1968.

SEGAL G.

74. *Configuration spaces and iterated loop spaces.* Invent. Math., **21** (1973), 213–221.

SERRE J.-P.

75. *Homologie singulière des espaces fibrés. Applications.* Ann. Math., (2) **54** (1951), 425–505.
76. *Groupes d'homotopie et classes de groupes abéliens.* Ann. Math., **58** (1953), 258–294.
77. *Cohomologie modulo 2 des complexes d'Eilenberg–MacLane.* Comment. Math. Helv., **27** (1953), 198–232.

SHOKUROV A.V.

78. *The relationships between the Chern numbers of quasicomplex manifolds.* Math Notes, **26** (1979), 560–566.

SPANIER E.H.

79. *Algebraic Topology.* Springer, 1966.

STEENROD N.E.

80. *Products of cocycles and extension of mappings.* Ann. Math., **48** (1947), 290–320.
81. *The Topology of Fibre Bundles.* Princeton Univ. Press, 1951.

STONG R.E.

82. *Notes on Cobordism Theory.* Princeton Univ. Press, 1968.

SULLIVAN D.

83. *Geomeric Topology: Localization, Periodicity, and Galois Symmetry. The 1970 MIT Notes.* Springer, 2005.

THOM R.

84. *Quelques propriétés globales des variétés différentiables.* Comment. Math. Helv., **28** (1954), 17–86.
85. *Les classes caracteristique de Pontryagin des variétés triangulées.* Sympos. Intern. Topol. Algebr., Mexico City, 1958.

TODA H.

86. *Composition Methods in Homotopy Groups of Spheres.* Princeton Univ. Press, 1962.

TURAEV V.G.

87. *The Milnor invariants ans Massey products* (Russian) Zap. Sem. LOMI, **66** (1976), 189–203; English translation: J. Soviet Mathematics, **12** (1979), 128–137.

WHITEHEAD J.H.C.

88. *A generalization of the Hopf invariant.* Ann. Math. (2), **51** (1950), 192–237.

WHITNEY H.

89. *Geometric Integration Theory.* Dover, 1956.

Name Index

A. Fomenko, D. Fuchs, *Homotopical Topology*, Graduate Texts in Mathematics 273,
DOI 10.1007/978-3-319-23488-5

Subject Index

A. Fomenko, D. Fuchs, *Homotopical Topology*, Graduate Texts in Mathematics 273,
DOI 10.1007/978-3-319-23488-5

Printed in the United States
By Bookmasters